Highway
Engineering

Pavements, Materials and Control of Quality

Highway
Engineering

Pavements, Materials and Control of Quality

Athanassios Nikolaides

CRC Press
Taylor & Francis Group
Boca Raton London New York

CRC Press is an imprint of the
Taylor & Francis Group, an **informa** business

CRC Press
Taylor & Francis Group
6000 Broken Sound Parkway NW, Suite 300
Boca Raton, FL 33487-2742

First issued in paperback 2017

ISBN-13: 978-1-4665-7996-5 (hbk)
ISBN-13: 978-1-138-89376-4 (pbk)

Library of Congress Cataloging-in-Publication Data

Nikolaides, Athanassios.
  Highway engineering : pavements, materials and control of quality / author, Athanassios Nikolaides.
    pages cm
  Includes bibliographical references and index.
  ISBN 978-1-4665-7996-5 (hardback)
  1. Pavements. I. Title.

TE250.N54 2014
625.7--dc23
                                                  2014039171

Visit the Taylor & Francis Web site at
http://www.taylorandfrancis.com

and the CRC Press Web site at
http://www.crcpress.com

This book is dedicated to my wife, Eleni, and my son, Giannis.

'Road gives life, highway improves it, we are obliged to
construct and maintain pavements effectively to the benefit of
the community, the economy and the environment'.

Contents

10   Layers of flexible pavement   453

Preface

Highway engineering is the term that replaced the traditional term *road engineering* used in the past, after the introduction of modern highways. Highway engineering is a vast subject that involves planning, design, construction, maintenance and management of roads, bridges and tunnels for the safe and effective transportation of people and goods.

This book concentrates on design, construction, maintenance and management of pavements for roads/highways. It also includes pavement materials since they are an integral part of pavements. It has been written for graduates, postgraduates as well as practicing engineers and laboratory staff and incorporates the author's 30 years of involvement in teaching, researching and practicing the subject of highway engineering.

Advancements in pavement materials, design, construction, maintenance and pavement management and the globalisation of the market make it imperative for the highway engineer to be aware of the techniques and standards applied globally.

One of the objectives of the book is to provide integrated information on the abovementioned disciplines of highway engineering.

Another objective is to include in one book both European and American standards and practices (CEN EN, ASTM, AASHTO and Asphalt Institute). This would result in a more useful reference textbook to pavement engineering courses taught in European and American educational establishments.

Another objective of this book is to provide a reference textbook to practicing pavement engineers and materials testing laboratory staff, working in countries employing European or American standards and techniques.

Apart from information regarding European and American practices, the reader can also find some specific information on practices employed in countries such as the United Kingdom, France and Greece, as well as Australia.

In addition, this book also aims to provide integrated information related to pavement materials (soil, aggregates, bitumen, asphalts and reclaimed material), material testing for acceptability and quality assurance, asphalt mix design, flexible and rigid pavement design, construction, maintenance and strengthening procedures, quality control of production and acceptance of asphalts, pavement evaluation, asphalt plants and pavement recycling. It also covers the basic principles of pavement management.

The book in its 18 chapters contains many tables, graphs, charts and photographs to assist the reader in learning and understanding the subject of pavement engineering and materials. It also contains a great number of references, a valuable tool to help the reader seek more information and enhance his or her knowledge.

The short description of all pavement material testing procedures, required by European and American standards, as well as pavement design and maintenance procedures covered, does not, by any means, substitute or replace the standards and procedures developed by the various organisations and agencies. The reader is advised to always consult the standards or manuals developed when engaged in testing, design, construction or maintenance works.

It is hoped that this textbook will not only contribute to the understanding of the wide and challenging subject of pavement engineering but also enable a more effective and economical design, construction and maintenance of pavements by employing updated standards, practices and techniques.

Prof. A.F. Nikolaides
Aristotle University of Thessaloniki, Greece
March 2014

Acknowledgements

I would like to express my deepest thanks to all those who encouraged me and contributed to the completion of this huge task. I sincerely thank Dr. E. Manthos, Emeritus Prof. Alan Woodside, Prof. Waheed Uddin, Dr. Dave Woodward, Dr. Cliff Nicholls and last but not least Emeritus Prof. George Tsohos. Since most of the text in this book is based on a similar Highway Engineering book (published in Greek) that I authored (2011), I also thank Miss Dimitra Kremida for assisting with the translation.

I also like to thank all organisations, institutions and private companies for providing and allowing me to use useful materials such as tables, figures and photographs. The organisations and institutions I would like to thank are AASHTO, the Asphalt Institute, ASTM, Austroads, BSI, Caltrans, CEN EN, FAA, Highways Agency (UK), TRB (USA), TRL (UK), EAPA, Energy Institute, EUROBITUME, ICE (UK), NAPA, OECD and PIARC. The private companies I would like to thank are Ames Engineering Inc., AMMANN Group, Anton-Paar ProveTec GmbH, APR Consultants, ARRB Group Ltd., Atlas Copco, Cooper Research Technology Ltd., Douglas Equipment, DYNAPAC, Dynatest International A/S, Euroconsult, FACE Companies, Findlay Irvine Ltd., Fugro Roadware, GSSI Inc., Impact Test Equipment Ltd., Interlaken Technology Corp., International Cybernetics Corp., MALÅ, Moventor Oy Inc., Nippo Sangyo Co. Ltd., Pavement Technology Inc., PipeHawk Plc., Roadtec Inc. (ASTEC), ROMDAS, Sarsys AB, Surface Systems & Instruments Inc., T&J Farnell Ltd., Vectra (France), VTI, WayLink Corp., WDM Ltd., Wirtgen GmbH, ARRA and Greenwood Engineering A/S.

Finally, I would like to thank all those who supported me (at a close distance) during the countless hours of writing this book. Special thanks go to my wife, son, mother, brother, sister and close friends. Their patience in tolerating my unsociable behaviour at times and their words of encouragement were the main factors that helped me write and finish this book. I will never forget their contribution and I thank them all from the bottom of my heart.

Author

Dr. Athanassios Nikolaides is a professor at the Aristotle University of Thessaloniki (AUTh), Greece, and director of the Highway Engineering Laboratory of the Department of Civil Engineering. He has extensive experience in the study of soil materials, unbound and bound aggregate materials, bituminous materials, asphalt mix design, pavement design, construction, maintenance, non-destructive testing, recycling and pavement management. From 1984 to 1989, he worked as a freelance engineer and project manager and since 1989 has been at AUTh; in 2000, he became full professor in the Department of Civil Engineering of AUTh. He does research and teaches Highway Engineering and Airport Engineering at the undergraduate and postgraduate levels. He has completed several research projects funded by various organisations or agencies and numerous technical studies funded by private companies. He has published more than 70 papers in journals and conference proceedings and presented more than 40 additional papers at seminars in various countries. He has also published three books in Greek under the following titles: *Highway Engineering* (2011, 3rd edition), *Airport Engineering* (2002) and *Flexible Pavements* (2005). He is a member of professional institutions and has been a member of scientific/technical committees of various international conferences; currently, he is a member of the editorial advisory panel of the ICE (UK) *Journal of Construction Materials*. He is the founder and president of the International Conference 'Bituminous Mixtures and Pavements' organised every 4 years since 1992. Prof. Nikolaides has served as a consultant to several organisations and private corporations in Greece, Europe and Indonesia.

Chapter 1

Soils

1.1 THE FORMATION OF SOILS

Soil is the natural material over which the pavement is going to be constructed. According to CEN EN ISO 14688-1 (2013), soil is defined as 'assemblage of mineral particles and/or organic matter in the form of a deposit but sometimes of organic origin, which can be separated by gentle mechanical means and which includes variable amount of water and air (and sometimes other gases)'. The term also applies to ground consisting of replaced soil or man-made materials exhibiting similar behaviour, for example, crushed rock, blast furnace slag and fly ash.

All types of soil derive from the disintegration of rocks and decomposition of vegetation. Disintegration and decomposition were caused by physical and chemical action. The major ones are wind, water, temperature variations and chemical reactions.

Soils are characterised by the way they were created. They are defined as *residual, sedimentary, aeolian* and *glacial.*

Residual soils are those formed by rocks located just below them. Climatologic conditions (temperature and rainfall) were the main reasons of disintegration of the parent materials. Those soil types consist of inorganic grainy materials, in the form of fine particles in upper layers and in the form of more coarse particles in lower layers. These soils can be used as pavement foundation layer, provided that no drastic chemical disintegration has occurred (tropical climatologic conditions).

Sedimentary soils are those formed by the deposition of materials that were in suspension in aqueous environments, such as lakes, rivers and oceans. The sedimentary soils vary from clean sand to flocculent clay of marine origin.

From the sedimentary soils, the alluvial soils are generally suitable as pavement foundation material.

The marine soils, those created by ocean erosion of the materials transferred from rivers to the sea, should be treated with care in pavement engineering, particularly when they contain a high percentage of fine particles.

Aeolian soils are those formed by aeolian action, that is, materials transported, eroded and deposited by winds. These soils appear as sand dunes or calcitic silt. Pavement construction on sand dunes, which are not protected by topsoil, appears to be problematic. A cut on calcitic silt may have very high gradient (slope) owing to the cohesive properties of calcium. The usage, however, of disturbed calcitic silt on embankment is problematic because cohesion has been lost.

Glacial soils are soils formed during the glacier era. These soils may extend to a depth of many kilometres; they consist of boulders, cobbles, gravel, sand, silt and clay, and they are widely found in the Northern Hemisphere.

Glacial soils may be characterised with respect to their content in inorganic materials. Soils in which the inorganic ingredients of mineral materials outclass the organic substances are called inorganic soils. Otherwise, they are called organic soils, which are characterised by their dark brown colour and their characteristic smell.

Regardless of the way they have been formed, the lack of homogeneity is a feature of soils. Soils appear to vary from loose to very well compacted, with or without cohesion, with continuous or non-continuous particle size distribution. The above heterogeneity appears in both horizontal and vertical levels. The highway engineer has to deal with many kilometres and in most cases has to use the existing soil without any adjustment. This fact makes the determination of its representative mechanical behaviour more difficult and tricky.

1.2 SOIL FRACTIONS

Natural soils, despite their lack of homogeneity, can be assorted in fractions or subfractions by reference to their particle size. The definition of 'particle size' in this case refers to the maximum size of the particle that is incorporated in the soil. This type of assortment is quite useful to engineers, since it is directly connected to the mechanical behaviour of the soil material. The basic fractions of soils are boulders, cobbles, gravels, sand, silt and clay.

The representative size of the particles for the above groups and subgroups differs slightly from one specification to another. The representative size of the particles for the above fractions and subfractions in accordance with CEN EN ISO 14688-1 (2013), AASHTO M 146 (2012) and ASTM D 2487 (2011) standards are shown in Table 1.1.

Boulders and cobbles, gravels and sands are granular soils. Their particles do not have any or almost any cohesion. They are easily recognizable and they are distinguished for their high permeability and good stability under the influence of axial load. The term *gravel* is

Table 1.1 Soil fractions

	CEN EN ISO 14688-1 (2013)	AASHTO M 146 (2012)	ASTM D 2487 (2011)
Groups		Particle size (mm)	
Boulders		>300	>300
Large boulder (LBo)	>630	—	—
Boulder (Bo)	>200–630	—	—
Cobble (Co)	>63–200	>75–300	>75–300
Gravel (Gr)	>2.0–63	>2.0–75	>4.75–75
Coarse (CGr)	>20–63	>25–75	>19–75
Medium (MGr)	>6.3–20	>9.5–25	—
Fine (FGr)	>2.0–6.3	>2.0–9.5	>4.75–4.75
Sand (Sa)	>0.063–2	>0.075–2.0	>0.075–4.75
Coarse (CSa)	>0.63–2.0	>0.425–2.0	>2.0–4.75
Medium (MSa)	>0.2–0.63	—	>0.475–2.0
Fine (FSa)	>0.063–0.2	>0.075–0.425	>0.075–0.475
Silt (Si)	>0.002–0.063	0.075–0.002	<0.075–PI <4
Coarse silt (CSi)	>0.02–0.063	—	—
Medium silt (MSi)	>0.0063–0.02	—	—
Fine silt (FSi)	>0.002–0.0063	—	—
Clay	≤0.002	<0.002	<0.075–PI >4
Peat		Organic soil	

used for natural granular materials of rivers, mines or other deposits, which in their majority have a spherical shape. Reversely, the term *crushed gravel* refers to a material having the same dimensions as gravel but derived from crushing natural gravels and has at least one broken and crushed surface. This is similar to the distinction between natural sand and crushed sand.

Silt is soil consisting of very fine particles, which, in contrast with the above groups, have some cohesion. It is rather difficult to visually recognise unless it is dried, broken and sieved using a 0.075 mm sieve (No. 200) or a 0.063 mm sieve. Thus, in this case, silt appears to be in the form of a powder. The silt particles range from 0.002 to 0.063 mm (0.075 mm), that is, larger than clay but smaller than sand particles. The shape of silt particles is mainly spherical.

Silts are characterised by low to modest plasticity, very low permeability and the fact that they are subject to substantial shrinkage and expansion owing to moisture change. Shrinkage and expansion are particularly obvious when the particle dimensions approach the dimensions of clay. The stability of the layer consists of silt under the influence of axial loads depending mainly on the existence or nonexistence of decomposed organic substances as well as plate-shaped inorganic particles, such as mica. Organic silts are unstable and present high compressibility. It is also very likely for silts, which include a high percentage of mica, to present high compressibility as well as elasticity.

Clay is the finest soil material with a particle size less than 0.002 mm. When dispersed in water, it gives a colloid in which the particles are in suspension for a very long time. In contrast to silts and sands, the shape of clay particles is flattened and elongated. Because of the size and nature of particles, a particular mass of clay has the largest specific surface of any other equivalent soil mass. Moreover, the surface of particles is more chemically active as well as unstable than any other soil material. Characteristically, it is reported that in 1 g of clay, there are approximately 90 billion particles, whereas in 1 g of silt and coarse sand (0.5–1.0 mm), there are 5.5 million particles and 700 particles, respectively (Millar et al. 1962).

The chemical activity of the clay is due to its surface electrical charge, which causes the attraction and adsorption of mainly positive ions, such as hydrogen ions, exiting in water, or calcium, or sodium ions. The swelling of clay in the presence of water is due to the attraction of hydrogen ions. The stabilisation of clay, however, is due to the attraction of calcium or sodium ions.

Clays are distinguished by their medium to very high plasticity, relatively high strength in dry state, high volume fluctuation with respect to moisture changes and their impermeability. By increasing the moisture content, the bearing capacity of the clay layer is dramatically decreased. Clayey soils appear to be highly problematic and should be treated with extra care. Clayey soils are often considered as inappropriate for pavement foundation layers, without any other treatment (stabilisation) or use of geotextiles.

Clayey soils, which include montmorillonite, appear to be the most problematic of all because of their unstable structure (two tetrahedral sheets sandwiching a central octahedral sheet). They also show high swelling and shrinkage, as well as very high plasticity. Less problematic soils are those that contain kaolinite. Their structure is rather stable. Kaloninite clay is a layered silicate mineral, with one tetrahedral sheet linked through oxygen atoms to one octahedral sheet of alumina octahedral. As a result, swelling, shrinkage and plasticity are rather lower in relation to montmorillonite. In between, there are clayey soils containing illite or chrorite – the most common minerals found in clayey soils (O'Flaherty 2002).

It is known that many soils do not belong exclusively to one of the abovementioned groups but simultaneously to two or three of them. For instance, there are soils that consist of an amount of sand and an amount of silt or an amount of sand and gravel, and so on. In these cases, the designation of such soils derives from the two basic ingredients, for example, 'silty sand' and 'sand-gravel'.

They are also found in soils under the designation 'loam', 'loess' and 'peat'. The first one is another expression for clayey soil. The second is aeolian soil with properties mentioned in Section 1.1. The third consists mainly of vegetable substances in decomposition and it contains a high percentage of moisture content (or absorbs a lot of moisture) and is rather compressive and absolutely inappropriate for the foundation layer of pavements. It should be noted that the knowledge of particle size distribution is not sufficient to fully judge the soil's properties and its behaviour under loading. For instance, in terms of shear resistance, natural spherical sand performs in a manner completely different from that of angular sand despite the fact that they may have the same particle size distribution. The behaviour between silty and clayey soils is more complicated. Thus, in any project (highways, roadworks, etc.), the engineer should thoroughly investigate the behaviour of soils in order to ensure the stability of the construction.

More instructive classification systems for soils have been developed by the American Association of State Highway and Transportation Officials (AASHTO) and the American Society for Testing and Materials (ASTM) standards. These two classification systems are described in Section 1.5.

1.3 PHYSICAL PROPERTIES OF SOILS

Natural soil can be assumed to be a large number of solid particles with scattered voids or pores. Some of these voids, or all of them, or even none of them contain water. The combination of the three phases (solid, liquid and air) gives the soil its basic properties, which affect its mechanical behaviour. For a better explanation and understanding of the properties of a soil, the three phases for a unit volume (V) or mass (M) have been separated in a simplified way as shown in Figure 1.1.

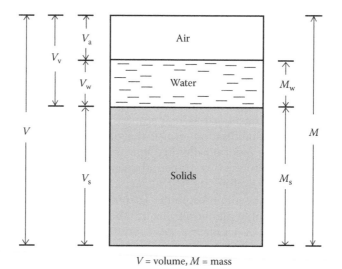

V = volume, M = mass

Figure 1.1 Diagram showing the three phases of soil.

1.3.1 Moisture content

Moisture (water) content of soil (w), often referred to as natural moisture, is defined as the ratio of the water mass contained in the soil (M_w) to the mass of solids (M_s) and is expressed as a percentage by mass of solids. The equation used is simply

$$w = (M_w/M_s) \times 100(\%).$$

The natural moisture of the soil is a crucial factor for the highway engineer, since the bearing capacity of the pavement's foundation layer depends directly from it.

More information regarding the determination of water content can be found in AASHTO T 265 (2012), ASTM D 4959 (2007), ASTM D 4643 (2008) and CEN ISO/TS 17892-1/AC (2005).

1.3.2 Void ratio or index

Void ratio or index (e) is defined as the ratio of the volume of air voids (V_a) and water voids (V_w) to the volume of solids (V_s). The equation used to determine void ratio or index is as follows:

$$e = (V_a + V_w)/V_s.$$

In case the soil is saturated, all existing voids are filled with water. In other words, $V_v = V_w$, and thus, the void ratio may be calculated using the relationship

$$e = V_w/V_s = (M_w/M_s)/(\rho_s/\rho_w) = w \times (\rho_s/\rho_w),$$

where ρ_s is the density of the soil, ρ_w is the density of water and w is the moisture content.

1.3.3 Porosity

Porosity (n) is defined as the ratio of the volume of all voids (V_v) (air voids and voids filled with water) to the total volume (V) and is expressed as a percentage of total volume. The equation used is simply

$$n = (V_v/V) \times 100(\%).$$

The porosity of the soil can also be calculated from the porosity index (e) using the following relationship:

$$n = [e/(1 + e)] \times 100(\%).$$

The porosity index and the porosity are parameters that characterise the soil whether it is loose or dense. However, in soil mechanics, the term *relative density* (D_r) is also used, since it defines soil densification. A value of relative density of approximately 0.15 signifies a very loose soil, while a value of 1.0 signifies a very dense soil. The equation used to determine the relative density of a soil can be found in any soil mechanics textbook.

1.3.4 Percentage of air voids

The percentage of air voids (n_a) is defined as the ratio of air voids (V_a) to the total volume (V) and is determined from the equation

$$n_a = (V_a/V) \times 100(\%).$$

1.3.5 Percentage of water voids

The percentage of water voids (n_w) is defined as the ratio of volume of voids filled with water (V_w) to the total volume (V) and is determined from the equation

$$n = (V_w/V) \times 100(\%).$$

Knowing the percentage of water voids (n_w) and the percentage of air voids (n_a), porosity (n) can also be calculated using the equation $n = n_w + n_a$.

1.3.6 Degree of saturation

The degree of saturation (S) is defined as the ratio of the volume of the water to the volume of all voids (V_v) (volume of air voids and voids filled with water). The equation to determine the degree of saturation is

$$S = (V_w/V_v) \times 100(\%).$$

1.3.7 Density and specific gravity of soil particles

The density of soil particles (ρ_s) is defined as the ratio of the mass of soil particles solids (M_s) to the corresponding volume of soil particles (V_s). In other words:

$$\rho_s = M_s/V_s \ (kg/m^3).$$

The specific gravity of a soil $(g$ or $\gamma)$ is the ratio of the weight in air of a given volume of soil particles at a stated temperature to the weight in air of an equal volume of distilled water at the stated temperature.

The term *density* is used by EN specifications while the term *specific gravity* is usually used by US specifications.

The density or specific gravity of soils can be determined in the laboratory by using a pyknometer or flasks of various sizes. The procedure is similar to the one followed for the determination of density or specific gravity of aggregate materials (see Section 2.12).

However, details regarding laboratory investigation of density and specific gravity of soils can be found in CEN ISO/TS 17892-3 (2005), ASTM D 854 (2010) or AASHTO T 100 (2010).

1.3.8 Density and unit weight of soil in place (in situ)

The soil density (ρ) in situ is defined as the ratio of the mass of solids (M_s) and the mass of water (M_w) to the total volume (V) of the soil. In other words:

$$\rho = (M_s + M_w)/V \ (kg/m^3).$$

In some cases, the soil density can also be called 'apparent soil density' or 'wet density of soil'.

The density of the undisturbed or compacted soil in situ is determined by the 'sand-cone' method (ASTM D 1556 2007 or AASHTO T 191 2013), the rubber balloon method (ASTM D 2167 2008) or nuclear methods (ASTM D 5195 2008 or AASHTO T 310 2013). All three methods are outlined in Section 1.11.

The density of the soil in its undisturbed state depends on various factors, the main factors being the soil classification group, the moisture content and the degree of natural compaction. Indicatively, in situ soil density values vary from 1200 to 1700 kg/m^3 for loose soils and 1500 to 2200 kg/m^3 for dense soils.

1.3.9 Density of saturated soil

The density of saturated soil (ρ_{sat}) is defined as the ratio of the mass of solids (M_s) and the mass of water, which occupies all pores (V_v), to the total volume (V). In other words:

$$\rho_{sat} = (M_s + V_v \times \rho_w)/V \text{ (kg/m}^3),$$

where ρ_w is the water density (kg/m^3).

1.3.10 Other useful relationships

The following useful relationships arise from some of the above soil properties:

$$w = (\rho - \rho_d)/\rho_d \text{ or } \rho_d = \rho/(1 + w)$$

$$e = (\rho_s - \rho_d)/\rho_d \text{ or } \rho_d = \rho_s/(1 + e)$$

$$\rho = (\rho_s + S \times e \times \rho_w)/(1 + e)$$

$$\rho_{sat} = (\rho_s + e \times \rho_w)/(1 + e),$$

where w is the moisture content (%), e is the void ratio, ρ is the soil density in situ (kg/m^3), S is the degree of saturation (%), ρ_d is the dry density (kg/m^3), ρ_s is the density of soil particles (kg/m^3), ρ_w is the density of water (kg/m^3) and ρ_{sat} is the density of saturated soil (kg/m^3).

It is noted that density can be replaced with the respective specific gravity in all the above-mentioned equations.

1.3.11 Laboratory maximum density of soils

The laboratory density of the disturbed and recompacted soil, along with the corresponding moisture content, is defined by the Proctor method (see Section 1.4.6). The determination of the optimum moisture is considered to be of great importance for roadworks, since pavement must be sited on sufficiently compacted soil material so that settlement does not occur. Sufficient compaction is achieved when the density of the soil material, after compaction, is greater than or equal to a certain required percentage of the laboratory optimum density determined. The required percentage to be obtained, depending on the type of soil, varies from 92% to 98%.

1.4 BASIC SOIL TESTS

1.4.1 Particle size analysis

Particle size analysis is the determination of the particle size distribution of a soil sample. It is one of the most important physical characteristics of soil since classification of soils is based on that and many geotechnical and geohydrological properties of soil are related to the particle size distribution.

The determination of the particle size distribution may involve two processes: sieving and sedimentation. Sieving is the process whereby the soil is separated in particle classes by the use of test sieves. Sedimentation is the process of the setting of soil particles in a liquid.

Conventional sieving is applicable to soils with less than 10% fines (materials passing through 0.063 or 0.075 mm sieves). Soils with more than 10% fines can be analysed by a combination of sieving and sedimentation (EN ISO/TS 17892-4). According to ASTM D 422 (2007) or AASHTO T 88 (2013), sieving by sedimentation is required normally when the percentage of fines is more than 20%.

A rather crucial factor for the credibility of the results is the representativeness of sampling and the mass of test portion. The mass of test portion required depends on the maximum size of the soil particles. For example, the minimum mass of test portion normally used for a maximum soil particle size of 9.5 mm is 0.5 kg and that for a maximum soil particle size of 75 mm is 5 kg.

1.4.1.1 Particle size distribution by sieving

Sieving is carried out using appropriate diameter and size sieves. The procedure is similar to aggregate sieving described in Section 2.11.

More specific information can be found in CEN ISO/TS 17892-4 (2005), ASTM D 422 (2007) or AASHTO T 88 (2013).

1.4.1.2 Particle size distribution by sedimentation

According to ASTM D 422 (2007) (AASHTO T 88 2013), materials passing through a 2.0 mm sieve are separated. A mass of test portion of approximately 100 g, if sandy soil is used, or 50 g, if clay and silt size, is placed in a 250 ml beaker. The sample is covered with 125 ml of stock solution (sodium hexametaphosphate solution, 40 g/L), stirred until is thoroughly wetted and allowed to soak for at least 12 or 16 h. At the end of soaking, the contents of the beaker are washed into a dispersion cup using distilled or demineralised water and dispersed for a period of 60 s in a mechanical stirring apparatus. An alternative air-jet method may be used for dispersion.

Immediately after dispersion, the soil–water slurry is transferred to the sedimentation cylinder, water is added until the total volume is 1000 ml and the contents are shaken by hand for approximately 60 s.

At the end, the cylinder is placed in a convenient location with stable temperature, and hydrometer readings are taken at intervals of 5, 15, 30, 60, 250 and 1440 min or as many as needed. In fact, the depth (d) of the hydrometer is recorded (see Figure 1.2).

After the last hydrometer reading is taken, transfer the suspension to a 0.075 mm sieve and wash with tap water until the wash water is clear. The material retained on the 0.075 mm sieve is transferred into a suitable container and dried in an oven at 110°C. The dried material, together with the portion retained on the 2.0 mm sieve, is sieved using the desired set of sieves. This sieving provides the gradation curve of the soil fraction >0.075 mm.

The gradation curve of the material passing through the 0.075 mm sieve, that is, the particle size distribution, is determined from the hydrometer readings according to Stock's law. Details for determining the particles' diameter can be found in ASTM D 422 (2007) or AASHTO T 88 (2013).

The two gradations obtained are combined and presented as a final result.

Similar to the above procedure is the procedure described in CEN ISO/TS 17892-4 (2005).

The conjunction of the three gradation curves, in other words, the curve of the fraction retained on the 2.00 mm sieve, the curve of fraction from 2.00 to 0.075 mm, and the curve of fraction passing through the 0.075 mm sieve, gives the final gradation curve of soil material. It should be mentioned here that bearing out the sieve analysis with hydrometer is only

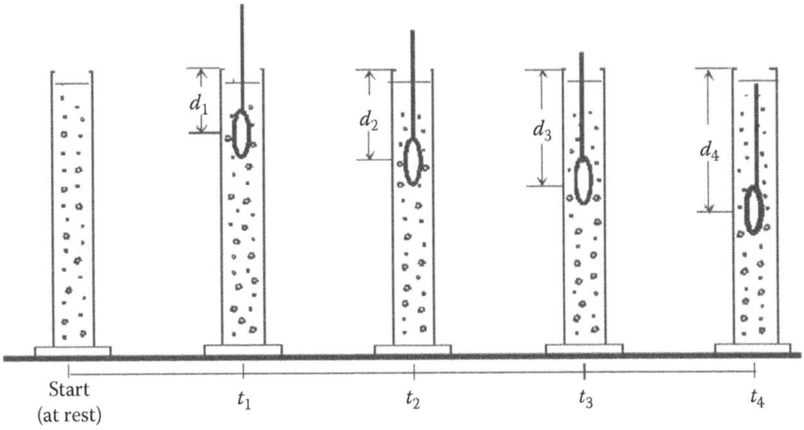

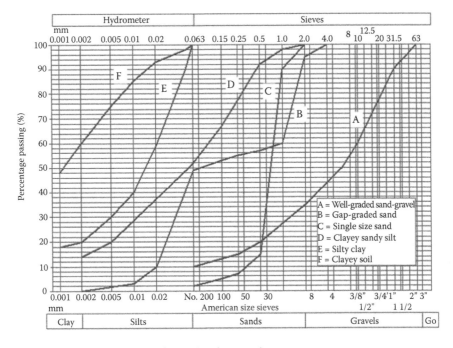

Figure 1.2 Schematic representation of the hydrometer test.

necessary for very fine soils (clayey soils, silty sands, clayey sands, etc.) or soil materials containing a great amount of material passing through the 0.075 mm sieve (higher than 20%).

1.4.1.3 The usefulness of aggregate gradation

The position, the shape and the slope of the gradation curve provide useful information for the soil material. The position of the gradation curve determines more or less the soil classification group, while the shape and slope determine the distribution of the particles and the deficiency or sufficiency of certain size particles.

Typical gradation curves of soil materials are given in Figure 1.3.

Figure 1.3 Indicative gradation curves of typical soil materials.

The particle size of the soil material affects its mechanical behaviour. Coarse soil materials have a better mechanical behaviour than fine ones in terms of strength and resistance to loading. Additionally, the particle size determines the ability or possibility of water retention or absorption, as well as the possibility of frost damage or swelling. Very fine soils (particularly silt and clay) absorb and retain water. As a result, they swell and they are also susceptible to frost damage. Coarse materials do not have any of the abovementioned deficiencies; hence, they are preferable and more suitable as foundation layer materials. More details regarding the suitability of soil materials, in general, can be found in Table 1.8.

The relatively moderate slope of curves A, C, E and F indicates a uniform particle distribution and consequently the existence of all sizes particles. The almost vertical 'S'-shaped curve (curve D) indicates a material with almost single-size particles. The double 'S'-shaped curve (curve B) indicates a material in which certain particle sizes are absent. This type of curve is known as a gap graded curve.

The uniformity of the soil material, desirable in most cases, is quantified by reference to the uniformity coefficient and the coefficient of curvature. The uniformity coefficient (C_U) and the coefficient of curvature (C_C) are determined by the following equations:

$$C_U = d_{60}/d_{10}$$

and

$$C_C = (d_{30})^2/(d_{10} \times d_{60}),$$

where d_{10}, d_{30} and d_{60} are the particle sizes corresponding to the ordinates 60% and 10% by mass of the percentage passing.

According to CEN EN ISO 14688-2 (2013), the shape of the grading curve with respect to C_U and C_C is characterised as shown in Table 1.2.

Similar but not the same values for determining the shape of the grading have been proposed in ASTM D 2487 (2011).

It should be noted that the shape of the grading influences the shear strength and the permeability of the soil material. Soils with non-uniformly distributed particles possess higher shear strength, while they have lower permeability in comparison to soils with uniformly distributed particles.

1.4.2 Liquid limit, plastic limit and plasticity index

The liquid limit (w_L or LL) of a material is the water content at which the soil passes from the liquid to plastic condition and is determined by the liquid limit test.

The plastic limit (w_P or PL) of a material is the water content at which the soil becomes too dry to be in a plastic condition and is determined by the plastic limit test.

Table 1.2 Shape of grading curve

Shape of grading curve	C_U	C_C
Multi-graded	>15	$1 < C_C < 3$
Medium graded	6–15	<1
Even graded	< 6	<1
Gap graded	Usually high	Any (usually <0.5)

Source: Reproduced from CEN EN ISO 14688-2, *Geotechnical Investigation and Testing – Identification and Classification of Soil. Part 2: Principles for a Classification*, Brussels: CEN, 2013. With permission (© CEN).

The liquid and plastic limits are known as Atterberg limits.

The numerical deference between the liquid limit and the plastic limit is defined as the plasticity index (I_P); that is, $I_P = w_L - w_P$, or PI = LL – PL.

The Atterberg limits and plasticity index are used in several engineering classification systems to characterise the fine-grained fractions of soils and to specify the fine-grained fraction of construction materials. They are also used to correlate with compressibility, permeability, compactibility, shrinkage and shear strength.

The liquid limit test is carried out using a liquid limit device also known as a Casagrande apparatus. The test is performed on a sample passing through a 0.425 mm sieve and in accordance with ASTM D 4318 (2010) or AASHTO T 89 (2013).

The soil material containing a certain amount of water is spread in a brass cup and divided into two portions by a grooving tool. Then, by repeatedly dropping the cup, the material flows close to the groove. The number of blows (required to be between 15 and 35) with the corresponding water content is recorded and plotted on a semilogarithmic graph with the water content being a coordinate on the arithmetic scale. From the straight line obtained, the liquid limit is the water content corresponding to the 25-drop abscissa.

The values for liquid limit vary widely. However, for clay material, the liquid limit values are expected to be within 40% to 50%, while for clay-slit, values are expected to be between within 25% to 50%.

The plastic limit, according to ASTM D 4318 (2010) or AASHTO T 90 (2008), is performed on a mass of a 20 g sample prepared for the liquid limit test. The water content in the sample is reduced so that it can be rolled without sticking to the hands. A smaller mass of approximately 2 g is rolled between the palm or fingers and a ground-glass plate to roll the mass into a thread of uniform diameter throughout its length. The thread is further rolled so its diameter reaches 3.2 mm within 2 min. Then, break the thread into several pieces. Squeeze the pieces together, knead the thumb and the finger and re-form into an ellipsoidal mass and reroll. Continue this alternate rolling to a thread of 3.2 mm diameter and repeat the procedure until no thread can be formed. Collect the portions of the crumbled thread and determine the water content. The average value of two water content determinations is reported as the plastic limit of the material tested.

Typical values of plastic limit for silts and clays are between 5% and 30%, with the silty soils having lower values. Materials that cannot be rolled to a 3.2 mm thread are characterised as non-plastic materials.

The plasticity index (I_P or PI) is the most commonly used parameter in pavement engineering. Soil materials with a high plasticity index value are unsuitable for pavement foundation. Examples of such materials include all clayey, silty and sand-silt materials. Table 1.3 gives indicative values of the plasticity index that are usually used, in relation to the degree of plasticity of the soil material.

Table 1.3 Plasticity of soils related to plastic index

Soil material	I_P (PI)	Characteristics of soil in a dry stage
Very high plasticity	>35 or >40	High cohesiveness, almost impossible to break lumps by hand when dry
High plasticity	16–35 or 20–40	Medium to high cohesiveness, difficult to break lumps by hand
Medium plasticity	7–15 or 10–20	Low to medium cohesiveness, lumps break with low hand pressure
Low plasticity	4–6 or 5–10	Low cohesiveness, easy to break lumps by hand
Slight plasticity to non-plastic	0–3 or 0–5	Very low to noncohesiveness, lumps break by contact

Table 1.4 Variation of characteristic soil properties with respect to PI and LL

Characteristic	Comparing soils of equal w_L with I_P increasing	Comparing soils of equal I_P with w_L increasing
Compressibility	Approximately the same	Increases
Permeability	Decreases	Increases
Rate of volume change	Increases	—
Dry strength	Increases	Increases

Source: O'Flaherty C.A., *Highways: The Location, Design, Construction and Maintenance of Pavements*, 4th Edition. Burlington, MA: Butterworth-Heinemann, 2002.

The plasticity index, apart from being used to classify soils into groups or categories, in relation to the liquid limit, makes it possible to comparatively assess soils in terms of compressibility, permeability, volume change and dry strength (cohesiveness in dry conditions). Table 1.4 gives the expected changes on these properties when either I_P (PI) or w_L (LL) varies.

Apart from the above commonly used soil characteristic parameters, there are also the shrinkage factors, such as shrinkage limit (S), shrinkage ratio (R), the volumetric change (VC) and linear shrinkage (LS). These characteristic parameters are not often used in pavement engineering projects. A detailed description of the test procedure to determine the above parameters can be found in AASHTO T 92 (2009).

1.4.3 Relationships between I_P, w_L and w_P

Two other characteristic parameters that are rarely used in pavement engineering are the liquidity index (I_L) and the consistency index (I_C). These indexes are defined by the following equations:

$$I_L = (w - w_P)/I_P$$

and

$$I_C = (w_L - w)/I_P,$$

where w is the natural water content, w_P is the plastic limit, w_L is the liquid limit and I_P is the plasticity index.

The above indexes, especially I_C, are mostly applied to clayey soils in order to estimate cohesiveness. Clayey soils at their natural water content may be in solid condition, if $I_C > 1$; in plastic condition, if I_C is between 0 and 1; or in liquid condition, if $I_C < 0$.

1.4.4 Alternative method to determine the liquid limit – cone penetrometer

In the early part of 1970, another apparatus to determine the liquid limit – the cone penetrometer – was developed and is still being used in some countries. Determining the liquid limit using this apparatus seems to be rather easier, because it does not depend on the experience of the person executing the liquid test and the results obtained have better repeatability.

The test consists of the measurement of the penetration depth of a standard cone, with a mass of 80 g, forced into the soil material. By carrying out measurements at different moisture contents, a diagram similar to the number of blows versus moisture content is obtained. The liquid limit is determined as the moisture content at which the penetration depth is equal to 20 mm.

The results obtained using this apparatus might be slightly different from those determined using the Casagrande device. However, the differences are not significant. A detailed description of the test is given in BS 1377: Part 2 (1990).

1.4.5 Moisture–laboratory density relationship

Pavements are constructed on soil materials that, prior to the construction of the overlying layers, must be compacted sufficiently. Insufficient compaction will cause settlement of the subgrade after a certain period.

The compaction of the soil material is directly related to the moisture content of the subgrade soil at compaction. Unless the moisture content is at its optimum value during compaction, maximum density cannot be achieved. The determination of the optimum water content is carried out in the laboratory using various methods, the most common of which is the proctor method (compaction).

1.4.6 Proctor compaction test (modified) by American standards

The proctor compaction test defines the relationship between soil density and moisture content with the aim of determining the maximum density at a certain moisture content, known as the optimum moisture content.

According to ASTM D 1557 (2012) or AASHTO T 180 (2010), a sufficient mass of soil material is mixed well with water, is transferred to a mould and is compacted by a dropping rammer. The mass of the rammer is 4.54 kg and the height of the drop is 457.2 mm. The term *modified test* is used since the standard mass of 2.5 kg and the drop height have been modified from those originally assigned.

Modification was necessary since the development and use of heavier rollers resulted in higher site densities than laboratory densities at even lower optimum water content. The phenomenon is observed mostly in cohesive soils. Table 1.5 shows the differences observed between a standard and a modified proctor when specimens were compacted in the laboratory using either compaction method.

The sizes of the moulds used are 101.6 or 152.4 mm (diameter) by 177.8 mm (height). The 101.6 mm mould is used only when the percentage retained on the 4.75 mm sieve is less than 7%. However, the 152.4 mm mould may be used in all cases.

The soil sample should have particles all passing through the 19.0 mm sieve. In case the soil contains particles retained on the 19.0 mm sieve and its percentage is 10% to 30%, the corresponding percentage retained shall be replaced by an equivalent mass of material

Table 1.5 Typical values of optimum moisture content obtained by the modified proctor compaction method

Type of soil material	Modified proctor compaction method		Standard proctor compaction method	
	Maximum dry density (kg/m³)	Optimum moisture content (%)	Maximum dry density (kg/m³)	Optimum moisture content (%)
Clayey soil	1875	18	1555	28
Silty clay	1945	12	1670	21
Sandy clay	2055	11	1840	14
Sand	2085	9	1940	11
Gravels, sand and clay	2200	8	2070	9

Source: O'Flaherty C.A., *Highways: The Location, Design, Construction and Maintenance of Pavements*, 4th Edition. Burlington, MA: Butterworth-Heinemann, 2002.

passing through the 19.0 mm sieve and retained on the 4.75 mm sieve. When the percentage retained on the 19.0 mm sieve is higher than 30%, follow the test procedure in accordance with ASTM D 4718 (2007).

The compaction of specimens is carried out in five layers of approximately 127 mm thickness. Each layer is compacted by 56 well-distributed blows. In the case where the small mould is used, the layers shall be five, but each layer is compacted by 25 blows. Once the compaction has been completed, the surface is flattened with a straight edge and the material with the mould is weighed. Then, a representative sample is collected for moisture content determination. The moisture content of the sample (w), as well as the dry specific gravity (apparent) (ρ_d) of the compacted soil, is calculated using the following equations:

$$w = [(A - B)/(B - C)] \times 100$$

and

$$\rho_d = \rho_m/(1 + w/100),$$

where w is the moisture content (%), A is the mass of the container and wet specimen (g), B is the mass of the container and oven dry specimen (g), C is the mass of the container (g), ρ_d is the dry specific gravity of the compacted specimen (kg/m^3) and ρ_m is the moist specific gravity of the compacted specimen (kg/m^3).

The moist specific gravity of the compacted soil specimen (ρ_m) is calculated using the relationship

$$\rho_m = M/V,$$

where M is the mass of the moist soil compacted specimen in the mould (kg) and V is the volume of the compaction mould (see Annex A1, ASTM D 1557 2012).

Once the above procedure is repeated for different moisture contents (two specimens per moisture content), the pair of values (moisture content, dry density) are plotted on linear-scale x–y axes. Joining the points, a curve like the one presented in Figure 1.4, curve A, is obtained.

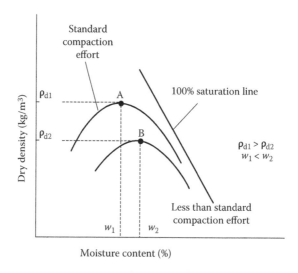

Figure 1.4 Graphical representation of dry density versus moist content.

The moisture content that gives the maximum dry density is defined as 'optimum moisture' for maximum compaction. Needless to say, the use of less compaction effort would result in different optimum moisture contents and, of course, less dry compacted density (see curve B, Figure 1.4).

Instead of dropping rammer, a vibration rammer may be used, which perhaps better simulates the compaction applied on site (vibration rollers). This type of compaction is suggested in a similar procedure proposed in BS 1377: Part 4 (1990).

1.4.7 Proctor compaction test by European standards

The modified proctor compaction test procedure in accordance with CEN EN 13286-2 (2012) slightly differs from the one described in ASTM D 1557 (2012) or AASHTO T 180 (2010).

The key differences include the maximum size of the particles of the soil to be tested, the sizes of the moulds and the mass and diameter of the rammer and its dropping height, in case a large-sized mould (250 mm) is used.

The recommended particle sizes, the percentage passing and the type of mould to be used are given in Table 1.6.

The standard mould size is the one with 150 mm diameter. However, any mould can be used provided its diameter is four times the maximum size (D) of the particles in the soil sample. Finally, when the percentage of particles retained in the 63 mm sieve is greater than 25%, the test is inappropriate.

The number of specimens required is at least five or at least three when the material is known.

When the 250 mm diameter mould is used, the mass and diameter of the rammer are 15 kg and 125 mm, respectively. The drop height differs and it is 600 mm instead of 457 mm when a 100 mm or a 150 mm diameter mould is used.

Additionally, in the case of the 100 mm diameter mould, the number of blows per layer is 25, instead of 56 when a 150 mm diameter mould is used. In the case of the 250 mm diameter mould, the number of blows is 98 and the number of layers is decreased to 3.

As for the height of the moulds, it is 120 mm for the 150 mm or 100 mm diameter moulds and 200 mm for the 250 mm diameter mould.

1.4.8 Moisture condition value test

As mentioned earlier, there is a strong relationship between compaction effort, moisture content and density.

Table 1.6 Size of mould and required mass of material for proctor test

Percentage passing (%), per sieve			Mass required (kg)	Mould diameter, mm (type)
16 mm	31.6 mm	63 mm		
100	—	—	15	100 (A)
			40	150 (B)
75–100	100	—	40	150 (B)
<100	75–100	100	40	150 (B)
—	<75	75–100	200	250 (C)

Source: Reproduced from CEN EN 13286-2, *Unbound and Hydraulically Bound Mixtures, Part 2: Test Methods for the Determination of the Laboratory Reference Density and Water Content – Proctor Compaction*, Brussels: CEN, 2012. With permission (© CEN).

The moisture condition value (MCV) test involves testing a soil material at fixed moisture content but by different compaction methods (number of blows or by a vibratory rammer). Thus, the compaction effort beyond which there is no further increase in density is determined. This test is much newer than the proctor test; it has been developed in the United Kingdom (Parsons 1992) and is also used by some other countries.

In fact, the test is carried out for quick acceptance control of materials (soils or stabilised) regarding their natural moisture content for effective compaction.

The device used is similar to the proctor compaction device, but it is smaller, manually operated and easily transported.

The device consists of a metal mould (100 mm in diameter and 200 mm in height) and a rammer (97 mm in diameter and 7 kg in mass), which drops from a standard height of 250 mm on the soil specimen surface.

After the initial load application, the rammer falls freely on the soil specimen, and after a predetermined number of blows, the penetration of the rammer is measured. The process is carried out repeatedly until no additional rammer penetration is observed. On completion, the soil sample is taken for redetermination of the moisture content.

The rammer penetration for any n blows is compared with a penetration for $4n$ blows, and their difference, which in fact constitutes an indication of change in density, is recorded as change in penetration. The data recorded, the number of blows and the change in penetration are represented in a semilogarithmic coordinate system (x-axis is the logarithmic axis, which represents the number of blows). The MCV is defined by the following equation:

$$MCV = 10 \times \log B,$$

where B is the number of blows, where the change in penetration is 5 mm.

The test is repeated for a number of specimens with different moisture content. Thus, each time, a different MCV value is determined. When the MCV values and the moisture content are plotted, a straight line is obtained. This is useful for the quick determination of the natural moisture of the material in situ provided that the relatively quick MCV test is executed on site (almost half an hour is required). The test is described in detail in BS 1377: Part 4 (1990).

More related details can be found in Parsons (1976), SDD AG 1 (1989), BS 1377: Part 4 (1990) and Parsons (1992).

As a general guide, an MCV of 8.5 is recommended to be the lower limit of acceptability of a soil material compacted in its natural moisture content. Specific conditions may require that the limit of 8.5 be lowered or raised marginally (SDD AG 1 1989).

1.5 SOIL CLASSIFICATION

The soil fractions as described in Section 1.2 along with the description of natural properties (such as origin, colour, shape, etc.) did not help engineers to easily recognise the soil's suitability for roadworks. As early as 1928, the AASHTO developed a soil classification system for highway engineering. The system has been revised several times, and the 1945 version formed the basis of the existing soil classification system by AASHTO.

A necessity was also recognised, more than 60 years ago, for runway pavements where the construction had to be more meticulous because of heavier loads applied to the pavement. As a consequence, in the 1940s, Casagrande developed another classification system for airfields (Taylor 1948). It was then named as Unified Soil Classification System

(USCS) and started to be used in all civil engineering works. In 1966, it was adopted by the ASTM, and today, it is also known as the ASTM classification system (ASTM D 2487 2011).

Since then, other countries adopted or developed their own soil classification systems.

In this book, the AASHTO, the ASTM, the European and the UK soil classification systems will be presented in brief.

All the abovementioned systems are based on particle size distribution and plasticity (plasticity index and liquid or plastic limit). The main variation among them lies on the designation of coarse (or very coarse) and fine soils.

The AASHTO soil classification system considers coarse soil materials (boulders, cobbles, gravels and sands) as those in which ≥65% of their mass is retained on a 0.0075 mm sieve. Fine soil materials (silts and clays) are those in which >35% of their mass passes through a 0.075 mm sieve.

The ASTM soil classification system considers coarse soil materials (boulders, cobbles, gravels and sands) as those in which >50% of their mass is retained on a 0.075 mm sieve. Fine soil materials are those in which ≥50% of their mass passes through a 0.075 mm sieve.

The European soil classification system distinguishes very coarse materials (boulders and cobbles) as those in which most particles are retained on a 63 mm sieve, coarse materials (gravels and sands) as those in which most particles are retained on a 0.063 mm sieve and fine soil materials as those with low plasticity and are dilatant (silts) or those that are plastic and non-dilatant (clays).

The basic UK soil classification system is similar to the ASTM classification system.

1.5.1 AASHTO soil classification system

The soil classification system is described in detail in AASHTO M 145 (2012).

The coarse soil material is divided into three major groups (A-1, A-2 and A-3) and seven subgroups (A-1-a, A-1-b, A-3, A-2-4, A-2-5, A-2-6 and A-2-7), depending on the retained percentage in certain sieves, the liquid limit and the plasticity index. The A-1 group is considered to be the coarsest, whereas the A-3 group is the least coarse.

Fine soil material is divided into four major groups (A-4, A-5, A-6 and A-7) and only A-7 is further divided into two subgroups (A-7-5 and A-7-6). The A-4 and A-5 groups refer to silts, while the A-6 and A-7 groups refer to clays.

The analytical soil classification table is shown in Table 1.7.

To classify soil material in groups, the percentage passing through a particular sieve, the liquid limit and the plasticity index must be known. In certain cases, the group index is also determined, which is notified at the end of the group categorisation. For instance, soil material under category A-2-6(3) means that the fine fraction (passing through the 0.075 mm sieve) is ≤35, the liquid limit is ≤40, the plasticity index is >10 (≥11) and the group index was found to be 3.

1.5.1.1 Group index of soils

The group index of soils (GI) indirectly expresses the bearing capacity of soil material. The values obtained are higher than zero (0). Low values from 0 to 4 indicate a soil material with good bearing capacity, whereas values >8 indicate a soil material with bearing capacity varying from low to bad.

Under no circumstances does the determination of group index exempt the engineer from not determining the bearing capacity of the soil material by other laboratory or in situ tests.

Table 1.7 AASHTO soil classification system

General classification	Coarse soil material, percentage passing through the 0.075 mm sieve <35%							Fine soil material (silt–clayey), percentage passing through the 0.075 mm sieve >35%				
Classification group	A-1		A-3	A-2				A-4	A-5	A-6	A-7	
Subgroup	A-1-a	A-1-b	A-3	A-2-4	A-2-5	A-2-6	A-2-7	A-4	A-5	A-6	A-7-5	A-7-6
% Passing from sieve												
2.00 mm	50 max	—	—	—	—	—	—	—	—	—	—	—
0.425 mm	30 max	50 max	51 min	—	—	—	—	—	—	—	—	—
0.075 mm	15 max	25 max	10 max	35 max	35 max	35 max	35 max	36 min	36 min	36 min	36 min	36 min
Characteristics (passing through the 0.425 mm sieve)												
Liquid limit	—	—	—	40 max	41 min	40 max	41 min	40 max	41 min	40 max	41 min	41 min
Plasticity index	6 max		N.P.	10 max	10 max	11 min	11 min	10 max	10 max	11 min	11 min[a]	11 min[b]
Typical types of soils	Gravels, sand and sand gravels		Fine sand	Silty or clayey sand gravels				Silty soils			Clayey soils	
Suitability as subgrade	Excellent to good							Fair to poor				

Source: AASHTO M 145, Classification of Soils and Soil-Aggregate Mixtures for Highway Construction Purposes, Washington, DC: AASHTO, 2012. With permission.

[a] The plasticity index of the materials in subgroup A-7-5 is equal to or less than (LL − 30), where LL is the liquid limit.
[b] The plasticity index of the materials in subgroup A-7-6 is greater than (LL − 30), where LL is the liquid limit.

The group index of a soil material depends on the liquid limit, plasticity index and the percentage passing through the 0.075 mm sieve. The relationship used to calculate the GI is as follows:

$$GI = (F - 35) \times [0.2 + 0.005 \times (LL - 40)] + 0.01 \times (F - 15) \times (PI - 10),$$

where F is the percentage passing through the 0.075 mm sieve (expressed as an integer), LL (or w_L) is the liquid limit and PI (or I_p) is the plasticity index.

The value of group index is expressed as an integer; if negative values are obtained, GI is zero.

1.5.2 Unified soil classification system (ASTM system)

The USCS or ASTM soil classification system is described in detail in ASTM D 2417.

In this system, the soils are divided into groups: (a) coarse soils, (b) fine soils (silts and clay) and (c) highly organic soils (peat).

As mentioned, coarse soils are those in which a mass of >50% is retained on a 0.075 mm sieve and fines are those in which ≥50% of its mass passes through a 0.075 mm sieve.

Coarse soils (gravels and sands) are further divided into eight subgroups (GW, GP, GM, GC, SW, SP, SM and SC), starting with very coarse (GW) to least coarse (SC). The fine soil materials are divided into six subgroups (ML, CL, OL, MH, CH and OH), ranked in order of decreasing fine particle size. Highly organic fine soils, peat, are classified in one group under the notification (P_t).

The two letters used for notifying the subgroup derive from a combination of the following letters: 'G' for gravel, 'S' for sand, 'M' for silt, 'C' for clay, 'W' for well graded, 'P' for poorly graded (mainly single-size particles), 'H' for high plasticity index, 'L' for low plasticity index and 'O' for organic. As a consequence, 'GW' symbolises well-graded gravel, whereas 'SP' denotes poorly graded sand (almost single sized). Further explanation is given in Table 1.8.

Awareness of attributes of particle size distribution (i.e. percentage passing through a 0.075 mm sieve, d_{10}, d_{60} and d_{30}), either coefficient of uniformity, C_U, or coefficient of curvature, C_C, as well as plasticity index (PI), is required for the classification of soils in some subgroups.

The well-graded materials, which generally plot as a smooth and regular curve, do not lack of any size of aggregate. The uniformity coefficients (C_U) of well-graded gravels are greater than 4 and those of well-graded sands are greater than 6.

The coefficient of curvature (C_C) ensures that the particle size distribution will have a concave curvature within relatively narrow limits for a given d_{60} and d_{10} combination. As for well-graded gravels, C_C values should range from 1 to 3.

The analytical table of the USCS or ASTM soil classification system (ASTM D 2487 2011) is given in Table 1.8 (see also Figure 1.5).

Other tables related to USCS have been developed to ease the work of engineers. One of them is shown in Table 1.9. A similar table has been published in BS 6031 (1981).

1.5.3 The European classification system

The European classification system is prescribed in EN ISO 14688-2 (2013) in conjunction with CEN EN ISO 14688-1 (2013). According to EN ISO 14688-2 (2013), soil classification is the assignment of soil into groups on the basis of certain characteristics, criteria and genesis.

The classification principles established are applicable to natural soils as well as similar man-made material in situ and re-deposited, to be used in all engineering works (roads, ground improvements, embankments, drainage systems, foundations and dams).

Table 1.8 USCS chart

Type of soil material	Criteria for classification (based on materials passing through the 75 mm [3-inch] sieve)			Group symbol	Group name[a]
Coarse soils, >50% retained on the 0.075 mm (No. 200) sieve	Gravels >50% retained on the 4.76 mm (No. 4) sieve	Clean gravels	$C_U \geq 4$ and $1 \leq C_C \leq 3^b$	GW	Well-graded gravel[c]
		With <5% fines[d]	$C_U > 4$ and/or $1 > C_C > 3^b$	GP	Poorly graded gravel[c]
		Gravels with fines	Fines: ML or MH	GM	Silty gravel[c,e,f]
		With >12% fines[d]	Fines: CL or CH	GC	Clayey gravel[c,e,f]
	Sands ≥50% passing through the 4.76 mm (No. 4) sieve	Clean sand	$C_U \geq 6$ and $1 \leq C_C \leq 3^b$	SW	Well-graded sand[g]
		With <5% fines	$C_U < 6$ and/or $1 > C_C > 3^b$	SP	Poorly graded sand[g]
		Sand with fines	Fines: ML ⇒ MH	SM	Silty sand[e,f,g]
		With >12% fines[h]	Fines: CL ⇒ CH	SC	Clayey sand[e,f,g]
Fine soils, ≥50% passing through the 0.075 mm (No. 200) sieve	Silts and clays with liquid limit <50	Inorganic	PI > 7 and plots on or above the 'A' line[i]	CL	Lean clay[j,k,l]
			PI < 4 or plots below the 'A' line[i]	ML	Silt[j,k,l]
		Organic	LL (dried)/LL (not dried) <0.75	OL	Organic clay[j,k,l,m] or organic silt[j,k,l,n]
	Silts and clays with liquid limit ≥50	Inorganic	PI plots on or above the 'A' line	CH	Fat clay[j,k,l]
			PI plots below the 'A' line	MH	Elastic silt[j,k,l]
		Organic	LL (dried)/LL (not dried) <0.75	OH	Organic clay[j,k,l,o] Organic silt[j,k,l,p]
Highly organic soils	Primarily organic matter, dark in colour and organic odor			PT	Peat

Source: Reprinted from ASTM D 2487, *Standard Practice for Classification of Soils for Engineering Purposes (Unified Soil Classification System)*, West Conshohocken, PA: ASTM International, 2011. With permission (© ASTM International).

a   If samples contain cobbles or boulders, or both, add 'with cobbles or boulders, or both' to the group name.
b $C_U = D_{60}/D_{10}, C_C = (D_{30})2/D_{10} \times D_{60}$.
c   If soil contains >15% sand, add 'with sand' to group name.
d   Gravels with 5% to 12% fine require dual symbols, such as GW–GM, well-graded gravel with silt; GW–GC, well-graded gravel with clay; GP–GM, poorly graded gravel with silt; GP–GC, poorly graded gravel with clay.
e   If fines classify as CL–ML, use dual symbol GC–GM or SC–SM.
f   If fines are organic, add 'with organic fines' to group name.
g   If soils contains ≥15% gravel, add 'with gravel' to group name.
h   Sands with 5% to 12% fines require dual symbols, such as SW–SM, well-graded sand with silt; SW–SC, well-graded sand with clay; SP–SM, poorly graded sand with silt; SP–SC, poorly graded sand with clay.
i   If Atterberg limits plot in hatched area, soil is a CL–ML, silty clay.
j   If soil contains 15% to 29% retained on sieve No. 200, add 'with sand' or 'with gravel', whichever is predominant.
k   If soil contains ≥30% retained on sieve No. 200, predominately sand, add 'sand' to group name.
l   If soil contains ≥30% retained on sieve No. 200, predominately gravel, add 'gravelly' to group name.
m   PI ≥ 4 and plots on or above 'A' line (see Figure 1.5).
n   PI < 4 or plots below 'A' line (see Figure 1.5).
o   PI plots on or above 'A' line (see Figure 1.5).
p   PI plots below 'A' line (see Figure 1.5).

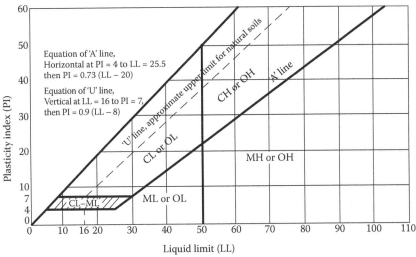

Figure 1.5 Plasticity index (PI) graph with liquid limit (LL). (Reprinted from ASTM D 2487, *Standard Practice for Classification of Soils for Engineering Purposes [Unified Soil Classification System]*, West Conshohocken, PA: ASTM International, 2011. With permission [© ASTM International].)

The European soil classification system distinguishes the following groups:

a. Very coarse materials, subdivided into boulders (Bo), those in which most particles are retained on the 200 mm sieve, and cobbles (Co), those in which most of the materials are retained on the 63 mm sieve.
b. Coarse materials, subdivided into gravels (Gr), those in which most particles are retained on the 2 mm sieve, and sands (Sa), those in which most particles are retained on the 0.063 mm sieve.
c. Fine materials, subdivided into silts (Si), those with low plasticity and that are dilatant, and clays (Cl), those with intermediate to high plasticity and that are non-dilatant.

The denomination of the materials into groups with similar properties is carried out using the principal component (Bo, Co, Gr, Sa, Si and Cl) at the end and the secondary or tertiary component before. As an example, the denomination 'sasiGr' stands for sand–silt–gravel.

Based on the above, Table 1.10 has been proposed. Table 1.10 shows further subdivision of the main soil subgroups.

If the soil materials are to be distinguished on the basis of the contents of the main fractions (gravel, sands, silts and clays) alone, Table 1.11 can be used.

Other quantifying terms that can be used to describe soils are also suggested and these are as follows: for sands and gravels, the density index (I_D); for fine soils, the untrained shear strength (c_u); for silts and clays, the consistency index (I_C).

Finally, some other parameters may be used for soil classification for specific purposes, namely: dry density, clay activity, mineralogical nature, saturation index, permeability, compressibility index (C_C), swelling index and carbonate index.

More details on all the above can be found in CEN EN ISO 14688-2 (2013) and CEN EN ISO 14688-1 (2013).

Table 1.9 Characteristics of soils for construction works in roads and airfields

Soil groups			Key attributes						
			Suitability as subgrade	Suitability subbase	Potential frost action	Compressibility and expansion	Drainage characteristics	Typical values[a] CBR (k value)[b]
Gravel and gravelly soils	GW		Well-graded gravels or gravel-sand mixtures, little or no fines	Excellent	Excellent	None to very slight	Almost none	Excellent	40–80 (80–135)
	GP		Poorly graded gravels or gravel-sand mixtures, little or no fines	Good to excellent	Good	None to very slight	Almost none	Excellent	30–60 (80–135)
	GM	d	Silty gravels, silty sand-gravels	Good to excellent	Good	Slight to medium	Very slight	Fair to poor	40–60 (80–135)
		u		Good	Fair	Slight to medium	Slight	Poor to practically impervious	20–30 (55–135)
	GC		Clayey gravels, gravel-sand clay mixtures	Good	Fair	Slight to medium	Slight	Poor to practically impervious	20–40 (55–135)
Sands and sandy soils	SW		Well-graded sands or gravelly sands, little or no fines	Good	Fair to good	None to very slight	Almost none	Excellent	20–40 (55–110)
	SP		Poorly graded sands or gravelly sands, little or no fines	Fair to good	Fair	None to very slight	Almost none	Excellent	10–40 (40–110)
	SM	d	Silty sands, sand and silt mixture	Fair to good	Fair to good	Slight to high	Very slight	Fair to poor	15–40 (40–110)
		u		Fair	Poor to fair	Slight to high	Slight to medium	Poor to practically impervious	10–20 (27–80)
	SC		Clayey sands, sand-clay mixture	Poor to fair	Poor	Slight to high	Slight to medium	Poor to practically impervious	5–20 (27–80)

Major division		Symbol	Description						k value[b]
Silts and clays	LL <50	ML	Inorganic silts and very fine sands, silty or clayey fine sands or clayey silts with slight plasticity	Poor to fair	Not suitable	Medium to very high	Slight to medium	Fair to poor	15 (27–55)
		CL	Inorganic clays of medium to low PI, gravelly clays, sandy clays, silty clays, lean clays	Poor to fair	Not suitable	Medium to high	Medium	Practically impervious	15 (14–40)
		OL	Organic silts and organic silt-clays of low plasticity	Poor	Not suitable	Medium to high	Medium to high	Poor	5 (14–40)
	LL >50	MH	Inorganic silts, micaceous or diatomaceous fine sandy or silty soils, elastic silts	Poor	Not suitable	Medium to very high	High	Fair to poor	10 (14–40)
		CH	Inorganic clays of high plasticity, fat clays	Poor to fair	Not suitable	Medium	High	Practically impervious	15 (14–40)
		OH	Organic clays of medium to high plasticity, organic silts	Poor to very poor	Not suitable	Medium	High	Practically impervious	5 (7–27)
Org.		PT	Peat and other highly organic soils	Not suitable	Not suitable	Slight	Very high	Fair to poor	—

Source: Adapted from Asphalt Institute MS-10, *Soils Manual. Manual Series No. 10*. Lexington, KY: Asphalt Institute. With permission.

[a] Design values.
[b] k value in N/cm²/cm (N/cm³).

Table 1.10 Principles of classification of soil in accordance with the european specification

Criterion	Soil group	Quantification	Denomination into groups of similar properties			Further subdivision as appropriate by
Wet soil does not stick together	Very coarse	Most particles >200 mm	Bo	xᵃBo boCo	coBo	Requires special consideration
		Most particles >63 mm	Co	saCo, grCo	sagrGr	
	Coarse	Most particles >2 mm	Gr	coGr saGr, grSa	cosaGr sasiGr, grsiSa	Particle size (grading)
		Most particles >0.063 mm	Sa	siGr, clGr orSa	siSa, clSa, saclGr	Shape of grading curve Relative density Permeability (Mineralogy and shape)
Wet soil sticks together	Fine	Low plasticity Dilatant	Si	saSi clSi, siCl	sagrSi saclSi	Plasticity Water content
		Plastic Non-dilatant	Cl	orSi, orCl	sagrCl	Strength, sensitivity Compressibility, stiffness (Clay mineralogy)
Dark colour, low density	Organic		Or	soar, siOr	clOr	Requires special consideration
Not naturally	Made ground	Deposited	Mg	xᵃMgᵇ	Man-made material	Requires special consideration
					Relaid natural materials	As for natural soils

Source: Reproduced from CEN EN ISO 14688-2, *Geotechnical Investigation and Testing – Identification and Classification of Soil. Part 2: Principles for a Classification*, Brussels: CEN, 2013. With permission (© CEN).

ᵃ x, any combination of components.
ᵇ Mg, made ground components.

Table 1.11 Division of mineral soils based on contents of fractions

Fraction	Content of fraction in wt% of material ≤63 mm	Content of fraction in wt% of material ≤0.063 mm	Name of soil	
			Modifying term	Main term
Gravel	20 to 40		Gravelly	
	>40			Gravel
Sand	20 to 40		Sandy	
	>40			Sand
Silt + clay (fine soil)	5 to 15	<20	Slightly silty	
		≥20	Slightly clayey	
	15 to 40	<20	Silty	
		≥20	Clayey	
	>40	<10		Silt
		10 to 20	Clayey	Silt
		20 to 40	Silt	Clay
		>40		Clay

Source: Reproduced from CEN EN ISO 14688-2, *Geotechnical Investigation and Testing – Identification and Classification of Soil. Part 2: Principles for a Classification*, Brussels: CEN, 2013. With permission (© CEN).

1.5.4 The UK soil classification systems

The basic UK soil classification system is based on and very similar to USCS. Details can be found in Dumbleton (1981).

The Highways Agency in the United Kingdom has developed another classification system of materials for highway earthworks. This system, apart from soil materials, also includes stabilised soils or any combination of materials (Highways Agency 2009a).

The proposed Highways Agency classification system also provides information for typical use of materials and property requirements, together with acceptability limits and compaction requirements.

The soil materials, as well as the stabilised soil materials or any other material or combination of materials for highway earthworks, are grouped into nine general categories denoted by numbers from 1 to 9. Each number corresponds to the following general category:

1. Category of materials for general granular fill
2. Category of materials for general cohesive fill
3. Category of materials for general chalk fill
4. Category of materials for landscape fill
5. Category of materials for topsoil
6. Category of materials for selected granular fill
7. Category of materials for selected cohesive fill
8. Category of materials for miscellaneous fill
9. Category of materials for stabilised materials

The first level of subclassification is notified by a letter from A to S, and the second level of subclassification, where needed, is notified by a number from 1 to 5. For example, class 6F2 refers to a selected granular fill material, coarse grading, for capping.

The classification table developed is known as Table 6/1 and consists of 32 pages. Details regarding Table 6/1 and many more can be found in the manual MCHW, Vol. 1-Series 600 and MCHW Vol. 2-Series NG 600 (Highways Agency 2009 a,c).

1.6 SOIL BEARING CAPACITY TESTS

The bearing capacity or the strength of the soil material over which the pavement is going to be constructed is expressed in pavement engineering by reference either to the California bearing ratio (CBR) or to the modulus of subgrade reaction, k. These parameters are very useful since they are utilised by many pavement design methodologies.

Apart from the abovementioned parameters, the resistance R value is also used. However, the resistance R value is indicative of performance when untreated or treated soil material is used.

1.6.1 CBR laboratory test

The method was developed by the California Road Service in the 1930s and has since been adopted by many administrations/organisations around the world. The original test method has been slightly modified by some organisations. The modifications were mainly concerned on the moisture at compaction and on the compaction effort. The procedure, however, in determining the CBR value remained unchanged.

The test determines the soil's bearing capacity from laboratory-compacted specimen, expressed as CBR.

CBR is defined as the ratio of the load required to cause a certain penetration of the plunger into the soil material to the load required to obtain the same penetration on a specimen of standard material.

The CBR laboratory test can be carried out by ASTM D 1883 (2007), AASHTO T 193 (2013) or CEN EN 13286-47 (2012); other equivalent national specifications may have been developed.

1.6.1.1 CBR test procedure according to ASTM D 1883 (2007)

1.6.1.1.1 Compaction of specimen

The soil material, after being pulverised so as no aggregations (lumps) exist, passed through a 19.0 mm sieve. In case there are remaining aggregates on the 19.0 mm sieve, an equivalent mass is replaced from the same soil material passing through the 19.0 mm sieve but was retained on the 4.75 mm sieve. This replacement is necessary for the validity of the result when coarse soil material is tested.

A sufficient amount of moist soil material is prepared at the desired moisture content (usually at the optimum moisture content). The total amount of material required to perform the test is approximately 35 kg.

A test portion is placed in cylindrical metal moulds (152.4 mm in diameter and 177.8 mm in height) and is compacted with a rammer similar to the one used in the modified proctor compaction method (weight of rammer, 4.54 kg; diameter, 50.8 mm; height of drop, 457 mm).

The compaction of specimen is carried out in three approximately equal layers of 125 mm compacted thicknesses. Each layer receives a certain number of blows. This number of blows, which is the same for each layer, is determined in such a way that the density reached per specimen is either a little lower or a little higher than the maximum density, determined by the proctor method. In most cases 10, 30 and 65 blows are used for three different CBR determinations.

1.6.1.1.2 Specimen saturation

After compaction and once the metal extension collar is removed, the soil material is flattened by means of a straightedge (knife or spatula) until it is even with the top of the mould. Afterwards, the soil specimen with the steel mould is inverted and clamped on a perforated base plate (a coarse filter paper is added to the interface of the perforated base plate and soil specimen). Sufficient annular weights (usually three) are added on the compacted soil material in the mould in order to produce a surcharge equal to the weight of the overlying layers. Soil material, mould and surcharge weights are immersed in water (water level shall be 25 mm above the specimen's surface). At the same time, a tripod with a dial indicator is placed over the mould and its initial indication is recorded. The whole system remains undisturbed for 96 h.

At the end of the 96 h and after the last indication has been recorded, the whole system is removed from the water bath. The tripod and the surcharge weights are removed and the specimen is allowed to drain downwards for 15 min. After this time, the compacted soil specimen is ready for the CBR determination test. From the measurements of the dial indicator (first and last indication), the swell (as a percentage of the initial height of the specimen) of the soil material is determined.

1.6.1.1.3 Loading–penetration measurement

After placing annular weights on the specimen (slotted metal plates of 2.26 kg weight each, with a bigger slot diameter than the diameter of the loading plunger), the compacted and wet specimen is placed on the CBR apparatus. The device imposes a steadily increased load

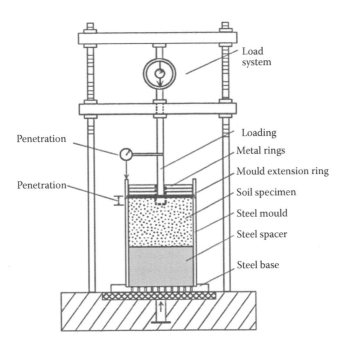

Figure 1.6 CBR apparatus.

through the cylindrical loading plunger (49.63 mm in diameter) with a penetration velocity of 1.3 mm/min. The loading plunger comes in contact with the specimen's surface; the penetration and load measuring gauges are set to zero and loading begins. Load indications are marked at regular intervals of depth of penetration. Once a penetration of approximately 8.0 mm is achieved, the loading stops. A typical layout of a CBR apparatus is given in Figure 1.6.

1.6.1.1.4 CBR determination

The pairs of value penetration of plunger and applied load or penetration of plunger and applied stress are placed in linear coordinates and a curve of form A or B is received (see Figure 1.7). The applied stress or load is related to the resistance in penetration.

If a type A curve is obtained, there is no need for correction before CBR determination. In case a type B curve is obtained, that is, the curve concaves upwards after the initial loading, correction of the curve is required. This correction consists of redetermination of the coordinate's origin.

The new position of the coordinate's origin is defined by the point where the extension of the linear section of the curve intersects with the ordinate's axis (see Figure 1.7).

After correction (if needed), the applied stresses are determined for penetration of 2.50 and 5.00 mm, respectively. The CBR is calculated using the following relationship:

CBR = [(stress for 2.5 or 5.0 mm penetration)/(6.89 or 10.34)] × 100,

where 6.89 or 10.34 is the applied stress (in MPa), which causes the plunger's penetration in a standardised material of 2.5 or 5.0 mm, respectively.

The CBR value is the value that corresponds to a penetration of 2.5 mm, provided that the value received is higher than the one corresponding to a penetration of 5.0 mm. If not, the test is repeated. If the repeated test gives the same outcome, then the value corresponding to a penetration of 5.0 mm is taken as the CBR of the material tested.

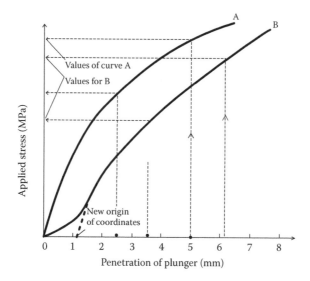

Figure 1.7 Diagram for CBR determination.

Results are usually expressed as an integer, provided that CBR values are higher than 10%. If values are lower than 10%, one decimal point is sufficient.

If the determination of the CBR value is based on the applied load, the values in the denominator of the relationship above are 13.45 or 20.02 kN, corresponding to 2.5 or 5.0 mm penetration, respectively.

This is similar to the determination of the CBR value in AASHTO T 193 (2013).

1.6.1.1.5 Determination of design CBR

Design CBR is usually defined by the relationship between CBR and the dry density of compacted soil material. The results of the three specimens compacted at different compaction efforts (number of blows), and thus having different densities, are presented in Figure 1.8.

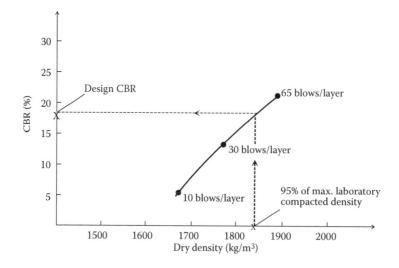

Figure 1.8 Determination of design CBR.

From the resulting curve in Figure 1.8, the design CBR is determined. Usually, the design CBR is the value obtained for density equal to 95% of the maximum density according to the (modified) proctor method. Additional information for the design CBR is given in Section 10.2.2.

1.6.1.2 CBR test procedure according to CEN EN 13286-47 (2012)

The test in accordance with CEN EN 13286-47 (2012) is carried out in materials passing through a 22.4 mm sieve. By carrying out the test, the CBR and the immediate bearing index are determined, in case the latter one is required.

The difference between CBR and immediate bearing index determination is the use or non-use of annular weights. Annular weights on soil material are used for the determination of CBR, regardless of how the specimen is cured. However, this is not the case when it comes to the determination of the immediate bearing index. Apart from the above difference, the rest of the procedure is the same for both tests.

If specimen saturation (of 96 h or more) is required as part of the curing of the specimen, the European standard also includes swelling measurement during specimen saturation as the respective American standard.

The specimen's compaction is carried out using the proctor method in accordance with CEN EN 13286-2 (2012) (normal or modified compaction, whichever is chosen), whereas the diameter of circular piston (50 ± 5 mm), the piston's penetration velocity (1.27 mm/min) and the rest of the procedure are similar to the American standard.

The CBR, as well as the immediate bearing index, is calculated by expressing the force on the piston for a given penetration (2.5 or 5 mm) as a percentage of a reference force (13.2 or 20 kN, for 2.5 or 5 mm penetration, respectively).

The final value is rounded to the nearest 0.5% if the value varies from 0 to 9, to the nearest integer if the value varies from 10 to 29 and to the nearest 5% value if the value is higher than 29.

Additional information is provided in CEN EN 13286-47 (2012).

1.6.1.3 Variations on determining the CBR value

Some countries do not fully abide by the abovementioned procedure (methodology) of CBR determination. The main deviations are the moisture content at compaction, the non-use of specimen saturation before penetration and the manner of compaction (static as against dynamic compaction).

In these cases, moisture content at compaction is not the optimum; the optimum value is the expected moisture content of the soil material during a pavement's service life. This results in not subjecting the specimens to saturation before testing. In most cases, the moisture content at compaction is the maximum expected moisture of the material and the CBR is determined at that moisture content.

Using static instead of dynamic compaction is advantageous since specimens can be compacted in any desirable dry density precisely. It also requires less manual effort.

As a consequence, all compaction and curing procedures for determining the CBR and even more for the determination of the design CBR value must be clearly stated in the contractual requirements.

1.6.2 CBR in situ

The CBR of compacted soil material can also be measured 'in situ'. This provides the engineer quick information as regards the material he has to handle at a specific natural moisture content.

Table 1.12 Comparison between laboratorial and in situ CBR values

Type of soil	Dry density (kg/m³)	Moisture content (%)	CBR values (%) Laboratorial	In situ
Clay (w_L = 69, w_p = 27)	1522	24.8	8.9	7.9
Clay (w_L = 59, w_p = 23)	1538	25.1	3.9	3.0
Silty clay (w_L = 37, w_p = 23)	1746	19.5	2.0	1.2
	1714	19.0	5.0	1.1
	1666	16.1	2.2	2.2
Sandy clay (w_L = 30, w_p = 18)	1522	19.2	2.2	3.1
	1858	12.2	14	7.0
Clayey sand	1826	12.6	10	9.0
	1746	10.0	12	18.0
Well-graded sand	1750	8.0	24	7.5
Crushed slag	2243	4.8	41	44

Source: Adapted from Croney, D., *The Design and Performance of Road Pavements*. Transport and Road Research Laboratory (TRRL), now TRL. London: HMSO, 1977. With permission.

The apparatus used is similar to the one used in the laboratory, with the only difference being the manual application of load. This requires the use of a means for providing reaction/resistance so that the load is applied on the surface. The rest of the procedures (i.e. test execution and eliciting results) are exactly the same as the procedure executed in the laboratory.

A key observation is that CBR values determined in situ always differ from values obtained in the laboratory, for the same moisture content and degree of compaction. This is primarily due to the confinement conditions. The restriction imposed by steel mould in soil material is higher than that of the compacted layer. Hence, the values determined in the laboratory are usually higher than the values obtained in situ. The differences observed are more intense in coarse soils. Table 1.12 shows some typical values obtained in a variety of soil materials.

Moreover, it should be mentioned that the in situ determination of CBR of coarse soils should be avoided, since particle dimensions in relation to the diameter of the plunger also affect the result.

1.6.3 Dynamic cone penetrometer

Dynamic cone penetrometer (DCP) is used to measure quickly the in situ bearing capacity or strength of the subgrade material as well as the weakly bound material. The strength is expressed in penetration rate or index and may be related to in situ CBR value.

One advantage of the DCP over the static cone penetrometer is that coarser and stronger material can be tested (with CBR values up to 100%). Additionally, the layer's thickness and its strength can be identified down to a depth of, normally, 1 m or more.

The device has a very basic construction and can be easily used. It uses a standard 8 kg rammer (4.6 kg optional for weaker materials) dropping through a height of 575 mm and a steel drive rod with a replaceable point or disposable 60° cone tip. During the test, the depth of penetration is recorded every time the load is applied and results are expressed in millimetres per blow, known as the DCP index or rate.

To convert the DCP index to CBR values, correlation equations are used. According to ASTM D 6951 (2009), the correlation equations used are as follows:

for all soils except for CL soils with CBR < 10 and CH soils,

$$CBR = 292/DCP^{1.12};$$

for CL soils with CBR < 10,

$$CBR = 1/(0.07019 \times DCP)^2;$$

and for CH soils,

$$CBR = 1/(0.002871 \times DCP),$$

where DCP is the dynamic cone penetration index (mm/blow).

The Highways Agency in the United Kingdom uses a different correlation equation developed by TRL (Highways Agency 2008), which is as follows:

$$CBR = 10^{(2.48-1.057 \times Log\ P)},$$

where P is the penetration index (or rate) (mm/blow).

It is stated that the accuracy of the above equation reduces for CBR values below 10%.

More details regarding the apparatus and execution of the test can be found in relevant standards such as ASTM D 6951 (2009), CEN EN ISO 22476-2 (2011) and DMRB HD29/08 (Highways Agency 2008).

1.6.4 Correlations between CBR and index properties of soil material

Correlations between CBR and index properties of the soil material or DCP value have been established (National Cooperative Highway Research Program [NCHRP] 2001, 2004) and proposed to be used with the new pavement design procedures of AASHTO known as Mechanistic–Empirical Pavement Design Guide (MEPDG) (AASHTO MEPDG-1 2008).

The correlation between CBR and index properties of soil material is as follows:

For coarse, clean material, typically non-plastic material such as GW, GP, SW and SP, for which wPI = 0:

$$CBR = 28.09\ (D_{60})^{0.358}.$$

For soils containing more than 12% fines and exhibiting some plasticity, such as GM, GC, SM, SC, ML, MH, CL and CH, for which wPI other than 0:

$$CBR = 75/[1 + 0.728 \times (wPI)],$$

where CBR is the California bearing ratio (%), D_{60} is the diameter at 60% passing from particle size distribution (mm) and wPI is the weighted plasticity index (= $P_{200} \times$ PI), where P_{200} is percentage passing through sieve No. 200, used as a decimal, and PI is the plasticity index.

1.7 PLATE BEARING TEST – MODULUS OF REACTION (k)

The plate bearing test is used for the determination of soil bearing capacity with respect to the modulus of surface reaction (k value). The subgrade bearing capacity in terms of k value is used, mainly, in rigid pavement design methodologies.

The determination of modulus of reaction is based on the Westergaard theory for stresses and deformations (deflections) for loading concrete plates, where the subgrade elastic reaction in a vertically loaded plate with pressure (p) is considered to be vertical and proportional to the vertical deformation (deflection) of subgrade, at all levels of loading. The constant of proportionality between the applied pressure (p) and the respective deflection (δ_z) is defined as the modulus of subgrade reaction (k); that is, $p = k \times \delta_z$.

The test is carried out on compacted material with certain moisture, using a steel circular plate and a load application system. The steel plate can be of various diameters, but the plate normally used has a 762 mm diameter. For increasing plate's rigidity, two additional circular plates of smaller diameter (approximately 650 and 550 mm) are placed on top of it. Load application is usually carried out with hydraulic jack assembly, which is properly adjusted to a fixed reaction beam of a load vehicle. Plate bearing test arrangement is shown in Figure 1.9.

When the load is applied, the pressure induced on the layer's surface results in a corresponding deflection, which is measured by two, three or even four gages. The average value of deflection measured at different magnitudes of pressure determines the pressure (kPa) versus deflection (mm) curve. Theoretically, the curve should be linear. However, this is not the case because of the nonelastic behaviour of the soil or unbound material. As a result, the k value is determined from a point of the curve as defined by the test procedure adopted. In most cases, either the pressure corresponding to 1.25 mm deflection or the deflection caused by 68.94 kPa pressure, using a 762 mm diameter plate, is used.

Thus, the modulus of reaction (k value) is calculated with one of the following equations:

$$k_{762} = p/1.25 \text{ (kPa/mm)}$$

or

$$k_{762} = 68.94/\delta_z \text{ (kPa/mm)},$$

where k_{762} is the modulus of reaction using a 762 mm diameter plate, p is the applied pressure (kPa) and δ_z is the deflection (mm).

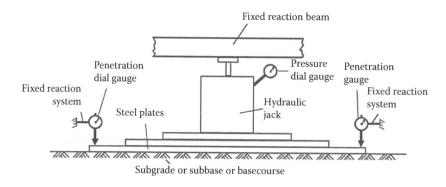

Figure 1.9 Schematic representation of the plate bearing test.

The above test is described in detail in ASTM D 1195 (2009) or BS 1377: Part 9 (1990). It should be noted that the test procedure according to ASTM D 1195 (2009) differs from the abovementioned, which is the one prescribed in BS 1377: Part 9 (1990). The difference is in the way the loading is applied and in the determination of the applied pressure.

It should be mentioned that during the test, the moisture content of the layer tested should be determined and reported.

1.7.1 Correlation between CBR and *k* value

Provided that composition and thickness uniformity of soil layer exists, it has been found that the following relationship (HA Vol. 7 IAN 73/06) between CBR and *k* stands:

$$CBR = 6.1 \times 10^{-8} \times (k_{762})^{1.733}(\%),$$

where k_{762} is the modulus of reaction measured with a 762 mm diameter plate for plate penetration, which is usually 1.25 mm.

In case a smaller plate is used, k_{762} can be calculated with the following relationship:

$$k_{762} = F \times k_{mm},$$

where F is the coefficient calculated with the following relationship:

$$F = 0.00124D + 0.0848,$$

where D is the diameter of loading plate (mm), and k_{mm} is the modulus of reaction determined by the plate bearing test, using a certain millimetre diameter plate.

Using the above relationships, the plate bearing test is used by some organisations, as well as for the estimation of CBR of soil layer, which contains a high percentage of coarse soils.

1.8 RESISTANCE *R* VALUE TEST

This test method is used to measure the potential strength of subgrade, subbase or even base course soil materials in pavements. The *R* value is also used by some agencies as an acceptance criterion of the material for subgrade or even subbase/base course.

The use of the test is limited nowadays, and for this reason, no detailed description will be given here. However, the reader can find all the details in the relevant standards (ASTM D 2844 2013 or AASHTO T 190 2013).

1.9 ELASTIC MODULUS AND RESILIENT MODULUS OF SOILS

The bearing capacity of the subgrade, expressed with the abovementioned parameters, constitutes a rather useful parameter and provides the engineer with the chance of designing a pavement with semianalytical methods. However, when the elastic theory for pavement structural analysis and design is to be used, it is necessary to know the fundamental elastic parameters, such as modulus of elasticity (E) and Poisson ratio (μ).

Modulus of elasticity of the subgrade for static load could be calculated with the Boussinesq theory in conjunction with the plate bearing test. The relationship, which can be used for defining modulus of elasticity, assuming that the soil material behaves elastically under static load, is as follows:

$$E = [\pi \times P \times r \times (1 - \mu^2)]/2 \times \delta_z,$$

where P is the applied pressure (kPa), r is the plate radius (mm), μ is the Poisson ratio and δ_z is the deflection (mm).

Depending on the Poisson ratio value, the above relationship is simplified as follows:

$$E = 1.45 \times (P \times r/\delta_z), \text{ for } \mu = 0.3$$

$$E = 1.50 \times (P \times r/\delta_z), \text{ for } \mu = 0.2.$$

The application of static loading instead of dynamic loading, which really exists on site, in conjunction with the difficulty of executing the test under different moisture content or density, and given the fact that soils and aggregates are nonelastic materials (load duration dependent), led to laboratory determination of the stiffness modulus (E) or resilient modulus (M_R).

The triaxial test with a repeated loading was found to be the most acceptable test for determining a soil material's stiffness.

1.9.1 Repeated triaxial test – resilient modulus test

The repeated triaxial test applies a repeated axial cyclic stress of fixed magnitude, load duration and cycle duration to a cylindrical specimen. While the specimen is subjected to the dynamic cyclic stress, it is also subjected to a static confining stress provided by a pressure chamber. The cyclic load application, though, is to better simulate the actual traffic loading.

The triaxial test under repeated axial loading is performed using a similar apparatus to the one used for the typical triaxial test in which a static axial loading is used (ASTM D 2850 2007).

The apparatus used for the determination of the resilient modulus is as shown in Figure 1.10.

The deviator stress σ_d ($=\sigma_1 - \sigma_2$) is repeated, at a fixed magnitude and frequency. The loading of the soil specimen, under the influence of deviator stress, results in a deformation, part of which is recoverable during the stage of unloading. This recoverable strain along with the deviator stress determines the resilient modulus (M_R), or the modulus of elasticity (E), of the material tested, using the equation

$$M_R(\text{or } E) = \sigma_d/\varepsilon_r \text{ (kPa)},$$

where ε_r is the recoverable strain (mm/mm).

A detailed description of the test procedure can be found in the Asphalt Institute MS-10 and AASHTO T 307 (2007).

The compaction of the soil specimens is carried out by either the double plunger method or the kneading compactor and the specimen is tested at maximum density. The test can also be carried out in undisturbed specimens extracted from the site.

The dimensions of the cylindrical specimens depend on maximum particle size. The length of the specimen should not be shorter than twice its diameter and its minimum diameter shall not be less than 71 mm or six times the maximum size particle. The cylindrical test samples are normally 100 mm in diameter by 200 mm in height.

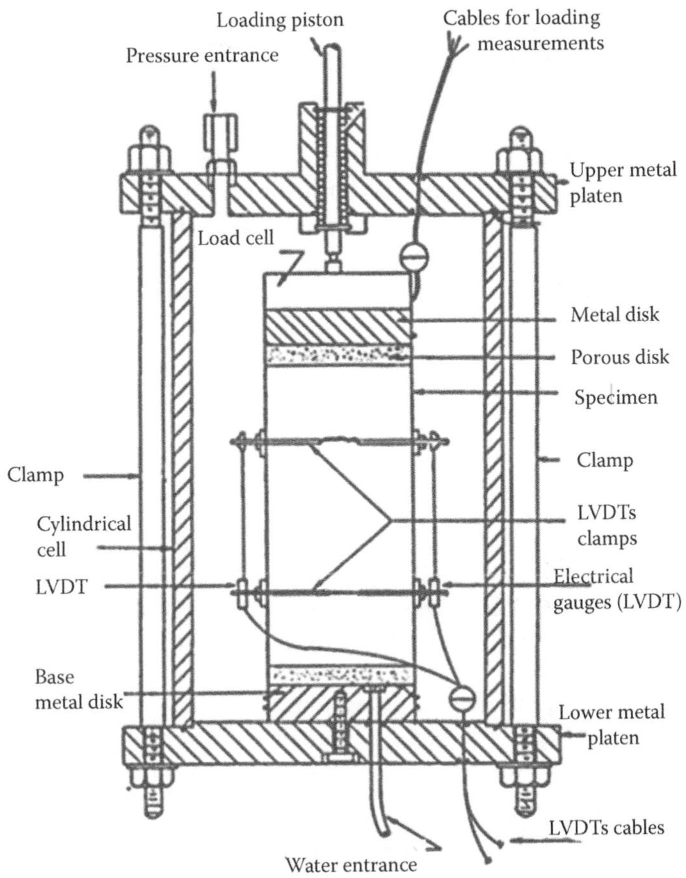

Figure 1.10 Schematic representation of apparatus determining M_R. (From Asphalt Institute MS-10, *Soils Manual. Manual Series No. 10*. Lexington, KY: Asphalt Institute. With permission.)

After compaction, the specimen is covered with a plastic membrane and is placed in the triaxial apparatus (chamber), where it is subjected to loadings of 200 repetitions. In each loading cycle of 200 repetitions, a different deviator stress is used by increasing or decreasing the radial stress, σ_3. At the end of each loading cycle of 200 repetitions, the recovered deformation is measured, from which both the recovered strain and the corresponding resilient modulus (M_R) are calculated using the equation above. The number of loadings at different levels of deviator stress is defined by the specification.

For a pavement design study, the above procedure may be repeated for different percentages of moisture contents; hence, the curve of resilient modulus versus moisture content is obtained. Upon knowing the representative moisture content of the layer during a pavement's service life, a more representative resilient modulus (M_R) or stiffness modulus (E) can be determined for further calculations.

1.9.2 Correlation between resilient modulus and CBR

The analytical and some semianalytical pavement design methodologies use a fundamental mechanical property for expressing the bearing capacity or strength of the subgrade soil and not the empirical CBR value. This property is the resilient modulus (M_R) or the stiffness modulus (E) of the material. The determination of the material's stiffness requires the use of more expensive and complex apparatuses than the CBR apparatus. On the other hand, there is great experience and a large amount of data collected from CBR measurements. The above two factors advocate that the existence of a correlation between CBR and stiffness would be very useful to the engineers.

After extensive work carried out by researchers on behalf of organisations/laboratories, reliable correlation equations have been established for estimating stiffness from CBR values. The most widely used equations for fine-grained soil material are as follows:

$$E = 17.6 \times CBR^{0.64} \text{ (MPa)} \tag{1.1}$$

$$M_R = 10.3 \times CBR \text{ (MPa)} \tag{1.2}$$

$$M_R = 8.0 + 3.8 \times R \text{ (MPa)} \tag{1.3}$$

$$M_R = 206.84 \times (a_i/0.14) \text{ (MPa)} \tag{1.4}$$

where E is the stiffness modulus (or surface stiffness modulus) (MPa), M_R is the resilient modulus (AASHTO T 307 2007) (MPa), CBR is the California bearing ratio (ASTM D 1883 2007 or AASHTO T 193 2013, rammer compaction) (%), R is the resistance R value (ASTM D 2844 2013 or AASHTO T 190 2013) and a_i is the AASHTO structural layer coefficients (AASHTO 1993).

In order to express the unit of MN/m² in psi unit, the following conversion factor can be used: 1 psi = 0.006895 MN/m².

Equation 1.1 proposed by Powel et al. (1984) is based on the work carried out by Jones (1958) and is used in the current UK pavement design methodology (Highways Agency 2006). The validity of the equation is restricted to fine soil material with laboratory CBR values from 2% to 12%.

Equations 1.2 and 1.3 are proposed to be used by the Asphalt Institute of the United States (Asphalt Institute MS-1). They are applicable to materials classified as A-7, A-6, A-5, A-4 and finer A-2 soils (AASHTO designation) or CL, CH, ML, SC and SP (USCS), or in general for materials that are estimated to have a resilient modulus of 207 MPa or less.

Equation 1.4 is used in the AASHTO guide for the design of pavement structures (AASHTO 1993).

It must be pointed out that the above relationships are valid for fine soil material. For very fine soil material with a plasticity index (PI or I_p) greater than 30, the resilient modulus is recommended to be determined by laboratory testing.

As for the coarse soil material (gravels and sands) or coarse crushed material, relevant information regarding the resilient modulus can be found in Section 10.5.6.

A broad estimation of resilient modulus, M_R (=E), as well as CBR and R value, may be carried out using Figure 1.14.

Finally, the resilient modulus (M_R) may be predicted from correlation equations where soil index properties are used, that is, LL, PI, percentage of clay, percentage of silt, percentage passing through No. 200 sieve, and so on (George 2004).

1.9.3 Dynamic plate test

The dynamic plate test is carried out in situ and determines the stiffness of the subgrade or unbound subbase/base course in terms of a stiffness modulus by using dynamic plate loading. The particular modulus is usually called surface modulus (E).

Figure 1.11 Light weight deflectometer. (a) Typical LWD. (b) LWD with two additional geophones. (Courtesy of Dynatest International A/S.)

The apparatus used is portable and is known as light weight deflectometer (LWD) (see Figure 1.11).

The apparatus consists of a steel handle with a steel circular plate of 300 mm (standard size) or 150 mm in diameter in its end.

During the test, a certain load (typically 10 kg) drops from a certain height to the steel plate, to produce a certain peak stress ranging from 30 to 150 kPa (typical value used, 100 kPa) with a load pulse time ranging from 8 to 20 ms. The applied load on each impact is measured with a load cell. The vertical displacement, deflection, is measured by a geophone (velocity transducer) positioned at the centre of the loaded area.

The surface stiffness modulus is calculated using the following general equation (Highways Agency 2009b):

$$E = (2 \times (1 - \mu^2) \times R \times P)/D,$$

where E is the surface stiffness modulus (MPa), μ is the Poisson ratio (normally 0.35), R is the radius of load plate (mm), P is the applied pressure (kPa) and D is the deflection under the centre of the plate (μm).

The standard LWD may be equipped with two additional geophones, if the thickness of the underlying layer is to be estimated (see Figure 1.11b).

All data are stored and can be transferred to a portable data processing unit or to a portable computer.

General instructions for executing the dynamic plate test with an LWD can be found in ASTM E 2583 (2011). More information about LWD and field measurements can be found in Nazzal et al. (2007), Hossain and Apeagyei (2010) and Marradi et al. (2011).

A device similar to the LWD is known as the German Dynamic Plate. The only differences are that the deflection is measured with an accelerometer, it operates with a constant applied stress of 100 kPa with a load pulse type of 18 ms and the stiffness range measured is limited to 10–200 MPa. More information can be found in Fleming et al. (2002, 2007).

It should be noted that the fundamental principles of the LWD are similar to those of a larger device known as falling weight deflectometer. This particular device is mainly used for determining the bearing capacity of pavements, as well as to determine the layers' stiffness modulus. ASTM D 4694 (2009) constitutes the relative standard for executing the test.

Other portable dynamic plate devices developed and used in the past were the TRL foundation tester and the Prima and Humboldt soil stiffness gauge (Fleming et al. 2000, 2002).

1.9.4 Springbox equipment

The Springbox equipment (see Figures 1.12 and 1.13) is a new device for laboratory testing of unbound granular and some weak hydraulically bound mixtures (Edwards et al. 2005). The Springbox uses the standard Nottingham Asphalt Tester loading frame and software.

It consists of a steel box containing a cubical sample of material, with an edge dimension of 170 mm, to which a repeated load is applied over the full upper surface. One pair of the box sides is fully restrained and the other is restrained through elastic springs, giving a wall stiffness of 10–20 kN/mm. The above arrangement simulates the lateral support of the material within the compacted layer.

The material to be tested is compacted with a vibrating hammer; the assembly provides means to introduce water to the sample or drain the water from its underside.

Loading takes the form of repeated vertical load applications of controlled magnitude at a frequency of at least 1 Hz and no greater than 5 Hz. The load capacity is equivalent to a vertical stress of at least 150 kPa.

Measurements of both vertical and horizontal (spring restrained) deflection can be made, with two measurement transducers for each measure. In the case of vertical deflection measurement, the equipment allows the transducers to make direct contact with the specimen, via holes in the loading platen.

The stiffness modulus of the material can be calculated from the averaged deflections measured over a series of loading patterns (Highways Agency 2009b).

Additional information of the Springbox test can be found in Edwards et al. (2004).

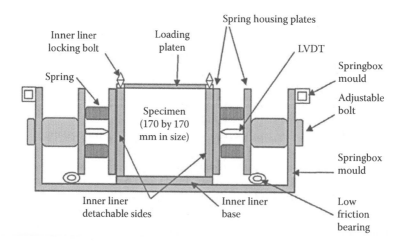

Figure 1.12 Cross section of springbox. (From Edwards J.P. et al., Development of a simplified test for unbound aggregates and weak hydraulically bound materials utilizing the NAT. *Proceedings of the 6th International Symposium on Pavements Unbound*, A.R. Dawson [ed.], pp. 3–11. Rotterdam, The Netherlands: A. A. Balkema, 2004.)

Figure 1.13 Springbox equipment. (Courtesy of Cooper Research Technology Ltd.)

1.9.5 Correlation between mechanical parameters of soils and soil classification groups

Correlations between mechanical parameters (M_R, CBR, k value and R value) and soil classification groups have been developed and presented in tabular–graphical form (Shell International 1985). The table shown in Figure 1.14 may be used to estimate the mechanical properties of a soil material from its classification group. The value obtained may be used in a pavement design calculation but only at its preliminary stage.

1.10 COMPACTION OF SOIL MATERIALS ON SITE

Sufficient compaction of the soil material constitutes a key prerequisite in avoiding premature settlement and other negative consequences. Sufficient compaction is ensured by applying necessary compaction energy at optimum, or almost optimum, moisture content. The application of the necessary compaction energy is obtained with the use of appropriate compaction plants applying a sufficient number of passes.

The thickness of the layer to be compacted constitutes another key factor for effective and uniform compaction with respect to depth. It is again related to the type of compaction plant used as well as the type of soil material. Vibratory rollers are able to compact thicker layers than static rollers. Pneumatic tyre rollers are not suitable for uniformly graded granular materials (single sized) or silty cohesive materials.

Rolling velocity and tyre pressure, in case of pneumatic tyre rollers, constitute additional factors affecting compaction. With respect to the velocity, it should not be higher than 3 km/h, in all cases. As for the tyre pressure, it is determined by the manufacturer and should not fluctuate during compaction.

The engineer, before compaction, for a given thickness and type of soil material, should determine the number of passes, so that sufficient compaction is achieved by using the

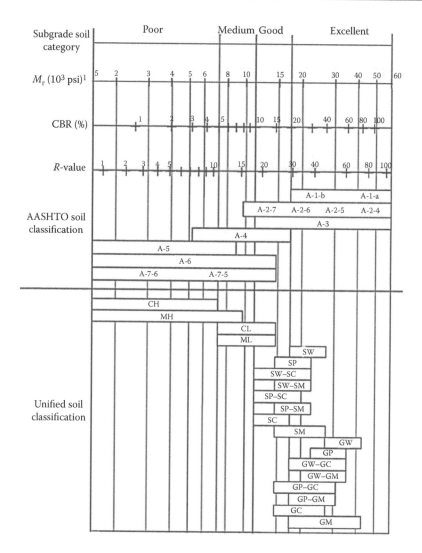

Figure 1.14 Correlation of mechanical parameters and soil classification groups. (From NCHRP, *Guide for Mechanistic–Empirical Design of New and Rehabilitated Structures, Final Report, Appendices CC-1*. Washington, DC: National Cooperative Highway Research Program, Transportation Research Board, National Research Council, 2001. With permission.)

compaction plant. This is achieved after trials on site by taking density measurements after a certain number of passes. The number of passes is sufficient when the achieved dry density satisfies the requirements. Usually, a certain percentage of the maximum dry density determined in the laboratory is required, which varies from 92% to 100%.

Adequacy or acceptability of compaction may also be based on MCV determined in situ (Highways Agency 2009b; Matheson and Winter 1997).

In some countries, the satisfactory compaction is judged by running the plate bearing test twice, with two different numbers of passes. The surface stiffness modulus is determined in both cases (E_1 and E_2) from the k value obtained. If the E_2/E_1 ratio is less than 2, the compaction is considered satisfactory.

Table 1.13 provides very useful information for compacting sufficiently different types of soil materials with alternative compaction plants (machines). The suggested minimum

Table 1.13 Compaction plants for soil materials (minimum number of passes per maximum compacted depth)

Type of compaction plant	Category	Wet cohesive materials		Well-graded granular, dry or stony cohesive materials		Uniformly graded granular or silty cohesive materials	
		D	N	D	N	D	N
Smoothed wheeled roller (or vibratory roller operating without vibration)	kg/m width of roll:						
	>2100–2700	125	8	125	10	125	10[a]
	>2700–5400	125	6	125	8	125	8[a]
	>5400	150	4	150	8	Unsuitable	
Grid roller	>2700–5400	150	10	Unsuitable		150	10
	>5400–8000	150	8	125	12	Unsuitable	
	>8000	150	4	150	12	Unsuitable	
Deadweight tamping roller	>4000–6000	225	4	150	12	250	4
	>6000	300	5	200	12	300	3
Pneumatic tyre roller	kg per wheel:						
	>1000–1500	125	6	Unsuitable		150	10[a]
	>1500–2000	150	5	Unsuitable		Unsuitable	
	>2000–2500	175	4	125	12	Unsuitable	
	>2500–4000	225	4	125	10	Unsuitable	
	>4000–6000	300	4	125	10	Unsuitable	
	>6000–8000	350	4	150	8	Unsuitable	
	>8000–12,000	400	4	150	8	Unsuitable	
	>12,000	450	4	175	6	Unsuitable	
Vibratory roller (rolling velocity 1.5 to 2.5 km/h)	kg/m width of a vibratory roll:						
	>270–450	Unsuitable		75	16	150	16
	>450–700	Unsuitable		75	12	150	12
	>700–1300	100	12	125	10	150	6
	>1300–1800	125	8	150	8	200	10[a]
	>1800–2300	150	4	150	4	225	12[a]
	>2300–2900	175	4	175	4	250	10[a]
	>2900–3600	220	4	200	4	275	8[a]
	>3600–4300	225	4	225	4	300	8[a]
	>4300–5000	250	4	250	4	300	6[a]
	>5000	275	4	275	4	300	4[a]
Vibrating plate compactor	kg/m² of base plate:						
	>880–1100	Unsuitable		Unsuitable		75	6
	>1100–1200	Unsuitable		75	10	100	6
	>1200–1400	Unsuitable		75	6	150	6
	>1400–1800	100	6	125	6	150	4
	>1800–2100	150	6	150	5	200	4
	>2100	200	6	200	5	250	4
Vibro-tamper	Mass:						
	>50–65 kg	100	3	100	3	150	3
	>65–75 kg	125	3	125	3	200	3
	>75–100 kg	150	3	150	3	225	3
	>100 kg	225	3	200	3	225	3

(continued)

Table 1.13 Compaction plants for soil materials (minimum number of passes per maximum compacted depth) (Continued)

Type of compaction plant	Category	Wet cohesive materials		Well-graded granular, dry or stony cohesive materials		Uniformly graded granular or silty cohesive materials	
		D	N	D	N	D	N
Power rammer	>100–500 kg	150	4	150	6	Unsuitable	
	>500 kg	275	8	275	12	Unsuitable	
Dropping weight compactor	Mass of rammer >500 kg weight drop:						
	>1–2 m	600	4	600	8	450	8
	>2 m	600	2	600	8	Unsuitable	

Source: Adapted from Highways Agency, *The Manual of Contract Documents for Highway Works (MCHW),Volume 1: Specification for Highway Works, Series 600: Earthworks.* London: Department for Transport, Highways Agency, 2009a.

Note: D, maximum thickness of compacted layer; N, minimum number of passes.

[a] Roller shall be towed by track-laying tractors. Self-propelled rollers are unsuitable.

number of passes (N) should ensure the achievement of the minimum required degree of compaction (Highways Agency 2009a).

1.11 DENSITY TESTS IN SITU

Compaction density obtained on site is measured in situ by one of the following methods: the sand-cone method, the rubber balloon method, nuclear methods or the drive-cylinder method. The complex impedance method may also be used.

1.11.1 Sand-cone method

The sand-cone method used for the determination of compacted layer density is perhaps the most popular of all tests available. It consists of creating an almost cylindrical hole, weighing the extracted material and measuring the volume of the hole created. The volume is measured by filling the hole with fine sand. Provided that the mass of the sand used for filling the hole and the apparent specific gravity are known, the volume of the hole as well as the dry density of the compacted layers can be calculated.

For the determination of the mass used to fill the cylindrical hole created, a (normally) metal cone having a plastic bottle in its upper part with an outflow valve was used (see Figure 1.15). The bottle contains a preweighted mass of fine sand.

After placing the cone on the hole, the sand outflow valve is opened; the sand then occupies the hole's space and the cone's space. Once sand outflow stops, the valve is closed and the remaining quantity of the sand in the plastic bottle is being weighed. The weight of the sand used for filling the cone and the hole arises from the difference in weight of the initial amount of sand and the one remaining in the bottle.

Knowing in advance the mass of sand required for filling the cone, the volume of the hole opened (V) is calculated by the following relationship:

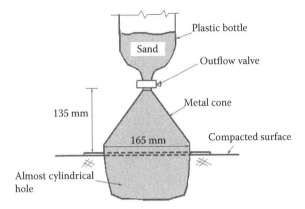

Figure 1.15 Sand-cone apparatus.

$$V = (m_1 - m_2)/G_{ap},$$

where m_1 is the mass of the sand used (g), m_2 is the mass of the sand in the cone (g) and G_{ap} is the apparent specific gravity of sand (g/cm^3).

The mass of soil material that came out of the hole is weighed immediately and a representative sample is received in order to determine its moisture content. As a result, the dry compacted density (ρ_d) is calculated by the following relationship:

$$\rho_d = [m_3/(1 + w)/V],$$

where m_3 is the weight of soil material with moisture (g), w is the moisture content (%) and V is the volume (cm^3) from the previous equation.

The sand-cone method is recommended for soil materials that contain particles not bigger than 50 mm in diameter. The volume of the hole created is by reference to the maximum particle contained in the soil material. The minimum volume for soil materials that do not contain particles larger than 4.75 mm is suggested to be 700 cm^3, whereas for soil materials containing particles up to 50 mm in diameter, the minimum suggested volume is 2830 cm^3. Further information and analytical description can be found in the standard ASTM D 1556 (2007) or AASHTO T 191 (2013).

1.11.2 Rubber balloon method

The rubber balloon method is used only by a few agencies and organisations. In this method, the volume of the hole created is measured by water, which is secluded by a rubber membrane, creating a balloon.

A specific device is used for performing the test, known as the rubber balloon apparatus. It consists of a volumetric cylinder, one end of which is made of a thin elastic membrane. The volumetric cylinder is filled with a known mass of water before testing. The device is placed over the hole, and with the aid of low air pressure, the water occupies the available space (Figure 1.16).

The volume of the hole is calculated by volume difference, which appears in the volume counter tube. Once the volume has been determined, the dry compacted density is calculated the same way as in the previous method.

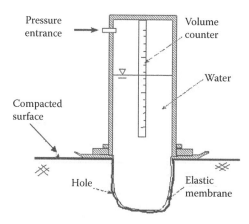

Figure 1.16 Schematic presentation of the apparatus for the rubber balloon test.

This method is simpler than the sand-cone method. However, it is not suggested for very soft soil materials, which are deformed easily, and the walls of the hole are not stable. Additionally, this test method may not be suitable for soils containing crushed rock fragments or sharp edge materials that may puncture the rubber membrane. Regarding the restriction on maximum particle size contained in the soil material, the maximum shall not be more than 63 mm.

A detailed description of the method can be found in ASTM D 2167 (2008).

1.11.3 Nuclear method

This is a rapid non-destructive method for in situ measurements of wet density, water content and the determination of dry density of soil and soil aggregate mixtures.

The rapid nondestructive determination of the compacted density (less than 5 min) allows the almost immediate detection of areas that compacted insufficiently. This is of great advantage since if measurements are taken during construction, corrections can be applied within hours.

The nuclear density devices are much more expensive than conventional devices (cone-sand or rubber balloon) but they are very popular in highway projects.

The density may be measured by direct transmission, backscatter or backscatter/air-gap method of gamma radiation. The measurements of moisture content are taken, in most devices available, in backscatter mode, regardless of the mode being used for density.

In the direct transmission method, the gamma source extends through the base of the gauge into a preformed hole to a desired depth. In the backscatter or backscatter method, the gamma and neutron sources and the detectors are in the same plane (see Figure 1.17).

The depth of measurements is normally up to 300 mm; hence, the devices are known as shallow-depth measurements. However, nuclear devices capable of taking measurements at a desired greater depth also exist. They are called devices for measurements at depths below surface.

The direct transmission gauge provides more representative results, especially in depths from 150 to 300 mm.

The backscatter devices are usually used for layer thickness of 100 mm to 150 mm, they are more practical and the results derived quicker than the direct transmission mode. These devices seem to be more popular than others.

The backscatter/air-gap devices constitute a variant of the previous ones in order to eliminate the possibility of errors attributed to the surface roughness.

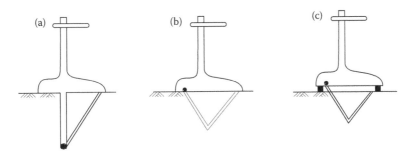

Figure 1.17 Types and representative nuclear gauges for density determination. (a) Direct transmission, (b) backscatter and (c) backscatter/air gap method.

All devices should be calibrated for the type of soil to be used. The precision of the measurements is affected by the uniformity of the layer and the chemical composition of the soil material. The results obtained by nuclear method are comparable to those of conventional methods.

The nuclear density devices are safe to operate because they possess an automatic cutoff safety system. However, because of the use of a radioactive element, every appropriate safety measure must be taken.

More details regarding the procedure of the test for shallow depths can be found in ASTM D 6938 (2010) or AASHTO T 310 (2013); with regard to measurements below surface, details can be found in ASTM D 5195 (2008).

It is mentioned that nuclear devices may also determine the compacted density of asphalt layers. In this particular case, the method of determination is specified by ASTM D 2950 (2011).

1.11.4 Drive-cylinder method

This test method is used to determine the in situ density of soils that do not contain significant amounts of particles coarser than 4.75 mm and which can be readily retained in the drive cylinder.

This test method is not recommended for use in organic or friable soils. This test method may not be applicable for soft, highly plastic, non-cohesive, saturated or other soils that are easily deformed, compress during sampling or may not be retained in the drive cylinder.

The test method involves obtaining a relatively intact soil sample by driving a thin-walled cylinder and the subsequent activities for the determination of in-place density. When sampling or in situ density is required at depth, test Method D 1587 should be used.

More information regarding the test is given in ASTM D 2937 (2010).

1.11.5 Complex impedance method

The test method is a procedure for estimating in place the density and water content of soils and soil aggregates on the basis of electrical measurements, using a Complex-Impedance Measuring Instrument.

The electrical properties of a soil are measured using a radiofrequency voltage applied to soil electrical probes driven into the soils and soil aggregates to be tested, in a prescribed pattern and depth. Certain algorithms of these properties are related to wet density and water content. This correlation between electrical measurements and density and water content is accomplished using a calibration methodology. In the calibration methodology, density and water content are determined by other ASTM test standards that measure soil density and water content, thereafter correlating the corresponding measured electrical properties to the soil physical properties.

Although this test method causes minimal disturbance, it may not be applicable to all types of soil.

More information regarding the test is given in ASTM D 7698 (2011).

REFERENCES

AASHTO. 1993. *Guide for design of pavement structures.* Washington, DC: American Association of State Highway and Transportation Officials.

AASHTO M 145. 2012. *Classification of soils and soil-aggregate mixtures for highway construction purposes.* Washington, DC: American Association of State Highway and Transportation Officials.

AASHTO M 146. 2012. *Terms relating to subgrade, soil-aggregate and fill materials.* Washington, DC: American Association of State Highway and Transportation Officials.

AASHTO MEPDG-1. 2008. *Mechanistic-empirical pavement design guide: A manual of practice.* Washington, DC: American Association of State Highway and Transportation Officials.

AASHTO T 88. 2013. *Particle size analysis of soils.* Washington, DC: American Association of State Highway and Transportation Officials.

AASHTO T 89. 2013. *Determining the liquid limit of soils.* Washington, DC: American Association of State Highway and Transportation Officials.

AASHTO T 90. 2008. *Determining the plastic limit and plasticity index of soils.* Washington, DC: American Association of State Highway and Transportation Officials.

AASHTO T 92. 2009. *Determining the shrinkage factors of soils.* Washington, DC: American Association of State Highway and Transportation Officials.

AASHTO T 100. 2010. *Specific gravity of soils.* Washington, DC: American Association of State Highway and Transportation Officials.

AASHTO T 180. 2010. *Moisture-density relations of soils using a 4.54kg rammer and a 457mm drop.* Washington, DC: American Association of State Highway and Transportation Officials.

AASHTO T 190. 2013. *Resistance R-value and expansion pressure of compacted soils.* Washington, DC: American Association of State Highway and Transportation Officials.

AASHTO T 191. 2013. *Density of soil in-place by the sand-cone method.* Washington, DC: American Association of State Highway and Transportation Officials.

AASHTO T 193. 2013. *The California bearing ratio.* Washington, DC: American Association of State Highway and Transportation Officials.

AASHTO T 265. 2012. *Laboratory determination of moisture content of soils.* Washington, DC: American Association of State Highway and Transportation Officials.

AASHTO T 307. 2007. *Determining the resilient modulus of soils and aggregate materials.* Washington, DC: American Association of State Highway and Transportation Officials.

AASHTO T 310. 2013. *In-place density and moisture content of soil and soil-aggregate by nuclear methods (shallow depth).* Washington, DC: American Association of State Highway and Transportation Officials.

Asphalt Institute MS-10. *Soils Manual. Manual Series No. 10,* 5th edition. Lexington, KY: Asphalt Institute.

ASTM D 422. 2007. *Standard test method for particle-size analysis of soils.* West Conshohocken, PA: ASTM International.

ASTM D 854. 2010. *Standard test method for specific gravity of soil solids by water pycnometer.* West Conshohocken, PA: ASTM International.

ASTM D 1195/D 1195M. 2009. *Standard test method for repetitive static plate load tests for soils and flexible pavement components for use in evaluation and design of airport and highway pavements.* West Conshohocken, PA: ASTM International.

ASTM D 1556. 2007. *Standard test method for density and unit weight of soil in place by the sand-cone method.* West Conshohocken, PA: ASTM International.

ASTM D 1557. 2012. *Standard test methods for laboratory compaction characteristics of soil using modified effort (56,000 ft-lbf/ft)*. West Conshohocken, PA: ASTM International.

ASTM D 1883-07e2. 2007. *Standard test method for CBR (California bearing ratio) of laboratory-compacted soils*. West Conshohocken, PA: ASTM International.

ASTM D 2167. 2008. *Test method for density and unit weight of soil in place by the rubber balloon method*. West Conshohocken, PA: ASTM International.

ASTM D 2487. 2011. *Standard practice for classification of soils for engineering purposes (Unified Soil Classification System)*. West Conshohocken, PA: ASTM International.

ASTM D 2844/D 2844M. 2013. *Standard test method for resistance R-value and expansion pressure of compacted soils*. West Conshohocken, PA: American Society for Testing and Materials.

ASTM D 2850. 2007. *Standard test method for unconsolidated-undrained triaxial compression test on cohesive soils*. West Conshohocken, PA: ASTM International.

ASTM D 2937. 2010. *Standard test method for density of soil in place by the drive-cylinder method*. West Conshohocken, PA: ASTM International.

ASTM D 2950/D 2950M. 2011. *Standard test method for density of bituminous concrete in place by nuclear methods*. West Conshohocken, PA: ASTM International.

ASTM D 4318. 2010. *Standard test method for liquid limit, plastic limit and plasticity index of soils*. West Conshohocken, PA: ASTM International.

ASTM D 4643. 2008. *Standard test method for determination of water (moisture) content of soil by microwave oven heating*. West Conshohocken, PA: ASTM International.

ASTM D 4694. 2009. *Standard test method for deflections with a falling-weight-type impulse load device*. West Conshohocken, PA: ASTM International.

ASTM D 4718. 2007. *Standard practice for correction of unit weight and water content for soils containing oversize particles*. West Conshohocken, PA: ASTM International.

ASTM D 4959. 2007. *Standard test method for determination of water (moisture) content of soil by direct heating*. West Conshohocken, PA: ASTM International.

ASTM D 5195. 2008. *Test method for density of soil and rock in-place at depths below surface by nuclear methods*. West Conshohocken, PA: ASTM International.

ASTM D 6938. 2010. *Standard test method for in-place density and water content of soil and soil-aggregate by nuclear methods (shallow depth)*. West Conshohocken, PA: ASTM International.

ASTM D 6951/D 6951M. 2009. *Standard test method for use of the dynamic cone penetrometer in shallow pavement applications*. West Conshohocken, PA: ASTM International.

ASTM D 7698-11a. 2011. *Standard test method for in-place estimation of density and water content of soil and aggregate by correlation with complex impedance method*. West Conshohocken, PA: ASTM International.

ASTM E 2583. 2011. *Standard test method for measuring deflections with a light weight deflectometer (LWD)*. West Conshohocken, PA: ASTM International.

BS 1377: Part 2. 1990. *Methods of test for soils for civil engineering purposes, Part 2: Classification tests*. London: British Standards Institution.

BS 1377: Part 4. 1990. *Methods of test for soils for civil engineering purposes, Part 4: Compaction-related tests*. London: British Standard Institution.

BS 1377: Part 9. 1990. *Methods of test for soils for civil engineering purposes, Part 9: In-situ testing*. London: British Standard Institution.

BS 6031. 1981. *Code of practice for earthworks*. London: British Standards Institution.

CEN EN 13286-2:2010/AC. 2012. *Unbound and hydraulically bound mixtures, Part 2: Test methods for the determination of the laboratory reference density and water content – Proctor compaction*. Brussels: CEN.

CEN EN 13286-47. 2012. *Unbound and hydraulically bound mixtures, Part 47: Test methods for the determination of the California bearing ratio, immediate bearing index and linear swelling*. Brussels: CEN.

CEN EN ISO 14688-1:2002/A1. 2013. *Geotechnical investigation and testing – Identification and classification of soil – Part 1: Identification and description*. Brussels: CEN.

CEN EN ISO 14688-2:2004/A1. 2013. *Geotechnical investigation and testing – Identification and classification of soil – Part 2: Principles for a classification*. Brussels: CEN.

CEN EN ISO 22476-2:2005/A1. 2011. *Geotechnical investigation and testing. Field testing. Part 2: Dynamic probing.* Brussels: CEN.

CEN ISO/TS 17892-1:2004/AC. 2005. *Geotechnical investigation and testing – Laboratory testing of soil – Part 1: Determination of water content.* Brussels: CEN.

CEN ISO/TS 17892-3:2004/AC. 2005. *Geotechnical investigation and testing – Laboratory testing of soil – Part 3: Determination of particle density – Pycnometer method.* Brussels: CEN.

CEN ISO/TS 17892-4:2004/AC. 2005. *Geotechnical investigation and testing – Laboratory testing of soil – Part 4: Determination of particle size distribution.* Brussels: CEN.

Cooper Research Technology Ltd. 2014. Available at http://www.cooper.co.uk.

Croney D. 1977. *The Design and Performance of Road Pavements.* Transport and Road Research Laboratory (TRRL), now TRL. London: HMSO.

Dumbleton M.J. 1981. *The British Soil Classification Systems for Engineering Purposes: Its Development and Relation to Other Comparable Systems.* TRRL Report LR 1030. Crowthorne, UK: Transport Research Laboratory.

Dynatest International A/S. 2014. Available at http://www.dynatest.com.

Edwards J.P., N.H. Thom, and P.R. Fleming. 2004. Development of a simplified test for unbound aggregates and weak hydraulically bound materials utilizing the NAT. *Proceedings of the 6th International Symposium on Pavements Unbound*, A.R. Dawson (ed.), pp. 3–11. Rotterdam, The Netherlands: A. A. Balkema.

Edwards P., N.H. Thom, P.R. Fleming, and J. Williams. 2005. Testing of unbound materials in the Nottingham asphalt tester springbox. *Transportation Research Record: Journal of the Transportation Research Board*, No. 1913, pp. 32–40. Washington, DC: Transportation Research Board of the National Academies.

Fleming P.R., M.W. Frost, and J.P. Lambert. 2007. A review of lightweight deflectometer for routine in situ assessment of pavement material stiffness. *Transportation Research Record: Journal of the Transportation Research Board*, No. 2004, pp. 80–87. Washington, DC: Transportation Research Board of the National Academies.

Fleming P.R., M.W. Frost, and C.D.F. Rogers. 2000. A comparison of devices for measuring stiffness in situ. *Proceedings of the Fifth International Symposium on Unbound Aggregates in Roads, UNBAR 5*, pp. 193–200. Nottingham, UK.

Fleming P.R., J.P. Lambert, M.W. Frost, and C.D. Rogers. 2002. In-situ assessment of stiffness modulus for highway foundations during construction. *Proceedings of the 9th International Conference on Asphalt Pavements*, pp. 1–2. Copenhagen, Denmark.

George K.P. 2004. *Prediction of Resilient Modulus from Soil Index Properties.* Report No. FHWA/MS-DOT-RD-04-172. Washington, DC: Federal Highway Administration.

Highways Agency. 2006. *Design Manual for Roads and Bridges (DMRB), Volume 7: Pavement Design and Maintenance*, Section 2, Part 3, HD 26/06. London: Department for Transport, Highways Agency.

Highways Agency. 2008. *Design Manual for Roads and Bridges (DMRB), Volume 7: Pavement Design and Maintenance*, Section 3, Part 2, HD29/08. London: Department for Transport, Highways Agency.

Highways Agency. 2009a. *The Manual of Contract Documents for Highway Works (MCHW), Volume 1: Specification for Highway Works*, Series 600: Earthworks. London: Department for Transport, Highways Agency.

Highways Agency. 2009b. *Design Manual for Roads and Bridges (DMRB), Volume 7: Pavement Design and Maintenance*, Section 2, Part 2, IAN 73/06 Rev. 1. London: Department for Transport, Highways Agency.

Highways Agency. 2009c. *The Manual of Contract Documents for Highway Works (MCHW), Volume 2: Specification for Highway Works*, Series NG 600: Earthworks. London: Department for Transport, Highways Agency.

Hossain M.S. and A.K. Apeagyei. 2010. *Evaluation of the Lightweight Deflectometer for In-Situ Determination of Pavement Layer Moduli.* Final Report VTRC 10-R. Charlottesville, VA: Virginia Transportation Research Council.

Jones R. 1958. In situ measurements of the dynamic properties of soil by vibration methods. *Geotechnique*, Vol. 8, No. 1, p. 1.

Marradi A., G. Betti, C. Sangiorgi, and C. Lantieri. 2011. Comparing light weight deflectometers to standardize their use in the compaction control. *Proceedings of the 5th International Conference 'Bituminous Mixtures and Pavements'*. Thessaloniki, Greece: Aristotle University of Thessaloniki.

Matheson G.D. and Winter M.G. 1997. *Use and Application of the MCA with Particular Reference to Glacial Tills*. TRL Report 273. Wokingham, UK: Transport Research Laboratory.

Millar C.E., L.M. Turk, and H.D. Forth. 1962. *Fundamentals of Soil Science*. New York: John Willey.

Nazzal M.D., M.Y. Abu-Farsakh, K. Alshibli, and L. Mohammad. 2007. Evaluating the light falling weight deflectometer device for in situ measurement of elastic modulus of pavement layers. *Transportation Research Record: Journal of the Transportation Research Board*, No. 2016, pp. 13–22. Washington, DC: Transportation Research Board of the National Academies.

NCHRP. 2001. *Guide for Mechanistic–Empirical Design of New and Rehabilitated Structures*. Final Report, Appendices CC-1. Washington, DC: National Cooperative Highway Research Program, Transportation Research Board, National Research Council.

NCHRP. 2004. Project 1-37A, *Guide for Mechanistic-Empirical Design of New and Rehabilitated Structures*, Part 2, Chapter 2. Washington, DC: National Cooperative Highway Research Program, Transportation Research Board, National Research Council.

O'Flaherty C.A. 2002. *Highways: The Location, Design, Construction and Maintenance of Pavements*, 4th Edition. Burlington, MA: Butterworth-Heinemann.

Parsons A.W. 1976. *The Rapid Determination of the Moisture Condition of Earthwork Material*. TRRL Report LR 750. Crowthorne, UK: Transport Research Laboratory.

Parsons A.W. 1992. *Compaction of Soils and Granular Materials: A Review of Research Performed at the Transport Research Laboratory*. London: HMSO.

Powell W.D., J.F. Potter, H.C. Mayhew, and M.E. Nunn. 1984. *The Structural Design of Bituminous Roads*, TRRL Report LR1132. Crowthorne, UK: Transport Research Laboratory.

SDD AG 1. 1989. *The Use and Application of the Moisture Condition Apparatus in Testing Soil Suitability for Earthworking, Application Guide No.1*. London: The Scottish Development Department.

Shell International. 1985. *Shell Pavement Design Manual*. London: Shell International Petroleum Company Limited.

Taylor D.W. 1948. *Fundamentals of Soil Mechanics*. New York: John Wiley & Son.

Chapter 2

Aggregates

2.1 GENERAL

Aggregates used in pavement construction can be crushed aggregates, natural aggregates, slags, mine waste, demolition materials, artificial aggregates, recycled materials or any other material, which meets the required mechanical, natural and chemical properties for the layer to be used. The first two aggregates are defined as conventional or 'primary aggregates', whereas the others can be defined as 'secondary aggregates'.

2.2 CRUSHED AGGREGATES

Crushed aggregates are produced in quarries by various rocks with proper mechanical and chemical properties. Rocks are classified into three general groups depending on how they were formed: igneous rocks, sedimentary rocks and metamorphic rocks.

Rocks formed by cooling molten elements (materials) are defined as *igneous rocks*. Depending on cooling rate, they are characterised as 'coarse' textured, such as granite rocks (slow cooling rate), or 'fine' textured, such as basalt (quicker cooling rate).

Igneous rocks are classified according to their chemical or mineral composition, in particular their silica content (SiO_2). They can be classified as (a) 'felsic', that is, containing a high silica content of approximately >65%, for example, granite; (b) 'intermediate', if the silica content is between approximately 55% and 65%, for example, andesite; (c) 'mafic', if the silica content is low, approximately 45%–55%, the iron–magnesium content is typically high, for example, basalt and gabbro; and (d) 'ultramafic', if the silica content is <45%, for example, peridotite.

Igneous rocks are distinguished into four main categories: gabbros, basalts, granites and porphyries.

Gabbros are mafic with a coarse-grained texture, basalts are also mafic but are finer grained, granites are felsic coarse grained and porphyries are mainly felsic (or intermediate) finely grained with large, clearly discernible crystals embedded. The main rock types used by engineers are given in Table 2.1.

Sedimentary rocks are classified depending on how they were formed. They are divided into four groups: (a) 'clastic sedimentary rocks', (b) biochemical (or biogenic) sedimentary rocks, (c) chemical sedimentary rocks and (d) 'other' sedimentary rocks formed by impact, volcanism and other processes.

Clastic sedimentary rocks are mainly composed of quartz, feldspar, rock fragments (lithic), clay minerals and mica. Clastic sedimentary rocks are further subdivided into conglomerates (mainly composed of rounded gravels), breccias (mainly composed of rounded

Table 2.1 Igneous rocks

Gabbros	Basalts	Granites	Porphyries
Gabbro	Basalt	Granite	Porphyry
Diorite	Andesite	Granodiorite	Rhyolite
Gneiss	Diabase/dolerite	Syenite	Obsidian
Norite	Quartzite dolerite	Quartzite diorite	Quartzite porphyry
Peridotite			

Table 2.2 Sedimentary rocks

Limestones	Sandstones	Quartzites	Siliceous
Limestone	Sandstone	Quartzitic	Chert
	Gritstone	sandstone	Flint
Dolomite (older name, dolostone)	Greywacke	Ganister	
	Tuff		

gravels), sandstones (composed mostly of sand-sized grains) and mudrocks (composed of at least 50% silt- and clay-sized particles).

Biochemical or biogenic sedimentary rocks are formed from calcareous skeletons of organisms. The main constituent mineral is calcite ($CaCO_3$). Limestone is the main rock type in this category.

Chemical sedimentary rocks are formed as precipitates from supersaturated solutions precipitating out around. Oolitic limestone is an example of rock in this category.

Sedimentary rocks may be distinguished into four categories for engineering purposes: limestones, gritstones, quartzites and pyrites (siliceous). Some of the sedimentary rocks that are classified under these categories are given in Table 2.2.

Metamorphic rocks are rocks formed by the transformation of other types of rocks such as igneous or sedimentary rocks under a high-temperature or high-pressure environment. They are divided into schists, hornfels, marbles and gneisses. Phyllite, schist, slate and all rocks that appear to be layered or foliated are examples of schist. These are all shortened along one axis during recrystallisation (metamorphism). Marble is transformed limestone, whereas gneiss is derived from transformed granite.

The majority of igneous rocks, sandstones from sedimentary rocks and hornfels from metamorphic rocks are usually hard rocks. They typically produce durable crushed aggregates that can be used for all pavement layers. Special care should be given to the adhesion of bitumen to these rocks.

Crushed aggregates derived from sedimentary rocks, such as limestone and dolomite, have very good to excellent properties and can be used in all pavement layers except for surface courses, because they have low resistance to polishing.

2.3 NATURAL AGGREGATES

Natural aggregates derived from natural deposits are known as gravels, sand-gravels or natural sand. They can be found in a slightly consolidated form in old streambeds or stream banks, in plateaus created during the postglacial era and in estuaries of rivers, streams or beaches. Natural deposit materials can be used in a wide range of applications such as subbase materials and in certain cases as base materials. They can also be used for producing bituminous mixtures, provided they are first processed (by presieving and washing) to remove any soil or silt that they might contain. Then, they are crushed and sieved to produce the desired particle size distribution.

Natural deposit materials are termed 'loose'. Thus, they are extracted with the aid of an excavator or even a 'strong' loader.

Natural deposit aggregates are mixtures of various rocks, the majority of which are mainly limestones, sandstones and granites. Because of their particular composition, crushed gravels have better mechanical properties than limestone rocks. For this reason, they should be preferred to limestones for wearing courses, when there is a lack of harder rocks.

2.4 SLAGS

Slags are by-products that are produced during the production process of metals such as iron and nickel. Slags vary depending on their chemical composition, specific weight and porosity. The use of slags in highway engineering is usually limited to works that are carried out at a relatively close distance to the production factories. Slags are used mainly as a substitute for aggregates not only in bases or subbases but also for producing bituminous mixtures. Only in a few cases are they used as a substitute for filler.

2.5 MINE WASTE

Mine waste is rock with a low metal content, rejected during the enrichment process. These materials are mainly used as base/subbase materials.

2.6 DEMOLITION MATERIALS

Demolition materials are used in the same way as mine waste, usually in subbase or base layers after preselection and crushing. Demolition materials may also be used in subbase and base mixtures (Woodside et al. 2011).

2.7 ARTIFICIAL AGGREGATES

Artificial aggregates are mainly produced from the calcination of rocks such as bauxite. Calcined bauxite has good antiskid properties. Other types of aggregates are designated by their low density or specific gravity (unit weight) and are used mainly in producing light-weight concrete.

2.8 RECYCLED (PULVERISED) AGGREGATES

Pulverised pavement materials are also known as recycled asphalt plannings. They are produced by crushed and screening old asphalt materials during reconstruction projects. They can be used as an aggregate replacement in new asphalt materials.

2.9 AGGREGATE SIZES

Aggregates are divided into coarse aggregates, fine aggregates, fines and filler aggregates or fillers.

According to European standards, coarse aggregates for bituminous mixtures (asphalts) are defined as aggregates whose particles are retained on a 2 mm sieve and pass through a 45 mm sieve. In the case of aggregates for unbound and hydraulically bound layers, the aggregate particles retained on a 2 mm sieve are defined as coarse aggregate without specifying an upper size sieve.

Fine aggregate for bituminous mixtures are defined as aggregates whose particles pass through a 2 mm sieve and retained on a 0.063 mm sieve. In the case of aggregates for unbound and hydraulically bound layers, aggregates whose particles pass through the 6.3 mm sieve are defined as fine.

Fines, in all cases, are defined as the particle size fraction of an aggregate that passes through the 0.063 mm sieve.

Filler aggregates or fillers are aggregates most of which pass through a 0.063 mm sieve. They are added to construction materials to provide certain properties.

According to American Association of State Highway and Transportation Officials (AASHTO) and American Society for Testing and Materials (ASTM) standards, coarse aggregate is retained on the 4.75 mm sieve (also known as the No. 4 sieve); fine aggregate passes through the 4.75 mm sieve and is retained on the 0.075 mm sieve (or No. 200 sieve); for fines, all particles pass through the 0.075 mm sieve (or No. 200 sieve).

2.10 AGGREGATE TESTS

Because of the variability of aggregate sources, aggregates should be tested to determine their suitability for use in asphalt (bituminous mixtures) or as material in base and subbase layers. Their suitability is defined in reference to geometrical, physical and chemical properties/requirements. Table 2.3 summarises the main test methods.

2.11 GEOMETRICAL PROPERTIES DETERMINATION TESTS

Geometrical properties such as particle size and particle size distribution, particle shape and angularity (angled surfaces caused by crushing or friction) affect the mechanical properties of layers with or without binder. Well-graded aggregates having a cubic shape, a small portion of flaky-shaped aggregates and angled surfaces caused by crushing are expected to have the best mechanical behaviour.

Particles with size smaller than 0.002 mm, that is, clay, might also affect cohesion of the asphalt (bituminous) mixture, as well as the mechanical behaviour of both asphalt and unbound layer in the presence of water.

A brief description of the main tests for defining the geometrical properties of aggregates is given in the following paragraphs.

2.11.1 Particle size distribution – sieving method

The distribution of aggregate sizes is determined by sieving or sieve analysis. The aim of sieve analysis is to determine the aggregate gradation curve.

Table 2.3 Tests required by the European standards CEN EN 13043 (2004) and CEN EN 13242 (2007) on aggregates for bituminous mixtures (BM) and unbound (UBM) or hydraulically bound mixtures (HBM)

			Requirement	
			BM	*UBM and*
A/A	*Tests*	*Test standard*	*(asphalts)*	*HBM*
1.	Geometrical properties			
1.1	Particle size distribution	CEN EN 933-1 (2005)	Yes	Yes
1.2	Grading of filler aggregates, if filler content is >10%	CEN EN 933-10 (2009)	Yes	—
	Particle shape of coarse aggregates:			
1.3	Flakiness index	CEN EN 933-3 (2012)	Yes	Yes
1.4	Shape index	CEN EN 933-4 (2008)	Yes	Yes
1.5	Percentage of crushed and broken surfaces in coarse aggregate particles	CEN EN 933-5 (2004)	Yes	Yes
1.6	Flow coefficient of aggregates (*angularity of fine aggregates*)	CEN EN 933-6 (2004)	Yes	—
	Cleanliness and quality of fines, if fines content >3%:			
1.7	Fines content	CEN EN 933-1 (2005)	Yes	Yes
1.8	Methylene blue, (MB_F) on 0/0.125 mm fraction	CEN EN 933-9 (2009)	Yes	—
1.9	Methylene blue (MB) on 0/2 mm fraction	CEN EN 933-9 (2009)	—	Yes
1.10	Sand equivalent	CEN EN 933-8 (2012)	—	Yes
2.	Physical properties			
	Resistance to fragmentation/crushing			
2.1	Resistance to fragmentation (Los Angeles test)	CEN EN 1097-2 (2010) §5	Yes	Yes
2.2	Resistance to impact (SZ)	CEN EN 1097-2 (2010) §6	Yes	Yes
	Resistance to polishing/abrasion/wear/ attrition			
2.3	Resistance to polishing of coarse aggregate (PSV test)	CEN EN 1097-8 (2009)	Yes	—
2.4	Resistance to surface abrasion (AAV test)	CEN EN 1097-8 (2009)	Yes	—
2.5	Resistance to wear of coarse aggregate (micro-Deval test)	CEN EN 1097-1 (2011)	Yes	Yes
2.6	Resistance to wear by abrasion from studded tyres (Nordic test)	CEN EN 1097-9 (2008)	Yes[a]	—
2.7	Classification of the constituents of coarse recycled aggregates	CEN EN 933-11 (2009)	—	Yes
	Particle density and water absorption			
2.8	Particle density and water absorption	CEN EN 1097-6 (2005)	Yes	Yes
2.9	Loose bulk density and voids	CEN EN 1097-3 (1998)	Yes	—
	Durability against freeze/thaw, weathering and thermal shock			
2.10	Magnesium sulfate test	CEN EN 1367-2 (2009)	Yes	Yes
2.11	Resistance to freezing and thawing	CEN EN 1367-1 (2007)	Yes[a]	Yes[a]
2.12	Water absorption as a screening test for freeze/thaw resistance	CEN EN 1097-6 (2005)	Yes	Yes

(continued)

Table 2.3 Tests required by the European standards CEN EN 13043 (2004) and CEN EN 13242 (2007) on aggregates for bituminous mixtures (BM) and unbound (UBM) or hydraulically bound mixtures (HBM) (Continued)

A/A	Tests	Test standard	Requirement BM (asphalts)	Requirement UBM and HBM
2.13	Boiling test for signs of 'Sonnenbrand'[b] in basalts	CEN EN 1367-3 (2004)	Yes	Yes
2.14	Resistance to thermal shock	CEN EN 1367-5 (2011)	Yes	—
2.15	Affinity of coarse aggregates to bituminous binders	CEN EN 12697-11 (2012)	Yes	—
2.16	Water content of aggregates	CEN EN 1097-5 (2008)	Yes	—
2.17	Particle density of filler	CEN EN 1097-7 (2008)	Yes	—
	Stiffening properties			
2.18	Voids of dry compacted filler	CEN EN 1097-4 (2008)	Yes	—
2.19	Delta ring and ball test of filler aggregates for bituminous mixtures	CEN EN 13179-1 (2000)	Yes	—
3.	Chemical properties			
3.1	*Chemical composition*			
	Petrographic description of aggregates	CEN EN 932-3 (2003)	Yes	Yes
3.2	Determination of lightweight contaminators	CEN EN 1744-1 (2009)	Yes	—
3.3	Determination of acid-soluble sulfate	CEN EN 1744-1 (2009)	—	Yes
3.4	Determination of total sulfur	CEN EN 1744-1 (2009)	—	Yes
3.5	Determination of water-soluble sulfate	CEN EN 1744-1 (2009)	—	Yes
	Determination of unsoundness of blast-furnace and steel slags	CEN EN 1744-1 (2009),		
3.6	Dicalcium silicate disintegration of air-cooled blast-furnace slag	§19.1	Yes	Yes
3.7	Iron disintegration of air-cooled blast-furnace slag	§19.2	Yes	Yes
3.8	Volume stability (expansion) of steel slag	§19.3	Yes	Yes
3.9	Water solubility of filler aggregate	CEN EN 1744-1 (2009) §16	Yes	—
3.10	Water susceptibility of fillers for bituminous mixtures	CEN EN 1744-4 (2005)	Yes	—
3.11	Water-soluble constituents	CEN EN 1744-3 (2002)	—	Yes
3.12	Calcium carbonate content of limestone filler aggregate	CEN EN 196-2 (2013)	Yes	—
3.13	Calcium hydroxide content of mixed filler	CEN EN 459-2 (2011)	Yes	—

[a] Usually in countries with cold climate.
[b] Type of weathering of basalts.

For bituminous mixtures or unbound/hydraulically bound layers, the size and number of sieves are selected from a basic sieve set and from one of two alternative sieve sets (set 1 or set 2). The basic set of sieves and the basic set of sieves plus set 1 or set 2 are shown in Table 2.4. Each European Union country selects the set it is going to use. The sieves from one set should never be mixed with the series of sieves of the other.

Between sieves 0.063 and 1 mm, European standards recommend using one or more sieves having a size between 0.063 and 1 mm. These sieves are 0.150, 0.250 and 0.500 mm regardless of the sieve set selected.

Table 2.4 Series of sieves recommended by European and American standards

| European standards (CEN EN 13043 2004 and 13242 2007) | | | American standards ASTM D 2940, ASTM D 3515, AASHTO M 92 (2010) |
Basic set (mm)	Basic set plus set 1 (mm)	Basic set plus set 2 (mm)	
0.063	**0.063**	0.063	0.075 mm (No. 200)
1	**1**	1	0.150 mm (No. 100)
2	**2**	2	0.300 mm (No. 50)
4	**4**	4	0.425 mm (No. 40)
—	5.6 (5)[a]	—	0.600 mm (No. 30)
—	—	6.3 (6)	1.18 mm (No. 16)
8	**8**	8	2.36 mm (No. 8)
—	—	10	4.75 mm (No. 4)
—	11.2 (11)	—	9.5 mm (3/8 in.)
—	—	12.5 (12)	12.5 mm (1/2 in.)
16	**16**	**16**	19.0 mm (3/4 in.)
—	—	20	25.0 mm (1 in.)
—	22.4 (22)	—	37.5 mm (1 1/2 in.)
31.5	**31.5 (32)**	**31.5 (32)**	50 mm (2 in.)
—	—	40	63 mm (2 1/2 in.)
—	45	—	75 mm (3 in.)
63	**63**	**63**	
—	—	80[b]	
—	90[b]	—	

Note: Bold characters denote the sieves of the basic series.

[a] The rounded value given may be used only for simplification of the size of the aggregate.
[b] This sieve may be used if needed. In special cases, the 125 mm sieve may also be used.

Table 2.4 also shows the equivalent sieves used in American standards.

Each sieve is designated by a number that corresponds to the aperture of the sieve mesh. Sieves used for sieve analysis have apertures of square shape, as specified by CEN EN 933-2 (1995) or ASTM E 11 (2013). They are designated by the edge lengths of the square apertures.

A basic condition of sieve analysis is to first obtain a representative bulk sample of aggregate and then to use a proper quantity of this for sieve analysis. The coarser the aggregate is, the larger the required quantity of sampled aggregate. A representative sample of the bulk sample is prepared in the laboratory by quartering or by using a splitting vessel. The minimum amount of aggregate required by European standards for sieve analysis, depending on the aggregate particle size, is shown in Table 2.5.

Table 2.5 Minimum mass of aggregates for sieving, per batch

Aggregate size, D (mm)	Minimum mass of aggregates for particle size distribution by sieving method (kg)
63	40
32	10
16	2.6
8	0.6
≤4	0.2

Source: Reproduced from CEN EN 933-1, *Tests for geometrical properties of aggregates – Part 1: Determination of particle size distribution – Sieve analysis*, Brussels: CEN, 2005. With permission (© CEN).

Table 2.6 Minimum masses of aggregate for sieving by American standards

Aggregate size (mm)	Minimum field sample mass of aggregates[a,b] (kg)	Minimum mass for sieving, per batch[c] (kg)
2.36	10	0.3
4.75	10	0.3
9.5	10	1
12.5	15	2
19	25	5
25	50	10
37.5	75	15
50	100	20
63	125	35
75	150	60
90	175	100

[a]   ASTM D 75 (2009) (AASHTO T 2 2010).
[b]   Regarding combined coarse and fine aggregates, the minimum weight shall be that of the coarse aggregate plus 10 kg.
[c]   ASTM C 136 (2006) (AASHTO T 27 2011).

According to American standards, the minimum mass of aggregate for sieving, as well as for initial sampling, is presented in Table 2.6.

During sieving, the 'overloading' of each sieve should be avoided. According to CEN EN 933-1 (2005), to avoid 'overloading', the portion retained on each sieve at the end shall not be over the specified proportion, expressed in grams, which is determined by the following equation:

$$(A \times \sqrt{d})/200$$

where A is the surface of the sieve (mm^2) and d is the aperture size of the sieve (mm).

If the retained portion on a sieve is larger than the one determined by the above equation, the portion is divided into two or three smaller parts and further sieving is carried out by hand. Finally, to determine the final portion retained, the subportions retained on a sieve are combined for the particular sieve.

The respective maximum retained acceptable proportions are not specified by American standards. According to American standards, effective sieving is ensured by extending the sieving time for 1 min. During this extension of time, the extra portion passing through the particular sieve shall not be more than 0.5% of the total sample weight.

2.11.1.1 Aggregate size

According to European standards, aggregate size designates an aggregate in terms of lower (d) and upper (D) sieve sizes expressed as d/D. This designation accepts the presence of some particles that are retained on the upper sieve (oversize) and some that pass through the lower sieve (undersize). The lower sieve size (d) can be zero.

In the case of a coarse aggregate for bituminous mixtures, the upper sieve size (D) is designated as the sieve from which 90% up to 99%, or 85% up to 99% or even 80% to 99% of the aggregate mass passes through the upper sieve. In the case of coarse aggregate for unbound or hydraulically bound layers, the above percentage might vary from 85% up to 99% or 80% up to 99%. The lower sieve size (d) for bituminous mixtures is designated as the sieve from which 0% to 10%, or 0% to 15%, or 0% to 20% and even 0% to 35% of the aggregate mass passes through the lower sieve. For unbound and hydraulically bound materials, the above percentages may be 0% to 15% or 0% to 20%. The coarse aggregate is categorised by the letters 'G_C' followed by two numbers that correspond to the percentage passing though the upper sieve 'D' and the percentage passing through the characteristic lower sieve 'd'. For example, coarse aggregate category G_C90/15 means that at least 90% of its mass passes through the upper sieve and not more than 15% passes through the lower sieve.

In the case of fine aggregate for bituminous mixtures, the percentage passing through the upper sieve 'D' (D ≤ 2 mm) can range from 85% up to 99%, while for unbound and hydraulically bound materials, the allowable ranges are 80% to 99% or 85% to 99%. The fine aggregate is categorised by the letters G_F followed by a number that corresponds to the percentage passing through sieve D. For example, G_F85 means that at least 85% passes the upper sieve.

In the case of an 'all-in aggregate' (aggregate mix), the aggregate is categorised by the letters G_A followed by a number that corresponds to the percentage passing through the upper sieve 'D' (D ≤ 45 mm for bituminous mixtures and D > 6.3 mm for unbound and hydraulically bound materials).

The designation or classification of the aggregate (coarse or fine) sizes according to American standards (ASTM D 448 2012) is based on the size number designation and nominal size range. The size number designation ranges from 1 to 10 with some intermediate numbering. The nominal size range corresponds to the upper and lower sieve. The upper sieve, unlike the European specification, is the one from which 100% of the mass passes through the upper sieve. The lower sieve is the one from which 0% to 10% or 0% to 15% of the aggregate mass passes through.

2.11.1.2 Sieving procedure and aggregate gradation curve determination

The sieving procedure is carried out on dry aggregates using the appropriate sieve series and a sieve shaker. If there is no suitable sieving apparatus, sieving can be carried out by hand.

According to the European standard CEN EN 933-1 (2005), before sieving, the aggregate sample is washed thoroughly and dried until it obtains a stable mass. This allows the percentage of fines to be determined. A detailed description of the sieving procedure according to the European and American standards is given in CEN EN 933-1 (2005), ASTM C 136 (2006), ASTM C 117 (2013), ASTM D 546 (2010), AASHTO T 11 (2009), AASHTO T 27 (2011) and AASHTO T 37 (2011).

After sieving for a sufficient period, the mass retained on each sieve is weighed and expressed initially as a retained percentage (RP) of the total mass (see the examples in Table 2.7, column 3).

The mass retained on each sieve is then expressed as a cumulative percentage passing through each sieve (see column 4 of Table 2.7). The cumulative percentage passing through each sieve is calculated by the following formulas:

For the first sieve, $(CPP)_1 = 100 - (RP)_1$;
for the second sieve, $(CPP)_2 = 100 - (RP)_1 - (RP)_2$;
for the third sieve, $(CPP)_2 = 100 - (RP)_1 - (RP)_2 - (RP)_3$;
for the fourth sieve, $(CPP)_2 = 100 - (RP)_1 - (RP)_2 - (RP)_3 - (RP)_4$; and so on,

where RP_1, RP_2, RP_3, RP_4, and so on, are retained percentages on the respective sieves.

Table 2.7 Sieving results

Sieves (mm)	Retained mass (g)	Retained percentage	Cumulative percentage passing
14	0	0.0	100
10	100	4.3	95.7
6.3	650	27.8	67.9
4	714	30.6	37.3
2	614	26.3	11.0
0.5	85	3.6	7.4
0.25	36	1.5	5.9
0.063	79	3.4	2.5
Filler	58	2.5	—
Total	2336	100.0	

The cumulative percentage with the corresponding sieve is plotted on graph paper and hence the aggregate gradation is obtained. The European specification suggests that both axes be linear, unlike the American specification that requires the x-axis to be logarithmic.

Determination of the filler portion is important as its presence or absence in the mixture directly affects the behaviour of either the bituminous mixture or the unbound aggregate mixture. In general, too much filler is more harmful than the lack of it. However, the latter is unusual.

High filler content in aggregates (usually >10% for aggregates used in bituminous mixtures or >15% for aggregates used in unbound layers) contributes to (a) increase of the bitumen content and increase of the water content for optimum compaction density, (b) decrease of the mixture workability, (c) the aggregate coating, (d) stiffness of the bituminous mix and (e) the 'balling' effect in either cold or hot bituminous mixtures.

2.11.2 Particle shape tests of coarse aggregates

Tests for determining the particle shape or form of aggregates are the Flakiness Index and Shape Index. They are carried out mainly on crushed aggregates. The shape of a crushed aggregate is affected by its petrological composition and the crushing process (number of crushing stages, type of crusher and size of rock used to feed the initial crusher).

2.11.2.1 Flakiness index test

According to the European specification CEN EN 933-3 (2012), the aggregates used for the Flakiness Index test have a nominal diameter smaller than 80 mm and bigger than 4 mm. The test consists of two sieving processes. First, the sample is divided into the various aggregate fractions depending on their size, that is, passing through an upper sieve D_i and retained on the lower sieve d_i. The fraction size is expressed as d_i/D_i. A portion from each fraction size is then sieved using a special Flakiness Index sieve. This sieve consists of stainless steel bars that have a specific space between them; hence, they are called bar sieves. The typical EN series of sieves with the corresponding opening between the bars is shown in Table 2.8.

The total Flakiness Index is designated as the total aggregate mass passing through each bar sieve, expressed as the percentage of the total sample mass of dry aggregates tested.

If the Flakiness Index for each aggregate size d_i/D_i is required, it is calculated by the mass passing through the bar sieve expressed as a percentage of mass of the respective aggregate size.

Table 2.8 Bar sieves according to CEN EN 933-3 (2012)

Particle size fraction d_i/D_i (mm)	Width of slot in bar sieve (mm)
63/80	40 ± 0.3
50/63	31.5 ± 0.3
40/50	25 ± 0.2
31.5/40	20 ± 0.2
25/31.5	16 ± 0.2
20/25	12.5 ± 0.2
16/20	10 ± 0.1
12.5/16	8 ± 0.1
10/12.5	6.3 ± 0.1
8/10	5 ± 0.1
6.3/8	4 ± 0.1
5/6.3	3.15 ± 0.1
4/5	2.5 ± 0.1

Source: Reproduced from CEN EN 933-3, *Tests for geometrical properties of aggregates – Part 3: Determination of particle shape – Flakiness index*, Brussels: CEN, 2012. With permission (© CEN).

More information, as well as a calculation example, is presented in CEN EN 933-3 (2012).

The Flakiness Index value determined in accordance with the European specification differs from the one determined by BS 812-Part 105, which was used in the past by many countries. Thus, care must be taken to the limiting values that may have been set.

2.11.2.2 Shape Index test (CEN EN 933-4 2008)

According to the European standard CEN EN 933-4 (2008), single particles of aggregate are classified according to their length L and their thickness E using a slide gauge (caliper). The Shape Index is calculated as the mass of aggregate with a ratio of dimension $L/E \geq 3$ expressed as a percentage of the total dry mass of the sample tested.

The test is conducted on each aggregate fraction d_i/D_i, where D (maximum particle size) < 2d (minimum particle size).

More information about conducting the test and the minimum quantities of aggregate required for the test is given in CEN EN 933-4 (2008).

The Shape Index and Flakiness Index test methods are essential when the aggregate is to be used in bituminous mixtures particularly those incorporating single-sized aggregates such as porous asphalt, surface dressing or precoated chippings for HRA. In general, the particle shape of coarse aggregates influences the mechanical properties of some construction materials and may affect placement, compaction and consolidation.

2.11.3 Flat particles, elongated particles or flat and elongated particles test

This test proposed by the American standard ASTM D 4791 (2010) is equivalent to the Flakiness and Shape Index test methods proposed by the European standards. It determines the percentage of flat particles, elongated particles or flat and elongated particles in coarse aggregates (retained on 9.5 mm or 4.75 mm sieve depending on the requirements).

Individual aggregate particles of specific sieve size are measured to determine the ratios of width to thickness, length to width or length to thickness.

Two procedures are detailed, that is, Method A and Method B. Method B is intended for use with Superpave specifications.

More information about conducting the test is given in the relevant standard ASTM D 4791 (2010).

2.11.4 Crushed and broken surfaces test

This is carried out in accordance with the European standard CEN EN 933-5 (2004) using coarse aggregate d_i/D_i, where $D_i < 63$ mm and $d_i > 4$ mm. The test is a manual segregation of the aggregate sample into two groups: (a) crushed and broken particles including totally crushed or broken particles and (b) rounded particle group including totally rounded aggregates.

The mass of each group is recorded and expressed as a percentage of the total sample mass.

A totally crushed or broken particle is defined as a particle with more than 90% of its surface crushed or broken and is expressed with the index C_{tc}.

A crushed or broken particle is defined as a particle with more than 50% of its surface crushed or broken and is expressed with the index C_c.

A rounded particle is defined as a particle with 50% or less of its surface crushed or broken and is expressed with the index C_r.

A totally rounded particle is defined as a particle with more than 90% of its surface rounded and is expressed with the index C_{tr}.

In conclusion, crushed or broken surfaces are defined as facets of gravel produced by crushing or breaking by natural forces and bound by sharp edges.

More information about the test method, minimum quantities of aggregate for testing, and so on, are given in CEN EN 933-5 (2004).

A similar method is given in the American standard ASTM D 5821 (2006). Particles with n-fractured (crushed) surfaces are separated from the aggregate sample. The percentage of fractured particles with the specified number of fractured faces is determined by mass or count. More detail about the test method is given in ASTM D 5821 (2006).

The percentage of crushed and broken surfaces affects inter particle friction and hence the shear strength of a mixture. It also affects friction and surface texture for aggregates used in pavement surface courses.

2.11.5 Flow coefficient test

This test is conducted using coarse aggregates (4–20 mm) and fine aggregates (0–4 mm) according to CEN EN 933-6 (2004). Angularity is expressed in reference to flow coefficient. For coarse aggregate, the flow coefficient is linked to the percentage of crushed and broken surfaces. As a result, this test can be used in association with the test specified in CEN EN 933-5 (2004). Shape and surface texture characteristics also influence the results.

The flow coefficient (E_c) of an aggregate is the time, expressed in seconds, for a specified volume of aggregate to flow through a given opening, under specified conditions using an apparatus consisting of bar sieves conforming to CEN EN 933-3 (2012) and a vibrating plate.

More details regarding the mass of test portion, equipment required, the procedure and how to determine flow coefficient (E_c) is given in CEN EN 933-6 (2004).

2.11.6 Assessment of cleanness of fine aggregates and fines

Assessment of cleanness of fine aggregates and fines is carried out to ascertain the existence of harmful fines, that is, material that passes through the 0.063 mm sieve (silt or clay particles). The presence of this material affects the behaviour of the bituminous mixture, causing swelling of unbound layers or asphalt layers, and may be the reason for ravelling.

The tests used are (a) the determination of fines content, (b) the sand equivalent test conducted on fine aggregates and (c) the methylene blue test conducted when the fines content is greater than 3%.

The fines content is determined during the procedure for determination of particle size distribution in accordance with CEN EN 933-1 (2005).

2.11.6.1 Sand equivalent test

This test is conducted to determine quickly the relative proportion of the fine clay-like material in fine aggregate and granular soils. A low sand equivalent value indicates the presence of clay-like proportion. This is detrimental to the quality of the aggregate and characterises the aggregates as 'non-clean'.

The sand equivalent test may be conducted in accordance with either CEN EN 933-8 (2012) or ASTM D 2419 (2009) (AASHTO T 176 2008).

According to CEN EN 933-8 (2012), the test is conducted on aggregates that pass through the 2 mm sieve. It is noted that, if lumps are retained on the 2 mm sieve, they should be broken up so that it is included in the test.

From a representative sample, a mass of 120 g of dried material is placed into a graduated transparent cylinder. A portion of calcium chloride solution is added to the cylinder until it reaches about 100 mm. The contents of the cylinder are left undisturbed for about 10 min and then, after loosening the material from the bottom and placing a stopper, it is shaken manually (more than 90 times) or in a mechanical shaker for 30 ± 1 s. Then, more calcium chloride solution is added to the cylinder until the cylinder is filled to the 380 ± 0.25 mm graduation mark. The cylinder and its content are then left undisturbed for 20 min. During this period, the material settles out from suspension to form two distinctive layers. The height of the clay suspension (h_c) and the height of the sand reading (h_s) are taken.

The sand equivalent (SE) is calculated by the following formula:

$$SE = h_s/h_c \times 100.$$

The average of two results is rounded to the nearest whole number. In case the difference between the two results is greater than 4 units, the test is repeated. More information about the test is given in CEN EN 933-8 (2012) and ASTM D 2419 (2009).

Differences between European and American standard methods can be found in the fine aggregate fraction, the mass of sample taken, the number of tests carried out at the same time and the expression of results. In the European standard, the fine aggregate assessed passes through the 2 mm sieve in contrast to the 4.75 mm sieve according to the American standard. In the European method, the test portion quantity is designated by mass (120 g), whereas in the American method, it is designated by volume (approximately 85 ml). The European sand equivalent value is determined by averaging two results rounded to the nearest whole number, in contrast to averaging three results rounded to the nearest 0.1% according to the American standard.

The key difference that affects the result is the aggregate size. The rest of the differences are believed to have a minor effect.

Research carried out by Nikolaides and Manthos (2007) on aggregates derived from various rocks (limestones, gabbros, basalts, crushed sand-gravel and steel slags) found that the sand equivalent values determined in accordance with the European standard $(SE_{0/2})$ were always lower than those determined in accordance with the American standard $(SE_{0/4.75})$. The results obtained are shown in Figure 2.1.

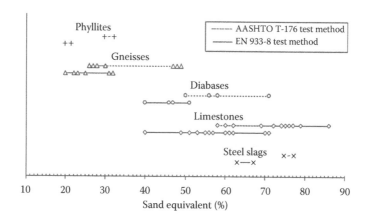

Figure 2.1 Sand equivalent values of sands from various rocks. (From Nikolaides A. and E. Manthos, *Sand equivalent of road aggregates tested with European and American standards and methylene blue results. Proceedings of the 4th International Conference, 'Bituminous Mixtures and Pavements'*, Vol. 1, p. 199. Thessaloniki, Greece: Aristotle University, 2007.)

From the same study, a good correlation was also found among the results, which is expressed by the following formula:

$$SE_{0/2} = 0.842 \times SE_{0/4.75} - 3.547 \ (R^2 = 0.91),$$

where $SE_{0/2}$ is the sand equivalent value determined by the CEN EN 933-8 (2012) procedure (%) and $SE_{0/4.75}$ is the sand equivalent value determined by the ASTM D 2419 (2009) procedure (%).

The difference observed is important as the limiting SE values set by many national specifications suggest that the test is conducted in accordance to the American standard. Hence, if a material has an acceptable but close to the limit SE value when derived using the American standard procedure, it will not have an acceptable SE value if tested using the European standard procedure. This necessitates altering the limiting value set if the European standard is specified for determining the SE value.

Generally, surface courses containing aggregates with a lower SE value than the limiting value are expected to suffer from ravelling and, ultimately, potholes. Subbase and base layers constructed from aggregates with low SE aggregates are prone to swelling. Finally, soils on which a pavement is to be constructed with a very low SE value (lower than 10%) will develop excessive swelling and have very low bearing capacity in wet/saturated conditions.

2.11.6.2 Methylene blue test

The methylene blue test is used to determine the quality of fines by ascertaining the existence of active clay minerals. In contrast with inactive clay minerals, active clays tend to swell depending on their water content. This swelling has a detrimental impact on both the bituminous mixture and the unbound layers of the pavement.

The test is based on the adsorption principle of clay minerals using methylene blue dye. During the test, the quantity of methylene blue required to cover all clayey ingredients is measured. The quantity of methylene blue dye adsorbed is related to the specific surface of the clay mineral (montmorillonite, illite and kaolinite). Active clay minerals have a large specific surface in contrast to inactive clay materials. As a result, the required methylene blue quantity will be proportional to the quantity and the type of clay minerals.

The methylene blue test is supplementary to the sand equivalent test, since the latter determines only the existence of clay particles and not the presence of active clay minerals. In some countries, the methylene blue test has replaced the sand equivalent test.

The test, according to CEN EN 933-9 (2009), is executed on either the 0/2 mm or 0/0.125 fine aggregate fraction. At least 200 g for the 0/2 mm particle size or 30 ± 0.1 g for the 0/0.125 mm particle size is added to 500 ± 5 ml of distilled water. The solution is stirred for 5 min and then a dose of 5 ml of methylene blue dye is added. Further stirring is carried out for at least 1 min and then the methylene blue test is performed by taking one drop of suspension by means of the glass rod and depositing it on a filter paper. The stain formed consists of a solid blue colour surrounded by a colourless wet zone. The test is considered to be positive if, in the wet zone, a halo is formed consisting of a persistent light blue ring of about 1 mm. The end point of the test is confirmed by repeating the stain test at 1 min intervals for 5 min, without adding more dye solution.

The methylene blue value (MB) of the 0/2 mm fraction, expressed in grams of dye per kilogram, is determined from the following relationship:

$$MB = (V_1/M_1) \times 10 \ (g/kg),$$

where V_1 is the total volume of dye solution injected (ml) and M_1 is the mass of the test portion (g).

If the test is carried out by adding kaolinite, the above equation becomes

$$MB = ((V_1 - V')/M_1) \times 10 \ (g/kg),$$

where V' is the volume of dye solution absorbed by the kaolinite (ml).

In the case a fraction 0/0.125 mm is used, the result is marked as MB_F. In both cases, the methylene blue value is recorded to the nearest 0.1 g/kg.

More information about the test is given in CEN EN 933-9 (2009). The test may also be carried out in accordance with the American standard ASTM C 837 (2009).

Methylene blue tests for determining both MB and MB_F values have been carried out on various rocks (Nikolaides and Manthos 2007). The results obtained are shown in Figures 2.2 and 2.3.

From Figure 2.2, it can be seen that when a 0.2 mm fraction is used, some samples from certain rocks gave MB values that are above the limiting value of 3.0 g/kg normally used.

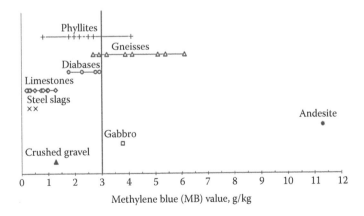

Figure 2.2 Graphical representation of methylene blue (MB), 0/2 mm fraction, from various types of rocks. (From Nikolaides A. and E. Manthos, Sand equivalent of road aggregates tested with European and American standards and methylene blue results. *Proceedings of the 4th International Conference, 'Bituminous Mixtures and Pavements'*, Vol. I, p. 199. Thessaloniki, Greece: Aristotle University, 2007.)

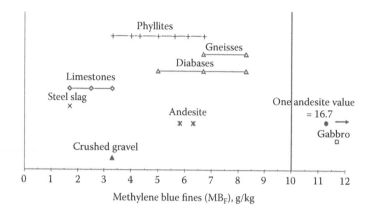

Figure 2.3 Graphical representation of methylene blue (MB$_F$) 0.0125 mm fraction for various types of rocks. (From Nikolaides A. and E. Manthos, Sand equivalent of road aggregates tested with European and American standards and methylene blue results. *Proceedings of the 4th International Conference, 'Bituminous Mixtures and Pavements'*, Vol. 1, p. 199. Thessaloniki, Greece: Aristotle University, 2007.)

However, all samples except gabbro and some andesite aggregates, when a 0/0.125 fraction was used, gave MB$_F$ values below the limiting value of 10 g/kg normally used.

The limestone, the crushed gravel and the diabase aggregates gave methylene blue values below the limiting values regardless of fraction size tested.

Prior to CEN EN 933-9 (2009), the methylene test was carried out in accordance to an ISSA specification (ISSA TB 145 1989). The aggregate fraction used was 0/0.075 mm and the sample mass was 1 g added to 30 ml of distilled water. The methylene blue dye was added in steps of 0.5 ml. The methylene blue value (MB$_f$) was expressed in grams of dye solution per 100 g of fines (ISSA TB 145 1989). The magnitude of the MB$_f$ value was one-tenth of the MB$_F$ value.

2.11.6.3 Grading of fines

The particle size distribution of fines should be determined in case their content in the mixture of aggregates is greater than 10%. In the case where a filler has been used, its particle size distribution should be determined regardless of the percentage added.

The grading of the fines is determined in accordance with CEN EN 933-10 (2009), using an air jet sieving apparatus.

A dried test portion is placed on a 0.063 mm sieve fitted on the air jet sieving apparatus and is sieved until sieving is completed (minimum, 3 min). The mass retained is recorded and placed on a 0.125 mm sieve for further sieving. By recording the mass retained again, the cumulative percentages of the original dry mass passing through the 0.125 and 0.063 mm test sieves are determined and compared with the requirements.

More details about the sieving procedure, sample preparation and the apparatus are given in CEN EN 933-10 (2009).

2.12 PHYSICAL PROPERTIES DETERMINATION TESTS

2.12.1 Resistance to fragmentation and polishing/abrasion tests

Resistance of fragmentation/crushing and polishing/abrasion are specified for determining the mechanical behaviour of aggregate under the disruptive action of traffic, as well as the wear experienced during production, laying and compaction of the materials.

The tests to determine resistance to fragmentation of aggregates are the following:

* Resistance to fragmentation by the Los Angeles test
* Resistance to impact test

With regard to the resistance to polishing/abrasion of the aggregates, the tests are as follows:

* Polished stone value (PSV) test
* Aggregate abrasion value (AAV) test
* Resistance to wear by micro-Deval test

In countries where studded tyres are used or permitted in the winter, the resistance to wear by abrasion from studded tyres test also needs to be conducted.

2.12.1.1 Resistance to fragmentation by the Los Angeles test

This test is perhaps the oldest and best known of all the aggregate tests. It determines the wear on aggregates under the influence of crushing and abrasion forces. The forces are developed during rotation of the aggregate and steel spheres in an apparatus known as the Los Angeles machine (see Figure 2.4). The machine consists of a rotating drum with internal dimensions 508 mm (length) by 711 mm (diameter).

The standard size of aggregate tested in accordance with CEN EN 1097-2 (2010) is 10/14 mm size aggregate.

The mass of the test sample used is 5000 ± 5 g and is obtained from a sampled mass of at least 15 kg of aggregate. It is required that 60% to 70% of the test sample mass should pass through the 12.5 mm sieve. If an 11.2 mm sieve is used, the percentage changes to 30% to 40%.

Along with the dry aggregate, 11 steel spheres 45 to 49 mm in diameter and mass between 400 and 445 g each are also added. The drum is then sealed and is rotated for 500 revolutions

Figure 2.4 Los Angeles apparatus. (Courtesy of Controls Srl.)

Table 2.9 Aggregate sizes for the Los Angeles test

Sieve size (mm)		Mass of indicated sizes (g)			
		Grading			
Percentage passing	Percentage retained	A	B	C	D
37.5	25	1250 ± 25	—	—	—
25	19	1250 ± 25	—	—	—
19	12.5	1250 ± 10	2500 ± 10	—	—
12.5	9.5	1250 ± 10	2500 ± 10	—	—
9.5	6.3	—	—	2500 ± 10	—
6.3	4.75	—	—	2500 ± 10	—
4.75	2.36	—	—	—	5000 ± 10
Total mass		5000 ± 10	5000 ± 10	5000 ± 10	5000 ± 10
Number of spheres		12	11	8	6

Source: Reprinted from ASTM C 131, *Standard test method for resistance to degradation of small-size coarse aggregate by abrasion and impact in the Los Angeles machine*, West Conshohocken, Pennsylvania, US: ASTM International, 2006. With permission (© ASTM International).

at 31 to 33 revolutions per minute. After completion of the test, the aggregates are sieved (wet sieving) to determine the mass of material that is retained on a 1.6 mm sieve.

The Los Angeles coefficient (LA) is calculated as

$$LA = (5000 - m)/50,$$

where m is the mass retained on a 1.6 sieve (g).

The lower the Los Angeles coefficient, the more durable and resistant the aggregate is to fragmentation.

When the test is carried out in accordance with the American ASTM C 131 (or AASHTO T 96 2010), the size of aggregate tested can be of maximum nominal size 37.5 mm, 19 mm, 9.5 mm or even 4.75 mm. Coarser than 37.5 mm aggregate can also be tested; details can be found in ASTM C 535 (2012).

The mass of the test sample is 5000 ± 10 g. The test sample required is shown in Table 2.9. The steel spheres used are similar in size and mass to the European CEN EN 1097-2 (2010) method. The number of revolutions is also the same but the number of steel spheres added to the drum depends on the size of the aggregate (grading) tested (see Table 2.9).

The American LA coefficient is calculated as the difference between the initial and final mass retained on a 1.7 mm sieve and is expressed as a percentage of the initial mass of the aggregate tested.

More details regarding the test are given in the relevant standards.

2.12.1.2 Resistance to impact test

The resistance impact test is an alternative test to the resistance to fragmentation by the Los Angeles test and is conducted in accordance with CEN EN 1097-2 (2010). It is also known as the German Schlagversuch test.

During the test, an aggregate of size 8/12.5 mm is added to a metal mould and is crushed under the influence of a dropped load (10 blows from a height of 370 mm). After crushing, the aggregate is sieved through the 0.2, 0.63, 2, 5 and 8 mm test sieves.

The impact crushing value (SZ) is calculated as

$$SZ = M/5 \ (\%),$$

where M is the sum of the percentage of the mass passing through each of the five above-mentioned sieves.

More information about the test accompanied by a calculation example is given in CEN EN 1097-2 (2010).

2.12.1.3 Polished stone value test

This test is conducted on coarse aggregates that are used for surfacing mixes. The PSV test determines the resistance of the coarse aggregate to the polishing action of vehicle tyres. The test consists of two parts: in the first part, test specimens are subjected to a polishing action in an accelerated polishing machine. In the second part of the test, the state of polish reached by each specimen is measured using the Pendulum tester apparatus. The PSV is then calculated from the friction determinations.

The test is carried out in accordance with CEN EN 1097-8 (2009) or with ASTM D 3319 (2011) (AASHTO T 279 2012). The key differences between the two standards are the nominal size of aggregate used, the polishing medium used, the number and the type of test wheels, the way the aggregates are subjected to polishing and the way the PSV index is calculated.

According to the European standard, the aggregate should pass through a 10 mm sieve and be retained on a 7.2 mm grid sieve. The polishing medium is emery (in two grades – emery corn and emery flour), the number of wheels are two (one for each type of emery) and the polishing time is 6 h (3 h +3 h for each polishing medium). With respect to the American standard, the aggregate should pass through a 12.5 mm sieve and be retained on a 9.5 mm sieve, the polishing medium is sand, only one wheel is used and the total polishing time may reach up to 10 h. The final result obtained by either standard cannot be the same. The following short description refers to the test carried out in accordance with CEN EN 1097-8 (2009).

Approximately 36 to 46 aggregate particles of size 7.2/10 mm are carefully placed in a single layer in a mould with their flattest surface lying on the bottom. The interstices between the aggregates are filled with fine sand. The quantity of sand used is such that only three quarters of the depth of the interstices is filled. Then, the mould is in-filled with an epoxy resin and any surplus is removed with a spatula. After the resin has hardened, the specimen is cleaned thoroughly to remove any sand and is transferred to the polishing machine.

The polishing machine (see Figure 2.5) has a metal wheel called the road wheel. This is 406 mm in diameter and holds the test specimens and the stone control specimens around its rim. The number of specimens placed around the wheel is 14 in total. The wheel rotates at 320 revolutions per minute during the test. The rubber-tyred wheel has a static contact force with the moulds of 725 N. For the first 3 h of the test, corn emery is fed onto the wheel at a rate of 27 ± 7 g/min together with a sufficient amount of water. The corn emery has a gradation with 98%–100% of the particles passing through the 0.600 mm sieve.

After the first 3 h, the second rubber-tyred wheel is fitted and the test is repeated for another 3 h using flour emery. The emery flour rate is 3 ± 1 g/min with the rate of water approximately twice that of the emery flour.

On completion of the test after 6 h, the test specimens are cleaned and tested for friction using the Pendulum friction tester (see Figure 2.6).

Figure 2.5 Typical accelerated polishing machine. (Courtesy of Controls Srl.)

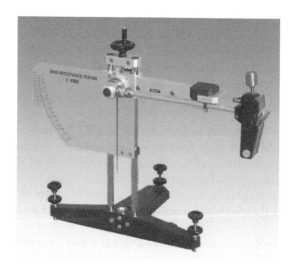

Figure 2.6 British pendulum friction tester. (Courtesy of Controls Srl.)

The PSV is determined using the following equation:

PSV = S + 52.5 − C,

where S is the mean value of skid resistance for the four aggregate test specimens and C is the mean value of skid resistance for the four PSV control stone specimens (this value should range from 49.5 to 55.5 units).

The determination of PSV is considered necessary for determining the suitability of aggregate for any type of surfacing asphalt and surface dressing. Aggregates are considered to be suitable when their PSV is greater than or equal to the value given by the national specification. These values are related to traffic level and the type of site.

The most detailed specification regarding PSV requirements is considered to be the one used in the United Kingdom. Table 2.10 shows the minimum PSV required for coarse aggregates in hot applied thin surface course systems used in the United Kingdom. A similar table exists with regard to chippings or coarse aggregates in other bituminous surfacings (Highways Agency 2012).

2.12.1.4 Aggregate abrasion value test

The resistance of the aggregate to surface wear by abrasion under traffic is determined in the laboratory by the AAV test. According to the European standard CEN EN 1097-8 (2009), the test is carried out on coarse aggregate passing through a 14 mm sieve and retained on a 10.2 mm grid sieve. The test is considered complementary to the PSV test. It is suggested that it should be used for high skid-resistant aggregates (typically those with a PSV of 60 or greater) to determine whether they are susceptible to abrasion under traffic.

At least 24 aggregate particles are placed in a steel mould in a single layer with their flattest surface down. The interstices between the aggregates are in-filled with fine sand to approximately three quarters of their depth. The rest of the mould is in-filled with epoxy resin, and a metal plate, which has been precoated with grease, is placed on top. After the resin hardens, the specimen is removed from the mould, cleaned thoroughly from any sand and transferred to the abrasion machine.

The aggregate abrasion machine (Figure 2.7) consists of a machined flat circular cast iron or steel grinding lap wheel not less than 600 mm in diameter. Two specimens are pressed against the surface of the grinding lap using a mass of 2000 ± 10 g. This gives a resultant force of 0.365 N/cm^2 acting on the specimen. The grinding lap then rotates horizontally for 500 revolutions at a speed of 28 to 31 min^{-1}. The interface of the aggregates and the abrasion head is continuously fed with sand at a rate of 800 ± 100 g/min for each specimen.

After completion of the test, the specimens are cleaned thoroughly and weighed with an accuracy of 0.1 g.

The AAV index is calculated in reference to the loss of weight of the specimen using the following equation:

$$AAV = 3(A - B)/\rho_{ssd},$$

where A is the mass of the specimen before abrasion (g), B is the mass of the specimen after abrasion (g) and ρ_{ssd} is the particle density of the aggregate (on a saturated surface dry basis) (Mg/m^3).

AAV limiting values relate to traffic volume. The limiting values proposed by the UK Highways Agency are given in Table 2.11.

2.12.1.5 Resistance to wear by the micro-Deval test

The test was developed in France by an engineer, Deval, and is similar to the resistance to fragmentation Los Angeles test. The test according to CEN EN 1097-1 (2011) uses 14/10 mm single-sized aggregates (passing through a 14 mm sieve and retained on a 10 mm sieve). This is similar to the Los Angeles test. The key differences with the Los Angeles test are the following: (a) the amount of aggregates used is smaller (500 ± 2 g); (b) the drum is smaller

Table 2.10 Minimum PSV of coarse aggregates in hot applied thin course system

Site category	Site description	IL[a]	Min PSV for given IL[a], traffic level[b] and type of site									
			<250	251–500	501–750	751–1000	1001–2000	2001–3000	3001–4000	4001–5000	5001–6000	>6000
A1	Motorways where traffic is generally free-flowing on a relatively straight line	0.30	50	50	50	50	50	50	50	63	63	63
		0.35	50	50	50	50	50	53	53	53	63	63
A2	Motorways where some braking regularly occurs	0.35	50	50	50	55	55	60	60	65	65	65
B1	Dual carriageways where traffic is generally free-flowing on a relatively straight line	0.30	50	50	50	50	50	50	50	53	63	63
		0.35	50	50	50	50	50	53	53	53	63	63
		0.40	50	50	50	50	53	58	58	58	63	68+
B2	Dual carriageways where some braking regularly occurs	0.35	50	50	50	55	55	60	60	65	65	65
		0.40	55	60	60	65	65	68+	68+	68+	68+	68+
C	Single carriageways where traffic is generally free-flowing on a relatively straight line	0.35	50	50	50	50	50	53	53	58	63	63
		0.40	50	53	53	58	58	63	63	63	68+	68+
		0.45	53	53	58	58	63	63	63	63	68+	68+
G1/G2	Gradients >5% longer than 50 m	0.45	55	60	60	65	65	68+	68+	68+	68+	68+
		0.50	60	68+	68+	HFS[c]	HFS	HFS	HFS	HFS	HFS	HFS
		0.55	68+	HFS[c]	HFS	HFS	HFS	HFS	HFS	HFS	HFS	HFS
K	Approaches to pedestrian crossings and other high-risk situations	0.50	65	65	65	68+	68+	68+	68+	68+	68+	HFS
		0.55	68+	68+	HFS	HFS	HFS	HFS	HFS	HFS	HFS	HFS

Site category	Description	IL[a]									
Q	Approaches to major and minor junctions on dual carriageways and single carriageways where frequent or sudden braking occurs but in a generally straight line	0.45	60	65	65	68+	68+	68+	68+	68+	HFS
		0.50	65	65	65	68+	68+	68+	HFS	HFS	HFS
		0.55	68+	68+	HFS	HFS	HFS	HFS	HFS	HFS	HFS
R	Roundabout circulation areas	0.45	50	55	60	60	65	65	68+	68+	68+
		0.50	68+	68+	68+	68+	68+	68+	68+	68+	68+
		0.55	HFS	HFS	HFS	HFS	HFS	HFS	HFS	HFS	HFS
S1/S2	Bends (radius <500 m) on all types of road, including motorway link roads; other hazards that require combined braking and cornering	0.45	68+	68+	68+	68+	68+	HFS[c]	HFS	HFS	HFS
		0.50	HFS	HFS	HFS	HFS	HFS	HFS	HFS	HFS	HFS
		0.55	HFS	HFS	HFS	HFS	HFS	HFS	HFS	HFS	HFS

Source: Highways Agency, *Interim Advice Note 156/12, Revision of Aggregate Specification for Pavement Surfacing*, London: Department for Transport, Highways Agency, 2012 (© Highways Agency).

Note: Where '68+' material is listed in this table, none of the three most recent results from consecutive PSV tests relating to the aggregate to be supplied must fall below 68.

For site categories G1/G2, S1/S2 and R, any PSV in the range given for each traffic level may be used for any IL and should be chosen on the basis of local experience of material performance. In the absence of this information, the values given for the appropriate IL and traffic level must be used.

a IL, investigatory level, defined in HD 28 (Highways Agency 2004).
b The traffic level is expressed in terms of the expected commercial vehicles (cv) per lane, per day (cv/lane/day).
c HFS, specialised high friction surfacing, incorporating calcined bauxite aggregate.

Figure 2.7 Aggregate abrasion machine. (Courtesy of Cooper Research Technology Ltd.)

Table 2.11 Maximum AAV by British specifications

Traffic (cv/lane/day) at design life	<250	251–1000	1001–1750	1751–2500	2501–3250	>3250
Max. AAV for chippings for hot rolled asphalt and surface dressing, and for aggregate in slurry and microsurfacing systems	14	12	12	10	10	10
Max. AAV for aggregate in thin surface course systems, exposed aggregate concrete surfacing and coated macadam surface course	16	16	14	14	12	12

Source: Highways Agency, *Design Manual for Road and Bridges (DMRB), Volume 7: Pavement design and maintenance, Section 5, Part 1*, HD 36/06. London: Department for Transport, Highways Agency, 2006 (© Highways Agency).

Note: For roads carrying less than 1750 cv/lane/day, aggregates of higher AAV may be used where experience has shown that satisfactory performance is achieved by an aggregate from a particular source.

with 200 ± 1 mm internal diameter and 154 ± 1 mm length; (c) the rotation time is longer, 2 h (12,000 revolutions); (d) the rotation speed is higher (100 revolutions per minute); (e) the diameter of the steel spheres added is smaller (10 ± 0.5 mm); and (f) 2.5 ± 0.05 l of water is added to the drum. During the micro-Deval test, the aggregates are subject to higher distress than that in the Los Angeles test.

After completion of the test, the aggregates are dried and sieved through the 1.6 mm sieve. The micro-Deval coefficient is calculated as

$$M_{DE} = (500 - m)/5,$$

where M_{DE} is the micro-Deval coefficient and m is the mass of aggregate retained on the 1.6 mm sieve (g).

The reported coefficient M_{DE} of the aggregates tested (the value is rounded to the nearest whole number) is the mean value of two samples tested. The test may also be conducted without the addition of water. The reported value is designated M_{DS}. A detailed description of the test procedure is given in CEN EN 1097-1 (2011).

The micro-Deval test is not as popular as the Los Angeles test. It is optional for some countries, whereas it is obligatory for a few such as France. The relationship between micro-Deval and Los Angeles coefficients following the French standard (XP P 18-545 2003) is normally –5 units of the LA value obtained. In certain cases, according to the French standard, the micro-Deval test is supplementary to the Los Angeles test. The maximum permissible M_{DE} and LA values for roadwork aggregates according to the French practice are given in XP P 18-545 (2003).

The micro-Deval test is also described by the American standard ASTM D 6928 (2010). The main differences from the European standard are as follows: (a) the aggregate fractions tested 19/9.5, 12.5/4.75 and 9.6/4.75 mm and (b) the proportion of sample tested, which is 1500 g in all cases.

2.12.1.6 Resistance to wear by abrasion from studded tyres – Nordic test

The test to determine resistance to wear of aggregate by abrasion from studded tyres test was developed in Finland, Norway and Sweden. In these countries, studded tyres are permitted and used during the winter period.

The test simulates the polishing/abrasion effect of these specific tyres on the surfacing coarse aggregates. The abrasion machine is known as the Nordic abrasion machine (see Figure 2.8).

The test is conducted normally on 11.2/16 mm or, alternatively, on 8/12 mm size aggregates, using an apparatus similar to the micro-Deval test. The only difference is three ribs are mounted on the interior of the drum to give extra abrasion. The test is performed in the presence of water.

Figure 2.8 Nordic abrasion machine. (Courtesy of Cooper Research Technology Ltd.)

After rotating the aggregates and steel spheres for the specified period, the aggregates are dried and then sieved to determine the mass retained on a 14 mm test sieve. The 14 mm sieve is nested with a 2.0 mm sieve protected by an 8.0 mm guard sieve. Their sum is deducted from the initial mass of the aggregates and expressed as a percentage of the initial mass of the aggregate. The value obtained is designated as the Nordic abrasion value.

More information about the test procedure is given in CEN EN 1097-9 (2008).

2.12.2 Particle density and water absorption tests – general

Determining the density or the specific gravity of aggregates is important as the value is used in calculations such as composition of aggregate in bituminous mixtures, determination of voids in a bituminous mixture and calculations of weight/volume.

The ratio of the mass of aggregate to the volume the aggregate occupies in water is designated as density and its units of measurement are g/cm^3, Mg/m^3 or kg/m^3. The mass of the volume of aggregates to the mass of an equal volume of water, normally at 25°C, or alternatively 23°C or 20°C, is designated as specific gravity. The value determined is dimensionless.

In either case, what is taken as volume of the aggregate affects the result.

In an ideal situation in which the aggregates do not have any internally enclosed or surface micro- or macropores, the measured volume (V) coincides with the absolute volume of aggregate. In this case, the *absolute density* (p) or the *absolute specific gravity of aggregates* (G) is determined.

However, in almost all cases, there are some impervious micropores enclosed in the mass of the aggregate (see Figure 2.9). As a result, the measured volume is not the real, but the apparent one. Thus, the density or the specific gravity derived is the *apparent density* (ρ_a) or the *apparent specific gravity* (G_a) of the aggregate.

Aggregates also have surface pores or cavities (see Figure 2.8). Hence, the actual volume determined is relative; that is, the aggregate volume is greater than the apparent volume.

In this case, the density or the specific gravity reported is designated as *relevant density* (ρ_{rd}) or *bulk specific gravity* (G) of the aggregate. These terms are simply referred to as density or specific gravity of aggregate.

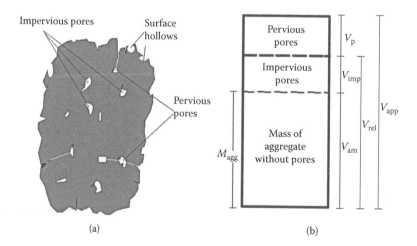

(a)   (b)

Figure 2.9 Schematically explanation of pores and respective volumes in the aggregate mass (*M*, mass; *V*, volume). (a) Aggregate particle. (b) Schematic presentation.

The European standard describing the procedure for the determination of aggregate particle density as well as the determination of water absorption is CEN EN 1097-6 (2005). The corresponding American standards are ASTM C 127 (2012) (AASHTO T 85 2013) and ASTM C 128 (2012) (AASHTO T 84 2013).

Both standards use wire baskets or pyknometers for the determination of the volume of the aggregate depending on its size.

According to CEN EN 1097-6 (2005), the density determination of aggregates passing through the 63 mm sieve and retained on the 31.5 mm sieve or retained on the 63 mm sieve in general is carried out with the use of wire baskets.

The testing of aggregates passing through the 31.5 mm sieve and retained on the 0.063 mm sieve is carried out with the use of pyknometers.

It should be noted that the density determination of aggregate particles from 4 to 31.5 mm may also be carried out with the use of wire baskets. However, in case of dispute, the pyknometer constitutes the reference method.

2.12.2.1 Determination of density of aggregate particles between 31.5 and 63 mm by wire-basket method

The basic equipment for the determination of density of this coarse aggregate are as follows: a watertight tank, a balance (weighing capacity of 0.1 g) and a wire basket of suitable size that is suspended from the balance.

After sampling the aggregate in accordance with CEN EN 932-1 (1996) and reducing the amount in accordance with CEN EN 932-2 (1999), a representative portion of dry coarse aggregate is placed in the wire basket, which is then immersed in a water tank for 24 ± 0.5 h, at 22°C ± 3°C. The minimum mass per test portion required is given in Table 2.12.

After 24 h, the basket containing the aggregate is weighed (M_2) in water at 22°C ± 3°C. The water temperature is recorded when the mass is determined.

The aggregate is then carefully removed from the wire basket and the empty basket is weighed in water (M_3).

The aggregate is placed on absorbent paper or cloth and dried until any surplus water is removed from its surface. The aggregate is then weighed (M_1).

Finally, the aggregate is placed in an oven at 110°C ± 5°C until it reaches constant mass (M_4).

The apparent density (ρ_a), oven dry density (ρ_{rd}) and saturated surface dry density (ρ_{ssd}), in Mg/m³, are calculated as follows:

$$\rho_a = \rho_w \times M_4/[M_4 - (M_2 - M_3)]$$

$$\rho_{rd} = \rho_w \times M_4/[M_1 - (M_2 - M_3)]$$

$$\rho_{ssd} = \rho_w \times M_1/[M_1 - (M_2 - M_3)]$$

Table 2.12 Minimum mass of mest portion (wire-basket method) by CEN EN 1097-6 (2005)

Maximum size of aggregates (mm)	Minimum mass of test portions (kg)
63.0	15
≤45	7

Source: Reproduced from CEN EN 1097-6, Tests for mechanical and physical properties of aggregates – Part 6: Determination of particle density and water absorption, Brussels: CEN, 2005. With permission (© CEN).

The water absorption (WA_{24}) after 24 h immersion in water is calculated as

$$WA_{24} = 100 \times (M_1 - M_4)/M_4,$$

where WA_{24} is water absorption (% by mass of dry sample), ρ_w is the density of water at the temperature recorded when the mass M_2 was determined (Mg/m³), M_1 is the mass of saturated surface dry aggregates in air (g), M_2 is the apparent mass in water of the basket containing the sample of saturated aggregate (g), M_3 is the apparent mass in water of the empty basket (g) and M_4 is the mass of the oven-dried test portion in air (g).

The calculated density values are rounded to the nearest 0.01 Mg/m³ and the value of water absorption is rounded to the nearest 0.1%.

More information about the test can be found in CEN EN 1097-6 (2005).

2.12.2.2 Determination of density of aggregate particles between 4 and 31.5 mm – pyknometer method

Apart from the basic equipment as before, pyknometers of appropriate capacity (1000 to 5000 ml) are required for the determination of the density of aggregate particles from 4 to 31.5 mm.

A representative portion of dry aggregate is placed in a pyknometer, which is then filled with water. The minimum mass per test portion required is given in Table 2.13. The pyknometer is gently rolled and jolted in a tipped position to remove any entrapped air. The entrapped air can also be removed by applying a vacuum. The pyknometer is then placed in a water bath set at 22°C ± 3°C for 24 ± 0.5 h.

After 24 h, the pyknometer containing the aggregate is overfilled with water, and after placing its cover, it is dried on the outside and weighed (M_2). At this point, the water temperature is also measured.

The aggregates are then removed and the pyknometer is refilled with water, dried on the outside and weighed with its cover (M_3). The temperature of water is again recorded. The two temperatures during determining mass M_2 and M_3 shall not differ by more than 2°C.

The wet aggregates are placed in absorption paper or cloth and gently surface-dried until no free-water content exists on their surfaces. The saturated surface dry aggregates are transferred to a tray and weighed (M_1). The aggregates are then dried in an oven set at 110°C ± 5°C, until a stable mass is reached (M_4).

Table 2.13 Minimum mass of test portions (pyknometer method) by CEN EN 1097-6 (2005)

Maximum size of aggregates[a] (mm)	Minimum mass of test portions[a] (kg)
31.5	5
16	2
8	1

Source: Reproduced from CEN EN 1097-6, *Tests for mechanical and physical properties of aggregates – Part 6: Determination of particle density and water absorption*, Brussels: CEN, 2005. With permission (© CEN).

[a] For other sizes, the minimum mass of test portion may be interpolated from the masses given in Table 2.12.

The apparent density (ρ_a), oven dry density (ρ_{rd}) and the saturated surface dry density (ρ_{ssd}) in Mg/m^3 is calculated as

$$\rho_a = \rho_w \times M_4/[M_4 - (M_2 - M_3)]$$

$$\rho_{rd} = M_4/[M_1 - (M_2 - M_3)]$$

$$\rho_{ssd} = M_1/[M_1 - (M_2 - M_3)]$$

The water absorption (WA$_{24}$), as a percentage of dry mass of sample, after being immersed in water for 24 h is calculated by the following formula:

$$WA_{24} = 100 \times (M_1 - M_4)/M_4,$$

where M_1 is the mass of saturated surface dry aggregates in air (g), M_2 is the mass of the pyknometer containing the sample of saturated aggregate and water (g), M_3 is the mass of the pyknometer filled with water only (g), M_4 is the mass of the oven-dried test portion in air (g) and ρ_w is the density of water at the test temperature (Mg/m^3).

The calculated density values are rounded to the nearest 0.01 Mg/m^3 and the water absorption value is rounded to the nearest 0.1%.

More information about the test procedure is given in CEN EN 1097-6 (2005).

Knowing the percentage of water absorption is important when the aggregates are to be used in unbound layers and in bituminous mixtures. The mass of aggregates is increased when they absorb water. They require higher water content for optimum compaction and a higher asphalt percentage for optimum behaviour in comparison with aggregates having low or almost zero water absorption. When water absorption is greater than 2%, the aggregates may be susceptible to freezing.

The water absorption of aggregates used in unbound layers may range from 0.1% to 4.0%. Limestones, sand-gravels and some igneous rocks have a low percentage, usually 0.1% to 0.8%. In contrast, slags or some igneous rocks of volcanic origin may have values of 2.0% to 4.0%. Other types of rock such as basalt and sandstone can have water absorption ranging from 0.5% to 2%.

The maximum permissible water absorption value is normally 1.75% or 2%. Aggregates with a higher absorption percentage can be also permitted, under certain circumstances, provided that the rest of the required physical and chemical properties are satisfied.

2.12.2.3 Determination of density of aggregate particles between 0.063 and 4 mm – pyknometer method

The determination of aggregate particle density between 0.063 and 4 mm using the pyknometer method is similar to the method for aggregate particles between 4 and 31.5 mm. The key difference is the amount of aggregate tested, which should not be less than 1 kg.

Before starting the test, the aggregate is sieved so that all the particles pass through the 4 mm sieve and are retained on the 0.063 mm sieve.

Following the procedure described in Section 2.12.2.2 and using the same equations, the density of aggregate particles between 4 and 0.063 mm is determined.

More information about the test is given in CEN EN 1097-6 (2005).

2.12.2.4 Determination of predried particle density of aggregates (normative methods)

The method for determining the density of predried aggregate in an oven at 110°C ± 5°C is described in CEN EN 1097-6 (2005), Annex A.

The density of very coarse aggregate between 31.5 and 63 mm is determined using the wire-basket method. For aggregates between 0.063 and 31.5 mm, the pyknometer method is used.

The volume of aggregate is determined either by immersing it for at most 10 min in water or by immersing it in a pyknometer of known volume.

The required test portions for very coarse aggregates are given in Table 2.11. For aggregates with particles smaller than 31.5 mm, the test portions are smaller than those given in Table 2.15; for details, see CEN EN 1097-6 (2005)/Annex A.

The density of predried aggregates (ρ_p) using the wire-basket method is calculated as

$$\rho_p = \rho_w \times M_1/[M_1 - (M_2 - M_3)],$$

where ρ_w is the density of water at the test temperature (Mg/m³), M_1 is the mass of the oven-dried test specimen (g), M_2 is the mass in water of the basket containing the test sample under water (g) and M_3 is the mass in water of the empty basket (g).

Using the pyknometer method, the density of predried aggregates (ρ_p) is calculated as

$$\rho_p = \rho_w \times (M_2 - M_1)/[V - (M_3 - M_2)],$$

where ρ_w is the density of water at the test temperature (Mg/m³); M_1 is the mass of the pyknometer and funnel (g); M_2 is the mass of the pyknometer, funnel and test specimen; M_3 is the mass of the pyknometer, funnel, test specimen and water (g) and V is the volume of the pyknometer (ml).

More information about the tests is given in CEN EN 1097-6 (2005). Method density determination of lightweight aggregates is also described in this standard.

The determination of predried density is not usual. It is only used when aggregates do not have a high water absorption value, they are dry and a quick determination of density is required.

2.12.2.5 Particle density of filler – pyknometer method

The determination of filler particle density is carried out using the pyknometer method. The pyknometers used are smaller than those used for determining the density of aggregates. The capacity of the pyknometer used is usually 50 ml.

According to CEN EN 1097-7 (2008), a small amount of dry filler, namely, 10 ± 1 g, is placed in the 50 ml pyknometer along with a suitable liquid. The suitability of liquid is determined by its ability to detach fine grains. Liquids such as water, denatured ethanol, redistilled kerosene or toluene have been found to be suitable for many types of fillers. Significant attention during the test should be paid to the determination of the volume of the pyknometer with liquid, as well as to the absence of enclosed air in the mass of filler located inside the pyknometer.

The particle density of filler (ρ_f) is calculated (Mg/m³) as follows:

$$\rho_f = (m_1 - m_0)/[V - (m_2 - m_1)/\rho_1],$$

where m_0 is the mass of the empty pyknometer with stopper (g), m_1 is the mass of the pyknometer with the filler test portion (g), m_2 is the mass of the pyknometer with the filler test portion, topped up with liquid (g), V is the volume of the pyknometer (ml) and ρ_l is the density of the liquid at 25°C (Mg/m^3).

The masses are determined with an accuracy of 0.001 g. The final result is the mean of three test values and is expressed in 0.01 Mg/m^3.

More information about calculating the volume of pyknometer and performing the test is given in CEN EN 1097-7 (2008).

The equivalent American standard test method is carried out according to AASHTO T 100 (2010) (ASTM D 854 2010). The density of particles smaller than 4.75 mm is determined using volumetric flasks of >100 ml. More information about the test is given in the relevant standards.

2.12.2.6 Determination of density of aggregate mix

When the determination of aggregate density is carried out separately, that is, for each aggregate fraction, the density of the aggregate mixture is calculated from the general formula:

$$\rho_x = 100/[(p_1/\rho_{x1}) + (p_2/\rho_{x2}) + \ldots + (p_n/\rho_{xn})],$$

where ρ_x is the density of aggregate mixture of any expression ρ_a, ρ_{rd}, ρ_{ssd} (Mg/m^3); $p_1, p_2, \ldots p_n$ is the percentage of aggregate fraction in the mixture (%); $\rho_{x1}, \rho_{x2}, \ldots \rho_{xn}$ is the respective density of aggregate fraction (Mg/m^3) and n is the number of aggregate fractions in the mixture.

2.12.2.7 Determination of loose bulk density and voids

Loose bulk density is the mass of dry aggregates that fills a container of a particular volume without any compaction divided by the volume of the container.

The test is conducted according to CEN EN 1097-3 (1998) and applies to natural, crushed and artificial aggregates.

The container of a particular capacity (see Table 2.14) is filled with dry aggregate. The amount of sample should be between 120% and 150% more than the amount required in order for the vessel to be filled without any compaction or vibration to occur.

The loose bulk density (ρ_b) is calculated by the following formula:

$$\rho_b = (m_2 - m_1)/V,$$

where m_1 is the mass of the empty container (kg), m_2 is the mass of the container and test specimen (kg) and V is the volume of the container (l).

Table 2.14 Minimum capacity of container depending on the aggregate size by CEN EN 1097-3 (1998)

Upper size of aggregate, D (mm)	capacity (l)
Up to 4	1
Up to 16	5
Up to 31.5	10
Up to 63	20

Source: Reproduced from CEN EN 1097-3, *Tests for mechanical and physical properties of aggregates – Part 3: Determination of loose bulk density and voids*, Brussels: CEN, 1998. With permission (© CEN).

The representative loose bulk density is the mean of three test values and is expressed in 0.01 Mg/m³.

The voids (*u*) in loose aggregates are calculated from the following formula:

$$u = [(\rho_p - \rho_b)/\rho_p] \times 100,$$

where *u* is the percentage of voids, ρ_b is the loose bulk density (Mg/m³) and ρ_p is the oven-dried or predried particle density (Mg/m³).

More information about the test is given in CEN EN 1097-3 (1998).

2.12.3 Thermal and weathering tests

2.12.3.1 Magnesium sulfate test

This test determines the resistance to weathering or disintegration of aggregate when subjected to the cyclic action of immersion in magnesium sulfate, followed by oven drying. It simulates the long-term resistance to weathering due to volume changes caused by the impact of alternating seasonal temperatures. If the aggregate is not resistant to ambient temperature change, its disintegration may cause loss of bearing capacity of the layer, cracks, potholes and pavement surface disintegration.

The simulation of volume change is carried out with the crystallisation and hydration of magnesium sulfate (Mg_2SO_4) within the pores of the aggregates during the drying stage of saturated aggregates. A mass of 500 g of aggregate particles between 10 and 14 mm, according to CEN EN 1367-2 (2009), is immersed for 17 ± 0.5 h in a magnesium sulfate solution. The aggregate is left to drain for a period of 2 ± 0.25 h and then dried at 110°C $\pm$ 5°C for 24 ± 1 h. This process is repeated five times.

The aggregates are then washed with tap water to remove any magnesium sulfate. The aggregates are then dried at 110°C $\pm$ 5°C and sieved manually using the 10 mm sieve. The loss of weight attributed to wear is recorded and expressed as a percentage of the original sample mass. The mean value of two tests, rounded to the nearest integer, is reported as the magnesium sulfate (MS) value.

More information about the test performance is given in CEN EN 1367-2 (2009).

The maximum permissible MS value normally allowed is 18% or 25% (MS_{18} or MS_{25}).

The American standard test is conducted in accordance with ASTM C 88 (2013) (AASHTO T 104 2011).

According to this standard, fine aggregates (<4.75 mm), coarse aggregates (4.75 to 63 mm) or even a mixture of fine and coarse aggregates can be tested. As a consequence, the result obtained differs depending on the size of the aggregate even for the same rock. More information can be found in ASTM C 88 (2013) or AASHTO T 104 (2011).

2.12.3.2 Determination of resistance to freezing and thawing

The resistance to freezing and thawing test provides information about the behaviour of an aggregate subjected to the cyclic action of freezing and thawing. The test is usually conducted in countries having subzero temperatures.

The test according to CEN EN 1367-1 (2007) is conducted on aggregate particles from 4 to 63 mm.

A predefined mass of dry aggregate is first immersed in a water bath at 20°C $\pm$ 5°C for 24 h and is then subjected to gradual freezing to –20°C for a predefined period. The sample is then thawed at 20°C $\pm$ 3°C. This freezing and thawing procedure is repeated 10 times.

The sample is then sieved through a sieve smaller than that used to determine the test sample (half the lower size sieve used). The loss of weight due to freezing and thawing is recorded and expressed as a percentage of the initial mass of the test sample. The representative freezing and thawing value is the mean of three test values.

More information about the test is given in CEN EN 1367-1 (2007).

2.12.3.3 Determination of resistance to thermal shock

This test determines the resistance of aggregate to thermal shock attributed to heating and drying during the production of hot bituminous mixtures. This test is used to condition aggregates for testing using the Los Angeles or impact test methods.

According to CEN EN 1367-5 (2011), test samples of aggregate particles 14 mm/10 mm and about 1000 g in mass are immersed in a water bath at room temperature for 2 ± 0.5 h. They are then placed onto a pan and then in an oven after they have been surface dried with the aid of absorbent paper or cloth.

The sample of aggregate is left in the oven at $700°C \pm 50°C$ for 180 ± 5 s.

After removing the aggregates from the oven, they are allowed to cool at room temperature.

The procedure is repeated in 1000 g increments until a sufficient amount of aggregate for the fragmentation test according to CEN EN 1097-2 (2010) is obtained.

Once a sufficient amount is obtained, the whole mass of aggregate is sieved through a 5 mm sieve. The mass passing through the sieve is recorded (M_2).

The percentage of undersized aggregates passing through the 5 mm sieve due to exposure to thermal shock is determined using the following equation:

$$I = (M_1 - M_2)/M_1 \times 100,$$

where I is the percentage of undersized aggregate due to thermal shock, M_1 is the initial mass of the test portion (g) and M_2 is the mass of the undersized aggregate passing through the 5 mm sieve (g).

Then, the Los Angeles fragmentation test or the impact test is carried out on two test samples in accordance to CEN EN 1097-2 (2010). The first test uses aggregates that underwent thermal shock and the second one uses aggregates that did not undergo thermal shock.

The thermal shock resistance value (V_{LA} or V_{SZ}), depending on which test is conducted (Los Angeles test or impact test), is determined by the following formula:

$$V_{LA} \text{ (or } V_{SZ}) = LA_2 \text{ (or } SZ_2) - LA_1 \text{ (or } SZ_1),$$

where LA_1 is the Los Angeles coefficient according to CEN EN 1097-2 (2010), paragraph 5.3; LA_2 is the Los Angeles coefficient for aggregates that underwent thermal shock, according to CEN EN 1097-2 (2010), paragraph 5.3; SZ_1 is the impact value according to CEN EN 1097-2 (2010), paragraph 6.3; and SZ_2 is the impact value for aggregates that underwent thermal shock according to CEN EN 1097-2 (2010), paragraph 6.3.

The percentage of undersized aggregates is expressed to 0.1% and the loss of resistance is rounded to the nearest whole number.

More information about the test is given in EN 1376-5.

2.12.3.4 Boiling test for 'Sonnenbrand' signs on basalts

The test is conducted for the determination of Sonnenbrand and is used to test basalt aggregate.

The test is conducted according to CEN EN 1367-3 (2004) and applies to pieces of rock and graded coarse aggregate.

The Sonnenbrand effect starts with the appearance of grey/white star-shaped spots. Hairline cracks are then usually generated. They reduce the strength of the minerals and the rock decays into smaller particles.

The required amount of basalt particles is placed in a steel can containing water, which is boiled for 36 ± 1 h. Sonnenbrand signs, as well as the development of cracks and possible cracking of aggregate particles, are visually determined. The mass loss after boiling, as well as the loss of resistance to strength after boiling, is determined.

The loss of mass (M_1) is determined as a percentage by weight of the initial sample after sieving through a sieve having an aperture of half the lower nominal size.

The loss of strength (S_{LA}) attributed to boiling is determined on aggregates that have and have not undergone the boiling test. More details and information about the test are given in CEN EN 1367-3 (2004).

2.12.4 Water content test

The water content or moisture test for aggregates is determined by drying them in a ventilation oven at 110°C ± 5°C.

According to CEN EN 1097-5 (2008), the required amount of aggregate is placed in a suitable tray, weighed with the tray and then placed in an oven set at 110°C ± 5°C.

The amount of test sample depends on the maximum aggregate particle (D). If $D \geq$ 1.0 mm, the minimum sample proportion in kilograms is $0.2D$. If $D < 1.0$ mm, the minimum proportion is 0.2 kg.

The mass of aggregate (M_1) is determined by subtracting the tray mass (M_2). The aggregates are left in the oven until constant mass (M_3) is achieved (the plate mass [M_2] is subtracted). Constant mass is achieved when the difference between successive weighings, after drying for at least 1 h and cooling at room temperature, is 0.1% or less of the previous dry mass.

The water content in the aggregate (w) is expressed as a percentage of the dry sample mass. This percentage is determined using the following formula:

$$w = [(M_1 - M_3)/M_3] \times 100 \ (\%),$$

where M_1 is the mass of the test portion (g) and M_3 is the constant mass of the dried test portion (g).

The result is expressed in 0.1%.

More details and information for the test are given in CEN EN 1097-5 (2008).

2.12.5 Voids of dry compacted filler (Rigden test)

This test is conducted on both natural and artificial fillers to determine their ability to retain bitumen. It is related to the space availability within the compacted filler.

The voids of the dry compacted filler are determined using the Rigden apparatus.

According to CEN EN 1097-4 (2008), 10 ± 1 g of filler is placed in the Rigden apparatus and compacted by applying 100 blows of a dropping load weighing 875 ± 25 g. After compaction, the height of the compacted filler inside the cylinder is determined.

Using the particle density of filler, the air void content of the compacted filler (v) is calculated from the following formula:

$$v = [1 - (4 \times 10^3 \times m_2)/\pi \times \alpha^2 \times \rho_f \times h)] \times 100,$$

where v represents voids (%), m_2 is the mass of compacted filler (g), α is the inner cylinder diameter (kg), ρ_f is the filler particle density (kg/m^3) and h is the height of the compacted filler (mm).

The void percentage of the filler is the mean value of three tests, rounded to the nearest 1%.

More information about the test is given in CEN EN 1097-4 (2008).

2.12.6 Delta ring and ball test of filler for bituminous mixtures

This test determines the stiffening effect of filler when mixed with bitumen. More information about the test is given in CEN EN 13179-1 (2000).

The test is normally performed on fillers that will be used in mastic asphalt applications. However, useful information may be provided for other asphalt performance criteria (Vansteenkiste and Vanelstraete 2008).

2.13 CHEMICAL PROPERTIES TESTS

2.13.1 Petrographic description of aggregates

The petrographic composition of both natural and artificial aggregates, if required, is determined according to CEN EN 932-3 (2003).

This standard specifies basic procedures for the petrographic examination of aggregates for purposes of general classification.

2.13.2 Determination of lightweight contamination

This test is a method for estimating the mass percentage of lightweight particles in fine aggregates (or in coarse aggregates), such as lignite and coal. These substances are critical in concrete and mortars as they may cause pop-outs or staining.

More details can be found in CEN EN 1744-1 (2009), paragraph 14.2.

2.13.3 Determination of acid-soluble sulfates

The determination of sulfate ions is applied to both natural aggregates and air-cooled blast-furnace slags according to CEN EN 1744-1 (2009), paragraph 12.

Sulfates are extracted from a representative mass of aggregates (about 2 g, passing through the 0.125 mm sieve) by dilute hydrochloric acid and determined gravitationally (from subsidence of sediment). Sulfate ion content is expressed as a percentage by mass of aggregate. More information can be found in CEN EN 1744-1 (2009).

2.13.4 Determination of total sulfur content

The determination of total sulfur content is applied to both natural aggregates and air-cooled blast-furnace slags and is carried out according to CEN EN 1744-1 (2009), paragraph 11.

A representative test portion of aggregates (about 1 g, passing through the 0.125 mm sieve) is treated with bromine and nitric acid to convert any sulfur compounds into sulfates. Sulfates in the form of $BaSO_4$ sink and then weighed. The total sulfur content is expressed as a percentage by mass of aggregate.

More information about the test is given in CEN EN 1744-1 (2009).

2.13.5 Unsoundness tests for blast-furnace and steel slags

2.13.5.1 Dicalcium silicate disintegration of air-cooled blast-furnace slag

This test is used on crushed air-cooled blast-furnace slags to determine their susceptibility to disintegration resulting from the inversion of dicalcium silicate from the 'b' form to the 'g' form. This is sometimes called 'lime disintegration'.

According to CEN EN 1744-1 (2009), paragraph 19, the surface of crushed slag is fluoresced under the influence of ultraviolet light in the field of visible spectrum. The aspect and colour of fluorescence enable the detection of slags that are suspect to silicate disintegration.

At the end of the test, observations made on the appearance of freshly broken surfaces are recorded. Slags that exhibit numerous or clustered large and small bright spots of a yellow, bronze or cinnamon colour on a violet background are recorded as suspect to disintegration.

Slags with a uniform shine in various shades of violet and those exhibiting a limited number of bright spots uniformly distributed are deemed sound.

More details and information about the test are given in CEN EN 1744-1 (2009).

2.13.5.2 Iron disintegration of air-cooled blast-furnace slag

This test is used on crushed air-cooled blast-furnace slags for determining their sensitivity to disintegration as a consequence of iron and manganese sulfide hydrolysis.

According to CEN EN 1744-1 (2009), paragraph 19.2, the disintegration of iron occurs by ageing in a humid atmosphere, in rain or more rapidly in water. It is observed by examination of slag particles that have been immersed in water.

Thirty slag particles of size 40 to 150 mm are immersed in water at 20°C ± 2°C for 2 days. If one or two particles crack or disintegrate, the test is repeated for another 30 particles. If a particle cracked or disintegrated again, the slag does not pass the test.

More details and information about the test are given in CEN EN 1744-1 (2009).

2.13.5.3 Volume stability (expansion) of steel slags

This test is used for steel slags to determine their susceptibility to expansion as a consequence of the hydration of free lime or free magnesium oxide.

According to CEN EN 1744-1 (2009), paragraph 19.3, a compacted slag specimen is subjected to a flow of steam at 100°C in a steam unit at ambient pressure for 24 or 168 h (7 days) depending on slag type.

Any change in volume, owing to the reaction of moisture with free lime or magnesium oxide, is recorded with the aid of a dial gauge. The result is expressed as a percentage increase of the volume in relation to the initial volume of the compacted sample.

More details and information are given in CEN EN 1744-1 (2009).

2.13.6 Water solubility of filler and aggregates

According to CEN EN 1744-1 (2009), paragraph 16, an appropriate aggregate test portion is extracted with an amount of water twice the mass of aggregates. The test portion is as follows: coarse aggregate, 2 ± 0.3 kg; fine aggregate, 500 ± 75 g; and filler, 10 ± 0.2 g. After 24 h extraction, the aggregate is dried and weighed.

The water solubility (WS) is calculated from the following formula:

$$WS = [(m_{11} - m_{12})/m_{11}] \times 100 \ (\%),$$

where m_{11} is the mass of aggregates before extraction (g) and m_{12} is the mass of aggregates after extraction (g).

More details and information about the test are given in CEN EN 1744-1 (2009).

2.13.7 Water susceptibility of fillers for bituminous mixtures

The water susceptibility of fillers for bituminous mixtures is determined by separation of the filler from the bitumen–filler mixture. The test is conducted according to CEN EN 1744-4 (2005).

The bitumen and filler mixture, which contains 50 ± 0.5 g of low-viscosity bitumen solution (50/70 bitumen and kerosene) and 10 ± 0.1 g of filler passing through the 0.125 mm sieve, is stirred with hot water at $60°C \pm 1°C$ for 300 ± 5 s and left to rest for 300 ± 5 s. The water is then emptied into a conical flask and stirred again for a further 300 ± 5 s.

The mixture at the end of 300 s is visually examined to determine any uncoated filler (separation is indicated by the turbidity of the water). If not, the filler is considered nonsusceptible to water. Otherwise, the water is collected and filtered to recover the filler.

The susceptibility of filler to water (W_s) is calculated from the following formula:

$$W_s = [(m_2 - m_1)/m_o] \times 100 \ (\%),$$

where m_o is the mass of filler test portion (g), m_1 is the mass of the filter paper (g) and m_2 is the mass of filter paper with filler retained (g).

The result is rounded to the nearest 1% by mass.

More details and information about the test are given in CEN EN 1744-4 (2005).

An alternative method for determining the susceptibility of filler is based on volume increase and Marshall stability loss. This gives a measure of the influence of fillers on the durability of asphalt in the presence of water.

An aggregate mixture 0/8 mm with a predetermined mass of aggregate fractions 5/8, 2/5, 0.125/2 and 0/0.125 mm (test filler) is mixed with 160/220 bitumen. The compacted Marshall specimens are placed in a water bath at $40°C \pm 1°C$ and allowed to soak for 48 h.

The percentage of volume increase (Q) and the percentage of stability loss (S_{MA}) are determined by comparing the two properties derived from samples tested dry and after being immersed in water.

More details about the test are given in CEN EN 1744-4 (2005), Annex A.

2.13.8 Water-soluble constituents in filler

The determination of water-soluble constituents in filler for paving mixtures is conducted in accordance with CEN EN 1744-3 (2002).

2.13.9 Calcium carbonate content of limestone filler aggregate

Determination of the calcium carbonate content of limestone filler aggregate for bituminous mixtures is conducted in accordance with CEN EN 196-2 (2013).

2.13.10 Calcium hydroxide content of mixed filler

Determination of the calcium hydroxide content of mixed filler for bituminous mixtures is conducted in accordance with CEN EN 459-2 (2011).

2.14 BLENDING TWO OR MORE AGGREGATES

Aggregates that are used in unbound, hydraulically bound and bituminous layers almost always consist of an admixture of two or more aggregate fractions or fine aggregate, medium and coarse aggregate and perhaps extra filler. These materials are proportioned in the mixture in such a way that the final gradation is within the specification limits. The procedure to determine the optimum proportions is known as aggregate composition.

The aggregate composition can be carried out by trial-and-error, mathematical (linear equation or least squares) or graphical methods.

2.14.1 Trial-and-error method

The trial-and-error method is used only by experienced engineers. This method is based on estimation of the approximate optimum proportions and verification of the estimation. In the case that the original estimation does not give a gradation within the limits specified, the process is repeated with different values.

2.14.2 Mathematical methods

2.14.2.1 Linear equation method

The linear equation method is the most common method used to blend two or more aggregates. It is based on linear equations of the form

$$a \times P_{Ai} + b \times P_{Bi} + c \times P_{Ci} + d \times P_{Di} + \ldots = P_X,$$

where $a, b, c, d \ldots$ are percentage proportions of individual aggregates A, B, C, D and so on, used in the composition; $P_{Ai}, P_{Bi}, P_{Ci}, P_{Di}, \ldots$ are the percentage of material passing through a given sieve (i) for the individual aggregate A, B, C, D and so on; and P_X is the percentage of combined aggregates passing through a given sieve (initially, this percentage is equal to the mid value determined from the range of limiting values given in the specification).

The linear equation system may be constructed to have as many equations as the number of sieves used.

Additionally, the system includes the basic equation

$$a + b + c + d + \ldots = 1 \text{ (or 100)}.$$

Determination of the percentages a, b, c, d and so on, is carried out by solving the system of linear equations. The disadvantage of this method is that more than one solution or combination can be found. To find the optimum or desired solution, successive approximations are needed, having determined an acceptable solution.

The method is best explained with the following example.

Example

Blend an aggregate mixture consisting of three aggregates A, B and C, such that the gradation of the final mix is within the specified limits. The percentage passing the particular size for each aggregate as well as the specified limits is shown in Table 2.15.

Solution

Let a, b and c be the percentages of aggregates A, B and C from which the final mixture will be derived and be within the specification limits.

Table 2.15 Results from sieve analysis and specification limits for aggregate composition

Sieve size (mm)	Aggregate			Specification limits	Limits' mid value
	A	B	C		
	Percentage passing (cumulative), by weight				
19.0	100	100	100	100	100
12.5	90.0	100	100	90–100	95
4.75	40.0	100	100	60–75	67.5
2.36	6.5	98.1	100	40–55	47.5
0.600	3.0	20.7	93.2	20–35	27.5
0.300	1.2	12.2	58.7	12–22	17.0
0.075	0.5	3.3	27.4	5–10	7.5

It is known that $a + b + c = 1$ (2.1)

Using successively the general equation (Equation 2.1), the following equations can be derived:

For the 12.5 mm sieve: $90 \times a + 100 \times b + 100 \times c = 95$ (2.2)

For the 4.75 mm sieve: $40 \times a + 100 \times b + 100 \times c = 67.5$ (2.3)

For the 2.36 mm sieve: $6.5 \times a + 98.1 \times b + 100 \times c = 47.5$ (2.4)

For the 0.60 mm sieve: $3.0 \times a + 20.7 \times b + 93.2 \times c = 27.5$ (2.5)

For the 0.30 mm sieve: $1.2 \times a + 12.2 \times b + 58.7 \times c = 17.0$ (2.6)

For the 0.075 mm sieve: $0.5 \times a + 3.3 \times b + 27.4 \times c = 7.5$ (2.7)

The solution to the above system of equations can be done in various ways. One is to subtract Equation 2.3 from Equation 2.2, which gives

$50 \times a = 27.5$; hence, $a = 0.55$.

Then replacing the value of a found into Equation 2.1, the equation becomes

$b + c = 1 - 0.55$ or $b = 0.45 - c$.

By replacing the values a and b into Equation 2.4, the equation becomes

$6.5 \times 0.55 + 98.1 \times (0.45 - c) + 100 \times c = 47.5$; hence, $c = -0.12$.

The resulting negative value means that the curve of the combined mix does not pass through the midpoint determined from the limiting values specified for the 2.36 mm sieve. However, by changing slightly the value 47.5 say to 48, a positive value can be obtained. This is

$6.5 \times 0.55 + 98.1 \times (0.45 - c) + 100 \times c = 48$; hence, $c = 0.15$.

Using the basic equation (Equation 2.1), the value of b can be determined:

$$0.55 + b + 0.15 = 1; \text{ hence, } b = 0.30.$$

Hence, the aggregate mixture could consist of 55% of aggregate A, 30% of aggregate B and 15% of aggregate C. This result is not necessarily the optimum or the desired one. To decide whether this proportion is the one that will be used, it is advised to determine the gradation of the mix and plot it against limiting values.

This is done by constructing Table 2.16 and plotting the result as in Figure 2.10.

The gradation of the final mix (curve X) deviates from the curve determined from the mid values of the limits specified (specification curve). If this mid curve was the desirable one, adjustment of the result and curve X is needed. This is carried out by altering the percentages derived so as to bring curve X as close as possible to the mid curve.

By successive adjustments, the best result was found to be when the proportion was 55% for aggregate A, 22% for aggregate B and 23% for aggregate C. This gives curve Y

Table 2.16 Evaluation of determined proportion

Sieve size (mm)	Aggregate			Aggregate mix (A + B + C)
	A = 55%	B = 30%	C = 15%	
	Percentage passing, by mass			
19.0	100 × 0.55 = 55	100 × 0.30 = 30	100 × 0.15 = 15	100.0
12.5	90.0 × 0.55 = 49.5	100 × 0.30 = 30	100 × 0.15 = 15	94.5
4.75	40.0 × 0.55 = 22	100 × 0.30 = 30	100 × 0.15 = 15	67.0
2.36	6.5 × 0.55 = 3.6	98.1 × 0.30 = 29.4	100 × 0.15 = 15	48.0
0.600	3.0 × 0.55 = 1.6	20.7 × 0.30 = 6.2	93.2 × 0.15 = 14	21.8
0.300	1.2 × 0.55 = 0.7	12.2 × 0.30 = 3.7	58.7 × 0.15 = 8.8	13.2
0.075	0.5 × 0.55 = 0.3	3.3 × 0.30 = 1.0	27.4 × 0.15 = 4.1	5.4

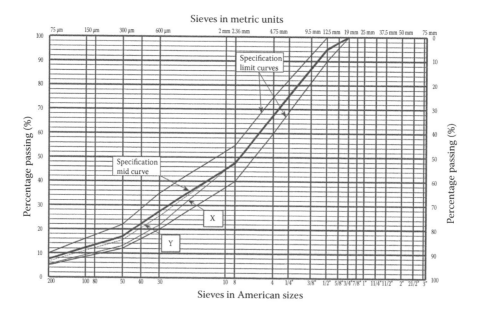

Figure 2.10 Gradation curves of aggregate mix after blending.

as shown in Figure 2.10. This is the final result of the procedure followed for blending the three aggregates, known as the job-mix formula.

However, in some cases, it may not be possible to get a proportion with all percentages being positive. This means that using all three sizes of aggregate, it is impossible to combine a mixture to have a gradation within the specified limits. In this case, examination to combine only two of the aggregates must be carried out. It may be possible to obtain a gradation within the specified limits; otherwise, with the given aggregates, it is impossible to blend a mix within the specified limits.

2.14.2.2 Least squares method

The equations derived in the example above can be solved by the least squares method. To solve the system of linear equations with the least squares method, the reader is referred to a suitable handbook. However, this method is nowadays included in most computer software readily available (Microsoft Excel, etc.).

It should be stated that the advantage of using the least squares method is the fact that one and only one solution – the optimum one – is obtained.

Hence, the reader in encouraged to use this method.

2.14.3 Graphical method

The graphical method of aggregate composition was used in the past by some engineers as it was considered simple and did not require solving mathematical equations. It was easy to combine two aggregates and a bit more complicated for three or more. Nowadays, the graphical method has been phased out. If the reader prefers to combine two aggregates with the graphical method, he can find relevant information in Asphalt Institute MS-2.

REFERENCES

AASHTO M 92. 2010. *Wire-cloth sieves for testing purposes*. Washington, DC: American Association of State Highway and Transportation Officials.

AASHTO T 2. 2010. *Sampling of aggregates*. Washington, DC: American Association of State Highway and Transportation Officials.

AASHTO T 11. 2009. *Materials finer than 75-μm (no. 200) sieve in mineral aggregates by washing*. Washington, DC: American Association of State Highway and Transportation Officials.

AASHTO T 27. 2011. *Sieve analysis of fine and coarse aggregates*. Washington, DC: American Association of State Highway and Transportation Officials.

AASHTO T 37. 2011. *Sieve analysis of mineral filler for hot mix asphalt (HMA)*. Washington, DC: American Association of State Highway and Transportation Officials.

AASHTO T 84. 2013. *Specific gravity and absorption of fine aggregate*. Washington, DC: American Association of State Highway and Transportation Officials.

AASHTO T 85. 2013. *Specific gravity and absorption of coarse aggregate*. Washington, DC: American Association of State Highway and Transportation Officials.

AASHTO T 96. 2010. *Resistance to degradation of small-size coarse aggregate by abrasion and impact in the Los Angeles machine*. Washington, DC: American Association of State Highway and Transportation Officials.

AASHTO T 100. 2010. *Specific gravity of soils*. Washington, DC: American Association of State Highway and Transportation Officials.

AASHTO T 104. 2011. *Soundness of aggregate by use of sodium sulfate or magnesium sulphate*. Washington, DC: American Association of State Highway and Transportation Officials.

AASHTO T 176. 2008. *Plastic fines in graded aggregates and soils by use of the sand equivalent test.* Washington, DC: American Association of State Highway and Transportation Officials.

AASHTO T 279. 2012. *Accelerated polishing of aggregates using the British wheel.* Washington, DC: American Association of State Highway and Transportation Officials.

Asphalt Institute MS-2. *Mix Design Methods for Asphalt Concrete and Other Hot-Mix Types*, 6th Edition. Lexington, KY: Asphalt Institute.

ASTM C 88. 2013. *Standard test method for soundness of aggregates by use of sodium sulfate or magnesium sulfate.* West Conshohocken, PA: ASTM International.

ASTM C 117. 2013. *Standard test method for materials finer than 75-μm (No. 200) sieve in mineral aggregates by washing.* West Conshohocken, PA: ASTM International.

ASTM C 127. 2012. *Standard test method for density, relative density (specific gravity), and absorption of coarse aggregate.* West Conshohocken, PA: ASTM International.

ASTM C 128. 2012. *Standard test method for density, relative density (specific gravity), and absorption of fine aggregate.* West Conshohocken, PA: ASTM International.

ASTM C 131. 2006. *Standard test method for resistance to degradation of small-size coarse aggregate by abrasion and impact in the Los Angeles machine.* West Conshohocken, PA: ASTM International.

ASTM C 136. 2006. *Standard test method for sieve analysis of fine and coarse aggregates.* West Conshohocken, PA: ASTM International.

ASTM C 535. 2012. *Standard test method for resistance to degradation of large-size coarse aggregate by abrasion and impact in the Los Angeles machine.* West Conshohocken, PA: ASTM International.

ASTM C 837. 2009. *Standard test method for methylene blue index of clay.* West Conshohocken, PA: ASTM International.

ASTM D 75/D 75M. 2009. *Standard practice for sampling aggregates.* West Conshohocken, PA: ASTM International.

ASTM D 448. 2012. *Standard classification for sizes of aggregate for road and bridge construction.* West Conshohocken, PA: ASTM International.

ASTM D 546. 2010. *Standard test method for sieve analysis of mineral filler for bituminous paving mixtures.* West Conshohocken, PA: ASTM International.

ASTM D 854. 2010. *Standard test methods for specific gravity of soil solids by water pycnometer.* West Conshohocken, PA: ASTM International.

ASTM D 2419. 2009. *Standard test method for sand equivalent value of soils and fine aggregate.* West Conshohocken, PA: ASTM International.

ASTM D 3319. 2011. *Standard practice for the accelerated polishing of aggregates using the British wheel.* West Conshohocken, PA: ASTM International.

ASTM D 4791. 2010. *Standard test method for flat particles, elongated particles, or flat and elongated particles in coarse aggregate.* West Conshohocken, PA: ASTM International.

ASTM D 5821. 2006. *Standard test method for determining the percentage of fractured particles in coarse aggregate.* West Conshohocken, PA: ASTM International.

ASTM D 6928. 2010. *Standard test method for resistance of coarse aggregate to degradation by abrasion in the micro-Deval apparatus.* West Conshohocken, PA: ASTM International.

ASTM E 11. 2013. *Standard specification for woven wire test sieve cloth and test sieves.* West Conshohocken, PA: ASTM International.

CEN EN 196-2. 2013. *Methods of testing cement – Part 2: Chemical analysis of cement.* Brussels: CEN.

CEN EN 459-2. 2011. *Building lime – Part – 2: Test methods.* Brussels: CEN.

CEN EN 932-1. 1996. *Tests for general properties of aggregates – Part 1: Methods for sampling.* Brussels: CEN.

CEN EN 932-2. 1999. *Tests for general properties of aggregates – Part 2: Methods for reducing laboratory samples.* Brussels: CEN.

CEN EN 932-3/A1. 2003. *Tests for general properties of aggregates – Part 2: Procedure and terminology for simplified petrographic description.* Brussels: CEN.

CEN EN 933-1/A1. 2005. *Tests for geometrical properties of aggregates – Part 1: Determination of particle size distribution-Sieve analysis.* Brussels: CEN.

CEN EN 933-2. 1995. *Tests for geometrical properties of aggregates – Part 2: Determination of particle size distribution-test sieves, nominal size of apertures.* Brussels: CEN.

CEN EN 933-3. 2012. *Tests for geometrical properties of aggregates – Part 3: Determination of particle shape-Flakiness index.* Brussels: CEN.

CEN EN 933-4. 2008. *Tests for geometrical properties of aggregates – Part 4: Determination of particle shape-Shape index.* Brussels: CEN.

CEN EN 933-5:2008/A1. 2004. *Tests for geometrical properties of aggregates – Part 5: Determination of percentage of crushed and broken surfaces in coarse aggregate particles.* Brussels: CEN.

CEN EN 933-6/AC. 2004. *Tests for geometrical properties of aggregates – Part 6: Flow coefficient of aggregates.* Brussels: CEN.

CEN EN 933-8. 2012. *Tests for geometrical properties of aggregates – Part 8: Assessment of fines-Sand equivalent test.* Brussels: CEN.

CEN EN 933-9. 2009. *Tests for geometrical properties of aggregates – Part 9: Assessment of fines-Methylene blue test.* Brussels: CEN.

CEN EN 933-10. 2009. *Tests for geometrical properties of aggregates – Part 10: Assessment of fines-Grading of filler aggregates (air jet sieving).* Brussels: CEN.

CEN EN 933-11. 2009. *Tests for geometrical properties of aggregates – Part 11: Classification test for the constituents of coarse recycled aggregate.* Brussels: CEN.

CEN EN 1097-1. 2011. *Tests for mechanical and physical properties of aggregates – Part 1: Determination of the resistance to wear (micro-Deval).* Brussels: CEN.

CEN EN 1097-2. 2010. *Tests for mechanical and physical properties of aggregates – Part 2: Methods for the determination of the resistance to fragmentation.* Brussels: CEN.

CEN EN 1097-3. 1998. *Tests for mechanical and physical properties of aggregates – Part 3: Determination of loose bulk density and voids.* Brussels: CEN.

CEN EN 1097-4. 2008. *Tests for mechanical and physical properties of aggregates – Part 5: Determination of the voids of dry compacted filler.* Brussels: CEN.

CEN EN 1097-5. 2008. *Tests for mechanical and physical properties of aggregates – Part 5: Determination of the water content determination by drying in a ventilated oven.* Brussels: CEN.

CEN EN 1097-6/A1. 2005. *Tests for mechanical and physical properties of aggregates – Part 6: Determination of particle density and water absorption.* Brussels: CEN.

CEN EN 1097-7. 2008. *Tests for mechanical and physical properties of aggregates – Part 7: Determination of the particle density of filler-Pyknometer method.* Brussels: CEN.

CEN EN 1097-8. 2009. *Tests for mechanical and physical properties of aggregates – Part 8: Determination of the polished stone value.* Brussels: CEN.

CEN EN 1097-9/A1. 2008. *Tests for mechanical and physical properties of aggregates – Part 9: Determination of the resistance to wear by abrasion from studded tyres-Nordic test.* Brussels: CEN.

CEN EN 1367-1. 2007. *Tests for thermal and weathering properties of aggregates – Part 1: Determination of resistance to freezing and thawing.* Brussels: CEN.

CEN EN 1367-2. 2009. *Tests for thermal and weathering properties of aggregates – Part 2: Magnesium sulfate test.* Brussels: CEN.

CEN EN 1367-3/AC. 2004. *Tests for thermal and weathering properties of aggregates – Part 3: Boiling test for 'Sonnenbrand' basalt.* Brussels: CEN.

CEN EN 1367-5. 2011. *Tests for thermal and weathering properties of aggregates – Part 1: Determination of resistance to thermal shock.* Brussels: CEN.

CEN EN 1744-1. 2009. *Tests for chemical properties of aggregates – Part 1: Chemical analysis.* Brussels: CEN.

CEN EN 1744-3. 2002. *Tests for chemical properties of aggregates – Part 3: Preparation of eluates by leaching of aggregates.* Brussels: CEN.

CEN EN 1744-4. 2005. *Tests for chemical properties of aggregates – Part 4: Determination of water susceptibility of fillers for bituminous mixtures.* Brussels: CEN.

CEN EN 12697-11. 2012. *Bituminous mixtures. Test methods for hot mix asphalt. Part 11: Determination of the affinity between aggregate and bitumen.* Brussels: CEN.

CEN EN 13043/AC. 2004. *Aggregates for bituminous mixtures and surface treatments for roads, airfields and other trafficked areas.* Brussels: CEN.

CEN EN 13179-1. 2000. *Tests for filler aggregate used in bituminous mixtures – Part 1: Delta ring and ball test.* Brussels: CEN.

CEN EN 13242+A1. 2007. *Aggregates for unbound and hydraulically bound materials for use in civil engineering work and road construction.* Brussels: CEN.

Controls Srl. 2014. Available at http://www.controls-group.com.

Cooper Research Technology Ltd. 2014. Available at http://www.cooper.co.uk.

Highways Agency. 2004. *Design Manual for Road and Bridges (DMRB), Volume 7: Pavement Design and Maintenance.* Section 3, Part 1, HD 28/04. London: Department for Transport, Highways Agency.

Highways Agency. 2006. *Design Manual for Road and Bridges (DMRB), Volume 7: Pavement Design and Maintenance.* Section 5, Part 1, HD 36/06. London: Department for Transport, Highways Agency.

Highways Agency. 2012. *Interim Advice Note 156/12, Revision of Aggregate Specification for Pavement Surfacing.* London: Department for Transport, Highways Agency.

ISSA TB 145. 1989. *Test method of methylene blue test adsorption value (MBV) of mineral aggregate fillers and fines.* Washington, DC: ISSA.

Nikolaides A. and E. Manthos. 2007. Sand equivalent of road aggregates tested with European and American standards and methylene blue results. *Proceedings of 4th International Conference 'Bituminous Mixtures and Pavements',* Vol. 1, p. 199. Thessaloniki, Greece: Aristotle University.

Vansteenkiste S. and A. Vanelstraete. 2008. Properties of fillers: Relationship with laboratory performance in hot mix asphalt. *Journal of the Association of Asphalt Paving Technologists,* Vol. 77, pp. 361–394. Philadelphia, PA: Association of Asphalt Paving Technologists.

Woodside A.R., W.D.H. Woodward, and J. McElhinney. 2011. The potential use of construction and demolition waste in road construction in Ireland. *Proceeding, 5th International Conference 'Bituminous Mixtures and Pavements',* Vol. 1, p. 13. Thessaloniki, Greece: Aristotle University of Thessaloniki.

XP P 18-545. 2003. *Granulats: Éléments de définition, conformité et codification.* Paris: AFNOR.

Chapter 3

Bitumen, bituminous binders and anti-stripping agents

3.1 GENERAL

Bitumen or asphalt is well known and used since ancient times, because it is the oldest and widely accepted structural material. It is used since 6000 BC as a waterproofing and binder material of great quality. A prominent example of bitumen use is cited in the Old Testament, since it was used as coating for Noah's Ark. The Sumerians used to use it in the prosperous shipbuilding industry, whereas the Babylonians used it as a binder in the mixture production for castle construction (Babel Tower). Asphalt was also used by the Egyptians both to mummify the dead bodies and to waterproof tanks. Around 3000 BC, the Persians also used bitumen for road construction. Finally, Herodotus and Plinius describe bitumen's export and use.

The Greek word *asphaltos* was used during Homeric times, which means a stable or solid substance. Afterwards, it was used by the Romans (*asphaltus*); hence, the term *asphaltic*, or even its root, exists until now in all modern languages.

Until the beginning of the 20th century, the asphalt or bitumen used was a natural product. The first natural deposits were found at the Dead Sea (or Salt Sea) where bitumen used to emerge from the bottom of the sea, floated to the surface and discharged into the banks. This was the reason why the ancient Greeks called this lake 'Lake Asphaltites'. Surface discharge of asphalt also existed in other parts of the Middle East. Later on, around 13th to 14th century AD, the largest surface deposits of natural asphalt in the world were discovered in Trinidad island (Lake Asphalt of Trinidad), as well as in the coasts of Venezuela.

Natural asphalt also exists in the form of rocks, that is, rocks (mainly limestones or sandstones) enriched with bitumen. This type of asphalt is known as rock asphalt or tar sand. Such kinds of rocks, in deposit size order, were found in Alberta in Canada (the area of Athabasca having the largest deposits in the world); East Venezuela; Malagasy; Utah, California, New Mexico and Kentucky in the United States; Buton Island in Indonesia; Albania (area of Selenitza); Romania (Derna area); Kazakhstan; as well as in various other areas, but in smaller deposits, such as in France (Gard and Thann), Switzerland (Traver Valley), Italy (Ragusa) and Greece (Paxos/Antipaxos and Marathoupoli).

Apart from Trinidad and Venezuela asphalt, all other natural asphalt sources were used only occasionally in pavement construction. Today, only a few of the abovementioned deposits are currently exploited (mainly deposits of Albania, Romania and Kazakhstan). The reason is the high cost of recovering the asphalt from the rock.

Apart from the natural asphalt, there is also the 'artificial asphalt', which is a residue of fractional distillation of crude oil (petroleum oil), simply called bitumen or asphalt nowadays.

Bitumen (or asphalt), apart from being used in the production of bituminous mixtures for road construction, also has a wide range of applications, such as waterproofing, protective coatings and a range of industrial products (more than 250; Asphalt Institute MS-4 1989).

3.1.1 Terminology today

At this stage, in order to avoid confusion, it is necessary to clarify the meaning of the terms *bitumen* and *asphalt*, as well as their adjectives *bituminous* and *bituminous binder*. The reason is that, nowadays, the above terms have different meanings, certainly among European and American specifications and practising engineers.

In Europe and according to CEN EN 12597 (2000), *bitumen* is virtually an involatile, adhesive and waterproofing material derived from crude petroleum or present in natural asphalt, which is completely or almost completely soluble in toluene and very viscous or almost solid at ambient temperatures.

According to the same specification, *asphalt* is a mixture of mineral aggregate and bituminous binder.

In addition, *bituminous* is the adjective applicable to binders and mixtures of binders and aggregates containing bitumen. Hence, *bituminous binder* is the adhesive material containing bitumen.

Furthermore, *natural asphalt* is a relatively hard bitumen found in natural deposits, often mixed with fine or very fine mineral matter, which is virtually solid at 25°C and viscous fluid at 175°C at the same time.

In American English and according to American specifications, more often the term *asphalt* is used rather than the term *bitumen*. According to ASTM D 8 (2013), 'asphalt' or 'bitumen' is 'a dark brown to black cement-like residuum obtained from the distillation of suitable crude oils'. Another term also used in North America is *asphalt cement*, which has the same meaning as asphalt.

The term *asphalt binder* has been utilised in the Superpave mix design method to classify the grade of asphalt cement used in an asphalt mix based on expected performance under specific environmental conditions and anticipated traffic loading. Asphalt binder according to ASTM D 8 (2013) is 'an asphalt which may or may not contain an asphalt modifier'.

Finally, in American specifications, the term *native asphalt* is used instead of the term *natural asphalt* used in European standards. The term *native asphalt* is defined as 'the asphalt occurring as such in nature'.

In order to avoid confusion to the readers of this book, the terms *bitumen* and *asphalt*, as defined by CEN EN 12597 (2000) will be adopted throughout the text, unless reference is made to the American standards or specifications.

3.2 NATURAL ASPHALT

Natural asphalt is derived from the natural mutation of petroleum. This mutation happened a million years ago under the influence of bacteria. This bacterial decomposition of petroleum resulted in reduced content of saturated hydrocarbons, mainly *n*-alkylenes, and of light oils, which became heavier and more viscous, taking the form of asphalt.

According to CEN EN 12597 (2000), natural asphalt is defined as 'a relatively hard bitumen found in natural deposits, often mixed with fine or very fine mineral matter, which is virtually solid at 25°C but which is a viscous fluid at 175°C'.

3.2.1 Trinidad asphalt

The asphalt that comes from Trinidad Island is the most widely known natural asphalt, in semi-solid or very viscous form, which is currently used in highway engineering constructions. Deposits are superficial and form 'lakes', which is the reason that this natural asphalt is also known as 'Pitch lake or Lake asphalt'. The largest lake asphalt can be found in the south part of Trinidad Island, near La Brea. The total surface of the lake is approximately 500,000 m² (50 ha), with a maximum depth of approximately 90 m, and it is estimated that there are approximately 15 million tonnes of asphalt. This constitutes one of the largest deposits of very good quality natural asphalt in the world (Chilingarian and Yen 1978).

The paradox is that although the asphalt is mined, the deposits remain almost the same. Surface mining is possible since the lake asphalt surface is hard enough to withstand the loads of the mining machinery.

Trinidad natural asphalt contains a small amount of water and foreign materials, mainly topsoil and stones, which are removed once the asphalt is heated, at 160°C, and 'sieved'. The final product, known as Trinidad Lake Asphalt (TLA) or Epuré, has the following typical composition: 54% asphalt, 36% mineral matter and 10% organic matter. Its specific gravity varies from 1.39 to 1.44 g/cm². The TLA's typical chemical composition, together with the typical chemical composition of other natural asphalts, is given in Table 3.1.

The pure asphalt (Trinidad Epuré) is very hard (penetration, 0–2 dmm; softening point, 93°C–99°C) to be used alone in the production of bituminous mixtures. Hence, it is mixed with soft crude petroleum bitumen (usually 200 pen [dmm]) in order to produce suitable paving asphalt (a proportion of approximately 50/50 results in a 50-pen asphalt/bitumen).

Table 3.1 Typical chemical compositions of some natural asphalts

Place/country of appearance	Average bitumen content (%)	Chemical composition (average values)				
		Saturated hydrocarbons (%)	Aromatic hydrocarbons (%)	Resins (%)	Asphaltenes (%)	Sulfur content (%)
Trinidad Island	54	5.7	24.8	38.5	31.0	6.0–8.0
Venezuela	64	—	—	—	—	5.9
Alberta (Athabasca) Canada	2–8	17.2	38.2	29.3	15.3	3.54
		13.6	39.8	40.6	6.0	5.30
		14.1	20.3	53.3	12.3	—
		14.2	20.6	51.6	13.7	—
Tham region, SE of France	—	14.6	30.6	47.9	6.9	—
		14.0	25.1	52.0	8.9	—
		10.8	31.8	30.6	26.8	—
Travers valley, Switzerland	—	20.5	12.4	53.6	13.5	0.77
		22.6	27.9	42.3	7.2	1.10
Selenizza, Albania	8–14	—	—	—	—	6.1
Derna, Romania	15–22	—	—	—	—	0.7
SW of France (shelf at depths of 1900–2600 m)	—	1.2–4.6	14.4–35.3	34.6–48.7	19.4–46.8	7.9–10.6

Source: Reprinted from *Bitumens, Asphalts, and Tar Sands*, Chilingarian, G.V. and Yen, T.F., Copyright 1978, with permission from Elsevier.

3.2.2 Venezuela and other natural asphalts (pitch lake asphalts)

The natural asphalt of Venezuela, from the Guanoco area, is also superficially deposited and forms a lake. The covered area is larger than Trinidad Lake, approximately 4,000,000 m^2, but it has lesser depth (maximum depth, 3 m). The quantities that exist are almost the same as those in Trinidad Island and the asphalt content is approximately 64%. This natural asphalt is not as widespread as the Trinidad asphalt.

Other asphalt lakes can be found in California (La Brea tar pits in urban Los Angeles, McKittrick tar pits in Kern County and Carpinteria tar pits in Santa Barbara County) and are mostly used as tourist attraction areas.

3.2.3 Rock asphalt and gilsonite

Rock asphalt is natural asphalt found in solid form. It was formed millions of years ago when layers of petroleum harden after heavier components settled while lighter components evaporated. The hardening took place, in most of the cases, within the pores of rocks (mainly limestone or sandstones); thus, a number of mineral materials are normally found within rock asphalt.

Rock asphalt, as natural asphalt in liquid-viscous form, was used historically in engineering works but today its use in road works is rare. Rock asphalt nowadays is used only as an additive to petroleum bitumen. Since it is a very hard material, its use is to harden petroleum bitumen.

The largest natural surface deposits of rock asphalt that are commercially exploited are found in Utah in the United States (28 km^2) and Kermanshah in Iran.

The rock asphalt in the Uintah Basin in Utah has the trade name gilsonite, from S.H. Gilson, the founder of a mining company in 1888.

Gilsonite today is a well-known additive for hardening petroleum bitumen.

3.2.4 Other natural rock asphalts

All the other natural asphalts in rock formation can be found at a depth ranging from 3 to 1000 m. The typical asphalt content of these deposits range from 5% to 20%. Because of this low asphalt content and the costly and time-consuming purifying procedure, the exploitation of these natural asphalts turns out to be unprofitable. As a result, their extensive usage in both highway engineering and other industrial applications is limited.

The only exception seems to be the rock asphalt of Buton Island (south of Sulawesi Island) in Indonesia. In this island, the rock asphalt forms hills and mountains. This porous rock contains hard natural asphalt (penetration of 5–10 dmm) at 15%–30% and soft limestone with fossilised shell impurities. The extraction is very simple and cheap, because of low local labour cost; thus, it is used locally for the production of bituminous mixtures.

The rock asphalt is grinded into fine grains 0–2 mm in diameter, and this constitutes the binder under the trade name Asbuton. Asbuton is mixed with aggregates and solvents and thus cold bituminous mixtures are produced for pavements with medium to low traffic volume. Because of the binder's specificity, to ensure the quality of the mixture and of the construction, a specific mix design methodology has been developed (MoPW 1989). In the early part of the 1990s, several pavements had been constructed or maintained successfully using the mixture designed with the specially developed mix design methodology (Nikolaides 1990, 1991a,b). Lasbutag and Latasbushir, trade names for coarse and fine bituminous mixtures, respectively, are considered to be alternative low-cost bituminous mixtures for maintenance and construction of pavements in the remote areas of Polynesia.

Natural rock asphalt can also be found in small deposits in Greece, particularly in Marathos, Paxos and Antipaxos islands; Divri; Suli; Prouso; Fteri; Kalarrytes; Dremissa village in Giona; Zakynthos; and Epirus.

Natural rock asphalt in larger deposits can be found in the Jordan Valley, Dead Sea banks, France, Switzerland, Antilles, Venezuela and Cuba.

3.3 TAR

Tar is a product of fractional distillation of primary tar produced by carbonation of natural organic matters, such as coal or wood. When the primary tar is derived from coal, the product is called pitch tar. Similarly, when it is derived from wood, it is called wood tar. The pitch tar is further mixed with oil distillates, to produce the processed tar, commonly known as tar. The tar differs in chemical composition and odour to bitumen. Tars consist of variable mixtures of phenols, polycyclic aromatic hydrocarbons (PAHs) and heterocyclic compounds, and their odour is characteristically more aromatic than the odour of the asphalt.

The classification of the tars is carried out mainly in reference to the equiviscous temperature (EVT). According to BS 76 (1974), there are eight different tar types, from 30°C to 58°C EVT. EVT is the temperature in which 50 ml of tar has a flow time of 50 s when passing through the 10 mm hole of the tar viscometer. As a consequence, the higher the EVT, the more viscous the tar is. Tar with 50°C–58°C EVT is suitable for dense bituminous mixtures in heavy traffic volume sites, whereas tar with 30°C–38°C EVT is used in open graded bituminous mixtures for low traffic volume sites. Tar with 34°C–46°C EVT is usually used for surface dressings.

The tar was widely used for many years both in highway engineering and for the production of insulators. Over the last years, the usage of tar has been minimised mainly for environmental and health reasons. In some countries, such as Germany and the Netherlands, and some US states, the use of tar in highway engineering works has been prohibited. Nowadays, tar is used almost exclusively for the production of specific bituminous mixtures, not affected by petrol or oils, known as fuel-resistant mixtures. It may also mix with petroleum bitumen to produce better-quality binder for surface dressing works. Better coating of aggregates, as well as better adhesion of bitumen to aggregate, is achieved with the use of tar/bitumen binder. Tar/bitumen binder is used almost exclusively used in the United Kingdom.

Apart from the abovementioned advantages, the tar, or tar/bitumen, does not offer anything else. On the contrary, it is more sensitive to temperature variations, and as a result, it softens easier and it hardens and becomes brittle quicker than bitumen of similar viscosity.

3.4 MANUFACTURED BITUMEN

Manufactured bitumen, or bitumen as it is always called, is a product of fractional distillation of crude oil. Crude oil is derived from organic matter (vegetable matter, marine organisms), which, millions of years ago, deposited in very thick layers together with mud and rocks at the bottom of oceans. Under the action of overlying pressure, waterborne rocks were formed. The saline environment disintegrated the organic matter, which, under the influence of high pressure, temperature, bacterial activity and probably radiation, transformed into hydrocarbons in the form of crude oil. Further rock deposits in later years forced the crude oil to rise towards the earth surface through the pores of the rocks. In places where the rocks were impermeable, the crude oil with gases formed underground reservoirs and remains there until it is extracted.

Depending on the initial composition of the organic matter, as well as the prevailing conditions, crude oil obtained its physical properties and chemical composition. As a consequence, crude oil differs from one oil field to another and varies from a black, viscous to a tawny, low-viscous liquid.

The major oil-producing areas are the Middle East, the United States, countries around the Caribbean, ex-Soviet Union countries, North Sea and Indonesia. Nowadays, there are more than 1500 types of crude oil worldwide. Only certain types of crude oil can produce bitumen suitable for the asphalt paving industry.

The bitumen from crude oil (petroleum) is produced from fractional distillation of crude oil under high-temperature vacuum conditions. An additional process (separation) is also used in a solvent deasphalting unit, usually placed after the distillation tower. The solvent deasphalting unit separates aliphatic compounds from asphaltenes also producing high-quality deasphalted oil. Further processing may also be carried out by 'blowing' (oxidation) if harder and more viscous bitumen is to be produced. The products are then called oxidised bitumens. A schematic representation of bitumen production is shown in Figure 3.1.

The type of bitumen produced is determined by both the origin of crude oil and the vacuum (10–100 mm Hg) and temperature conditions (350°C–400°C) exist in the distillation column.

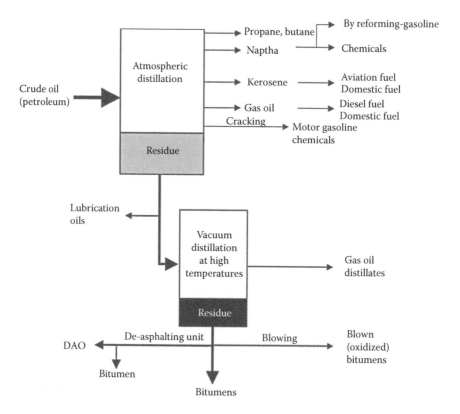

Figure 3.1 Simplified representation of crude oil distillation process for bitumen production and other materials.

Bitumen can be subject to further processing apart from blowing, such as emulsification or dissolution with solvents. In these cases, bituminous emulsions and cut-back and flashed bitumens are produced, respectively.

Finally, various chemical additives may be added to the bitumen, producing the modified bitumens.

3.4.1 Groups and grades of bitumen

The bitumen and bituminous binders may be divided into three general groups: (a) the paving grade bitumens, (b) the hard paving grade bitumens and (c) the oxidised bitumens.

Paving bitumens are bitumens produced from the residue distillation of crude oil, whereas oxidised bitumens are produced from bitumen subjected to a blowing process during which oxidation of bitumen occurs.

3.4.1.1 Paving grade bitumens

Paving grade bitumens in Europe are designated by the nominal penetration or viscosity ranges as appropriate.

Paving grade bitumens according to CEN EN 12591 (2009) are divided into three groups: (a) bitumen from 20 to 220 dmm in penetration, at 25°C; (b) bitumen from 250 to 900 dmm in penetration, at 25°C; and (c) soft bitumen, which are designated by kinematic viscosity, at 60°C.

The paving grades used in the construction and maintenance of roads, airfields and other paved areas are the bitumens with penetration values from 20 to 220 dmm. Within this range of values, eight different grades are distinguished, with the first one being 20/30 and the last being 160/220. The selection of the most suitable grade is based on the climatic and traffic conditions encountered.

Paving grade bitumens are characterised by their consistency at intermediate (determined by penetration test), consistency at elevated service temperatures (determined by softening point or viscosity test) and durability (determined by resistance to hardening test). Their brittleness at low service temperature (determined by the Fraass breaking test) and temperature dependence of consistency (determined by the penetration index) may also need to be determined to meet regional requirements for specific conditions such as extreme cold or wide ambient temperature variations. Flash point is also determined as well as, optionally, the density.

The requirements for the above properties of paving grade bitumens specified by CEN EN 12591 (2009) for grades from 20 to 220 dmm, per subgrades, are given in Tables 3.2 and 3.3.

The requirements for grades 250 to 900 dmm or for soft bitumens can be found in the same specification, CEN EN 12591 (2009).

In the United States, paving bitumens are graded in reference to either penetration or viscosity value. Since 1980, bitumens have been graded mainly in reference to their viscosity (kinematic), at 60°C, hence the term *viscosity-graded asphalts* (bitumens). Other properties such as penetration, flash point and kinematic viscosity at 135°C are also considered.

The requirements of the viscosity-graded bitumens (asphalts or asphalt cements) in accordance with ASTM D 3381 (2012) (or AASHTO M 226 2012) are based on original bitumen or on residue after rolling thin-film test. In the first case, the viscosity grade bitumens are notified by the letters AC – in the second case, by the letters AR – followed by numbers related to viscosity values.

Table 3.2 Paving grade bitumen specifications for grades from 20 to 220 dmm penetration

Property	Test method	Paving grade bitumen							
		20/30	30/45	35/50	40/60	50/70	70/100	100/150	160/220
Penetration at 25°C (dmm)	CEN EN 1426 (2007)	20–30	30–45	35–50	40–60	50–70	70–100	100–150	160–220
Softening point (°C)	CEN EN 1427 (2007)	55–63	52–60	50–58	48–56	46–54	43–51	39–47	35–43
Softening point (°C)	CEN EN ISO 2592 (2001)	≥240	≥240	≥240	≥230	≥230	≥230	≥230	≥220
Solubility (%)	CEN EN 12592 (2007)	≥99.0	≥99.0	≥99.0	≥99.0	≥99.0	≥99.0	≥99.0	≥99.0
Resistance to hardening at 163°C	CEN EN 12607-1 (2007)								
Retained penetration (%)		≥55	≥53	≥53	≥50	≥50	≥46	≥43	≥37
Increase in softening point (°C)									
Severity 1, or		≤8 or	≤8 or	≤8 or	≤9 or	≤9 or	≤9 or	≤10 or	≤11 or
Severity 2[a]		≤10	≤11	≤11	≤11	≤11	≤11	≤12	≤12
Change of mass[b] (absolute value) (%)		≤0.5	≤0.5	≤0.5	≤0.5	≤0.5	≤0.8	≤0.8	≤1.0

Source: Reproduced from CEN EN 12591, *Bitumen and bituminous binders – Specifications for paving grade bitumens*, Brussels: CEN, 2009. With permission (© CEN).

[a] When Severity 2 is selected, it shall be associated with the requirement for Fraass breaking point or penetration index or both measured on the unaged binder (see Table 3.3).

[b] Change in mass can be either positive or negative.

Table 3.4 gives the requirements of the AC viscosity-graded bitumens (asphalts) in accordance with ASTM D 3381 (2012). There are six distinct grades, with the first being AC-2.5 and last being AC-40.

Additionally, Table 3.5 gives the requirements of penetration-graded bitumens (asphalts) in accordance with ASTM D 946 (2009). In this case, there are five distinct grades, with the first being 40/50 and last being 200/300.

3.4.1.2 Hard paving grade bitumens

Hard paving bitumens have very high stiffness modulus values and are used for the construction and maintenance of road and airport pavements or other kinds of bitumen surfaces. Hard paving bitumens constitute an extension of common paving bitumens. They are usually used in locations with very high daily traffic flow, when annual ambient temperatures are intermediate or high.

Hard bitumens, according to CEN EN 13924 (2006), are designated by a range of penetration values, at 25°C. Sometimes, they also bear the discreet letter 'H' or 'HB'.

Table 3.3 Paving grade bitumen specifications for grades from 20 to 220 dmm penetration – properties associated with regulatory or other regional requirements

Property	Test method	Paving grade bitumen							
		20/30	30/45	35/50	40/60	50/70	70/100	100/150	160/220
Penetration index[a] (P.I.)	CEN EN 12591 (2009), Annex A[b]				−1.5 to +0.7 or NR[c]				
Dynamic viscosity at 60°C (Pa·s)	CEN EN 12596 (2007)	≥440 or NR	≥260 or NR	≥225 or NR	≥175 or NR	≥145 or NR	≥90 or NR	≥55 or NR	≥30 or NR
Fraass breaking point[a] (°C)	CEN EN 12593 (2007)		≤−5 or NR	≤−5 or NR	≤−7 or NR	≤−8 or NR	≤−10 or NR	≤−12 or NR	≤−15 or NR
Kinematic viscosity at 135°C (mm²/s)	CEN EN 12595 (2007)	≥530 or NR	≥400 or NR	≥370 or NR	≥325 or NR	≥295 or NR	≥230 or NR	≥175 or NR	≥135 or NR

Source: Reproduced from CEN EN 12591, *Bitumen and bituminous binders – Specifications for paving grade bitumens*, Brussels: CEN, 2009. With permission (© CEN).

[a] When Severity 2 is selected, it shall be associated with the requirement for Fraass breaking point or penetration index or both measured on the unaged binder.
[b] Reference to normative Annex A of CEN EN 12591 (2009) dealing with the calculation of penetration index.
[c] NR, no requirement may be used when there are no regulations or other regional requirements for the property in the territory of intended use.

Table 3.4 Requirements for viscosity-graded asphalt cement

Property	Test method (ASTM)	Viscosity grade (grading based on original asphalt)					
		AC-2.5	AC-5	AC-10	AC-20	AC-30	AC-40
Viscosity at 60°C (Pa·s)	D 2171	25 ± 5	50 ± 10	100 ± 20	200 ± 40	300 ± 60	400 ± 80
Viscosity at 135°C (mm²/s)	D 2170 (2010)	≥125	≥175	≥250	≥300	≥350	≥400
Penetration at 25°C (pen)	D 5 (2013)	≥220	≥140	≥80	≥60	≥50	≥40
Flash point (Cleveland open cup) (°C)	D 92 (2012)	≥165	≥175	≥220	≥230	≥230	≥230
Solubility in trichloroethylene (%)	D 2042 (2009)	≥99.0	≥99.0	≥99.0	≥99.0	≥99.0	≥99.0
After thin-oven test at 163°C/5 h	D 1754						
Viscosity at 60°C (Pa·s)	D 2171	≤125	≤252	≤500	≤1000	≤1500	≤2000
Viscosity at 25°C (Pa·s)	D 113 (2007)	≥100[a]	≥100	≥75	≥50	≥40	≥25

Source: Reprinted from ASTM D 3381, *Standard specification for viscosity-graded asphalt cement for use in pavement construction*, West Conshohocken, Pennsylvania, US: ASTM International, 2012. With permission (© ASTM International).

[a] If ductility is <100 cm, material will be accepted if ductility at 15°C is ≥100 cm at a pull rate of 5 cm/min.

Table 3.5 Requirements for penetration-graded asphalt cement

Property	Test method (ASTM)	Penetration grade				
		40/50	60/70	85/100	120/150	200/300
Penetration at 25°C (pen)	D 5 (2013)	40–50	60–70	85–100	120–150	200–300
Softening point (°C)	D 36	≥49	≥46	≥42	≥38	≥32
Flash point (Cleveland open cup) (°C)	D 92 (2012)	≥230	≥230	≥230	≥220	≥175
Ductility at 25°C (cm)	D 113 (2007)	≥100	≥100	≥100	≥100	≥100[a]
Solubility in trichloroethylene (%)	D 2042 (2009)	≥99.0	≥99.0	≥99.0	≥99.0	≥99.0
After thin-film oven test at 163°C/ 5 h:	D 1754					
Retained penetration (%)	D 5 (2013)	≥55	≥52	≥47	≥42	≥37
- Ductility at 25°C	D 113 (2007)	—	≥50	≥75	≥10	≥100[a]

Source: Reprinted from ASTM D 946, *Standard specification for penetration-graded asphalt cement for use in pavement construction*, West Conshohocken, Pennsylvania, US: ASTM International, 2009. With permission (© ASTM International).

[a] If ductility at 25°C is less than 100 cm, material will be accepted if ductility at 15.0°C is ≥100 cm at a pull rate of 5 cm/min.

Hence, as in paving grade bitumen, a 10/20 dmm hard paving bitumen denotes that the penetration is expected to range between 10 and 20 dmm.

As with common paving bitumens, the tests conducted on hard paving bitumens aim at determining their consistency at intermediate service temperatures (penetration test) and at elevated service temperatures (softening point and dynamic viscosity test) and their durability (resistance to hardening test). Kinematic viscosity, Fraass breaking point, flash point and solubility are also properties considered useful in the specification of hard paving bitumens.

The limiting values of hard paving bitumen specified by CEN EN 13924 (2006) are shown in Table 3.6. It should be noted that after choosing the grade of hard paving bitumen, 15/25 or 10/20, all the other properties can satisfy the requirements of any of the technical classes provided per property. For example, the hard paving grade bitumen 10/20 may have softening point class 4 (60°C–76°C), dynamic viscosity class 3 (≥700), change in mass class 2 (≤0.5%), increase in softening point class 3 (≤10°C) and so on. More details can be found in CEN EN 13924 (2006).

3.4.1.3 Oxidised bitumen

Oxidised bitumen is mainly used in roofing, waterproofing, adhesives and insulations.

The oxidation process is conducted after bitumen production and consists of blowing air through heated bitumen (temperatures between 240°C and 320°C). The oxidation process is conducted per bitumen batch or in a continuous bitumen feed. Oxidation 'dehydrates' and 'polymerises' bitumen. As a consequence, the molecular weight of asphaltenes is increased, additional asphaltenes are created from the continuous oil phase (maltenes) and thus bitumen with a higher molecular weight is produced. Because of these changes, bitumen becomes harder and it becomes less susceptible to temperature changes (the penetration index increased). However, because of hardening, oxidised bitumen is cracked easier at temperatures below 0°C.

The extent and magnitude of change in the rheological properties of bitumen that take place during oxidation are affected by the viscosity of the initial bitumen, the origin of crude oil, the duration of oxidation and the temperature at which the oxidation is carried out. In general, prolonged oxidation combined with high temperatures results in the production

Table 3.6 Specifications for hard paving grade bitumens

Essential requirement	Surrogate characteristic	Test method	Unit	Classes				
				0	1	2	3	4
Consistency at intermediate service temperature	Penetration at 25°C	CEN EN 1426 (2007)	dmm	NPD	TBR	15–25[e]	10–20	
Consistency at elevated service temperature	Softening point[c]	CEN EN 1427 (2007)	°C	NPD	TBR	55–71[c,e]	58–78[c]	60–76[c]
	Dynamic viscosity at 60°C	CEN EN 12596 (2007)	Pa·s	NPD	TBR	≥550[e]	≥700	
Durability (resistance to hardening at 163°C, EN 12607-1)[a]	Change of mass	CEN EN 12607-1 or 3	%	NPD	TBR	≤0.5		
	Retained penetration (%)	CEN EN 1426 (2007)	%	NPD	TBR	≥55		
	Softening after hardening	CEN EN 1427 (2007)	°C	NPD	TBR	≥Original min + 2[f]		
	Increase in softening point	CEN EN 1427 (2007)	°C	NPD	TBR	≤8	≤10	
	Increase in softening point and penetration index (of original bitumen)	CEN EN 1427 (2007) I_p (see Annex A)	°C (unitless)	NR	TBR	≤10 −1.5 to +0.7	≤10 ≤−1.5	
Other properties[b]	Kinematic viscosity at 135°C	CEN EN 12595 (2007)	mm²/s	NR	TBR	≥600[e]	≥700	
	Fraass breaking point	CEN EN 12593 (2007)	°C	NR	TBR	≤0[e]	≤3	
	Flash point[d]	CEN EN ISO 2592 (2001)	°C	NR	TBR	≥235	≥245	
	Solubility	CEN EN 12592 (2007)	%	NR	TBR	≥99.0		

Source: Reproduced from CEN EN 13924, *Bitumen and bituminous binders – Specifications for hard paving grade bitumens*, Brussels: CEN, 2006. With permission (© CEN).

Note: NPD, no performance determined; NR, no requirement; TBR, level or range to be reported by the supplier (it is not used for the purpose of regulatory marking).

a Only RTFOT shall be used for referee purposes.
b These additional properties are not part of the mandated essential characteristics but have been considered useful in specification of hard paving grade bitumens in some cases.
c Important – a restricted softening point range, of ±5°C about a midpoint, shall be declared by the supplier; the overall range shall be within the range in the table.
d By Cleveland open cup method.
e In selecting combinations of classes, it is intended that values marked 'e', if selected, shall only be used with the softer grade, 15/25 pen.
f The softening point after treatment shall be at least 2°C above the selected minimum value for the original bitumen (see note 'd' above).

Table 3.7 Properties and test methods of oxidised bitumen

Property	Test method	Unit	Limits and tolerances
Softening point[a]	EN 1427 (2007)	°C	±5 of midpoint value[e]
Penetration at 25°C	CEN EN 1426 (2007)	dmm	±5 of midpoint value[e]
Solubility in toluene[b]	CEN EN 12592 (2007)	%	≥99.0
Loss in mass after heating	CEN EN 13303	%	≤0.5
Flash point	CEN EN ISO 2592 (2001)	°C	>250
Fraass breaking point	CEN EN 12593 (2007)	°C	NR[c]
Staining properties	CEN EN 13301	mm	NR[c]
Dynamic viscosity[d]	CEN EN 13302 (2010)	Pa·s	NR[c]
Density	CEN EN 15326 (2009)	kg/m³	NR[c]

Source: Reproduced from CEN EN 13304, *Bitumen and bituminous binders – Framework for specification of oxidised bitumens*, Brussels: CEN, 2009. With permission (© CEN).

[a] Ring-and-ball softening point testing for oxidised bitumens are carried out in glycerol, as the values typically are above 80°C.
[b] If other solvents are used, it shall be stated in the test report.
[c] NR, no requirements. Values can be agreed upon between the client and the supplier.
[d] The type of instrument used to determine viscosity should be agreed upon between supplier and client.
[e] Midpoint value: value that defines the ring-and-ball softening point class or the penetration class.

of harder oxidised bitumens. However, both parameters affect production cost and productivity. In a research, it was found that the presence of β-diketone with dicarboxylic or hydroxylic acids reduces the oxidation period but with the same desired result (Economou et al. 1990).

In contrast to paving bitumens, the oxidised bitumen products are graded in reference to a combination of the softening point values, and penetration at 25°C, expressed in multiples of 5. Hence, an oxidised bitumen 85/25 means that the softening point of the product is between 80°C and 90°C and the penetration is between 20 and 30 dmm.

Typical grades for oxidised bitumens are as follows: 85/25, 85/40, 95/25, 95/35, 100/40, 105/35, 110/30 and 115/15.

Per EN 13304 (2009), oxidised bitumens are specified by the rules given in Table 3.7.

With regard to American specifications for oxidised bitumen (asphalt), information can be found in standards related to particular use, such as ASTM D 2521 (2014) for canal, ditches, pond lining and so on.

3.4.2 Performance graded asphalt binders – superpave

The extensive Strategic Highways Research Program, completed in the United States in 1993, concluded in performance graded (PG) specifications for bitumens (asphalt binders) known as Superpave (superior performing pavements) specification. These specifications were first introduced by Asphalt Institute (Asphalt Institute SP-1) and soon became an AASHTO standard (AASHTO M 320) and ASTM standard (ASTM D 6373 2007).

The grading designations are related to the high pavement design temperature at a depth 20 mm below the pavement surface and the low pavement design temperature at the pavement surface.

The high pavement design temperature at a depth 20 mm below the pavement surface is determined using the following equation (Asphalt Institute SP-1 2003):

$$T_{20mm} = (T_{air} - 0.00618 \times Lat^2 + 0.2289 \times Lat + 42.2) \times 0.9545 - 17.78,$$

where T_{20mm} is the high pavement design temperature at a depth of 20 mm, T_{air} is the 7-day average high air temperature (°C) and Lat is the geographical latitude of the project (°).

The low pavement design temperature at the pavement surface is calculated using the following equation (Asphalt Institute SP-1 2003):

$$T_{pav} = -1.56 + 0.72 \times T_{air} - 0.004 \times Lat^2 + 6.26 \log_{10}$$
$$\times (H + 25) - Z \times (4.4 + 0.52 \times \sigma_{air})^{1/2}$$

where T_{pav} is the low pavement design temperature at the pavement surface (°C), T_{air} is the low air temperature (°C), Lat is the latitude of the project (°), H is the depth to surface (mm), σ_{air} is the standard deviation of the mean low air temperature (°C) and Z is 2.055 for 98% reliability (from normal distribution table).

More details for calculating the above temperatures can be found in an example given in Asphalt Institute (2003). A Superpave computer program that performs all these calculations based on minimal user input is provided and may also be used.

The PG asphalt binder specification is intended to reduce bitumen's contribution to permanent deformation, fatigue and cracking at low temperatures of the pavement.

The specified properties per performance grade asphalt binders (bitumens) are given in Table 3.8. A similar table is also provided by ASTM D 6373 (2007), which incorporates practice D 6816 (2011) for determining the critical low cracking temperature using a combination of test method D 6648 (2008) and test method D 6723 (2012) test procedures. According to ASTM D 6373 (2007), if no table is specified, the default is Table 3.8.

The selection of the suitable performance grade of binder as described is for typical highway loading conditions (fast, transient loads). For slow-moving design loads, the binder should be selected one higher temperature grade to the right (one grade 'warmer'). For example, PG 64 instead of the PG 58 determined should be used.

Also, an additional shift is proposed for an extraordinary high number of traffic loads. These are locations where design lane traffic is expected to be $>10 \times 10^6$ equivalent single axle loads (ESAL). An ESAL is defined as one 80 kN dual tyre axle (Asphalt Institute SP-1 2003).

When the continuous grading temperatures and the continuous grades for PG asphalt binders need to be calculated, it is carried out in accordance to ASTM D 7643 (2010).

Continuous grading temperatures, T_c, are the high, intermediate and low temperatures at which the specification requirements given in Tables 1 or 2 of Specification D 6373 are met. The continuous grade is a grade defined by upper and lower continuous grading temperatures.

3.4.3 Chemical composition of bitumen

Bitumen (asphalt) is a complex chemical compound composed predominately of carbon and hydrogen (hydrocarbon), with a small amount of heterocyclic compounds containing sulfur, nitrogen and oxygen (Traxler 1936). Bitumen also contains traces of metals including nickel, magnesium, iron, vanadium and calcium in the form of inorganic salts and oxides. Elementary analysis of bitumens produced from a variety of crude oil showed that most bitumen contains carbon, 82%–88%; hydrogen, 8%–11%; oxygen, 0%–1.5%; and nitrogen, 0%–1%. The exact composition of bitumen differs, and it depends on both the source of the crude oil and the modification during its fractional distillation. It also depends on the oncoming ageing in service (Shell Bitumen 2003).

Table 3.8 PG asphalt binder specification by superpave

Performance grade	PG 46			PG 52							PG 58					PG 64					
	-34	-40	-46	-10	-16	-22	-28	-34	-40	-46	-16	-22	-28	-34	-40	-10	-16	-22	-28	-34	-40
Average 7-day, pavement design temperature (°C)	<46			<52							<58					<64					
Minimum pavement design temperature (°C)	>-34	>-40	>-46	>-10	>-16	>-22	>-28	>-34	>-40	>-46	>-16	>-22	>-28	>-34	>-40	>-10	>-16	>-22	>-28	>-34	>-40
Original binder																					
Flash point, D 92 (2012) (°C)					>230																
Viscosity, D 4402 (2013): Max. 3 Pa·s, test temperature (°C)					135																
Dynamic shear, D 7175: $G'/\sin\delta$, min 1.0 kPa; Test temperature at 10 rad/s (°C)	46			52							58					64					
After ageing by rolling thin film oven (Test Method D 2872 2012)																					
Mass loss, max. (%)					1.00																
Dynamic shear, D7175: $G'/\sin\delta$, min. 2.2 kPa; Test temperature at 10 rad/s (°C)	46			52							58					64					
After ageing in PAV (Practice D 6521 2008)																					
PAV ageing temperature (°C)	90			90							100					100					
Dynamic shear, D7175: $G'/\sin\delta$, 5000 kPa; Test temperature at 10 rad/s (°C)	10	7	4	25	22	19	16	13	10	7	25	22	19	16	13	31	28	25	22	19	16
Creep stiffness, D 6648 (2008): S_c max. 300 MPa, m-value, min. 0.3; Test temperature at 60 s (°C)	-24	-30	-36	0	-6	-12	-18	-24	-30	-36	-6	-12	-18	-24	-30	0	-6	-12	-18	-24	-30
Direct tension, D 6723 (2012): Failure strain, min. 1.0%; Test temperature at 1.0 mm/min (°C)	-24	-30	-36	0	-6	-12	-18	-24	-30	-36	-6	-12	-18	-24	-30	0	-6	-12	-18	-24	-30

	PG 70						PG 76					PG 82			
Performance grade	-10	-16	-22	-28	-34	-40	-10	-16	-22	-28	-34	-16	-22	-28	-34
Average 7-day, pavement design temperature (°C)	<70						<76					<82			
Minimum pavement design temperature (°C)	>-10	>-16	>-22	>-28	>-34	>-40	>-10	>-16	>-22	>-28	>-34	>-16	>-22	>-28	>-34
Original binder															
Flash point, D 92 (2012) (°C)								>230							
Viscosity, D 4402 (2013): Max. 3 Pa·s, test temperature (°C)								135							
Dynamic shear, D 7175: $G^*/\sin\delta$, min 1.0 kPa, Test temperature at 10 rad/s (°C)	70						76					82			
After ageing by rolling thin film oven (Test Method D 2872 2012)															
Mass loss, max. (%)							1.00								
Dynamic shear: $G^*/\sin\delta$, min. 2.2 kPa, Test temperature at 10 rad/s (°C)	70						76					82			
After ageing in PAV (Practice D 6521 2008)															
PAV ageing temperature (°C)	100						100					100			
Dynamic shear: $G^*/\sin\delta$, 5000 kPa, Test temperature at 10 rad/s (°C)	34	31	28	25	22	19	37	34	31	28	22	34	31	28	
Creep stiffness, D 6648 (2008): S, max. 300 MPa, m-value, min. 0.3, Test temperature at 60 s (°C)	0	-6	-12	-18	-24	-30	0	-6	-12	-18	-24	-6	-12	-18	-24
Direct tension, D 6723 (2012): Failure strain, min. 1.0%, Test temperature at 1.0 mm/min (°C)	0	-6	-12	-18	-24	-30	0	-6	-12	-18	-24	-6	-12	-18	-24

Source: Reprinted from ASTM D 6373, *Standard specification for performance graded asphalt binder*, West Conshohocken, Pennsylvania, US: ASTM International, 2007. With permission (© ASTM International).

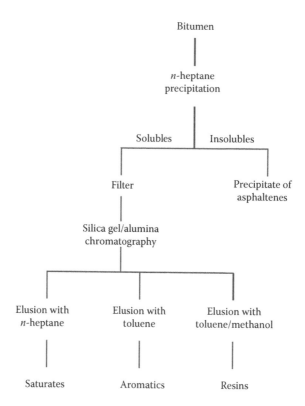

Figure 3.2 Schematic representation of the analysis for broad chemical composition of bitumen. (From Shell Bitumen, *The Shell Bitumen Handbook*. Surrey, UK: Shell Bitumen UK, 1990.)

Despite the complexity of bitumen's chemical composition, it is possible to be separated into two broad chemical groups, the asphaltenes and the maltenes. Maltenes can be further subdivided into saturated hydrocarbons, aromatic hydrocarbons and resins.

This separation of bitumen into the abovementioned fractions can be carried out using four methods: (a) solvent extraction, (b) chromatography, (c) adsorption by finely divided solids and removal of unabsorbed solution by filtration and (d) molecular distillation used in conjunction with one of the other techniques.

The first two of the abovementioned methods are mostly used. The solvent extraction is relatively simple and quick, but the separation is poorer than the one resulting from chromatography. Chromatography is the most widely used method for the detection of asphaltenes. Asphaltenes are insoluble to an *n*-heptane solution; thus, they are precipitated as sediment. In contrast to asphaltenes, maltenes are soluble in *n*-heptane, as well as to other solvents. Figure 3.2 provides a schematic representation of bitumen separation.

3.4.3.1 Asphaltenes

Asphaltenes are complex polar aromatic compounds, black or dark brown solids, insoluble in *n*-heptane, of high molecular weight, containing in addition to carbon and hydrogen some nitrogen, sulfur and oxygen. The asphaltene content directly affects the rheological properties of the bitumen. When asphaltene content increases, the bitumen is harder (low penetration and high softening point) and more viscous (high viscosity). The percentage of asphaltenes in bitumen usually ranges from 5% to 28%.

3.4.3.2 Maltenes

3.4.3.2.1 Resins

Resins have similar components to asphaltenes but they are soluble in *n*-heptane. They are solid or semi-solid, dark brown in colour and strongly adhesive. Resins are dispersing agents to asphaltenes and their proportion to asphaltenes control the gel/sol type of character of bitumen. Their molecular weight is lower than asphaltenes.

3.4.3.2.2 Aromatics

Aromatics are naphthenic aromatic hydrocarbons, have the lowest molecular weight of the compounds in the bitumen and represent the main dispersion medium of asphaltenes. They are viscous fluids of dark brown colour and they can be found at 40% to 65% in bitumen.

3.4.3.2.3 Saturates

Saturates are aliphatic hydrocarbons together with alkyl naphthenes and alkyl aromatics. Their molecular weight is similar to the molecular weight of aromatics and their components contain both waxy and non-waxy saturates. Saturates are light yellow to white in colour and its content ranges from 5% to 20% in bitumens.

Bitumen is generally considered to be a colloidal system consisting of high-molecular-weight micelles dispersed or dissolved in a dispersed oil medium of lower molecular weight, namely, maltenes (Girdler 1965). Any fluctuation in the percentage of asphaltenes and maltenes, particularly of resins and saturates, influences the viscosity and the temperature sensitivity of bitumen. The fluctuation of the abovementioned substances takes place mainly during production of bitumen.

3.4.4 Changes in bitumen composition during distillation, mixing, laying and time in service

During fractional distillations (atmospheric or with vacuum), the lightest volatile ingredients of the bitumen are removed, which results in an increase in asphaltene concentration. Asphaltene concentration also increases during the oxidation (air-blowing) process. This fact makes the bitumen harder and less susceptible to temperature variations (increase of the penetration index value).

The observed change in bitumen composition during distillation and air-blowing is demonstrated in Shell Bitumen (2003).

The chemical composition of the bitumen also changes during its usage, namely, from the very moment it is mixed with aggregates until the end of the pavement's service life. In engineering terminology, this change is known as ageing and results in hardening of the bitumen. Typical changes in composition with respect to mixing, laying and time in service are presented in Figure 3.3.

As can be seen in Figure 3.3, the major changes take place during the mixing and compaction stages. Consequently, great attention should be given to the recommended temperatures during these stages.

After compaction of the bituminous mixture, natural hardening of the bitumen takes place owing to temperature drop. From this moment, chemical hardening gradually starts to occur.

The natural hardening is caused by reorientation of the molecules owing to the temperature drop and the volatilisation of some volatile ingredients. Hardening attributed to molecule reorientation is reversible, whereas hardening attributed to volatilisation is a non-reversible phenomenon.

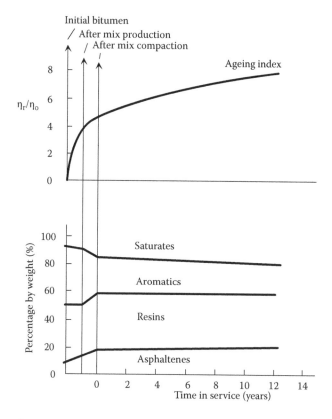

Figure 3.3 Changes in bitumen composition and its ageing index (ratio of recovered viscosity, η_r, over initial viscosity, η_o) during mixing, laying and in service. (From Chipperfield E.H. et al., Asphalt characteristics in relation to road performance. *Proceedings of the Association of Asphalt Paving Technologists*, Vol. 39, p. 575. Seattle, WA, 1970.)

Chemical hardening of the bitumen with time in service is also known as ageing of the bitumen. It occurs because of the oxidation of organic compounds and because of further volatilisation of volatile ingredients of the bitumen. This stage is also called bitumen oxidation.

Oxidation or ageing of the bitumen affects the mechanical behaviour of the bitumen and usually reduces the pavement's service life. The changes that occur are as follows: reduction of penetration, increase of softening point, reduction of elasticity and adhesion ability and increase of friability. Oxidation and ageing can be decelerated with the use of chemical additives.

Oxidation and, consequently, ageing of bitumen after construction are affected by climatic conditions (ambient temperatures and periods of sunshine) and voids in the bituminous mixture. Greater oxidation of bitumen is expected in geographical locations where ambient temperatures are high and sunshine periods are longer in a year. A similar result is expected between open and dense graded bituminous mixtures.

Measurements have shown that an 80/10 penetration grade bitumen extracted from an asphalt concrete surface layer with air voids of 8%, after 2 years in service in the latitude of Thessaloniki, Greece, showed a reduction in penetration by 20%. Bitumen (80/100) extracted from a similar bituminous mixture (cold asphalt concrete surface layer with 9% air voids) but lay in the latitude of Jakarta, Indonesia, after 1 year in service, showed a reduction in penetration of 35% to 45%.

Finally, it should be mentioned that bitumen oxidation is more intense at the surface (surface layer) than within the bituminous mixture (layers beneath).

The hardening of the bitumen during mixing, laying and compaction can be simulated in the laboratory by either the rolling thin film oven test (RTFOT) or the thin film oven test (TFOT), and its ageing can be simulated by the pressurised ageing vessel test (see Chapter 4).

Bitumen also hardens during hot storage; hence, prolonged storage should be avoided.

3.5 CUT-BACK AND FLUXED BITUMINOUS BINDERS

Cut-back bitumen constitutes bitumen whose viscosity has been reduced by the addition of a relatively volatile flux. In case the volatile flux is derived from petroleum, the material is called petroleum cut-back bitumen. Typical petroleum-derived fluxes used are white spirit and kerosene. Petroleum cut-back bitumens are almost exclusively used, and, for simplicity, they are called cut-back bitumens (or asphalts).

When the viscosity of the bitumen is reduced by the addition of a flux oil (a relatively involatile oil), the bitumen is called fluxed bitumen. Flux oils can be derived from petrochemical, carbochemical or petroleum origin materials, or a mixture of these materials. In this case, the product is called petroleum fluxed bitumen or mineral fluxed bitumen. Typical flux oil is the gas oil of various boiling ranges.

Additionally, flux oils can be derived from plant-based (vegetal) products. In this case, the product is called vegetal fluxed bitumen.

The viscosity of the bitumen is significantly reduced that cut-back bitumens or fluxed bitumens are considered as 'liquids' and require far less heating energy during application in comparison to 'solid' (at ambient temperatures) bitumen.

Cut-back and fluxed bituminous binders are suitable for use in the construction and maintenance of roads, airfields and other paved areas. The bitumen used may be paving grade bitumen (in most cases) or by addition of polymer.

However, cut-back bitumens and, to a lesser extent, fluxed bitumens used to be extensively used in the past, mainly for the production of open or semi-dense graded bituminous mixtures. Today, their use is limited to prime coating, to in situ production of cold/semi-warm bituminous mixtures particularly in remote areas away from a hot mix plant and to the production of ready-mixed bituminous mix for repairing–filling works (potholes, utility cuts, local depressions, etc.).

Cut-back bitumens, as well as the fluxed bitumens, are designated by the viscosity and setting ability or viscosity and curing time. They may also be designated by the bitumen type (paving grade or with addition of polymer).

3.5.1 Cut-back and fluxed bitumens according to CEN EN 15322 (2013)

According to CEN EN 15322 (2013), cut-back bitumens and fluxed bitumens are designated by two letters ('Fm' or 'Fv'), one number, followed by one or two letters ('B' or 'BP') and a number.

The letters 'Fm' or 'Fv' describe the flux material; if it is relatively volatile flux or flux oil, the notification 'Fm' is used; otherwise, if it is vegetal, the notification 'Fv' is used.

The number after the letters corresponds to a viscosity class, the letters following notify the type of bitumen, 'B' for paving grade bitumen and 'BP' if with addition of polymer, whereas the last number corresponds to the class of setting ability.

The recommended classes of cut-backs and fluxed bituminous binders with the technical requirements specified by CEN EN 15322 (2013) are as shown in Table 3.9.

According to the European standard, there are three classes (2, 3 and 4) of cut-back bitumens and fluxed bitumen with respect to low and medium viscosity and three classes of high viscosity (5, 6 and 7).

Table 3.9 Specification framework for technical requirements and performance classes of cut-back and fluxed bituminous binders

Technical requirements	CEN EN standard	Unit	Classes										
			0	1ª	2	3	4	5	6	7	8	9	10
Viscosity													
Efflux time 4 mm at 25°C[b]	12846-2 (2011)	s			<200[c]								
Efflux time 10 mm at 25°C[b]		s				15–500[c]							
Efflux time 10 mm at 40°C[b]		s					50–500[c]						
Efflux time 10 mm at 60°C[b]		s						20–300[c]	250–500[c]				
Dynamic viscosity at 60°C[b,d]	13302 (2010)	Pa·s								<10	10–50	30–100	>80
Solubility	12592 (2007)	%	NR[e]	TRB	>99.0								
Flash point	ISO 13736 (2013)	°C		TRB	≤23	>23	>35	>45	>45	>60	>65	>160	>200
	ISO 2719 (2002)	°C											
	ISO 2592 (2001)	°C											
Adaptivity with reference to aggregate	13358 (2010)	l	NR[e]	TRB	≥75	≥90							
Fm grades setting ability by distillation test													
Total distillate at 360°C	13358 (2010)	%	NR[e]	TRB	<5	<10	<15	<20	<32	<55			
% of total distillate fraction, 190°C		%	NR[e]	TRB	<5	2–15	10–25	>20					
% of total distillate fraction, 225°C		%	NR[e]	TRB	<15	10–25	20–40	35–60	>55				
% of total distillate fraction, 260°C		%	NR[e]	TRB	<20	15–40	35–60	>55					
% of total distillate fraction, 190°C		%	NR[e]	TRB	<40	35–70	65–90	>80					
Fv grades setting ability by softening point of recovered binder	13074-1 (2011)	°C	NR[e]	TRB	≤35	>35	>39	>43	>50	>55			

Source: Reproduced from CEN EN 15322, *Bitumen and bituminous binders – Framework for specifying cut-back and fluxed bituminous binders*, Brussels: CEN, 2013. With permission (© CEN).

[a] Class 1, TBR, may not be used for regulatory declaration and marking purposes.

[b] For marking purposes of each product, only one class of viscosity, from Class 2 to Class 10, may be used.

[c] In addition to the appropriate selected class (Class 2 to Class 4), a restricted efflux time range of ±35%, with a minimum of ±10 s about a midpoint value, shall be defined. The restricted range shall be within the class limits defined in the table.

[d] A restricted dynamic viscosity range of ±35% around midpoint values shall be defined, which shall be within the specified class limit.

[e] NR, no requirement may be used when there are no regulations for the property in the territory of intended use.

The low and medium viscosity is expressed as efflux time, while the high viscosity is expressed by dynamic viscosity (see Table 3.9).

Hence, using Table 3.9, the characterisation Fm 4 B 6 indicates a medium-viscosity cut-back or flux oil material based on paving grade bitumen that contains relatively volatile petroleum-based flux of which more than 55% distils at 225°C.

Similarly, the characterisation Fv 7 B 6 indicates a high-viscosity polymer-modified bituminous binder cut-back or fluxed material that contains a flux of vegetable origin, in which the recovered binder has a softening point higher than 50°C.

Apart from the technical properties/requirements of the cut-back bitumens and fluxed bitumens, the standard also defines the technical properties of the stabilised (recovered) bitumen. The properties/requirements, which are specified for stabilised bitumens with penetration below 330 dmm, are such as presented in Table 3.10. There is also a similar table for softer bitumens with penetration higher than 330 dmm (see CEN EN 15322 2013).

Table 3.10 Specification framework for the technical requirements and performance classes of stabilised binders, when penetration at 25°C after stabilisation is ≤330 dmm

			Stabilisation procedure: CEN EN 13074-1 (2011) followed by CEN EN 13074-2 (2011)							
			Classes							
Technical requirements	CEN EN standard	Unit	0	1ª	2	3	4	5	6	7
Consistency at intermediate temperature Penetration at 25°C	1426 (2007)	0.1 mm	NRᶜ	TBR	≤50	≤100	≤150	≤220	≤330	
Consistency at elevated temperature Softening point	1427 (2007)	°C	NRᶜ	TBR	≥55	≥50	≥43	≥39	≥35	≥30
Cohesion (modified binders only)			**0**	**1**	**2**	**3**	**4**	**5**	**6**	
Cohesion energy by tensile test at 5°C (100 mm/min traction)ᵇ	13587 (2010) 13703 (2003)	J/cm²	NRᶜ	TBR	≥1	≥2	≥3			
Cohesion energy by force ductility at 5°C (50 mm/min traction)ᵇ	13589 (2008) 13703 (2003)	J/cm²	NRᶜ	TBR	≥1	≥2	≥3			
Cohesion energy by pendulum testᵇ	13588 (2008)	J/cm²	NRᶜ	TBR	≥0.5	≥0.7	≥1.0	≥1.2	≥1.4	
Elastic recovery at 10°C (for elastic polymer binders)	13398 (2010)	%	NRᶜ	TBR	≥30	≥40	≥50	≥75		
Elastic recovery at 25°C (for elastic polymer binders)	13398 (2010)	%	NRᶜ	TBR	≥30	≥40	≥50	≥75		

Source: Reproduced from CEN EN 15322, *Bitumen and bituminous binders – Framework for specifying cut-back and fluxed bituminous binders*, Brussels: CEN, 2013. With permission (© CEN).

a   Class 1, TBR, may not be used for regulatory declaration and marking purposes.
b   The cohesion of stabilised binder from polymer-modified cut-back and fluxed bituminous binders, which are used for surface dressings, shall be determined in accordance with CEN EN 13588 (2008). For binders used in asphalt mixes, the test methods given in either CEN EN 13587 (2010) or CEN EN 13589 (2008) may be used. For binders used in other applications, any one of the three methods listed above, CEN EN 13587 (2010), CEN EN 13589 (2008) or CEN EN 13588 (2008), may be used.
c   NR, no requirement may be used when there are no regulations for the property in the territory of intended use.

Furthermore, the standard also requires defining the properties of the stabilised bitumen after long-term accelerated ageing conditioning by a pressure ageing vessel (PAV).

For more information, please refer to CEN EN 15322 (2013).

3.5.2 Cut-back asphalts according to American standards

The American standards recognise only the petroleum cut-back bitumens (asphalts). Depending on the rate of solvent evaporation or rate of curing, petroleum cut-back asphalts (bitumens) are divided into three categories. These categories are slow curing (SC), medium curing (MC) or rapid curing (RC) cut-back asphalts. MC cut-back asphalts are perhaps the type mostly used in pavement engineering.

Petroleum cut-back asphalts apart from curing rate are graded with reference to their minimum kinematic viscosity at 60°C.

According to ASTM D 2027 (2010) (AASHTO M 82 2012), there are five different grades of cut-back asphalts of MC (MC-30, MC-70, MC-250, MC-800 and MC-3000). All of their characteristic properties are shown in Table 3.11.

Table 3.11 Requirements for cut-back asphalt – medium-curing type

Property	Test method (ASTM)	MC-30 Min	MC-30 Max	MC-70 Min	MC-70 Max	MC-250 Min	MC-250 Max	MC-800 Min	MC-800 Max	MC-3000 Min	MC-3000 Max
Kinematic viscosity at 60°C (mm²/s)	D 2170 (2010)	30	60	70	140	250	500	800	1600	3000	6000
Flash point (°C)	D 3143 (2008)	38	—	38	—	66	—	66	—	66	—
Distillate test: Distillate, volume % of total distillate to 360°C:	D 402 (2008)										
– to 225°C		—	25	0	20	0	10	—	—	—	—
– to 260°C		40	70	20	60	15	55	0	35	0	15
– to 316°C		75	93	65	90	60	87	45	80	15	75
Residue from distillation to 360°C, per cent volume by difference	D 402 (2008)	50	—	55	—	67	—	75	—	80	—
Tests on residue from distillation: Viscosity at 60°C (Pa·s)[a]	D 2170 (2010)	30	120	30	120	30	120	30	120	30	120
Ductility at 25°C (cm)	D 113 (2007)	100	—	100	—	100	—	100	—	100	—
Solubility in trichloroethylene (%)	D 2042 (2009)	99.0	—	99.0	—	99.0	—	99.0	—	99.0	—
Water (%)	D 95 (2013)	—	0.2	—	0.2	—	0.2	—	0.2	—	0.2

Source: Reprinted from ASTM D 2027, Standard specification for cutback asphalt (medium-curing type), West Conshohocken, Pennsylvania, US: ASTM International, 2010. With permission (© ASTM International).

[a] Instead of viscosity of the residue, the specifying agency may specify penetration (100 g/5 s) at 25°C of 120–250 for all MC grades. However, in no case will both be required.

Table 3.12 Energy requirement in the production of bituminous binders

Energy required to produce 1 gal bituminous binder (Btu)		
Cut-back asphalt medium curing	Cationic emulsified asphalt	Asphalt cement (bitumen)
MC-30: 70,000	CRS-1: 2640	80/100 pen (dmm): 2500
MC-70: 63,200	CRS-2: 2715	
MC-250: 47,000	CMS-2: 2715	
MC-800: 36,200	CMS-2h: 2595	
MC-3000: 29,500	CSS-1: 2595	
	CSS-1: 2595	

Source: From Asphalt Institute, *Energy requirements for roadway pavements.* IS-173, USA: Asphalt Institute, 1979. With permission.

With respect to SC and RC, there are four different grades in each category (MS or MR-70, MS or MR-250, MS or MR-800 and MS or MR-3000). More details can be found in ASTM D 2026 (2010) and ASTM D 2028 (2010) (AASHTO M 81 2012).

It should be noted that cut-back bitumens are energetically unadvisable and are environmentally harmful. During production, normally costly solvents are used, which evaporate to the environment after being mixed with the aggregates or sprayed to the surface. The evaporation of the solvent contributes to waste of energy and to air pollution. It has been estimated (Asphalt Institute IS-173 1979) that to produce a gallon of an MC cut-back asphalt, 12 to 18 times more energy is required than to produce 1 gallon of bitumen, depending on the cut-back bitumen grade. Comparative energy required values between cut-back asphalts, bituminous emulsions and paving grade bitumen are presented in Table 3.12.

According to the abovementioned data, the usage of petroleum cut-back bitumens, as well as the fluxed bitumens, has been limited, and in some countries, their usage has been prohibited (Asphalt Institute MS-19; Chipperfield and Leonard 1976; Nikolaides and Oikonomou 1987; Transportation Research Board No. 30 1975).

3.6 BITUMEN EMULSIONS

Bitumen emulsions are emulsions in which the dispersed phase (discontinuous phase) is bitumen and the water or aqueous solution is the continuous phase.

The suspension of bitumen particles is achieved by charging their surface with the same electric charge after the addition of the emulsifier, or emulsifying agent, during the emulsification process.

Apart from the above three basic ingredients, a bitumen emulsion may contain, in minute amounts, one or more of the following: stabilisers, break control agents, acidity regulators, coating improvers or anti-stripping agents.

When a modified bitumen emulsion is to be produced, polymer is also added.

The size of the dispersed bitumen particles ranges from approximately 0.0001 to approximately 0.015 mm. Figure 3.4 shows a microscopic view of a bitumen emulsion.

In an unmodified bituminous emulsion, the proportion by mass of bitumen, water, emulsifier and other additives typically range within the values: 40%–70% bitumen, 58%–28% water, 0.5%–1.5% emulsifier and less than 1% other additives.

Depending on the surface charge of the particles, bitumen emulsions are distinguished into two basic categories: cationic (or acid), where particles are electro-positively charged, and anionic (or alkaline), where bitumen particles are electro-negatively charged. The term *cationic* is derived from the fact that, when static electric current passes through the emulsion,

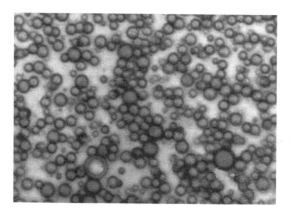

Figure 3.4 Bitumen emulsion, photo taken by microscope.

the positively charged bitumen particles are superimposed to the cathode. Similarly, the term *anionic* is derived from the negatively charged particles superimposed to the anode. The alternative terms *acid* and *alkaline* are due to the fact that bitumen particles are found to be suspended in acid or alkaline water environment, respectively, mainly caused by the surplus of emulsifier.

There are also non-ionic bitumen emulsions, where bitumen particles are in suspension of neutral water environment (pH approximately 7). These emulsions are rarely used in highway engineering.

Finally, there are also emulsions stabilised with clay, known as clay-stabilised emulsions. They are mostly used for industrial applications such as roofing and sealing and not in highway engineering works. They are essentially anionic emulsions.

The anionic bitumen emulsions were first to be used in highway engineering works in the beginning of the 20th century (Barth 1962). In the early 1960s, the cationic emulsions were developed, solving the inherent problem of re-emulsification of the anionic bitumen emulsions. Today, almost exclusively cationic emulsions are used.

Bitumen emulsions are used in a wide range of highway engineering works, such as production of cold bituminous mixtures for the pavement construction or maintenance, surface dressings, prime coating, tack coating, pre-coating of chippings, slurry sealing, micro-surfacing and slope stabilisation.

The European country with the highest annual consumption of bitumen emulsion in the year 2011 is France, 0.96 million tonnes, followed by Romania, 0.28 million tonnes, and Spain, 0.12 million tonnes (EAPA 2011).

The use of bitumen emulsions compared to cut-back bitumens and paving grade bitumens has the following advantages:

- They require less energy for production and application.
- They contribute to the atmospheric pollution reduction.
- They make bituminous (asphalt) works safer.
- They are able to coat successfully wet aggregates or wet surfaces with bitumen.
- They accelerate the progress of bituminous works and improve the construction productivity.
- They prevent some construction failures, such as surface layer slippage and bleeding, and eliminate bitumen ageing owing to prolonged heating.
- They show better results in prime coating and tack coating works.

The above advantages result from the fact that, because of low viscosity compared with bitumen and cut-back bitumens, the usage of bitumen emulsions does not presuppose the usage of heating at any stage of application. As a consequence, bituminous works are safer and can be carried out at lower ambient temperatures than those required for hot bituminous mixtures. Furthermore, low viscosity is achieved with the addition of water and not a solvent. This way, there is evaporation of only water and not evaporation of an expensive substance, which requires high thermal energy input. Moreover, in contrast with solvents, the evaporation of water does not add to the atmospheric pollution. Finally, emulsifiers used for the production of cationic emulsions have, by nature, adhesive properties, which further increase the adhesive ability of bitumen.

3.6.1 Types and classifications of bitumen emulsions

Bitumen emulsions, apart from the basic distinction (anionic or cationic) based on the polarity of the surface charge of the bitumen particles, are further distinguished into grades depending on the rate or setting or breaking value. American standards distinguish three bitumen emulsion grades: slow, medium and rapid setting.

The European standard distinguishes bitumen emulsions into a higher number of grades depending on the breaking behaviour.

Additionally, bitumen emulsions are designated by the percentage of bitumen content, the type of bitumen or its hardness.

According to CEN EN 13808 (2013), a combination of letters and numbers is used to describe a bitumen emulsion. The cationic emulsions are designated by the letter 'C'; two numbers indicating the nominal binder content; one or up to three letters such as B (for paving grade bitumen), P (for polymer-modified bitumen) and F (for bitumen with more than 3% [m/m] flux oil); and a number indicating the class of breaking behaviour.

As an example, the designation of bitumen emulsion C 69 B 2 means that the emulsion is cationic, its nominal binder content is 69%, the bitumen contained is paving grade bitumen and its breaking value belongs to class 2 (rapid rate of breaking).

The designation of bitumen emulsion C 65 BP 3 means that the emulsion is cationic, its nominal binder content is 65%, produced from bitumen and contains polymer, and its breaking value belongs to class 3 (almost rapid rate of breaking).

Additionally, the designation of bitumen emulsion C 69 BF 3 means that the emulsion is cationic, its nominal binder content is 69%, produced from bitumen and contains more than 3% flux, and its breaking value belongs to class 3.

As for the anionic type of bitumen emulsions, there was no EN specification available at the time of writing. This is probably due to the rare use or abandonment of use of anionic bitumen emulsion.

American specifications for cationic emulsified asphalts (cationic bitumen emulsions), namely, ASTM D 2397 (2012) (AASHTO M 208 2009), use the letter C for cationic emulsions, and ASTM D 977 (2012) (AASHTO M 140 2013) does not use a letter for anionic emulsified asphalts. They both use two letters indicating setting rate (RS for rapid, MS for medium and SS for slow), a number (1 or 2) and sometimes the letter 'h'. For cationic emulsions, number 1 indicates that the minimum bitumen content in the emulsion, for slow- and rapid-setting emulsions, is 60% and 57%, respectively. Number 2, which is only used for rapid- and medium-setting emulsions, indicates that the minimum bitumen content is 65%. Letter 'h', when used, indicates that the bitumen contained is harder than the usual one, 40 to 90 pen grade instead of 100 to 250 pen grade. Something similar is used for the anionic emulsions with the difference that the designation begins with the letters RS, MS or SS.

Thus, according to American standards, CMS-2 indicates that the emulsion is a cationic emulsion of medium setting with minimum bitumen content, 65%. Furthermore, CSS-1h indicates that the emulsion is cationic, of slow setting with minimum bitumen percentage, 57%, and the bitumen is harder than the 80/100 pen bitumen.

It should be noted that in ASTM D 977 (2012), there are also emulsions designated by the initials HF (high flow). This emulsion grade includes particular anionic emulsions, which help the formation of thicker bituminous films to prevent draining of bitumen from the aggregates.

Each emulsion, depending on its grade, should also meet the requirements of other properties, such as viscosity, storage stability, coating ability, oversized particles (residue on sieving) and so on. Additionally, the recovered bitumen should satisfy the penetration, softening point, ductility, solubility and viscosity requirements, to name a few, depending on the specification employed.

According to CEN EN 13808 (2013), the required properties of a cationic emulsion suitable for use in the construction and maintenance of roads, airfields and other paved areas are selected from the classes shown in Table 3.13. Additionally, the technical requirements and performance of the residual, recovered, stabilised and aged binder from the cationic bituminous emulsion is selected from the classes shown in Table 3.14.

The initials NR and DV used in Tables 3.13 and 3.14 are abbreviations of the following: NR for 'no requirement' (this class has been included for countries where the characteristic property is not subject to regulatory requirements) and DV for 'declared value', which means that the manufacturer is required to provide a value or value range for the product.

It is noted that the CEN EN 13808 (2013) standard applies to emulsions of paving bitumen, fluxed bitumen or cut-back bitumen and to emulsions of polymer-modified bitumen, polymer-modified fluxed bitumen or polymer-modified cut-back bitumen, which also includes latex-modified bituminous emulsions.

The grades and required properties of cationic emulsions according to ASTM D 2397 (2012) are presented in Table 3.15. It should be noted that the specification includes a grade of cationic emulsion, 'quick setting', exclusively used in slurry seal systems.

For anionic bitumen emulsions, a relevant specification table and other information are given in ASTM D 977 (2012).

3.6.2 Usage of bitumen emulsions with respect to setting rate

Applications of rapid-, medium- or slow-setting emulsions mainly depend on the gradation of aggregate mix, effectively on the relative surface area of the aggregates.

Rapid-setting emulsions are always used with open graded aggregate mixtures (low relative surface area). On the other hand, slow-setting emulsions are always used with dense aggregate mixtures (high relative surface area). Regarding intermediate mixtures, such as semi-dense graded mixtures, medium-setting emulsions may be used. However, this demarcation is not that strict. When it comes to other applications, such as tack coating, rapid- or medium-setting emulsions are used, whereas only slow-setting emulsions are used for prime coating.

3.6.3 Emulsifiers

Emulsifiers for anionic emulsions are fatty acids, which saponify when they react with sodium or potassium hydroxide. For instance,

$$\text{R-COOH (fatty acid)} + \text{NaOH} \rightarrow \text{RCOO}^- + \text{Na}^+ + \text{H}_2\text{O}.$$

Table 3.13 Specification framework for cationic bituminous emulsions – properties of the emulsion

Technical requirements	CEN EN document	Unit	0	1ᵃ	2	3	4	5	6	7	8	9	10	11	12
							Performance classes for the technical requirements of cationic bituminous emulsions								
Binder content	1428 (2012)[a]	% (m/m)			<38	38–42	48–52	53–57	58–62	63–67	65–69	67–71	≥69	≥71	
or Residual binder	1431 (2009)[b]	% (m/m)			<38 (C35)	≥38 (C40)	≥48 (C50)	≥53 (C55)	≥58 (C60)	≥63 (C65)	≥65 (C67)	≥67 (C69)	≥69 (C70)	≥71 (C72)	
Breaking behaviour															
Breaking value (Forsh. filler)	13075-1 (2009)	None			<100	70–155	110–195	>170	—	—	—	—	—	—	
or Fines mixing time, or	13075-2 (2009)	s			—	—	—	—	>90	≥180	≥300	—	—	—	
or Mixing stability with cement	12848 (2009)	g			—	—	—	—	—	—	—	>2	≤2	—	
Residue on sieving – 0.5 mm sieve	1429 (2013)	% (m/m)			≤0.1	≤0.2	≤0.5	—	—	—	—	—	—	—	
Viscosity															
Efflux time 2 mm at 40°C	12846-1 (2011)	s	NR		≤20	15–730	40–130	—	—	—	—	—	—	—	
or Efflux time 4 mm at 40°C	12846-1 (2011)	s	NR		—	—	—	5–70	40–100	—	—	—	—	—	
or Efflux time 4 mm at 50°C	12846-1 (2011)	s	NR		—	—	—	—	—	5–30	≥25	—	—	—	

(continued)

Table 3.13 Specification framework for cationic bituminous emulsions – properties of the emulsion (Continued)

Technical requirements	CEN EN document	Unit	Performance classes for the technical requirements of cationic bituminous emulsions												
			0	1[a]	2	3	4	5	6	7	8	9	10	11	12
or Dynamic viscosity at 40°C[c]	13302 (2010)	mPa·s	NR	—	—	—	—	—	—	—	—	≤30	20–300	100–1000	>1000
Adhesivity with ref. aggregate	13614 (2011)	None	NR	—	—	—	—	—	—	—	—	—	—	—	—
Penetration power	12849 (2009)	min	NR	DV	—	—	—	—	—	—	—	—	—	—	—
Oil distillate content[d]	1431 (2009)	% (m/m)	NR		≤2.0	≤3.0	≤5.0	≤8.0	≤10.0	5–15	>15	—	—	—	—
Residue on sieving – 0.16 mm sieve	1429 (2013)	% (m/m)	NR		≤0.25	≤0.5	—	—	—	—	—	—	—	—	—
Efflux time, 85°C	16345 (2012)	s	NR		25–45	20–100	—	—	—	—	—	—	—	—	—
Storage stability by sieving (7 days storage) – 0.5 mm sieve	1429 (2013)	% (m/m)	NR		≤0.1	≤0.2	≤0.5	—	—	—	—	—	—	—	—
Setting tendency (7 days storage)	12847 (2009)	% (m/m)	NR		≤5	≤10	—	—	—	—	—	—	—	—	—

Source: Reproduced from CEN EN 13808, *Bitumen and bituminous binders – Framework for specifying cationic bituminous emulsions*, Brussels: CEN, 2013. With permission (© CEN).

a   The binder content shall be defined as [100 – water content]; it may also be determined by the method described in EN 1431 (2009), in which case, it shall be defined as [per cent by mass of the residual binder + per cent by mass of oil distillate].

b   The residual binder content determined is the binder residue from a bituminous emulsion after distillation of water and oil distillate.

c   Dynamic viscosity at a shear rate of 50 s⁻¹. If the shear rate of 50 s⁻¹ does not produce the level of accuracy specified by EN 13302 (2010), another shear rate can be used.

d   The per cent by mass of oil distillate can be determined on bulk quantities of flux following the determination of the density by EN ISO 3838 and the determination of per cent by volume of distillate obtained by EN 1431 (2009).

Table 3.14 Specification framework for the technical requirements and performance classes for residual, recovered, stabilised and aged binders from cationic bituminous emulsions

Technical requirements	CEN EN document	Unit	Performance classes for the technical requirements of cationic bituminous emulsions[a]										
			1	2	3	4	5	6	7	8	9	10	11
Consistency at intermediate service temperature													
Penetration at 25°C	1426 (2007)	0.1 mm	DV	≤50	≤100	≤150	≤220	≤270	≤330	—	—	—	—
or Penetration at 15°C		mm		—	—	—	—	—	—	90–170	140–260	180–360	—
Consistency at elevated service temperature													
Softening point	1427 (2007)	°C	DV	≥60	≥55	≥50	≥46	≥43	≥39	≥35	<35	—	—
or Dynamic viscosity at 60°C	12596 (2007)	Pa·s	DV	≥18	≥12	≥7	≥4.5	<4.5	—	—	—	—	—
or Kinematic viscosity at 60°C	13302 (2010)	mm²/s	DV	≥16,000	≥8000	≥6000	≥4000	≥2000	≥2000	—	—	—	—
Cohesion (modified binders only)													
Cohesion energy by tensile test (100 mm/min traction)[a]	13587 (2010) 13703 (2003)	J/cm²	DV	≥3 at 5°C	≥2 at 5°C	≥1 at 5°C	≥2 at 10°C	≥1 at 10°C	≥0.5 at 10°C	≥1 at 15°C	≥0.5 at 15°C	≥0.5 at 20°C	≥0.5 at 25°C
or Cohesion energy by force ductility (50 mm/min traction)[a]	13589 (2008) 13703 (2003)	J/cm²		≥3 at 5°C	≥2 at 5°C	≥1 at 5°C	≥0.5 at 5°C	≥2 at 10°C	≥1 at 10°C	≥0.5 at 10°C	≥0.5 at 15°C	≥0.5 at 20°C	—
or Cohesion by pendulum test[a]	13588 (2008)	J/cm²		≥1.4	≥1.2	≥1.0	≥0.7	≥0.5	—	—	—	—	—
Brittleness at low service temperature (Fraass breaking point)	12593 (2007)	°C	DV	≤−25	≤−20	≤−15	≤−10	≤−5	≤0	≤5	—	—	—
Elastic recovery (for elastomeric polymer binders)	13398 (2010)	%	DV										
– at 10°C				≥75	≥50	—	—	—	—	—	—	—	—
– at 25°C				—	—	≥75	≥50	—	—	—	—	—	—

Source: Reproduced from CEN EN 13808, *Bitumen and bituminous binders – Framework for specifying cationic bituminous emulsions*, Brussels: CEN, 2013. With permission (© CEN).

[a] One cohesion method shall be chosen based on end application. The cohesion of residual binders from polymer-modified emulsions that are used for surface dressings shall be determined in accordance with EN 13588 (2008). For binders used in other applications, the method used shall be any one of EN 13587 (2010), EN 13589 (2008) or EN 13588 (2008).

Table 3.15 Types and requirements for cationic emulsified asphalt

Property	Test method (ASTM)	Rapid setting		Medium setting		Slow setting		Quick setting
		CRS-1	CRS-2	CMS-2	CMS-2h	CSS-1	CSS-1h	CQS-1H
Test on emulsions:								
Viscosity, Saybolt Furol, 25°C (s)	D 244 (2009)	—	—	—	—	20–100	20–100	20–100
Viscosity, Saybolt Furol, 50°C (s)	D 244 (2009)	20–100	100–400	50–450	50–450	—	—	—
Storage stability[a], 24 h (%)	D 6930 (2010)	<1	<1	<1	<1	<1	<1	—
Demulsibility, 35 ml, 0.8% (%)	D 6936 (2009)	>40				—	—	—
Coating ability and water resistance:								
Coating, dry aggregate	D 244 (2009)	—	—	>Good	>Good	—	—	—
Coating, after spraying		—	—	>Fair	>Fair	—	—	—
Coating, wet aggregate		—	—	>Fair	>Fair	—	—	—
Coating, after spraying		—	—	>Fair	>Fair	—	—	—
Particle charge test		Positive	Positive	Positive	Positive	Positive	Positive	Positive
Sieve test[a] (%)	D 6933 (2008)	<0.1	<0.1	<0.1	<0.1	<0.1	<0.1	<0.1
Cement mixing test (%)	D 6935 (2011)	—	—	—	—	<2.0	<2.0	—
Distillation:								
Oil distillate, by vol. emuls. (%)	D 6997 (2012)	<3	<3	<12	<12	—	—	—
Residue (%)	D 244 (2009)	>60	>65	>65	>65	>57	>57	>57
Tests on residue from distillation:								
Penetration, 25°C, 100 g/5 s (pen)	D 5 (2013)	100–250	100–250	100–250	40–90	100–250	40–90	40–90
Ductility, 25°C, 5 cm/min (cm)	D 113 (2007)	>40	>40	>40	>40	>40	>40	>40
Solubility in trichloroethylene (%)	D 2042 (2009)	>97.5	>97.5	>97.5	>97.5	>97.5	>97.5	>97.5

[a] This test requirement on representative samples is waived if successful application of the material has been achieved in the field.

Regarding cationic emulsions, emulsifiers are mostly monoamines, diamines, amido-amines, polyamines or imidazolines (Hoiberg 1965), which are dissolved with acids, mostly hydrochloric or acetic, before emulsification. The chemical reaction with addition of hydro-chloric acid is of the following formula:

$$R\text{-}NH_2(\text{amine}) + HCL \rightarrow RNH_3^+ + CL^-.$$

Compounds of quaternary ammonium are also used, which are water soluble, and they do not require acid or oxyethylene addition (Oikonomou and Nikolaides 1985).

Each molecule of the emulsifier has two groups, the hydrophobic and the hydrophilic. The hydrophobic or lipophilic is non-polar and soluble in bitumen (R) and the hydrophobic one is polar and soluble in water (COO^- or NH_3^+). During emulsification, the molecules of the emulsi-fier settle onto the microscopic globules (micelles) of bitumen. As a consequence, their surface, because of the polarity of the hydrophilic part, is charged uniformly. As a result, repulsive forces are developed between the globules, and this ensures their stable suspension in the continuous aqueous phase; hence, a bituminous emulsion is produced. This phenomenon is explained in Figure 3.5. Depending on the polarity of the hydrophilic part, the emulsion is designated as cat-ionic or anionic. Particle polarity may be determined according to CEN EN 3/18 1430 (2009).

3.6.4 Breaking mechanism of bitumen emulsions

When cationic or anionic emulsions come in contact with the aggregates or with the sur-face of the pavement, they begin to 'break'. In other words, the bitumen particles, owing to the disturbance of the balance of the emulsion–aggregate system, cannot be in suspension anymore and the bitumen particles (micelles) begin to settle on the aggregate's surface. Imbalance is caused by chemical reaction or water loss from the system (emulsion dehydra-tion). Furthermore, the reduction of the emulsifier content below a crucial level may also disturb the balance of the system. Emulsion breaking is indicated by the colour change of the emulsion from brown to black.

The breaking process of an anionic emulsion differs from the breaking process of a cat-ionic one. Regarding the anionic (alkaline) emulsion, breaking is mainly due to the dehydra-tion of the emulsion (amount of water loss, initially attributed to water absorption by the aggregates and then to water evaporation) and not to chemical reaction (Bohn 1965). Water loss is followed by coalescence, natural settlement, agglomeration, adhesion of micelles to the surface of aggregates and finally by emulsion breaking. Full breaking of anionic bitu-men emulsions can occur only after almost complete removal of the water from the system (bitumen emulsion and aggregate mixture). Otherwise, the system can be 're-emulsified' with the presence of rainwater. Re-emulsification is undesirable since it leaves the aggregates uncoated.

Regarding cationic bitumen emulsions, breaking is caused initially by a chemical reac-tion and secondarily by dehydration of the bitumen emulsion. In particular, breaking of a cationic emulsion takes place at three stages (Gaestel 1967; Scott 1974). At the first stage, absorption of redundant molecules of the emulsifier and of the bitumen electro-charged particles on the aggregate surface occurs, owing to electrostatic attraction. At the second stage, violent coalescence and natural settlement of particles occur owing to the imbalance of the system. At the third stage, dehydration of the emulsion takes place which causes fur-ther coalescence and finally breaking of the emulsion. The breaking process of a cationic emulsion with a siliceous aggregate particle is explained in Figure 3.6. These three breaking

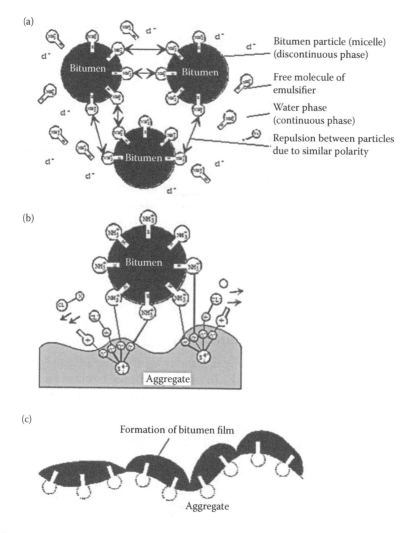

Bitumen particle (micelle)
(discontinuous phase)

Free molecule of
emulsifier

Water phase
(continuous phase)

Repulsion between particles
due to similar polarity

Formation of bitumen film

Aggregate

Figure 3.5 Breaking mechanism and adhesion of cationic bitumen emulsion on siliceous aggregate particle. (a) Suspension of bitumen droplets in the water – continuous. (b) Initiation of breaking and adhesion of bitumen particles to the aggregate surface. (c) Aggregation and deposition of bitumen particles.

stages take place within a very short period; thus, the danger of re-emulsification of the system is nullified. This fact shows a major advantage of cationic emulsions over anionic emulsions.

In case a cationic emulsion with limestone aggregates is used, the breaking process of the emulsion is the same, since limestone used in bituminous mixtures is not absolutely electro-positively charged. The total surface charge of limestone aggregates is designated as a mixture of electro-positive and electro-negative charges (Mertens and Wright 1959). As a consequence, the first stage of the emulsion breaking, which is due to the electrostatic attraction, also occurs. Hence, cationic emulsions can be used with all aggregates, and because they reach complete breaking much quicker than the anionic ones, they have almost phased out the anionic bitumen emulsions.

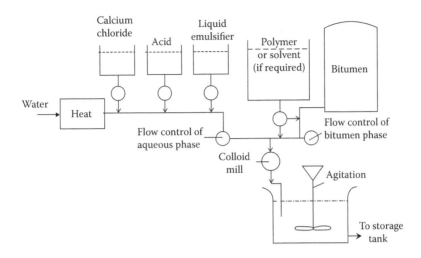

Figure 3.6 Schematic presentation of cationic bitumen emulsion production – continuous plant.

3.6.5 Production of bitumen emulsions

Bitumen emulsions are produced in specific stationary or mobile unit plants, which consist of storage tanks for raw materials, pumps and piping network, emulsification system (colloid mill or high-speed mixer) and tanks with stirring ability to store the final product. A schematic representation of cationic bitumen emulsion production of continuous flow is shown in Figure 3.6.

Bitumen (usually 80/100 or 60/70 grade), water, emulsifier, chemicals used for the water solubility of the emulsifier, stabilisers and particular oils in a small proportion and solvents are used as raw materials.

The bitumen is heated at a suitable temperature (110°C–140°C), so as to acquire low viscosity, and is pumped to a colloid mill. The emulsifier, the acid, calcium chlorite or any other water-soluble additives are added to the heated water (normally less than 85°C), and all lead to the colloid mill. Other additives such as solvent or polymer in liquid form are also added to the colloid mill.

The high-speed mill (up to 6000 revolutions per minute) consists of a conic rotor and its respective stator. The gap between rotor and stator is adjustable and is normally approximately 0.25–0.50 mm. Any variation of the gap affects the sizes of the bitumen globules (particles) and, hence, the quality of the bitumen emulsion.

The bitumen and the aqueous solution of the emulsifier and other additives as they enter the colloid mill are subjected to high shear stresses, which divide the bitumen into microscopic particles (0.5 to 0.015 mm in diameter). These particles, because of the presence of the emulsifier, are uniformly charged to the same polarity. The produced bitumen emulsion, having a temperature of approximately 80°C–90°C, is pumped into storage tanks to cool down.

Apart from for the continuous flow production units, there are also batch production units. In this case, the predetermined proportions of water, emulsifier and all the other chemicals for each batch are mixed with water in a separate tank before the aqueous solution is pumped into the colloid mill. The hourly output of this system is always lower than the output of the continuous flow system.

3.6.6 Properties of bitumen emulsion

The produced bitumen emulsion must be stable, so not to break during transportation or/and storage for a certain period of time. Additionally, it should have a suitable viscosity, so to be sprayed at ambient temperatures without the need of too much heating. Finally, it should adhere to the aggregates properly and break at a desired breaking rate. Obviously, certain properties of the emulsion, such as stability and breaking, seem to be contrary. This is true, but the abovementioned properties are required at different stages of emulsion usage. The first one is required during storage, whereas the second one is needed during application.

3.6.6.1 Storage stability of bitumen emulsion

Storage stability of bitumen emulsion is of great importance, because otherwise the produced and stored emulsion could not be properly used or it would be absolutely unsuitable for use. Storage stability is mainly affected by the size of the bitumen globules being in suspension in the aqueous phase as well as by their size distribution. The larger the globule size, the bigger their mass, and as a result, they cannot remain in suspension for a long period. They start to fall through and accumulate at the bottom of the tank or the barrel. The sedimentation velocity is higher when the size distribution of the globules is non-uniform. While falling down, these particles are likely to draw other particles at the bottom, leading to a bigger volume agglomeration of globules. As a consequence, the area at the lower part of the tank/barrel is rich in bitumen, while the area at the upper part is bitumen deficient. At the initial stage of settlement, simple stirring or shaking may recall the system to its initial form. Thus, it can be said that this agglomeration phenomenon called flocculation is reversible.

Further flocculation of bitumen globules, attributed to the weight of the mass of globules formed and the dominant hydrostatic pressure, causes fusion of the globules and the formation of a unified bituminous mass. This phenomenon is called coalescence. Coalescence is an irreversible process. Thus, after coalescence begins, the bitumen content of the emulsion is drastically reduced and the emulsion is unsuitable for use.

The size of bitumen globules, which should normally be between 0.5 and 0.015 mm for at least 85% of the globules, is mainly affected by the distance of the gap between the rotor and the stator. If for some reason (mainly metal wear) the gap becomes bigger, the emulsion produced will be more 'coarse'. The size of the globules can also be affected by the low temperature of the bitumen or the aqueous solution during the emulsification process.

The storage stability of the emulsion may also be affected by the viscosity of the emulsion, the specific gravity of the bitumen and the emulsifier content of the emulsion.

Emulsions with relatively low viscosity and emulsions produced by bitumen with relatively high specific gravity are more prone to settlement than emulsions with high viscosity and bitumen of lower specific gravity. The use of an increased emulsifier quantity in the system decelerates the settlement rate. However, this solution should not be preferred, since the emulsifier constitutes the most expensive ingredient and it will have further effects, such as increase, mainly, of the emulsion break time.

Emulsion storage stability is quickly determined by the 24 h settlement test (see Section 4.20.13). The size of bitumen globules, as well as their size distribution, can be determined using electronic microscopes or electronic apparatuses for sieve analysis. A quick and far less costly method to determine the suitability of the bitumen emulsion in road works with respect to its bitumen globule (particle) size is the sieving test (see Section 4.20.12).

During storage or transport of bitumen emulsions, the cleanness of tanks or barrels should not be overlooked. In case of residues or foreign solid particles, it is almost certain that the stability and emulsion's general behaviour will be affected.

At this point, it must be stated that regardless of the cleanness of the storage area, the emulsion should not be exposed to ambient air for a long period. A partial evaporation of the water causes surface breaking, which appears as a 'crust' of bitumen in the surface of the tank, container or barrel. Thus, barrels or tanks should be sealed properly and, if possible, be completely filled. Furthermore, it is preferable that barrels or tanks are placed vertically.

It must be emphasised that an anionic emulsion should never be mixed with a cationic emulsion, since as a result of their opposite electrostatic charge, instant breaking occurs. Barrels or tanks that were used in the past for the opposite emulsion type should be cleaned thoroughly with plenty of water or solution in case they are to be used again. It should be ensured that no trace of emulsion of opposite type exists. Tanks that were used for bitumen storage in the past can be used for emulsion storage, once they are checked for foreign bodies or detached parts of old bitumen.

3.6.6.2 Viscosity of bitumen emulsion

Bitumen emulsions are used at ambient temperatures, and as a result, an essential change in their viscosity can possibly have negative impacts or requirements for partial heating. Emulsion viscosity mainly affects the emulsion's ability to be successfully sprayed at ambient temperatures and to successfully coat the aggregates with bitumen. It also affects the workability of the cold bituminous mixture at the mixing and compaction stages.

The viscosity of the emulsion mainly changes with the fluctuation of bitumen content. As the bitumen content increases, the viscosity of the emulsion also increases. The effect of the increase is more profound when the bitumen content exceeds 65% (see Figure 3.7).

The same figure also presents the effect of temperature on the emulsion viscosity. As the temperature is reduced, the emulsion viscosity increases. This increase is more distinct to emulsions having a bitumen content of more than 65%. Moreover, the emulsion viscosity is affected by the bitumen viscosity during emulsification, the acidity of the aqueous solution and the emulsifier content (Shell Bitumen 2003). When the viscosity of the bitumen entering the colloid mill is reduced, the average diameter of particles is reduced and the emulsion viscosity slightly increases. Additionally, a relative increase of the viscosity is observed when the acidity of the aqueous solution is reduced and emulsifier content is increased.

The grade of bitumen (hardness) used for emulsion production also affects the viscosity of the emulsion but to a lesser extent. The extent of influence depends on the type of emulsifier (Nikolaides 1983). The type of emulsifier also affects the Newtonian or non-Newtonian behaviour of the emulsion.

Lyttleton and Traxler (1948) examined a great number of anionic emulsions and found that almost all of them had non-Newtonian behaviour, with the majority of them being thixotropic. Cationic emulsions examined by other researchers exhibited the same behaviour, but as a whole, their viscosity was lower than the viscosity of the anionic ones, for the same bitumen content. Clear non-Newtonian behaviour has been found (Nikolaides 1983) in cationic emulsions when the bitumen content exceeded 60% (see Figure 3.8).

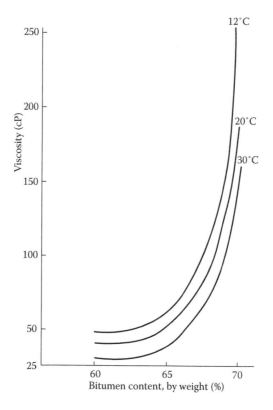

Figure 3.7 Viscosity variation attributed to bitumen content and temperature variation.

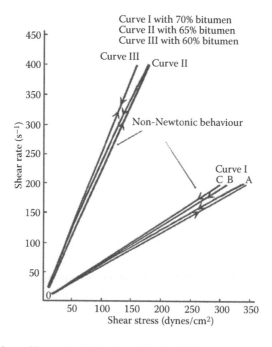

Figure 3.8 Newtonian and non-Newtonian behaviour of cationic bitumen emulsion.

3.6.6.3 Adhesiveness of bitumen emulsion

In all cases where bitumen is used, the need for good adhesiveness of bitumen with aggregate surface is of the greatest importance.

The factors that affect the adhesiveness of the emulsion and thus of the bitumen with the aggregates are as follows: the emulsifier type and quantity, the bitumen quality and grade, the pH of the aqueous solution of the emulsifier and the type of aggregates.

A sufficient quantity of the selected emulsifier should be included in the emulsion so that a sufficient number of free molecules of the emulsifier exist. The pH of the aqueous solution of the emulsifier should fluctuate depending, mainly, on the chemical composition of the bitumen and the type of aggregates. Regarding the effect of the aggregate type on the adhesiveness of the emulsion/bitumen, it should be stated that cationic emulsions adhere very well to all types of aggregates for the production of bituminous mixtures. In contrast, anionic ones adhere better only to limestone aggregates.

3.6.6.4 Breaking rate of bitumen emulsion

The breaking rate of an emulsion is one of the most important factors to help achieve a good performance of the emulsion and satisfactory mixing or spraying. The breaking rate should be such, in order to provide the required time for the uniform and full dispersion of the bitumen on the surface of aggregates, to offer sufficient workability during laying and, at the same time, to accelerate the development of the cohesion of the mixture. This may seem difficult, but in practice, it is achievable. The breaking rate of a bitumen emulsion for spraying applications, apart from prime coating, should be higher than that for mixing applications.

The breaking rate depends on the following factors:

a. The composition of the bitumen emulsion
b. The rate of water evaporation—weather conditions (temperature and wind velocity)
c. The absorption of aggregates
d. The physical and chemical characteristics of aggregates
e. The disturbance of the emulsion/aggregate system (cold mix) during mixing and laying

The breaking rate can be changed by altering the emulsion composition, namely, the amount of emulsifier, bitumen and additives. In a cationic bitumen emulsion, if the amount of emulsifier and the bitumen content is reduced, or if the amount of acid or the acid/emulsifier ratio is increased, the rate of breaking is reduced. Additionally, a reduction on the size of the bitumen globules and its size distribution also reduces the breaking rate. In case an increase of breaking rate is required, fluctuation of the above parameters should be reversed (Nikolaides 1983).

The evaporation rate of water has a direct impact on the rate of emulsion breaking. Since the evaporation rate is related to the climatic conditions, the rate of emulsion breaking increases when the ambient temperature and the wind velocity increases or when the relative humidity is reduced. Hence, the emulsion used should be susceptible as little as possible to the abovementioned factors. Cationic emulsions seem to be more susceptible to the above factors.

The absorption of aggregates is related to the loss of water from the bitumen emulsion. Thus, for the same emulsion, when porous aggregates are used, the rate of emulsion breaking is expected to be faster. The increase of breaking rate attributed to aggregate porosity can be eliminated by pre-moistening the aggregates (3% to 5% of water by mass of aggregates may be used).

Physical and chemical characteristics of aggregates affecting the breaking rate are the particle size distribution and the maximum particle size of the aggregate mixture (both related to the aggregate specific surface area, the relative moisture content, the surface texture, the origin of parent rock material and the amount and type of filler).

When the specific surface area of aggregates increases, which means that more fine aggregates are in the mixture, the bitumen emulsion breaks faster. Likewise, an increase of the breaking rate is observed when the filler percentage increases. However, the effect on the breaking rate differs if the filler is cement, limestone or lime, for a given bitumen emulsion. Nevertheless, the impact of the amount of filler is greater than the type of filler used.

As the relative moisture of aggregates increases, the rate of emulsion breaking decreases. Even when using slow-setting emulsions, pre-wetting of the aggregates with a small amount of water (approximately 1%–3%) is almost always the case during the production of a cold bituminous mixture. The pre-wetting of the aggregates results in neutralising the surface ions, to a certain extent.

Finally, the rate of emulsion breaking is affected by the mixing and spreading device used (i.e. mixing unit with a spreader box or motor grader). In the first case, the rate increases because higher shear forces are applied to the emulsion/aggregate system. Further breaking of the emulsion is accelerated by rolling the mat for a few minutes after laying (20–30 min).

At this point, it must be stated that the rate of emulsion breaking can also be modified by using chemical substances. In case of asphalt dressing, the rate of emulsion breaking may be accelerated by spraying a chemical substance on the already sprayed surface with emulsion, shortly before the aggregate dispersion. These substances usually play a dual role: they accelerate the emulsion setting and improve further bitumen and aggregate adhesion.

In the beginning of the 1990s, an additive that fully controlled the breaking rate of the emulsion, depending on the type of application, was developed. The additive was an aqueous solution (neutraliser) in oil (reverse emulsion) and it was added to the bitumen emulsion shortly before its use. This additive gave excellent results both on surface dressings and on cold bituminous mixtures of any type with conventional or modified bitumen (Redilius 1993).

3.7 ANTI-STRIPPING AGENTS

Anti-stripping agents, or adhesion agents, are mainly organic compounds used for improving the adhesion of bitumen on the aggregate surface, reducing or eliminating the danger of bitumen stripping from the aggregate surface, in the presence of water.

The use of anti-stripping agents applies both to surface dressings (Cawsey and Gourlet 1989; Woodside and Macoal 1989), where excellent and permanent adhesion is required, and to hot or cold bituminous mixtures (Anderson et al. 1982; Christensen and Anderson 1985). In both cases, the bitumen–aggregate adhesion is improved and the mixture resistance to the disastrous impact of water is increased, in case hydrophilic aggregates are used.

As a consequence, the cohesion, stability and tensile strength of the bituminous mixture are preserved or even improved when it is subjected to a water attack. This is due to the insertion of anti-stripping agent between the hydrophilic aggregates and the oleophilic bitumen, which develops a strong bond between bitumen and aggregate surface. For every aggregate category, namely, siliceous, silico-calcareous and calcareous, there is an appropriate type of anti-stripping agent that should be used.

3.7.1 Types of anti-stripping agents

Anti-stripping agents are chemical substances that have a similar composition to emulsifiers and are divided into (a) anionic type, such as organic acids (creosote) and fatty acids (oleic acid, stearic acid, pine pitch, etc.), and (b) cationic type, such as amines (simple amines, diamines, tertiary amines, polyamines and imidazolines) and salts of quaternary ammonium.

Organic acids are usually used with aggregates bearing a positive electrostatic charge on their surface, such as limestone, dolomites, laterites and so on. Fatty acids are usually used with aggregates that are not electrostatically charged, silico-calcareous aggregates, such as quartzites, rhyolites and so on. Finally, all anti-stripping agents of cationic type are used with aggregates, which are negatively surface charged, as the majority of silica rocks.

The operating procedure of anti-stripping agents is the same as that of the emulsifiers. In other words, the hydrophobic part of their molecules is dissolved in the bitumen and the electrostatically charged free end forms a strong bond with the aggregate surface. This bond is stronger than the one that would be formed between bitumen and aggregates under normal conditions, even if aggregates were positively charged. In case an amount of water appears in the system, the hydrophilic end of molecules, which have already formed a bond with the aggregate surface, does not permit detachment of bitumen from the aggregate surface.

3.7.2 Usage of anti-stripping agents

3.7.2.1 Anti-stripping agents in surface dressings

As it was mentioned, the anti-stripping agents are used in surface dressings. They may be incorporated using one of the following ways:

i. Dissolved (usually in tar oil) and sprayed onto the aggregates before aggregate spreading
ii. Dissolved in tar oil and sprayed on the surface already coated with bitumen, just before spreading the aggregates
iii. Dissolved into bitumen before spraying begins

The first way is fairly time-consuming and thus is rarely used. The second one significantly reduces the viscosity of the bitumen film and delays the bond development. The third way has an advantage over the other ways, since a good adhesion is achieved not only between aggregates but also with the old surface. Thus, the third way has been established and is used because of more positive results observed.

A key element is the determination of the necessary quantity of anti-stripping agent to be used. This quantity depends on the type of anti-stripping agent; the type, texture and porosity of the aggregate; and the composition, viscosity and surface tension of the bitumen and the ambient temperature. In general terms, the required quantity of anti-stripping agent usually ranges between 0.5% and 2.5%. A small increase of the quantity is observed when the viscosity of the bitumen increases and ambient temperatures decrease.

3.7.2.2 Anti-stripping agents in asphalts

Anti-stripping agents are also used in the production of hot or warm asphalts (bituminous mixtures) in order to improve the adhesion between bitumen and hydrophilic aggregates in the presence of water. The addition of anti-stripping agent not only preserves the cohesion, stability and stiffness of the asphalts when wet but also prevents ravelling.

The anti-stripping agent is added to the bitumen shortly before mixing the bitumen and the aggregates for the production of asphalts. Adding the anti-stripping agent to bitumen shortly before mixing is due to the fact that usually most anti-stripping agents are not stable and break up when being at high temperatures for a long period.

Asphalts treated with anti-stripping agent, apart from increasing their retained stability, showed better behaviour in bitumen ageing (Anderson et al. 1982; Christensen and Anderson 1985). Furthermore, some anti-stripping agents decrease the susceptibility of the bitumen to temperature variations (Anderson et al. 1982).

Finally, it should be stated that in all cases, the effectiveness and suitability of each anti-stripping agent differ between aggregates and also between bitumens. Thus, the anti-stripping agent–bitumen–aggregate system should be examined in order to determine the optimum usage of the anti-stripping agent for a given type of aggregate and origin of bitumen.

3.7.3 Test methods for determining susceptibility to stripping or affinity between aggregate and bitumen

The determination of susceptibility to stripping in the presence of water or affinity between aggregate and bitumen and thus the necessity of anti-stripping agent usage is carried out by various laboratory tests. In fact, by carrying out these tests, the hydrophilia of the aggregate is determined.

The tests may be conducted on loose or on compacted bituminous mixture. In the first case, the affinity between aggregates and bitumen or aggregate's hydrophilia is determined directly. In the second case, the moisture susceptibility of the compacted bituminous mixture (specimen) is specified and the susceptibility to stripping or aggregate/bitumen affinity or aggregate's hydrophilia is determined indirectly.

The tests that belong to the first category are (a) the rolling bottle method, (b) the static method and (c) the boiling water stripping method. These tests are described in EN 12697-11 (2012).

In the second category, fracture tests on specimens after immersing them into water are used for the determination of (a) the ratio of indirect tensile strength (ITSR) (CEN EN 12697-12 2008) or (b) the index of the retained compressive strength (ASTM D 1075 2011) and (c) the ratio of retained Marshall stability.

Finally, for surface dressings, the Vialit plate shock test method is used (CEN EN 12272-3 2003).

3.7.3.1 Rolling bottle test method

In the rolling bottle test method, the affinity is determined visually by recording the degree of the bitumen coverage of the aggregate particles after hand mixing and rotating the loose bituminous mixture in the presence of water.

The rolling bottle test is a simple but subjective test, suitable for routine testing. This test is not appropriate for highly abrasive aggregates.

In accordance with CEN EN 12697-11 (2012), Part A, a minimum aggregate quantity of 600 g passing through an 11.2 mm sieve and retained on an 8 mm sieve (or passing through an 8 mm sieve and retained on a 5.6 mm sieve, respectively) is washed using an 8 mm or a 5.6 mm sieve and is dried. From the initial aggregate quantity, 510 g aggregates are heated and mixed well (by means of a spatula) with a certain quantity of heated bitumen, namely, 16 ± 0.2 g or 18 ± 0.2 g, depending on the aggregate size. This quantity corresponds to approximately 3% bitumen per weight of mixture. After mixing, the loose bituminous mixture is left onto a flat metal lid or silicone-coated paper at ambient temperatures 20°C $\pm$ 5°C for 12 to 64 h.

The loose bituminous mixture is split into three equal parts and placed into specific glass bottles, which are filled to the middle with distilled water. After sealing the bottles with a screw cap, the bottles are placed on a rolling machine for 6 h ± 15 min. The rotation speed is proportional to the hardness of the bitumen used. The temperature during rolling is the ambient temperature, namely, 15°C to 25°C.

After rolling is finished, the water is removed and the content is transferred into a bowl or flat surface, where the percentage of bitumen coverage on aggregates is estimated (to the nearest 5%) by one or two skilled operators (in case of two estimations, the average value is recorded).

The aggregate particles are left for 6 ± 1 h and 24 ± 1 h and returned to the bottles filled with fresh distilled water for further rolling (6 h ± 15 min rolling time). Optionally, 48 h or even 72 h resting periods may be used.

Each time, not only the percentage of the aggregate coverage is recorded, but also the case where lumps of particles are created. If lumps exceed 10% of the total number of particles, the test result shall be discarded.

More information is given in CEN EN 12697-11 (2012).

3.7.3.2 Static test method

In the static test method, the affinity is also determined visually by recording the degree of bitumen coverage of the aggregate particles after leaving the coated aggregates in water for a certain period.

This static test is simple but subjective and suitable for routine testing for all types of aggregates with respect to their hardness.

According to CEN EN 12697-11 (2012), Part B, a sufficient aggregate quantity passing through a 10 mm sieve and retained on a 6 mm sieve is washed using a 6 mm sieve and then dried. From the dried aggregate, a mass of 150 g is heated at 135°C ± 5°C and mixed well (by hand) for approximately 5 min with a certain quantity of heated bitumen at 135°C ± 5°C. The amount of bitumen is equal to 4% by weight of aggregates and similar to the one going to be used.

If any particles are not coated after 5 min of mixing, a new aggregate proportion is mixed again with the bitumen proportion increased in steps of +0.5% by weight of aggregates until a mix that gives a complete coating of the aggregate is obtained.

Once full aggregate coating is achieved, a glycerol and dextrine mixture (50%/50%) is sprayed on the mix and the loose mixture is left for 1 h ± 5 min in one or two trays.

After the above time has passed, the loose bituminous mixture is covered with distilled water at a stable temperature of 19°C ± 1°C for 48 ± 1 h. After immersion, the water is removed and the loose mixture is dried at 19°C ± 9°C. Then, each particle is examined for incomplete coating by the binder. If more than three aggregate particles in a sample have an incomplete binder coating, the test is repeated on three more samples.

During testing the number of particles with an incomplete coating of binder is recorded each time and the average value is determined and reported as the final result. The quantity of bitumen used is also recorded and reported. More information is given in CEN EN 12697-11 (2012).

It is stated that a similar test method used to be executed using AASHTO T 182 or ASTM D 1664 is now withdrawn.

3.7.3.3 Boiling water stripping test method

The boiling water stripping test is used to express the affinity in reference to the degree of bitumen coverage on loose aggregates after immersion in boiling water containing certain reagents (hydrochloric acid or hydrofluoric acid and phenolphthalein [as indicator]).

It is designated as an objective test with high precision. However, it requires a greater amount of operative skills from the laboratory staff and the use of chemicals. As a consequence, hygiene and safety measures are required.

The boiling water stripping test can be used for any bitumen–aggregate combination, which might be calcareous, silico-calcareous or siliceous aggregates.

According to CEN EN 12697-11 (2012), Part C, a minimum aggregate quantity of 2000 g passing through a 14 mm sieve and retained on a 7 mm sieve (another aggregate fraction may also be used) is washed using the 7 mm sieve and then dried. From the dried aggregate, a mass of 1500 ± 52 g is heated and mixed well (by a spatula) with a certain quantity of heated bitumen (31.5 ± 0.2 g). This quantity corresponds to approximately 2.1% bitumen by weight of total mix (the quantity of bitumen is modified if aggregate fraction other than 7/14 mm is used). After mixing, the loose bituminous mixture is transferred into a vessel and then into cold water to cool off (conditioning stage).

Two samples of coated aggregates (approximately 200 g each) are boiled in 600 ml demineralised water for 10 min and then removed to dry and cool off. The coated aggregate samples are weighed again and each was put in 5 min contact with approximately 200 g of 0.1 N hydrochloric acid in an 800 ml glass beaker, if calcareous aggregates are tested. In case silico-calcareous or siliceous aggregates are tested, use 1 h $\pm$ 1 min contact period and 0.1 N hydrofluoric acid.

After the contact time has elapsed, separate the hydrochloric acid from the aggregate by pouring the solution gently into a 250 ml graduated cylinder. Titrate a 25 ml aliquot portion with 0.1 N sodium hydroxide or N potassium hydroxide, depending on the type of aggregate used, in the presence of phenolphthalein. Repeat the titration on a second aliquot of 25 ml. Determine the average volume in millilitres of 0.1 N NaOH or KOH required for the titration to the nearest 0.05 ml.

Then the volume of acid consumed is calculated using a formula, and from that, the percentage (%) of stripped aggregates is determined using a prepared calibration curve. The degree of bitumen coverage (%) is calculated as follows: 100-% of stripping.

More information is given in CEN EN 12697-11 (2012).

3.7.3.4 Water sensitivity of bituminous specimens by determination of the indirect tensile

The water sensitivity of a bituminous mixture is evaluated by determining the effect of saturation and accelerated water conditioning on the indirect tensile strength of cylindrical specimens of a bituminous mixture. The test can be used for evaluating the effect of moisture with or without anti-stripping additives.

According to CEN EN 12697-12 (2008), a set of not less than six cylindrical specimens is prepared. The specimens shall have a diameter of 100 ± 3 mm when the nominal maximum particle size of the bituminous mixture is 22 mm, and 150 ± 3 mm or 160 ± 3 mm for bigger nominal maximum particle size.

The specimens are compacted using one of the approved compaction methods, specified in the relevant standard (impact, gyratory, vibratory compaction or using a slab compactor).

The specimens are divided into two subgroups of three or more specimens, having approximately the same average length and average bulk density. The difference of the average length of the two subgroups must not exceed 5 mm, whereas the difference of the average bulk density shall not exceed 30 kg/m³.

One of the subgroups of specimens is stored in a flat laboratory surface at 20°C $\pm$ 5°C. This subgroup is designated as the 'dry' subgroup of specimens.

The other subgroup is placed on a perforated shelf in a vacuum container filled with distilled water at 20°C ± 5°C, to a level at least 20 mm above the upper surface of the test specimens. A vacuum of 6.7 kPa (50 mm Hg) is applied for 30 ± 5 min, and then for another 30 ± 5 min, the specimens remain submerged in water under atmospheric pressure. This subgroup is designated as the 'wet' subgroup of specimens. The dimensions of specimens are measured and their volume is calculated. Any specimen whose volume has increased for more than 2% is rejected.

Then, the wet subgroup of specimens is stored in a water bath at 40°C ± 1°C for a period of 68 to 72 h.

After that, all specimens are brought to a test temperature of 25°C ± 2°C for at least 2 h for 100 mm diameter specimens or 4 h for specimens with a diameter of 150 mm or larger.

The indirect tensile strength of all specimens is measured using the apparatus and the methodology specified in CEN EN 12697-23 (2003).

The ratio of the indirect tensile strength (ITSR) is calculated by the following formula:

$$ITSR = (ITS_w/ITS_d) \times 100,$$

where ITSR is the ratio of indirect tensile strength (%) to the nearest whole number, ITS_w is the average indirect tensile strength of the wet specimen group (kPa) and ITS_d is the average indirect tensile strength of the dry specimen group (kPa).

The average value of tensile strengths is rounded to three significant figures.

More information about the test is given in CEN EN 12697-12 (2008) and CEN EN 12697-23 (2003). The test is similar to the one specified in ASTM D 4867 (2009), ASTM D 6931 (2012) or AASHTO T 283 (2011).

The ratio of indirect tensile strength should be higher than the value determined by the contractual requirements. If the ratio is lower than the required value, the use of a suitable anti-stripping agent is required and the test is repeated.

3.7.3.5 Retained compressive strength index test

The retained compressive strength index test is conducted in accordance with ASTM D 1075 (2011). The determination of the impact of water on the bituminous mixture cohesion is expressed as the percentage of the retained compressive strength.

The test requires cylindrical specimens of usually 101.6 mm in diameter and 101.6 ± 2.5 mm height to be produced. Specimens are produced and tested on compressive strength according to the standard ASTM D 1074 (2009) (AASHTO T 167 2010). The compaction is carried out with a compression testing machine, which applies an initial compressive load of 1 MPa (150 psi) to the bituminous mixture for a few seconds and then a full load of 20.7 MPa (3000 psi) for 2 min.

If the specimens are required to be compacted at a certain void percentage, usually 6%, the abovementioned full compressive load varies accordingly.

The specimens are extracted from the moulds and left to room temperature for 24 h. Once bulk specific gravity is determined, half of them (at least three specimens) are tested in axial compression, without lateral support, at a uniform rate of vertical deformation (5.08 mm/min for the 101.6 mm specimen) until crushing.

The rest of the specimens (at least three) are placed into a water bath at 60°C for 24 h or into a water bath at 49°C for 4 days. After 4 days, they are transferred into another water bath at 25°C for 2 h, so that the test temperature is obtained. Then, they are subject to compression, as the first set of specimens.

Compressive loads of both wet and dry specimens are recorded, so that the average compressive strength can be calculated. The index of the retained strength is determined from the ratio of the average compressive strength of wet specimens to the average compressive strength of dry specimens, multiplied by 100 (i.e. expressed as a percentage).

More details can be found in ASTD D 1075 (2011) and ASTM D 1074 (2009).

3.7.3.6 Retained marshall stability test method

The retained Marshall stability test is conducted by some organisations, mainly in the United States, as an alternative to determine the effect of moisture to the bituminous mixtures. The test consists of preparing Marshall specimens and testing them dry and wet (after certain hours of saturation) using the Marshall apparatus. Saturation conditions have not yet been standardised. The retained Marshall stability is calculated as the ratio of the average Marshall stability of the wet specimens to the average Marshall stability of the dry specimens, expressed in percentage.

3.7.3.7 Vialit plate shock test method

The Vialit plate test, according to CEN EN 12272-3 (2003), determines the binder aggregate adhesivity and the influence of adhesion agents or interfacial dopes in adhesion characteristics as an aid to design binder aggregate systems for surface dressing.

In particular, CEN EN 12272-3 (2003) specifies methods of measurements of (a) the mechanical adhesion of the binder to the surface of aggregates, (b) the active adhesivity of the binder to chippings, (c) the improvement of the mechanical adhesion and active adhesivity by adding an adhesion agent either into the mass of the binder or by spraying the interface between binder and chippings, (d) the wetting temperature and (e) the fragility temperature.

The adhesivity value is determined as the sum of number of chippings remaining bonded to the plate and the number of fallen chippings that are stained by the binder.

The wetting temperature is determined as the lowest binder temperature used when spreading the chippings on the plate, which results in 90% of the chippings being stained by the binder.

The fragility temperature is the lowest test temperature at which 90% of the aggregates remain bonded to the plate.

The test is suitable for all bituminous binders for surface dressings, such as conventional or polymer-modified bitumen, cut-back bitumens, fluxed bitumen or bituminous emulsions with conventional or modified bitumen.

During testing, a binder quantity is spread out on a 200 ± 1 mm $\times$ 200 ± 1 mm steel plate and is then cooled or stabilised at $5°C \pm 1°C$. The binder rate of spread ranges from 0.7 to 1.3 kg/m² depending on the chipping size. Then, 100 graded chippings (4/6 or 6/10 mm size) or 50 graded chippings (10/14 mm size) are laid down on the binder and rolled using a manual rubber cylinder.

After the light rolling, the plate is placed in a heating chamber at $5°C \pm 1°C$ for 20 ± 2 min. Then, it is removed and placed reversely in the Vialit apparatus. Within up to 1 min, an impact load of dropping metal ball is applied to the plate, three times within 10 s. After loading, the number of fallen chippings unstained by the binder is recorded as 'a'.

The adhesivity value ($b + c$) constitutes the number of the rest of the chippings, namely, those fallen but stained by the binder, recorded as 'b', and those bonded to the plate after testing, recorded as 'c'.

More information about the test and the way to determine wetting temperature and fragility temperature is given in CEN EN 12697-3.

Figure 3.9 Pendulum (Vialit) device. (Courtesy of Cooper Research Technology Ltd.)

3.7.3.8 Determination of cohesion of bituminous binders with the pendulum (Vialit) test

The pendulum (Vialit) test, according to CEN EN 13588 (2008), determines the cohesion of any bituminous binder (pure, modified or fluxed). Cohesion is one of the measures of the performance of the bituminous binder, particularly modified bitumen or bitumen recovered from bitumen emulsion. It is important to use binders that have a sufficient level of cohesion according to the level of traffic to be supported.

The test method measures the cohesion of bituminous binders at temperatures ranging from −10°C to +80°C and for expressing the relationship between cohesion and temperature.

Cohesion is defined as energy per unit area required to fully detach a cube from the support, with the previously bonded faces of the cube and support remaining fully covered by binder (CEN EN 13888).

A 10 mm side steel cube is fixed to a steel support by a film of binder of 1 mm thickness. The assembly is brought to the test temperature and the cube is dislodged by the impact of a swinging pendulum. The energy absorbed by rupture of the binder is calculated from the angle (α) of swing of the pendulum. The determination is performed over a range of at least six temperatures covering the cohesion peak of the binder. A pendulum (Vialit) device is shown in Figure 3.9.

More details about the test can be found in CEN EN 13588 (2008).

3.8 MODIFIED BITUMENS AND SPECIAL BITUMENS

Modified bitumens are bitumens whose rheological properties have been modified during manufacture by the use of one or more chemical agents (modifiers). The modification alters and improves certain bitumen properties, which results in the improvement of the respective bituminous mixture or application and therefore improved construction quality.

The necessity of using modified bitumens initially emerged during the 1970s when, due to a large number of bitumen origin and production sources, variability in bitumen properties was noticed. Additionally, the price of bitumen was increased during the same period owing to the energy crisis.

The first resulted in several pavement failures associated to the quality of bitumen, whereas the second one resulted in an increase of the total construction cost. As a consequence, the need to improve bitumen quality emerged in order to ensure the qualitative stability and improvement of the construction quality.

Factors such as the rapid increase of traffic volume and axial loading, the higher demands of users for better and constant ride quality and the users' disturbance during maintenance works made the need for improvement of bitumen properties and consequently of the bituminous works imperative.

The required improvements refer to temperature susceptibility, stiffness, elasticity, adhesivity and ageing of the bitumen; in other words, for bitumen to not become too soft at high temperatures, to not become brittle or fractured at subzero temperatures, to deform less under loading, to adhere better to aggregates and to age at a slower rate. The chemical industry met the abovementioned demands and presented a wide rage of chemical additives (modifiers), each one satisfying some or almost all of the above requirements.

Special bitumens, according to CEN EN 12597 (2000), are bitumens manufactured by processes and from feedstocks chosen to confer special properties that meet stringent requirements for paving or industrial applications.

3.8.1 Usage and role of modified bitumen

Modified bitumen can be used in the whole range of bitumen works, namely, for producing bituminous mixtures (hot to cold), in surface dressings, surface sealing, water insulation and so on. Modified bitumens may be employed 'directly' or in the form of cut-backs or emulsions, or blended with, for example, natural asphalt.

The role of modified bitumen in paving is mainly fourfold: to increase asphalt's resistance to permanent deformation, to improve asphalt's fatigue life, to increase asphalt's stiffness modulus and to improve adhesion between bitumen and aggregate particles. All the above tasks are to be fulfilled without affecting, if possible, the workability of the mixture. Only certain modifiers are able to fully accomplish these rather complicated and difficult tasks.

The increase of asphalt's permanent deformation resistance solves the problem of premature rutting typically developed in areas with high–medium traffic volume and normal–high ambient temperatures.

The improvement of asphalt's fatigue life delays the development of the pavement's fatigue cracking.

The increase of asphalt's stiffness modulus improves the load spreading ability of the asphalt layer; hence, lower stresses are transferred to the subgrade. This could be interpreted as the ability to decrease the asphalt layer's thickness, and hence the pavement's thickness, for a given subgrade strength and under the given traffic conditions.

Finally, improvement of the adhesion between bitumen and aggregate particles positively affects the life of surface dressings, as well as open graded and porous asphalts, and eliminates the development of ravelling.

Modified bitumens are characterised by the relatively high production and supply cost. However, this should not be taken as a deterrent factor, since their cost must not be compared with the cost of graded bitumen, but with the total pavement construction cost inclusive of future maintenance cost.

As stated, modified bitumen can be used in all bituminous mixtures or bitumen works. However, for optimising cost/benefit, it is recommended to be used in asphalts for highly stressed areas (highways, airports, bridge decks, etc.) or asphalts with high-quality/cost hard aggregates or surface dressings with high-quality/cost chippings.

3.8.2 Bitumen modifiers, methods of modification and main changes to the properties of the bitumen

Bitumen modifiers can be synthetic polymers, natural rubber (latex) and some chemical additives such as sulfur and certain organo-metallic compounds. Fibres and fillers (inorganic powders) are not considered to be bitumen modifiers. Table 3.16 gives some typical bitumen modifiers, as well as significant improvements to asphalts. Polymers are the most common type of bitumen modifiers, with thermoplastic elastomers being the most popular polymer.

Table 3.16 also lists some compounds (additives), such as fillers or fibres, occasionally added to asphalts to influence their mechanical properties rather than to modify the bitumen.

Most bitumen modifiers and other additives are found in solid form (mainly fine particles) while others are in liquid form (mainly oleic solution).

The thermoplastic modifiers are added to the bitumen in a separate mixing–modification procedure before mix production or spray application, and the product (modified bitumen) may be stored. The thermoplastic polymers may be added to the aggregate mixer during

Table 3.16 Some bitumen modifiers/additives and supervened improvements

Type of additive		Example	Main improvements
Polymers	Thermoplastic elastomers	Styrene–butadiene–styrene (SBS), styrene–butadiene–rubber (SBR), styrene–isoprene–styrene (SIS), styrene–ethyl–butadiene–styrene (SEBS), ethyl-propyl-dien tetropolymer (EPDM), isobutene–isoprene copolymer (IIR), polybutadiene (PBD), natural rubber	(1), (2), (3), (4), [8], [9], [10], [11], [13]
		Crumb rubber	(2), [8], [9], [11]
	Thermoplastic polymers (plastics)	Ethylene–vinyl acetate (EVA), ethylene–methyl acrylate (EMA), ethylene–butyl acrylate (EBA), polyethylene (PE), polypropylene (PP), polyvinyl chloride (PVC), polystyrene (PS)	(2), (3), [8], [9], [10]
	Thermosetting polymers	Resins: epoxy resin, acrylic resin, polyurethane resin, phenolic resin	(2), (3), (4), (6), [8], [9], [10]
Chemical modifiers		Sulfur, lignin and certain organo-metallic compounds	(2), (5), (6), [8], [9], [12]
Natural asphalts		Trinidad lake asphalt, rock asphalt, gilsonite	(2), (4), (6), [8], [9]
Fillers		Hydrated lime, lime, carbon black, fly ash fillers	(4), (6), [8], [9]
Fibres		Cellulose, mineral, plastic, glass, asbestos fibres	[9], [11]
Hydrocarbons		Recycled or rejuvenating oils	(5), (7), [12]

Improvements to bitumen:

(1) Improves elastic behaviour

(2) Improves thermal susceptibility

(3) Improves binding ability

(4) Ageing retardation

(5) Viscosity reduction

(6) Hardens the bitumen

(7) Rejuvenates the bitumen

Improvements to asphalts:

[8] Stiffness increases

[9] Increases resistance to permanent deformation

[10] Cohesion improvement

[11] Better behaviour to fatigue cracking

[12] Workability increases

[13] Better behaviour to thermal cracking

mixing. However, this procedure is not recommended since optimal modification results cannot be achieved.

The thermosetting modifiers – two liquid compounds (resin and hardener) – are first blended together and then with the bitumen a few seconds before application as surface coating/surface dressing (main use) or before mixing with aggregates for the production of hot asphalts (occasional use).

The crumb rubber is either blended with the bitumen or added to the aggregates in the asphalt plant mixer before the bitumen is charged to the mixer. The first process is known as the wet process and effectively modifies the bitumen. The second process is known as the dry process and effectively modifies the properties of the asphalt. Better results are expected when the wet process is employed.

The fibres and filler additives are added to the asphalt plant mixer during the production of hot or cold asphalts.

The modification of the bitumen with thermoplastic elastomers or polymers in a purposely built plant ensures proper dispersion of the modifier, full polymerisation and homogeneity of the final product (modified bitumen). Homogeneity and optimum modification results are affected by the duration and method of mixing (high-shear mixing or slow mixing) and the relaxation (or 'digestion') period in addition to the origin of the bitumen and its compatibility to the modifier used. Relaxation period is the period needed for full polymerisation to take place and all cross-links to be developed.

During the modification process of the bitumen, chemical and physico-chemical changes, depending on the type of modifier/additive used, occur. These changes affect the characteristic properties of the bitumen such as penetration, softening point, viscosity, cohesion, resistance to hardening, elastic recovery and Fraass breaking point.

The modification of a given bitumen with crumb rubber (wet process) depends on the particle size of the crumb rubber, the quantity of crumb rubber blended, the mixing temperature and the duration of mixing.

3.8.3 Modification by thermoplastic elastomers

Bitumen modification with elastomers is currently the most common method for producing paving modified bitumen. The bitumen produced is often called elastomer bitumen.

Thermoplastic elastomers or elastomers are designated as elastic polymers, which, after polymerisation, can be extended under the influence of a tensile force, and by removing the force, they can quickly revert to their initial length. The elastic behaviour of these polymers is due to the fact that parts of the macromolecular chains at ambient temperatures not only can be moved under the influence of a force but also revert back to their initial position after removing the force, provided the glass transition temperature (T_g) is lower than the usual ambient temperatures.

According to the International Union of Pure and Applied Chemistry (IUPAC), the polymer is defined as 'a substance composed of molecules characterized by the multiple repetition of one or more species of atoms or groups of atoms (constitutional units) linked to each other in amounts sufficient to provide a set of properties that do not vary markedly with the addition or removal of one or a few of the constitutional units' (IUPAC 1974).

The category of elastomers includes a wide range of products, such as natural rubber (NR), styrene–butadiene rubber (SBR), styrene–butadiene–styrene copolymer (SBS; known as thermoplastic rubber), styrene–isoprene–styrene copolymer (SIS), polyurethane rubber, polyether–polyester copolymer, olefinic copolymers, ethylene–propylene rubber (EPR) and so on (see also Table 3.16).

From the abovementioned elastomers, those widely used today for the production of elastomer bitumen are SBS, SBR elastomer, SIS and, to a smaller extent, NR. It should be stated here that crumb rubber from used tyres is also indirectly classified into this category.

During polymerisation, linear as well as multi-armed copolymers are produced, as shown in Figure 3.10. During dispersion of the elastomer in the hot bitumen, the elastomer absorbs components from the bitumen (maltenes) and the polymer extends (swells) (Vonk and Van Gooswilligen 1989).

A typical example of the structure of copolymer SBS in bitumen in three-dimensional space is given in Figure 3.11. Thermoplastic elastomers derive their strength and elasticity

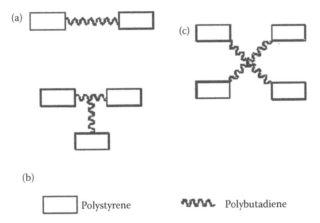

Figure 3.10 Linear and branched linked thermoplastic elastomers. (a) Linear link. (b) Branched link. (c) Radial link. (From Bull A.L. and W.C. Vonk, *Thermoplastic Rubber/Bitumen Blends for Roof and Road*. Thermoplastic Rubbers Technical Manual, TR 8.15. London: Shell International Petroleum Company Ltd., 1984; Shell Bitumen, *The Shell Bitumen Handbook*. Surrey, UK: Shell Bitumen UK, 1990.)

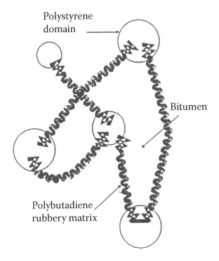

Figure 3.11 Three-dimensional SBS thermoplastic elastomer network. (From Vonk W.C. and G. Van Gooswilligen, Improvement of paving grade bitumens with SBS polymers. *Proceedings of the 4th Eurobitume*, Vol. I, p. 299. Madrid, 1989.)

from the physical molecule cross-linking in space. Strength is provided by the polystyrene and elasticity is attributed to the polybutadiene (Vonk and Van Gooswilligen 1989).

Compatibility of polymers with the bitumen to be modified is required for producing homogeneous and stable modified bitumen. The bitumen chemical composition is of great importance. Key factors related to the bitumen chemical composition affecting the compatibility, for a given elastomer, are the asphaltene and aromatic content. Generally, phase segregation (bitumen and elastomer) is expected if the asphaltene content is high and the aromatic content is relatively normal. Similarly, incompatibility exists when the content of aromatic compounds is high or the percentage of asphaltenes is very low (Van Gooswilligen and Vonk 1986).

For a given compatible system, the quality of polymer dispersion achieved is also influenced by the concentration of the polymer and the shear rate applied by the mixer (Shell Bitumen 2003).

Testing the compatibility of the bitumen with the modifier is a rather complicated procedure; however, it should be carried out whenever the origin (source) of the bitumen changes or the bitumen chemical composition changes significantly. The advantage of using elastomers is that the possibility of its compatibility with the paving grade bitumen is very high.

The compatibility of the bitumen and elastomer system can be detected on the final product, the modified bitumen, microscopically. In a compatible system, a homogeneous and continuous sponge-like structure will appear, whereas in a non-compatible system, a non-continuous structure with coarse particles will be shown.

The compatibility of the system or the successful bitumen modification can also be ascertained with a simple laboratory test, known as 'storage stability test', originally suggested by Shell and standardised by CEN EN 13399 (2010).

The storage stability test consists of storing the modified bitumen for a certain period at high temperature and then testing the bitumen of the top and bottom part of the storage vessel, in terms of softening point test. Typical storage stability results of SBS polymer-modified bitumens stored for up to 6 days at 180°C ± 2°C, obtained in the Highway Engineering Laboratory of the Aristotle University, are shown in Figure 3.12.

The non-compatible bitumen/polymer system (curves with a solid line) shows large numerical differences in softening point of the top and bottom samples after 3 and 6 days'

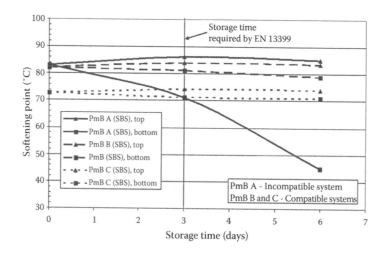

Figure 3.12 Effect of compatibility of the system on the storage stability of three bitumen/SBS blends.

storage at 180°C, when compared to softening point before the storage test. The compatible system shows very small differences in softening point after the same storage period. The CEN EN 14023 (2010) storage stability requirement is that the softening point difference between the top and the bottom sample should be ≤5°C, after 3-day storage at 180°C.

In a non-compatible system, almost the entire polymer is separated and accumulated at the top of a storage tank. The separation and accumulation of the polymer are often quite distinct from an expert's naked eye (change of appearance or colour of the top surface).

3.8.3.1 Characteristic properties of elastomer-modified bitumen

The addition of elastomer to bitumen alters its characteristic properties, and the final product – the modified bitumen – becomes more elastic and less susceptible to temperature variations. Additionally, the modified bitumen with elastomers has better cohesion properties and higher resistance to ageing compared with the unmodified paving grade bitumen.

In particular, the elastic recovery of the bitumen has remarkably improved when modified with elastomers. Typical values of 25% to 35%, at 25°C, for conventional bitumen can be increased, so as to reach elastic recovery values higher than 95%. As a consequence, the strain developed when bitumen is loaded is recovered to a large extent and plastic (permanent) deformation is minimised.

By adding an elastomer to the bitumen, the penetration value decreases, the softening point increases and the penetration index increases. These changes indicate that the bitumen becomes harder and less susceptible to temperature variations.

The addition of elastomer also lowers the Fraass breaking point. As a result, the modified bitumen is expected to crack at lower subzero temperatures compared to unmodified paving grade bitumen.

Finally, modification with elastomers increases bitumen's viscosity. This implies that an increase of working temperatures is required in order to achieve the same viscosity values as the unmodified bitumen, to carry out mixing, laying and compaction.

It was also found that SBS-modified bitumen does not age as much as conventional paving bitumen (Nikolaides and Tsochos 1992; Nikolaides et al. 1992).

With respect to crumb rubber modification, a recent study has shown that crumb rubber (size 0/0.8 mm and quantity 5%, 10% and 15% by mass of bitumen), when blended to 70/100 and 50/70 graded bitumens, results in improved penetration, softening point, flexural creep stiffness and shear modulus G^*. The quantity of the crumb rubber and the provenance of the bitumen directly influence the above properties of the modified bitumen (Neutag and Beckedahl 2011).

In another study using crumb rubber (size 0/0.7 mm and quantity 15% and 18%), it was also found that the characteristic and rheological properties of the bitumen improved. However, the storage stability requirement of the rubberised modified bitumen (RmB) was not satisfied. When wet and dry processes were compared on extracted RmB from an asphalt concrete mixture (AC 8 mm), it was found that the effect of crumb rubber was more profound when the wet process was used (Lukac and Valant 2011).

3.8.3.2 Characteristic properties of asphalts with elastomer-modified bitumen

The asphalts (bituminous mixtures) produced with elastomer-modified bitumen have, above all, improved elastic properties. As a result, the asphalt layer with elastomer-modified bitumen has better permanent deformation and fatigue performance when compared to the asphalt layer with conventional bitumen.

Dynamic and static stiffness moduli of the asphalt with modified bitumen also increase in comparison to the asphalts with conventional bitumen. However, the latter does not constitute the most characteristic changes in asphalts produced with SBS-modified bitumen.

Permanent deformation behaviour of asphalts is determined by the wheel tracking test (see Sections 7.6.5 and 7.6.6). From an extensive study, it was found that the wheel tracking parameters such as rate or the wheel tracking slope and the rut depth or the proportional rut depth of the asphalt concrete mixtures (AC 20 mm) with SBS-modified bitumen were, in all cases, lower than those obtained for the AC 20 mm mixtures with 50/70 grade bitumen (Nikolaides and Manthos 2009).

Permanent deformation behaviour can also be determined by using the triaxial cyclic compression test (see Section 7.6.2). Figure 3.13 shows results obtained when asphalt concrete mixtures were used (AC 20 mm) with and without polymer-modified bitumen.

Cracking or fatigue performance of an asphalt is represented by its fatigue curve (see Section 7.7.1). Typical results of fatigue curves of an asphalt concrete (AC 20 mm) with unmodified and SBS-modified bitumen are shown in Figure 3.14. From Figure 3.14, it is obvious that for the same tensile strain developed, the asphalt concrete with SBS-modified bitumen will sustain more repetitive loading before fatigue cracking develops.

Finally, the use of elastomer bitumen increases Marshall stability, but it may also increase the Marshall flow above the limiting value. This increase cannot be taken as a drawback, since the nature of the test is such that it does not take into account the recovered (elastic) deformation during unloading and in effect does not simulate dynamic loading. Regarding the optimum binder content of the asphalt with or without modified bitumen, it was found that, when asphalt concrete mixtures were tested, the optimum binder content as determined by the Marshall procedure was not significantly different (Nikolaides and Tsochos 1992).

The addition of crumb rubber from recycled tyres with the aim of improving the permanent deformation of the asphalts has been examined by various researchers. The results obtained are encouraging; however, difficulties related to the homogeneity of the mixture during mixing (development of 'balling' effect), when the dry process is used, have been reported. In dry process mixing, only a small amount of very fine crumb rubber is integrated

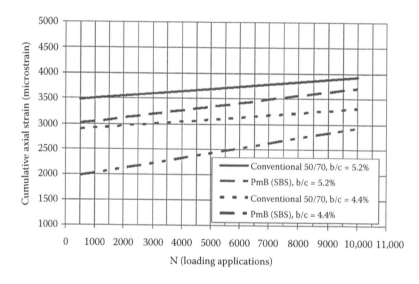

Figure 3.13 Triaxial cyclic compression results of AC-20 mixtures with 50/70 grade and elastomer-modified bitumen.

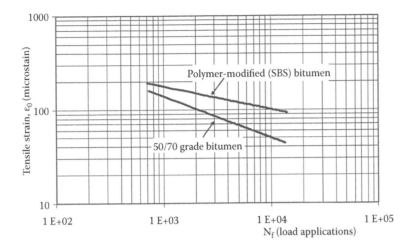

Figure 3.14 Fatigue lines of elastomer-modified bitumen and 50/70 grade bitumen.

into the bitumen. Most particles remained unchanged within the bitumen mass, acting as elastic solid particles. The percentages of added crumb rubber examined varied widely from 5% to 40% by weight of bitumen.

When the wet process is employed, the homogeneity of mixture is ensured and the positive effect of the crumb rubber to the permanent deformation performance of the asphalts is more distinct. It has been reported that the permanent deformation characteristics, rut depth and the rut rate, of an asphalt concrete (AC 8 mm) with rubberised bitumen decreased in comparison to the AC 8 mm with Venezuela 50/70 grade bitumen, when the wet process was used (Neutag and Beckedahl 2011). The stiffness modulus determined by the indirect tensile test was found to increase.

3.8.4 Modification of bitumen by thermoplastic polymers (plastics)

Bitumen modification with thermoplastic polymers, also known as thermo-softening plastics or plastics, is not as common as bitumen modification with elastomers for paving works.

Thermoplastic polymers are polymers that, when heated above a certain temperature (glass transition temperature, T_g), become soft and pliable or mouldable. However, when temperature drops, they harden and preserve their shape. The ability of the macromolecular chains to move and recover their original position at ambient temperatures is limited. The intermolecular bonds of the polymeric chain can absorb only vibrations and torsions and are less resistant to environmental stress cracking. As a consequence, thermoplastic polymers do not present elastic properties at ambient temperatures.

The main thermoplastics used for the modification of the bitumen are the following: ethylene vinyl acetate (EVA), polyvinyl chloride (PVC), polyethylene (PE), polypropylene (PP), polystyrene (PS), ethylene methyl acrylate (EMA) and ethylene butyl acrylate (EBA).

The effectiveness of thermoplastic polymers and, hence, the resulting characteristic properties of the modified bitumen depend on the type and quantity used, the chemical composition of the bitumen and, to a lesser extent, the manner and duration of mixing. The crystalline structure of the ethylene, responsible for the provision of hardness, constitutes the determinant factor of the copolymer behaviour (Gilby 1985; Woolley 1986).

The addition of a thermoplastic polymer to bitumen increases bitumen viscosity and the bitumen becomes harder at ambient temperatures. Bitumen elasticity is not affected significantly. The main drawbacks of the thermoplastic bitumen are that the bitumen's elasticity is not improved substantially and that separation of the thermoplastic polymer may occur during heating. The latter is restored to a great extent by further mixing or recirculation of the modified bitumen before using it.

3.8.4.1 Characteristic properties of modified bitumen and asphalts with thermoplastic polymers

Some of the characteristic properties of thermoplastic polymers are modified when they are added to bitumen. Penetration decreases and softening point increases in particular. These changes indicate that the bitumen becomes harder and less susceptible to temperature variations. Additionally, bitumen's binding ability increases. However, the Fraass breaking point and bitumen elasticity did not significantly improve (Brule and Lebourlot 1993).

The use of modified bitumen thermoplastic polymers in asphalts significantly increases asphalts' dynamic and static stiffness modulus.

Early investigation on the effect of polymers on rolled asphalt surfacing found that the most effective and simplest polymer to use was the EVA copolymer (Denning and Carswell 1981a).

However, it does not improve the elastic behaviour of the asphalt (bituminous mixture). Thus, asphalts with thermoplastic polymer-modified bitumen present good resistance to deformation at high ambient temperatures, but generally their fatigue cracking resistance is similar to or even less than that of asphalts with unmodified bitumen.

In an evaluation study on asphalts with modified bitumens, it was found that asphalt layers with thermoplastic polymer-modified bitumen, when laid over cracked and uncracked surfaces, have more intensive cracking within a very short period, in contrast to the asphalt layers where elastomer-modified bitumen was used (Anderson et al. 1999).

Because of the abovementioned properties and results, the use of thermoplastic polymer-modified bitumen is recommended in areas with high ambient temperatures where the asphalt layers are subjected to high traffic loads but not over cracked pavement surfaces.

Regarding the asphalt's characteristic Marshall properties, the use of thermoplastic polymer-modified bitumen increases stability and decreases when compared to the same asphalt with the same binder content.

3.8.5 Modified bitumen with thermosetting polymers (resins)

The use of thermosetting polymers, known as resins, is not as widespread as other polymers in the modified bitumen production, even though bitumen properties have improved in all aspects and the produced modified bitumen is the best that can be presented by the chemical industry. The reason for this is probably their high cost.

The asphalts produced with thermosetting polymer-modified bitumen have excellent adhesive ability, excellent resistance to deformation, excellent fatigue performance and high stiffness modulus.

Thermosetting polymers or resins are produced by blending an epoxy resin and a hardener, both in liquid form, and when they chemically react, they form a tough, thermoset elastomer characterised by no flow or deformation at high temperature or under heavy axle loads, high flexural fatigue resistance, high solvent resistance and no fatting up. When blended with bitumen, the modified bitumen displays the properties of the epoxy rather than those of the bitumen (Dinnen 1981).

Since the epoxy bitumen presents the properties of the thermosetting polymer, the available time to use the modified bitumen is limited and depends to a great extent on the mixing temperature. The higher the temperature, the shorter is the time of use.

The thermosetting bitumen begins to 'cure' and increase its strength (hardens), once it is applied. The curing time depends on ambient temperatures. The higher the ambient temperature, the longer is the curing time. Once curing is completed, future temperature increase, which would soften the conventional bitumen, does not affect the hardness of the thermosetting bitumen at all. The completely cured bitumen is designated as an elastic material having no viscous behaviour, and it is very resistant to various chemical substances including solvents, diesel and oils.

3.8.5.1 Characteristic properties of asphalts with thermosetting polymers

Asphalts produced with thermosetting bitumen are far superior to asphalts produced with conventional bitumen. These asphalts have exceptionally high dynamic and static stiffness modulus at moderate and high ambient temperatures, high Marshall stability, excellent resistance to permanent deformation and high resistance to fatigue, and they present high cohesion. Typical comparative results of hot rolled asphalt with resin-modified and conventional bitumen are given Table 3.17.

Although the use of thermosetting polymers has the most positive results of all polymers on asphalt, their use is limited. The reason is the high purchase cost of the modified bitumen. As a consequence, the modified bitumen with thermosetting polymers is currently used in certain cases where the construction cost is not the determinant factor. Such cases include those aimed at reducing or eliminating repetitive accidents caused exclusively by the lack of surface skid resistance (black spots) and those that entail constructing a flexible pavement or an asphalt layer that is subjected to extreme loading, taking into account the fact that a long-lasting, maintenance-free service is required.

In some countries, such as the United Kingdom, for high-quality surface dressings with calcined bauxite chippings, the thermosetting polymer-modified bitumen is the only binder recommended to be used.

Table 3.17 Results of HRA with resin-modified bitumen and conventional bitumen

Mix property	HRA with epoxy-modified bitumen	HRA with conventional bitumen
Marshall stability (kN)	45	7.5
Marshall flow (mm)	4.0	4.0
Marshall quotient (kN/mm)	11.2	1.9
Rut rate, at 45°C (mm/h)	0	3.2
Stiffness[a] (kN/m^2)		
at 0°C	2.0×10^{10}	1.5×10^{10}
at 20°C	1.2×10^{10}	3.0×10^9
at 40°C	3.3×10^9	4.0×10^8
at 60°C	9.5×10^8	—
Flexular fatigue resistance (cycles to failure)		
with 6% binder content	1.0×10^6	3.0×10^4
with 7% binder content	$>20 \times 10^6$	2.0×10^5

Source: Dinnen, A., Epoxy bitumen binders for critical road conditions. *Proceedings of the 2nd Eurobitume*, p. 294, 1981.

[a] Determined from prismoidal specimen under constant deformation of 1.0 mm at 25°C.

3.8.6 Modification by chemical modifiers

The chemical modifiers, such as sulfur, lignin and certain organo-metallic compounds, do not really modify the bitumen but, rather, the properties of the asphalt mixture. The sulfur or the lignin extend the properties of the bitumen and modify the properties of the asphalt. Hence, the sulfur and the lignin are called bitumen extenders. The organo-metallic compounds modify the asphalt mixture by their catalytic action.

The bitumen extenders replace the bitumen needed in an asphalt mixture while the organo-metallic compounds are added to the bitumen. The typical range of bitumen replacement seems to be 30% to 40%, when sulfur is used, and 10% to 12% when lignin is used. When organo-metallic compounds are used, approximately 2% to 3%, by mass of bitumen, is added.

The concept of using a sulfur extender/modifier is very old, but it was not until the 1970s, during the advent of sulfur-extended asphalts, that a process was developed and started to be used on project sites (Beatty et al. 1987; Denning and Carswell 1981b; Kennepohl et al. 1975).

The high sulfur cost in comparison to bitumen during that period, the blending and storage problems encountered and the health/safety concerns had affected the use of sulfur in asphalt works. The development of solid flour pellets, added directly to the asphalt mixing process, and changes in sulfur/bitumen cost eliminated the abovementioned problems.

The sulfur pellets developed, initially called Shell SEAM and now Shell Thiopave, have been tested and used successfully in various projects (Nicholls 2009). The primary advantages of Shell Thiopave–modified asphalt are an increase in strength, stability and, possibly, durability. The economic advantages brought about by reduced bitumen consumption, reduced pavement thicknesses and lower energy consumption should propel its use on commercial projects (Nicholls 2011). Results from another experimental program have shown that the rutting performance of sulfur-modified warm mix asphalt (using Thiopave) was comparable or superior to conventional mixtures, but more susceptible to cracking given its stiffness characteristics (Elseifi et al. 2011).

Lignin, a biomass by-product (wood, corn, etc.), may also be used as an extender to bitumen. First laboratory test applications were carried out in the late 1970s with positive results (Sundstrom et al. 1983; Terrel and Rimsritong 1979). Recent research has shown that the use of lignin-containing co-products to asphalt binders causes a slight stiffening effect depending upon the percentage and type of lignin used and that only high-temperature properties were positively affected (McCready and Williams 2007). Further future increase in the price of bitumen in conjunction with future surplus of biomass by-products may lead to a higher (certain) percentage replacement of bitumen by lignin. If a high percentage of biomass compound is used as a bitumen replacement, then it will certainly reduce CO_2 emissions. Extensive work on biomass by-products has been carried out in Australia by Ecopave (Ecopave 2014).

Another bitumen extender product based on crumb rubber, the so-called elastomeric asphalt extender, has been recently introduced in the market under the trade name RuBind.

RuBind is added directly to the pug mill or dryer drum, replacing part of the bitumen. Extensive research and development during 2011 and 2012 showed that mixtures produced with RuBind outperform conventional hot mix asphalts and even common modified and asphalt rubber mixes (RuBind 2013).

The organo-metallic compounds (organo-magnesium, organo-copper, organo-cobalt, etc.) having been pre-mixed with oleic chemical compounds (carrier) are blended to the bitumen, which is then mixed with the aggregates. The properties of the asphalt produced are modified by a catalytic action, after mixing, laying and compacting.

The first multi-metallic catalyst, Chemcrete, came out in the late 1970s and was found to improve the stability, stiffness and rutting behaviour of the asphalt, after a period of weeks (Chemcrete Co. 1987; Daines et al. 1985; Nikolaides 1992). This delayed catalytic action was found to cause early rutting under certain conditions. IntegraBase, the successor product to Chemcrete, has a slightly modified formulation aimed at improving certain performance metrics such as blending optimisation and cure time reductions (Resperion 2013).

3.8.7 Natural asphalts as additives

Natural asphalts such as Trinidad asphalt, rock asphalt or gilsonite are blended with bitumen derived from crude petroleum in order to improve its properties. Since the above additives have very low penetration and very high softening point, the resulting bitumen becomes harder. Since the chemical composition of the natural asphalt is similar to the bitumen from crude petroleum, performing the blending procedure is far easier and homogeneity of the final modified bitumen is ensured.

Because of its chemical composition, TLA is considered as a thermoplastic material. Apart from hardening the bitumen and making it less temperature susceptible, the addition of TLA to petroleum bitumen improves the asphalt's durability, stability, stiffness, anti-rutting performance and fatigue characteristics. TLA is available in semi-solid state or in pellets. Recently, a new product, in pellet form, was introduced and was found to be suitable for warm asphalt mixtures (Lake Asphalt of Trinidad and Tobago Ltd 2014).

The proportion blended with the petroleum binder varies and depends on the original bitumen and the desired final properties. With regard to pellets, the appropriate proportions range from 25% to 40%, by mass of bitumen.

The rock asphalt or the gilsonite comes in powder form of certain particle size and is graded in softening point (140°C to 190°C). They are normally added to the aggregate/bitumen mixture during the mixing process. A typical dosage is 3%, by mass of bitumen, which can be adjusted according to the type of bitumen and asphalt mixture used. The addition of rock asphalt or gilsonite primarily improves the rutting resistance of the asphalt mix.

3.8.8 Addition of fillers to bitumen

The addition of fillers to bitumen does not modify the properties of the bitumen. However, it is as an old and very economical way to increase the stability and stiffness of the asphalt (bituminous mixture) so as to improve resistance to deformation. Today, with the development of other modifiers and additives, their use is limited to the production of bituminous materials for water insulation, acting as a substance to increase volume.

Cement, lime, limestone filler and filler from other rocks, such as a mixture of diabase and dolomite, as well as sulfur filler, were used in the past (Piber and Pichler 1993). A substantial effect on the permanent deformation behaviour was seen only with the use of sulfur (Denning and Carswell 1981b; Fromm and Kennepohl 1979).

3.8.9 Addition of fibres to bitumen

Adding fibres to bitumen does not modify any characteristic property of the bitumen. Their function is to enhance the strength of the bituminous mixture (asphalt). Hence, when the term *modified bitumen with fibres* is used, it refers to the modification of bituminous mixture.

Fibres added to the bituminous mixtures are natural, synthetic or regenerated fibres, such as cellulose, mineral (asbestos), metallic (iron) (Gottschall and Hollnsteiner 1985) and carbon, fibreglass or polymer fibres.

The addition of fibres takes place at the stage of mix production or directly after spraying the bitumen or the bituminous emulsion on the pavement's surface. The latter creates a stress-absorbing membrane to inhibit reflective cracking (Yeates 1994).

Another function of the fibres is to increase the relative surface area of the mixture so as to incorporate higher percentages of bitumen and thus reduce the rate of oxidation of bitumen. Reduction of the rate of oxidation increases the pavement's service life (improves fatigue life and slows appearance of ravelling). From all fibre types, cellulose fibres have been found to allow the higher increase of bitumen content in the mixture (Peltonen 1989).

Mixtures to which fibres are added are the SMA and porous asphalt and occasionally in micro-surfacing. Dense asphalt mixtures with fibres were also found to have high resistance to fatigue (Samanos and Serfass 1993). The addition of plastic or iron fibres was found to increase the resistance to permanent deformation (Courard and Rigo 1993). The percentage of iron fibres added was 0.4% to 1.5%, while for the plastic fibres, it was 0.1 to 0.7%, by mass of bitumen.

In all cases where fibres are used, besides the appropriate amount to be used, the following should be ensured: (a) uniform distribution in the mixture; (b) optimum length, which depends on the type of fibres used; and (c) resistance to high temperatures (higher than the temperatures of heated bitumen).

3.8.10 Addition of hydrocarbons

Various oils of high molecular weight, normally derived from recycled oils after rejuvenating process, or reductant agents are sometimes added to bitumen. They are intended to improve the viscosity (usually to decrease it) or to rejuvenate bitumen's properties. They are normally used in recycling or in countries with very low ambient temperatures.

In the first case, along with heavy oils, virgin graded bitumen is also added to improve the adhesivity of the aged bitumen. The percentage and type of hydrocarbon used are determined by the chemical industry, which supplies these materials taking into account the properties of the aged bitumen. These materials are also known as bitumen rejuvenators.

In the second case, the addition of heavy oils aims exclusively to decrease bitumen's viscosity so as to improve the workability of the mixture during laying and compaction. This policy is very common in countries with cold climates, such as Scandinavian countries; thus, these oils, along with some percentage of solvent, are called Scandinavian oils.

3.8.11 Special bitumens

3.8.11.1 Multigrade bitumens

Multigrade bitumen is a special binder that, for the same level of penetration as conventional bitumen, is not as susceptible to temperature changes. This means that it is more consistent at high temperatures and less fragile at low temperatures. In fact, multigrade bitumens have higher penetration index values than conventional bitumen.

The asphalts produced with this type of bitumen provide better resistance to rutting and better performance to low-temperature fatigue cracking, in comparison to those produced with the same-grade conventional bitumen.

Multigrade bitumens are manufactured at the refinery by means of a special refining process. The process involves the chemical reaction, under tightly controlled conditions, of oxygen and hydrocarbons. Additives, aromatic components and others (not polymers) may be used depending on the feedstocks.

Multigrade bitumens come with trade names such as Multiphate, Multibit, Bitrex, Biturox, KNB-MGB, to name a few. They are normally used for the production of high-stiffness bituminous mixtures to be used in highly stressed pavements susceptible to deformation owing to heavy traffic volumes and high, particularly, ambient temperature.

Experimental measurements carried out in 2009–2011 on the use of multigrade bitumens in chip sealing (surface dressing) did not show any advantage for multigrade bitumens compared to standard binders with a similar 25°C penetration value (Herrington et al. 2011).

Multigrade bitumens do not replace polymer-modified bitumens.

A European specification is going to be developed for multigrade bitumen (prEN 13924-2 under approval).

3.8.11.2 Fuel-resistant bitumens

Bitumens that are resistant to fuels and oils are specially modified bitumens and are available under patented trade names. With the modification, the bitumen does not lose any of the required properties for the production of a satisfactory bituminous mixture. It should be stated that today's fuel-resistant bitumens are not tar-based bitumens.

Asphalts with fuel-resistant bitumen are used in places where fuel leaks are expected (airport aprons, gas stations, parking areas, etc.).

3.8.11.3 Coloured bitumen

Coloured or pigmentable bitumens are those whose colour is different from black, and when mixed with aggregates, the pavement's surface is coloured.

The majority of coloured bitumens are produced by the addition of a pigment during mixing process with aggregates. The most common pigment used is the iron oxide, which results in the production of the most stable coloured (red) asphalt.

The disadvantages of coloured bitumens/asphalts are as follows: (a) steep price owing to the high cost of the pigment and (b) the fact that the only truly stable colour that can be achieved is red, which, depending on the traffic, type of asphalt and ambient temperatures, gets 'dirty' (due to rubber tyre traces left on the surface).

Over the last few years, the industry has developed synthetic binders that can be coloured in any colour with relatively less cost. Synthetic bitumens have almost the same properties as common paving bitumens. More information can be found in Shell Bitumen (2003) and the Web.

To preserve the colour on the pavement's surface, it is recommended that aggregates of the same colour as that of the pigmentable bitumen be used.

It should be stated that coloured surfaces can also be produced with slurries or micro-surfacing using coloured emulsions.

Coloured asphalts are used in bus lanes, leisure paths or roads, tunnels where a lighter than black colour may be desirable and so on. Another application of the coloured binder/thin asphalt layer is on the construction of the waterproofing layer in bridges. Its purpose is to make the layer where the bridge waterproofing system starts clearly visible for it not to be damaged during future maintenance works of the overlying bituminous layers.

3.8.11.4 Bitumens for joint and crack filling

Bitumens for filling joints and cracks in pavements of roads, airports, parking places and other trafficked areas are specially modified bitumens, so as to resist thermal tensile and compressive stresses developed as a result of temperature changes. At the same time, they possess excellent resistance to permanent deformation. Depending on the additives incorporated in the bitumen, these modified bitumens can also be fuel resistant.

Bitumens for filling joints are hot-applied (160°C–180°C) or cold-applied at ambient temperatures.

More information on the specified properties of hot-applied and cold-applied bitumens for filling joints and cracks according to European standards is given in CEN EN 14188-1 (2004) and CEN EN 14188-2 (2004), respectively.

The corresponding American standards on bitumens for filling joints and cracks are ASTM D 3406 (2006), ASTM D 5329-07 (2009), ASTM D 5893-04 (2010) and ASTM D 7116 (2005).

3.8.12 Specifications of polymer-modified bitumens

To ensure construction quality and since there is a great variety of polymer-modified bitumens in the market, the properties of the polymer-modified bitumen must be specified precisely to satisfy the needs.

In the European Union, the grades of polymer-modified bitumens are selected/specified in accordance with CEN EN 14023 (2010).

The properties specified for all polymer-modified bitumens ensure (a) consistency at intermediate service temperatures (with reference to penetration at 25°C), (b) consistency at elevated service temperatures (with reference to softening point), (c) cohesion (with respect to force ductility, tensile test or Vialit pendulum), (d) resistance to hardening, (e) brittleness at low service temperature (with respect to Fraass breaking point) and (f) stain recovery (with respect to elastic recovery).

Other informative characteristic properties may also be specified, such as for flash point, plasticity range, homogeneity, storage stability and density.

The framework specifications for polymer-modified bitumens by CEN EN 14023 (2010) are given in Tables 3.18 through 3.20.

The properties in Table 3.18 shall be specified for all polymer-modified bitumens and are considered essential properties. The properties in Table 3.19 are required so as to meet specific regional conditions and are characterised as special for regional requirements. The properties in Table 3.20 are additional properties; they are non-mandated but have been found useful to describe better the polymer-modified bitumens.

According to the standard, the modified bitumen is graded by the penetration range and the softening point. It is noted that having specified the penetration range, the rest of the properties are selected from any of the classes provided.

For instance, a polymer-modified bitumen (pmb) designated as 25/55–70 is a bitumen with penetration varying from 25 to 55 dmm, which belongs to class 3 of Table 3.18, and with a softening point $\geq 7°C$, belonging to class 4. The rest of the essential, special and additional properties are chosen in such a way that the modified bitumen can satisfy the project and regional requirements (traffic volume, ambient temperature, etc.).

In the United States, properties of chemically modified bitumens are specified by ASTM D 6154 (2009). Regarding Trinidad Lake modified asphalt, ASTM D 5710 (2005) or ASTM D 6626 (2009) is used.

3.8.13 Proposed grades of modified bitumen

Undoubtedly, the engineer has the freedom to choose/specify any combination of properties for the modified bitumen that will be used. This introduces a problem to the producers since they have to use different recopies and provide different storage facilities per grade requested. After many years of experience in pavement construction projects, the author proposes three representative grades of polymer-modified bitumens (elastomer modified)

Table 3.18 Framework specifications for polymer-modified bitumens – properties that apply to all polymer-modified bitumens

Essential requirement and property		Test method	Unit	Classes for all polymer-modified bitumens									
				2	3	4	5	6	7	8	9	10	11
Penetration at 25°C		CEN EN 1426 (2007)	dmm	10–40	25–55	45–80	40–100	65–105	75–130	90–150	120–200	200–300	
Softening point		CEN EN 1427 (2007)	°C	≥80	≥75	≥70	≥65	≥60	≥55	≥50	≥45	≥40	
Cohesion	Force ductility (50 mm/min)[a], or	CEN EN 13589 (2008) and CEN EN 13703 (2003)	J/cm²	≥3 at 5°C	≥2 at 5°C	≥1 at 5°C	≥2 at 0°C	≥2 at 10°C	≥3 at 10°C	≥0.5 at 15°C	≥2 at 15°C	≥0.5 at 20°C	≥0.5 at 25°C
	Tensile test (100 mm/ min)[a], or	CEN EN 13587 (2010) and CEN EN 13703 (2003)	J/cm²	≥3 at 5°C	≥2 at 5°C	≥1 at 5°C	≥3 at 0°C	≥3 at 10°C					
	Vialit pendulum[a]	CEN EN 13588 (2008)	J/cm²	≥0.7									
Resistance to hardening[b]	Retained penetration	CEN EN 12607-1 CEN EN 1426 (2007)	%	≥35	≥40	≥45	≥50	≥55	≥60				
	Increase in softening point	CEN EN 12607-1 CEN EN 1427 (2007)	°C	≤8	≤10	≤12							
	Change of mass (positive or negative)		%	≤0.3	≤0.5	≤0.8	≤0.8						
Flash point		CEN EN ISO 2592 (2001)	°C	≥250	≥235	≥220							

Sources: Reproduced from CEN EN 14023, *Bitumen and bituminous binders – Framework specification for polymer modified bitumens*, Brussels: CEN, 2010; and from CEN EN 13808, *Bitumen and bituminous binders – Framework for specifying cationic bituminous emulsions*, Brussels: CEN, 2013. With permission (© CEN).

a One cohesion method shall be chosen on the basis of end application. Vialit cohesion (CEN EN 13588 2008) shall only be used for surface dressing binders.

b The main test is the RTFOT at 163°C. For some highly viscous polymer-modified bitumens where the viscosity is too high to provide a moving film, it is not possible to carry out the RTFOT at the reference temperature of 163°C. In such cases, the procedure shall be carried out at 180°C in accordance with CEN EN 12607-1.

Table 3.19 Framework specifications for polymer-modified bitumens – properties associated with regulatory or other regional requirements

Property	Test method	Unit	Classes for regional requirements										
			0	1	2	3	4	5	6	7	8	9	10
Fraass breaking point	CEN EN 12593 (2007)	°C	NR[a]	TBR[b]	≤0	≤−5	≤−7	≤−10	≤−12	≤−15	≤−18	≤−20	≤−22
Elastic recovery at 25°C (or at 10°C)	CEN EN 13398 (2010)	%	NR[a]	TBR[b]	≥80 (≥75)	≥70 (≥50)	≥60 —	≥50 —					

Source: Reproduced from CEN EN 14023, *Bitumen and bituminous binders – Framework specification for polymer modified bitumens*, Brussels: CEN, 2010. With permission (© CEN).

[a] NR, no requirement may be used when there are no regulations or other regional requirements for the property in the territory of intended use.

[b] TBR, to be reported may be used when there are no regulations or other regional requirements for the property in the territory of intended use, but the property has been found useful to describe polymer-modified bitumens.

Table 3.20 Framework specifications for polymer-modified bitumens – additional properties

Property	Test method	Unit	Classes for the additional properties of polymer-modified bitumens							
			0	1	2	3	4	5	6	7
Plasticity range (S.P.–Fraass point)	CEN EN 14023 (2010), para. 5.2.8.4	°C	NR[a]	TBR[b]	≥85	≥80	≥75	≥70	≥65	≥60
Storage stability[c], Difference in:	CEN EN 13399 (2010)									
- softening point	CEN EN 1427 (2007)	°C	NR[a]	TBR[b,c]	≤5	—	—	—		
- penetration	EN 1426 (2007)	dmm	NR[a]	TBR[b,c]	≤9	≤13	≤19	≤26		
After hardening test, EN 12607-1										
Drop in softening point	CEN EN 1427 (2007)	°C	NR[a]	TBR[b]	≤2	≤5				
Elastic recovery:	CEN EN 13398 (2010)									
- at 25°C		%	NR[a]	TBR[b]	≥70	≥60	≥50			
- at 10°C		%	NR[a]	TBR[b]	≥50	—	—			

Source: Reproduced from CEN EN 14023, *Bitumen and bituminous binders – Framework specification for polymer modified bitumens*, Brussels: CEN, 2010. With permission (© CEN).

Note: The following data may be given by the supplier of the polymer-modified bitumen in the product data sheet:

- Polymer dispersion (see CEN EN 13632)
- Solubility (see CEN EN 12592 2007), using the appropriate solvent declared by the supplier
- Handling temperatures
- Minimum storage and pumping temperatures
- Maximum and minimum mixing temperatures; for comparison purposes, CEN EN 13302 (2010) or CEN EN 13702 (2010) should be used
- Density (see CEN EN 15326 2009)

[a] NR, no requirement may be used when there are no requirements for the property in the territory of intended use.

[b] TBR, to be reported may be used when there are no regulations or other regional requirements for the property in the territory of intended use, but the property has been found useful to describe polymer-modified bitumens.

[c] Storage conditions of the polymer-modified binder shall be given by the supplier. Homogeneity is necessary for polymer-modified bitumens. The tendency of polymer-modified bitumens to separate during storage may be assessed by the storage stability test (see CEN EN 13399 2010). If the product does not fulfil the properties in Table 3.20, Classes 2 to 5, information shall be given by the supplier regarding storage conditions for the polymer-modified bitumen to avoid separation of the components and to ensure the homogeneity of the product.

Table 3.21 Proposed grades of modified bitumens for locations with AAAT >12°C

Property	Standard	Unit of measurement	PMB Type		
			10/40–80	25/55–70	25/50–60
Penetration at 25°C	CEN EN 1426 (2007)	dmm	10–40	25–55	25–55
Softening point	CEN EN 1427 (2007)	°C	≥80	≥70	≥60
Force ductility (50 mm/ min traction)	CEN EN 13589 (2008) and CEN EN 13703 (2003)	J/cm²	≥3 at 5°C	≥ 2 at 5°C	≥2 at 5°C
Elastic recovery at 25°C	CEN EN 13398 (2010)	%	≥80	≥80	≥70
Storage stability, Difference in:					
- softening point	CEN EN 13399 (2010)	dmm	≤5	≤5	≤5
- penetration	CEN EN 1427 (2007)	°C	≤9	≤9	≤9
Fraass breaking point	CEN EN 12593 (2007)	°C	≤−10	≤−10	≤−10
Flash point	CEN EN ISO 2592 (2001)	°C	≥235	≥235	≥235
Density	CEN EN 15326 (2009)	g/cm³		Declare	
After hardening, EN 12607-1					
Elastic recovery at 25°C	CEN EN 13398 (2010)	%	≥70	≥70	≥60
Retained penetration	CEN EN 1426 (2007)	%	≥60	≥60	≥60
Increase in softening point	CEN EN 1427 (2007)	°C	≤8	≤8	≤10
Change of mass	CEN EN 12607-3	%	≤0.5	≤0.5	≤0.5

for use in project locations with average annual air temperatures (AAAT) above 12°C. The three proposed grades of polymer-modified bitumens are as follows: pmb 10/40–80, pmb 25/55–70 and pmb 25/55–60. The proposed grades of polymer-modified bitumens and their characteristic properties are shown in Table 3.21.

pmb 10/40–80 is proposed to be appropriate for pavements with high traffic volume (ESAL > 5×10^7 and AAAT ≥ 17°C) and for aprons of international airports (regardless of ambient temperatures).

pmb 25/55–70 is proposed to be appropriate for pavements with medium traffic volume (ESAL = 1×10^7 to 5×10^7) and for aprons of all other airport categories apart from international airports, regardless of AAAT.

Finally, pmb 25/50–60 is proposed to be appropriate for pavements with low traffic volume (ESAL < 1×10^7), regardless of AAAT.

3.9 HANDLING OF BITUMINOUS BINDERS

3.9.1 Transportation delivery

The transportation or delivery of the bituminous binders is carried out by delivery vehicles having thermo-insulated tanks of various capacities, typically 20 to 40 m³. The delivery vehicles should possess an accurate temperature measuring system/sensor and a sufficient pumping system with circulation facility.

If bitumen is to be transported to relatively long distances, the vehicles should also possess a heating system. The provision of a heating system is absolutely necessary when transporting modified bitumen.

The tank of the delivery vehicle should always be clean, before loading bituminous materials. Small proportions of foreign matter, usually petrol, diesel, oils and so on, directly influence bitumen's properties and, hence, the bituminous mixture and the quality of pavement. Furthermore, safety is affected since the presence of these substances may lower the flash point of the bitumen substantially. It has been found that 0.1% of diesel in bitumen can decrease the flash point up to 28°C and increase the bitumen penetration by 10 units (Asphalt Institute MS-4 2007). In case of transporting bitumen emulsions, the presence of foreign substances will certainly affect the breaking time of the emulsion.

Frequent alternations of transported material should be avoided. If not possible, emptying and scrupulous cleaning of the tank should be carried out every time the type of material to be transported changes. Scrupulous cleaning must also be carried out in case cationic or anionic material is to be transported alternately.

3.9.2 Storage

The bitumens in asphalt plants are stored in preferably vertical or horizontal thermo-insulated heated tanks of various capacities (typical range, 40 to 100 m³), depending on production demand. Seprentine heating coils are the preferred heating system rather than direct fire from a gas burner.

Long storage periods at working temperatures should be avoided so as not to harden the bitumen. In the case of polymer-modified bitumen, extended storage at working temperatures may cause separation of the polymer.

During storage, overheating should also be avoided for hardening and safety (self-ignition) reasons. The maximum temperatures for safe use should always be lower than bitumen's flash point (normally at least 30°C lower).

Storage tanks should be kept full of bitumen, if the bitumen is to be stored for a prolonged period at working temperatures, thereby reducing the air and the oxygen inside them. Oxygen may cause oxidation of the bitumen, while in the case of storing bituminous emulsion, a prolonged storage period will certainly cause surface breaking of the emulsion.

Storage tanks should have a recirculation or agitation facility, so as to achieve the desired temperature quicker and to avoid local overheating of the bitumen. The recirculation or agitation facility may also restore the disturbed homogeneity of the bituminous material.

Finally, storage tanks should have a reliable temperature and volume measuring system.

3.9.3 Temperatures of bituminous binders at stages of usage

In every stage of usage (pumping, mixing, spraying of the bitumen and compaction of the bituminous mixture), bituminous binders must be at the appropriate temperature to possess the appropriate viscosity.

During pumping, the bituminous binder should have such a viscosity so as to allow easy flow in the pipelines, avoid pipeline blockage, cause no problem to the pumps and ensure that the vehicles can be completely discharged. For all the abovementioned, bitumen's viscosity should not exceed a maximum value during pumping. It is recognised that the maximum viscosity value of the bitumen at pumping should not exceed 2 Pa·s (Shell Bitumen 1990).

During the mixing stage, the hot bitumen should also have an appropriate viscosity for satisfactory coating to be achieved. If the viscosity is too high during mixing, the aggregates will not be coated uniformly and properly. If the viscosity is too low, the aggregates may look coated but most of the bitumen will drain off during transportation or storage. A viscosity value approximately 0.2 Pa·s is considered to provide satisfactory coating (Shell Bitumen 2003).

Similarly, during spraying, the bitumen should have an appropriate viscosity to carry out the work successfully. If the viscosity is too high, there will be a blockage problem with the nozzles, affecting the uniformity and the quantity of bitumen sprayed. If the viscosity is too low, there is a risk that the sprayed bitumen will be in a 'foggy spray' form and be blown away by the wind, resulting in the sprayed surface not receiving the required amount of bitumen. The optimum viscosity value for spraying is considered to be 0.06 Pa·s for penetration grades and 0.03 Pa·s for cut-back grades (Shell Bitumen 1990). For bitumen emulsions with up to 60% bitumen content, ambient temperatures above 10°C are considered sufficient for proper spraying.

Finally, during compaction, the bitumen should also have an appropriate viscosity, so that effective compaction is ensured. If the viscosity is too high, the bituminous mixture cannot be compacted properly. If the viscosity of the bitumen is too low, the bituminous mixture becomes too soft that it cannot sustain the dead load of the roller and compaction is delayed. It is widely recognised that the optimal bitumen viscosity for compaction is between 2 and 20 Pa·s (Shell Bitumen 2003).

Considering all the above, the working temperatures can be determined by measuring the viscosity of the bitumen at three different temperatures (low, medium and high). Then by plotting the values in a semi-logarithmic scale, it is possible to determine the optimum or critical temperatures for spraying, mixing, compaction or pumping (see Figure 3.15). Figure 3.15 has been created as an example using three different indicative typical grade bitumens.

For simplicity, typical handling temperatures have been recommended by various institutes and organisations. Table 3.22 gives the handling and storage temperatures recommended by the Institute of Petroleum (UK).

A similar table that gives the recommended handling and storage temperatures of the PG bitumens is provided by the Asphalt Institute (2007). With respect to the Asphalt Institute's table, as an example, the recommended range of storage and mixing temperatures for the softer bitumen PG 46-28 is 127°C–143°C and 115°C–146°C, respectively. On the other hand, for the hardest bitumen PG 82-22, the corresponding temperatures are 157°C–168°C and 143°C–171°C, respectively.

For modified bitumens' handling and storage temperatures, it is recommended that the ones provided by the supplier be used.

However, a recent study has concluded that the steady shear flow test method and the phase angle test method can be used successfully for determining the mixing and compaction temperatures of modified bitumen, as well as of the unmodified bitumen, for laboratory use (Randy et al. 2010).

3.9.4 Health, safety and environmental issues of bitumens

Even though the bitumen from crude petroleum is not a particularly dangerous product, all health and safety guidelines and regulations must be followed and respected.

Most potential hazards of bitumen arise from handling the bitumen at elevated temperatures (more than 100°C). At elevated temperatures, skin burns and inhalation of vapour and fume emissions are the most common hazards that occur. However, there is always a danger associated with water coming into contact with hot bitumen and its self-ignition–combustion.

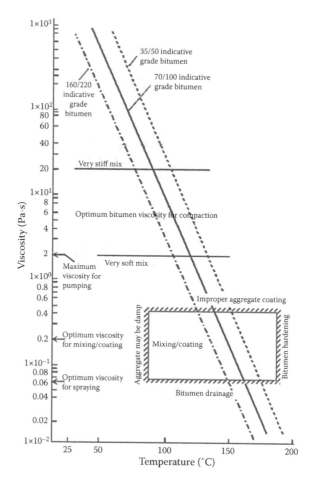

Figure 3.15 Schematic representation of the relation between temperature and viscosity, at different stages of asphalt use.

3.9.4.1 Skin burns

Skin burns revolve around the wearing of protective gear (clothing, etc.) and avoiding any skin contact with hot bitumen. Protective gear such as heat-resistant gloves, face shields, heat-resistant material overalls and safety boots should always be worn, and all relevant precautions published should be strictly followed.

3.9.4.2 Inhalation of vapour and fumes

A variety of vapours and fumes are emitted during handling bitumens at elevated temperatures. Visible emissions or fumes normally start to develop at approximately 150°C and the amount of fume generated doubles for each 10°C to 12°C increase in temperature (Shell Bitumen 2003).

The 2000 National Institute for Occupational Safety and Health (NIOSH) hazard review continues to support the assessment of the 1977 NIOSH criteria document on asphalt fumes (NIOSH 1977), which associated exposure to asphalt fumes from paving and other uses of asphalts with irritations to eyes, nose and throat (NIOSH 2000). In the same review, it was also stated that some studies reported evidence of acute lower respiratory track symptoms among workers exposed to asphalt fumes.

Table 3.22 Recommended bitumen handling and storage temperatures

| Grade | Minimum pumping temperature[a] (°C) | Typical bitumen temperature at time of application (°C) | | Max. handling and storage temperature (°C) | Typical long-term storage temperature[d] (°C) |
		Mixing and short-term storage[b]	Spraying[c]		
Paving grades (CEN EN 12591 2009)					
250/330	100	135	165	190	60
160/220	110	140	175	190	65
100/150	110	150	190	190	70
70/100	120	155	—	190	75
50/70	125	160	—	190	80
40/60	125	165	—	200	80
35/50	130	165	—	200	85
30/45	130	170	—	200	85
20/30	140	175	—	200	90
Hard paving grades (CEN EN 13924 2006)					
15/25	145		—	200	90
10/20	150		—	200	90
Hard industrial grades (CEN EN 13305)					
H80/90	160	200	—	230	120
H/100/120	190	220	—	230	130
Oxidised grades (CEN EN 13304 2009)					
75/30	150	195	—	230	110
85/25	165	210	—	230	110
85/40	165	210	—	230	110
95/25	175	215	—	230	120
105/35	190	220	—	230	130
115/15	205	225	—	230	130

Source: Institute of Petroleum, Bitumen safety code, Model code of safe practice in the petroleum industry, Part 11, 4th Edition. London: Energy Institute, 2005.

[a] Maximum pumping viscosity: approximately 2000 cSt (all grades).
[b] Mixing/coating viscosity: approximately 200 cSt (all grades).
[c] Spraying viscosity: approximately 60 cSt (penetration grades).
[d] Based on protracted storage period without addition of fresh binder. For bulk bitumens, the temperature should not fluctuate above and below 100°C as this increases the risk of condensation leading to boilover.

3.9.4.3 Toxicity of bitumen

With respect to toxicity of hydrocarbon components of bitumen (PAHs) identified in asphalt fumes at various work sites, the measured concentrations and the frequency of occurrence have been low (NIOSH 2000).

To minimise possible acute or chronic health effects from exposure to asphalt, asphalt fumes and vapours and asphalt-based paints, NIOSH recommends an exposure limit of 5 mg/m^3 during any 15 min period and implementation of the following practices: (a) prevent thermal exposure, (b) keep the application temperature of heated asphalt as low as possible, (c) use engineering controls and good work practices at all work sites to minimise worker exposure to asphalt fumes and asphalt-based paint aerosols and (d) use appropriate respiratory protection (NIOSH 2000).

An extensive study analysed 433 lung cancer cases and 1253 controls from Denmark, Finland, France, Germany, Netherlands, Norway and Israel and concluded the following: (a) no evidence found of an association between lung cancer and exposure to bitumen fumes, (b) factors identified as the likely contributors to the slightly elevated incidence of cancer mortalities were tobacco smoking and previous exposure to coal tar, (c) other occupational exposures do not seem to play a part and (d) the study further acknowledged the importance of the continuing trend towards minimising inhalation and dermal exposures (Eurobitume 2010; Olsson et al. 2010).

A substantial amount of information on health, safety and environmental aspects of bituminous materials can also be found in the *Bitumen safety code* (Institute of Petroleum 2005) and *Bitumens and bitumen derivatives* report (CONCAWE 1992).

3.9.4.4 Contact with water

When water comes into contact with hot water, the water undergoes a very sudden expansion causing a foaming affect. This foaming effect is dangerous to the personnel and could cause fire and explosion. For this reason, care must always be taken when handling bitumen emulsion followed by bitumen.

3.9.4.5 Combustion

Under conditions of high temperature and in the presence of oxygen, an exothermic reaction can occur, leading to the risk of fire or explosion of the tanks. Hence, manholes in bitumen tanks should be kept closed and access to tank roofs should be restricted (Shell Bitumen 2003).

3.9.4.6 Skin and eye contamination

The hazards associated with skin contact of bitumens other than burns are negligible. However, cut-back bitumens and bituminous emulsions, because they are handled at lower temperatures, increase the chance of skin contact. Studies carried out by Shell demonstrated that the bitumen is unlikely to penetrate the skin and the bitumens diluted with solvents are unlikely to present a carcinogenic risk. Nevertheless, bituminous emulsions can cause irritation to the skin and eyes and can produce allergic responses in some individuals (Shell Bitumen 2003).

A hot bitumen splash may cause serious eye injury. Direct contact with cut-back, emulsions and small particles of cold hard bitumens may cause eye irritation (CONCAWE 1992).

Proper body and eye protective gear (clothing, glasses, etc.) should always be used in order to minimise skin and eye contamination by bituminous materials.

Coal tar pitch contains a large proportion of carcinogenic compounds (Wallcave et al. 1971) and thus its use has been prohibited in many countries.

3.9.4.7 Environmental aspects of bitumens

The assessment of bitumen's impacts on the environment should be carried out by Life Cycle Assessment (LCA).

LCA covers the entire life cycle including extraction of the raw material, manufacturing, transport and distribution, product use, service and maintenance, and disposal (recycling, incineration or landfill). LCA can be divided into two distinct parts, the life cycle inventory (LCI) and life cycle impact (Shell Bitumen 2003).

Eurobitume has carried out an LCI on bitumen, polymer-modified bitumen and bitumen emulsion (Eurobitume 2012). The results obtained may be used in an LCA study.

Table 3.23 Energy consumed and greenhouse gases emitted for the production of 1 t of material

Product	Energy (MJ/t)	CO_{2eq} (kg/m²)
Bitumen	4900	285
Emulsion 60%	3490	221
Cement (Portland cement)	4976	980
Crushed aggregates	40	10
Steel	25,100	3540
Quicklime	9240	2500
Plastic	7890	1100
Production of hot mixed asphalt	275	22
Production of warm mix asphalt	234	20
Production of high-modulus asphalt	289	23
Production of cold mix asphalt	14	1.0
Laying of hot mix asphalt	9	0.6
Laying of cold mix materials	6	0.4
Cement concrete road paving	2.2	0.2

Source: Adapted from Chappat M. and J. Bilal, *The Environmental Road of the Future: Life Cycle Analysis, Energy Consumption and Greenhouse Gas Emissions*, Colas Group, http://www.colas.com/FRONT/COLAS/upload/com/pdf /route-future-english.pdf, 2003.

Table 3.23 gives values of energy consumed and greenhouse gases (CO_{2eq}) emitted during the manufacture of 1 tonne of some finished products from extraction (quarry, oil deposit, etc.) until the sale at the production unit (refinery, cement plant, etc.). These data may be used in an LCA to be carried out. CO_{2eq} is the amount of greenhouse gas emitted, expressed in CO_2 equivalent. The main greenhouse gases are CO_2, N_2O and CH_4.

Table 3.24 Total energy consumption and total GHGe for some pavement construction materials

Product	Total[a]	
	EC[b] (MJ/t)	GHGe[c] (kg/t)
Asphalt concrete	680	54
Road base asphalt concrete	591	47
High-modulus asphalt concrete	699	55
Warm mix asphalt concrete	654	53
Emulsion-bound aggregates	365	30
Cold mix asphalt	457	36
Untreated granular material	113	15
Cement-bound material	319	51
Cement concrete slabs without dowels	738	200
Continuous reinforced concrete	1226	15
Asphalt concrete with 10% RAP	642	51
Road base asphalt concrete with 20% RAP	538	44

Source: Adapted from Chappat M. and J. Bilal, *The Environmental Road of the Future: Life Cycle Analysis, Energy Consumption and Greenhouse Gas Emissions*, Colas Group, http://www.colas.com/FRONT/COLAS/upload/com /pdf/route-future-english.pdf, 2003.

[a] Includes energy consumption for binders, aggregates, manufacture, transport and laying.
[b] EC, energy consumption.
[c] GHGe, greenhouse gas emissions, expressed in CO_2 equivalence.

From Table 3.23, it can be seen that the highest energy consumed is in producing steel, and although the energy consumed to produce bitumen and cement is the same, the CO_{2eq} emissions during the manufacture of bitumen is approximately one-third of the cement.

Additionally, Table 3.24 gives the total energy consumption and the total greenhouse gas emissions (GHGe) for some pavement construction materials or products, determined in the same analysis (Chappat and Bilal 2003). It must be noted that total energy or GHGe is the sum of energy consumed or GHG emitted at all stages: production of aggregates and binder required, manufacture of the material, transport to work site and laying.

As can be seen in Table 3.24, the energy consumption for the construction of a concrete layer is higher than that of the asphalt layer with virgin materials. Similarly, when 20% reclaimed material is used, the energy consumed for an asphalt base layer is less than that for an asphalt base layer with virgin materials. The processes that use unheated aggregate and cold applied binders utilise the least amount of energy per tonne.

For the construction of a concrete layer, the highest energy demand is required for the manufacture of cement, while for the construction of an asphalt layer, most of the energy is required for the manufacture of asphalt cement and heating during the hot mix production process (Chappat and Bilal 2003).

An LCA concluded that, on an annualised basis (based on life expectancy), different asphalt works require differing amounts of energy per year of pavement life. New construction, major rehabilitation, thin hot mix asphalt overlay and hot in-place recycling have the highest energy consumption and range from 6.3 to 12.6 MJ/m²/year. Chip seals, slurry seals, micro-surfacing and crack filling utilise lower amounts of energy per year of extended pavement life and range from 1.3 to 3.3 MJ/m²/year. Crack seals and fog seals consume the least amount of energy per year of extended pavement life, less than 1.3 MJ/m²/year (Chehovits and Galehouse 2010).

Finally, regarding atmospheric pollution, asphalt works pollute the environment less when compared with other sources such as car engines or central heating burners (Hangebrauck et al. 1967).

REFERENCES

AASHTO M 81. 2012. *Cutback asphalt (rapid-curing type)*. Washington, DC: American Association of State Highway and Transportation Officials.

AASHTO M 82. 2012. *Cutback asphalt (medium-curing type)*. Washington, DC: American Association of State Highway and Transportation Officials.

AASHTO M 140. 2013. *Emulsified asphalt*. Washington, DC: American Association of State Highway and Transportation Officials.

AASHTO M 208. 2009. *Cationic emulsified asphalt*. Washington, DC: American Association of State Highway and Transportation Officials.

AASHTO M 226-80. 2012. *Viscosity-graded asphalt cement*. Washington, DC: American Association of State Highway and Transportation Officials.

AASHTO T 167. 2010. *Compressive strength of hot mix asphalt*. Washington, DC: American Association of State Highway and Transportation Officials.

AASHTO T 283. 2011. *Resistance of compacted hot mix asphalt (HMA) to moisture-induced damage*. Washington, DC: American Association of State Highway and Transportation Officials.

Anderson D.A., D. Maurer, T. Ramirez, D.W. Christensen, M.O. Marasteanu, and Y. Mehta. 1999. Field performance of modified asphalt binders evaluated with Superpave test methods: I-80 Test project. *Transportation Research Record: Journal of the Transportation Research Board*, No. 1661, pp. 60–68. Washington, DC: Transportation Research Board of the National Academies.

Anderson D.A., E.L. Dukaz, and J.C. Peterson. 1982. The effect of antistrip additives on the properties of asphalt cement. *Proceedings of the Association of Asphalt Paving Technologists*, Vol. 51, p. 298. Kansas City, MO.

Asphalt Institute IS-173. 1979. *Energy Requirements for Roadway Pavements*. Lexington, KY: Asphalt Institute.

Asphalt Institute MS-4. 1989. *The Asphalt Handbook*, 1989 Edition. Manual Series No. 4. Lexington, KY: Asphalt Institute.

Asphalt Institute MS-4. 2007. *The Asphalt Handbook*, 7th Edition. Manual Series No. 4. Lexington, KY: Asphalt Institute.

Asphalt Institute SP-1. 2003. *Superpave: Performance Graded Asphalt, Binder Specification and Testing*, 3rd Edition. Lexington, KY: Asphalt Institute.

ASTM D 5/D 5M. 2013. *Standard test method for penetration of bituminous materials*. West Conshohocken, PA: ASTM International.

ASTM D 8. 2013. *Standard terminology relating to materials for roads and pavements*. West Conshohocken, PA: ASTM International.

ASTM D 92-12b. 2012. *Standard test method for flash and fire points by Cleveland open cup tester*. West Conshohocken, PA: ASTM International.

ASTM D 95-13e1. 2013. *Standard test method for water in petroleum products and bituminous materials by distillation*, West Conshohocken, PA: ASTM International.

ASTM D 113. 2007. *Standard test method for ductility of bituminous materials*. West Conshohocken, PA: ASTM International.

ASTM D 244. 2009. *Standard test methods and practices for emulsified asphalts*. West Conshohocken, PA: ASTM International.

ASTM D 402. 2008. *Standard test method for distillation of cutback asphaltic (bituminous) products*. West Conshohocken, PA: ASTM International.

ASTM D 946/D 946M-09a. 2009. *Standard specification for penetration-graded asphalt cement for use in pavement construction*. West Conshohocken, PA: ASTM International.

ASTM D 977-12b. 2012. *Standard specification for emulsified asphalt*. West Conshohocken, PA: ASTM International.

ASTM D 1074. 2009. *Standard test method for compressive strength of bituminous mixtures*. West Conshohocken, PA: ASTM International.

ASTM D 1075. 2011. *Standard test method for effect of water on compressive strength of compacted bituminous mixtures*. West Conshohocken, PA: ASTM International.

ASTM D 2026/D 2026M-97e1. 2010. *Standard specification for cutback asphalt (slow-curing type)*. West Conshohocken, PA: ASTM International.

ASTM D 2027/D 2027M. 2010. *Standard specification for cutback asphalt (medium-curing type)*. West Conshohocken, PA: ASTM International.

ASTM D 2028/D 2028M. 2010. *Standard specification for cutback asphalt (rapid-curing type)*. West Conshohocken, PA: ASTM International.

ASTM D 2042. 2009. *Standard test method for solubility of asphalt materials in trichloroethylene*. West Conshohocken, PA: ASTM International.

ASTM D 2170/D 2170M. 2010. *Standard test method for kinematic viscosity of asphalts (bitumens)*. West Conshohocken, PA: ASTM International.

ASTM D 2397. 2012. *Standard specification for cationic emulsified asphalt*. West Conshohocken, PA: ASTM International.

ASTM D 2521/2521M-76e1. 2014. *Standard specification for asphalt used in canal, ditch, and pond lining*. West Conshohocken, PA: ASTM International.

ASTM D 2872-12e1. 2012. *Standard test method for effect of heat and air on a moving film of asphalt (rolling thin-film oven test)*. West Conshohocken, PA: ASTM International.

ASTM D 3143. 2008. *Standard test method for flash point of cutback asphalt with tag open-cup apparatus*. West Conshohocken, PA: ASTM International.

ASTM D 3381/D 3381M. 2012. *Standard specification for viscosity-graded asphalt cement for use in pavement construction*. West Conshohocken, PA: ASTM International.

ASTM D 3406. 2006. *Standard specification for joint sealant, hot-applied, elastomeric-type, for Portland cement concrete pavements*. West Conshohocken, PA: ASTM International.

ASTM D 4402/D 4402M. 2013. *Standard test method for viscosity determination of asphalt at elevated temperatures using a rotational viscometer*. West Conshohocken, PA: ASTM International.

ASTM D 4867/D 4867M. 2009. *Standard test method for effect of moisture on asphalt concrete paving mixtures.* West Conshohocken, PA: ASTM International.

ASTM D 5329. 2009. *Test methods for sealants and fillers, hot-applied, for joints and cracks in asphaltic and Portland cement concrete pavements.* West Conshohocken, PA: ASTM International.

ASTM D 5710. 2005. *Specification for Trinidad lake modified asphalt.* West Conshohocken, PA: ASTM International.

ASTM D 5893/D 5893M. 2010. *Standard specification for cold applied, single component, chemically curing silicone joint sealant for Portland cement concrete pavements.* West Conshohocken, PA: ASTM International.

ASTM D 6154. 2009. *Standard specification for chemically modified asphalt cement for use in pavement construction.* West Conshohocken, PA: ASTM International.

ASTM D 6373-07e1. 2007. *Standard specification for performance graded asphalt binder.* West Conshohocken, PA: ASTM International.

ASTM D 6521. 2008. *Standard practice for accelerated aging of asphalt binder using a pressurized aging vessel (PAV).* West Conshohocken, PA: ASTM International.

ASTM D 6626. 2009. *Standard specification for performance graded Trinidad lake modified asphalt binder.* West Conshohocken, PA: ASTM International.

ASTM D 6648. 2008. *Standard test method for determining the flexural creep stiffness of asphalt binder using the bending beam rheometer (BBR).* West Conshohocken, PA: ASTM International.

ASTM D 6723. 2012. *Standard test method for determining the fracture properties of asphalt binder in direct tension (DT).* West Conshohocken, PA: ASTM International.

ASTM D 6816. 2011. *Standard practice for determining low-temperature performance grade (PG) of asphalt binders.* West Conshohocken, PA: ASTM International.

ASTM D 6930. 2010. *Standard test method for settlement and storage stability of emulsified asphalts.* West Conshohocken, PA: ASTM International.

ASTM D 6931. 2012. *Standard test method for indirect tensile (IDT) strength of bituminous mixtures.* West Conshohocken, PA: ASTM International.

ASTM D 6933. 2008. *Standard test method for oversized particles in emulsified asphalts (sieve test).* West Conshohocken, PA: ASTM International.

ASTM D 6935. 2011. *Standard test method for determining cement mixing of emulsified asphalt.* West Conshohocken, PA: ASTM International.

ASTM D 6936. 2009. *Standard test method for determining demulsibility of emulsified asphalt.* West Conshohocken, PA: ASTM International.

ASTM D 6997. 2012. *Standard test method for distillation of emulsified asphalt.* West Conshohocken, PA: ASTM International.

ASTM D 7116. 2005. *Standard specification for joint sealants, hot applied, jet fuel resistant types, for Portland cement concrete pavements.* West Conshohocken, PA: ASTM International.

ASTM D 7643. 2010. *Standard practice for determining the continuous grading temperatures and continuous grades for PG graded asphalt binders.* West Conshohocken, PA: ASTM International.

Barth E.J. 1962. *Science and Technology,* Chapter 7. New York: Gordon & Breach Publication.

Beatty T.L., K. Dunn, E.T. Harrigan, K. Stuart, and H. Weber. 1987. Field evaluation of sulfur-extended asphalt pavements. *Transportation Research Record No. 1115.* Washington, DC: Transportation Research Board.

Bohn A.O. 1965. Chemistry of breaking of asphalt emulsions. *Highway Research Record,* Vol. 67, p. 195. Washington, DC: Transportation Research Board of the National Academies.

Brule B. and F. Lebourlot. 1993. Choix de bitumes pour melanges bitumes – EVA. *Proceedings of the 5th Eurobitume,* Vol. 1, p. 91. Stockholm.

BS 76. 1974. *Specifications for tars for road purposes.* London: British Standards Institution.

Bull A.L. and W.C. Vonk. 1984. *Thermoplastic Rubber/Bitumen Blends for Roof and Road.* Thermoplastic Rubbers Technical Manual, TR 8.15. London: Shell International Petroleum Company Ltd.

Cawsey D.C. and C.S. Gourlet. 1989. Bitumen–aggregate adhesion: Development of a stripping index. *Proceedings of the 4th Eurobitume, Madrid,* Vol. 1, p. 225. Madrid.

CEN EN 1426. 2007. *Bitumen and bituminous binders. Determination of needle penetration.* Brussels: CEN.

CEN EN 1427. 2007. *Bitumen and bituminous binders – Determination of the softening point – Ring and Ball method.* Brussels: CEN.

CEN EN 1428. 2012. *Bitumen and bituminous binders – Determination of water content in bitumen emulsions – Azeotropic distillation method.* Brussels: CEN.

CEN EN 1429. 2013. *Bitumen and bituminous binders – Determination of residue on sieving of bituminous emulsions, and determination of storage stability by sieving.* Brussels: CEN.

CEN EN 1430. 2009. *Bitumen and bituminous binders – Determination of particle polarity of bituminous emulsions.* Brussels: CEN.

CEN EN 1431. 2009. *Bitumen and bituminous binders – Determination of residual binder and oil distillate from bitumen emulsions by distillation.* Brussels: CEN.

CEN EN 12272-3. 2003. *Surface dressing – Test methods – Part 3: Determination of binder aggregate adhesivity by the Vialit plate shock test method.* Brussels: CEN.

CEN EN 12591. 2009. *Bitumen and bituminous binders – Specifications for paving grade bitumens.* Brussels: CEN.

CEN EN 12592. 2007. *Bitumen and bituminous binders – Determination of solubility.* Brussels: CEN.

CEN EN 12593. 2007. *Bitumen and bituminous binders – Determination of the Fraass breaking point.* Brussels: CEN.

CEN EN 12595. 2007. *Bitumen and bituminous binders – Determination of kinematic viscosity.* Brussels: CEN.

CEN EN 12596. 2007. *Bitumen and bituminous binders – Determination of dynamic viscosity by vacuum capillary.* Brussels: CEN.

CEN EN 12597. 2000. *Bitumen and bituminous binders – Terminology.* Brussels: CEN.

CEN EN 12607-1. 2007. *Bitumen and bituminous binders – Determination of the resistance to hardening under the influence of heat and air – Part 1: RTFOT method.* Brussels: CEN.

CEN EN 12697-11. 2012. *Bituminous mixtures – Test methods for hot mix asphalts – Part 11: Determination of the affinity between aggregate and bitumen.* Brussels: CEN.

CEN EN 12697-12. 2008. *Bituminous mixtures – Test methods for hot mix asphalts – Part 12: Determination of the water sensitivity of bituminous specimens.* Brussels: CEN.

CEN EN 12697-23. 2003. *Bituminous mixtures – Test methods for hot mix asphalts – Part 23: Determination of the indirect tensile strength of bituminous specimens.* Brussels: CEN.

CEN EN 12846-1. 2011. *Bitumen and bituminous binders – Determination of efflux time by the efflux viscometer-Part 1: Bituminous emulsions.* Brussels: CEN.

CEN EN 12846-2. 2011. *Bitumen and bituminous binders – Determination of efflux time by the efflux viscometer – Part 2: Cut-back and fluxed bituminous binders.* Brussels: CEN.

CEN EN 12847. 2009. *Bitumen and bituminous binders – Determination of settling tendency of bituminous emulsions.* Brussels: CEN.

CEN EN 12848. 2009. *Bitumen and bituminous binders – Determination of mixing stability with cement of bituminous emulsions.* Brussels: CEN.

CEN EN 12849. 2009. *Bitumen and bituminous binders – Determination of penetration power of bituminous emulsions.* Brussels: CEN.

CEN EN 13074-1. 2011. *Bitumen and bituminous binders – Recovery of binder from bituminous emulsion or cut-back or fluxed bituminous binders – Part 1: Recovery by evaporation.* Brussels: CEN.

CEN EN 13074-2. 2011. *Bitumen and bituminous binders. Recovery of binder from bituminous emulsion or cut-back or fluxed bituminous binders. Part 2: Stabilisation after recovery by evaporation.* Brussels: CEN.

CEN EN 13075-1. 2009. *Bitumen and bituminous binders – Determination of breaking behaviour – Part 1: Determination of breaking value of cationic bituminous emulsions, mineral filler method.* Brussels: CEN.

CEN EN 13075-2. 2009. *Bitumen and bituminous binders – Determination of breaking behaviour – Part 2: Determination of fines mixing time of cationic bituminous emulsions.* Brussels: CEN.

CEN EN 13302. 2010. *Bitumen and bituminous binders – Determination of dynamic viscosity of bituminous binder using a rotating spindle apparatus.* Brussels: CEN.

CEN EN 13304. 2009. *Bitumen and bituminous binders – Framework for specification of oxidised bitumens.* Brussels: CEN.

CEN EN 13358. 2010. *Bitumen and bituminous binders – Determination of the distillation characteristics of cut-back and fluxed bituminous binders made with mineral fluxes.* Brussels: CEN.

CEN EN 13398. 2010. *Bitumen and bituminous binders – Determination of the elastic recovery of modified bitumen.* Brussels: CEN.

CEN EN 13399. 2010. *Bitumen and bituminous binders – Determination of storage stability of modified bitumen.* Brussels: CEN.

CEN EN 13587. 2010. *Bitumen and bituminous binders – Determination of the tensile properties of bituminous binders by the tensile test method.* Brussels: CEN.

CEN EN 13588. 2008. *Bitumen and bituminous binders – Determination of cohesion of bituminous binders with pendulum test.* Brussels: CEN.

CEN EN 13589. 2008. *Bitumen and bituminous binders – Determination of the tensile properties of modified bitumen by the force ductility method.* Brussels: CEN.

CEN EN 13614. 2011. *Bitumen and bituminous binders – Determination of adhesivity of bituminous emulsions by water immersion test.* Brussels: CEN.

CEN EN 13702. 2010. *Bitumen and bituminous binders – Determination of dynamic viscosity of modified bitumen by cone and plate method.* Brussels: CEN.

CEN EN 13703. 2003. *Bitumen and bituminous binders – Determination of deformation energy.* Brussels: CEN.

CEN EN 13808. 2013. *Bitumen and bituminous binders – Framework for specifying cationic bituminous emulsions.* Brussels: CEN.

CEN EN 13924/AC. 2006. *Bitumen and bituminous binders – Specifications for hard paving grade bitumens.* Brussels: CEN.

CEN EN 14023. 2010. *Bitumen and bituminous binders – Framework specification for polymer modified bitumens.* Brussels: CEN.

CEN EN 14188-1. 2004. *Joint fillers and sealants – Part 1: Specifications for hot applied sealants.* Brussels: CEN.

CEN EN 14188-2. 2004. *Joint fillers and sealants – Part 2: Specifications for cold applied sealant.* Brussels: CEN.

CEN EN 15322. 2013. *Bitumen and bituminous binders – Framework for specifying cut-back and fluxed bituminous binders.* Brussels: CEN.

CEN EN 15326:2007+A1. 2009. *Bitumen and bituminous binders. Measurement of density and specific gravity. Capillary-stoppered pyknometer method.* Brussels: CEN.

CEN EN 16345. 2012. *Bitumen and bituminous binders – Determination of efflux time of bituminous emulsions using the Redwood No. II viscometer.* Brussels: CEN.

CEN EN ISO 2592. 2001. *Determination of flash and fire points. Cleveland open cup method.* Brussels: CEN.

CEN EN ISO 2719. 2002. *Determination of flash point. Pensky-Martens closed cup method.* Brussels: CEN.

CEN EN ISO 13736. 2013. *Determination of flash point – Abel closed-cup method.* Brussels: CEN.

Chappat M. and Bilal J. 2003. *The Environmental Road of the Future: Life Cycle Analysis, Energy Consumption and Greenhouse Gas Emissions.* Colas Group, Boulogne-Billancourt Cedex, France. Available at http://www.colas.com/FRONT/COLAS/upload/com/pdf/route-future-english.pdf.

Chehovits J. and L. Galehouse. 2010. Energy usage and green gas emissions of pavement preservation processes for asphalt concrete pavements. *1st International Conference on Pavements Preservation.* Newport Beach, CA.

Chemcrete Co. 1987. *Tougher Flexible Roads and Modified Binders.* Note 87, Internal Reports. Chemcrete Co., Richardson, TX, USA. Available at http://www.chem-crete.com.

Chilingarian G.V. and T.F. Yen. 1978. *Bitumens, asphalts, and tar sands.* Amsterdam, Oxford, New York: Elsevier Scientific Publishing Company.

Chipperfield E.H., J.L. Duthie, and R.B. Girdler. 1970. Asphalt characteristics in relation to road performance. *Proceedings of the Association of Asphalt Paving Technologists,* Vol. 39, p. 575. Seattle, WA.

Chipperfield E.H. and M.J. Leonard. 1976. How our European neighbours tackle with bitumen road construction. *Journal of the Institution of Highway Engineers,* Vol. XXIII, No. 12, p. 9. London: The Institution of Highway Engineers.

Christensen D.W. and D.A. Anderson. 1985. Effect of amine additives on the properties of asphalt cement. *Proceedings of the Association of Asphalt Paving Technologists*, Vol. 54, p. 593. San Antonio, TX.

CONCAWE. 1992. *Bitumens and Bitumen Derivatives*. Product Dossier No. 92/104. Brussels: CONCAWE.

Cooper Research Technology Ltd. 2014. Available at http://www.cooper.co.uk.

Courard L. and J.M. Rigo. 1993. The use of fibers in bituminous concrete as a solution to decrease rutting. *Proceedings of the 5th Eurobitume*, Session 3, p. 497. Stockholm.

Daines M.E., J. Carswell, and D.M. Colwill. 1985. *Assessment of 'Chem-crete' as an Additive for Binders for Wearing Courses and Roadbase*. TRL Research Report 54. Crowthorne, UK: Transport Research Laboratory.

Denning J.H. and J. Carswell. 1981a. *Improvements in Rolled Asphalt Surfacing by the Addition of Organic Polymers*. TRRL Report LR 989. Crowthorne, UK: Transport Research Laboratory.

Denning J.H. and J. Carswell. 1981b. *Improvements in Rolled Asphalt Surfacing by the Addition of Sulphur*. TRLL Report LR 963. Crowthorne, UK: Transport Research Laboratory.

Dinnen A. 1981. Epoxy bitumen binders for critical road conditions. *Proceedings of the 2nd Eurobitume*, p. 294. Cannes.

EAPA. 2011. *Asphalts in Figures 2011*. Brussels, Belgium: European Asphalt Pavement Association.

Economou N.D., V.P. Papageorgiou, and A.F. Nikolaides. 1990. New organoboron complexes as catalysts in asphalt blowing. *Chemical Chronika*, New Series, Vol. 19, pp. 233–241.

Ecopave. 2014. Available at http://www.ecopave.com.au.

Elseifi M.A., L.N. Mohammad, and S.B. Cooper. 2011. Laboratory evaluation of asphalt mixtures containing sustainable technologies. *Proceedings, Association of Asphalt Paving Technologists*, Vol. 80, pp. 227–254.

Eurobitume. 2010. *Health Study Update 2010*. Brussels: Eurobitume.

Eurobitume. 2012. *Life Cycle Inventory: Bitumen*, 2nd Edition. Brussels: Eurobitume.

Fromm H.J. and G.J.A. Kennepohl. 1979. Sulfur-asphaltic concrete on three Ontario test roads. *Proceedings of the Association of Asphalt Paving Technologists*, Vol. 48, p. 135.

Gaestel C. 1967. The breaking mechanism of asphalt emulsions. *Chemistry and Industry*, Vol. 2, p. 221.

Gilby G.W. 1985. Ethylene-vinylacetate (EVA) copolymers as modifiers for bitumen binders. *Asphalt Technology*, Vol. 36, No. 36, pp. 37–41.

Girdler R.B. 1965. Constitution of asphaltenes and related studies. *Proceedings of Association of Asphalt Paving Technologists*, Vol. 34, p. 34.

Gottschall A. and S. Hollnsteiner. 1985. Steel fiber reinforced asphalt. *Proceedings of the 3rd Eurobitume*, Session IV, p. 518. Hague.

Hangebrauck R.P., J.U. von Lehmden and J.E. Meeker. 1967. *Sources of Polynuclear Hydrocarbons in the Atmosphere*. U.S. Department of Health, Education and Welfare. Publication No. 999-AP-33, Washington, DC.

Herrington P., M. Gribble, and G. Bentley. 2011. *Multigrade Bitumen for Chipsealing Applications*. NZ Transport Agency Research Report 460. Wellington, New Zealand: NZ Transport Agency.

Hoiberg A.J. 1965. *Bituminous Materials: Asphalts, Tars and Pitches*, Vol. II, Chap. 10. New York: Interscience.

Institute of Petroleum. 2005. *Bitumen Safety Code, Model Code of Safe Practice in the Petroleum Industry*, Part 11, 4th Edition. London: Energy Institute.

IUPAC. 1974. *Basic Definitions of Terms Relating to Polymers*. London: Butterworths.

Kennepohl G.J.A., A. Logan, and D.C. Bean. 1975. Conventional paving mixes with sulfur-asphalt binders. *Proceedings, Association of Asphalt Paving Technologists*, Vol. 44, pp. 485–518.

Lake Asphalt of Trinidad and Tobago Ltd. 2014. Available at http://www.trinidadlakeasphalt.com.

Lukac B. and A.Z. Valant. 2011. Performance of crumb rubber modified binders and asphalts. *Proceedings of the 5th International Conference 'Bituminous Mixtures and Pavements'*, Vol. 1, p. 133. Thessaloniki, Greece: Aristotle University of Thessaloniki.

Lyttleton D.Y. and R.N. Traxler. 1948. Flow properties of asphaltic emulsions. *Industrial and Engineering Chemistry*, Vol. 40, p. 2115.

McCready N.S. and R.C. Williams. 2007. The utilization of agriculturally derived lignin as an anti-oxidant in asphalt binder. *Proceedings of the 2007 Mid-Continent Transportation Research Symposium.* Ames, IA: Iowa State University.

Mertens E.W. and J.R. Wright. 1959. Cationic asphalt emulsions: How they differ from conventional emulsions in theory and practice. *Proceedings of the Highway Research Board,* Vol. 38, p. 386.

MoPW. 1989. *Supervision Manual for Lasbutag.* Directorate General Bina Marga, Jakarta: Ministry of Public Works.

Neutag L. and H.J. Beckedahl. 2011. Performance of crumb rubber modified binders and asphalts. *Proceedings of the 5th International Conference 'Bituminous Mixtures and Pavements',* Vol. 1, p. 104. Thessaloniki, Greece: Aristotle University of Thessaloniki.

Nicholls J.C. 2009. *Review of Shell Thiopave™ Sulphur Extended Asphalt Modifier.* TRL Report TRL 672. Crowthorne, UK: Transport Research Laboratory.

Nicholls J.C. 2011. Shell THIOPAVE™ sulphur extended asphalt modifier. *Proceedings of the 5th International Conference 'Bituminous Mixtures and Pavements',* Vol. 1, p. 104. Thessaloniki, Greece: Aristotle University of Thessaloniki.

Nikolaides A. 1990. *ASDP Method: Evaluation of Performance of Lasbutag and Latasbusir Trial Sections.* Internal Report, Jakarta: Ministry of Public Works.

Nikolaides A. 1991a. *Evaluation of Performance of Lasbutag Pilot Projects in S.E. Sulawesi.* Internal Report, Jakarta: Ministry of Public Works.

Nikolaides A. 1991b. *Asbuton-Micro: Pilot Project in Bandung-Evaluation.* Internal Report, Jakarta: Ministry of Public Works.

Nikolaides A. 1992. *Use of Chemical Additives in the Production of Hot and Cold Bituminous Mixtures.* MoT, GSR&T Research Program. Highway Engineering Laboratory, Aristotle University of Thessaloniki (in Greek).

Nikolaides A. and G. Tsochos. 1992. Mechanical properties of asphalt concrete type A 265-B with elastomer modified bitumen. *Proceedings of 1st National Conference on Bituminous Mixtures,* p. 417. Thessaloniki, Greece (in Greek).

Nikolaides A., G. Tsochos, and N. Tressos. 1992. The effect of additives to the rheological properties of bitumen. *Proceedings of 1st National Conference on Bituminous Mixtures,* p. 394. Thessaloniki, Greece (in Greek).

Nikolaides A.F. 1983. Design of dense graded cold bituminous emulsion mixtures and evaluation of their engineering properties. Ph.D. Thesis, University of Leeds, Department of Civil Engineering.

Nikolaides A.F. and E. Manthos. 2009. Wheel tracking performance of asphalt concrete mixture with conventional and modified bitumen. *MAIREPAV 6, 6th International Conference on Maintenance and Rehabilitation of Pavements & Technological Control,* Vol. I, p. 213. Torino, Italy, July.

Nikolaides A.F. and N. Oikonomou. 1987. The necessary replacement of cutbacks with bitumen emulsion. *Technical Chronicals A,* Vol. 7, No. 1, p. 197. Athens: Technical Chamber of Greece (in Greek-English summary).

NIOSH. 1977. *Criteria For a Recommended Standard: Occupational Exposure to Asphalt Fumes.* NTIS Publication No. PB-277-333, DHEW (NIOSH) Publication No. 78-106. Department of Health, Education, and Welfare, Public Health Service, Center for Disease Control, National Institute for Occupational Safety and Health. Cincinnati, OH: NIOSH.

NIOSH. 2000. *Hazard Review: Health Effects of Occupational Exposure to Asphalt.* Publication No. 2001-110. U.S. Department of Health and Human Services (DHHS), National Institute for Occupational Safety and Health (NIOSH). Cincinnati, OH: NIOSH.

Oikonomou N. and A.F. Nikolaides. 1985. Cationic emulsions in antiskidding surface courses. *10th National Conference in Chemistry,* Vol. B, p. 1094. Patra, Greece (in Greek).

Olsson A., H. Kromhout, M. Agostini, J. Hansen, C.F. Lassen, C. Johansen, K. Kjaerheim, S. Langård, I. Stücker, W. Ahrens, T. Behrens, M.-L. Lindbohm, P. Heikkilä, D. Heederik, L. Portengen, J. Shaham, G. Ferro, F. de Vocht, I. Burstyn, and P. Boffetta. 2010. A case–control study of lung cancer nested in a cohort of European asphalt workers. *Environmental Health Perspectives,* Vol. 118, No. 10, p. 1418. Available at http://ehp03.niehs.nih.gov/article/info:doi/10.1289/ehp.0901800.

Peltonen P. 1989. Fibres as additives in bitumen. *Proceedings of the 4th Eurobitume,* Session IV, p. 938. Madrid.

Piber H. and W. Pichler. 1993. Mineral fiber for the modification of asphalt. *Proceedings of the 5th Eurobitume*, Session 2, p. 493. Stockholm.

prEN 13924-2. Under approval. *Bitumen and bituminous binders – Specification framework for special bitumen – Part 2: Multigrade bituminous binders*. Brussels: CEN.

Redilius P. 1993. A novel system for delayed breaking control of bituminous emulsions. *Proceedings of the 1st World Congress on Emulsions*, Vol. 1, Workshop 22, Order No. 147. Paris.

Resperion. 2013. Available at http://www.resperion.com.

RuBind. 2013. Available at http://www.rubind.com.

Samanos J. and J.P. Serfass. 1993. New developments in fiber-based bituminous mixes for road mainte-nance. *Proceedings of the 5th Eurobitume*, Session 3, p. 579. Stockholm.

Scott J.A. 1974. A general description of the breaking process of cationic emulsions in contact with mineral aggregate. *Proceedings of Symposium on Theory and Practice of Emulsion Technology*, p. 151. Brunel University.

Shell Bitumen. 1990. *The Shell Bitumen Handbook*. Surrey, UK: Shell Bitumen UK.

Shell Bitumen. 2003. *The Shell Bitumen Handbook*, 5th Edition. London: Thomas Telford Publishing for Shell Bitumen.

Sundstrom D.W., H.E. Klei, and J.E. Stephens. 1983. *The Addition of Lignin From Gasohol Plants to Asphalt*. Research Report, Project No. 80-3. Storrs, CT: Civil Engineering Department, University of Connecticut.

Terrel R.L. and S. Rimsritong. 1979. Wood lingins used as extenders for asphalt in bituminous pave-ments. *Proceedings, Association of Asphalt Paving Technologists*, Vol. 48, pp. 111–134.

Transportation Research Board No. 30. 1975. *Bituminous Emulsions for Highway Pavement*. NCHR Programme Synthesis of Highway Practice. Washington, DC: National Research Council.

Traxler R.N. 1936. The physical chemistry of asphaltic bitumen. *Chemical Review*, Vol. 19, No. 2, pp. 119–143.

Van Gooswilligen G. and W.C. Vonk. 1986. *The Role of Bitumen in Blends with Thermoplastic Rubbers for Roofing Applications*. Thermoplastic Rubbers Technical Manual TR 8.16. Chertsey, Surrey, UK: Shell Bitumen, UK.

Vonk W.C., M.C. Phillips, and M. Roele. 1993. Ageing resistance of bituminous road binders: The ben-efit of SBS modification. *Proceedings of the 5th Eurobitume*, Vol. 1, p. 156. Stockholm.

Vonk W.C. and G. Van Gooswilligen. 1989. Improvement of paving grade bitumens with SBS polymers. *Proceedings of the 4th Eurobitume*, Vol. 1, p. 299. Madrid.

Wallcave L., H. Garcia, R. Feldman, W. Lijinsky, and P. Shubik. 1971. Skin tumorigenesis in mice by petroleum asphalts and coal tar pitches of known polynuclear aromatic hydrocarbon content. *Toxicology and Applied Pharmacology*, Vol. 18, p. 41.

West C.R., D.E. Watson, P.A. Turner, and J.R. Casola. 2010. *Mixing and Compaction Temperatures of Asphalt Binders in Hot-Mix Asphalt*. NCHRP Report 648. Transportation Research Board. Washington, DC: National Academy of Sciences.

Woodside A. and P. Macoal. 1989. The use of adhesion agents and their effect. *Proceedings of the 4th Eurobitume Symposium*, Vol. 1, p. 200. Madrid.

Woolley K.G. 1986. Polymer modified bitumen for extra value asphalt. *Journal of the Institute of Asphalt Technology*, Vol. 38, pp. 45–51.

Yeates C. 1994. An evaluation of the use of a fiber-reinforced membrane to inhibit reflecting cracking. *Symposium on Performance and Durability of Bituminous Materials*. Leeds, UK.

Chapter 4

Laboratory tests and properties of bitumen and bitumen emulsion

4.1 GENERAL

The purpose of the laboratory tests performed on bitumen is to define its characteristic properties, so as to ascertain its suitability and predict its behaviour during the service life of the pavement. The term *characteristic properties* include all properties, such as technological, mechanical, rheological, physical and chemical. Technological properties are those defined by empirical and not fundamental tests, unlike the mechanical and rheological properties. The main technological properties are penetration, softening point, ductility, flash point, solubility, weight loss after heating and Fraass breaking point. All the abovementioned technological properties, along with viscosity (rheological property), constitute the main criteria for suitability or conformity of bitumen specified by international standards.

Apart from the normative tests, many laboratories perform additional tests for the determination of fundamental mechanical properties of bitumen such as stiffness modulus, tensile strength and resistance to fatigue. These properties are of key importance for predicting the mechanical behaviour of bitumen and thus of the bituminous mixture.

With the appearance of modified bitumen, there was a necessity to establish supplementary tests. Thus, cohesion tests, the elastic recovery test and storage stability tests have been added.

This chapter provides a description of all laboratory tests performed on both conventional and modified bitumen. Additionally, the laboratory tests performed on bitumen emulsions are also described in brief.

4.2 PENETRATION TEST

The penetration test is the most widely known test, developed at the end of the 19th century and still in use, for grading the paving bitumen. This test method also determines the consistency of the bitumen and bituminous binders at intermediate service temperature. Higher values of penetration indicate softer consistency.

The test involves the determination of the penetration depth of a standard penetration needle into the bituminous binder under standard conditions of temperature (25°C), applied load and loading time. The standardised needle first touches the surface of the bituminous specimen and then is allowed to penetrate into the mass of the specimen under the influence of its own weight and an additional mass so that the total load is 100 ± 0.1 g, for a period of 5 s. After loading, the penetration depth of the needle is measured in 0.1 mm or dmm. This unit is also called 'pen' (1 pen = 0.1 mm). Figure 4.1 shows an automatic penetration test apparatus.

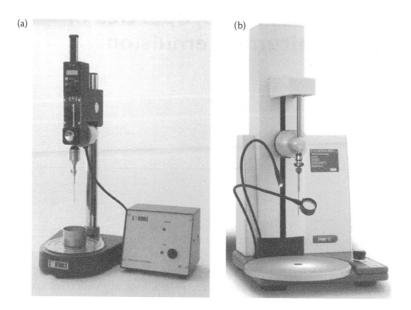

Figure 4.1 Penetration test apparatuses. (a) Controls penetrometer. (b) Anton-Paar penetrometer. (Courtesy of Controls Srl. and Anton Paar ProveTec GmbH.)

According to CEN EN 1426 (2007), the testing temperature for bituminous binders with penetration above 330 dmm is reduced to 15°C, while the load and duration of loading remain the same.

The test is described in CEN EN 1426 (2007) and in ASTM D 5 (2013) or AASHTO T 49 (2011).

According to CEN EN 1426 (2007), the recommended size of the test sample container for penetration <160 dmm is 35 mm internal depth and 55 mm internal diameter. When the penetration is ≥160 dmm and ≤330 dmm the recommended sizes are 45 and 70 mm, respectively.

At least three measurements are taken from different points of the specimen surface (distance apart ≥10 mm). The average value rounded to the nearest integer is reported as the penetration value of the bitumen. According to CEN EN 1426 (2007), the three consecutive measurements should not differ (minimum and maximum value) by more than 2, 4, 6 and 8 dmm, when the penetration is between 0 and 49 dmm, between 50 and 149 dmm, between 150 and 249 dmm or ≥250 dmm, respectively. The respective permitted differences according to ASTM D 5 (2013), for the same range of grades, are 2, 4, 12 and 20 dmm.

The above testing conditions are typical. Non-typical test conditions are those when the test is carried out at temperatures other than 25°C, such as 0°C, 4°C, 45°C or 46.1°C. When the testing temperatures is lower than 25°C, the total load and the loading time change to 200 g and 60 s, respectively. At 46.1°C, the total load is 50 g and the loading time is 5 s (ASTM D 5 2013).

It should be noted that maintaining the test temperature constant throughout the measurements is of key importance. This is achieved by the use of a water bath, in which water circulates through a heating/cooling system. Furthermore, special attention should be given to the preparation of the bitumen specimens, so that no trapped air is created within the mass of the bitumen and the specimen surface is free of dust or other foreign micro-particles. The above is ensured by allowing the hot bitumen to be poured into the container to cool down in air temperature (15°C to 30°C), while the container is loosely covered.

More details about the test can be found in CEN EN 1426 (2007), ASTM D 5 (2013) or AASHTO T 49 (2011).

4.3 SOFTENING POINT TEST (LEFT)

The consistency of the bitumen and the bituminous binders at elevated temperature is empirically determined with the softening point test, also known as a 'ring and ball' (R&B) test. Along with the penetration test, the softening point test is used for grading the oxidised bitumens.

Bitumen, a viscoelastic material, does not have a defined melting point. It gradually becomes softer and less viscous as the temperature rises.

Softening point is defined as the temperature at which material under standardised test conditions attains a specific consistency. The test is used to determine the softening point of bitumen and bituminous binders in the 28°C to 150°C range.

Two horizontal brass rings (approximately 19.8 mm internal diameters) filled with bituminous binder are placed in a ring holder and heated at a controlled rate in a liquid bath while each supports a steel ball. The softening point is the mean of the temperatures at which the two discs soften enough to allow each ball, enveloped in bituminous binder, to fall a distance of 25.0 mm.

The ring holder with the rings containing the bituminous binder is placed in a glass beaker filled with iced distilled or de-ionised water (approximately 5°C ± 1°C), before the placement of the steel balls and the initiation of the test. After 15 min, the steel balls are positioned on each ring and the water is heated at a uniform rate of 5°C ± 1°C, under continuous stirring, following the instructions of the specification. As the temperature rises, the bitumen softens and begins to form a meniscus, which constantly augments as the temperature rises. For each ring and ball, record the temperature indicated by the thermometer at the instant the bituminous binder surrounding the ball touches the bottom plate. If an automatic apparatus is used, record the temperature at which the meniscus interrupts the ray of light. An automatic softening point apparatus is shown in Figure 4.2.

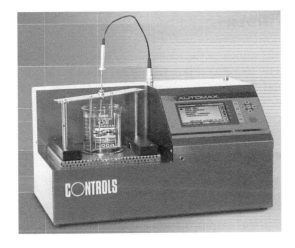

Figure 4.2 Ring and ball apparatus, automatic version. (Courtesy of Controls Srl.)

The average of the two measurements, rounded up to the nearest 0.2°C for bituminous binders with a softening point up to 80°C or up to the 0.5°C for bituminous binders with a softening point >80°C, is defined as the softening temperature or softening point ($t_{R\&B}$).

If the difference between the two temperatures exceeds 1°C for softening points below 80°C or exceeds 2°C for softening points above 80°C, the test is repeated.

When modified bitumens are tested, the test is repeated when (a) the difference between the two measurements is >2°C, (b) the metal ball breaks the surrounding bitumen film before touching the lower platform or (c) a partial detachment of the bitumen from the metal ball is observed.

It should be mentioned that the preparation of the test samples is of key importance. They should have a flat and smooth surface, which is achieved by placing the casting rings on a smooth metal or class surface and cutting the excess amount of binder with the use of warmed knife or blade. The smooth steel or glass surface prior to casting is covered with a thin layer of release agent (mixture of glycerol and dextrin or mineral talc, or another commercially available release agent) to prevent the bituminous binder adhering to the pouring plate.

The bath liquid when softening points are up to 80°C is, as mentioned, distilled or deionised water. For softening points above 80°C, the bath liquid is glycerol and the starting temperature of the test is 30°C instead of 5°C ± 1°C.

More details of the test can be found in CEN EN 1427 (2007), ASTM D 36/D 36M (2012) or AASHTO T 53 (2009).

The softening point may be estimated from the penetration value. After extensive laboratory tests by the author, the following equation was derived, for paving grade bitumens with penetration ranging from 40 to 100 dmm, with a very good correlation coefficient (0.98):

$$t_{R\&B} = 87.3 - 22.5 \times lgP \text{ (correlation coefficient = 0.983),}$$

where $t_{R\&B}$ is the softening point (°C) and lgP is the logarithm (base 10) of penetration at 25°C.

The above equation is slightly modified compared to the one derived earlier on (Nikolaides 1988) because of additional laboratory tests. It should be highlighted that the above equation does not in any case replace the softening point test.

4.4 PENETRATION INDEX

The penetration index (I_p) indicates the temperature susceptibility of the paving grade bitumens, for grades 20/30 to 160/220. The penetration index is calculated from the values of penetration, at 25°C, and the softening point determined.

The penetration index (I_p), according to CEN EN 12591 (2009), Annex A, is calculated from the following equation:

$$I_p = (20 \times t_{R\&B} + 500 \times lgP - 1952)/(t_{R\&B} - 50 \times lgP + 120),$$

where lgP is the logarithm (base 10) of penetration at 25°C and $t_{R\&B}$ is the softening point (°C).

The result is rounded to the nearest 0.1 unit.

The above equation, developed by Pfeiffer and Van Doormaal (1936), is based on the following hypotheses:

a. The penetration value of the bitumen at the temperature of the softening point is equal to 800 dmm.
b. When the logarithm (base 10) of penetration (P) is plotted against temperature (T), a straight line is obtained, for example, $\lg P = A \times T + B$, the slope (A) of which is determined by the relation

$$A = \lfloor (20 - I_p)/(10 + I_p) \rfloor \times (1/50).$$

Research work carried out by Lefebvre (1970) and Heukelom (1973) has demonstrated that the hypothesis of the penetration of all graded bitumens at a temperature equal to the softening point being equal to 800 dmm is not absolutely valid. For certain bitumens, especially for the hard bitumens with high softening point (>65°C) and high penetration index (+3.6), as well as for bitumens with high paraffin content (>2%), the deviation is significant. Similar results were also found at a research conducted on paving grade bitumens by Nikolaides (1988). In these cases, it is recommended to determine the penetration at two different temperatures and then determine the penetration index using the following general equation:

$$(20 - I_p)/(10 + I_p) = 50(\lg P_1 - \lg P_2)/(T_1 - T_2),$$

where $\lg P_1$ and $\lg P_2$ are the logarithm (base 10) of penetration at temperatures T_1 and T_2, respectively.

The limiting theoretical values of the penetration index are −10 for bitumens with very high susceptibility to temperature variations, up to +20 for bitumens almost independent of temperature variations. In practice, penetration index varies between −3 and +7 for paving grade bitumens and oxidised bitumens. The smaller the penetration index, the more sensitive the bitumen is to temperature variations.

It is noted that penetration index value equal to zero $(I_p = 0)$ is attributed to a bitumen with a penetration of 200 dmm at 25°C and a softening point of 40°C.

4.5 DUCTILITY TEST

The ductility test indirectly measures the tensile properties of the bituminous materials and may be used for specification requirements. Because most of the typically used paving bitumens meet the specification requirement at 25°C, its usefulness is questioned by many researchers.

During the ductility test, the bitumen specimen is pulled apart at a specified speed and temperature condition (50 mm/min, 25°C) until it ruptures or reaches the length limitations of the machine. The elongation length at rupture, measured in centimetres, is defined as the bitumen's ductility value. The test is performed on three briquette specimens and the average of the three values is determined. In the event of discrepancy of results, they should be within the acceptable range set by the specification (AASHTO T 51 2009; ASTM D 113 2007).

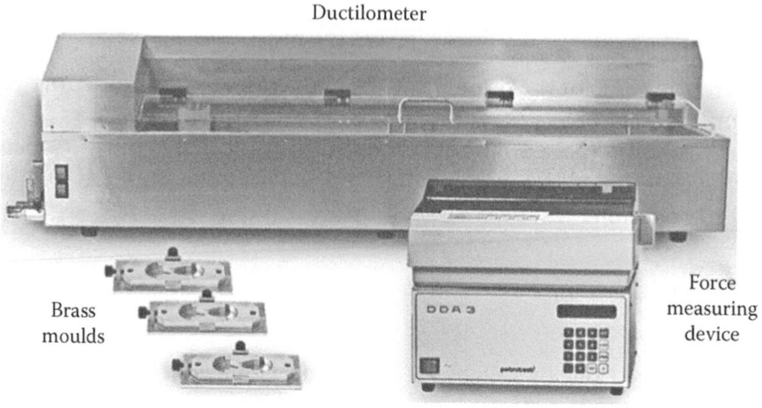

Figure 4.3 Ductility apparatus with capability to determine ductility force. (Courtesy of Anton Paar ProveTec GmbH.)

The specimens are prepared by pouring the heated bitumen into ductility brass moulds assembled on a brass or metallic plate. Before pouring the bitumen, the surface of the metallic plate is coated with a release agent (typically a mixture of glycerin and talc powder). This prevents specimen from sticking to the plate and to the detachable middle part of the mould (side pieces), after allowing the filled mould to cool to room temperature.

During preparation of the specimens, development of air bubbles should be avoided by allowing gradual cooling at room temperatures, and during cooling, protect the moulds from dust. After cooling, the specimens are placed in a water bath, at test temperature, for approximately half an hour and then removed for trimming the excess amount of bitumen may be present; to avoid excess bitumen, the moulds should be filled to the correct level. The trimmed specimens are placed in the apparatus' water bath for approximately 90 min.

During the testing, the water bath temperature must be kept constant and any water oscillations or undulations resulting in early fracture of the created asphalt 'thread' must be avoided. A typical ductility apparatus, known as a ductilometer, and moulds are shown in Figure 4.3.

It should be pointed out that the same apparatus is used for the elastic recovery test. For details, see Section 4.6.

4.6 FORCE DUCTILITY TEST

The force ductility test is performed on bituminous binders, in particular those of polymer-modified bitumens, for the determination of the conventional energy of bituminous binders from tensile characteristics.

The force ductility test is carried out in accordance to CEN EN 13589 (2008), using the ductilometer apparatus, which is equipped with an additional device capable of measuring the tensile force within the range of 1 to 300 N to an accuracy of ± 0.1 N (see Figure 4.3).

The specimens are prepared as in the ductility test and the traction rate is as in the ductility test, 50 ± 2.5 mm/min.

The test temperature is usually $5.0°C \pm 0.5°C$. However, in the case of soft bituminous binders, the test may be performed at a lower temperature ($0.0°C \pm 0.5°C$), whereas for hard polymer-modified bitumens, the test should be performed at $10.0°C \pm 0.5°C$ or even at $15.0°C \pm 0.5°C$ (CEN EN 13589 2008).

After the bitumen specimen is placed in the apparatus and the testing temperature is obtained, they are stretched to an elongation of 1333% (400 mm). If the specimen breaks prior to the desired elongation, the test is repeated. In case the second specimen also breaks prior to the desired elongation, the test is repeated by increasing the temperature in steps of 5°C, until the test is complete without brittle break.

For each test specimen, the deformation energy and finally the conventional energy are accomplished from the computerised data of coupled force/elongation.

The deformation energy, E_i, in joules (J), is the energy supplied by test pieces, until displacement, i, of the moving element is achieved.

The conventional energy, E'_s (in J/cm^2), is the quotient of deformation energy, E_i (in joules), and the initial cross section of the test pieces (in square centimetres).

For the force ductility test, the conventional energy (average of three specimens) corresponds to two elongation points (0.2 and 0.4 m). Details for calculating the deformation energy and the conventional energy, after performing the force ductility test in accordance to CEN EN 13589 (2008), are given in CEN EN 13703 (2003).

The force ductility test is also described in AASHTO T 300 (2011).

4.7 ELASTIC RECOVERY TEST

The elastic recovery test is used for the determination of the elastic recovery of the bituminous binders in a ductilometer at a given temperature. It is primarily applicable to modified bitumens with thermoplastic elastomers. However, it can also be used with conventional bitumen or other bituminous binders that generate only small recovery.

The test is carried out in accordance to CEN EN 13398 (2010) and ASTM D 6084 (2006), or AASHTO T 301 (2013), by using the same apparatus, mould and procedure as in the ductility test. The testing temperature is usually 25°C. The only difference is that the bituminous specimen is not stretched until rupture but only up to a predetermined elongation (200 mm by CEN EN 13398 2010).

The bitumen thread thus produced is cut in the middle to obtain two halves of thread. After a predetermined time for recovery has elapsed (30 min by CEN EN 13398 2010), the shortening of the half threads is measured (by a ruler) and expressed as the percentage of the elongation length.

The elastic recovery (R_E) is specified by the following relation:

$$R_E = (d/200) \times 100,$$

where d is the distance between the two shrinked parts of bitumen (mm) and R_E is the elastic recovery (%) rounded to full per cent.

The test is performed on two specimens, and their arithmetic mean, rounded up to 1%, is taken as the representative elastic recovery value of the bituminous binder. The difference between the two measurements should be less than 5% in absolute value. Otherwise, a third specimen is tested and the representative elastic recovery is the arithmetic mean of the two values that differ the least. If the third value differs by more than 5% in absolute value, the test is repeated with two new samples.

With regard to ASTM D 6084 (2006), the test may be carried out using two testing procedures, which differ between then on the length of elongation and elapsed time. Procedure A requires a 100 mm elongation length and 60 min elapsed time before taking measurements. Procedure B requires a 200 mm elongation length (maintaining the specimen in this stretched position for 5 min) and 60 min elapsed time.

The elastic recovery test, because of its similarity to the ductility test, is also sometimes called the modified ductility test.

4.8 VISCOSITY

Viscosity is a fundamental characteristic property of bitumen, since it determines the way it will behave in a specific temperature or at a range of temperatures. Viscosity is defined as a measure of fluid's resistance to flow. It describes the internal friction of a moving fluid. Broadly speaking, it could be said that viscosity is an expression of the coherence or fluidity that decreases when the temperature increases and increases when the temperature decreases.

In the fundamental way of measuring the viscosity, the gap between two parallel plates (one of which may move relative to the other) is filled with fluid, in this case, bitumen. The force that opposes the movement is developed solely because of the presence of bitumen. This force (F) is proportional to the surface (A) covered with fluid and the relative velocity of movement of one plate to the other (v) and inversely proportional to the distance (d) between the plates. If a constant (η) expressing the intermediate material is also introduced, which is the viscosity constant or, briefly, viscosity (η), then the following relation applies:

$$F = \eta \times A \times v/d \text{ or } \eta = (F \times d)/(A \times v).$$

By the abovementioned way of measuring the viscosity, that is, sliding plate, the absolute or otherwise known as *dynamic viscosity* (η_d or η) is measured. At this point, it should be mentioned that the relative movement of the two solid surfaces could also be rotary such as in a system of cylindrical container and rotating cylinder, or of a flat surface (plate) with a rotating cone, or of a flat surface (plate) with a rotationally oscillating circular plate. All these systems are basically the different types of viscosity measuring devices that measure dynamic or absolute viscosity (dynamic viscometers).

According to the International System of Units (SI), the dynamic viscosity measurement unit is the Pascal-second (Pa·s). This unit is the fundamental measurement unit of viscosity. Respectively, in the CGS system, the viscosity measurement unit is in dyne·s/cm (=1 g·s/cm). This unit is known as poise (P). The relation between Pa·s and poise is as follows: 1 Pa·s = 10 P. Sometimes, centipoise (cP) is also used, where 1 cP = 0.01 P. Centipoise may have been adopted due to the fact that water's viscosity at 30°C is 1 cP.

Viscosity can also be measured with viscometers where the movement or the developed force is due to gravity, by the fluid's own weight. Such a system is developed when the fluid flows in special glass tubes (capillary tubes). In this case, the *kinematic viscosity* (η_k or v) is measured, and the measurement unit is in square millimetres per second. This unit is also known as centistoke (cSt). However, when the fluid is 'forced' to flow under negative pressure (vacuum), then the dynamic viscosity (Pa·s) is measured.

Between the kinematic and the dynamic viscosity, the following relation applies:

Kinematic viscosity (η_k) = dynamic viscosity (η_d)/fluid density.

Finally, viscosity is often measured in relation to the time required for a specific mass of fluid to pass through an efflux orifice. The movement is, as in the previous case, due to the fluid's own weight. Such a system is developed when the fluid is placed in a special container and a specific volume is let to pass through an efflux orifice. This system is called cup viscometer and the unit of measurement, in this case, is in seconds. The viscosity measured in seconds may be converted to dynamic (η_d) or to kinematic viscosity (η_k) using the following relations:

$$\eta_d = \text{flow time} \times \text{fluid's density} \times C \ (\text{Pa·s})$$

$$\eta_k = \text{flow time} \times C \ (\text{mm}^2/\text{s}),$$

where C is the viscometer's conversion constant.

4.8.1 Types of viscometers

There is a great number of apparatuses (viscometers) for measuring viscosity of bitumen or bituminous binders. These viscometers could be classified as (a) rotational, (b) capillary, (c) cup or efflux and (d) sliding plate viscometers. The most common types of viscometers used for determining the viscosity of bitumen and bituminous binders are presented in Table 4.1. Table 4.1 also provides information on the type of viscometer used per bituminous binder.

The typical assemblies of some basic types of viscometers mentioned in Table 4.1 are provided in Figure 4.4.

Table 4.1 Classification of viscometers for bitumen and bituminous binders

Viscometer type	Viscosity (unit)	Main field of use per bituminous binder
Rotational		
Rotating spindle or coaxial (Figure 4.4a$_1$)	Dynamic (Pa·s)	Bitumen/modified bitumen, cut-backs and bitumen emulsions
Cone and plate (Figure 4.4a$_2$)		Bitumen/modified bitumen
Rotational paddle	Apparent (mPa·s)	Bitumen emulsions
Capillary (Figure 4.4b)		
Cannon–Fenske	Kinematic	Bitumens, cut-back and fluxed bitumens
Zeitfuchs Cross-arm	Kinematic	
Lantz-Zeitfuchs reverse flow	Kinematic	
BS/IP/RF U-tube reverse flow	Kinematic (mm^2/s)	
Cannon–Manning vacuum	Dynamic	Cut-backs and soft bitumens
Asphalt Institute vacuum	Dynamic	
Modified Koppers vacuum	Dynamic (Pa·s)	
Cup or efflux viscometers (Figure 4.4c$_1$ and c$_2$)		
Standard tar viscometer (STV)	Time of flow of bitumen (s)	Bitumen emulsions and cut-back or fluxed bitumens
Saybolt Furol		
Engler		
Redwood I and II		
Sliding plate		
Sliding plate viscometer (Figure 4.4d)	Dynamic (Pa·s)	Bitumen/modified bitumen
Oscillatory plate		
Dynamic shear rheometer (DSR) (Figure 4.7)	Dynamic (Pa.s)	Bitumen/modified bitumen

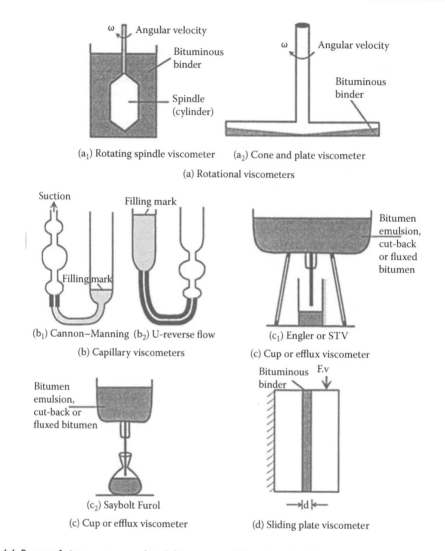

Figure 4.4 Range of viscometers used with bitumen and bituminous binders.

4.8.2 Viscosity tests by rotational viscometers

4.8.2.1 Rotating spindle viscometer

This rotating spindle viscometer is used for the determination of dynamic viscosity of a variety of bituminous binders: modified and unmodified bituminous binders, bituminous emulsions, cut-back and fluxed bituminous binders by means of rotating spindle (coaxial viscometer) viscometer, at typical test conditions (temperatures and rate of shear).

The principle of the test method is that the torque applied to a spindle (e.g. a cylinder), which is rotating in a special sample container containing the test sample, measures the relative resistance of the spindle to rotation and provides a measure of the dynamic viscosity of the sample. It may be necessary to apply a form factor to yield the actual dynamic viscosity at the test temperature.

The typical test temperatures for unmodified or modified bitumens range from 90°C to 180°C. The test temperatures usually used when unmodified bitumens are tested are 90°C,

105°C and 135°C, and when polymer-modified bitumens are tested, the test temperatures are 135°C, 150°C and 165°C.

The typical test temperature for bituminous emulsions is 40°C, and for cut-back and fluxed bituminous binders, it is 60°C. However, other temperatures may also be used, such as 90°C for cut-back and fluxed bituminous binders.

The rotational viscometer test is specified by ASTM D 6373 (2007) as a standard test for performance-graded asphalt binders and is carried out at 135°C.

In accordance with CEN EN 13302 (2010), for unmodified or modified bitumens, a small volume of heated sample (specified for the spindle to be used) is placed in the sample cylindrical container, which is then placed in the temperature-controlled device (environmental chamber). The sample together with the appropriate size spindle is left for a certain period to reach uniform testing temperature. Figure 4.5 shows a rotational viscometer.

Upon reaching the required test temperature, the spindle starts to rotate at a speed such that the desired shear rate is achieved with a precision of ±10%. Readings of torque, viscosity and shear rate are taken after the shear rate is stabilised for a period of 60 ± 5 s.

For bitumen emulsions, to break storage-induced thixotropy effects, the spindle rotates at a speed such that a shear rate of 10 ± 2 s^{-1} is achieved and is then left for 60 ± 5 s. After this period, the spindle is set to rotate at a speed such that a shear rate of 2.0 ± 0.2 s^{-1} is achieved. Readings are taken after the shear rate is stabilised for a period of 30 ± 3 s.

Similarly, for cut-back and fluxed bituminous binders to break possible storage-induced thixotropy effects, the spindle rotates initially at a speed such that a shear rate of 50 ± 5 s^{-1} is achieved and is then left for 60 ± 5 s. After this period, the spindle is set to rotate at a speed such that the shear rate of 2 s^{-1} is achieved. Readings are taken, as for bitumen emulsions, after the shear rate is stabilised for a period of 30 ± 3 s.

The dynamic viscosity is expressed in Pa·s or in millipascal-seconds (mPa·s) and is the mean of the two independent measurements, provided that the values do not differ by more than 10%. In case the two viscosity values differ by more than 10%, the individual values and their mean are reported.

By changing the size of the spindles, the rotational viscometer can measure the viscosity of any fluid substance. Additionally, due to the fact that the apparatus has the ability to fluctuate shear rate, the identification of a Newtonian (a fluid having a viscosity that is independent of the shear rate) or non-Newtonian fluid is also possible. Additionally, in the case

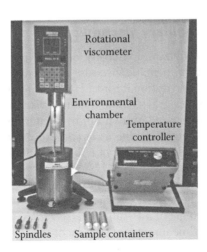

Figure 4.5 Rotational (coaxial) viscometer (Brookfield-type AUTh laboratory).

of a non-Newtonian fluid, the pseudoplastic or plastic behaviour, or even the thixotropic or rheopectic behaviour, can be detected.

A detailed description of the dynamic viscosity test by rotational viscometer can be found in CEN EN 13302 (2010), ASTM D 2196 (2010), ASTM D 4402 (2013) or AASHTO T 316 (2013).

4.8.2.2 Cone and plate viscometer

The cone and plate viscometer measures the dynamic viscosity of modified bituminous binders. Although the test method has been developed for modified bituminous binders, it may also be suitable for other bituminous binders.

According to CEN EN 13702 (2010), a small sample of bitumen is placed on a plate, a cone is pressed onto the sample, any surplus sample is removed and the system is brought to the test temperature. Then, a stress is applied to the sample by rotation and the torque is measured at the applied shear rate. The dynamic viscosity (η) (in Pa·s), which is the ratio of shear stress (τ) (in Pa) to shear rate (γ) (in s^{-1}), is calculated automatically by the apparatus (viscometer) used.

More information regarding the test can be found in CEN EN 13702 (2010).

This is similar to the test described in ASTM D 4287 (2010), which is used to determine the viscosity under high shear conditions, comparable to those encountered during spraying, brushing and so on. It is stated that the high shear cone and plate viscometer test method is suitable for paints and varnishes, whether they are Newtonian in behaviour or not, than for bituminous binders.

4.8.2.3 Rotational paddle viscometer

The rotational paddle viscometer measures the apparent viscosity of bitumen emulsions with viscosities between 30 and 1500 mPa·s (centipoises) at 50°C.

The apparatus consists of a paddle that rotates at 100 rpm and measures viscosity in centipoise. The preset temperature and rotational speed allow for an automated and consistent determination of an emulsified asphalt viscosity within a short time.

The rotational paddle viscometer is used when, rarely, the apparent viscosity of a bitumen emulsion needs to be determined. A detailed description of the viscosity test by rotational paddle viscometer can be found in ASTM D 7226 (2011).

4.8.3 Viscosity test by capillary viscometers

The kinematic as well as the dynamic viscosity of the bituminous binders can be measured by capillary viscometers. Kinematic viscosity is measured with capillary viscometers in which the flow is caused by gravitational force. Hence, kinematic viscosity is a measure of a liquid's resistance to flow under gravity. The dynamic viscosity is measured with vacuum viscometers, where the flow is assisted (drawn up) by vacuum pressure (40 kPa).

The *capillary viscometers* determine the *kinematic viscosity* of bituminous binders at 135°C or at 60°C (soft bitumen) with a range of 6 to 300,000 mm²/s (CEN EN 12595 2007) or up to 100,000 mm²/s (ASTM 2710).

The most common capillary viscometers used for the determination of kinematic viscosity are those listed in Table 4.1. The selected viscometer should give an efflux time greater than 60 s. Figure 4.4b₂ shows the shape of a U-tube reverse-type viscometer.

By performing the test, the time for a fixed volume of the liquid to flow through the capillary of a calibrated glass capillary viscometer under an accurately reproducible head and at a closely controlled temperature is determined, called efflux time. The kinematic viscosity (v) (in mm/s^2) is calculated by multiplying the efflux time (t) (in seconds) by the viscometer calibration factor (C) (in mm/s^2).

A detailed description of the test for the determination of the kinematic viscosity by capillary viscometers is given in CEN EN 12595 (2007), ASTM D 2170 (2010) or AASHTO T 201 (2010).

The *vacuum capillary viscometers* determine the *dynamic viscosity* of bituminous binders at 60°C with a range of 0.0036 to more than 580,000 Pa·s (CEN EN 12596 2007) or up to 20,000 Pa·s (ASTM D 2171 2010). Dynamic viscosity is a measure of the resistance to the flow of a liquid and is commonly called the viscosity of the liquid.

The most common capillary viscometers used for the determination of dynamic viscosity are those listed in Table 4.1. Figure 4.4b$_1$ shows the shape of the Cannon–Manning viscometer.

The purpose of the test is to determine the time for a fixed volume of the liquid to be drawn up through a capillary tube by means of a vacuum, under closely controlled conditions of vacuum and temperature.

The dynamic viscosity (η) (in Pa·s) is calculated by multiplying the efflux time (t) (in seconds) by the viscometer calibration factor (K) (in Pa).

A detailed description of the test for the determination of the dynamic viscosity test by vacuum capillary viscometers is given in CEN EN 12596 (2007), ASTM D 2171 (2010) or AASHTO T 202 (2010).

4.8.4 Viscosity test by efflux or cup viscometers

The efflux or cup viscometers are used to determine the viscosity of bitumen emulsions and cut-back or fluxed bituminous binders.

During the test, a viscometer cup is filled with a certain quantity of liquid bituminous binder and is left for a sufficient time until reaching the required temperature (usually 40°C, 50°C or 25°C). Then, the bituminous binder is effluxed through an orifice of various diameters (typically 2, 4 or 10 mm) and length (depending on the viscometer), until a required volume (50, 60, 100 mm or other depending on the viscometer) is collected in a graduated flask or cylinder (receiver) and the efflux time (t) is recorded in seconds.

The efflux time, indirectly, is a measure of viscosity. However, the efflux time (t) may be converted to dynamic viscosity (η) if the density of the liquid (ρ) and the constant of the cup viscometer (C) were known ($\eta = t \times \rho \times C$). Similarly, the kinematic viscosity (v) may be determined, if the constant of the cup viscometer (C) was known ($v = t \times C$).

There are various types of efflux or cup viscometers, two of which are shown in Table 4.1. The standard tar viscometer type (STV), also called efflux viscometer, is recommended to be used for bituminous emulsions (CEN EN 12846-1 2011) or for cut-back and fluxed bituminous binders (CEN EN 12846-2 2011). An efflux viscometer (STV) is shown in Figure 4.6.

The Saybolt Furol viscometer is mostly used in the United States and the testing procedure is described in ASTM D 88 (2013), ASTM D 7496 (2011) (for emulsified asphalts), ASTM E 102 (2009) (for bituminous materials at high temperatures) or AASHTO T 72 (2010).

The testing procedure when the Engler viscometer is used is described in ASTM D 1665 (2009).

The Redwood viscometer is considered to be the first efflux viscometer developed, designed in the late 1800s by Redwood, and, after some modification, is still in use. The test procedure may be carried out according to CEN EN 16345 (2012).

Figure 4.6 Efflux viscometer (STV). (Courtesy of Controls Srl.)

4.8.5 Viscosity tests by shear plate viscometers

4.8.5.1 Sliding plate viscometer

The sliding plate viscometer, also known as sliding plate micro-viscometer (Figure 4.4a), measures absolute or dynamic viscosity. The apparatus comprises a loading system that applies a shear stress and a recording system of flow as a function of time. The bitumen sample is placed between two plates so as to create a very thin film of 5–50 μm. The apparatus can measure the viscosity only in the range of 10^5 to 10^9 Pa·s; thus, it is not suitable for low-viscosity measurements.

Today, the use of the sliding plate viscometer is very limited and mainly for comparative studies with other viscometers. For further information, see Griffin et al. (1957) and Shell Bitumen (2003).

4.8.5.2 Dynamic shear rheometer

Although the dynamic shear rheometer (DSR) can measure dynamic viscosity, its main use is to determine the viscous and elastic behaviour of bituminous binders at medium to high temperatures, particularly to determine the complex shear modulus (G^*) and the phase angle (δ) of bituminous binders when tested in dynamic (oscillatory) shear, using parallel plate geometry. Details for determining viscosity with DSR can be found in Tredrea (2007).

All types of bituminous binders can be tested with the DSR, such as unaged, aged and recovered bituminous binders, cut-backs and bituminous binders stabilised from emulsions.

In the DSR, the sliding plate is circular and oscillates on the horizontal plane at a preselected frequency or at a range of frequencies. The typical range of frequencies used is 0.1 to 10 Hz (0.62 to 62.83 rad/s). When testing bituminous binders for compliance to ASTM D 6373 (2007), the frequency used is 10 rad/s (1.59 Hz).

Figure 4.7 shows a schematic representation of the DSR. The centre line of the upper plate, represented by point A in Figure 4.7, rotationally oscillates between points B and C.

The test may be carried out under strain control conditions or under stress control conditions. ATSM D 7175 (2008) proposes target strain or stress values, depending on the bitumen tested (original, RTFO residue or pressurised ageing vessel [PAV] residue).

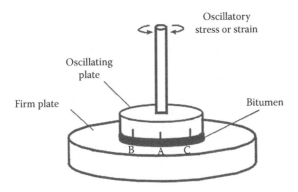

Figure 4.7 Schematic representation of the DSR.

The testing temperature (–*s*) is selected within the typical range 25°C to 85°C. When a bituminous binder is tested for compliance to ASTM D 6373 (2007), the temperature is selected from an appropriate table and is within the range 4°C to 88°C.

The size of the sample tested is very small, 25 or 8 mm in diameter with a typical thickness (gap between plates) of 1 or 2 mm. Special preparation of the samples is required which is described in the specification adopted.

The complex shear modulus, G^*, measured by the DSR is defined as the ratio calculated by dividing the absolute value of the peak-to-peak shear stress, τ, by the absolute value of the peak-to-peak shear strain, γ. The complex shear modulus is expressed in kilopascals or pascals, up to three significant figures.

Similarly, phase angle, δ, is the phase difference between stress and strain in sinusoidal harmonic oscillation, expressed in degrees, to the nearest 0.1°. Figure 4.8 shows a schematic representation of the phase angle under stress control conditions.

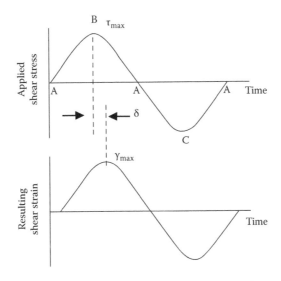

Figure 4.8 Schematic representation of phase angle (δ) between stress and strain in sinusoidal harmonic oscillation (stress control conditions).

The phase angle, δ, takes values from 0° to 90° (δ = 0° for elastic materials and δ = 90° for viscous materials).

After the determination of the above two variables, the ratio (G^*/sin δ) is also calculated, when compliance of the bituminous binder with ASTM D 6373 (2007) or AASHTO M 320 (2010) is required to be examined.

The complex shear modulus is an indicator of the stiffness or resistance of the bituminous binder to deformation under load. The complex shear modulus and the phase angle define the resistance to shear deformation of the bituminous binder in the linear viscoelastic region. Finally, the complex modulus and the phase angle are used to determine or calculate performance-related criteria in accordance to specifications.

Further information regarding the way the test is conducted, as well as the calculation of the above variables, is provided in CEN EN 14770 (2012), ASTM D 7175 (2008) or AASHTO T 315 (2012).

4.9 FRAASS BREAKING POINT

The Fraass breaking point provides a measure of brittleness of bitumen and bituminous binders at low (subzero) temperatures. The Fraass breaking point test was developed in 1937 (Fraass 1937) and began to be broadly used upon the advent of modified bitumens, since, by executing this test, the effect of the chemical additive on the behaviour of the modified bitumen could be determined. Nowadays, it has been adopted by many organisations and has been incorporated in European standards for specification requirements of paving grade bitumens (CEN EN 12591 2009) and of modified bitumens (CEN EN 14023 2010).

The Fraass breaking point, according to CEN EN 12593 (2007), is the temperature at which the bituminous binder of a specified and uniform thickness will break under defined loading conditions.

The Fraass apparatus consists of a bending apparatus that can apply repetitive bending stress by flexing the coated test plate (hence tensile strain on the film of bitumen), a glass cylindrical cooling apparatus, a graduated thermometer covering temperatures from –38°C to –30°C, with 0.5°C subdivision scale marks, and a sample preparation unit for creating the bitumen film on the steel plate–coated test plate. The temperature drop during application of bending stress is achieved by adding small quantities of solid carbon dioxide (dry ice) in alcohol, circulating the twin-wall glass tube. The Fraass apparatus is shown in Figure 4.9.

The test plates, made of tempered spring steel, have the following dimensions: 41 mm long, 20 mm wide and 0.15 mm thick. The quantity of bituminous binder placed on the test plate is $0.40 \times \rho_{25}$ (g), where ρ_{25} is the density of the binder at 25°C (in g/cm³), when the expected softening point of the binder is ≤100°C.

The coated plate is placed on the Fraass apparatus, at a starting temperature of at least 15°C above the expected breaking point, and is subject to repeated bending with gradual decrease of temperature at a rate of 1°C per minute. The temperature at which the first crack appears is reported as the Fraass breaking point. The average of the two results is taken as the final Fraass breaking value, provided that the two measured values do not differ by more than 2°C. Otherwise, two further tests are performed, and the average of the four values is taken as a representative value. In any case, the result is expressed in degrees Celsius rounded to the nearest whole number. An analytical description of the test is provided in CEN EN 12593 (2007).

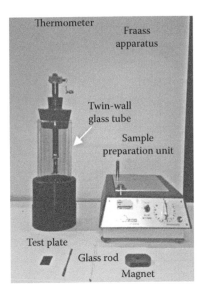

Figure 4.9 Fraass test apparatus.

The breaking point practically corresponds to an equi-stiffness temperature; a temperature at which the bitumen reaches critical stiffness that fractures. It has been shown that upon fracture, the bitumen has a stiffness value of 2.1×10^9, which approaches the maximum stiffness value of bitumen, 2.7×10^9 Pa (Thenoux et al. 1985).

The Fraass breaking point may be estimated by the penetration and the softening point using the Heukelom chart.

4.10 HEUKELOM CHART – BITUMEN TEST DATA CHART

The Heukelom chart, also known as bitumen test data chart (BTDC), was developed by Heukelom in the late 1960s and enables Fraass breaking point, penetration, softening point and viscosity to be plotted as a function of temperature in one chart (Heukelom 1969). The chart consists of a horizontal linear scale for temperature, and two logarithmic vertical scales for penetration and viscosity. The viscosity scale has been devised such that when plotting all test results, paving-graded bitumens give straight line relationship. Hence, the Heukelom chart provides the facility to predict the temperature-viscosity characteristics of paving grade bitumens by using only penetration and softening data. Other bitumens such as oxidised (blown) or waxy bitumens do not give a straight line relationship.

When data of paving grade bitumens with different grades manufactured from the same crude petroleum are plotted on the BTDC, the straight lines obtained will be almost parallel (see Figure 4.10). On the contrary, same-grade bitumens manufactured from different crude types will give non-parallel straight lines (Heukelom 1973).

The Heukelom chart also provides the facility to estimate the penetration index (PI) of paving grade bitumen. This is achieved by drawing a parallel line form the focal point, located on the chart, to the straight line obtained when plotting penetration and softening point. The penetration index is read from the intersection of the parallel line with the PI scale positioned on the chart.

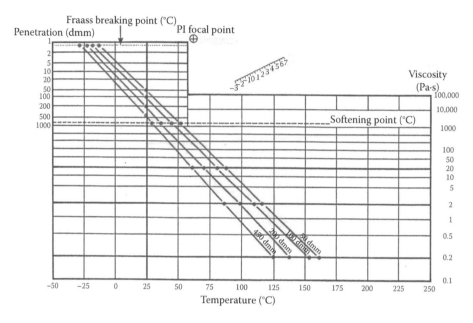

Figure 4.10 Heukelom chart and characteristic properties of penetration grade bitumens manufactured from one crude. (From Shell Bitumen, *The Shell Bitumen Handbook*. Surry: Shell Bitumen UK, 1990.)

4.11 RESISTANCE TO HARDENING TESTS

4.11.1 Rolling thin oven test (RTOT) and rotating flask test (RFT)

This test method is for measuring the combined effect of heat and air on a thin moving (rolling) film of bitumen or bituminous binder, simulating the hardening that a bituminous binder undergoes in an asphalt plant. By rolling the bituminous sample, the development of 'crust', which slows the evaporation of volatile substances, is avoided. The test is known as rolling thin film oven test (RTFOT) and is specified by CEN EN 12607-1 (2007), ASTM D 2872 (2012) or AASHTO T 240 (2013).

The oven has a tray rotating along the horizontal axis and a hot air blowing system over the specimens (see Figure 4.11).

The small quantity of bitumen, 35 ± 0.5 g, is poured in each special glass container, and when the oven attains the test temperature, 163°C ± 0.5°C, the samples are positioned in the vertical circular carriage. The carriage assembly starts to rotate at a rate of 15 revolutions per minute (rpm) by applying airflow at a rate of 4 l/min.

After rotating for 75 min, two samples are taken out, allowed to cool and weighed, in order to determine the change in mass. The rest of the samples are immediately poured in the same collecting vessel for penetration and softening testing and, if required, for determining dynamic viscosity (η).

The penetration, softening point and viscosity values after the RTFOT (hardened bitumen) are compared to the corresponding values before RTFOT.

A more detailed description of the test is provided in CEN EN 12607-1 (2007), ASTM D 2872 (2012) or AASHTO T 240 (2013).

Similar to the RTFOT method is the RFT method. The RFT method uses a rotating flask and 100 g bitumen sample. The test temperature is 165°C, the duration of rotation is

Figure 4.11 Oven for the RTFOT method. (Courtesy of Cooper Research Technology Ltd.)

150 min, the rotation rate is 20 rpm and the flow rate of air supply is 0.5 l/min. More details are given in CEN EN 12607-3 (2007).

4.11.2 Thin film oven test (TFOT) method

The TFOT method is similar to the RTFOT and measures the combined effects of heat and air on a film of bitumen or bituminous binder, simulating the hardening that a bituminous binder undergoes during mixing in an asphalt mixing plant.

The differences between the TFOT and the RTFOT method are the type of oven used (see Figure 4.12), the quantity of the bitumen sample, the type of containers, the duration of rotation and the absence of applying airflow on the samples.

The quantity of bitumen, 50 ± 0.5 g, is placed on a stainless steel or aluminium cylindrical pan of 140 mm diameter and 9.5 mm wall height, thus forming a film of approximately 3.2 mm thickness. The oven is ventilated and possesses a rotating metallic tray (of minimum 250 mm diameter) on the vertical axis, on which the bitumen specimens are positioned.

After 5 h, during which the specimens are constantly rotating at $163°C \pm 1°C$, the weight loss, the penetration, softening point and viscosity values of the hardened bitumen are

Figure 4.12 Oven for the TFOT method.

measured. The penetration, softening point and viscosity values after the TFOT test are compared to the initial values (before hardening). Analytical description of the test is provided in CEN EN 12607-2 (2007), ASTM D 1754 (2009) or AASHTO T 179 (2009).

4.11.3 Accelerated long-term ageing/conditioning by the rotating cylinder method (RCAT)

The accelerated ageing/conditioning test procedure involves rotating cylinder ageing (RCA), that is, binder ageing at moderate temperatures in a large cylinder rotating in an oven under oxygen flow conditions. Prior to long-term ageing with this method, samples are prepared under the condition they would be applied to the road.

The test method is applicable to bitumen, modified binders and bituminous mastics (homogenous mixture of filler and bituminous binder) and stabilised (recovered) bitumen from bituminous emulsions, cut-back or fluxed bitumen.

According to CEN EN 15323 (2007), the bituminous binder is first preconditioned as necessary to simulate the condition to which it would be applied to the road. Preconditioning of the sample is carried out by RTOT or by TFOT. Preconditioning may also be carried out by the procedure RCAT163 described in Annex A of CEN EN 15323 (2007).

After preconditioning, a sample of 525 to 550 g is poured into a preheated stainless steel testing cylinder and the cylinder is placed in the drive mechanism of the ageing system (see Figure 4.13). The cylinder containing the sample is left for 60 ± 5 min without rotation or inflow of oxygen, to enable the binder to reach the test temperature of 90°C.

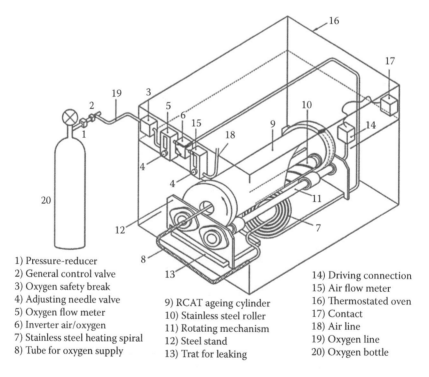

1) Pressure-reducer
2) General control valve
3) Oxygen safety break
4) Adjusting needle valve
5) Oxygen flow meter
6) Inverter air/oxygen
7) Stainless steel heating spiral
8) Tube for oxygen supply

9) RCAT ageing cylinder
10) Stainless steel roller
11) Rotating mechanism
12) Steel stand
13) Trat for leaking

14) Driving connection
15) Air flow meter
16) Thermostated oven
17) Contact
18) Air line
19) Oxygen line
20) Oxygen bottle

Figure 4.13 Schematic presentation of RCAT long-term ageing system. (Reproduced from CEN EN 15323, *Bitumen and bituminous binders – Accelerated long-term ageing/conditioning by the rotating cylinder method (RCAT)*, Brussels: CEN, 2007. With permission [© CEN].)

Then, the cylinder starts to rotate (1 rpm) and oxygen is fed at a flow rate of 4.5 ± 0.5 l/h (adjust the pressure gauge of the gas bottle that supplies the oxygen to 0.1 MPa, 1 bar).

After predetermined exposure times, typically 17 and 65 h, a portion of 25 to 30 g is taken from the sample to monitor the ageing process. The total mass of removed samples should not exceed 120 g.

The remaining sample is left for 140 h (±15 min) at 90°C and with oxygen supply. At ageing time completion (140 h ± 15 min), the heating and oxygen flow stops and the sample is emptied from the testing cylinder.

To empty the open testing cylinder, turn it upside down and place it with the opening down on a 0.5 l metal box against the inner wall of an oven preheated to 160°C ± 5°C. Allow it to stand for 30 to 35 min for penetration grade bitumen and for residues from emulsion or cut-back and for 40 to 45 min for binders containing polymers or for rubber-bitumen.

The recovered bituminous binder has undergone long-term ageing/conditioning and is ready for further testing.

Analytical description of the test is provided in CEN EN 15323 (2007).

4.12 FLASH AND FIRE POINT – CLEVELAND OPEN CUP METHOD

The flash and fire points of petroleum products such as bitumens and bituminous binders, having an open cup flash point above 79°C, are determined by the Cleveland open cup apparatus. For fuel oils, the most common test method used is the closed cup procedure that uses the Pensky–Martens apparatus (see ISO 2719 2002 or ASTM D 93 2013).

The flash point is determined as the lowest temperature at which application of test flame causes the vapour of the test portion to ignite and the flame to propagate across the surface of the liquid under the specified conditions of the test.

The fire point is determined as the lowest temperature at which application of a test flame causes the vapour of the test portion to ignite and sustain burning for a minimum of 5 s under the specified conditions of the test.

The flash and fire points are useful for safety and security reasons for the avoidance of accidents in case of overheating the bitumen.

According to CEN EN ISO 2592 (2001), a quantity of bitumen that has filled the test cup (Cleveland cup) is initially heated at an intense rate (14–17°C/min). When the temperature rises to approximately 56°C below the expected flash point, decrease the heat so that the rate of temperature rise for the last (23 ± 5)°C before the expected flash point is 5–6°C/min. At this heating stage, a test flame passes periodically, every 2°C, over the centre of the cup. The temperature at which flash of the vapours appears is defined as the flash point. The fire point is determined afterwards using the same device after continuation of heating.

The flash and fire temperatures calculated under ambient barometric pressure are corrected for normal atmospheric pressure, 101.3 kPa, using a correction equation provided by the specification. A detailed description of the test can be found in CEN EN ISO 2592 (2001), ASTM D 92 (2012) or AASHTO T 48 (2010).

The flash point of the paving grade bitumens as determined by the Cleveland open cup method may vary from ≥165°C to ≥245°C; for cut-backs of flashed bituminous binders, the flash point may be >160°C.

However, for cut-backs and fluxed bituminous binders, the flash point is usually determined by the Pensky–Martens closed-cup apparatus (ISO 2719 2002 or ASTM D 93 2013), by the tag open-cup apparatus (AASHTO T 79 2012; ASTM D 3143 2008) or by the Abel closed-cup apparatus (CEN EN ISO 13736 2013).

When the Pensky–Martens apparatus is used, the expected flash point of cut-backs and fluxed bituminous binders should be >60°C; when the tag open-cup apparatus is used, the expected flash point may vary from >38°C to >66°C; and when the Abel closed-cup apparatus is used, the flash point may vary from 23°C or less to >45°C.

4.13 SOLUBILITY TEST

The solubility test is performed to determine the degree of solubility of bituminous binders, having little or no organic impurities or rock salts (salts, free carbon etc.) other than recovered bituminous binders from asphalts, in a specific solvent. Toluene is used as the solvent for reference tests. It is noted that the bituminous binders will have different solubilities if different solvents are used.

According to CEN EN 12592 (2007), a mass of 2 g of bitumen is dissolved in 100 ml of solvent and then the solution is filtered. The amount of substances withheld by the filter is washed, dried, weighed and expressed as a percentage of the initial mass of bitumen. This percentage deducted from 100 determines the bitumen solubility. The degree of solubility should be above a certain specified value, normally 99%. A detailed description of the test is provided by CEN EN 12592 (2007), ASTM D 2042 (2009) or AASHTO T 44 (2013).

4.14 TENSILE TEST

The tensile test is an alternative to the force ductility test for the determination of the tensile properties and cohesion of bituminous binders, particularly polymer-modified binders.

The test is performed according to CEN EN 13587 (2010) at a temperature of 5°C using the ductilometer apparatus as in the force ductility test but at a constant traction rate of 100 mm/min.

For the tensile test, the conventional energy (average of three specimens) corresponds to an elongation of 0.2 m (400%); for example, $E'_s = E'_{0.2}$.

Details for calculating the deformation energy and the conventional energy, after performing the tensile test in accordance to CEN EN 13587 (2010), are given in CEN EN 13703 (2003).

4.15 COHESION WITH PENDULUM TEST

The cohesion with pendulum test is a method for measuring the cohesion of bituminous binders at temperatures in the range of –10°C to 80°C and for expressing the relationship between cohesion and temperature.

This test method is applicable for graded bitumen, modified bitumen and fluxed bitumen and is used only when the binders are going to be used in surface dressings.

In the case of fluxed bitumen, the test is performed on the binder containing fluxant or on the binder from which the solvent has been removed. For bitumen emulsions, the test is carried out on the residual binder obtained after recovery.

The cohesion in this test is determined as the energy per unit area (J/cm^2) required to fully detach a cube from the support, with the previously bonded faces of the cube and support remaining fully covered by binder.

The test is executed in accordance to CEN EN 13588 (2008), where more information can be found.

4.16 STORAGE STABILITY TEST

This test method measures the storage stability of the modified bitumen at high temperatures. The polymer-modified bitumens in particular are known to display phase separation under certain conditions, mainly storage.

According to CEN EN 13399 (2010), a homogeneous specimen of modified bitumen is placed in a special thin unvarnished aluminium tube and positioned vertically in an oven of 180°C ± 5°C for 3 days. The specimen is then allowed to cool and is cut into three equal parts. The material contained in the two tube ends (top and bottom) is further analysed for potential changes of its characteristic properties.

Usually, after the 3-day storage, the penetration and softening point are determined and the values obtained are compared to those before storage. The differences in values expressed as percentages or units must comply with limiting values required.

Further information is provided in CEN EN 13399 (2010).

4.17 MINERAL MATTER OR ASH IN ASPHALT MATERIALS

This test is conducted to determine the inorganic residual percentage occurring after the bitumen combustion. During this test, a small quantity of bitumen or tar is burnt at very high temperatures until the ash is free of carbon. The remaining quantity of inorganic impurities is weighed and expressed as a percentage of the initial weight of the specimen.

This test is rarely used as the presence of inorganic impurities is not necessarily harmful to bitumen.

A detailed description of the test is provided in AASHTO T 111 (2011).

4.18 CAPILLARY-STOPPERED PYCNOMETER TEST FOR DETERMINATION OF DENSITY AND SPECIFIC GRAVITY OF BITUMEN

This test method determines the specific gravity and density of bituminous binders, apart from bitumen emulsions, at 25°C ± 0.2°C using capillary-stoppered pycnometers. The method may be performed at other temperatures, but when doing so, the density values of the water or other liquid used should be determined.

According to CEN EN 15326 (2009), a quantity of heated binder is poured in the pycnometer, filling approximately 3/4 of its volume. After the pycnometer containing the bituminous binder is allowed to cool in ambient temperature for at least 40 min, the test sample is weighed (weight A), together with the pycnometer's glass stopper. The remaining volume is then fully completed with tested liquid (boiled or de-ionised water for bitumens or isopropanol for cut-back or fluxed bitumen); the pycnometer is placed in a water bath for at least 30 min in order to reach the temperature of 25°C and is weighted (with the stopper) (weight B). After the pycnometer is emptied and thoroughly cleaned, it is weighed in air (weight C). Then, the pycnometer is filled completely with tested liquid and is weighed (weight D). All weight measurements include the stopper.

The specific gravity and the density of bituminous binder are calculated by the following equations:

$$SG_{25} \text{ or } d_{25} = (A - C)/[(D - C) - (B - A)] \text{ and}$$

$$\rho = [(A - C)/[(D - C) - (B - A)]] \times \rho_T,$$

where SG_{25} or d_{25} is the specific gravity of bituminous binder at 25°C, A is the mass (g) of the pycnometer partially filled with bituminous binder sample, B is the mass (g) of the pycnometer plus bituminous binder sample plus test liquid, C is the mass (g) of the pycnometer including weight of the stopper, D is the mass (g) of the pycnometer filled with test liquid (water or isopropanol), ρ is the density of bituminous binder (kg/m³) and ρ_T is the density of tested liquid at test temperature (water density at 25°C is 997.0 kg/m³ and isopropanol density at 25°C is 782.7 kg/m³).

The average of at least two determinations is expressed in three decimal places, for specific gravity, and to the nearest 1 kg/m³, for density.

A detailed description of the test is provided in CEN EN 15326 (2009), ASTM D 70 (2009) or AASHTO T 228 (2009).

4.19 DETERMINATION OF WATER IN BITUMEN BY DISTILLATION METHOD

The determination of water in bitumen, when required to be determined, is carried out by distillation method. During the test, a quantity of bitumen is dissolved with a solvent (usually xylene) and the solution is then distilled. At the distillation, the mass of water is collected in a graduated glass tube.

The volume of the collected water divided by the initial weight of the bitumen specimen, multiplied by hundred, expresses the water content percentage in the bitumen. A detailed description of the test is provided in ASTM D 95 (2013) or CEN EN 1428 (2012) (see also Section 4.20.6).

4.20 BITUMEN EMULSION TESTS

The basic tests conducted on bitumen emulsions as required by CEN EN 13808 (2013), ASTM D 2397 (2012) or AASHTO M 208 (2009) are described in the following paragraphs. For the list of tests on bitumen emulsions, see also Tables 3.13 and 3.15.

Apart from the tests performed on bitumen emulsions, tests are also performed on the residual bitumen contained in the emulsion. The tests on the bitumen residue (see Tables 3.14 and 3.15) have been described in the previous paragraphs of this chapter.

The residual bitumen for properties determination is extracted by distillation procedure. The azeotropic distillation (by means of a carrier vapour from a water-immiscible solvent-carrier liquid) is used only for the determination of the water content in bitumen emulsion.

4.20.1 Particle polarity test

This test is performed in order to determine the bitumen particle polarity of the bitumen emulsion and, hence, the type of emulsion (cationic or anionic).

During the test, two steel plates (electrodes) are dipped into the bitumen emulsion and a direct electric current of 8–10 mA is passed through the plates. After approximately 30 min or when the current drops to 2 mA, whichever is achieved first, the bitumen deposition on the electrodes is observed. If bitumen deposition appears on the negative electrode (cathode), the emulsion is cationic. Otherwise, the emulsion is anionic.

A detailed description of the test is provided in CEN EN 1430 (2009), ASTM D 244 (2009) and AASHTO T 59 (2013).

4.20.2 Breaking value of cationic bitumen emulsion – mineral filler method

This test determines the breaking value of the cationic bitumen emulsion.

Breaking value is a dimensionless number that corresponds to the amount of reference filler, in grams, needed to coagulate 100 g of bitumen emulsion.

According to CEN EN 13075-1 (2009), reference filler (Sikaisol filler) is added at a steady rate of 0.25–0.45 g/s to 100 ± 1 g of cationic emulsion and the emulsion with the filler is constantly stirred by spatula at a rate of 1 rpm. Stirring may be carried out semi-mechanically.

The mixture becomes thicker as the filler is added. The emulsion is considered broken when the mix comes off completely (or substantially) from the enamelled or stainless steel mixing dish.

The quantity of filler used, in grams, divided by the quantity of emulsion used, in grams, multiplied by 100, and expressed to the nearest integer, gives the emulsion breaking value.

This value is multiplied by a coefficient of 1.4 to be converted to breaking value with the Forshammer filler, which was the reference filler initially used for this test. The limiting values specified in CEN EN 13808 (2013) are based on the use of the Forshammer filler.

Further information for the test method can be found in CEN EN 13075-1 (2009).

4.20.3 Mixing stability with cement of bitumen emulsions

The test determines the mixing stability of bitumen emulsions with cement. It applies to overstabilised cationic bitumen emulsions and to slow-setting and overstabilised anionic bitumen emulsions.

Mixing stability with cement is the mass of coagulated material (bitumen and cement), which is produced when a bituminous emulsion is mixed with cement under the conditions of the test.

According to CEN EN 12848 (2009), a mass of 50 g of cement, without lumps (passed through a 0.16 mm sieve), is added to 100 ml of bitumen emulsion and the mixture is mixed for 1 min. Immediately after that, another 150 ml of water is added and the stirring continues for another 3 min.

The mixture is then poured to a 2 mm sieve and is washed off until the washings are clear. Then, the sieve is placed on a pan and is left for approximately 1 h in an oven at 110°C ± 5°C to dry.

The quantity of material (cement and bitumen) in grams withheld in the sieve, along with that which may have fallen on the pan during drying, expresses the mixing stability with cement, S_c. The value is rounded to the nearest 0.1 g.

Further information is provided by CEN EN 12848 (2009). Similar is the test described by ASTM D 6935 (2011).

4.20.4 Determination of fines mixing time of cationic bitumen emulsions

This test determines the fines mixing time of diluted cationic bituminous emulsions, under standardised conditions.

The fines mixing time is the time, in seconds, for the mixability of a mixture of mineral filler and bitumen emulsion without noticeable breaking effect under the conditions specified.

According to CEN EN 13075-2 (2009), under normal laboratory conditions (18°C to 25°C), a mass of 100 ± 0.5 g of emulsion is diluted with 50 ± 0.5 g of water. Then, using a stopwatch, a specific mass of filler, 150 ± 1 g, is poured at a rate of 10 g per 5 s and mixed by spatula, at a rate of 1 r/s, so that the entire filler mass is added within 75 s. The filler may be Sikaisol or any other with the same properties.

The stirring/mixing continues until the emulsion breaks. An emulsion breaking is considered to occur when the bituminous emulsion and filler mixture becomes pasty and forms lumps, which do not adhere to the walls of the pan. This, combined with a noticeable increase in stirring power, indicates the end of mixability.

When the breaking occurs, the elapsed time is recorded. The arithmetic mean of two results of mixing time, expressed to the nearest integer, is the fines mixing time of the emulsion under test.

In the case the bituminous emulsion does not break within 300 s, the procedure stops and the result is reported as >300 s.

Further information is provided in CEN EN 13075-2 (2009).

4.20.5 Determination of penetration power of bitumen emulsions

This test determines the penetration power of bituminous emulsions, through a reference filler, and is applicable to low-viscosity bitumen emulsions and is conducted at normal laboratory temperatures (18°C to 28°C).

Penetration power is the ability of a bitumen emulsion to penetrate into a reference filler.

According to CEN EN 12849 (2009), a certain mass of bitumen emulsion (10.0 ± 0.1 g) is poured on a reference filler and the time required to penetrate is recorded. The reference filler material mixture consists of 50.0 ± 0.1 g silica sand and 50.0 ± 0.1 g of silicon filler, the characteristic properties of which are determined in Annexes A and B of CEN EN 12849 (2009).

The quantity of reference filler material is placed in a glass tube of 41.5 ± 0.5 cm diameter and >11.5 cm length, which, at its one end, has a fused-on glass filter disc with pore size between 160 and 250 μm.

After pouring the bitumen emulsion, determine the time for the emulsion to completely penetrate the filler mixture, that is, when the structure of the filler at its upper surface can be clearly recognised.

If penetration of the filler mixture is not completed within 20 min, the test is stopped. The test is repeated with new emulsion and filler material quantities. If the two results differ for more than 3 min, a third test is conducted. The average of the two nearest values is taken as the final result.

Further information is provided by CEN EN 12849 (2009).

4.20.6 Determination of water content in bitumen emulsions – azeotropic distillation method

This test determines the water content in a bitumen emulsion by azeotropic distillation method. Azeotropic distillation refers to the technique of adding another component to generate a new, lower-boiling azeotrope that is heterogeneous (e.g. producing two immiscible liquid phases), for example, xylene and water. Azeotrope is a mixture of two or more liquids in such a way that its components cannot be altered by simple distillation. The azeotropic distillation method, for simplicity, is referred to as distillation method.

The water contained in a bitumen emulsion or in the graded bitumen, if such a case arises, is distilled by means of a carrier vapour from a water-immiscible solvent-carrier liquid.

A sample of bitumen emulsion is placed in a round bottomed flask of 500 ml so that after distillation to receive 15 ml to 25 ml of water. An approximate amount of 100 ml of solvent carrier is added together with a number of anti-bumping granules to avoid the creation of excessive foaming during heating. The solvent carrier is usually xylene and in certain cases toluene.

The round-bottomed flask is assembled in the distillation apparatus, which includes a receiver (trap) and a condenser. The distillation continues until no increase of the water volume in its receiver is observed. The percentage of water (w) contained in the bitumen emulsion is calculated by the following relation:

$$w = m_{\mathrm{W}}/m_{\mathrm{E}} \times 100,$$

where m_{W} is the mass of the water distilled from the test material (g), which is equal to the volume of water (ml) collected in the graduated receiver, and m_{E} is the mass of bitumen emulsion used (g).

The result is expressed to the nearest 0.1%. Further information is provided in CEN EN 1428 (2012).

It is noted that the percentage of bitumen residue in the emulsion, if desired, may be determined from the deduction [100 − water percentage (w)].

A similar method to the above distillation method is described in ASTM D 95 (2013).

4.20.7 Determination of bitumen residue by evaporation of bitumen emulsion

This test does not require a specific apparatus and is used by many laboratories for the quick determination of the bitumen content in the emulsion. The residue from the evaporation may be tested as required.

According to ASTM D 6934 (2008), a quantity of 50 ± 0.1 g of emulsion sample is placed in a pre-weighed beaker and is left in an oven of $163°C \pm 3°C$ for 2 h. After the end of the 2 h, the sample is stirred with a pre-weighed metallic or glass rod and all – beaker, sample and rod – are placed again in the oven again for one more hour.

After cooling the beaker, the sample and the rod are weighed. The percentage of bitumen residue is specified as a percentage to the initial emulsion weight. Report the average of three determinations to the nearest 0.1% as the percentage of bitumen residue by evaporation.

This test method for residue by evaporation tends to give a bitumen residue lower in penetration and ductility than the distillation test method. If the residue from evaporation fails to meet the requirements for properties specified for residue from distillation, tests shall be re-run using the distillation test method.

More information on the test method can be found in ASTM D 6934 (2008). A similar method is described in CEN EN 13074 (2011).

4.20.8 Determination of bitumen residue by moisture analyser

This test method involves a rapid quantitative determination of the residue in bitumen emulsion using a moisture analyser. It is applicable to all non-solvent-containing emulsion types, anionic, cationic, non-polymer-modified or polymer-modified bitumen emulsions.

The residue obtained from this test method may also be subjected to further rheological characterisation tests.

A sample of bitumen emulsion, minimum 1 g and up to 3 g, is placed in a moisture analyser equipped with a heating element and capable of running either isothermally or in a programmable temperature-gradient mode.

The instrument calculates the residue automatically at the end of the run. More details can be found in ASTM D 7404 (2012).

4.20.9 Determination of residual binder and oil distillate from bitumen emulsions by distillation

This test method is for the quantitative determination of residual binder and oil distillate in bitumen emulsions for specification acceptance. The method is used to obtain residue and oil distillate for further testing.

Residual binder is the residue from a bitumen emulsion after distillation of water and oil distillate.

Oil distillate is the hydrocarbon fraction that is distilled and collected in the graduated cylinder under conditions specified.

According to CEN EN 1431 (2009), a bitumen emulsion sample of approximately 200 g is placed in a pre-weighted aluminium alloy still or iron still (including lid, clamp, thermometers and gasket, if gasket is used). The lid allows two thermometers to be inserted through a stopper and an outlet of suitable diameter to enable a connection tube to be connected also through a stopper.

The thermometers are placed so that the end of the bulb of one is 6.5 ± 1.0 mm from the bottom of the still and the bulb of the other is 165 ± 2 mm from the bottom of the still.

The content is heated with the ring burner initially placed at a distance of 152 cm from the bottom of the still. A schematic representation of the distillation apparatus is provided in Figure 4.14.

The content is heated until the reading of the lower thermometer reaches 215°C. At this point, the ring burner is lowered to a position until the reading of the thermometer is 260°C $\pm$ 5°C. This temperature is maintained for 15 min.

After 15 min, the content in the still is allowed to cool and is weighed. The entire distillation process should last, approximately, 60 min. If the residual binder is not to be determined, the content in the still is poured immediately through a 300 μm pre-heated sieve to suitable moulds or containers for carrying out any required further tests.

If determination of residual binder is to be carried out, allow the still with its content to cool to room temperature. The determination of the residual binder (r), as mass percentage, after distillation is carried out by using the following equation:

$$r = B_m/A_m \times 100,$$

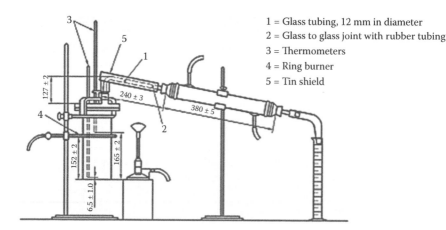

1 = Glass tubing, 12 mm in diameter
2 = Glass to glass joint with rubber tubing
3 = Thermometers
4 = Ring burner
5 = Tin shield

Figure 4.14 Apparatus for distillation test of bitumen emulsions. (Reprinted from ASTM D 6997, *Standard test method for distillation of emulsified asphalt*, West Conshohocken, Pennsylvania, US: ASTM International, 2012. With permission [© ASTM International].)

where A_m is the specimen mass of the bitumen emulsion (g) and B_m is the mass of the residue after distillation (g).

If the oil distillate (o) is also to be defined, by volume, the following equation is used:

$$o = D \times \rho/A_m \times 100,$$

where D is the volume of oil distillate in the graduated cylinder (ml), ρ is the bitumen emulsion density (g/ml) and A_m is the specimen mass of the bitumen emulsion (g).

The residual binder result is expressed as a percentage to the nearest 1%, while that of the oil distillate is expressed as a percentage at the nearest 0.1%.

The binder content of the emulsion by the distillation method is the sum of the mass percentage of the residual binder plus the mass percentage of oil distillate.

For the determination of the mass percentage of oil distillate, it is necessary to determine the density of the oils withheld by distillation. If this is not possible, the value of 0.850 may be used.

Further details on the test are provided in CEN EN 1431 (2009). Similar is the test method described in ASTM D 6997 (2012) or AASHTO T 59 (2013).

4.20.10 Determination of efflux time of bitumen emulsion by the efflux viscometer

This test determines the efflux time of a bitumen emulsion, using an efflux viscometer. The efflux time is the time of efflux of a given quantity of the emulsion through an orifice of a specified size at a specified temperature.

The test for bitumen emulsions is carried out in accordance with CEN EN 12846-1 (2011), when an STV efflux viscometer is used.

The Saybolt viscometer may also be used when testing bitumen emulsions and the test is carried out according to ASTM D 7496 (2011).

For additional information, see also Section 4.8.4.

4.20.11 Dynamic viscosity test

This test method determines the dynamic viscosity of bituminous emulsions by means of a rotating spindle viscometer.

The dynamic viscosity test is part of the framework for specifying requirements by EN 13308 and is performed in accordance with CEN EN 13302 (2010). A brief description of the test is given in Section 4.8.2.

The dynamic viscosity test determined by a rotating spindle viscometer (rotational viscometer test) is also covered by ASTM D 2196 (2010).

4.20.12 Determination of residue on sieving and storage stability by sieving

This test method determines the quantity of coarse particles of binder present in a bitumen emulsion by utilising sieving and, from that, storage stability.

Storage stability is defined as the ability of a bituminous emulsion not to form more coarse particles within a specified period (n-days).

The presence of coarse particles may affect the storage, pumping and handling of the emulsion and furthermore even the distribution of the bitumen within the aggregate mix.

When conducting the test in accordance with CEN EN 1429 (2013), approximately 1000 g of emulsion is filtered through a 0.5 mm sieve, after the latter has been wetted with sodium hydroxide (S_a) or hydrochloric acid (S_c) solution. The sieve is then washed with a solution of S_a or S_c and then with water and dried in an oven at 105°C. The residue retained on the 0.5 mm sieve is expressed as a percentage by mass of bitumen emulsion sieved through the 0.5 mm sieve, $R_{0.500}$, to the nearest 0.01%.

Part of the filtered emulsion through the 0.5 mm sieve (i.e. 50 g) is first diluted with 50 cm³ of the S_a or S_c solution and then filtered through a 0.160 sieve, which has wetted with the S_a or S_c solution. The sieve is then washed up to three times with a solution of S_a or S_c and then with water and dried in an oven at 105°C. The residue retained on the 0.160 mm sieve is expressed as a percentage by mass of bitumen emulsion sieved through the 0.160 mm sieve, $R_{0.160}$, to the nearest 0.01%.

In order to specify the storage stability after n-days, usually 7 days, a mass of 50 g of bitumen emulsion filtered through the 0.5 mm sieve is stored for n-days and is then filtered through a 0.5 mm sieve. The residue retained on the 0.5 mm sieve after n-days of storage is expressed as a percentage by mass of bitumen emulsion stored, $R_{n\text{-days}}$, to the nearest 0.01%.

In all cases, the tests are performed under normal laboratory ambient temperatures (between 18°C and 28°C). For routine production control, a single test is performed. Duplicate tests are required for referee purposes only.

For very viscous emulsions, the emulsion may be pre-heated and then test the samples at 60°C ± 5°C. Alternatively, the 50 g of diluted emulsion may be diluted further with either the S_a or the S_c solution, as appropriate.

Further details on the above tests are provided in CEN EN 1429 (2013).

The above test method is similar to that specified in ASTM D 6933 (2008). The main difference is that a 0.850 mm sieve instead of a 0.5 mm sieve is used. Additionally, the temperature at which the sieve test is performed is related to the viscosity of the bitumen emulsion. For those materials whose viscosity in Saybolt Furol seconds is 100 s or less at 25°C, the test is performed at ambient temperature. For those materials whose viscosity is greater than 100 s at 25°C and those whose viscosity is specified at 50°C, use a test temperature of 50°C ± 3°C.

4.20.13 Determination of settling tendency after *n*-days

The settling test is carried out to determine the settling tendency of bituminous emulsion stability during its storage.

Settling tendency is the difference in water content of the top layer and the bottom layer of a prescribed volume of sample after standing for a specified time at ambient temperature.

When conducting the test in accordance with CEN EN 12847 (2009), depending on the water content determination method, one or four 500 ml stoppered glass graduated cylinders are filled with emulsion and allowed to stand undisturbed for n-days, usually 7 days. The samples are tightly stoppered so that there is no water loss.

After the standing period of 7 days, a quantity of approximately 55 ml of emulsion is carefully drawn from the upper part of the cylinder. The quantity collected is placed in a beaker to determine the water content (A). Then, a quantity of approximately 390 ml is removed from the cylinder. The remaining quantity of emulsion, approximately 55 ml, is stirred and drained into a second beaker to obtain a second test portion for water content determination (B).

The water content is determined in accordance with CEN EN 1428 (2012) or CEN EN 1431 (2009) from the two emulsion specimens collected from the two different positions. The settling tendency, ST, of the test sample, expressed in mass percentage, is determined by the difference ($A - B$). The settling tendency is often also called long-term storage stability.

It is noted that negative values obtained for the settling tendency indicate that the bituminous phase rises to the surface. Further details for the above test are provided in CEN EN 12847 (2009).

Similar is the test described in ASTM D 6930 (2010). However, the testing procedure according to ASTM D 6930 (2010) uses the percentages of residue from the top and bottom sample. Additionally, the determination of the difference of the percentages of residue after 24 h is called storage stability (after 24 h) and the determination of the difference of the percentages of residue after 5 days is called storage stability (5 days).

4.20.14 Adhesivity of bituminous emulsions by water immersion test

This test determines the adhesion of a cationic bitumen emulsion coated onto aggregate when immersed in water. The adhesion is defined as the ability of a binder to wet the surface of an aggregate and to remain bonded over time. Adhesivity is the qualitative assessment of the measurement of the adhesion.

According to CEN EN 13614 (2011), two procedures are distinguished, one for emulsions with limited storage stability and one for emulsions that can be stored. In both procedures, the bitumen emulsion comes into contact with the washed reference aggregates, and after its breaking, under specific conditions, the coated aggregates are placed in a 400 ml beaker and covered with approximately 300 ml of water. After a certain period and under specified conditions, the surface coated with the film of binder is assessed visually.

The reference aggregate, as light in colour as possible, or aggregates from a specific job site pass through a 10 mm sieve and retained on a 6.3 mm sieve (or alternatively 11 and 8 mm). Each country should define petrographically its own reference aggregate.

4.20.14.1 With emulsion of limited storage stability (breaking index <120)

The test is carried out with 100 ± 2 g of aggregates and 150 ± 2 g of emulsion. The aggregates are poured into a 15 to 20 cm diameter enamel dish containing the emulsion and allowed for 60 ± 5 s, without stirring.

After 60 s, the excess amount of emulsion is removed; the aggregates are washed, placed in a 400 ml glass beaker and covered with approximately 300 ml at room temperature. Immediately afterwards, the surface coated with the film of binder is observed and assessed, using the grades 100 (all surface coated), 90 (more than approximately 90% of the surface is coated), 75 (approximately 75% to 90% of the surface is coated), 50 (approximately 50% to 75% of the surface is coated), <50 (less than 50% of the surface is coated) and 0 (the binder is separate from the aggregates).

If the grade is ≥ 90, then the beaker covered with a glass watch is put in a $60°C \pm 3°C$ oven for 20 ± 4 h and then the aggregate surface is re-assessed.

4.20.14.2 With emulsions that can be stored (breaking index >120)

In this case, the test is carried out with 200 ± 2 g of reference aggregates and an amount of emulsion corresponding to 10 ± 1 g of bitumen. The aggregates are poured into the enamel dish containing the emulsion. After good stirring, the mixture is spread on a glass watch and left for 20 ± 4 h in a ventilated oven of $60°C \pm 3°C$.

After the end of the 20 h period, the surface coated with the film of binder is assessed according to the grading scheme given above.

For emulsions showing more than 90% of surface coated for the first test, provide the result for the second part of the test. For emulsions that can be stored, report one grading.

Further details on the above test are provided in CEN EN 13614 (2011).

4.20.15 Coating ability and water resistance test

This test method determines the ability of a bitumen emulsion to coat an aggregate, to withstand a mixing action while remaining as a film on the aggregate and to resist the washing action of water after completion of the mixing. Its intention is to identify a suitable bitumen emulsion for mixing with coarse-graded calcareous aggregates. However, it can be applied to other aggregates also.

The test method is specified in ASTM D 244 (2009) or AASHTO T 59 (2013) and is similar to the adhesivity test described in the above paragraph. The main differences compared to the adhesivity test (CEN EN 13614 2011) are as follows:

a. The same testing procedure is followed regardless of storage stability of the bitumen emulsion.
b. Different quantities of reference aggregates and bitumen emulsion are used.
c. Calcium carbonate is added as dust to the reference aggregate.
d. Wet- and dry-coated aggregates, after spraying with tap water in both cases, are assessed for degree of coating.
e. Assessment is carried out using only three grades: good (for fully coated aggregates), fair (for coated area in excess of uncoated area) or poor (for uncoated area in excess of coated area).

A mass of 461 g of referenced aggregate (passing through a 19 mm sieve and retained on a 4.75 mm sieve) is first mixed thoroughly with 4 g of calcium carbonate ($CaCO_3$) dust. Then, 35 g of bitumen emulsion is added into the aggregate and mixed vigorously with a mixing blade for 5 min. At the end of the mixing period, the excess amount of emulsion is drained and the mixture is split into two equal parts. One part is sprayed with water until overflow water runs clear. The other part is surface air-dried in the laboratory at room temperature.

Assessment of the coating of the total aggregate surface area by bitumen emulsion is carried out in both test samples, using the grades good, fair or poor.

The above procedure is repeated, but on wet aggregates, by adding 9.3 ml of water to the aggregate and calcium carbonate dust mix.

More details for the above test method are provided in ASTM D 244 (2009) or AASHTO T 59 (2013).

4.20.16 Demulsibility test

The demulsibility test is used to identify or classify a bitumen emulsion as rapid setting (RS) or medium setting (MS) by measuring the amount of bitumen that is broken from the bitumen emulsion. This is obtained by utilising specified amounts and concentrations of calcium chloride solution for anionic bitumen emulsions and dioctyl sodium sulfosuccinate for cationic bitumen emulsions.

The demulsibility test, carried out in accordance to ASTM D 6936 (2009), is appropriate both for anionic- and for cationic-type RS and MS emulsions.

Sample conditioning is required before testing. Bitumen emulsions with viscosity testing requirements of 50°C are heated to 50°C ± 3°C and stirred, before testing, to achieve homogeneity. Bitumen emulsions with viscosity testing requirements of 25°C are stirred, before testing, at 25°C ± 1°C.

After thoroughly mixing 100 g of bitumen emulsion with the appropriate solution, the product passes through a 1.4 mm wire cloth as well as the rinse water for cleaning the beaker and the rod. Washing is continued until the rinse water drains clear. Then, the wire cloth enclosing the bitumen residue is placed in an oven at 163°C to dry completely (for 1 h or longer).

The mass of demulsibility residue (A) in grams divided by the mass of residue in the sample of 100 g of bitumen emulsion and multiplied by 100 expresses the demulsibility percentage.

More details of the test method are provided in ASTM D 6936 (2009).

4.20.17 Identification test for **RS cationic bitumen emulsion**

The identification test is conducted for identifying RS cationic bitumen emulsion. This is achieved by the inability to coat specific silica sand particles.

The test method differs from the usual coating test because the material passes the requirement when it fails to coat the specified silica sand.

According to ASTM D 244 (2009), a 35 g sample of bitumen emulsion is added to 465 g of pure silica sand, which has been treated with 5% hydrochloric acid solution and washed and is mixed vigorously with a spatula for approximately 2 min. Any emulsion excess is then removed and the mixture is spread in an absorbent paper for visual estimation of coated or uncoated area in the mixture. An excess of uncoated area over coated area is considered as a passing rating for identification of rapid-setting cationic bitumen emulsion.

More details for the test can be found in ASTM 244 (2009).

4.21 MECHANICAL PROPERTIES OF BITUMEN

Unlike many traditional construction materials such as iron, cement and so on, whose mechanical behaviour is almost independent of temperature and duration of loading, the behaviour of bitumen is not and it changes from elastic to viscous. When the load (stress) is applied for a very short time and at the same time the temperature is very low, the bitumen behaves as a purely elastic material. Conversely, when stress is applied for a very long time and the temperature is high, the behaviour of the bitumen is viscous.

Between the abovementioned two extreme conditions, bitumen behaves as a viscoelastic material and such conditions are those that exist on project sites. Therefore, in service, the bitumen behaves as a viscoelastic material and its mechanical properties depend on both temperature (T) and stress (σ) loading time (t).

4.21.1 Viscoelastic behaviour of bitumen

The knowledge of the viscoelastic behaviour and in general the mechanical behaviour of bitumen is of significant importance to the engineer.

Many researchers have attempted to display the viscoelastic behaviour of bitumen with mathematical equations. The equations developed were based on the combination of two mechanical elements that represent the elastic and viscous behaviour. These elements are the spring and the dashpot (see Figure 4.15a and b, respectively).

In the first element, the strain (ε_1) is analogous to the applied stress (σ), is related to the elastic modulus (E) of the spring and is always recoverable. The stress/strain relationship that exists is $\sigma = E \times \varepsilon$. This is a Newtonian behaviour and Hooke's law is applied.

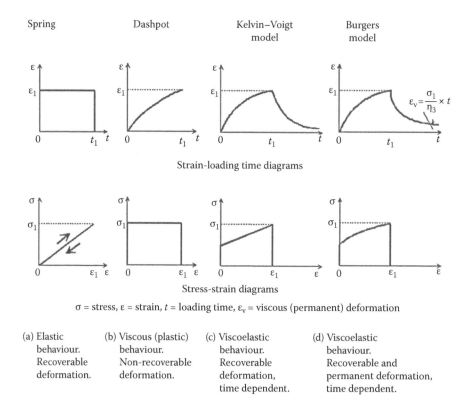

Strain-loading time diagrams

Stress-strain diagrams

σ = stress, ε = strain, t = loading time, ε_v = viscous (permanent) deformation

(a) Elastic behaviour. Recoverable deformation.	(b) Viscous (plastic) behaviour. Non-recoverable deformation.	(c) Viscoelastic behaviour. Recoverable deformation, time dependent.	(d) Viscoelastic behaviour. Recoverable and permanent deformation, time dependent.

Figure 4.15 Representation of elastic, viscous and viscoelastic materials.

In the second element, the strain is analogous to the viscosity of the material within the dashpot and is time dependent. The stress/strain relationship that exists is $\sigma = \eta/t$. This is a non-Newtonian behaviour.

Two of the many combinations of spring and dashpot elements developed, known as Kelvin–Voigt and Burgers models, are shown in Figure 4.15c and d. In the first model (Figure 4.15c), during loading, the strain gradually increases but at a decelerating rate as time lapses. After unloading ($\sigma = 0$), the strain gradually decreases and, after infinite time, is zeroed. This behaviour is essentially the typical behaviour of an ideal thermoplastic elastomer material.

The second model (Figure 4.15d) describes the complicated viscoelastic behaviour of bitumen. Upon application of stress, the model immediately presents elastic deformation and continues to deform at a non-linear rate. Thus, for a given temperature, if a constant stress (σ_1) is applied, the strain (ε) after time (t) could be calculated using the Burgers model by the following equation:

$$\varepsilon = \sigma_1/E_1 + (\sigma_1/E_2) \times (1 - e^{-k}) + (\sigma_1/\eta_3) \times t,$$

where ε is the strain developed, σ_1 is the applied constant stress, E_1 is the modulus of elasticity of the spring, E_2 is the modulus of elasticity of the Kelvin–Voigt element, η_3 is the viscosity of the dashpot material, $k = t \times (E_2/\eta_2)$, η_2 is the viscosity of Kelvin–Voigt element and t is the loading time.

When unloading ($\sigma_1 = 0$), the elastic deformation is instantly recovered, while deformation that is attributed to the viscoelastic behaviour of bitumen is recovered slowly. After a period of relaxation, some deformation cannot be recovered; this is the plastic or permanent deformation. The latter is due to the viscous behaviour of bitumen. The unrecoverable deformation is equal to $[(\sigma_1/\eta_3) \times t]$ and it causes the permanent deformation (rutting) of the bituminous layers.

Apart from the Kelvin–Voigt and Burgers models, many other more complicated models have been developed, aiming at a better simulation of the recovery curve. All models developed constitute combinations of the Kelvin–Voigt and Burgers models. More information can be found in Meyers and Chawla (1999) and Ward (1983).

4.21.2 Stiffness of bitumen

Bitumen is a viscous material unlike other construction materials such as steel or concrete, which are elastic materials. In 1954, Van der Poel introduced the concept of stiffness modulus as the fundamental parameter to describe the mechanical properties of bitumens by analogy to the elastic modulus of solids (Van der Poel 1954).

Stiffness modulus of bitumen (S_{bit} or often S_b), by analogy to the elastic modulus (E) (Young's modulus), is the ratio of stress (σ) to the strain (ε). However, the stiffness modulus of a viscous material depends on the loading time (t) and the temperature (T). Thus, the stiffness modulus of the bitumen at a given loading time and temperature can be determined by the following equation:

$$(S_{bit})_{t,T} = \sigma/\varepsilon_{t,T}.$$

The stiffness modulus variability of a viscous material as a function of loading time and temperature is demonstrated in Figure 4.16.

Figure 4.16 shows that, for the given bitumen, at very short loading times, the stiffness modulus is visually constant, asymptotic toward 2.5 to 3.0×10^9 Pa, and is, in this region, largely independent of temperature and loading time; that is, $S_b = E$, elastic behaviour (Shell Bitumen 1990). At higher loading times, the stiffness modulus changes (decreases) distinctly as the loading time increases. The same is valid when the temperature changes. At high loading times or temperatures, the behaviour of the bitumen is viscous. At intermediate values of loading time for a range of temperatures, the behaviour of the bitumen is viscoelastic.

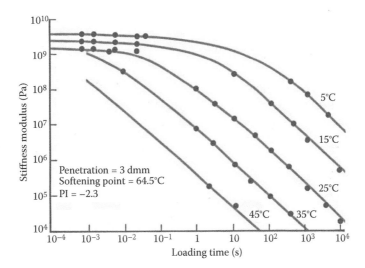

Figure 4.16 The effect of temperature and loading time on the stiffness of low PI bitumen. (From Shell Bitumen, *The Shell Bitumen Handbook*. Surry: Shell Bitumen UK, 1990.)

The stiffness modulus is also affected when the grade of bitumen and its penetration index change. This has been demonstrated in Shell Bitumen (1990, 2003).

The fluctuation of stiffness modulus of the bitumen with respect to loading time and temperature has a direct effect on the stiffness of the bituminous mixtures and performance of bituminous layers.

4.21.3 Determination of stiffness modulus of bitumen

The stiffness modulus of bitumen (S) can be determined in the laboratory by apparatuses applying static (creep) or dynamic loading. In both cases, the shear modulus, G, may be determined, which is defined as

G = shear stress (τ)/shear strain (γ).

For a homogeneous, isotropic and linear elastic material, it is known that the elastic modulus, E, and the shear modulus, G, are related by the following equation:

$E = 2 \times (1 + \mu) \times G,$

where μ is Poisson's ratio.

Poisson's ratio value for the almost uncompressible bitumen may be assumed to be equal to 0.5. Then, the above relationship becomes

$E \approx 3 \times G.$

When static load is applied, the shear stress is applied for a long period of time, t, usually for up to 10^5 s. When dynamic load is applied, the shear stress is applied for much shorter duration, depending on the frequency, f, used.

Hence, at a given temperature, the shear modulus at a loading time t, or at frequency f, is given by the ratio of the amplitude of shear stress, τ, and shear strain, γ, according to

$G_{t \text{ or } f} = (\tau/\gamma)_{t \text{ or } f}.$

It follows that the stiffness modulus under static (creep) or dynamic conditions, S, is

$S_t = 3 \times G_t \text{ or } S_f = 3 \times G_f.$

By combining static-creep and dynamic tests, a range of stiffness modulus can be obtained. When measurements of stiffness modulus as a function at time, at various temperatures, are carried out and the results are plotted in logarithmic scales, a graph of the type shown in Figure 4.16 is obtained.

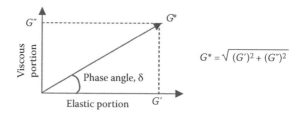

Figure 4.17 Graphical representation of complex shear modulus.

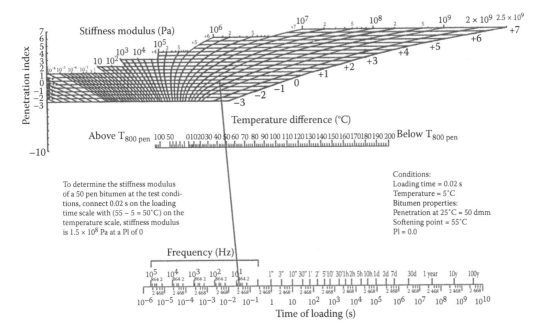

Figure 4.18 Van der Poel nomograph for the determination of the stiffness modulus (S_{bit}) of bitumens. (From Shell Bitumen, *The Shell Bitumen Handbook*. Surry: Shell Bitumen UK, 1990; Van der Poel C, *Journal of Applied Chemistry*, Vol. 4, p. 221, 1954.)

The device may be used when the test is carried out under static-creep loading, which is the one developed by Shell laboratory at the end of the 1960s (Fenijn and Krooshof 1970), similar to the sliding plate viscometer or the improved version known as AARB elastomer tester (Tredrea 2007).

When the test is carried out under dynamic loading, the DSR is used, where the complex shear modulus, G^*, may be utilised (see Figure 4.17).

4.21.4 Estimation of stiffness modulus

The laboratory determination of the stiffness modulus within a range of temperatures and times of loading is a time-consuming process and requires the existence of relatively expensive equipment. If not available, the stiffness modulus of conventional bitumen can be predicted using the Van der Poel nomograph (Van der Poel 1954).

Using the Van der Poel nomograph, the stiffness modulus of a bitumen can be predicted at any conditions of temperature and loading time by using only penetration and softening point data.

Figure 4.18 shows the Van der Poel nomograph and a worked example.

It is noted that Van der Poel's nomograph cannot be used for determining the stiffness modulus of modified bitumen or of waxy bitumen.

4.21.5 Determination of tensile strength of bitumen

The tensile properties of a bituminous binder, unmodified or polymer-modified, is determined by the direct tensile test.

Tensile properties are considered the tensile stress, the tensile strain, the elongation and energy at yield point and on fracture. Some properties are used as criteria for assessing the quality of the bituminous binder.

During the test, a specimen, held by its ends between two jaws, is extended in a chamber at constant speed until fracture or a given per cent elongation is achieved. The temperature in the chamber is regulated at the desired test temperature.

In general, stress and per cent elongation are noted at the maximum point of the stress–strain curve (flowing threshold), at breaking point and at a per cent elongation of 400%.

According to CEN EN 13587 (2010), the general values of test temperature are –20°C, –10°C, –5°C, 0°C, 5°C, 10°C, 15°C and 20°C, and the general values of test speed (elongation rate) are 1, 10, 50, 100 and 500 mm/min.

According to ASTM D 6723 (2012) or AASHTO T 314 (2012), the test temperatures range between 0°C and –36°C and are selected from ASTM D 6373 (2007) or AASHTO M 320 (2010) according to the grade of the bituminous binder. As for the elongation rate, it is fixed to 1 mm/min.

The samples are prepared in accordance to the specification adopted and after pouring the bituminous material in suitable moulds. The shape of the resulting specimen is as shown in Figure 4.19.

The specimen is carefully placed in the specimen attachment device (specimen gripping system) and is tested until fracture, after a short period for stabilisation of the test temperature (±0.5°C).

According to CEN EN 13587 (2010), the stress and the per cent elongation at flowing threshold, at fracture, at a per cent elongation of 400% and at maximum per cent elongation, if fracture is not reached, are calculated. The final result is the mean of the three values.

According to ASTM D 6723 (2012) or AASHTO T 314 (2012), only the stress and the strain at failure is computed, from at least two samples.

For further details, refer to CEN EN 13587 (2010) and ASTM D 6723 (2012) or AASHTO T 314 (2012).

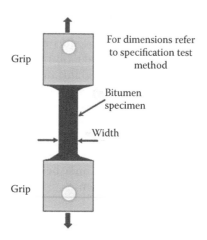

Figure 4.19 Shape of specimen for direct tension test.

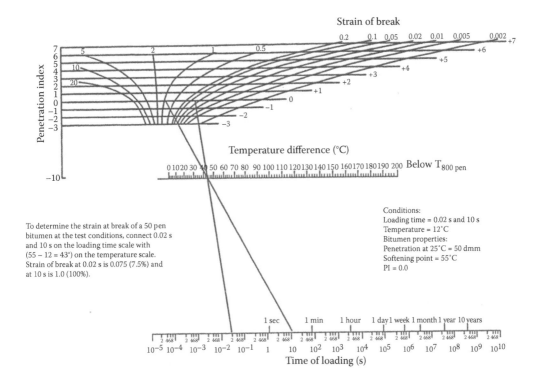

Figure 4.20 Nomograph for determining strain at break of bitumens. (From Heukelom W., *Proceedings of the Association of Asphalt Paving Technologists*, Vol. 35, p. 358, 1966; Shell Bitumen, *The Shell Bitumen Handbook*. Surry: Shell Bitumen UK, 1990.)

4.21.6 Prediction of strain at break

The strain at break of graded bitumen can be predicted using the Heukelom nomograph (Heukelom 1966). The nomograph is shown in Figure 4.20 together with two examples of bitumen with a softening point of 53°C and a PI of zero. At 10°C, the strain at break is 0.075 (7.5%), when the loading time is 0.02 s, and 1.0 (100%), when the loading time is 10 s.

From the nomograph, it can also easily be seen that, for a given loading time, the strain at break at low temperatures is low (hence stiffness modulus is high), while at high temperatures, it is high (hence stiffness modulus is low). This means that, when strain at break is low, the bitumen fractures easily. When the strain at break is high, the bitumen does not fracture but deforms.

4.21.7 Flexural creep stiffness

The flexural creep stiffness of bituminous binders in the range of 30 MPa to 1 GPa is determined by the bending beam rheometer test. Apart from the flexural stiffness, the *m*-value and the flexural creep compliance are also determined. The flexural creep stiffness test is performed on unaged or aged bituminous binders and at low to very low temperatures, ranging from 0°C to −36°C.

The flexural creep stiffness, or the flexural creep compliance, describes the low-temperature stress–strain time response of bituminous binder at the test temperature within the range of linear viscoelastic response.

The low-temperature thermal cracking performance of asphalt pavements is related to the creep stiffness and the *m*-value of the asphalt binder contained in the mix.

The creep stiffness and the *m*-value, according to ASTM D 6373 (2007) or AASHTO M 320 (2010), are used as performance-based specification criteria for bituminous binders.

The flexural creep stiffness at loading time t, $S(t)$ [or measured flexural creep stiffness, $S_m(t)$], is the ratio obtained by dividing the bending stress by the bending strain.

The *m*-value is the absolute value of the slope of the curve of the logarithm of the stiffness versus the logarithm of time.

To obtain the *m*-value, the calculated stiffness, $S_c(t)$ [or $S_e(t)$], should be determined. The calculated stiffness is obtained by fitting a second-order polynomial to the logarithm of the measured stiffness, usually at 8.0, 15.0, 30.0 60.0, 120.0 and 240.0 s, and the logarithm of time.

The flexural creep compliance, $D(t)$, if needed to be determined, is the inverse of $S_m(t)$, that is, the ratio obtained by dividing the bending shear by the bending stress.

The bending beam rheometer (Figure 4.21) is composed of a loading frame with test specimen supports, a controlled low/very low temperature liquid bath that maintains the test specimen at the test temperature and provides a buoyant force to counterbalance the force resulting from the mass of the test specimen and a computer-controlled data acquisition system for the execution of the test and the processing of the results.

According to CEN EN 14771 (2012), after the prismatic beam of bituminous binder has been prepared, by using appropriate moulds, it is placed in the liquid bath, is left for 60 min until a uniform temperature has been achieved to its entire mass and is pre-loaded with 35 ± 10 mN load, and the displacement is measured, The beam is inverted and loaded again with 35 ± 10 mN load and the displacement is measured. If the two readings agree within 1.0 mm, calculate the average and proceed. If the two readings differ by more than 1.0 mm, the flatness of the test specimen is suspect and it should be discarded.

The asphalt beam is then loaded automatically for 1 s with 980 ± 50 mN load, and then by reducing the load to 35 ± 10 mN, it is left to recover for 20 s. After this recovery period, the test practically starts, by applying a load of 980 ± 50 mN for 240 s.

Measurements for which the midpoint deflection of the test specimen is greater than 4.0 mm are suspect. Strains in excess of this value may exceed the linear response of the binder.

Measurements for which the midpoint deflection of the test specimen is less than 0.08 mm are also suspect. When the midpoint deflection is too low, the test system resolution may not be sufficient to produce reliable results.

The deflection over time are automatically recorded, and the flexural creep stiffness is calculated (S_t) at times of t = 8, 15, 30, 60, 120 and 240 s, as well as the *m*-value (*m* is the slope of the straight line from the $\lg S_t$ and $\lg t$ value pairs, where t is loading time).

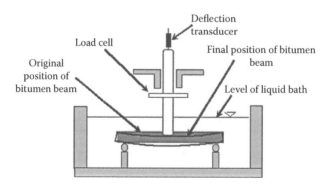

Figure 4.21 Schematic representation of the bending beam rheometer.

ASTM D 6373 (2007), for performance-graded bituminous binders, requires that the creep stiffness [$S_c(t)$ or $S_e(t)$] does not exceed 300 MPa at 60 s and the m-value is ≥0.300, at any test temperature.

A more detailed description and further details on the test are provided in CEN EN 14771 (2012) and ASTM D 6648 (2008) or AASHTO T 313 (2012).

4.21.8 Fatigue resistance

The strength of bitumen, and of many other materials, is reduced by repeated loading. A study contacted by Heukelom on the tensile strength of bitumen proved that the bitumen's resistance to fatigue is solely dependent on its stiffness. From the nomograph developed, the fatigue strength of the bitumen may be determined by each stiffness value (Heukelom 1966; Shell Bitumen 2003).

4.22 PERFORMANCE-GRADED ASPHALT BINDER TESTS

This section summarises the tests required by ASTM 6373 (2007) or AASHTO M 320 (2010) for performance-graded bituminous binders, also known as Superpave tests (Asphalt Institute SP-1 2003).

Table 4.2 summarises all tests required together with their purpose and their significance of use.

Performance-related specifications do not exist in European specifications. A position paper by Eurobitume has made proposals towards performance-related specification framework (Eurobitume 2012).

4.22.1 Flash and fire test points by Cleveland open cup

This test is performed in accordance to ASTM D 92 (2012) or AASHTO T 48 (2010). A brief description of the test is provided in Section 4.12.

4.22.2 Rotational viscometer for viscosity determination

This test is performed only on unaged bitumen and in accordance to ASTM D 4402 (2013) or AASHTO T 316 (2013). A brief description of the test is provided in Section 4.8.2.1.

4.22.3 DSR test

This test is performed in accordance with ASTM D 7175 (2008) or AASHTO T 315 (2012) on unaged and aged bitumen after hardening by RTFO test and PAV test. A brief description of the test is provided in Section 4.8.5.2.

4.22.4 Rolling thin film oven test

The test, or bitumen hardening process, is performed in accordance with ASTM D 2872 (2012) or AASHTO T 240 (2013). A brief description of the hardening test is provided in Section 4.11.1.

4.22.5 Accelerated ageing by PAV

The accelerated ageing by pressurised vessel, or bitumen ageing process, is performed in accordance with ASTM D 6521 (2008) or AASHTO R 28 (2012).

Table 4.2 Performance-graded asphalt binder tests

Test	Purpose	Significance of use
Flash and fire point (ASTM D 92 2012)	To determine the flash point and fire point (temperature) of bituminous binder.	Safety, to assess the overall flammability hazard of the bitumen and tendency to support combustion.
Rotational viscometer (ASTM D 4402 2013)	To determine the viscosity of bitumen at elevated temperatures.	To establish handling, mixing or application temperatures.
Dynamic shear rheometer (DSR) (ASTM D 7175 2008)	To determine the dynamic shear modulus and phase angle of unaged and aged asphalt binders when tested in dynamic (oscillatory) shear using parallel plate geometry.	To determine the resistance to deformation under loading, the resistance to shear deformation in the linear viscoelastic region and performance-related criteria of the binder.
Rolling thin film oven (RTFO) (ASTM D 2872 2012)	To measure the effect of heat and air on a moving film of semi-solid asphalt binder.	To indicate approximate changes in properties of bitumen during conventional hot mixing (approximately 150°C) as determined by viscosity and other rheological measurements.
Accelerated ageing by pressurised ageing vessel (PAV) (ASTM D 6521 2008)	To accelerate ageing (oxidation) of asphalt binders so as to simulate the changes in rheology during in-service oxidative ageing. It is not intended to accurately simulate the relative rates of ageing.	To simulate the in-service oxidative ageing that occurs in asphalt binders during pavement service. Residue from this conditioning practice is used to estimate the physical or chemical properties of asphalt binders after several years of in-service ageing in the field.
Bending beam rheometer (BBR) test (D 6648 2008)	To determine the characteristic properties of aged or unaged bitumen such as flexural creep stiffness and m-value at low to very low temperatures (0°C to −36°C).	To measure how much the binder creeps under constant load at lowest service temperature. The creep stiffness and m-value of the asphalt binder contained in the mix are used as performance-based specification criteria.
Direct tension (DT) (ASTM D 6723 2012)	To determine the fracture properties, failure strain and failure stress of aged or unaged bitumen under direct tension at a low to very low temperature range (+6°C to −36°C).	To measure the amount of stress and strain binder exhibited. From this, the low grade of asphalt binder is specified.

During the test, the bitumen is subjected to accelerated ageing (oxidation) by means of pressurised air and elevated temperature for 20 h, in order to simulate in-service bitumen ageing attributed to environmental conditions.

The test is performed in a special high-pressure cylindrical vessel and using a bituminous sample hardened after RTFOT. The control test conditions are as follows: 2.10 ± 0.1 MPa pressure and 90°C, 100°C or 110°C temperature, depending on the grade of bituminous binder.

The cylindrical pressure vessel has the ability to receive 10 pans with bitumen (approximately 50 g of bitumen in each pan). After the 20 h has passed, the specimens are placed in an oven at 163°C for half an hour and then they are ready to be used for further testing.

From this test, only the ageing temperature is recorded.

More information can be found in ASTM D 6521 (2008) or AASHTO R 28 (2012).

4.22.6 Bending beam rheometer test for flexural creep stiffness

This test is carried out in accordance to ASTM D 6648 (2008) or AASHTO T 313 (2012). A brief description of the hardening test is provided in Section 4.21.7.

4.22.7 Direct tension test for fracture properties

This test is performed on aged bitumen by PAV test and in accordance to ASTM 6723 (2012) or AASHTO T 314 (2012). A brief description of the hardening test is provided in Section 4.21.5.

REFERENCES

AASHTO M 208. 2009. *Cationic emulsified asphalt*. Washington, DC: American Association of State Highway and Transportation Officials.

AASHTO M 320. 2010. *Performance-graded asphalt binder*. Washington, DC: American Association of State Highway and Transportation Officials.

AASHTO R 28. 2012. *Accelerated aging of asphalt binder using a pressurized aging vessel (PAV)*. Washington, DC: American Association of State Highway and Transportation Officials.

AASHTO T 44. 2013. *Solubility of bituminous materials*. Washington, DC: American Association of State Highway and Transportation Officials.

AASHTO T 48. 2010. *Flash and fire points by Cleveland open cup*. Washington, DC: American Association of State Highway and Transportation Officials.

AASHTO T 49. 2011. *Penetration of bituminous materials*. Washington, DC: American Association of State Highway and Transportation Officials.

AASHTO T 51. 2009. *Ductility of asphalt material*. Washington, DC: American Association of State Highway and Transportation Officials.

AASHTO T 53. 2009. *Softening point of bitumen (ring-and-ball apparatus)*. Washington, DC: American Association of State Highway and Transportation Officials.

AASHTO T 59. 2013. *Emulsified asphalts*. Washington, DC: American Association of State Highway and Transportation Officials.

AASHTO T 72. 2010. *Saybolt viscosity*. Washington, DC: American Association of State Highway and Transportation Officials.

AASHTO T 79. 2012. *Flash point with tag open-cup apparatus for use with material having a flash point less than 93.3°C*. Washington, DC: American Association of State Highway and Transportation Officials.

AASHTO T 111. 2011. *Mineral matter or ash in asphalt materials*. Washington, DC: American Association of State Highway and Transportation Officials.

AASHTO T 179. 2009. *Effect of heat and air on asphalt materials (thin-film oven test)*. Washington, DC: American Association of State Highway and Transportation Officials.

AASHTO T 201. 2010. *Kinematic viscosity of asphalts (bitumens)*. Washington, DC: American Association of State Highway and Transportation Officials.

AASHTO T 202. 2010. *Viscosity of asphalts by vacuum capillary viscometer*. Washington, DC: AASHTO. American Association of State Highway and Transportation Officials.

AASHTO T 228. 2009. *Specific gravity of semi-solid asphalt materials*. Washington, DC: American Association of State Highway and Transportation Officials.

AASHTO T 240. 2013. *Effect of heat and air on a moving film of asphalt binder (rolling thin-film oven test)*. Washington, DC: American Association of State Highway and Transportation Officials.

AASHTO T 300. 2011. *Force ductility test of asphalt materials*. Washington, DC: American Association of State Highway and Transportation Officials.

AASHTO T 301. 2013. *Elastic recovery test of asphalt materials by means of a ductilometer*. Washington, DC: American Association of State Highway and Transportation Officials.

AASHTO T 313. 2012. *Determining the flexural creep stiffness of asphalt binder using the bending beam rheometer (BBR)*. Washington, DC: American Association of State Highway and Transportation Officials.

AASHTO T 314. 2012. *Determining the fracture properties of asphalt binder in direct tension (DT)*. Washington, DC: American Association of State Highway and Transportation Officials.

AASHTO T 315. 2012. *Determining the rheological properties of asphalt binder using a dynamic shear rheometer (DSR)*. Washington, DC: American Association of State Highway and Transportation Officials.

AASHTO T 316. 2013. *Viscosity determination of asphalt binder using rotational viscometer*. Washington, DC: American Association of State Highway and Transportation Officials.

Anton Paar ProveTec GmbH. 2014. Available at http://www.anton-paar.com.

Asphalt Institute SP-1. 2003. *Superpave: Performance Graded Asphalt, Binder Specification and Testing*, 3rd Edition. Superpave Series No. 1. Lexington, KY: Asphalt Institute.

ASTM D 5/5M. 2013. *Standard test method for penetration of bituminous materials*. West Conshohocken, PA: ASTM International.

ASTM D 36/D 36M. 2012. *Standard test method for softening point of bitumen (ring-and-ball apparatus)*. West Conshohocken, PA: ASTM International.

ASTM D 70-09e1. 2009. *Standard test method for density of semi-solid bituminous materials (pycnometer method)*. West Conshohocken, PA: ASTM International.

ASTM D 88. 2013. *Standard test method for Saybolt viscosity*. West Conshohocken, PA: ASTM International.

ASTM D 92-12b. 2012. *Standard method of test for flash and fire points by Cleveland open cup tester*. West Conshohocken, PA: ASTM International.

ASTM D 93. 2013. *Standard test methods for flash point by Pensky–Martens closed cup tester*. West Conshohocken, PA: ASTM International.

ASTM D 95-13e1. 2013. *Standard test method for water in petroleum products and bituminous materials by distillation*. West Conshohocken, PA: ASTM International.

ASTM D 113. 2007. *Standard test method for ductility of bituminous materials*. West Conshohocken, PA: ASTM International.

ASTM D 244. 2009. *Standard test methods for emulsified asphalt*. West Conshohocken, PA: ASTM International.

ASTM D 1665. 2009. *Standard test method for Engler specific viscosity of tar products*. West Conshohocken, PA: ASTM International.

ASTM D 1754/D 1754M. 2009. *Standard test method for effect of heat and air on asphaltic materials (thin-film oven test)*. West Conshohocken, PA: ASTM International.

ASTM D 2042. 2009. *Standard test method for solubility of asphalt materials in trichloroethylene*. West Conshohocken, PA: ASTM International.

ASTM D 2170/D 2170M-10. 2010. *Standard test method for kinematic viscosity of asphalts (bitumens)*. West Conshohocken, PA: ASTM International.

ASTM D 2171/D 2171M. 2010. *Standard test method for viscosity of asphalts by vacuum capillary viscometer*. West Conshohocken, PA: ASTM International.

ASTM D 2196. 2010. *Standard test methods for rheological properties of non-newtonian materials by rotational (Brookfield type) viscometer*. West Conshohocken, PA: ASTM International.

ASTM D 2397. 2012. *Standard specification for cationic emulsified asphalt*. West Conshohocken, PA: ASTM International.

ASTM D 2872-12e1. 2012. *Standard test method for effect of heat and air on a moving film of asphalt (rolling thin-film oven test)*. West Conshohocken, PA: ASTM International.

ASTM D 3143. 2008. *Standard test method for flash point of cutback asphalt with tag open-cup apparatus*. West Conshohocken, PA: ASTM International.

ASTM D 4287. 2010. *Standard test method for high-shear viscosity using a cone/plate viscometer*. West Conshohocken, PA: ASTM International.

ASTM D 4402/D 4402M. 2013. *Standard test method for viscosity determination of asphalt at elevated temperatures using a rotational viscometer*. West Conshohocken, PA: ASTM International.

ASTM D 6084. 2006. *Standard test method for elastic recovery of bituminous materials by ductilometer.* West Conshohocken, PA: ASTM International.

ASTM D 6373-07e1. 2007. *Standard specification for performance graded asphalt binder.* West Conshohocken, PA: ASTM International.

ASTM D 6521. 2008. *Standard practice for accelerated aging of asphalt binder using a pressurized aging vessel (PAV).* West Conshohocken, PA: ASTM International.

ASTM D 6648. 2008. *Standard test method for determining the flexural creep stiffness of asphalt binder using the bending beam rheometer (BBR).* West Conshohocken, PA: ASTM International.

ASTM D 6723. 2012. *Standard test method for determining the fracture properties of asphalt binder in direct tension (DT).* West Conshohocken, PA: ASTM International.

ASTM D 6930. 2010. *Standard test method for settlement and storage stability of emulsified asphalts.* West Conshohocken, PA: ASTM International.

ASTM D 6933. 2008. *Standard test method for oversized particles in emulsified asphalts (sieve test).* West Conshohocken, PA: ASTM International.

ASTM D 6934. 2008. *Standard test method for residue by evaporation of emulsified asphalt.* West Conshohocken, PA: ASTM International.

ASTM D 6935. 2011. *Standard test method for determining cement mixing of emulsified asphalt.* West Conshohocken, PA: ASTM International.

ASTM D 6936. 2009. *Standard test method for determining demulsibility of emulsified asphalt.* West Conshohocken, PA: ASTM International.

ASTM D 6997. 2012. *Standard test method for distillation of emulsified asphalt.* West Conshohocken, PA: ASTM International.

ASTM D 7175. 2008. *Standard test method for determining the rheological properties of asphalt binder using a dynamic shear rheometer.* West Conshohocken, PA: ASTM International.

ASTM D 7226. 2011. *Test method for determining the viscosity of emulsified asphalts using a rotational paddle viscometer.* West Conshohocken, PA: ASTM International.

ASTM D 7404. 2012. *Standard test method for determination of emulsified asphalt residue by moisture analyzer.* West Conshohocken, PA: ASTM International.

ASTM D 7496. 2011. *Standard test method for viscosity of emulsified asphalt by Saybolt Furol viscometer.* West Conshohocken, PA: ASTM International.

ASTM E 102/E 102M. 2009. *Standard test method for Saybolt Furol viscosity of bituminous materials at high temperatures.* West Conshohocken, PA: ASTM International.

CEN EN 1426. 2007. *Bitumen and bituminous binders – Determination of needle penetration.* Brussels: CEN.

CEN EN 1427. 2007. *Bitumen and bituminous binders – Determination of softening point-ring and ball method.* Brussels: CEN.

CEN EN 1428. 2012. *Bitumen and bituminous binders – Determination of water content in bitumen emulsions – Azeotropic distillation method.* Brussels: CEN.

CEN EN 1429. 2013. *Bitumen and bituminous binders – Determination of residue on sieving of bituminous emulsions, and determination of storage stability by sieving.* Brussels: CEN.

CEN EN 1430. 2009. *Bitumen and bituminous binders – Determination of particle polarity of bituminous emulsions.* Brussels: CEN.

CEN EN 1431. 2009. *Bitumen and bituminous binders – Determination of residual binder and oil distillate from bitumen emulsions by distillation.* Brussels: CEN.

CEN EN 12591. 2009. *Bitumen and bituminous binders – Specifications for paving grade bitumens.* Brussels: CEN.

CEN EN 12592. 2007. *Bitumen and bituminous binders – Determination of solubility.* Brussels: CEN.

CEN EN 12593. 2007. *Methods of test for petroleum and its products-bitumen and bituminous binders – Determination of the Fraass breaking point.* Brussels: CEN.

CEN EN 12595. 2007. *Bitumen and bituminous binders – Determination of kinematic viscosity.* Brussels: CEN.

CEN EN 12596. 2007. *Bitumen and bituminous binders – Determination of dynamic viscosity by vacuum capillary.* Brussels: CEN.

CEN EN 12607-1. 2007. *Bitumen and bituminous binders – Determination of the resistance to hardening under the influence of heat and air – Part 1: RTFOT method*. Brussels: CEN.

CEN EN 12607-2. 2007. *Bitumen and bituminous binders – Determination of the resistance to hardening under the influence of heat and air – Part 2: TFOT Method*. Brussels: CEN.

CEN EN 12607-3. 2007. *Bitumen and bituminous binders – Determination of the resistance to hardening under the influence of heat and air – Part 3: RFT Method*. Brussels: CEN.

CEN EN 12846-1. 2011. *Bitumen and bituminous binders – Determination of efflux time by the efflux viscometer – Part 1: Bituminous emulsions*. Brussels: CEN.

CEN EN 12846-2. 2011. *Bitumen and bituminous binders – Determination of efflux time by the efflux viscometer – Part 2: Cut-back and fluxed bituminous binders*. Brussels: CEN.

CEN EN 12847. 2009. *Bitumen and bituminous binders – Determination of settling tendency of bituminous emulsions*. Brussels: CEN.

CEN EN 12848. 2009. *Bitumen and bituminous binders – Determination of mixing stability with cement of bituminous emulsions*. Brussels: CEN.

CEN EN 12849. 2009. *Bitumen and bituminous binders – Determination of penetration power of bituminous emulsions*. Brussels: CEN.

CEN EN 13074-2. 2011. *Bitumen and bituminous binders – Recovery of binder from bituminous emulsion or cut-back or fluxed bituminous binders – Part 2: Stabilisation after recovery by evaporation*. Brussels: CEN.

CEN EN 13075-1. 2009. *Bitumen and bituminous binders – Determination of breaking behaviour – Part 1: Determination of breaking value of cationic bituminous emulsions, mineral filler method*. Brussels: CEN.

CEN EN 13075-2. 2009. *Bitumen and bituminous binders – Determination of breaking behaviour – Part 2: Determination of fines mixing time of cationic bituminous emulsions*. Brussels: CEN.

CEN EN 13302. 2010. *Bitumen and bituminous binders – Determination of dynamic viscosity of bituminous binder using a rotating spindle apparatus*. Brussels: CEN.

CEN EN 13398. 2010. *Bitumen and bituminous binders – Determination of the elastic recovery of modified bitumen*. Brussels: CEN.

CEN EN 13399. 2010. *Bitumen and bituminous binders – Determination of storage stability of modified bitumen*. Brussels: CEN.

CEN EN 13587. 2010. *Bitumen and bituminous binders – Determination of the tensile properties of bituminous binders by the tensile test method*. Brussels: CEN.

CEN EN 13588. 2008. *Bitumen and bituminous binders – Determination of cohesion of bituminous binders with pendulum test*. Brussels: CEN.

CEN EN 13589. 2008. *Bitumen and bituminous binders – Determination of the tensile properties of modified bitumen by the force ductility method*. Brussels: CEN.

CEN EN 13614. 2011. *Bitumen and bituminous binders – Determination of adhesivity of bitumen emulsions by water immersion test – Aggregate method*. Brussels: CEN.

CEN EN 13702. 2010. *Bitumen and bituminous binders – Determination of dynamic viscosity of modified bitumen by cone and plate method*. Brussels: CEN.

CEN EN 13703. 2003. *Bitumen and bituminous binders – Determination of deformation energy*. Brussels: CEN.

CEN EN 13808. 2013. *Bitumen and bituminous binders – Framework for specifying cationic bituminous emulsions*. Brussels: CEN.

CEN EN 14023. 2010. *Bitumen and bituminous binders – Framework specification for polymer modified bitumens*. Brussels: CEN.

CEN EN 14770. 2012. *Bitumen and bituminous binders – Determination of complex shear modulus and phase angle – Dynamic shear rheometer (DSR)*. Brussels: CEN.

CEN EN 14771. 2012. *Bitumen and bituminous binders – Determination of the flexural creep stiffness-Bending beam rheometer (BBR)*. Brussels: CEN.

CEN EN 15323. 2007. *Bitumen and bituminous binders – Accelerated long-term ageing/conditioning by the rotating cylinder method (RCAT)*. Brussels: CEN.

CEN EN 15326: 2007+A1. 2009. *Bitumen and bituminous binders – Measurement of density and specific gravity – Capillary-stoppered pyknometer method*. Brussels: CEN.

CEN EN 16345. 2012. *Bitumen and bituminous binders – Determination of efflux time of bituminous emulsions using the Redwood No. II Viscometer.* Brussels: CEN.

CEN EN ISO 2592. 2001. *Determination of flash and fire points – Cleveland open cup method.* Brussels: CEN.

CEN EN ISO 13736. 2013. *Determination of flash point – Abel closed-cup method.* Brussels: CEN.

Controls Srl. 2014. Available at http://www.controls-group.com.

Cooper Research Technology Ltd. 2014. Available at http://www.cooper.co.uk.

Eurobitume. 2012. *Position Paper: Performance Related Specification for Bituminous Binders.* Brussels: European Bitumen Association.

Fenijn J. and R.C. Krooshof. 1970. The sliding plate rheometer. A simple instrument for measuring the visco-elastic behaviour of bitumens and related substances in absolute units. *Proceedings of the 9th Annual Conference CTAA*, Vol. 15, p. 123.

Fraass A. 1937. Test method for bitumen and bituminous mixture with specific reference to low temperature. *Bitumen*, Vol. 7, p. 152.

Griffin R.L., T.K. Miles, C.J. Penther, and W.C. Simpson. 1957. *Sliding Plate Microviscometer for Rapid Measurement of Asphalt Viscosity in Absolute Units.* Emeryville, CA: Shell Development Co.

Heukelom W. 1966. Observations on the rheology and fracture of the bitumens and asphalt mixes. *Proceedings of the Association of Asphalt Paving Technologists*, Vol. 35, p. 358.

Heukelom W. 1973. An improved method of characterizing asphaltic bitumens with the aid of their mechanical properties. *Proceedings of the Association of Asphalt Paving Technologists*, Vol. 42, p. 67.

Heukelom W.A. 1969. A bitumen test data chart for showing the effect of temperature on the mechanical behaviour of asphaltic bitumens. *Journal of the Institute of Petroleum*, Vol. 55, p. 404. London: Institute of Petroleum.

ISO 2719. 2002. *Determination of flash point, Pensky–Martens closed cup method.* Geneva: International Organization for Standardization.

Lefebvre J.A. 1970. A modified penetration index for Canadian asphalt. *Proceedings of the Association of Asphalt Paving Technologists*, Vol. 39, p. 443.

Meyers M.A. and K.K. Chawla. 1999. *Mechanical Behavior of Materials.* New Jersey, USA: Prentice Hall, Inc.

Nikolaides A. 1988. Mechanical behavior of paving asphalt binders used in Greece. *Technical Chronicals, Scientific Area A*, Vol. 8, Part. 3: Athens, Greece: Technical Chamber of Greece (TEE).

Pfeiffer J. and P. Van Doormaal. 1936. The rheological properties of asphaltic bitumens. *Journal of the Institution of Petroleum*, Vol. 22, p. 414.

Shell Bitumen. 1990. *The Shell Bitumen Handbook.* Surry: Shell Bitumen UK.

Shell Bitumen. 2003. *The Shell Bitumen Handbook*, 5th Edition. London: Thomas Telford Publishing.

Thenoux G., G. Lees, and C.A. Bell. 1985. Laboratory investigation of the Fraass brittle test. *Journal of the Association of Asphalt Paving Technologists*, Vol. 54, p. 529.

Tredrea P.F. 2007. Superpave binder properties and the role of viscosity. *2007 AAPA Pavements Industry Conference: Innovation to Implementation.* Sydney: Australian Asphalt Pavement Association.

Van der Poel C. 1954. A general system describing the visco-elastic properties of bitumen and its relation to routine test data. *Journal of Applied Chemistry*, Vol. 4, p. 221.

Ward I.M. 1983. *Mechanical Properties of Solid Polymers.* Weinheim: John Wiley & Sons.

Chapter 5

Hot asphalts

5.1 GENERAL

Various types of hot asphalts (mixture of mineral aggregate and bituminous binder–bituminous mixture) are used in the construction of flexible pavements, depending on the project requirements, to ensure optimal use of the asphalt. From all available asphalts, each country uses a combination of those that are considered to be the most appropriate to their climatic and traffic conditions.

This chapter considers all types of hot asphalts specified by European and American standards. Throughout the text, more often the term *asphalt* rather than *bituminous mixture* will be used, as adopted by European specifications. Additionally, the adjective *hot* will also not be used since this chapter refers only to hot asphalts.

5.2 DETERMINATION AND ROLE OF ASPHALTS

Asphalts have to fulfil a wide range of requirements. Particularly, asphalts must

- Be resistant to permanent deformation
- Be resistant to fatigue cracking
- Contribute to the pavement's bearing capacity
- Be impervious to water (hence, the underlying layers are protected)
- Present good workability and easiness of compaction
- Be maintained easily
- Be cost-effective

Additionally, asphalts for surface courses should also

- Be resistant to the polishing action of tyres
- Be resistant to the catastrophic effect of the weather
- Provide an even surface for comfortable and safe driving
- Provide a surface so that the noise created by vehicle wheels is tolerable
- Provide a surface that requires as little maintenance as possible

The above requirements prove the necessity of proper mix design and utilisation of all available materials and technologies, so as to ensure the long-lasting performance of the pavement.

5.3 CHARACTERISTIC TYPES OF ASPHALTS

Asphalts are characterised by the particle size distribution of the aggregate mixture. Theoretically, there are unlimited types of asphalts, namely, from asphalts consisting only of almost single-sized coarse aggregates to mixtures consisting only of fine aggregates (sand). All types of asphalts used range between these two extreme cases.

Some countries, such as the United States, England, Germany and France, pioneers in invention and production of asphalts, for many years now, have developed asphalts that are used worldwide. These mixtures were asphalt concrete (AC), macadam, hot rolled asphalt (HRA), mastic asphalt (MA) and gussasphalt.

Over the last 40 years or more, new mixtures were developed in order to improve some features, such as drainage ability, noise reduction, durability and the ability to be laid in less than a 4 cm thick layer, hence economising on materials. The mixtures developed were porous asphalt (PA) in the United Kingdom, stone mastic asphalt (SMA) in Germany and AC for very thin layers (BBTM) in France.

The basic feature of all asphalts is the gradation curve of the aggregate mixture, which may be continuously graded or gap graded. In the first case, all particle sizes exist in the mix at an appreciable proportion. In the second case, certain sizes are in a small proportion, hence creating a kind of a gap.

Depending on the portion of each particle size, asphalts are designated as porous, open, semi-open or dense-graded mixtures. Thus, their voids vary from high to low content, and as a consequence, some are more permeable than others.

Indicative aggregate gradation curves of the most commonly used asphalts are given in Figure 5.1. From the asphalts shown in Figure 5.1, the asphalt with most voids is the PA, whereas the asphalt with the least air voids is the MA. The rest of the mixtures in descending air void order are BBTM, AC (dense), SMA and HRA.

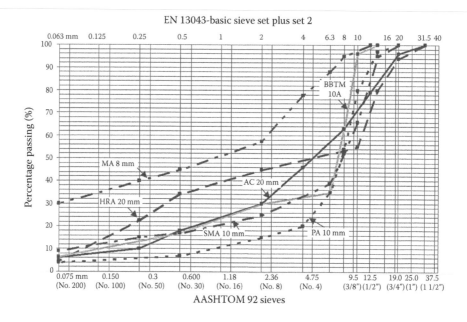

Figure 5.1 Aggregate grading curves of typical asphalts.

The HRA, as well as the BBTM, is a gap-graded mixture. Their aggregate gradation lacks certain size aggregates; for the given HRA mixture, the sizes from 2 to 10 mm are in small percentage, and for the BBTM, the sizes from 2 to 6.3 mm are also in small percentage.

Another characteristic difference among the asphalts is the way their strength is developed. In general, asphalts owe their strength to two key factors: the stiffness of the bituminous mortar (bitumen/sand/filler) and the interlocking and contact of the aggregate particles. In turn, the stiffness of the mortar is, almost exclusively, due to the hardness (viscosity) of the bitumen.

MA and HRA owe their strength almost exclusively to the stiffness of the mortar, whereas all other asphalts primarily owe their strength to the mechanical interlocking and contact of the aggregate particles.

Typical characteristic properties and a comparison of the abovementioned asphalts, as well as gussasphalt and macadam, in increasing void percentage order are given in Table 5.1.

European standards distinguish seven types of asphalts:

- Asphalt concrete (AC)
- Asphalt concrete for very thin layers (BBTM or AC-VTL)
- Porous asphalt (PA)
- Hot rolled asphalt (HRA)
- Stone mastic asphalt (SMA)

Table 5.1 Comparison of composition characteristics and properties of typical and non-typical hot asphalts

	Type of asphalt								
	MA	Guss*	HRA	SMA	AC	DMac*	BBTM	OMac*	PA
Composition (%)									
Coarse aggregate	Lo	Me	Me	Hi	Hi	Hi	Hi	Hi	Hi
Fine aggregate	Hi	Hi	Hi	Lo	Me	Me	Lo	Lo	Lo
Filler	VHi	Hi	Me	Hi	Me	Me	Hi	Lo	Lo
Bitumen	VHi	Hi	Hi	Hi	Me	Me	Hi	Lo	Lo
Hardness of bitumen	VHa	Ha	Ha	Me	Me/Ha	Me/Ha	Me	Me/So	Me
Properties									
% Air voids	0–1	1–2	2–3	3.5–4.5	3–5	5–10	6–15	>15	>18
Resist. to deformation	Me/Hi	Hi	Me	Hi	Me/Hi	Me	Hi	Me	Me
Durability	Hi	Hi	VHi	VHi	Hi	Hi	Hi	Lo	Lo
Structural contribution	Hi	Hi	Hi	Hi	Hi	Hi	Me	Lo	Lo
Text. depth/skid resist.	—	Me	Hi**	Me	Me	Me	Hi	Hi	Hi
Reduction of noise	No	No	No	No	No	No	Yes	Yes	Yes
Reduction of 'spray'	No	No	No	No	No	No	Yes	Yes	Yes
Workability	Lo	Lo	Me	Me	Me	Me	Me/Hi	Hi	Hi/Me
Cost (per unit area)	VHi	Hi	Hi	Hi+	Me	Me	Lo	Me–	Hi

Note: AC, asphalt concrete; BBTM, asphalt concrete for very thin layers; DMac, dense-graded macadam; Guss, gussasphalt; Ha, hard; Hi, high; HRA, hot rolled asphalt; Lo, low; MA, mastic asphalt; Me, medium; OMac, open-graded macadam; PA, porous asphalt; SMA, stone mastic asphalt; So, soft; VHa, very hard; VHi, very high. *Not distinguished by EN standards, **With the use of coated chippings.

- Mastic asphalt (MA)
- Soft asphalt (SA)

Macadam and gussasphalt are no longer distinguished as asphalts by European standards. Macadam is very similar to AC. Gussasphalt, almost exclusively used in Germany, is being phased out for practical and economic reasons. It requires plant modifications, the use of special vehicles for transporting it to the site and a high percentage of bitumen.

The American standards basically distinguish three asphalts for road construction: the AC, the open (porous) asphalt and the SMA.

5.4 ASPHALT CONCRETE

AC was initially developed in the United States to meet the need for an asphalt resistant to heavy traffic loads and aircraft loads. It is the most well-known type of asphalt, and it is used by almost all countries around the world. Many countries use gradation curves as they are specified by ASTM D 3515 (2010) (withdrawn in 2010), whereas European countries are obliged to compose gradation curves in accordance with the requirements of CEN EN 13108-1 (2008).

AC consists of coarse, fine aggregates and fines (filler), in various proportions, producing dense- and up to open-graded mixtures. The strength and the stability of this mixture derive from the interlocking of the aggregate particles, assisted by the hardness of the bitumen.

The AC is used in all layers of flexible pavements. However, its use should be limited or excluded from surface layers in roads with a speed limit above 70 km/h. There are more effective and, in some cases, more economic mixtures for use in surface layers than AC.

5.4.1 AC in accordance to European standards

AC in the European Union (EU) is designed in accordance with CEN EN 13108-1 (2008). The standard provides the general requirements of the constituent materials and of the AC mixture, as well as evaluation of conformity requirements. It does not provide a composition method (mix design).

The standard provides two ways of specifying AC.

The first, known as the 'empirical' approach, specifies AC in terms of compositional recipes and requirements for constituent materials with additional requirements based on performance-related tests. These requirements are given in Sections 5.4.1.3 and 5.4.1.4.

The second, known as the 'fundamental' approach, specifies AC in terms of performance-based requirements linked to a limited prescription of composition and constituent materials, offering a greater degree of freedom. These requirements are given in Sections 5.4.1.3 and 5.4.1.5.

As experience is gained with performance-based testing, the aim of the specification is to specify AC in terms of only fundamental performance-based properties.

The standard CEN EN 13108-1 (2008) covers all types of AC from dense- to open-graded mixtures.

5.4.1.1 Constituent material requirements

5.4.1.1.1 Binder

The binder material for empirically or fundamentally specified AC mixtures can be conventional paving grade bitumen, modified bitumen or hard bitumen. The conventional paving bitumen should conform to CEN EN 12591 (2009), the modified bitumen should conform to CEN EN 14023 (2010) and the hard bitumen should conform to CEN EN 13924 (2006). Natural asphalt, conforming to CEN EN 13108-4 (2008), Annex B, may also be added.

In an empirically specified mixture, the grade of bitumen, the type and the grade of modified bitumen or the amount and category of natural asphalt (if used) is always selected and specified.

For selecting the paving grade of added bitumen in surface courses with reclaimed asphalt, when more than 10% by mass of the total mixture of reclaimed asphalt is used, the resulting binder should meet the penetration, or softening point, requirements of the selected grade.

The penetration, or the softening point, of the binder in the resulting mixture is calculated using an equation provided by CEN EN 13108-1 (2008), Annex A. These equations are also given in Section 17.8.1.

The same applies for selecting the paving grade of added bitumen in binder courses and regulating courses and bases with reclaimed asphalt, when more than 20% by mass of the total mixture of reclaimed asphalt is used.

It is noted that the above applies only when the reclaimed asphalt has been produced with paving grade bitumen.

5.4.1.1.2 Aggregates

The aggregate, coarse, fine aggregates and added filler can be any natural, manufactured or recycled material conforming to CEN EN 13043 (2004). The added filler is of mineral origin, but cement or lime may also be used. More information with regard to aggregates is given in Chapter 2.

In case of using reclaimed asphalt, the upper sieve size D of the aggregate in the reclaimed asphalt should not exceed the upper sieve size D of the mixture, which is to be composed. Furthermore, aggregate properties in the reclaimed asphalt should fulfil the requirements specified for the aggregates in the composed mixture.

5.4.1.1.3 Additives

Other additives, such as fibres, colouring agents and so on can be used provided they conform to the requirements of the respective standard.

5.4.1.2 Requirements for the target composition of AC

The mixture at target composition, when empirical requirements are used, should fulfil general requirements plus empirical requirements selected from Sections 5.4.1.3 and 5.4.1.4.

Alternatively, when fundamental requirements are used, the mixture at the target composition should fulfil general requirements plus fundamental requirements selected from Sections 5.4.1.3 and 5.4.1.5.

According to CEN EN 13108-1 (2008), empirical and fundamental requirements should not be combined, avoiding over-specification.

5.4.1.3 General requirements

At the target mixture, the following should be determined: the composition, the grading of aggregates, the void content, the resistance to permanent deformation and the water sensitivity.

In case the AC is going to be applied in airfields, its resistance to fuel and its resistance to de-icing fluid should also be determined and declared.

Finally, there may be a need to determine its resistance to abrasion by studded tyres.

As a general requirement, the temperature of the mixture at any place in the plant should be stated and the coating/homogeneity should be ensured.

Finally, if the manufacturer declares a Euroclass for reaction to fire (i.e. it is subject to regulatory requirements), the AC should be tested and classified in accordance to CEN EN 13501-1 (2009).

5.4.1.3.1 Grading of aggregates

The requirements of the aggregate grading are expressed in terms of maximum and minimum percentages passing through some characteristic sieves: 1.4D, D, 2 mm and 0.063 mm.

D, the upper sieve size of the aggregate in millimetres, determines the grade of the AC. A certain mass aggregate may be retained in this sieve.

The 1.4D sieve is designated as the sieve is derived from the multiplication of D with 1.4 rounded to the nearest sieve. All aggregate mass passes through this sieve.

D sieves, as well as sieves between D and 2 mm, should be selected from the basic sieve set and the two alternative sieve sets, set 1 or set 2, specified. Information about the basic sieve test and sieve sets 1 and 2 is given in Section 2.11.1.

The overall grading limits at the characteristic sieves, as specified by CEN EN 13108-1 (2008) for AC using the basic sieve set plus set 2, are given in Table 5.2. It is stated that these limits should never be exceeded. The respective overall grading limits using the basic sieve set plus set 1 can be found in the reference source.

5.4.1.3.2 Void content

The void content of the compacted specimens should lie within the minimum and the maximum values selected from the categories given in CEN EN 13108-1 (2008).

The minimum void content should range from 0.5% to 6% voids, assigned to 12 categories, $V_{min0.5}$ to $V_{min6.0}$ (a step of 0.5% is used). Similarly, the maximum void content should range from 2% to 14%, assigned to 17 categories, $V_{max2.0}$ to $V_{max14.0}$.

The compaction of the specimens may be carried out by an impact (Marshall) compactor according to CEN EN 12697-30 (2012), by a gyratory compactor according to CEN

Table 5.2 Overall limits of target composition – basic sieve set plus set 2

D	4	6 (6.3)	8	10	12 (12.5)	14	16	20	32 (31.5)
Sieve (mm)	Passing sieve (%) by mass of aggregates								
1.4D[a]	100	100	100	100	100	100	100	100	100
D	90–100	90–100	90–100	90–100	90–100	90–100	90–100	90–100	90–100
2	50–85	15–72	10–72	10–60	10–55	10–50[b]	10–50[b]	10–50[b]	10–50
0.063	5.0–17.0	2.0–15.0	2.0–13.0	2.0–12.0	2.0–12.0	0–12.0	0–12.0	0–11.0	0–11.0

Source: Reproduced from CEN EN 13108-1, *Bituminous mixtures – Material specifications – Part 1: Asphalt Concrete*, Brussels: CEN, 2008. With permission (© CEN).

[a] Where the sieve calculated as 1.4D is not an exact number in the ISO 565/R20 series, then the next nearest sieve in the set should be adopted.
[b] For application on airfields, the maximum percentage passing 2 mm may be increased to 60%.

(a)   (b)

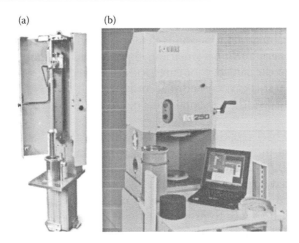

Figure 5.2 (a) Marshall and (b) gyratory compactor. (Courtesy of Controls Srl.)

EN 12697-31 (2007) or by a vibratory compactor according to CEN EN 12697-32 (2007). Marshall and gyratory compactor devices are shown in Figure 5.2.

Specimens may also be carried out by a roller compactor, according to CEN EN 12697-33 (2007).

In all cases, the compaction energy applied is selected and declared by the designer or the producer.

The air void content (V_m) is calculated according to the following equation:

$$V_m = (\rho_m - \rho_b)/\rho_m \times 100,$$

where V_m is the air void content of the mixture (0.1% [v/v]), ρ_m is the maximum density of the bituminous mixture (kg/m^3) and ρ_b is the bulk density of the specimen (kg/m^3).

The maximum density of the bituminous mixture is determined volumetrically according to Procedure A of CEN EN 12697-5 (2012). This procedure is similar to the procedure used to determine the maximum density by Rise, following the American standard.

Specimen bulk density is determined with the respective method depending on the expected percentage of air voids.

Particularly, if the percentage of air voids is expected to be ≤7%, the specimen bulk density is determined following procedure B of CEN EN 12697-6 (2012). During this procedure, the mass of the saturated surface dry specimen, in grams, is taken into account to calculate the specimen volume.

If the percentage of voids is expected to range from 7% to 10%, the specimen bulk density is determined following Procedure C of CEN EN 12697-6 (2012). During this procedure, the mass of the sealed specimen, in grams, is taken into account to calculate the specimen volume. The specimen sealing is usually carried out with paraffin.

Finally, if the percentage of voids is expected to be ≥10%, the specimen bulk density is determined following Procedure D of CEN EN 12697-6 (2012). During this procedure, the calculation of the specimen bulk density is conducted geometrically (diameter – surface area multiplied by height).

It is stated that the specimen's volume required for the determination of other volumetric properties, such as voids in the mineral aggregate (VMA) and voids filled with bitumen (VFB), is determined by the same way as explained above.

5.4.1.3.3 Resistance to permanent deformation

The resistance to permanent deformation is determined with respect to the wheel-tracking test. The resistance to permanent deformation, expressed accordingly, should be lower than the maximum values of the category (or categories) specified.

The wheel-tracking test is carried out in accordance to CEN EN 12697-22 (2007), using deferent devices. For an AC, the test conditions per device used are determined in CEN EN 13108-20 (2008).

Depending on the wheel-tracking device used, the categories for maximum permissible values and the units of measurements vary. The wheel-tracking devices recommended by CEN EN 13108-20 (2008) to be used when AC is tested are the large-sized wheel-tracking device or the small-sized device, procedure B.

When the large-sized wheel-tracking device is used, the resistance to permanent deformation is expressed in reference to proportional rut depth (P), expressed in per cent. The maximum values of the proportional rut depth range from 5% to 20%, assigned to five categories, P_5, $P_{7.5}$, P_{10}, P_{15} and P_{20}.

When the small device, procedure B, is used, the resistance to permanent deformation is expressed in reference to the wheel-tracking slope, expressed as millimetres per 1000 loading cycles (WTS_{AIR}) and proportional rut depth (PRD_{AIR}), expressed in per cent. The maximum values of WTS_{AIR} range from 0.03 to 1.0 mm/1000 cycles, assigned to 11 categories from $WTS_{AIR\,0.03}$ to $WTS_{AIR\,1.0}$. The maximum values of PRD_{AIR} range from 1.0% to 9.0%, assigned to seven categories from $PRD_{AIR\,1.0}$ to $PRD_{AIR\,9.0}$. More details can be found in CEN EN 13108-1 (2008).

5.4.1.3.4 Water sensitivity

The water sensitivity of AC compacted cylindrical specimens is determined with respect to the indirect tensile strength ratio (ITSR), and a minimum value should be achieved as selected, specified or declared.

The minimum ITSR, expressed in per cent, should be either 60%, 70%, 80% or 90%, assigned to categories $ITSR_{60}$, $ITSR_{70}$, $ITSR_{80}$ or $ITSR_{90}$.

The indirect tensile strength ratio is determined in accordance to CEN EN 13697-12 (see also Section 3.7.3.4).

5.4.1.3.5 Resistance to fuel for application on airfields

AC's resistance to fuels is required to be determined only when it is to be used in airfields or fuel stations.

The resistance to fuel test of the cylindrical specimens, which are produced and compacted as the specimens for air voids determination, is carried out according to CEN EN 12697-43 (2005).

The resistance to fuel should be selected from the categories good, moderate or poor.

5.4.1.3.6 Resistance to de-icing fluid for application on airfields

Similar to the previous requirement, the mixture resistance to de-icing fluids is recommended to be determined only when the AC is to be used in airfields.

The resistance to de-icing fluids is determined by using cylindrical specimens, similar to those produced for the resistance to fluid test, according to CEN EN 12697-41 (2013).

The resistance to de-icing fluid, expressed as minimum retained strength, in per cent, should be 55%, 70%, 85% or 100%, assigned to categories β_{55}, β_{70}, β_{85} or β_{100}.

5.4.1.3.7 Resistance to abrasion by studded tyres

Resistance to abrasion by studded tyres, when required, is carried out on cylindrical specimens in accordance to CEN EN 12697-16 (2004). The specimens are prepared as for void content determination.

The resistance to abrasion by studded tyres, expressed as maximum abrasion value, in millilitres, should range from 20 to 60, assigned to 10 categories from Abr_{A20} to Abr_{A60} (see CEN EN 13108-1 2008).

This requirement is exclusively used by some northern European countries that impose the use of studded tyres during winter months.

5.4.1.3.8 Temperature of the mixture in the plant

According to CEN EN 13108-1 (2008), the temperatures at any place in the plant when paving grade bitumen is used should be within the limits given in the standard. For 35/40 or 40/60 paving grade bitumen, the temperature range is 150°C to 190°C, and for 50/70 or 70/100 paving grade bitumen, it is 140°C to 180°C. Measurements should be carried out in accordance to CEN EN 12697-13 (2001).

The minimum temperature at delivery is determined by the producer.

When using modified or hard bitumen, or additives, temperature limits different from the above may be applied and should be documented and declared.

5.4.1.3.9 Coating and homogeneity

The AC, when discharged from the mixer, should be homogeneous in appearance, with the aggregates completely coated with binder. What is more, there should be no indication of balling of fine aggregates.

5.4.1.4 Empirical requirements

Apart from the general requirements, the target composition should also fulfil some empirical requirements, such as grading, binder content, voids filled with bitumen, voids in the mineral aggregates and void content at 10 gyrations. When the AC is to be applied in airfields, the Marshall values should also be determined.

5.4.1.4.1 Grading

The grading envelope of the target composition mixture should be within maximum and minimum values for the percentages passing through the following sieves: 1.4D, D, a characteristic coarse sieve, 2 mm, a characteristic fine sieve and 0.063 mm. The grading envelope should be within the overall limits specified (see Table 5.2 for the basic sieve set plus set 2 or CEN EN 13108 for the basic sieve set plus set 1).

In addition, the grading envelope may include the percentage passing through one optional sieve between D and 2 mm and one optional fine sieve between 2 and 0.063 mm. The optional fine sieve can be one of the following sieves: 1, 0.5, 0.25 or 0.125 mm.

The ranges between the maximum and minimum values for the grading envelope should be selected as a single value within the given limits provided in Table 12 of CEN EN 13108-1 (2008).

5.4.1.4.2 Binder content

The minimum binder content of the target composition should range from 3.0% to 8.0%, by mass of total mixture, assigned to 26 categories from $B_{min3.0}$ to $B_{min8.0}$ (see CEN EN 13108-1 2008).

The above minimum binder content values specified are to be corrected when the apparent particle density of the aggregate deviates from 2650 Mg/m³ by multiplying them by the factor (*a*), which is determined by the following equation:

$$a = 2650/\rho_a,$$

where ρ_a is the apparent particle density of the aggregate (Mg/m³) determined by the weighted mean of the total mineral fraction according to CEN EN 1097-6 (2013).

5.4.1.4.3 Voids filled with bitumen

The minimum and maximum percentage of voids filled with bitumen (VFB_{min} and VFB_{max}) of the compacted specimens is selected from appropriate tables given in CEN EN 13108-1 (2008).

The minimum VFB values range from 50% to 78%, assigned to eight categories, VFB_{min50} to VFB_{min78}. The maximum VFB values range from 50% to 97%, assigned to 16 categories, VFB_{max50} to VFB_{max97}.

Voids filled with bitumen (VFB) are determined according to the following equation:

$$VFB = ((B \times \rho_b/\rho_B)/VMA) \times 100\ (\%),$$

where *B* is the percentage of binder in the total mix (in 0.1%), ρ_b is the bulk density of the specimen (kg/m³), ρ_B is the density of binder (kg/m³) and VMA denotes the voids in the mineral aggregate (in 0.1%).

More information is given in CEN EN 13108-20 (2008), D.2, and CEN EN 12697-8 (2003).

5.4.1.4.4 Voids in the mineral aggregates

The minimum percentage of voids in mineral aggregates (VMA) of the compacted specimens is selected from appropriate tables given in CEN EN 13108-1 (2008).

The minimum VMA values range from 8% to 16%, assigned to six categories, $VMA_{min8.0}$ to $VMA_{min16.0}$.

Voids in mineral aggregates are determined according to the following equation:

$$VMA = V_m + B \times \rho_b/\rho_B\ (\%),$$

where VMA denotes the voids in the mineral aggregate (in 0.1%), V_m represents the air void content of the specimen (in 0.1%), *B* is the binder content of the specimen (in 100% mix) (in 0.1%), ρ_b is the bulk density of the specimen (in kg/m³) and ρ_B is the density of the binder (in kg/m³).

More information is given in CEN EN 13108-20 (2008), D.2, and CEN EN 12697-8 (2003).

5.4.1.4.5 Void content at 10 gyrations

When the compaction is conducted with a gyratory compactor, the minimum void content after 10 gyrations should be either 9%, 11% or 14%, assigned to categories $V10G_{min9}$, $V10G_{min11}$ or $V10G_{min14}$.

Voids after 10 gyrations are determined according to CEN EN 13108-20 (2008), D.2, and CEN EN 13108-31 standards.

5.4.1.4.6 Marshall values for application on airfields

The Marshall values such as Marshall stability, flow and quotient (Q) are only determined when the AC is to be used in airfield projects.

The testing equipment used to measure stability and flow is shown in Figure 5.3.

The specification with respect to Marshall stability sets requirements not only on a minimum value but also on a maximum value.

The minimum and maximum Marshall stability value (S_{min} and S_{max}) is selected from appropriate tables provided in CEN EN 13108-1 (2008).

The minimum stability values range from 2.5 to 12.5 kN, assigned to five categories, $S_{min2.5}$, $S_{min5.0}$, $S_{min7.5}$, S_{min10} or $S_{min12.5}$.

The maximum stability values range from 7.5 to 15.0 kN, assigned to five categories, $S_{max7.5}$, $S_{max10.0}$, $S_{max12.5}$, S_{max15} or $S_{max17.5}$.

The flow (F) values range from 1 to 8 mm, assigned to seven flow categories, $F_{1.0}$ to $F_{8.0}$. From these values, a minimum and maximum are selected, with the range to be at least 2 mm.

Finally, the minimum quotient (Q_{min}) values range from 1.0 to 4.0 kN/mm, assigned to seven categories, $Q_{min1.0}$ to $Q_{min4.0}$.

The characteristic Marshall properties are determined according to CEN EN 13108-20 (2008), D.10, and CEN EN 12697-34 (2012).

Figure 5.3 Marshall stability tester. (Courtesy of Controls Srl.)

5.4.1.5 Fundamental requirements

When the target composition is determined by fundamental requirements, the AC should fulfil the general requirements, as outlined in Section 5.4.1.3, plus the fundamental requirements, which are grading, binder content, stiffness, resistance to permanent deformation in the triaxial compression test and resistance to fatigue.

5.4.1.5.1 Grading and binder content

Using fundamental requirements, a less prescriptive way of specifying a composition is required. At the target composition, the grading should conform to Section 5.4.1.3.1.

Similarly, for the binder content, only a minimum value is required, which is 3.0%, by mass.

5.4.1.5.2 Stiffness

The stiffness (S) of the cylindrical specimens should be between a minimum and a maximum value selected from values/categories provided by CEN EN 13108-1 (2008).

The minimum stiffness values range from 1500 to 21,000 MPa, assigned to 13 categories, $S_{min1500}$ to $S_{min21000}$.

The maximum stiffness values range from 7000 to 30,000 MPa, assigned to eight categories, $S_{max7000}$ to $S_{max30000}$.

The stiffness is determined with one of the methods/apparatuses described in EN 12697-26 (2012), and the test conditions (temperature and loading frequency) are given in Table D.3 of CEN EN 13108-20 (2008).

Specimens are prepared according to CEN EN 13108-20 (2008), and the type and conditions of compaction (impact, gyratory or roller compaction) are selected from Table C.1 of CEN EN 13108-20 (2008).

5.4.1.5.3 Resistance to permanent deformation in the triaxial compression test

The resistance to permanent deformation under the triaxial compression test of cylindrical specimens is expressed in reference to creep rate (f_c) and should be less than the maximum value specified.

The maximum values of creep rate range from 0.2 to 16 µstrain/loading cycle, assigned to 16 categories, $f_{cmax0.2}$ to f_{cmax16}.

The resistance to permanent deformation under the triaxial compression test is determined according to CEN EN 12697-25 (2005), Method B, and the test conditions are determined by Table D.2 of CEN EN 13108-20 (2008).

Specimens are prepared and compacted as those for determination of stiffness.

5.4.1.5.4 Resistance to fatigue

The resistance to fatigue of compacted specimens is expressed in reference to the tensile strain at 10^6 loading cycles (ε_6) and should be selected from values specified.

The values of tensile strain (ε_6) range from 50 to 310 µstrain, assigned to 13 categories, ε_{6-50} to ε_{6-310}.

The resistance to fatigue is determined according to CEN EN 12697-24 (2012), Annex A or B, and test conditions are determined by Table D.4 of CEN EN 13108-20 (2008).

Specimens are prepared and compacted as those for determination of stiffness.

Table 5.3 Tolerances of gradation and binder content about target composition

Percentage passing[a]	Individual samples		Mean of four samples	
	D < 16 mm	D < 16 mm	D < 16 mm	D < 16 mm
D	−8 ± 5	−9 ± 5	±4	±5
D/2 or other characteristic coarse sieve	±7	±9	±4	±4
2 mm	±6	±7	±3	±3
Characteristic fine sieve	±4	±5	±2	±2
0.063 mm	±2	±3	±1	±2
Soluble binder content	±0.5	±0.6	±0.3	±0.3

Source: Adapted from CEN EN 13108-21, *Bituminous mixtures – Material specifications – Part 21: Factory Production Control*, Brussels: CEN, 2008. With permission (© CEN).

[a] A tolerance of −2% should apply to the requirement of 100% passing 1.4D.

5.4.1.6 Tolerances and test frequencies for conformity assessment

Tolerances of AC, produced by any asphalt plant in terms of the target composition, should be within certain limits.

In accordance to CEN EN 13108-21 (2008), tolerances of the gradation curve and bitumen content of a single result, or of the mean of four results, for an AC, or any other asphalt apart from MA and HRA, are given in Table 5.3.

The frequency of samplings for gradation and binder content determination, once the compliance method has been chosen (by a single result or mean of four results), depends on the operating compliance level (OCL) of the plant. More information is given in CEN EN 13108-21 (2008), Annex A.

5.4.2 AC in accordance to American standards

In American standards, as in European standards, ACs are also designated by the maximum aggregate size. Mixture and gradation limits specified by ASTM 3515 (2010) and control points, shown in boldface, specified by Superpave mix methodology (Asphalt Institute SP-2 2001) for dense AC are given in Table 5.4. It is stated that ASTM 3515 (2010) has been withdrawn in 2009 and has not been replaced. However, to the author's opinion, ASTM 3515 (2010) gradation limits are still helpful to determine the gradation envelope of dense AC until a replacement comes out.

The control limits determined by Superpave mix methodology work in the same principle as the general requirements of CEN EN 13108-1 (2008).

Regarding the target mix gradation, the Superpave mix design methodology, apart from the control points, also suggests the use of the minimum and maximum boundaries on certain sieves. The zone determined is known as the restricted zone. Minimum and maximum limit values (boundaries) that determine the restricted zones, per designated mix of dense AC, are given in Table 5.5.

The criterion of restricted zone is obligatory and it is recommended that the target mix gradation should pass outside the restricted zone.

The restricted zone prevents a gradation from following the maximum density line in fine aggregate sieves. Gradations that follow the maximum density line often have inadequate VMA to allow room for sufficient binder for durability. These gradations are typically sensitive to asphalt content, which may easily become plastic with even minor variations in binder content (Asphalt Institute MS-2).

Table 5.4 Gradation specifications for dense asphalt concrete mixtures in accordance to American standards

Sieve (mm)	Dense-graded asphalt concrete mixtures				
	Mix designation				
	37.5 mm	25.0 mm	19.0 mm	12.5 mm	9.5 mm
	(%) Passing, by mass				
63	—	—	—	—	—
50	100	—	—	—	—
37.5	**90–100**[a]	100	—	—	—
25.0	—	**90–100**[a]	100	—	—
19.0	56–80	—	**90–100**[a]	100	—
12.5	—	56–80	—	**90–100**[a]	100
9.5	—	—	56–80	—	**90–100**[a]
4.75	23–53	29–59	35–65	44–74	55–85
2.36	**15–41**	**19–45**	**23–49**	**28–58**	**32–67**
1.18	—	—	—	—	—
0.600	—	—	—	—	—
0.300	4–16	5–17	5–19	5–21	7–23
0.150	—	—	—	—	—
0.075	**0–6**	**1–7**	**2–8**	**2–10**	**2–10**

Sources: Adapted from ASTM D 3515, *Standard specification for hot-mixed, hot-laid bituminous paving mixtures*, West Conshohocken, Pennsylvania, US: ASTM International, 2010; and Asphalt Institute, *Superpave mix design*, Superpave Series No. 2 (SP-2), 3rd Edition, Lexington, USA: Asphalt Institute, 2001.

[a] Figures in boldface refer to control points proposed by Superpave design methodology (Asphalt Institute SP-2 2001).

Table 5.5 Boundaries of aggregate restricted zone

Sieve (mm)	Type of asphalt concrete				
	AC 37.5	AC 25.0	AC 19	AC 12.5	AC 9.5
4.75	34.7–34.7	39.5–39.5	—	—	—
2.36	23.3–27.3	26.8–30.8	34.6–34.6	39.1–39.1	47.2–47.2
1.18	15.5–21.5	18.1–24.1	22.3–28.3	25.6–31.6	31.6–37.6
0.6	11.7–15.7	13.6–17.6	16.7–20.7	19.1–23.1	23.5–27.5
0.3	10.0–10.0	11.4–11.4	13.7–13.7	15.5–15.5	18.7–18.7

Source: Asphalt Institute, *Superpave mix design*, Superpave Series No. 2 (SP-2), 3rd Edition, Lexington, USA: Asphalt Institute, 2001. With permission.

Tolerances of the gradation curve, as well as of the binder content, from the target composition, are given in Table 5.14.

Bitumen selection is based on either penetration or viscosity-graded specifications determined by ASTM D 946 (2009) and ASTM D 3381 (2012) or AASHTO M 226 (2012), respectively, or, when Superpave mix design is followed, on performance-graded specifications according to ASTM D 6373 (2007) or AASHTO M 320 (2010).

Aggregates used should have the appropriate physical and mechanical properties required.

Table 5.6 Superpave aggregate consensus property requirements

Design ESALs[a]	Coarse aggregate angularity (%), minimum[b]		Uncompacted void content of fine aggregate (%), minimum		Sand equivalent (%), minimum	Flat and elongated (%), maximum
	≤100 mm	>100 mm	≤100 mm	>100 mm		
<0.3	55/—	—/—	—	—	40	—
0.3 to <3	75/—	50/—	40	40	40	10
3 to <10	85/80[b]	60/—	45	40	45	10
10 to <30	95/90	80/75	45	40	45	10
≥30	100/100	100/100	45	40	50	10

Source: Asphalt Institute, *Superpave mix design*, Superpave Series No. 2 (SP-2), 3rd Edition, Lexington, USA: Asphalt Institute, 2001. With permission.

[a] Design ESALs are the anticipated project traffic level expected on the design lane over a 20-year period.

[b] 85/80 denotes that 85% of the coarse aggregate has one fractured face and 80% has two or more fractured faces.

According to Superpave methodology, the properties of aggregates that were almost consensually considered to affect the good performance of AC are the following: the percentage of fractured faces of coarse aggregate (coarse aggregate angularity) (ASTM D 5821 2006 or AASHTO T 335 2009), the uncompacted void content of fine aggregate (fine aggregate angularity) (ASTM C 1252 2006 or AASHTO T 304 2011), the sand equivalent (ASTM D 2419 2009 or AASHTO T 176 2008) and flat and elongated particles in coarse aggregate (ASTM D 4791 2010). Information about the abovementioned tests is given in Chapter 2.

The required minimum values for coarse and fine aggregate limits proposed as a function of traffic level and layer thickness (≤100 mm or >100 mm) are given in Table 5.6.

Apart from the above consensually decided properties, aggregates should meet the requirements for resistance to degradation in the Los Angeles machine (ASTM C 131 2006 and C 535 2012 or AASHTO T 96 2010), soundness of aggregates (ASTM C 88 2013 or AASHTO T 104 2011) and contaminants such as clay lumps and friable particles in aggregates (ASTM C 142 2010 or AASHTO T 112 2012). A short description of tests for the determination of the above properties is given in Chapter 2.

Limit values, which are used by various organisations or states in the United States, per property, typically range from a maximum percentage between 35% and 45% for the degradation in the Los Angeles machine, a maximum percentage between 10% and 20% for the soundness of the aggregate (for five cycles) and between 0.2% and 10% for clay lumps and friable particles (depending on the exact composition of the contaminant).

5.4.2.1 Superpave mix design

The AC mix design in the United States is suggested to be conducted with the Superpave methodology.

According to the Superpave methodology (AASHTO M 323 2013; Asphalt Institute SP-2 2001), which sooner or later will completely replace Marshall mix design, the AC is currently designed with volumetric criteria, once the materials (bitumen and aggregates) have been chosen.

AC performance criteria will also be used in the future on the basis of fundamental properties. The criteria related to asphalt performance, as well as the determination of the respective tests to define fundamental properties, have not been determined yet.

The volumetric Superpave design criteria determined are given in Table 5.7.

The fundamental difference between the Superpave and the Marshall methodology is the way specimens are compacted. In the Superpave mix design methodology, the specimens are compacted with the gyratory compactor (Table 5.8) (ASTM D 4013 2009) and the relative density

Table 5.7 Superpave volumetric mixture design requirements

Design ESALs[a] (×10^6)	Required density (% of theoretical maximum specify gravity)			Voids in mineral aggregate (VMA) (%), minimum					VFA[b,g]	D/B[c]
				Nominal maximum aggregate size (mm)						
	N_{ini}[d]	N_{des}[e]	N_{max}[f]	37.5	25.0	19.0	12.5	9.5		
<0.3	≤91.5								70–80	0.6–1.2[h]
0.3 to <3	≤90.5	96.0	≤98.0	11.0	12.0	13.0	14.0	15.0	65–78	
3 to <10	≤89.0								65–75	
10 to <30										
≥30										

Source: Asphalt Institute, *Superpave mix design*, Superpave Series No. 2 (SP-2), 3rd Edition, Lexington, USA: Asphalt Institute, 2001. With permission.

[a] Design ESALs are the anticipated project traffic level, expected on the design lane over a 20-year period. Determine the design ESALs for 20 years and choose the appropriate N_{design} level.
[b] VFA, voids filled with asphalt.
[c] D/B, dust-to-bitumen ratio.
[d] N_{ini}, initial number of gyrations (see Table 5.8).
[e] N_{des}, design number of gyrations (see Table 5.8).
[f] N_{max}, maximum number of gyrations (see Table 5.8).
[g] For 9.5 mm nominal maximum size mixtures, the specified VFA range should be 73% to 76% for design traffic levels ≥3 × 10^6. For 25.0 mm nominal maximum size mixtures, the specified lower limit of the VFA should be 67% for design traffic levels <0.3 × 10^6. For 37.5 mm nominal maximum size mixtures, the specified lower limit of the VFA should be 64% for all design traffic levels.
[h] If the aggregate gradation passes beneath the boundaries of the aggregate restricted zone, consideration should be given to increasing the dust (filler) to binder ratio criteria from 0.6–1.2 to 0.8–1.6.

Table 5.8 Gyratory compaction effort for superpave hot mix design

Design ESALs[a] (×10^6)	Compaction parameters		
	N_{ini}	N_{des}	N_{max}
<0.3	6	50	75
0.3 to <3	7	75	115
3 to <30	8	100	160
≥30	9	125	205

Source: Asphalt Institute, *Superpave mix design*, Superpave Series No. 2 (SP-2), 3rd Edition, Lexington, USA: Asphalt Institute, 2001. With permission.

[a] Based on 20 years design life regardless of actual design life.

is determined by ASTM D 6925 (2009) or AASHTO T 312 (2012). In the Marshall design methodology, the specimens are compacted using an impact compactor (see Section 5.4.2.2).

The final step on the Superpave mix design procedure is the determination of the moisture sensitivity of the target mix.

This consists of the production of six cylindrical specimens and testing their indirect tensile strength according to ASTM D 4867 (2009) or AASHTO T 283 (2011). The specimens are compacted to 7% voids. One set of three specimens is partially saturated using a vacuum chamber, followed by optional freeze/thaw conditioning cycle and 24 h heating at 60°C. Then, all specimens are tested for determination of tensile strength.

Water sensitivity is designated as the ratio of the average tensile strength of saturated specimens to the average tensile strength of dry specimens. The result is expressed as a percentage. The minimum ratio value required is 80%.

It is stated that the water sensitivity test is similar to the test conducted by CEN EN 12697-23 (2003) and it constitutes a general requirement for AC composition following CEN EN 13108-1 (2008).

5.4.2.1.1 Steps of Superpave mix design – summarised

The steps followed in AC mix design by Superpave procedure are summarised below:

A. Selection of materials

    For the selection of binder, aggregate and possibly modifiers, see Section 5.4.2.

B. Selection of aggregate gradation

  1. Select three trial gradations taking into account the control points given in Table 5.6 and considering the restricted zones of Table 5.7.

  2. Select trial bitumen content for the production of trial asphalt mixtures (empirically or using an equation, see Asphalt Institute SP-2 2001).

  3. Determine the initial and design number of gyrations (N_{ini} and N_{des}), from Table 5.10.

  4. Mix (at an appropriate mixing temperature) and compact, using a gyratory compactor, the specimens (minimum of two specimens per trial mixture).

  5. Determine the theoretical maximum specific gravity of the bituminous mixtures (G_{mm}) (ASTM D 2041 2011) and the bulk specific gravity of compacted bituminous mixtures (G_{mb}) (ASTM D 2726 2011 or ASTM D 1188 2007).

  6. Evaluate the trial mixtures, using the average values for each trial gradation, by determining the following:

    a. The average percentage (%) of G_{mm} after compaction with N_{ini} and N_{des} number of gyrations

    b. The air voids (%) and VMA (%)

  7. Estimate the bitumen content to achieve 4% air voids (96% G_{mm} at N_{des}).

  8. Estimate the volumetric (VMA and VFA) and mixture compaction properties at the estimated bitumen content.

  9. Determine the D/B ratio.

  10. Compare the properties of the mixtures with the volumetric requirements of Table 5.9.

  11. Select the most promising trial blend (gradation) for further analysis. If none satisfy the requirements, further trial gradations/blends need to be evaluated.

C. Selection of design bitumen content

  1. Mix and compact the mixture with the selected aggregate gradation at five different bitumen contents (at the estimated content above, the estimated bitumen content ±0.5% and the estimated bitumen content +1.0%). During compaction, keep densification data.

  2. Determine compaction and volumetric properties with respect to bitumen content.

    a. Determination of compaction property %G_{mm} at N_{ini}, N_{des} and N_{max}

    b. Determination of volumetric properties (air voids, VMA and VFA)

    c. Determination of D/B ratios

    d. Draw volumetric property diagrams with respect to bitumen content

  3. Select the design bitumen content for 4% air voids.

  4. Determine VMA and VFA at the above bitumen content.

  5. Compare air voids properties, VMA, VFA, D/B and %G_{mm} with the design requirements of Table 5.9.

D. Evaluation of the effect of moisture on the design mix

  1. Execute the moisture sensitivity test, according to ASTM D 4867 (2009), using the design mix determined in (C) above. If the criterion of tensile strength is satisfied (tensile strength ratio ≥ 80%), this is the design mix.

Table 5.9 Marshall mix design criteria

Marshall method mix criteria	Light traffic ITA, <10^4	Medium traffic ITA, 10^4–10^6	Heavy traffic ITA, >10^6
	Surface and other layers		
Compaction, no. of blows	2 × 35	2 × 50	2 × 75
Minimum stability, kN (lb)	3.34 (750)	5.34 (1200)	8.01 (1800)
Flow, mm (0.01 in)	2.0–4.5 (8–18)	2.0–4.0 (8–16)	2.0–3.5 (8–14)
Air voids, %	3–5	3–5	3–5
Voids filled with asphalt, VFA, %	70–80	65–78	65–75
	For all cases		
Voids in mineral aggregate (VMA), %	Nominal maximum particle size[a] (mm)	Minimum VMA (%), for design air voids[b]	
		3%	4%	5%

		3%	4%	5%
	63 (2.5″)	90	100	110
	50 (2.0″)	9.5	10.5	11.5
	37.5 (1.5″)	100	110	120
	25.0 (1.0″)	110	120	130
	19.0 (3/4″)	120	130	140
	12.5 (1/2″)	130	140	150
	9.5 (3/8″)	140	150	160
	4.75 (No. 4)	160	170	180
	2.36 (No. 8)	190	200	210
	1.18 (No. 16)	21.5	22.5	23.5

Source: Asphalt Institute MS-2, *Mix design methods – For Asphalt concrete and other hot-mix types*, Manual Series 2 (MS-2), 6th Edition, Lexington, USA: Asphalt Institute. With permission.

[a] The nominal maximum particle size is one size larger than the first sieve to retain more than 10%.
[b] Interpolate minimum VMA for design air void values between those listed.

5.4.2.2 Marshall mix design

Although the Marshall mix design (Asphalt Institute MS-2) is relatively old, it is currently used by several agencies/organisations because it uses simpler and cheaper laboratory equipment. However, many engineers believe that the way specimens are compacted does not simulate compaction on site. Additionally, the Marshall stability and flow do not adequately assess the asphalt shear strength. These two parameters do not ensure the asphalt resistance in permanent deformation (resistance to rutting).

Following the Marshall methodology, the AC composition is based on the following Marshall criteria: Marshall stability and flow, percentage of air voids, percentage of voids in the mineral aggregate (VMA) and percentage of voids filled with asphalt (bitumen) (VFA). The limit values per property/criterion are given in Table 5.9.

5.4.2.2.1 Marshall mix design procedure

The Marshall mix design is used to determine the optimum bitumen content of dense-graded AC with 25.4 mm maximum nominal aggregate size. This method is specified by Asphalt Institute (MS-2), by ASTM D 6926-04 (2010) and by ASTM D 6927-06 (2006) or AASHTO T 245 (2013).

It consists of a specimen production 100 mm (or 101 mm) in diameter, compacted with an impact Marshall compactor, using a Marshall apparatus for stability and flow determination as well as for volumetric property determination (air voids, voids in the mineral aggregate [VMA] and voids filled with asphalt [bitumen] [VFA]).

Because they are empirical properties, Marshall stability and flow cannot be used in analytical pavement design calculations.

5.4.2.2.1.1 PREPARATION OF TEST SPECIMENS

The target aggregate mix is placed in a heated oven to dry and obtain the desired temperature. Adequate aggregate mass is transferred into the heated mixer and then the desired mass of binder is also added to the mixer. The mass of aggregates required for producing one Marshall specimen is approximately 1.2 kg. The mass of bitumen is such that it corresponds to the desired binder content (at least five different AC mixtures should be produced with different binder contents, using steps from 0.3% to 0.5%). Three or more specimens are usually produced from each mixture. Mixing temperature may be determined by the bitumen type used (Table 3.22).

After mixing, the hot AC is placed in a cylindrical metal mould 100 mm (or 101 mm) in diameter. The mass needed should be adequate to produce a specimen of approximately 63.5 mm in height. The mass of AC has two layers. Each layer is subject to slight pre-compaction with a heated metal rod (15 times around the perimeter and 10 times over the interior). Then, the mould containing the mixture is placed in a compaction apparatus (Marshall compactor, Figure 5.2a) and compacted.

During compaction, the asphalt is subject to 35, 50 or 75 blows on each side. The compacted specimen is allowed to cool in the mould and then extracted by an extraction device. The specimens, after being checked for evenness on both sides, are numbered and weighed in air and water. After 24 h and after measuring the height, they are tested for stability and flow.

Preparation of specimens is described in detail in ASTM D 6926 (2010) or AASHTO T 245 (2013).

5.4.2.2.1.2 SPECIMEN STABILITY AND FLOW TESTS

The specimens are placed in a water bath at 60°C $\pm$ 1°C for 30 to 40 min before testing using the Marshall apparatus (Figure 5.3). The Marshall apparatus compresses diametrically the specimen at a constant rate (50.8 mm/min) until breaking occurs. The applied load at breaking, measured in kilonewtons (or pounds) is the Marshall stability. The resulting deformation at breaking, measured in millimetres (or inches), is the flow.

The stability value obtained by specimens with a thickness other than 63.5 mm is corrected by multiplying the stability with the correlation ratios given in the Asphalt Institute manual (Asphalt Institute MS-2).

The procedure for determining stability and flow is described in detail in ASTM D 6927 (2006).

5.4.2.2.1.3 DETERMINATION OF THE DESIGN ASPHALT CONTENT OF THE MIX

For the determination of the optimum binder content in mix, apart from stability (corrected stability) and flow, the following volumetric properties should be calculated: bulk density (ρ_b) or bulk specific gravity (G_{mb}), the percentage of air voids (V_a), the percentage of voids in the mineral aggregate (VMA) and the percentage of voids filled with asphalt (bitumen) (VFA).

The above parameters at optimum binder content completely characterise the AC mix. Stability ensures sufficient strength, while flow ensures acceptable deformation over a period of service life.

The voids determine the expected voids in the bituminous layer after some years of usage.

If AC is designed with a lower void content than the one specified, it will certainly deform prematurely, presenting rutting and shoving.

VMA and VFA are additional parameters for ensuring good performance. VMA ensures that the space among aggregates is adequate to accommodate the mass of bitumen, whereas

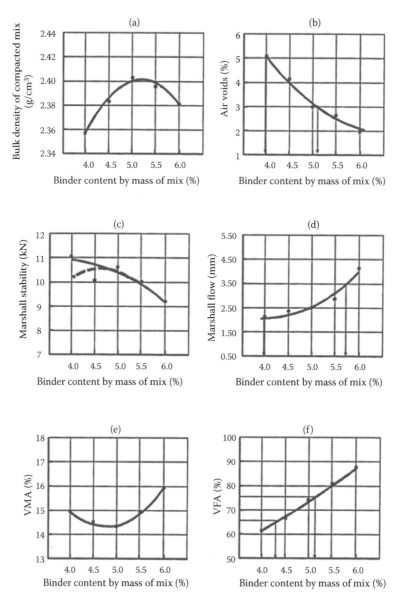

Figure 5.4 Typical graphs of Marshall characteristic properties.

VFA ensures that an adequate number of voids are filled with bitumen. The latter is of great importance because it indirectly determines the necessary binder quantity in the mixture. In other words, it determines the minimum bitumen quantity required for good mix cohesion and the maximum bitumen content to avoid premature deformation or bleeding. How to calculate the above parameters is explained in Section 5.4.2.3.

Once the above properties are calculated, six diagrams are plotted. They all have the horizontal axis representing the binder content and the vertical axis representing the above properties. A typical example of diagrams is shown in Figure 5.4.

As shown in Figure 5.4a, the density curve may present a maximum value as the binder content increases. The stability curve always presents a descending trend and sometimes a maximum value (see Figure 5.4c, two alternative curves).

Air voids continuously decrease as the binder content increases (Figure 5.4b). Correspondingly, the deformation increases as the binder content increases (Figure 5.4d), whereas the VMA initially decreases up to a minimum value and then increases (Figure 5.4e). Finally, the VFA continuously increases as the binder content increases (Figure 5.4f).

The mixture that meets all the requirements of Table 5.9 at the same time is selected as the target mixture. There is more than one such mixture. To eliminate and find the optimum, it is suggested that the diagram shown in Figure 5.5 be drawn.

VMA, VFA, V_a, stability and deformation criteria are placed in this diagram together with respective limit values of bitumen percentage in the mixture. The common area, in which the mixture meets all Marshall criteria for any bitumen content, determines the acceptable binder content. The optimum binder content is designated as the average binder content value corresponding to this area (4.7% in the example shown in Figure 5.5). The boundaries of the common area (4.3% and 5.1%) determine the acceptable variation of binder content for good performance.

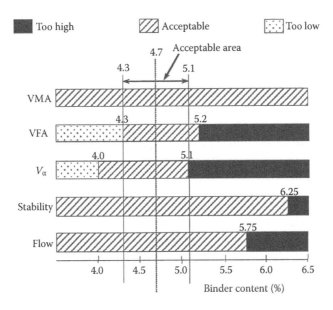

Figure 5.5 Graphical representation to narrow the range of acceptable binder content and find the optimum on the basis of Marshall criteria.

5.4.2.3 Modified Marshall mix design for mixtures with maximum nominal aggregate size >25 mm

Dense-graded ACs with a maximum nominal aggregate size >25 mm (usually for base layers) cannot be designed with the standard Marshall mix design outlined above.

The determination of the optimum binder content is achieved with the modified Marshall mix design procedure.

The procedure is the same as the standard Marshall mix design procedure described above, except for the following differences attributed to mixes composed of large aggregates, that is, >25 mm.

The size of the specimen is 152.4 mm (6 in.) (diameter) by 95.2 mm (5.88 in.) (height). The compaction hammer has a flat, circular tamping face 149.4 mm (5.88 in.) in diameter and a mass of 10.21 kg (22.50 lb).

The number of blows applied is 1.5 times higher than the number of blows used in the standard method, namely, 112 blows instead of 75 blows and 75 blows instead of 50 blows.

Design criteria are similar to those given in Table 5.11, with the only difference that minimum stability values increase by 2.25 times and flow limits increase by 1.5 times.

Because of the different sizes of specimens, different stability correlation ratios are used, which can be found in Asphalt Institute MS-2.

The rest of the procedure that is used to define the optimum binder content is exactly the same as the one mentioned in Section 5.4.2.2.

More details can be found in ASTM D 5581 (2007) and Asphalt Institute MS-2.

5.4.2.4 Tolerances from target mix

Tolerance limits for the aggregate grading and binder content of the produced AC in relation to the target mix for quality assurance are established by the agency, together with the procedure for verifying tests.

Relevant general information can be found in *The Asphalt Handbook* (Asphalt Institute MS-4 2007) and a mix design methods manual (Asphalt Institute MS-2).

5.4.2.5 Volumetric properties of compacted bituminous mixture

The volume of the compacted specimen of any bituminous mixture consists of the volume occupied by aggregates, the volume occupied by bitumen and the volume of air voids. The volume, which is occupied by bitumen and air voids, is known as volume in mineral aggregates (VMA). When bituminous binder is added, part of the volume of air voids is filled with bitumen (asphalt). The volume is known as voids filled with asphalt (VFA). The above volumetric characteristic properties are presented in Figure 5.6.

Aggregates attributed to the surface pores normally possess and absorb a certain quantity of bitumen. As a consequence, the remaining bitumen quantity is in fact the one that coats the aggregates, fills the voids and provides cohesion in the mixture. This quantity of bitumen is designated as 'effective' bitumen quantity, and it is always less than the initial quantity of bitumen added, unless the aggregate's absorption is zero (ideal case).

The surface pores, in the absence of bitumen, absorb water (surface voids permeable to water). Because of the lower viscosity of water in comparison to bitumen's viscosity, water absorption is always higher than bitumen absorption. The schematic representation of an aggregate-coated particle given in Figure 5.7 illustrates the above, as well as other concepts.

Provided that the bulk specific gravity (G_{sb}) and the effective specific gravity (G_{se}) of the total aggregate, as well as the specific gravity of the bitumen (G_b), are known, the volumetric properties of bituminous mixture (asphalt mixture) may be determined.

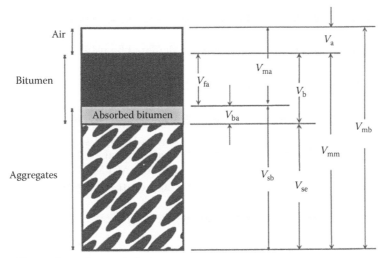

V_{mb} = volume of compacted mix
V_a = volume of air voids
V_b = volume of bitumen
V_{ma} = volume of voids in mineral aggregates (VMA)
V_{sb} = volume of mineral aggregate (by bulk specific gravity)
V_{se} = volume of mineral aggregate (by effective specific gravity)
V_{fa} = voids filled with asphalt (VFA)
V_{mm} = voidless volume of paving mix
V_{ba} = volume of absorbed bitumen

Figure 5.6 Representation of volumes in a compacted bituminous mixture. (Adapted from the Asphalt Institute, *The Asphalt Handbook*, Manual Series 4 [MS-4], 7th Edition, Lexington, USA: Asphalt Institute, 2007.)

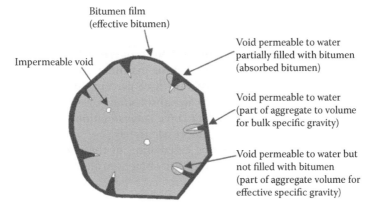

Figure 5.7 Coated aggregate particle illustrating voids related to the determination of specific gravities.

The *bulk specific gravity* of the total aggregate (G_{sb}) is calculated by the bulk specific gravities of aggregate fractions and their respective percentage, by weight, in the total aggregate mix.

The bulk specific gravity of the aggregates is calculated by using the following equation:

$$G_{sb} = 100/[(P_1/G_1) + (P_2/G_2) + \ldots + (P_n/G_n)],$$

where G_{sb} is the bulk specific gravity of the total aggregates; P_1, P_2, ... P_n are the percentages by weight of aggregates, 1, 2, ... n (%); G_1, G_2, ... G_n are the bulk specific gravities of aggregates 1, 2, ... n; and n is the number of different-sized aggregates.

The bulk specific gravities are determined according to ASTM C 127 (2012) or AASHTO T 85 (2013) and ASTM C 128 (2012), for coarse aggregate, or AASHTO T 84 (2013), for fine aggregate.

The *effective specific gravity* of the total aggregate (G_{se}) is calculated by using the following equation:

$$G_{se} = (100 - P_b)/[(100/G_{mm}) - (P_b/G_b)],$$

where G_{mm} is the maximum specific gravity of the bituminous mixture (ASTM D 2041 2011), P_b is the percentage of bitumen by total mass of mixture (%) and G_b is the specific gravity of bitumen.

The effective specific gravity of the total aggregate should have a value between the apparent specific gravity (G_{sa}) and the bulk specific gravity of the total aggregate (G_{sb}) (i.e. $G_{sa} < G_{se} < G_{sb}$). If not, the test determining the maximum specific gravity of loose bituminous mixture is repeated.

The *apparent specific gravity* (G_{sa}) is the ratio of the mass in air of a unit volume of an impermeable material at a stated temperature to the mass in air of equal density of an equal volume of gas-free distilled water at a stated temperature. In other words, the aggregate apparent specific gravity does not include the volume of water-permeable voids in the aggregate (Asphalt Institute MS-4 2007).

The *maximum specific gravity* (G_{mm}) of the loose hot bituminous mixture should be determined, at all binder contents, to calculate the volumetric properties of the compacted bituminous mixture.

However, given the fact that bitumen absorption in the range of bitumen contents used does not change significantly, it is suggested that the maximum specific gravity of the loose hot bituminous mixture is determined at two or three percentages of bitumen close to the expected optimum bitumen content. When water absorption is less than 1%, one determination of G_{mm} is considered sufficient. If two or three determinations of G_{mm} are carried out, the average value is taken to determine G_{se} more accurately.

The above G_{se} value can be used for the calculation of the maximum specific gravity (G_{mm}) of the bituminous mixture, for any bitumen content, using the following equation:

$$G_{mm} = 100/[(P_s/G_{se}) + (P_b/G_b)],$$

where P_s is the percentage of the aggregate by total mass of mixture (%), P_b is the percentage of bitumen by total mass of mixture (%) and G_{se} is the effective specific gravity of the aggregate.

5.4.2.5.1 Air voids of the compacted bituminous mixture

The air voids in a compacted bituminous mixture consists of the small air spaces among the coated aggregate particles. The percentage of air voids in the compacted bituminous mixture is calculated by the following equation:

$$P_a = [(G_{mm} - G_{mb})/G_{mm}] \times 100 \ (\%),$$

where P_a is the percentage of air voids (%), G_{mm} is the maximum specific gravity of the hot asphalt mix (ASTM D 2041 2011) and G_{mb} is the bulk specific gravity of the compacted mixture (ASTM D 2726 2011 or ASTM D 1188 2007, if coated samples are used).

More details for the determination of the per cent air voids in compacted dense or open bituminous paving mixtures can be found in ASTM D 3203 (2011) or AASHTO T 269 (2011).

5.4.2.5.2 Voids in the mineral aggregate (VMA)

The voids in the mineral aggregate (VMA) are defined as the void space between the aggregates in a compacted bituminous mixture that includes the air voids and the effective bitumen content, expressed as a percentage of the total mix volume.

The VMA is expressed as a percentage of the total volume of the compacted mixture (specimen) (V_{mb}) and is determined by the following equation:

$$\mathrm{VMA} = 100 - (G_{mb} \times P_s/G_{sb}) \ (\%),$$

where G_{mb} is the bulk specific gravity of the compacted mixture, G_{sb} is the bulk specific gravity of the aggregate and P_s is the percentage of the aggregate, by total mass of mixture (%).

More details on the determination of the voids in the mineral aggregate can be found in ASTM D 6995 (2005).

By definition, VMA is also equal to the algebraic sum of the volume of air voids ($V_a = P_a$) plus the volume that is occupied by the effective bitumen quantity (V_{be}) or the initial bitumen quantity volume (V_b), only in the case where there is no bitumen absorbed by the aggregates, namely:

$$\mathrm{VMA} = V_a + V_{be} \ (\text{or } V_b) \ (\%).$$

From the above equation, the volume of bitumen can be calculated provided the VMA and the volume of air voids are known.

The *volume of the bitumen* (V_{be} or V_b), as a percentage of the total mixture, can be independently calculated by the following equation:

$$V_{be} \ (\text{or } V_b) = [(G_{mb}/G_{mm}) - (G_{mb} \times P_s/G_{sb})] \times 100 \ (\%)$$

where G_{mb} is the bulk specific gravity of the compacted mixture, G_{mm} is the maximum specific gravity of the bituminous mixture (ASTM D 2041 2011), G_{sb} is the bulk specific gravity of the aggregate and P_s is the percentage of the aggregate, by total mass of mixture.

5.4.2.5.3 Void filled with asphalt (VFA)

The void filled with asphalt (bitumen) (VFA) is the percentage of the void space among the aggregate particles (VMA) that is filled with bitumen (effective bitumen, i.e. not including the absorbed bitumen). The VFA is determined using the equation below (Asphalt Institute MS-4 2007):

$$VFA = 100 \times [(VMA - V_a)/VMA] \text{ (\%)},$$

where VMA denotes the voids in the mineral aggregate (%) and V_a represents the air voids in the compacted mixture (%).

5.4.2.5.4 Bitumen absorption

Bitumen (asphalt) absorption is expressed as a percentage by mass of aggregate rather than as a percentage of total mass of the bituminous mixture. The absorbed asphalt is determined by the following equation (Asphalt Institute MS-4 2007):

$$P_{ba} = 100 \times [(G_{se} - G_{sb})/G_{sb} \times G_{se}] \times G_b \text{ (\%)},$$

where P_{ba} is the percentage of absorbed asphalt, by mass of aggregate; G_{se} is the effective specific gravity of the aggregate; G_{sb} is the bulk specific gravity of the aggregate; and G_b is the specific gravity of the asphalt (bitumen).

The percentage of bitumen absorption is always lower than the percentage of water absorption of aggregates. It is difficult to determine exactly how much lower the absorption in bitumen is. This percentage (absorption) mainly depends on the porosity of the aggregate (percentage and pore size), bitumen temperature and its viscosity, the time elapsed after mixing and the temperature at which the asphalt is conditioned until the above calculations are made. Research has shown that bitumen absorption may be up to 60% of the water absorption for porous paving aggregates.

5.4.2.5.5 Effective bitumen content

The bitumen quantity left after bitumen loss by absorption into the aggregate particles is designated as effective bitumen (bitumen) content. It is the effective bitumen quantity that will coat the aggregates, will provide cohesion in the asphalt mix, will fill some of the asphalt mix voids and will, in general, affect the bituminous mixture performance.

The effective bitumen (asphalt) content is expressed as a percentage of the total mixture, and it is calculated by the following equation (Asphalt Institute MS-4 2007):

$$P_{be} = P_b - (P_{ba}/100) \times P_s \text{ (\%)},$$

where P_{be} is the percentage of the effective bitumen (asphalt) content, by total mass of mixture; P_b is the percentage of bitumen (asphalt), by total mass of mixture (%); P_{ba} is the percentage of the absorbed bitumen (asphalt), by mass of aggregate (%); and P_s is the percentage of the aggregate, by total mass of mixture (%).

5.4.2.6 Test method for determination of the theoretical maximum specific gravity and density

As it was mentioned earlier, the theoretical maximum specific weight of the uncompacted (loose) bituminous mixture (G_{mm}) is determined in the laboratory according to ASTM D 2041 (2011). The test method is also known as the Rise method.

According to the abovementioned standard, a known mass of uncompacted bituminous mixture is placed in a vacuum bowl or vacuum flask and is completely covered by water, at 25°C. The air contained within the bituminous mixture is removed by agitation (or vibration) and by gradual increase of vacuum pressure. When the vacuum pressure reaches approximately 30 mm Hg (4.0 kPa), the sample is kept for an additional 15 min, approximately, while agitation continues. This facilitates the removal of trapped air among the particles coated with bitumen.

The theoretical maximum specific gravity (G_{mm}) of the bituminous mixture, at 25°C, is determined by the use of the following equation, when vacuum flasks are used:

$$G_{mm} = A/(A + D - E),$$

where A is the mass of dry bituminous mixture sample in air (g), D is the mass of flask and cover plate filled with water (at 25°C) (g) and E is the mass of flask, cover plate, sample and water (at 25°C) (g).

ASTM D 2041 (2011) also provides equations for the determination of maximum specific gravity of bituminous mixture when bowls are used.

The theoretical maximum density of the bituminous mixture, at 25°C, is expressed in either grams per cubic centimetre or kilograms per cubic metre.

5.4.2.7 Test method for determination of bulk specific gravity and density of compacted bituminous mixture

The bulk specific gravity of a dense compacted bituminous mixture is determined from the mass of the specimen in the air and the mass of the volume of water for the volume of the specimen. The latter, when the specimen has no open or interconnecting voids or does not absorb more than 2% of water by volume, is determined by submerging the specimen into the water bath without any other treatment of its surface. The procedure analytically is described in ASTM D 2726 (2011) or AASHTO T 166 (2013).

In brief, the specimen is placed into a basket suspended on a precision balance, so that it is automatically weighed. The specimen is completely submerged into a water bath at 25°C. After 3 to 5 min, the mass is indicated and it is designated as the mass of the specimen in water (the balance has been previously nullified with the basket being empty and completely submerged in water). Then, the specimen is removed from the water bath; it is dried by blotting quickly with a damp cloth towel and weighed. This mass is designated as the mass of the saturated surface dry specimen in air.

The bulk specific gravity of the compacted mixture (G_{mb}) is calculated by the following equation:

$$G_{mb} = A/(B - C),$$

where A is the mass of the dry specimen in air (g), $B - C$ is the mass of the volume of water for the volume of the specimen at 25°C, B is the mass of the saturated surface dry specimen in air (g) and C is the mass of the specimen in water (g).

When the specimen has open or interconnecting voids and absorbs more than 2% of water by volume, the bulk specific weight is determined after coating or sealing the specimen.

When the specimen is coated with paraffin, the determination is carried out according to AASHTO T 275 (2012), while when it is coated with parafilm (elastomeric film), the determination is carried out according to ASTM D 1188 (2007).

The test procedure is the same. The only difference is that the compacted asphalt (specimen or core) is coated with paraffin or parafilm in order to prevent water absorption and filling of the voids (pores) of the specimen with water.

The bulk specific gravity of the compacted bituminous mix specimen, in both cases, is determined by the following equation:

$$G_{mb} = A/[D - E - (D - A)/G_p],$$

where A is the mass of the compacted specimen in air (g); D is the mass of the dry, coated specimen in air (g); E is the mass of the dry, coated specimen underwater (g); and G_p is the specific gravity of coating material determined at 25°C.

Alternatively, the bulk specific gravity of a compacted bituminous mixture can be determined by using the automatic vacuum sealing test method (ASTM D 6752 2011 or AASHTO T 331 2013). This test method is applicable to all types of bituminous mixtures and requires the use of a vacuum chamber to seal the specimen with a plastic bag before submerging it into the water bath for mass determination. It is aimed at correcting inconsistencies in sample mass determinations resulting from drainage of water from samples and inaccuracy in saturated surface dry mass of absorptive coarse and open-graded mixes.

To determine the density of the compacted bituminous mixture specimen, the bulk specific gravity, in all cases, is multiplied by the density of water at the required temperature. At 25°C, the density of water is 0.9970 g/cm^3, while at 4°C, it is 1.000 g/cm^3.

5.4.2.8 Bitumen film thickness

The bitumen (asphalt) film thickness, although it is not a hot bituminous mix design criterion, is a useful parameter since the rate of oxidisation or ageing of the asphalt is very much affected by the bitumen (asphalt) film thickness. A bituminous mixture can satisfy all mix design requirements but aggregates may be coated by a thin bitumen film. A surface layer with aggregates coated with a thin bitumen film will be prone to ravelling and will be deficient in fatigue life. The range of bitumen (asphalt) film thickness considered for a satisfactory performance of the bituminous mixture is 8.0 to 15.0 μm.

The theoretical average bitumen film thickness, according to Roberts et al. (1996), can be calculated using the following equation:

$$T_F = 1000 \times (P_{be}/(SA \times P_s \times G_b)),$$

where T_F is the theoretical average bitumen (asphalt binder) film thickness (μm), P_{be} is the percentage (by weight) of effective bitumen (asphalt binder) in the mix (%), P_s is the percentage (by weight) of the aggregate in the mix (%), G_b is the specific gravity of the bitumen (asphalt binder) and SA is the surface area of the total aggregate (m^2/kg).

For the determination of the surface area of the total aggregate, SA, the procedure provided by the Asphalt Institute may be used (Asphalt Institute MS-2). The procedure utilises the aggregate grading and the surface area factors per sieve size (see Table 5.15). The calculations consist of multiplying the total percentage passing through a sieve by the 'surface

Table 5.10 Surface area factors and example of calculation of surface area

Sieve (mm)	Surface area factor (SAF) (m²/kg)	Example			
		Sieve	Passing %	SAF	Surface area
		19.0 mm	100 ×	0.41	= 0.41
>4.75	0.41[a]	9.5 mm	90 ×	—	—
4.75	0.41	4.75 mm	75 ×	0.41	= 0.31
2.36	0.82	2.36 mm	60 ×	0.82	= 0.49
1.18	1.64	1.18 mm	45 ×	1.64	= 0.74
0.600	2.87	0.600 mm	35 ×	2.87	= 1.00
0.300	6.14	0.300 mm	25 ×	6.14	= 1.54
0.150	12.29	0.150 mm	18 ×	12.29	= 2.21
0.075	32.77	0.075 mm	10 ×	32.77	= 3.28
		Surface area of total aggregate, SA			9.98 m²/kg

[a] Surface area factor is 0.41 m²/kg for any material retained above the 4.75 mm (No. 4) sieve.

area factor' corresponding to that sieve. The sum of these products is the surface area of the total aggregate mix (see the example in Table 5.10).

It must be noted that the surface area factors shown in Table 5.10 are applicable only to the listed sieves, which should be used during sieving.

The surface area factors have been determined assuming that the aggregate specific gravity is 2650. When the aggregate used has a significantly different specific gravity, the result of the surface area of the total aggregate mix should be multiplied by a correction factor 'α' ($= 2650/SG_a$).

Film thickness is associated with the richness modulus (k) used in France in mix design procedures to determine the minimum binder content. For more details, see Corte and Di Benedetto (2004).

Film thickness is also used in determining the minimum binder requirement in cold mix design described in Chapter 6.

5.4.3 AC for airport pavements

Although national specifications cover material specification and contractual requirement of AC for airport pavements, some aviation authorities provide their own standards. These particular specifications, in most of the cases, are not mandatory and they work in conjunction with the existing national specifications.

The Federal Aviation Authority (FAA) has its own recommendation and mix design requirements on dense AC. Since it is a globally recognised authority, brief information regarding the AC mixtures proposed is given here.

According to the FAA (2011a,c), the dense AC for surfacing course, base course, binder or levelling course is recommended to be one of the types listed in Table 5.11.

The aggregates should consist of crushed stone, crushed gravel or crushed slag (air cooled, blast furnaced) with or without natural sand or other inert, finely divided mineral aggregate, provided they meet the requirements summarised in Table 5.12.

The asphalt cement binder should conform to ASTM D 6373 (2007) (performance grade) or ASTM D 3381 (2012) (viscosity grade), to ASTM D 946 (2009) (AASHTO M 320 2010)

Table 5.11 Types and grading limits of asphalt concrete in accordance to FAA

Sieve (mm)	Type of asphalt concrete			
	37.5 mm	25.0 mm	19.0 mm	12.5 mm
37.5	100	—	—	—
25.0	86–98	100	—	—
19.0	68–93	76–98	100	—
12.5	57–81	66–86	79–99	100
9.5	49–69	57–77	68–88	79–99
4.75	34–54	40–60	48–68	58–78
2.36	22–42	26–46	33–53	39–59
1.18	13–33	17–37	20–40	26–46
0.600	8–24	11–27	14–30	19–35
0.300	6–18	7–19	9–21	12–24
0.150	4–12	6–16	6–16	7–17
0.075	3–6	3–6	3–6	3–6

Sources: From FAA, AC 150/5370-10F, Part 5, Item P-401, *Flexible Surface Courses, Item P-401, Plant Mix Bituminous Pavements*: FAA, 2011a; and FAA, AC 150/5370-10F, Part 5, Item 403, *Plant Mix Bituminous Pavements (Base, Levelling or Surface Course)*: FAA, 2011c.

Table 5.12 FAA's aggregate property requirements for AC

Requirement		Pavements designed for aircraft gross weight	
		>60,000 lb	<60,000 lb
Coarse aggregate			
Wear by LA test (%)	Surface course	≤40	≤50
(ASTM C 131 2006)	Base and other	≤50	
Fractured particles (%) (ASTM D 5821 2006)	All layers	70/85[a]	50/65
Flat and elongated particles (%) (ASTM D 4791 2010 – Method B)	All layers	≤8 (for ratio 5:1) ≤20 (for ratio 3:1)	
Soundness, sodium sulphate (%)	All layers	10 ≤ (13)[b]	
Fine aggregate			
Sand equivalent (%) (ASTM 2419 2009)	All layers	≥45	
P.I. and L.L. (AASHTO T 90 2008; ASTM D 4318 2010)	All layers	≤6 and ≤20	
Mineral filler: Shall meet the requirement of ASTM D 242 (2009)			

Sources: Adapted from FAA, AC 150/5370-10F, Part 5, Item P-401, *Flexible Surface Courses, Item P-401, Plant Mix Bituminous Pavements*: FAA, 2011a; and FAA, AC 150/5370-10F, Part 5, Item 403, *Plant Mix Bituminous Pavements (Base, Levelling or Surface Course)*: FAA, 2011c.

[a] The ratio 85/80 denotes that 85% of the coarse aggregate has one fractured face and 80% has two or more fractured faces.
[b] With magnesium sulfate.

Table 5.13 Marshall hot mix design criteria for dense AC mixtures by FAA

Marshall criteria		Pavements designed for aircraft gross weight	
		>60,000 lb	<60,000 lb
Compaction, blows		2 × 75	2 × 50
Min. stability, kN (lb)	Surface course	9.56 (2150)	6.00 (1350)
	Base and other	8.01 (1800)	4.45 (1000)
Flow, mm (0.01 in)	Surface course	2.5–3.5 (10–14)	2.5–4.5 (10–18)
	Base and other	2.0–4.0 (8–16)	2.0–4.0 (8–20)
Air voids, %	Surface course	2.8–4.2	2.8–4.2
	Base and other	2.0–5.0	2.0–5.0
VMA	—	(See Table 5.19)	
Tensile strength ratio, %		≥75 (ASTM D 4867 2009)	

Sources: Adapted from FAA, AC 150/5370-10F, Part 5, Item P-401, *Flexible Surface Courses, Item P-401, Plant Mix Bituminous Pavements:* FAA, 2011a; and FAA, AC 150/5370-10F, Part 5, *Item 403, Plant Mix Bituminous Pavements (Base, Levelling or Surface Course):* FAA, 2011c.

Table 5.14 Minimum VMA values for dense AC mixtures by FAA

Voids in mineral aggregate (VMA) (%)	For all cases	
	Maximum nominal aggregate size (mm) (%)	Minimum VMA (%)
	37.5 (1 1/4″)	13.0
	25.0 (1.0″)	14.0
	19.0 (3/4″)	15.0
	12.5 (1/2″)	16.0

Sources: From FAA, AC 150/5370-10F, Part 5, Item P-401, *Flexible Surface Courses, Item P-401, Plant Mix Bituminous Pavements:* FAA, 2011a; and FAA, AC 150/5370-10F, Part 5, *Item 403, Plant Mix Bituminous Pavements (Base, Levelling or Surface Course):* FAA, 2011c.

(penetration grade) and to the requirements and notes of Advisory Circular 150/5370-10E, Items 401 and 403.

The composition of the bituminous mixture to be used (job mix) is determined using the Marshall hot mix design method, as described in the previous sections. However, the design criteria used by the FAA are slightly different. FAA's criteria for the Marshall hot mix design are given in Tables 5.13 and 5.14.

Recycled AC is allowed to be used in all pavement layers except runway surface course. The recycled AC, consisting of reclaimed asphalt pavement (RAP), coarse aggregate, fine aggregate, mineral filler and asphalt cement, should meet the requirements set for an AC (see Tables 5.16 and 5.17). However, the amount of RAP to be incorporated in the mix is limited to 30%.

More detailed information regarding construction methods, materials acceptance, contractor quality control and basis of payment can be found in FAA AC 150/5370-10F, Part 5.

5.4.4 Comments on European and American composition methods of AC

Regarding the composition of ACs, the European standard for AC mixtures, CEN EN 13108-1 (2008), realising the weakness of the Marshall mix design, has broadened the requirements/properties of the target mixture. This ensures good performance in relation not only to permanent deformation but also to the mixture's fatigue life.

As a consequence, it is now possible to specify the target AC using an empirical or a fundamental specification. Both are based on a combination of requirements for composition and constituent materials with performance-related requirements. The difference between the two specifications is that the fundamental specification provides more degrees of freedom than the empirical specification.

Although the target mixture is now determined in Europe on performance-related requirements, European limit values have not yet been set. Until European limit values are set, each country specifies its own limit values.

The American Superpave design methodology was developed bearing in mind the use of performance-related requirements. At present, the Superpave design selects the bitumen on the basis of performance-related specifications but the target asphalt mix is still determined on the basis of volumetric properties.

Additionally, unlike the European specification CEN EN 13108-1 (2008), the compaction of the specimens under the Superpave design methodology is restricted to gyratory compaction only. This is a positive step since it simulates better the compaction carried out on site and eliminates variations on the mechanical properties of asphalt mixtures attributed to type of compaction.

A recent study has shown that the stiffness of AC (AC 12.5 mm) obtained from gyratory-compacted specimens was higher than that from Marshall-compacted samples. Fatigue performance of gyratory-compacted specimens was better than that of Marshall specimens. Statistical analysis showed that the difference between stiffness, VMA and VFA values was statistically significant (Nikolaides and Manthos 2014).

5.4.5 Open-graded AC

The use of open-graded AC for binder and base course is not as popular as dense-graded AC mixtures and, when used, its use is limited to surface course.

The European practice for surface course layer is to use BBTM and, when circumstances arise, such as ultimate reduction of traffic noise generated or maximising spray reduction, to use PA. If open-graded AC needs to be designed, CEN EN 13108-1 (2008) is used since it covers all types of AC mixtures.

With regard to the American practice for designing open-graded mixtures, ASTM D 7064 (2008) is used. These mixtures are recommended to be used in friction (surface) courses. Since these mixtures are similar to PA, they are described in detail in Section 5.6.2.

5.5 AC FOR VERY THIN LAYERS

ACs for thin or very thin layers were developed in France during the 1980s. The aim was to produce an AC for surface courses with better surface characteristics than traditional dense AC, with the capability of being laid at a thickness of less than 4 cm.

Two types of mixtures were developed (using their French names): BBTM (béton bitumineux très mince) and BBUM (béton bitumineux ultra mince).

The difference between BBTM and BBUM was that the first mixture could be laid in 20 to 30 mm thickness, whereas the second could be laid thinner (i.e. less than 20 mm thickness).

With the publication of the relevant European standard in 2006 (CEN EN 13108-2 2008), the BBTM type was adapted and named AC for very thin layers. To avoid confusion, the abbreviation to describe this type of mixture was kept the same, that is, BBTM.

BBUM mixtures remain 'proprietary' mixtures and thus are not covered by European specifications.

BBTMs are to be used for surface courses with a thickness of 20 to 30 mm in roads, airfields or any other trafficked surface.

The characterisation 'very thin layers' was used to distinguish layers laid in 30 to 40 mm thickness, which are designated as 'thin layers'.

BBTM consists of coarse aggregates of maximum nominal particle size (D) up to 10 mm (or 11.2 mm), fine aggregates 0/2 or 0/4 mm and binder. The binder can be paving grade bitumen. Natural asphalt and additives may also be added.

The aggregate particles are generally gap graded to form a stone-to-stone contact and provide an open surface texture.

The key advantage of BBTM compared with any other surface course asphalt laid at a thickness of 30 mm or more is its lower cost and, in general, it being economical, as a result of using a lower quantity of expensive and, in some countries, rare surfacing aggregates.

Compared with SMA, which is normally laid at a thickness of 30 or 40 mm, BBTM is much cheaper since, in addition, they do not require the addition of fibres and provides a layer with more superior surface characteristics.

5.5.1 BBTM in accordance to European standards

BBTM in Europe is designed in accordance to CEN EN 13108-2 (2008). The standard, apart from specifying requirements for constituent materials, also specifies empirical requirements for the particular bituminous mixtures.

5.5.2 Constituent materials

5.5.2.1 Binder

The binder of BBTM may be paving grade bitumen, conforming to CEN EN 12591 (2009), or modified bitumen, conforming to CEN EN 14023 (2010). Natural asphalt conforming to CEN EN 13108-4 (2008), Annex B, may also be added.

The selection of paving grade bitumen is based on regional climatic and traffic conditions. The author's opinion, after more than 10 years of experience with BBTM, is that a 35/50, 50/70 or 70/100 grade bitumen is sufficient in most cases.

Similarly, the use of polymer-modified bitumen (thermoplastic elastomer type) is almost inevitable when aggregates with no good affinity to bitumen (hydrophilic aggregates) are used, like most surfacing aggregates are in many regions. Apart from eliminating the premature ravelling, the use of polymer-modified bitumen enhances the good rutting performance of BBTM, the stiffness of the mixture and its fatigue performance.

According to CEN EN 13108-2 (2008), when using more than 10% by mass of the total mixture of reclaimed asphalt from mixtures in which mainly paving grade bitumen has been used and when the binder added to the mixture is a paving grade bitumen and the grade of the bitumen is specified, the binder should conform to the requirements outlined in Section 5.4.1.1.1.

5.5.2.2 Aggregates

Coarse and fine aggregates and added filler should conform to CEN EN 13043 (2004) as appropriate for the intended use.

The amount of filler, when added, should always be specified. Filler includes material such as cement or hydrated lime.

Reclaimed asphalt from recycling old bituminous layers may also be used to produce BBTM. The properties of reclaimed asphalt declared in accordance with CEN EN 13108-8 (2005) should conform to specified requirements appropriate to the intended use.

In case of using reclaimed asphalt, the upper sieve size D of aggregates that are contained in the reclaimed asphalt should not be larger than the upper sieve size D of the mixture to be composed. Furthermore, aggregate properties in the reclaimed asphalt should fulfil the requirements specified for the aggregates in the BBTM to be composed.

In general, the use of reclaimed asphalt is rare owing to the difficulty encountered to control efficiently the grading of the resulting mixture.

5.5.2.3 Additives

The nature and properties of all additives, when used, should be declared and should conform to the appropriate standards or specifications.

5.5.3 Mixture composition

5.5.3.1 Grading

The requirements for the grading of the aggregate mixture are expressed in terms of minimum and maximum percentage passing through some characteristic sieves, namely, 1.4D, D, 2 mm and 0.063 mm.

The minimum and maximum grading limits using the basic sieve set plus set 2, as specified by CEN EN 13108-2 (2008) for BBTM, are given in Table 5.15. The respective grading limits using the basic sieve set plus set 1 can be found in CEN EN 13108-2 (2008). Information about the basic sieve test and sieve sets 1 and 2 is given in Section 2.11.1.

One or two optional coarse sieves between D and 2 mm should be selected from 6.3, 8, 10 or 12.5 mm sieves, for the basic sieve set plus set 2; the corresponding optional coarse sieves for the basic sieve set plus set 1 are given in CEN EN 13108-2 (2008).

Similarly, one optional fine sieve between 2 and 0.063 mm should also be selected. The optional fine sieve should be selected from sieves 1, 0.5, 0.25 and 0.125 mm.

Table 5.15 Grading of target composition of BBTM, basic set plus set 2

D	6		8		10			
	6A	6B	8A	8B	10A	10B	10C	10D
Sieve (mm)	Percentage passing (%) by mass of aggregates							
1.4D^a	100							
D	90–100							
Optional course	Maximum and minimum value to be specified; range between maximum and minimum value to be selected from the following values: 10, 15 and 20							
2	25–35	15–25	25–35	15–25	25–35	15–25	25–35	27–33
Optional fine	Maximum and minimum value to be specified. Range between maximum and minimum value to be selected from the following values: 4, 5, 6, 7, 8, 9 and 10							
0.063	7–9^b	4–6	7–9	4–6	7–9	4–6	10–12^b	4.5–6.5

Source: Reproduced from CEN EN 13108-2, *Bituminous mixtures – Material specifications – Part 2: Asphalt concrete for very thin layers*, Brussels: CEN, 2008. With permission (© CEN).

[a] Where the sieve calculated as 1.4D is not an exact number, then the next nearest sieve should be adopted.
[b] For special mixtures, the filler content may be altered to 9–11.

The target composition of the mix should be within the resulting grading envelope.

For a layer thickness of 25 to 30 mm, the author proposes that BBTM grade 11 or 10 mm be chosen.

5.5.3.2 Binder content

The minimum binder content of the target composition should range from 5.0% to 6.4%, assigned to eight categories, $B_{min5.0}$ to $B_{min6.4}$.

The values of minimum percentages of binder content specified in CEN EN 13108-2 (2008) when aggregate apparent specific gravity is other than 2650 Mg/m^3 are corrected by multiplying by the factor (α), as in AC (see Section 5.4.1.4.2).

Needless to say, the bitumen content also includes the bitumen contained in the reclaimed asphalt, if used.

5.5.3.3 Additives

In case additives are used, their type, as well as their content in the mixture, is determined.

5.5.4 Mixture properties

At the target mixture, the following properties should be determined: void content, water sensitivity and mechanical stability.

In case the BBTM is going to be applied in airfields, its resistance to fuel and its resistance to de-icing fluid should also be determined and declared.

Finally, there may be a need to determine its resistance to abrasion by studded tyres.

5.5.4.1 Void content

The void content of the compacted specimens is selected to be within limit values, assigned to categories, depending on the method of compaction used.

Compaction may be carried out, as for AC specimens, by impact, gyratory, and vibratory compactors (see Section 5.4.1.3.3).

The void limits to be selected and the corresponding categories, per method of compaction, are given in Tables 5.16 and 5.17.

The void content is determined in accordance with all points referred to in Section 5.4.1.3.3.

The author proposes to choose categories V_{i7-10} or V_{i11-15}, when specimens are to be compacted by an impact compactor. He cannot propose void content categories when compaction is carried out by a gyratory compactor.

Table 5.16 Void content categories for compaction with impact or vibratory compactor

Void content (%)	Category V_i or V_v
3–6	V_{i3-6} or V_{v3-6}
7–10	V_{i7-10} or V_{v7-10}
11–15	V_{i11-15} or V_{v11-15}

Source: Reproduced from CEN EN 13108-2, *Bituminous mixtures – Material specifications – Part 2: Asphalt concrete for very thin layers*, Brussels: CEN, 2008. With permission (© CEN).

Table 5.17 Void content categories for compaction with gyratory compactor

Void content (%)	Category V_g
10–17	V_{g10-17}
12–19	V_{g12-19}
18–25	V_{g18-25}
20–25	V_{g20-25}

Source: Reproduced from CEN EN 13108-2, *Bituminous mixtures – Material specifications – Part 2: Asphalt concrete for very thin layers*, Brussels: CEN, 2008. With permission (© CEN).

5.5.4.2 Water sensitivity

The water sensitivity of BBTM-compacted cylindrical specimens is determined with respect to the indirect tensile strength ratio (ITSR) and a minimum value should be achieved as selected, specified or declared.

The minimum ITSR, expressed in per cent, should be either 75%, 90% or 100%, assigned to categories $ITSR_{75}$, $ITSR_{90}$ or $ITSR_{100}$.

The indirect tensile strength ratio is determined in accordance to CEN EN 13697-12 (see also Section 3.7.3.4).

5.5.4.3 Mechanical stability

The mechanical stability of BBTM is expressed in terms of rutting performance, determined by carrying out the wheel-tracking test (CEN EN 12697-22 2007).

Slab specimens of 30 or 50 cm in thickness are produced according to CEN EN 13108-20 (2008), paragraph 6.5, and the wheel-tracking device, the test conditions and test duration are selected from CEN EN 13108-20 (2008), Table D.1.

The mechanical stability, expressed in maximum proportional rut depth, should comply with the value, or category, selected from those given in CEN EN 13108-2 (2008). The maximum proportional rut depth, for example, for a BBTM 10 mm type can be 5, 7.5, 10 or 15, corresponding to categories P_5, $P_{7.5}$, P_{10} or P_{15}.

5.5.4.4 Resistance to fuel and de-icing fluid for application on airfields

As in the case of AC, when BBTM is to be used in airfields, its resistance to fuels by CEN EN 12697-43 (2005) as well as its resistance to de-icing fluids by CEN EN 12697-41 (2013) should be determined.

For resistance to fuels, see Section 5.4.1.3.5 and substitute AC with BBTM.

For resistance to de-icing fluids, see Section 5.4.1.3.6 and substitute AC with BBTM.

5.5.4.5 Resistance to abrasion by studded tyres

Resistance to abrasion by studded tyres, when required, is carried out on cylindrical specimens in accordance to CEN EN 12697-16 (2004). The specimens are prepared as for void content determination.

The resistance to abrasion by studded tyres, expressed as maximum abrasion value, in millilitres, should range from 20 to 60, assigned to 10 categories from Abr_{A20} to Abr_{A60} (see CEN EN 13108-2 2008).

This requirement is exclusively used by north European countries that allow or impose the use of studded tyres during winter months.

5.5.5 Other requirements

Other requirements to be stated are the temperature of the mixture at any place in the plant, coating/homogeneity and, if needed, reaction to fire.

5.5.5.1 Temperature of the mixture

BBTM temperatures at any place in the plant when produced with paving grade bitumen should be within the range of limits provided by CEN EN 13108-2 (2008). For paving grade bitumen 50/70 or 70/100, for example, the limit range is 140°C to 180°C. Measurements should be carried out in accordance to CEN EN 12697-13 (2001).

The minimum mixture temperature at delivery is determined by the producer.

When modified bitumen, hard bitumen or mixtures with additives are used, different temperature limits may be applied and should be documented and declared.

5.5.5.2 Coating and homogeneity

The same applies for coating and homogeneity as in AC (see Section 5.4.1.3.9).

5.5.6 Tolerance and test frequencies for conformity assessment

For tolerances from the target composition and test frequencies for conformity assessment, see Section 5.4.1.6.

5.6 POROUS ASPHALT

PA is a bituminous mixture with a high content of interconnecting voids that allow passage of water and air in order to provide the compacted mixture with drain and noise-reducing features. It is used for surface courses and it can be laid in one or more than one layer.

PAs were developed during the 1960s in England, initially for airfield surface courses in order to eliminate rainwater, thus preventing aquaplaning. Successful full-scale trials on roads application led PA to be used in highway pavements as well (Brown 1973; Daines 1992; Nicholls 1997). Since then, other countries started to apply PA as a surfacing course. PA can successfully be laid over concrete surfaces (Nicholls 2001).

The paradox is that although PA was developed in England, PA became more widespread in other countries. PA is widely used in Holland, Switzerland, Italy and Spain. The PA production percentage amounted to 13% of the annual production for hot and warm mixtures in 2010 in Holland. The corresponding figures for the rest of the abovementioned countries were 8%, 5% and 2% (EAPA 2011). From the same source, the corresponding figure for Japan was 11.2%, for the year 2010.

It is interesting to note that as of 2006 in England, PA is not recommended as a surface course material for new or old pavements. The main reasons were the premature failures observed in some projects and its high cost.

The service life of surface courses with PA was earlier found to be two-thirds of the service life of HRA (Nicholls 1997).

The typical property of PA is the high air void content (usually >18%), as well as the high number of interconnecting voids. Because of the above two characteristics, there is quick drainage of surface rainwater and a reduction of the tyre/pavement.

The surface of pavements with PA has very good macro-texture and a good anti-skidding coefficient, even if the surface is wet. What is more, because of its surface texture, driver glare by light beam reflection of vehicles on the pavement surface is reduced.

Quick removal of rainwater results in the elimination of aquaplaning and in the dramatic reduction of water spray by driving vehicles, causing a serious visibility problem during overtaking. In the 1960s, 10% of the accidents in the United Kingdom were caused by the wet surface of pavements (Maycock 1966).

The noise generated from moving vehicles on open-textured surfaces, similar to PA, has been found to be reduced when compared with conventional surfacing materials of the same macro-texture and skid resistance coefficient. The reduction measured was 3 to 4 dB(A), when the pavement surface was dry, and 7 to 8 dB(A), when the surface was wet (Nelson and Ross 1981). Similar results were also found by the Belgium Road Research Centre (Decoene 1989) and by Phillips et al. (1995). At a later study, resurfacing an old concrete pavement with PA led to wind-normalised reductions in noise exposure of the order of 4.5 to 6 dB(A) $L_{A10,18h}$ at villages approximately 0.5 km from the motorway (Baughan et al. 2002).

Weaknesses and drawbacks of PA reported over the years may be summarised as follows: (a) quicker oxidation of bitumen, (b) small mix tolerance to variations in bitumen content, (c) less service life when compared to dense-graded mixtures, (d) loss of effectiveness and functionality caused by clogging of pores with dirt and detritus, (e) requirement for good underlying sealed surface with adequate transverse slope, (f) reduction of bearing capacity of the layer compared with a layer with dense-graded bituminous mixture and (g) requirement for a higher quantity of salt during winter maintenance to avoid formation of ice.

The first three drawbacks can be solved to a great extent using modified bitumen, whereas the rest may not be considered to be so serious to decrease the effectiveness and usefulness of PA.

Twin-layer PA is being used nowadays by some countries to minimise the above weaknesses and drawbacks. In an extensive project report, it was concluded that although twin-PA costs more, when costs are expressed in terms of noise reduction achieved per unit spent, a twin-layer surface may be more cost-effective than single mitigation measures such as noise barriers or combined measures such as low-noise pavements and noise barriers (Morgan et al. 2007).

Finally, another point that must be considered when using PA is the arrangement of water drained from the porous layer. The pavement's concrete kerb should have openings extended to a depth at least equal to the thickness of the porous layer and placed in such a way to allow the side water to flow in and drain out. Similarly, drainage grids should also have openings on their sides to allow the water coming from the porous mat to enter the manhole. A very good relevant report that dealt with the above has been presented by Nicholls and Carswell (2001a).

5.6.1 PA in accordance to European standards

PA in Europe is designed in accordance to CEN EN 13108-7 (2008). The standard, except for requirements for constituent materials, specifies only empirical requirements for the bituminous mixtures. A mix design procedure for PA based on performance-related criteria has been proposed by Nicholls and Carswell (2001b).

5.6.1.1 Constituent materials

The requirements for the constituent materials, namely, binder and aggregates, and the use of recycled asphalt and other additives are the same as those for BBTM (see Section 5.5.2).

5.6.1.2 Mixture composition

The target PA composition should declare the materials by which the mixture is composed, the grading, the binder content and whether reclaimed asphalt or other additives are used. With respect to the use of reclaimed asphalt, it should not exceed, unless agreed upon, 10% by mass of total mixture if PA is for surface course and 20% by mass of total mixture if PA is for any other course.

5.6.1.2.1 Grading

The requirements for the grading of the aggregate mixture are expressed in terms of minimum and maximum percentage passing through the characteristic sieves 1.4D, D, 2 mm and 0.063 mm, one or two optional sieves between D and 2 mm, and one optional sieve between 2 and 0.063 mm.

Sieve D and the optional coarse sieves between D and 2 mm, if the basic sieve set plus set 2 is used, are selected from the following sieves: 4, 6.3, 8, 10, 12.5, 14, 16 and 20 mm; if the basic sieve set plus set 1 is used, see CEN EN 13108-7 (2008). Information about the basic sieve test and sieve sets 1 and 2 is given in Section 2.11.1.

The optional fine sieve between 2 and 0.063 mm, in all cases, is selected from the following sieves: 1, 0.5, 0.25 or 0.125 mm.

The overall grading limits for PA, as specified by CEN 13108-7 (2008), at the characteristic sieves are shown in Table 5.18, and the target composition of the mix should be within these limits.

The ranges between the maximum and minimum values for the grading envelope should be selected as a single value within the given limits provided in Table 2 of CEN EN 13108-7 (2008).

5.6.1.2.2 Binder content

The minimum binder content of the target composition should range from 3.0% to 7.0%, assigned to nine categories, $B_{min3.0}$ to $B_{min7.0}$.

The minimum binder contents specified are corrected if the density of the aggregate used is different from 2650 Mg/m^3. For correction factor, see Section 5.4.1.4.2.

5.6.1.2.3 Additives

When additives are used, each type and amount of additive constituent is determined precisely. The typical type of additive used to reduce binder drainage is cellulose or mineral fibre.

Table 5.18 Overall limits of target composition of porous asphalt

Sieve (mm)	Passing sieve (%), by mass
1.4D	100
D	90–100
2	5–25
0.063	2–10

Source: Reproduced from CEN EN 13108-7, *Bituminous mixtures – Material specifications – Part 7: Porous asphalt*, Brussels: CEN, 2008. With permission (© CEN).

5.6.1.3 Mixture properties

At the target mixture, the following empirical properties should be determined: drainage capacity, water sensitivity, particle loss and binder drainage.

In case PA is going to be applied in airfields, its resistance to fuel, its resistance to de-icing fluid, and the affinity between binder and aggregates should also be determined and declared.

5.6.1.3.1 Drainage capacity

The drainage capacity is determined by the requirement for *void content* or *minimum permeability* (horizontal or vertical permeability).

5.6.1.3.1.1 VOID CONTENT

The void content of the prepared specimens should lie between the minimum and maximum values specified.

The minimum void content should range from 14% to 28%, assigned to eight categories, V_{min14} to V_{min28}.

The maximum void content should range from 18% to 32%, assigned to eight categories, V_{max18} to V_{max28}.

The prepared specimens are compacted by am impact, gyratory or vibratory compactor, under conditions selected from CEN EN 13108-20 (2008), Table C.1.

The void content is determined in the usual way and in accordance to CEN EN 13108-20 (2008), D.2 (see also Section 5.4.1.3[c]).

5.6.1.3.1.2 PERMEABILITY – PERMEABILITY TEST

The minimum horizontal permeability, or vertical permeability, of cylindrical specimens of bituminous mixtures should range from 0.1×10^{-3} to 4.0×10^{-3} m/s, assigned to nine categories, $K_{v0.5}$ to $K_{v4.0}$, for vertical permeability, and $K_{h0.1}$ to $K_{h4.0}$, for horizontal permeability.

The cylindrical specimens are prepared in the laboratory using an appropriate compaction device or cored from slabs prepared in the laboratory or cored out of the road. The thickness of the specimens should not be less than 2.5 times the nominal size of the aggregate in the mixture. The nominal diameter size of specimens should be either 100 or 150 mm when the nominal maximum size of the aggregate is equal to or less than 22 mm, and it should be 150 mm when the nominal maximum size of the aggregate is greater than 22 mm.

The principle of the permeability test is that a column of water with a constant height is applied to a cylindrical specimen and is allowed to permeate through the specimen for a controlled time in either a vertical or a horizontal direction depending upon the parameter being measured.

The vertical permeability value, K_v, or the horizontal permeability value, K_h, is calculated by using Darcy's formula and the modified Darcy's formula, respectively, after measuring the resultant flow rate of water Q_v or Q_h.

The test is carried out at ambient temperature within the range 15°C to 25°C.

More about the permeability test can be found in CEN EN 12697-19.

5.6.1.3.2 Water sensitivity

The water sensitivity of PA-compacted cylindrical specimens is determined with respect to the indirect tensile strength ratio (ITSR) and a minimum value should be achieved as selected, specified or declared.

The minimum ITSR, expressed in per cent, should be either 50%, 60%, 70%, 80%, 90% or 100%, assigned to six categories, $ITSR_{50}$, $ITSR_{60}$, $ITSR_{70}$, $ITSR_{80}$, $ITSR_{90}$ or $ITSR_{100}$.

The indirect tensile strength ratio is determined in accordance to CEN EN 13697-12 (see also Section 3.7.3.4).

5.6.1.3.3 Particle loss – Particle loss test

The particle loss (PL) of PA-compacted cylindrical specimens should be less than the maximum value selected, specified or declared.

The maximum particle loss, expressed in per cent, should be either 10%, 15%, 20%, 30%, 40% or 50%, assigned to six categories, PL_{10}, PL_{15}, PL_{20}, PL_{30}, PL_{40} and PL_{50}.

5.6.1.3.3.1 PARTICLE LOSS TEST

The particle loss test is executed only on PA specimens to estimate the abrasiveness of PA, with a maximum particle size up to 25 mm.

Particle loss is assessed by the loss of mass of the specimens after certain turns in the Los Angeles machine. The test is carried out in accordance with CEN EN 12697-17 (2007).

At least five specimens 100 ± 3 mm in diameter and 63.5 ± 5 in height are compacted using 2 × 50 Marshall blows (as specified in CEN EN 13697-30) or 40 gyrations by a gyratory compactor (as specified in EN 13697-31). After the specimens have been compacted and removed from the moulds, their mass, density and void content are determined.

The specimens are left for at least 2 days at ambient temperature, not higher than 25°C, before placing them into the Los Angeles machine. One specimen at a time is placed in the drum, with the metal balls removed, and the drum turns at 30 to 33 rpm for 300 turns. Test ambient temperature should be 15°C to 25°C.

On completion, the specimen is removed, cleaned with a cloth by eliminating all particles that are clearly loose, and then weighed again.

The particle loss (PL) is calculated from the following equation:

$$PL = 100 \times (W_1 - W_2)/W_2,$$

where PL is particle loss (%), W_1 is the initial mass of the specimen (g) and W_2 is the final mass of the specimen (g).

The average value of at least five specimens is rounded to the nearest 1%.

More information about the test is given in CEN EN 12697-17 (2007).

5.6.1.3.4 Binder drainage – Binder drainage test

The binder drainage test is conducted on loose PA mixture according to CEN EN 12697-18 (2004).

The binder drainage (D), expressed in percentage, should be 0%, assigned to category D_0.

The binder drainage test of PA is carried out by using the basket method according to CEN EN 12697-18 (2004).

The binder drainage is defined as the binder, fine particles and additives, if any, separated from the mixture after the mixing process or during transport of the mixture to the site.

The principle of the test method is that the quantity of material lost by drainage, after 3 h at the test temperature, is measured in mixtures placed on baskets made out of perforated metal plates.

The test temperature is that of mixing temperature plus 25°C, when paving grade bitumen is used, or plus 15°C on top of the reference mixing temperature defined by the supplier.

The quantity of the mixture required for the test is in the region of 1140 to 1180 g, for aggregates with a density between 2.65 and 2.75 Mg/m^3.

The drained material (D) is determined by the following equation:

$$D = [(W_2 - W_1)/(1100 + B)] \times 100,$$

where D is the drained material (%), W_1 is the initial mass of the tray and foil (g), W_2 is the mass of the tray and foil finally with the drained material (g) and B is the initial mass of the binder in the mixture (g).

The average value of at least two single results of drained material is calculated, rounded to the closest 0.1%.

The basket method can be used for determining the binder drainage for different binder contents or with a single binder content, eliminating the successive repetitions. It also enables the effects of varying fine aggregate types or including any anti-draining additive to be quantified.

More information about the test is given in CEN EN 12697-18 (2004).

5.6.1.4 Complementary tests for application on airfields

When PA is to be used in airfields, it is required to also determine the following: (a) resistance to fuels, (b) resistance to de-icing fluids and (c) bitumen–aggregate affinity.

For resistance to fuels, see Section 5.4.1.3.5 and substitute AC with PA.

For resistance to de-icing fluids, see Section 5.4.1.3.6 and substitute AC with PA.

The bitumen–aggregate affinity, expressed in number of aggregate particles not completely coated by binder after immersion in water, should be less than the maximum value selected, specified or declared.

The maximum value should be either 1%, 3%, 5% or 10%, assigned to categories BBA_1, BBA_3, BBA_5 and BBA_{10}.

The determination of the affinity between aggregate and bitumen is conducted in accordance to EN 12697-11 (2012), Part B. For a short description of the test, see Section 3.7.3.2.

5.6.1.5 Other requirements

Other requirements to be stated are the temperature of the mixture at any place in the plant, coating/homogeneity and, if needed, reaction to fire.

5.6.1.5.1 Temperature of the mixture

PA temperatures at any place in the plant when produced with paving grade bitumen should be within the range of limits provided by CEN EN 13108-7 (2008). For paving grade bitumen 50/70 and 70/100, for example, the limit range is 140°C to 175°C for both. Measurements should be carried out in accordance to CEN EN 12697-13 (2001).

The minimum mixture temperature at delivery is determined by the producer.

When modified bitumen, hard bitumen or mixtures with additives are used, different temperature limits may be applied and should be documented and declared.

5.6.1.5.2 Coating and homogeneity

When PA is discharged from the mixer, the mixture should be homogeneous in appearance, with all aggregates completely coated with bitumen and there should be no indication of balling of fine aggregates.

Table 5.19 Permitted combinations of requirements

Requirements for:	Combinations		
	1	*2*	*3*
Binder content	x	x	x
Grading	x	x	x
Minimum void percentage	x	—	—
Maximum void percentage	x	x	x
Horizontal permeability	—	x	—
Vertical permeability	—	—	x
ITSR	x	x	x
Particle loss	x	x	x

Source: Reproduced from CEN EN 13108-7, *Bituminous mixtures – Material specifications – Part 7: Porous asphalt*, Brussels: CEN, 2008. With permission (© CEN).

5.6.1.6 Basic requirements of PA to avoid over-specification

According to CEN EN 13108-7 (2008), the overall quality of PA avoiding over-specification is ensured when one of the combinations of requirements given in Table 5.19 is selected. The combination of requirements selected is verified during the construction stage by carrying out the respective tests.

5.6.1.7 Tolerances, test frequencies for finished asphalt and conformity assessment

For tolerances from the target composition and test frequencies for conformity assessment, see Section 5.4.1.6.

5.6.2 PA in accordance to American standards

PA in the United States is similar to the open-graded friction course (OGFC) specified by ASTM D 7064 (2008). The terminology used is that 'an OGFC is a special type of hot mix asphalt (HMA) surface mixture used for reducing hydroplaning and potential for skidding, where the function of the mixture is to provide a free-draining layer that permits surface water to migrate laterally through the mixture to the edge of the pavement'.

OGFCs may also be placed to reduce the tyre–pavement interface noise and may also be placed to reduce the occurrence and severity of reflective cracking.

The mix design is based on the volumetric properties of the mix and particularly on air voids (void content) and the presence of stone-on-stone contact. The specimens are compacted by a gyratory compactor or another compactor providing equivalent compacted density.

According to ASTM D 6932 (2008), the binder content of an OGFC mixture typically ranges from 5% to 7% when paving grade bitumen is used, from 6% to 8% when modified bitumen is used or from 8% to 10.5% when bitumen–rubber is used.

5.6.2.1 Constituent materials

5.6.2.1.1 Aggregates

The coarse crushed aggregates should have resistance to fragmentation by Los Angeles lower than 30%, when tested in accordance with ASTM C 131 (2006).

In case crushed gravels are used, 90% of aggregates should have two crushed surfaces and 95% of aggregates should have one crushed surface, when the test is conducted according to ASTM D 5821 (2006).

The percentage of flat and elongated aggregates should not exceed 10%, with a maximum dimension-to-minimum dimension ratio of 5:1, in accordance with ASTM D 4791 (2010).

Fine aggregates should have an uncompacted void content of at least 40%, when tested according to ASTM C 1252 (2006). The sand equivalent of aggregates passing through the 2.36 mm sieve should be at least 45%, when tested in accordance with ASTM D 2419 (2009).

5.6.2.1.2 Binder

The binder should be selected on the basis of ambient temperature, traffic and the expected functional performance of the OGFC mixture. The selected bituminous (asphalt) binder should satisfy the requirements of ASTM D 6373 (2007) (performance-graded asphalt binder). However, other grades of asphalt binders, such as penetration grade (ASTM D 946 2009) or viscosity grade (ASTM D 3381 2012), may be suitable.

Modified bitumen may also be used. Mixtures with modified bitumens (asphalt cements) provide significant improvement in performance.

5.6.2.1.3 Additives

Additives, such as fibres of cellulose or inorganic fibres, may be used to prevent drainage. A typical range of fibre mass used is 0.2% to 0.5% by mass of total mixture. The precise amount of fibres to be used is determined after executing the drainage test in accordance with ASTM D 6390 (2011) (AASHTO T 305 2012).

5.6.2.2 Grading

The grading of the aggregate mix for the production of the trial mixtures, as well as of the target mixture, is recommended to be within the master limits given in Table 5.20 or within the limits given in the Appendix X1 of ASTM D 7064 (2008), which are limits used successfully by various states. Any other grading limits that have demonstrated good performance may also be used.

Table 5.20 Grading limits for OGFC mixtures

Sieve (mm)	Percent passing (%)
19.0 (3/4")	100
12.5 (1/2")	85–100
9.5 (3/8")	35–60
4.75 (No. 4)	10–25
2.36 (No. 8)	5–10
0.075 (No. 200)	2–4

Source: Reprinted from ASTM D 7064, *Standard practice for open-graded friction courses (OGFC) mix design*, West Conshohocken, Pennsylvania, US: ASTM International, 2008. With permission (© ASTM International).

5.6.2.3 Mix design of OGFC

The mix design of the OGFC mixture, in accordance to ASTM D 7064 (2008), after selecting the constituent materials consists of the following stages:

- a. Selection of trial gradings
- b. Selection of trial binder content
- c. Determination of VCA in the coarse aggregate fraction
- d. Selection of desired grading
- e. Selection of optimum binder content
- f. Evaluation of moisture susceptibility

5.6.2.3.1 Selection of trial gradings

Three trial aggregate gradation curves are selected and should be within the limits of Table 5.30 or within other limits that have been demonstrated to give gradings with good performance. The three trial gradings should generally fall along the coarse and fine limits of the grading envelope, with one falling in the middle.

5.6.2.3.2 Selection of trial binder content

For each trial aggregate grading, binder content between 6.0% and 6.5% by mass of bituminous mixture is initially selected. When using modified bitumen, higher binder contents should be selected.

5.6.2.3.3 Determination of VCA in the coarse aggregate fraction

For a better performance of the OGFC mixture, there should be stone-on-stone contact in the coarse aggregate skeleton (all coarse aggregate is retained on the 4.75 mm sieve).

The condition for stone-on-stone contact within an OGFC mixture is defined as the point at which the per cent of voids of the compacted mixture is less than the voids in the coarse aggregate (VCA) in the dry-rodded test. The dry-rodded test is carried out in accordance to ASTM C 29/C 29M (2009) from which the VCA_{DRC} is determined using the following equation:

$$VCA_{DRC} = (G_{CA} \times \gamma_w - \gamma_s)/(G_{CA} \times \gamma_w),$$

where G_{CA} is the bulk specific gravity of the coarse aggregate, γ_s is the bulk density of the coarse aggregate fraction in the dry-rodded condition (kg/m³) and γ_w is the density of water (998 kg/m³).

5.6.2.3.4 Selection of desired grading

After short-term ageing of the mixtures in accordance to AASHTO R 30 (2010), trial specimens (usually three per gradation curve) are compacted by a gyratory compactor (ASTM D 6925 2009) using 50 gyrations. Then, the specimens' bulk specific gravity is determined using geometric measurements of diameter and height, according to ASTM D 3203 (2011) (AASHTO T 269 2011). In addition, the theoretical maximum density of the loose mixture is also determined in accordance with ASTM D 2041 (2011) or ASTM D 6857 (2011).

Using the results of the abovementioned parameters, the void content (V_a) and the VCA of the compacted mix (VCA_{MIX}) are determined using the following equations:

$$V_a = 100 \times [1 - (G_{mb}/G_{mm})]$$

and

$$VCA_{MIX} = 100 - [(G_{mb}/G_{CA}) \times P_{CA},$$

where G_{mb} is the bulk specific gravity of the compacted mixture, G_{mm} is the theoretical maximum specific gravity of the mixture, G_{CA} is the bulk specific gravity of the coarse aggregate fraction and P_{CA} is the percentage of coarse aggregate fraction in total mixture.

Out of the three trial curves, the one that has the highest void content value (the minimum permissible void content is usually 18%) and a VCA_{MIX} equal to or lower than VCA_{DRC} is considered optimum and is selected as the desired grading.

5.6.2.3.5 Selection of optimum binder content

Once optimum grading has been selected, mixtures with at least three different binder contents are prepared so as to determine the optimum binder content (use increments of 0.5%).

The mass of the mixture should be enough to prepare eight specimens per binder content. Six compacted specimens are intended to be used for the Cantabro abrasion test (three of them without 'ageing' and the other three after 'ageing'), whereas the remaining two uncompacted specimens are intended for the draindown (drainage) test and the determination of theoretical maximum density. The void content of the compacted specimens is also determined.

The *drainage test* is conducted at 15°C higher than the expected production temperature and according to ASTM D 6390 (2011). The maximum draindown should not exceed 0.3% of the mass of total mixture.

The *void content* of the mixture to be chosen should be higher than or equal to 18%.

The *Cantabro abrasion test* is conducted using the Los Angeles apparatus (ASTM C 131 2006) and according to ASTM D 7064/D 7064M (2008), Appendix X2. This test is similar to the test described in Section 5.6.1.3.3.

The average weight loss value after the Cantabro abrasion test of non-'aged' specimens should not exceed 20%. Correspondingly, the average weight loss of 'aged' specimens should not exceed 30%, whereas no specimen should have a weight loss higher than 50%.

The OGFC mixture meeting all the above requirements is the one that determines the optimum binder content.

5.6.2.3.6 Evaluation of moisture susceptibility

The evaluation of moisture susceptibility is carried out in the mixture with the optimum bitumen content and is determined by the tensile strength test carried out in accordance to ASTM D 4867 (2009) (AASHTO T 283 2011). However, the relevant modification to the specimen compaction (use 50 gyrations) and specimen conditioning (use of five freeze/thaw cycles), as stated in ASTM D 7064/D 7064M (2008), is applied.

If the mixture with the optimum binder content does not meet the above requirement (a rare case when modified bitumen is used), anti-stripping liquid or hydrated lime, or both, are added and the test is repeated.

In case the above measures proved to be ineffective, the aggregate source or the type/grade of bitumen, or both, should be changed to obtain better aggregate/binder compatibility.

5.6.2.4 Production and placement

The production and placement of OGFC, as well as sampling and testing during construction, and the construction procedure are carried out in accordance to ASTM D 6932 (2008).

5.6.3 PA in accordance to FAA

As it was mentioned, PA or OGFC is used in surfacing airport pavements. FAA standards call this type of mixture porous friction course (PFC) mixture.

According to AC 150/5370-10F, Item P-402 (FAA 2011b), the binder should be viscosity-graded bitumen (asphalt cement) with a synthetic rubber (polymer) added at a percentage of not less than 2%, by mass of bitumen. The properties of the modified bitumen should comply with the requirements of Table 5.21.

The aggregates should consist of crushed stone, crushed gravel or crushed slag (air cooled, blast furnaced) with or without natural sand or other inert, finely divided mineral aggregate, provided they meet the requirements summarised in Table 5.22.

Table 5.21 Properties of polymer-modified bitumen for PFC

Mixture properties	Value
Viscosity at 60°C (140°F), poises (ASTM D 2171 2010)	1600–2400
Viscosity at 135°C (275°F), centistokes (ASTM D 2170 2010)	≥325
Flash point, °C (ASTM D 92 2012)	≥232
Ductility at 25°C (77°F), at 5 cm/min, cm (ASTM D 113 2007)	≥100
Ductility at 4°C (39.2°F), at 5 cm/min, cm (ASTM D 113 2007)	≥50
Toughness, inch-pounds (ASTM D 5801 2012)	≥110
Tenacity, inch-pounds (ASTM D 5801 2012)	≥75
After thin film oven test (test on residue)	
Viscosity at 60°C (140°F), poises (ASTM 2171 2010)	≤8000
Ductility at 25°C (77°F), at 5 cm/min, cm (ASTM D 113 2007)	≥100
Ductility at 4°C (39.2°F), at 5 cm/min, cm (ASTM D 113 2007)	≥25

Source: FAA, AC 150/5370-10F, Part 5, Item P-402, *Porous friction course*: FAA, 2011b.

Table 5.22 Aggregate property requirements for PFC

Requirement	Pavements designed for all types of aircraft gross weight
Coarse aggregate	
Fractured particles, % (ASTM D 5821 2006)	100/75[a]
Flat and elongated particles, % (ASTM D 4791 2010-Method B)	≤8 (for ratio 5:1)
Coarse and fine aggregate	
Wear by LA test, % (ASTM C 131 2006)	≤30
Soundness, sodium sulfate-5 cycles, %	12
Fine aggregate	
Sand equivalent, % (ASTM 2419 2009)	(Not specified) ≥45[b]
P.I. and L.L. (AASHTO T 90 2008; ASTM D 4318 2010)	≤6 and ≤20
Mineral filler: Shall meet the requirement of ASTM D 242 (2009)	

Source: FAA, AC 150/5370-10F, Part 5, Item P-402, *Porous friction course*: FAA, 2011b.

[a] The ratio 100/75 denotes that 100% of the coarse aggregate has one fractured face and 75% has two or more fractured faces.
[b] Typically required, unless local conditions require a lower value.

Table 5.23 Grading limits for PFC aggregates

Sieves (mm)	Gradations		Tolerances (Job mix)
	19.0 mm	12.5 mm	
19.0 (3/4″)	100	—	—
12.5 (1/2″)	70–90	100	±5%
9.5 (3/8″)	40–65	85–95	±5%
4.75 (No. 4)	15–25	30–45	±5%
2.36 (No. 8)	8–15	20–30	±2%
0.60 (No. 30)	5–9	9–17	±2%
0.075 (No. 200)	1–5	2–7	±2%
Binder content			±0.2%
Temperature of mixture			±7°C

Source: FAA, AC 150/5370-10F, Part 5, Item P-402, *Porous friction course:* FAA, 2011b.

The aggregate gradation should fall within the limits given in Table 5.23. Tolerances of the produced mixture from the design (target) mixture are also given in the same table.

More information regarding the mix design, the mix construction methods, testing and methods of measurement and payment are given in the relevant reference.

5.7 HOT ROLLED ASPHALT

HRA was developed in the United Kingdom aiming to produce a more economic and workable asphalt than MA, which was widely used until then. The incorporation of coarse aggregates in the mixture slightly increased the stiffness of the mixture. As a consequence, there was no need to use hard bitumen, such as in the case of MA. Furthermore, with the addition of coarse aggregates, the volume of the mixture increased and at the same time workability was improved.

The key feature of this mixture is that the aggregate particle size distribution curve presents a 'discontinuity' between certain sieves. For this reason, HRA is also known as gap-graded mixture.

Generally, the HRA is a dense, gap-graded asphalt, in which the mortar of fine aggregate, filler and high viscosity binder are the main contributors to the performance of the laid material and its stiffness.

HRA may be used for surface courses, binder courses, regulating courses and asphalt base courses, on roads, airfields and other trafficked areas. When HRA is used in surfacing, high-quality coated chippings (Green and Montgomery 1972) are used to provide the required surface texture. However, despite the good performance of HRA as a surfacing material (Nicholls 1998; Weston et al. 2001), its use as a surfacing material is limited nowadays.

HRA requires relatively higher compaction energy than AC.

The use of high-viscosity bitumen results in higher temperatures at compaction, which often are difficult to maintain, particularly during winter months. The production cost of HRA, owing to high binder content, is higher than that of AC.

The use of modified bitumen (polymers and sulfur) or Trinidad lake asphalt improves HRA's resistance to permanent deformation and its weathering properties (Carswell 1987; Denning and Carswell 1981a,b, 1983; Jacobs 1980).

Today, HRA is used almost exclusively in the United Kingdom.

5.7.1 HRA in accordance to European standards

The standards for HRA have traditionally been based on empirically compositional recipes combined with specifications for the constituent materials with additional requirements based on performance-related tests.

The HRA in Europe is designed in accordance to the CEN EN 13108-4 (2008) specification. There is no sufficient experience available with regard to fundamental testing of HRA. As a consequence, this European standard, for the moment, specifies empirical requirements combined with a stiffness requirement. However, the ultimate aim of the specification is to specify the required fundamental properties of the HRA.

The content of the following sections are based on the CEN EN 13108-4 (2008) specification.

5.7.2 Constituent materials

Regarding the constituent materials used in the production of HRAs, such as binder, aggregates and the use of reclaimed asphalt and additives, all are referred to Section 5.4.1.1 and Sections 5.4.1.1.1 through 5.4.1.1.3 apply.

In the case where coated chippings are used, their properties should be in accordance with Annex C of CEN EN 13108-4 (2008).

5.7.3 Mixture composition

5.7.3.1 General

The properties of HRA to be used should be determined by the designer and be satisfied by the target mix. At the same time, the producer should qualify that its product meets the requirements.

The composition of HRA, as mentioned, is based on empirical requirements (properties) and stiffness modulus.

The target composition should declare and document the materials, from which the mixture is composed, namely, the particle size distribution of the aggregate mix (grading), the bitumen content and whether reclaimed asphalt or other additives are used.

When reclaimed asphalt is used, in which modified bitumen and an additive modifier had been used, and when the produced mixture contains modified bitumen or a modifier, the amount of reclaimed asphalt should not, for surface courses, exceed 10% by mass of the total mixture. When the recycled HRA is to be used in binder, regulating or base courses, the reclaimed asphalt percentage can be up to 20%.

5.7.3.2 Grading

The requirements of the gradation curve of the aggregate mixture are expressed in reference to percentage passing through some characteristic sieves, namely, 1.4D, D, 2 mm and 0.063 mm.

The D sieve is designated as the sieve with an aperture size that determines the HRA grade type. In this sieve, a small mass of aggregate is allowed to be retained.

The 1.4D sieve is designated as the sieve that is derived from the multiplication of D with 1.4, rounding the result to the nearest sieve.

The maximum and minimum percentages (limits), specified by CEN EN 13108-4 (2008), within which the target aggregate composition for binder, regulating or base courses should be, are given in Table 5.24, for the basic sieve set plus set 2; a similar table but for the basic sieve test plus set 1 is provided by CEN EN 13108-4 (2008). Information about the basic sieve test and sieve sets 1 and 2 is given in Section 2.11.1.

Table 5.24 Grading of target composition of HRA for base and binder course mixtures for basic sieve set plus set 2

D	50/10	50/14	50/20	60/20	60/32
Sieve (mm)	Passing sieve (%) by mass of aggregates				
40	—	—	—	—	100
31.5	—	—	100	100	99–100
20	—	100	99–100	99–100	59–71
14	100	98–100	74–91	39–65[a]	39–65[a]
10	98–100	72–93	44–66	—	—
2[b]	40–50	40–50	40–50	37	37
0.5	17–51	17–51	18–50	13–39	13–39
0.25	14–31	14–31	15–30	10–25	10–25
0.063	3–6	3–6	4–5	4.0	4.0

Source: Reproduced from CEN EN 13108-4, *Bituminous mixtures – Material specifications – Part 4: Hot Rolled Asphalt*, Brussels: CEN, 2008. With permission (© CEN).

[a] The upper compliance value of 65 can be extended to 85 where evidence is available that the mixture so produced is suitable.
[b] For mixtures containing rock fine aggregate, and in some instances sands or blends of sand and crushed rock fines, the minimum binder content given may be reduced by up to 0.5%. Experience shows that this is advisable to avoid an over-rich mixture. Alternatively, a reduction in the target passing 2 mm of up to 5% can be permitted.

Tables for grading target composition are also provided for HRA to be used in surface courses (see Tables 3 and 4 of CEN EN 13108-4 2008).

5.7.3.3 Binder content

The minimum binder content of the target composition should range from 4.6% to 11%, assigned to 33 categories from $B_{min4.6}$ to $B_{min11.0}$ (a 0.2% step is used).

The minimum binder contents specified are corrected if the density of the aggregate used is different from 2650 Mg/m^3. For the correction factor, see Section 5.4.1.4.2.

It is stated that the binder content also includes the binder content contained in the reclaimed asphalt, when used.

5.7.3.4 Binder volume

The minimum binder volume of the target composition is an alternative requirement to binder content. The minimum binder volume requirements, when used, should range from 6.0% to 11.0%, assigned to 11 categories, $B_{vol6.0}$ to $B_{vol11.0}$ (a 0.5% step is used).

The binder volume is determined in accordance to CEN EN 12697-8 (2003). The density determination method is selected from CEN EN 13108-20 (2008), Table D.1.

5.7.3.5 Additives

When additives are also used, their type as well as their content in the mixture should be determined.

5.7.4 Mixture properties

At the target mixture, the following empirical properties should be determined: void content, water sensitivity, resistance to permanent deformation and stiffness.

In case HRA is going to be applied in airfields, its resistance to fuel and to de-icing fluid should also be determined and declared.

5.7.4.1 Void content

The void content of the compacted HRA specimens at the target composition should lie between minimum and maximum values. The minimum values range from 0.5% to 2.0% (minimum void category, $V_{min0.5}$ to $V_{min2.0}$), and the maximum void values range from 3.0% to 8.0% (maximum void category, $V_{max3.0}$ to $V_{max8.0}$).

The compaction of the test specimens is selected from CEN EN 13108-20 (2008), Table C.1, and the void content is determined in accordance to CEN EN 13108-20 (2008), paragraph D.2 (see also Section 5.4.1.3.3).

5.7.4.2 Water sensitivity

The water sensitivity (or susceptibility) of the compacted cylindrical specimens is expressed in indirect tensile strength ratio (ITSR). The minimum ITSR value should be either 60%, 70% or 80%, assigned to categories $ITSR_{60}$, $ITSR_{70}$ and $ITSR_{80}$.

The indirect tensile strength ratio is determined in accordance to CEN EN 13697-12 (see also Section 3.7.3.4).

5.7.4.3 Resistance to permanent deformation

The resistance to permanent deformation is determined with respect to maximum wheel-tracking rate and maximum rut depth, using a small device and procedure A of CEN EN 12697-22 (2007).

The maximum wheel-tracking rate should range from 5.0 to 20.0 μm/cycle, assigned to seven categories, $WTS_{Aair5.0}$ to $WTS_{Aair20.0}$.

The maximum rut depth should range from 3.0 to 16.0 mm, assigned to categories from $Rd_{Aair3.0}$ to $Rd_{Aair16.0}$.

Test conditions and compaction of specimens are as determined in CEN EN 13108-20 (2008).

5.7.4.4 Stiffness

The stiffness (S) of the specimens is determined by one of the methods/apparatuses described in CEN EN 12697-26 (2012) and the test conditions are as those given in Table D.3 of CEN EN 13108-20 (2008).

The stiffness should lie between the minimum and maximum values specified or declared.

The minimum stiffness should range between 1500 and 21,000 MPa, assigned to 13 categories, $S_{min1500}$ to $S_{min21,000}$.

The maximum stiffness should range between 7000 and 30,000 MPa, assigned to eight categories, $S_{max7000}$ to $S_{max30,000}$.

The compaction of the test specimens is as determined in CEN EN 13108-20 (2008).

5.7.4.5 Resistance to fuel and resistance to de-icing fluid for application on airfields

When HRA is to be used in airfields, its resistance to fuels by CEN EN 12697-43 (2005) as well as its resistance to de-icing fluids by CEN EN 12697-41 (2013) should be determined.

For resistance to fuels, see Section 5.4.1.3.5 and substitute AC with HRA.
For resistance to de-icing fluids, see Section 5.4.1.3.6 and substitute AC with HRA.

5.7.5 Other requirements

Other requirements to be stated are the temperature of the mixture at any place in the plant, coating/homogeneity and, if needed, reaction to fire.

5.7.5.1 Temperature of the mixture

HRA temperatures at any place in the plant when produced with paving grade bitumen should be within the range of limits provided by CEN EN 13108-4 (2008). For paving grade bitumen 50/70 and 70/100, for example, the limit range is 145°C to 185°C and 140°C to 180°C, respectively. Measurements should be carried out in accordance to CEN EN 12697-13 (2001).

The minimum mixture temperature at delivery is determined by the producer.

When using modified or hard bitumen, or additives, temperature limits different from the above may be applied and should be documented and declared.

5.7.5.2 Coating and homogeneity

When HRA is discharged from the mixer, the mixture should be homogeneous in appearance, with all aggregates completely coated with bitumen and there should be no indication of balling of fine aggregates.

5.7.6 Tolerances, test frequencies for finished asphalt and conformity assessment

Tolerances of HRA, produced by any asphalt plant in terms of the target composition, should be within certain limits.

In accordance to CEN EN 13108-21 (2008), tolerances of the gradation curve and bitumen content of individual samples, or of the mean of four samples, for an HRA mixture are given in Table 5.25.

The frequency of samplings for gradation and binder content determination, once the compliance method has been chosen (by a single result or the mean of four results), depends on the OCL of the plant. More information is given in CEN EN 13108-21 (2008), Annex A.

Table 5.25 Tolerances of gradation and binder content about target composition

Percentage passing[a]	Individual samples		Mean of four samples	
	D < 16 mm	D ≥ 16 mm	D < 16 mm	D ≥ 16 mm
D	−8 ± 5	−9 ± 5	±4	±5
D/2 or other characteristic coarse sieve	±7	±9	±3	±4
2 mm	±5	±7	±2	±3
Characteristic fine sieve	+4[b]	±5	±2	±3
0.063 mm	±2	±3	±2	±2
Soluble binder content	±0.6	±0.6	±0.25	±0.3

Source: Adapted from CEN EN 13108-21, *Bituminous mixtures – Material specifications – Part 21: Factory Production Control*, Brussels: CEN, 2008. With permission (© CEN).

[a]   A tolerance of −2% should apply to the requirement of 100% passing 1.4D.

[b]   For HRA mixtures with D = 4 mm and below, the tolerance for the characteristic fine sieve should be ±10%.

5.8 STONE MASTIC ASPHALT

SMA was developed in Germany and Scandinavian countries in the mid-1960s. The prime reason for the development of this mixture was the demand to produce a mixture that can sustain the destructive action of studded tyres, to have good resistance to permanent deformation and long service life.

Nowadays, although the first demand no longer exists (the use of studded tyres was prohibited in 1975), SMA mixtures are used, mainly as wearing courses, by many agencies as a long-lasting wearing course material. SMA can be applied in all road categories, as well as in airfields, bridges, bus stops and other paved areas.

The specificity of the SMA is that the aggregate skeleton consisting of coarse aggregates is bound with a mastic mortar (bitumen, sand, filler and fibres). SMA is a dense gap-graded mixture.

SMA differs from HRA, since the mortar fills the voids, whereas in HRA, it constitutes the base of the mixture to which coarse, almost single-sized aggregates are added.

The SMA mixture is characterised by the higher bitumen content used compared to AC. This provides a longer service life but increases the cost. The incorporation of higher bitumen content is achieved by the addition of inorganic or organic fibres. SMA is produced in typical asphalt plants, but small modification for the provision of the fibre dosage system is needed.

SMA is specified by European and American standards.

5.8.1 SMA in accordance to European standards

SMA in the EU is designed in accordance to CEN EN 13108-5 (2008). Although the aim of the standard is to specify the required fundamental properties of the SMA, at the moment, it specifies empirical requirements only. Needless to say, the standard sets requirements for the constituent materials.

5.8.1.1 Constituent materials

Regarding materials that compose SMA, namely, binder and aggregates, and the use of recycled asphalt and other additives, relevant points referred to in Section 5.4.1.1 for AC apply.

The only difference to the points referred to in Section 5.4.1.1 is the fact that hard bitumen is not recommended to be used in SMA.

5.8.1.2 Mixture composition

5.8.1.2.1 General

The target composition should declare and document the materials, from which the mixture is composed, namely, the grading, the bitumen content and whether reclaimed asphalt or other additives are used.

5.8.1.2.2 Grading

The requirements of the gradation curve of the aggregate mixture are expressed in reference to percentage passing through some characteristic sieves, namely, 1.4D, D, 2 mm and 0.063 mm.

The D sieve is designated as the sieve with an aperture size that determines the SMA grade type. In this sieve, a small mass of aggregate is allowed to be retained.

The 1.4D sieve is designated as the sieve that is derived from the multiplication of D with 1.4, rounding the result to the nearest sieve.

Table 5.26 Grading of target composition of SMA-basic sieve set plus set 2

D (mm)	4	6 (6.3)	8	10	12 (12.5)	14	16	20
Sieve	% passing by mass							
1.4D				100				
D				90–100				
2	25–45	20–40	20–40	20–35	20–35	15–35	15–30	15–30
0.063	5.0–14.0	5.0–14.0	5.0–14.0	5.0–13.0	5.0–13.0	5.0–12.0	5.0–12.0	5.0–12.0

Source: Reproduced from CEN EN 13108-5, *Bituminous mixtures – Material specifications – Part 5, Stone Mastic Asphalt*, Brussels: CEN, 2008. With permission (© CEN).

In addition, the grading envelope may include the percentages passing through one optional sieve between D and 2 mm and one optional fine sieve between 2 and 0.063 mm.

The D sieve, as well as sieves between D and 2 mm, should be selected from 4.0, 6.3, 8, 10, 12.5, 14, 16 or 20 mm sieves, for the basic sieve set plus set 2; the corresponding optional coarse sieves for the basic sieve set plus set 1 are given in CEN EN 13108-5 (2008). Information about the basic sieve test and sieve sets 1 and 2 is given in Section 2.11.1.

The optional fine sieve, in all cases, can be selected from the following sieves: 1, 0.5, 0.25 and 0.125 mm.

The maximum and minimum percentages (limits), specified by CEN EN 13108-5 (2008), within which the target aggregate composition should be, are given in Table 5.26, for the basic sieve set plus set 2.

The ranges between the maximum and minimum values for the grading envelope should be selected as a single value within the given limits provided in Table 3 of CEN EN 13108-5 (2008).

5.8.1.2.3 Binder content

The minimum binder content of the target composition should range between 5.0% and 7.6%, assigned to eight categories, $B_{min5.0}$ to $B_{min7.6}$ (a 0.2% step is used).

Regarding the requirement for correction of binder content when the aggregate density is different from 2650 Mg/m^3, relevant points referred to in Section 5.4.1.4.2 apply.

5.8.1.2.4 Additives

When additives are also used (mainly inorganic or organic fibres), their type as well as their content in the mixture should be determined and declared.

5.8.1.3 Mixture properties

At the target mixture, the following empirical properties should be determined: void content, voids filled with bitumen, binder drainage, water sensitivity, resistance to permanent deformation and, when required, resistance to abrasion by studded tyres.

In case SMA is going to be applied in airfields, its resistance to fuel and to de-icing fluid should also be determined and declared.

5.8.1.3.1 Air void content

The void content of the prepared specimens should lie between the minimum and maximum values specified.

The minimum void content should range from 1.5% to 6.0%, assigned to 10 categories, $V_{min1.5}$ to V_{min6} (a step of 0.5% is used).

The maximum void content should range from 3.0% to 6.0%, assigned to nine categories, V_{max3} to V_{max8}.

The prepared specimens are compacted by an impact, gyratory or vibratory compactor, under conditions selected from CEN EN 13108-20 (2008), Table C.1.

The void content is determined in the usual way and in accordance to CEN EN 13108-20 (2008), D.2 (see also Section 5.4.1.3.3).

5.8.1.3.2 Voids filled with bitumen

The voids filled with bitumen of the prepared specimens should lie between the minimum and maximum values specified.

The minimum voids filled with bitumen should range from 71% to 86%, assigned to six categories, VFB_{min71} to VFB_{min86} (steps of 3% are used).

The maximum void content should range from 77% to 92%, assigned to six categories, VFB_{max77} to VFB_{max92} (steps of 3% are used).

The prepared specimens are compacted by an impact, gyratory or vibratory compactor, under conditions selected from CEN EN 13108-20 (2008), Table C.1.

Voids filled with bitumen (VFB) are determined in accordance with all points referred to in Section 5.4.1.4.3.

5.8.1.3.3 Binder drainage

The binder drainage, expressed as the maximum percentage of drained material, should be either 0.3%, 0.6% or 1.0%, assigned to categories $D_{0.3}$, $D_{0.6}$ or $D_{1.0}$.

The binder drainage test is conducted on loose mixtures in accordance to CEN EN 12697-18 (2004) (see also Section 5.6.1.3.4).

5.8.1.3.4 Water sensitivity

The water sensitivity of SMA-compacted cylindrical specimens is determined with respect to the indirect tensile strength ratio (ITSR) and a minimum value should be achieved as selected, specified or declared.

The minimum ITSR, expressed in per cent, should be either 60%, 70%, 80% or 90%, assigned to categories $ITSR_{60}$, $ITSR_{70}$, $ITSR_{80}$ or $ITSR_{90}$.

The indirect tensile strength ratio is determined in accordance to CEN EN 13697-12 (see also Section 3.7.3.4).

5.8.1.3.5 Resistance to permanent deformation

The resistance to permanent deformation is determined with respect to maximum wheel-tracking slope (WTS_{AIR}) and maximum proportional rut depth (PRD_{AIR}), when a small device and procedure B of CEN EN 12697-22 (2007) are used.

Alternatively, the resistance to permanent deformation is determined with respect to maximum proportional rut depth (P), when a large device is used, in accordance to CEN EN 12697-22 (2007).

The maximum wheel-tracking slope should range from 0.33 to 1.0 mm per 10^3 load cycle, assigned to 11 categories, $WTS_{AIR0.33}$ to $WTS_{AIR1.0}$.

The maximum proportional rut depth should range from 1.0% to 5.0%, assigned to five categories, $PRD_{AIR1.0}$ to $PRD_{AIR5.0}$.

Alternatively, the maximum proportional rut depth (P) should range from 5.0% to 20.0%, assigned to five categories, P_5 to P_{20}.

Test conditions and compaction of specimens are as determined in CEN EN 13108-20 (2008).

5.8.1.3.6 Resistance to abrasion by studded tyres

Resistance to abrasion by studded tyres, when required, is carried out on cylindrical specimens in accordance to CEN EN 12697-16 (2004). The specimens are prepared as for void content determination.

The resistance to abrasion by studded tyres, expressed as the maximum abrasion value, in millilitres, should range from 20 to 60, assigned to 10 categories from Abr_{A20} to Abr_{A60} (see CEN EN 13108-5 2008).

This requirement is exclusively used by north European countries that impose the use of studded tyres during winter months.

5.8.1.3.7 Resistance to fuel and resistance to de-icing fluid for application on airfields

When SMA is to be used in airfields, its resistance to fuels by CEN EN 12697-43 (2005) as well as its resistance to de-icing fluids by CEN EN 12697-41 (2013) should be determined.

For resistance to fuels, see Section 5.4.1.3.5 and substitute AC with SMA.

For resistance to de-icing fluids, see Section 5.4.1.3.6 and substitute AC with SMA.

5.8.1.4 Other requirements

Other requirements to be stated are the temperature of the mixture at any place in the plant, coating/homogeneity and, if needed, reaction to fire.

5.8.1.4.1 Temperature of the mixture

SMA temperatures at any place in the plant when produced with paving grade bitumen should be within the range of limits provided by CEN EN 13108-5 (2008). For paving grade bitumen 50/70 and 70/100, for example, the limit range is 150°C to 190°C and 140°C to 180°C, respectively. Measurements should be carried out in accordance to CEN EN 12697-13 (2001).

The minimum mixture temperature at delivery is determined by the producer.

When modified bitumen or mixtures with additives are used, different temperature limits may be applied and should be documented and declared.

5.8.1.4.2 Coating and homogeneity

When SMA is discharged from the mixer, the mixture should be homogeneous in appearance, with all aggregates completely coated with bitumen and there should be no indication of balling of fine aggregates.

5.8.1.5 Tolerances, test frequencies for finished asphalt and conformity assessment

All points referred to in Section 5.4.1.6 for AC also apply to SMA.

5.8.2 SMA in accordance to American standards

The SMA according to American standards is called stone matrix asphalt and is also known as SMA. It is designed in accordance to AASHTO M 325 (2012).

The SMA design is based on the volumetric properties in terms of air voids (V_a), voids in mineral aggregate (VMA) and the presence of stone-on-stone contact.

In addition, the target mixture should also satisfy the requirements for water sensitivity determined in terms of tensile strength ratio (TSR).

The standard also specifies the minimum quality requirements for the constituent materials (bitumen–asphalt binder, aggregate, mineral filler and stabilising additives).

5.8.2.1 Constituent materials

Crushed aggregates, binder and additives (cellulose or mineral fibres) should be as specified by AASHTO M 325 (2012).

5.8.2.2 Grading

The grading of the target mix should be within the area determined by the values of Table 5.27.

5.8.2.3 Target composition of the bituminous mixture

The target mix should satisfy the requirements of Table 5.28.

The compaction of the specimens is carried out by a gyratory compactor in accordance to AASHTO T 312 (2012). The number of gyrations used is 100. However, when aggregate wear by Los Angeles is >30%, then the desired number of gyrations reduces to 75.

The tensile strength ratio (TSR) of SMA is determined from specimens with air voids 6.0% ± 1.0%, and the test is conducted in accordance to AASHTO T 283 (ASTM D 4867 2009).

The drainage test is conducted according to AASHTO T 305, at temperatures expected during production.

More information can be found in ASSHTO M 325 (2012). For production, laying and other directives, also see AASHTO R 46 (2012).

Table 5.27 Grading limits for SMA according to AASHTO M 325 (2012)

Sieve (mm)	Type of mixture		
	19 mm	12.5 mm	9.5 mm
25.0 (1″)	100		
19.0 (3/4″)	90–100	100	
12.5 (1/2″)	50–88	90–100	100
9.5 (3/8″)	25–60	50–80	70–95
4.75 (No. 4)	20–28	20–35	30–50
2.36 (No. 8)	16–24	16–24	20–30
1.18 (No. 16)	—	—	Max 21
0.60 (No. 30)	—	—	Max 18
0.30 (No. 50)	—	—	Max 15
0.075 (No. 200)	8.0–11.0	8.0–11.0	8.0–12.0

Source: From AASHTO M 325, *Stone Matrix Asphalt (SMA)*, American Association of State Highway and Transportation Officials, Washington, DC: AASHTO, 2012. With permission.

Table 5.28 Requirements for SMA according to AASHTO M 325 (2012)

Mixture properties	Requirements
Air voids (%)	4.0[a]
VMA (%)	≥17.0
VCA$_{mix}$ (%)	Less than VCA$_{DRC}$[b]
Binder content (%)	≥6.0[c]
Moisture sensitivity (TSR)	≥0.7
Drainage (%)	≤0.3

Source: From AASHTO M 325, *Stone Matrix Asphalt (SMA)*, American Association of State Highway and Transportation Officials, Washington, DC: AASHTO, 2012. With permission.

[a] Regarding road pavements with low traffic volume or locations with low ambient temperatures, the void content may be lower, but in no case less than 3%.
[b] For the determination of VCA$_{mix}$ and VCA$_{DRC}$, see AASHTO R 46 (2012) and Sections 5.6.2.3(c) and 5.6.2.3.4.
[c] Experience has shown that the binder content should be between 6.0% and 7.0%. A percentage of binder content less than 6.0% reduces service life.

5.9 MASTIC ASPHALT

MA is perhaps the oldest mixture, developed almost at the same time in England and in France more than 100 years ago.

The primary features of this mixture, in comparison to all the abovementioned asphalts, are its high filler content, its higher binder content, its almost zero permeability, its use of harder bitumen and its high cost primarily due to the high binder content incorporated.

MA is designated as a voidless bituminous mixture in which the filler volume and binder volume exceed the remaining volume of voids in the mix.

MA may be used in roads, airfields and other trafficked areas for surface courses or binder courses. MA is primarily used for protection layers and inter-layers for bridges, tunnels and roughs. It should not be confused with MA for waterproofing purposes in the construction and civil engineering fields specified in CEN/TC 314-MA for waterproofing.

MA is produced on a conventional mixing plant and can be laid mechanically or by hand. When MA is to be used for waterproofing bridge decks and small areas in roads, a two-stage manufacturing process may be used. This involves mixing and casting the MA into blocks at a factory and then the blocks are melted down at a depot or on site using small mixers (Mathews and Ferne 1970).

In all cases, chippings are added to paving grade MA at the end during the stage of laying.

5.9.1 MA in accordance to European standards

MA in the EU is designed in accordance to CEN EN 13108-6 (2008) based on empirical requirements. The standard also sets requirements for the constituent materials.

The content of the following sections is based on the CEN EN 13108-6 (2008) standard.

5.9.1.1 Constituent materials

Regarding the materials that make up MA, namely, binder and aggregates, and the possible use of recycled asphalt and other additives, all relevant points referred to in Section 5.4.1.1 also apply.

The only difference with points referred to in Section 5.4.1.1 is that, in MA, when paving grade bitumen is used, only grades between 20/30 and 70/100 are selected, and when a hard grade bitumen is used, it should be from grades 10/20 and 15/25.

5.9.1.2 Mixture composition

The target mix composition should declare and document the materials from which the mixture is composed, namely, the particle size distribution, the bitumen content and, when used, the reclaimed asphalt or other additives.

5.9.1.2.1 Grading

The requirements of gradation curve of the aggregate mixture are expressed in reference to the percentage passing through some characteristic sieves, namely, 1.4D, D, 2 mm and 0.063 mm. Explanations for the abovementioned typical sieves are given in Section 5.5.3.1.

In addition, the grading envelope may include the percentages passing through one optional sieve between D and 2 mm and one optional fine sieve between 2 and 0.063 mm.

The D sieve, as well as sieves between D and 2 mm, should be selected from 4.0, 6.3, 8, 10, 12.5, 14 or 16 mm sieves, for the basic sieve set plus set 2; the corresponding optional coarse sieves for the basic sieve set plus set 1 are given in CEN EN 13108-6 (2008). Information about the basic sieve test and sieve sets 1 and 2 is given in Section 2.11.1.

The optional fine sieve, in all cases, can be selected from the following sieves: 1, 0.5, 0.25 and 0.125 mm.

The overall grading limits of the target composition of the MA are specified in CEN EN 13108-6 (2008) and are given in Table 5.29, for the basic sieve set plus set 2.

The ranges between the maximum and minimum values for the grading envelope should be selected as a single value within the given limits provided in Table 3 of CEN EN 13108-6 (2008).

5.9.1.2.2 Binder content

The minimum binder content of the target composition should range from 6.0% to 9.5%, assigned to 10 categories, $B_{min6.0}$ to $B_{min9.5}$.

Regarding the requirement for correction of binder content when the aggregate density is different from 2650 Mg/m³, the points referred to in Section 5.4.1.4.2 apply.

Table 5.29 Overall limits of target composition of MA-basic sieve set plus set 2

D (mm)	4	6 (6.3)	8	10	12 (12.5)	14	16
Sieve				% passing by mass			
1.4D				100			
D				90–100			
2	50–80	45–70	45–70	40–70	40–70	35–55	35–55
0.063	20–45	20–40	20–40	18–35	18–35	18–28	18–28

Source: Reproduced from CEN EN 13108-6, *Bituminous mixtures – Material specifications – Part 6: Mastic Asphalt*, Brussels: CEN, 2008. With permission (© CEN).

5.9.1.2.3 Additives

When additives are also used, their type as well as their content in the mixture should be determined.

5.9.1.3 Mixture properties

At the target composition, the following empirical properties should be determined: (a) indentation (resistance to permanent deformation) and, when required, (b) resistance to abrasion by studded tyres, (c) resistance to fuel for application on airfields and (d) resistance to de-icing fluid for application on airfields.

5.9.1.3.1 Indentation (resistance to permanent deformation)

The resistance to permanent deformation of MA is determined by the indentation test conducted in accordance to CEN EN 12697-20 (using cube or Marshall specimens), CEN EN 12697-21, Test W or Test B (using plate specimens) or CEN EN 12697-25 (2005) (cyclic compression test).

At the target composition, the minimum indentation when determined in accordance to CEN EN 12697-20 or CEN EN 12697-21 should range from 1.0 to 3.0 mm, assigned to five categories, $I_{min1.0}$ to $I_{min5.0}$.

The maximum indentation when determined in accordance to CEN EN 12697-20 or CEN EN 12697-21 should range from 3.0 to 15.0 mm, assigned to 12 categories, $I_{max3.0}$ to $I_{max15.0}$.

When the indentation test is carried out according to CEN EN 12697-20, the maximum indentation increase after 30 min should range from 0.3 to 0.8 mm, assigned to five categories, $I_{nc0.3}$ to $I_{nc0.8}$.

For mixture with D ≤ 11 and indentation ≤2.5 mm when tested in accordance to CEN EN 12697-20, the maximum dynamic indentation should range from 1.0 to 4.5 mm, assigned to eight categories, $I_{dyn1.0}$ to $I_{dyn4.5}$.

5.9.1.3.2 Resistance to abrasion by studded tyres

The resistance to abrasion by studded tyres is determined in accordance to CEN EN 12697-16 (2004), procedure A.

The maximum abrasion value should range from 20 to 60 ml, assigned to 10 categories, Abr_{A20} to Abr_{A60}.

5.9.1.3.3 Resistance to fuel and resistance to de-icing fluid for application on airfields

When MA is to be used in airfields, its resistance to fuel is determined in accordance to CEN EN 12697-43 (2005) and its resistance to de-icing is determined in accordance to CEN EN 12697-41 (2013).

For resistance to fuel, all points referred to in Section 5.4.1.3.5 are also valid.

For resistance to de-icing fluids, all points referred to in Section 5.4.1.3.6 are also valid.

5.9.1.4 Other requirements

Other requirements to be stated are the temperature of the mixture at any place in the plant, coating/homogeneity and, if needed, reaction to fire.

5.9.1.4.1 Temperature of the mixture

MA temperatures at any place in the plant when produced with paving grade bitumen should be within the range of limits provided by CEN EN 13108-5 (2008).

When modified bitumen or hard bitumen is used, different temperature limits may be applied and should be documented and declared.

5.9.1.4.2 Coating and homogeneity

When MA is discharged from the mixer, the mixture should be homogeneous in appearance, with all aggregates completely coated with bitumen and there should be no indication of balling of fine aggregates.

5.9.1.5 Tolerances and conformity assessment

Tolerances of MA produced in terms of the target composition should be within certain limits, given in CEN EN 13108-21 (2008).

The frequency of samplings for gradation and binder content determination, once the compliance method has been chosen (by a single result or the mean of four results), depends on the OCL of the plant. More information is given in CEN EN 13108-21 (2008), Annex A.

5.9.2 MA in accordance to American standards

No relevant American standard currently exists for MA for roads, airfields and other trafficked areas.

5.10 SOFT ASPHALT

Soft asphalt (SA) is designed to be used for surface courses and can also be used for binder courses, regulating courses and bases on roads, airfields and other trafficked areas, especially in climates with low temperatures as in the Nordic countries.

SA is an asphalt consisting of paving aggregates and soft bitumen as specified in EN 12591/AC (2009) (Table 2 or Table 3) (see also Section 3.4.1).

SA is designed in accordance with CEN EN 13108-3 (2008), which employs empirical requirements.

No American standard exists for designing SA.

Because of their limited use, SAs are not going to be discussed in detail in this book.

5.11 HIGH-MODULUS ASPHALTS

High-modulus asphalts (HiMAs) are bituminous materials with high stiffness modulus, high resistance to rutting, good spreading ability and good durability. They are used only for base or binder courses in roads, airfields or other paved areas.

The higher resistance to rutting achieved make HiMAs more attractive to be used in heavily trafficked areas. Additionally, the superior load-spreading properties of HiMAs will allow either the same design life to be achieved by significantly reducing the thickness of the asphalt layers or a longer life to be obtained when using the same thickness determined by conventional asphalts.

HiMAs, known as enrobé à module élevé (EME), were introduced in France in the early 1980s as a measure to reduce the usage of oil-derived products by reducing asphalt layers' thickness. The high stiffness of this material enabled the base course material to be reduced by up to 40% when compared with conventional French asphalt, grave bitumen (Nunn and Smith 1994).

Later on, a similar mixture such as high-stiffness road base macadam, called heavy-duty macadam, was tested on road trials (Nunn and Smith 1997) and compared with EME (Sanders and Nunn 2005).

Over the last 10 years, various other countries carried out successful trials and evaluation of HiMAs on road pavements (Bańkowski et al. 2009; Capitão and Picado-Santos 2006; Nkgapele et al. 2012; Perret et al. 2004; Wu et al. 2011). HiMA with hard bitumen modified by an SBS additive has also been tried successfully over bridge deck pavement (Li et al. 2011).

The aggregate grading of HiMA complies with the requirements of AC; hence, they are also called HiMA concrete. Nevertheless, the French notification EME is also used, particularly in France. Hence, HiMAs are also known as AC-EME and, occasionally, EME.

The binder of HiMAs is hard paving grade bitumen 15/25 or 10/20, in accordance to CEN EN 13924 (2006), or very low paving grade bitumen 20/30, in accordance to CEN EN 12591 (2009). Modified bitumen may also be used provided the mixture's fundamental property requirements have been satisfied.

The target mix composition is determined by CEN EN 13108-1 (2008) using basic and fundamental property requirements.

The French practice for AC-EME designates three types of grading: AC10-EME, AC14-EME and AC20-EME. The recommended thickness to be laid is between 60 and 80 mm for AC10-EME, between 70 and 130 mm for AC14-EME and between 90 and 150 mm for AC20-EME (Delorme et al. 2007).

As for the composition of the aggregate mix, the target grading is recommended to fall, initially, within the initial limits given in Table 5.30.

With respect to binder content, the initial minimum binder content is recommended to be the values stated in Table 5.31.

Apart from compositional and constituent material requirements, the French mix design guide uses the following performance-based fundamental requirements: (a) void content,

Table 5.30 Limiting values for target gradings for AC-EME, LPC design guide

	AC of high stiffness (AC-EME)			
	D = 20 or 14 mm		D = 10 mm	
	Percentage passing			
Sieve (mm)	Range	Target value	Range	Target value
6.3	45–65 (50–70 for 0/14)	53	45–65	55
4	40–60	47	—	52
2	25–38	33	28–38	33
0.063	5.4–6.7	6.7	6.3–7.2	6.7

Source: Adapted from Delorme, J.-L. et al., *LPC Bituminous Mixtures Design Guide*, Paris Cedex 15, France: Laboratoire Central des Ponts et Chaussées (LCPC), 2007.

Table 5.31 Typical initial values of binder content of AC-EME mixtures, LPC design guide

	AC-EME 1		AC-EME 2	
Minimum binder content (%)	*10 or 14 mm*	*20 mm*	*10 or 14 mm*	*20 mm*
B_{min}, for ρ_a = 2.65 g/cm³ᵃ	4.0	4.0	5.4	5.3
B_{min}, for ρ_a = 2.75 g/cm³ᵃ	3.9	3.9	5.2	5.1

Source: Adapted from Delorme, J.-L. et al., *LPC Bituminous Mixtures Design Guide*, Paris Cedex 15, France: Laboratoire Central des Ponts et Chaussées (LCPC), 2007.

[a] ρ_a is the apparent particle density of the aggregate, determined on the weighted mean of the total mineral fraction.

(b) water sensitivity, (c) resistance to permanent deformation, (d) wheel tracking, (e) stiffness and (f) fatigue. Cylindrical specimens are compacted by a gyratory compactor.

The limiting values of the above fundamental property requirements for the two classes of mixtures distinguished, AC-EME 1 and AC-EME 2, are given in Table 5.32.

With regard to UK practice as well as those of other countries, a practice similar to that of the French is followed. However, as for the stiffness requirement, lower stiffness values are also adopted (>9000 MPa, at 15°C).

AC-EME mixtures, because of their high stiffness modulus, are laid and compacted at higher-than-AC temperatures. The typical minimum temperatures are as follows: >140°C for laying and >120°C for compaction. During compaction, heavy rollers are used (vibrating steel rollers and pneumatic tyre rollers).

More information related to these mixtures is given in Delorme et al. (2007), SETRA (Service of Technical Studies of the Roads and Expressways) (2008), BS PD 6691 (2007) and Sanders and Nunn (2005).

Table 5.32 Limiting values of fundamental properties for AC-EME, LPC design guide

	Requirements	
Property	*AC-EME 1*	*AC-EME 2*
Water sensitivity	ITSR ≥ 70%ᵃ	
Permanent deformation – wheel tracking, large device, after 30,000 cycles at 60°C	P ≤ 7.5%ᵇ	
Stiffness, at 15°C, 10 Hz or 0.02 s	S ≥ 14,000 MPaᶜ	
Fatigue, 2 points, at 10°C, 25 Hz	$\varepsilon \geq 100$ μstrainᵈ	$\varepsilon \geq 130$ μstrainᵈ
Compaction conditions	Voids	
Cylindrical specimens: - No. of gyrations: 80 for AC10-EME - No of gyrations: 100 for AC14-EME - No of gyrations: 120 for AC20-EME	V ≤ 10%	V ≤ 6%
Slab specimens: - Irrespective of mix type	7%–10%	3%–6%

Source: Adapted from Delorme, J.-L. et al., *LPC Bituminous Mixtures Design Guide*, Paris Cedex 15, France: Laboratoire Central des Ponts et Chaussées (LCPC), 2007.

[a] ITSR = indirect tensile ratio, determined in accordance to CEN EN 12697-12 (2008), in compression.
[b] P = proportional rut depth, determined in accordance to CEN EN 12697-22 (2007), large device.
[c] S = stiffness, determined by either the complex modulus test (trapezoidal or parallelepiped specimen) or the uniaxial tensile test (cylindrical or parallelepiped specimens), according to CEN EN 12697-26 (2012).
[d] ε = stain at 10^6 cycles, determined according to CEN EN 12697-24 (2012), Annex A.

5.12 WARM MIX ASPHALTS

Warm mix asphalts (WMAs) are asphalts produced, handled and compacted at temperatures lower than those of conventional hot asphalts. This is achieved by lowering the viscosity of the binder using additives or applying foaming processes, both known as WMA systems. The aggregate mix gradation is the same as for the hot mix variety.

Producing and using WMA have environmental and paving benefits.

In the environmental aspect, fuel energy savings, reduction of greenhouse gas emission and reduction of bitumen fumes, and hence, improvement of workers' welfare, are achieved.

The fuel energy savings may be 20% to 35% on the plant, depending on the WMA system, which is equivalent to approximately 1.5 to 2.0 L of fuel per tonne of material. The reduction in greenhouse gas emission is closely associated with the reduction in energy consumption, that is, 20% to 35% reduction of $CO_{2\text{-eq}}$, which translates to approximately 4.1 to 5.5 kg of CO_2 per tonne of mix. The reduction of bitumen fumes is in the magnitude of 30%, while more optimistic studies are indicating reductions of up to 90% behind the paver (Croteau and Tessier 2008).

In the paving aspect, compaction and transportation are facilitated, the paving season is extended into the winter/cold months, WMA facilitates higher rates of reclaimed asphalt (or RAP) in recycling, the usage of WMA allows quick return of traffic and multiple lifts with WMA can be overlaid sooner (Croteau and Tessier 2008).

Warm asphalt mix technologies are classified in relation to their mixing temperature.

In general, two types of lower-temperature asphalts are distinguished, the WMA and the Half (or semi) warm asphalt (HWMA). The WMA is produced at temperatures ranging from 100°C to 140°C, while the HWMA is produced at even lower temperatures ranging from 70°C to 100°C (D' Angelo et al. 2008). Nevertheless, the term *WMA* is often used when referring to either type of asphalts, which may confuse the matter.

Figure 5.8 shows how WMA and HWMA fit into the full range of techniques from cold to hot mix and the approximate fuel usage per type of mix.

WMA systems have been used with all types of asphalt mixtures, including dense-graded asphalt, stone matrix asphalt (SMA) and PA. They have also been used with polymer-modified binders and in mixes containing RAP (D' Angelo et al. 2008).

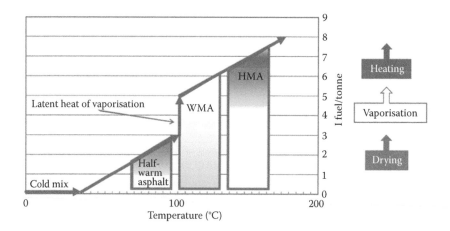

Figure 5.8 Classification by temperature range (temperature and fuel usage are approximations). (From D' Angelo J. et al., *Warm-Mix Asphalt: European Practice*. Technical Report FHWA-PL-08-007. Washington, DC: FHWA, Office of International Programs, Department of Transportation, 2008.)

The development of lower-temperature asphalts is relatively new. The necessity is believed to be triggered by Germany's consideration to review asphalt fumes' exposure limits, after United Nations' discussions in 1992. The first reported trials of WMA were undertaken in Germany and Norway between 1995 and 1999, and the first road trial took place in Germany in 1999 (Croteau and Tessier 2008). Road trials have been carried out ever since, including those in other European countries and the United States.

The argument of producing more economical asphalt at the end due to the fact that less fuel energy is required during mixing needs to be justified. The cost of the additive and of the investment for mixing plant modification (needed in some cases) may finally result in a more expensive mixture when compared to hot asphalts.

5.12.1 WMA technologies

The main objective of any WMA technology (system) is to modify the viscosity of the binder in a manner such that suitable mixing and field compaction is achieved at lower temperatures, while adequate viscosity is maintained at service temperatures. The latter is very important to avoid premature rutting.

The WMA technologies (systems) available today can be categorised into five, depending on the additive/technique used.

These technologies (when used) are as follows: (a) organic additives, (b) chemical additives (surfactants), (c) foaming techniques, (d) bitumen emulsion–based process (system), and (e) modification of the binder/aggregate mixing process.

Generally, the organic additive and the chemical additive systems produce WMA, the bitumen emulsion–based process produces HWMA and the other systems vary (Nicholls and James 2013).

The organic additives are waxes (Fischer-Tropsch processed synthetic wax or Montana wax) and fatty acid amides. They are added to the binder or directly to the mixture.

The organic additives affect the viscosity of the bitumen. They are often referred to as 'intelligent fillers' as they provide reduced viscosity at mixing/placing temperatures and increased viscosity at service temperatures (Croteau and Tessier 2008). On the basis of the above provision, the sulfur additive may be classified under this type of additive.

In general, the type of additive must be selected carefully so that its melting point is higher than the expected in-service temperatures (otherwise, permanent deformation may occur) and to minimise embrittlement of the asphalt at low temperatures (EAPA 2010).

Products currently available in the market are Asphaltan-B, BituTech PER, Ecoflex, isomerised paraffin, LEADCAP, Licomont BS 100, Sasobit, Shell Thiopave, SonneWarmix and Sübit (D' Angelo et al. 2008; NAPA 2011; Nicholls and James 2013; Prowell 2007).

The *chemical additives* are mainly surfactants, and they are a relatively new emerging group of additive for WMA (Anderson and May 2008). These additives do not reduce bitumen viscosity. As surfactants, they improve the ability of the bitumen to coat the aggregate particles at lower temperatures. Certain chemicals are added in a manner similar to anti-stripping agents at a concentration of 0.3% by mass of the bitumen (Croteau and Tessier 2008).

Products currently available in the market are Cecabase RT, Evotherm 3G, HyperTherm, Low-emission asphalt, Qualitherm, Rediset WMX and Revix (NAPA 2011; Nicholls and James 2013).

The *foaming techniques* rely on introducing small amount of water into hot bitumen. The water is introduced to the bitumen either directly by injection nozzles or indirectly by adding synthetic zeolites in powder form.

When the hot bitumen comes into contact with water, steam bubbles are forced into the continuous phase of the bitumen, which then expand until a thin film of bitumen holds the bubbles intact through their surface tension (Artamendi et al. 2011). When an amount of zeolite, typically 0.3% by weight of mixture, is added to the mixture shortly before or at the same time as the binder, it releases a small amount of water, creating a controlled foamed effect (D' Angelo et al. 2008).

In the foamed state, the viscosity of the bitumen is reduced, which allows the aggregate particles to be fully coated at lower mixing temperatures.

Products currently available in the market are Advera WMA, Aspha-min, Astec Green Systems, ECOMAC, LEAB, LT asphalt and Tri-Mix warm mix injection system (NAPA 2011; Nicholls and James 2013).

The *bitumen emulsion–based* processes were developed in North America and consist of mixing a specific high-residue bitumen emulsion with hot aggregate at reduced mixing temperatures (85°C to 115°C). As the emulsion is mixed with the hot aggregate, the water flashes off as steam. The bitumen emulsion is specifically designed for the WMA process and includes additives to improve coating, workability and adhesion (Croteau and Tessier 2008).

Products currently available in the market are Evotherm DAT and Warm recycling (Nicholls and James 2013).

The *modification of binder/aggregate mixing processes* are proprietary processes that are based on either mixing the binder (in foam or liquid state) with coarse and fine aggregates sequentially or mixing the aggregate with two different binders (again in foam or liquid state) sequentially. These processes are relatively inexpensive provided that plant modifications are minor (Croteau and Tessier 2008).

Products currently available in the market are Eco-Foam II, Accu-Shear Dual WMA System, Aquablack WMA, Double Barrel Green, Green Machine, Half-warm foamed bitumen process, HGrant Warm Mix System, Low-emission asphalt, Low-energy asphalt, Meeker warm mix, Terex WMA system, Ultrafoam GX and WAM foam (Nicholls and James 2013).

5.12.2 Performance of WMA and HWMA

On the basis of the laboratory and short-term (4 years or less) field performance data, WMA mixes appear to provide the same performance as, if not a better performance than, HMA (D' Angelo et al. 2008). A similar performance between WMA and HMA has been found on US sites after 2 years of service (Diefenderfer and Hearton 2010).

Recent laboratory and field evaluation of HWMA proved that their mechanical properties, such as stiffness, resistance to permanent deformation and resistance to water damage, were considered adequate and similar to those of the equivalent hot mix materials. Additionally, less hardening of the binders recovered from HWMA was found when compared to those recovered from the hot mixtures (Artamendi et al. 2011).

With respect to deformation, stiffness and fatigue resistance, Nicholls and James (2013) concluded the following.

The deformation resistance for WMA and HWMA is generally acceptable and often better than HMA mixtures, but the performance does depend on the system (Nicholls and James 2013).

The stiffness is likely to be lower for WMA and HWMA than for the equivalent HMA, but the difference should not be significant with a suitable choice of material.

In fatigue resistance, it appears that it decreases when moving from HMA to WMA and HWMA, although the situation will depend on the selected system.

The sensitivity to moisture of the WMA and HWMA mixtures was reported to be not as good as that of HMA (Austerman et al. 2009). However, the effect of using a WMA system on stripping has been shown to be dependent on the system used (Croteau and Tessier 2008), with some reports indicating a slight decrease in stripping resistance whereas others show no significant trend. Therefore, the addition of anti-stripping agent may be necessary when using certain WMA additives (Nicholls and James 2013).

The binder characteristics may be affected by certain additives, particularly the organic additives. Certain organic additives may have the tendency to stiffen the binder at lower temperatures, which may consequently increase the potential for thermal cracking. However, it was found that that the stiffening effect is related to binder and additive type (Croteau and Tessier 2008).

5.12.3 Specifications of WMA

At the time of writing this book, there was no specification published by European, American or any other standards.

Since WMA and HWMA systems can be used with all types of HMA (AC, AC for thin layers, SMA, PA and MA), the constituent materials to produce a WMA or an HWMA mixture are the same as for HMA, and because the expected performance of these mixtures should be the same as that of HMA mixtures, there is perhaps no need for a separate specification.

However, the variety of WMA additives/systems certainly needs to be addressed in a future specification regarding the batching, mixing, conditioning and compaction of the WMA mixture. The required additions/modifications could easily be implemented in Appendices or Annexes to the current HMA specifications. It is believed that the expected revision of the European specification for hot asphalts in 2015 will include WMA or HWMA mixtures. Such a proposal has already been made and carried out by NCHRP Report 691 (2011) for the AASHTO R 35 (2012) specification.

In 2006, Germany has issued a bulletin on the use of WMA, while in France, SETRA issues a product certificate after monitoring a test trial for a minimum of 3 years (D' Angelo et al. 2008).

At present, WMA mixtures are allowed to be used by implementing current specifications and performance requirements as for HMA, requiring a 4- or 5-year materials and workmanship warranty.

5.12.4 Mix design practices for WMA

The only mix design practice for WMA available at the period of writing this book was the one prepared by the NCHRP and proposed to be added as an appendix to AASHTO R 35 (2012) for Superpave volumetric design of HMA. The mix design practice for WMA is described in NCHRP Report 691 (2011), Appendix A, and its basic features are outlined below.

5.12.4.1 NCHRP mix design practice for WMA – Brief description

The NCHRP mix design for WMA refers to the AC type of mixtures that are produced at temperatures approximately 28°C (50°F) lower (or more) than temperatures typically used in the production of HMA. Is should be read in conjunction with AASHTO R 35 (2012).

5.12.4.1.1 Additional laboratory equipment

The proposed amendment to the specification AASHTO R 35 (2012) specifies additional laboratory equipment for mixing and for binder additive WMA processes.

5.12.4.1.2 WMA process selection

In selecting the WMA process, consideration should be given to the following factors: (1) available performance data, (2) the cost of the warm mix additives, (3) planned production and compaction temperatures, (4) planned production rates, (5) plant capabilities and (6) modifications required to successfully use the WMA process with available field and laboratory equipment.

5.12.4.1.3 Binder grade selection

A performance grade of binder should be selected in accordance with Section 5 of AASHTO M 323 (2013) considering the environment and the traffic volume at the project site.

5.12.4.1.4 Use of RAP

RAP can be used in the WMA after selection is made in accordance with Section 6 of AASHTO M 323 (2013) and the recommendations stated.

5.12.4.1.5 Batching, heating and preparation of WMA mixtures with different WMA additives/processes

Details are also given on batching, size of specimens and heating of the materials. A minimum of 24 gyratory specimens 150 mm in diameter are needed, in total, for the determination of optimum binder content and performance evaluation. The height of the specimens is 115, 95 or 175 mm, depending on the type of testing.

Details are also given for preparation of WMA mixtures with WMA additives added to the binder, added to the mixture, with a wet fraction of aggregate or preparation of foamed asphalt mixtures.

5.12.4.1.6 WMA mixture evaluation

At the optimum binder content determined in accordance with Section 10 of AASHTO R 35 (2012), mixture evaluation is required on (1) coating, (2) compactibility, (3) moisture sensitivity and (4) rutting resistance. The requirements set are as shown in Table 5.33.

The gyration ratio is determined using the following equation:

$$\text{Gyration ratio} = (N_{92})_{T-30}/(N_{92})_T,$$

Table 5.33 Summarised target WMA mixture requirements

Property	Test specification	Requirement	
Coating (%)	AASHTO T 195 (2011)	≥95	
Compactibility (gyration ratio)	(See below)	<1.25	
Moisture sensitivity (ratio)	AASHTO T 283 (2011)	≥0.80	
Rutting resistance (FN)[a]	AASHTO TP 79 (2013)	ESAL[b]	FN[a]
		<3	—
		3 to <10	≥30
		10 to <30	≥105
		≥30	≥415

Source: NCHRP Report 691, *Mix design practices for Warm Mix Asphalt*. Washington, DC: Transportation Research Board, 2011.

[a] FN, flow number.
[b] ESAL, traffic level in million equivalent standard axle loads.

where $(N_{92})_{T-30}$ represents gyrations to 92% relative density at 30°C below the planned field compaction temperature and $(N_{92})_T$ denotes gyrations to 92% relative density at the planned field compaction temperature.

The relative density for each gyration is determined by using the following equation:

$$G_{mmN} = [(G_{mb} \times h_{d})/G_{mm} \times h_{N}] \times 100 \ (\%),$$

where G_{mmN} is the relative density at N gyrations; G_{mb} is the bulk specific gravity of specimen compacted to N_{design} gyrations (AASHTO T 166 2013); G_{mm} is the theoretical maximum specific gravity (AASHTO T 209 2012); h_{d} is the height of specimen after N_{design} gyrations, from the Superpave gyratory compactor (mm); and h_{N} is the height of the specimen after N gyrations, from the Superpave gyratory compactor (mm).

5.13 OTHER TYPES OF ASPHALTS

In both European and American markets, the types of asphalt that are recognised are the ones mentioned and described in the previous paragraphs. Undoubtedly, there are asphalts with denominations different from the abovementioned that are used in the international market. Such asphalts include the following: Topekamixture, Gussasphalt, Asfaltbeton med Nedtro-mlede Skaever (ABS) mixture and so on.

These asphalts, except for Gussasphalt, are classified in one of the categories mentioned in the previous paragraphs and they are named as such for local or commercial needs. Gussasphalt is MA with added coarse aggregate particles, used almost exclusively in Germany and rarely used nowadays.

Given their local or restricted use of other types of asphalts, no further details will be given in this book.

REFERENCES

AASHTO M 226. 2012. *Viscosity-graded asphalt cement.* Washington, DC: American Association of State Highway and Transportation Officials.

AASHTO M 320. 2010. *Performance-graded asphalt binder.* Washington, DC: American Association of State Highway and Transportation Officials.

AASHTO M 323. 2013. *Superpave volumetric mix design.* Washington, DC: American Association of State Highway and Transportation Officials.

AASHTO M 325. 2012. *Stone matrix asphalt (SMA).* Washington, DC: American Association of State Highway and Transportation Officials.

AASHTO R 30. 2010. *Mixture conditioning of hot mix asphalt (HMA).* Washington, DC: American Association of State Highway and Transportation Officials.

AASHTO R 35. 2012. *Superpave volumetric design for hot mix asphalt (HMA).* Washington, DC: American Association of State Highway and Transportation Officials.

AASHTO R 46. 2012. *Designing stone matrix asphalt.* Washington, DC: American Association of State Highway and Transportation Officials.

AASHTO T 84. 2013. *Specific gravity and absorption of fine aggregate.* Washington, DC: American Association of State Highway and Transportation Officials.

AASHTO T 85. 2013. *Specific gravity and absorption of coarse aggregate.* Washington, DC: American Association of State Highway and Transportation Officials.

AASHTO T 90. 2008. *Determining the plastic limit and plasticity index of soils.* Washington, DC: American Association of State Highway and Transportation Officials.

AASHTO T 96. 2010. *Resistance to degradation of small-size coarse aggregate by abrasion and impact in the Los Angeles machine.* Washington, DC: American Association of State Highway and Transportation Officials.

AASHTO T 104. 2011. *Soundness of aggregate by use of sodium sulfate or magnesium sulfate.* Washington, DC: American Association of State Highway and Transportation Officials.

AASHTO T 112. 2012. *Clay lumps and friable particles in aggregate.* Washington, DC: American Association of State Highway and Transportation Officials.

AASHTO T 166. 2013. *Bulk specific gravity of compacted (Gmb) hot mix asphalt (HMA) using saturated surface-dry specimens.* Washington, DC: American Association of State Highway and Transportation Officials.

AASHTO T 176. 2008. *Plastic fines in graded aggregates and soils by use of the sand equivalent test.* Washington, DC: American Association of State Highway and Transportation Officials.

AASHTO T 195. 2011. *Determining degree of particle coating of asphalt mixtures.* Washington, DC: American Association of State Highway and Transportation Officials.

AASHTO T 209. 2012. *Theoretical maximum specific gravity (Gmm) and density of hot mix asphalt (HMA).* Washington, DC: American Association of State Highway and Transportation Officials.

AASHTO T 245. 2013. *Resistance to plastic flow of bituminous mixtures using Marshall apparatus.* Washington, DC: American Association of State Highway and Transportation Officials.

AASHTO T 269. 2011. *Percent air voids in compacted dense and open asphalt mixtures.* Washington, DC: American Association of State Highway and Transportation Officials.

AASHTO T 275. 2012. *Bulk specific gravity of compacted hot mix asphalt (HMA) using paraffin-coated specimens.* Washington, DC: American Association of State Highway and Transportation Officials.

AASHTO T 283. 2011. *Resistance of compacted hot mix asphalt (HMA) to moisture-induced damage.* Washington, DC: American Association of State Highway and Transportation Officials.

AASHTO T 304. 2011. *Uncompacted void content of fine aggregate.* Washington, DC: American Association of State Highway and Transportation Officials.

AASHTO T 305. 2012. *Determination of draindown characteristics in uncompacted asphalt mixtures.* Washington, DC: American Association of State Highway and Transportation Officials.

AASHTO T 312. 2012. *Preparing and determining the density of hot mix asphalt (HMA) specimens by means of the Superpave gyratory compactor.* Washington, DC: American Association of State Highway and Transportation Officials.

AASHTO T 331. 2013. *Bulk specific gravity (Gmb) and density of compacted hot mix asphalt (HMA) using automatic vacuum sealing method.* Washington, DC: American Association of State Highway and Transportation Officials.

AASHTO T 335. 2009. *Determining the percentage of fracture in coarse aggregate.* Washington, DC: American Association of State Highway and Transportation Officials.

AASHTO TP 79. 2013. *Determining the dynamic modulus and flow number for hot mix asphalt (HMA) using the asphalt mixture performance tester (AMPT).* Washington, DC: American Association of State Highway and Transportation Officials.

Anderson R.M. and R. May. 2008. *Engineering Properties, Emission and Field Performances of Warm Mix Asphalts.* Interim Report, NCHRP 9-47. Lexington, KY: Asphalt Institute.

Artamendi I., P. Phillips, and B. Allen. 2011. Laboratory and field evaluation of warm asphalt mixtures. *Proceedings of the 5th International Conference Bituminous Mixtures and Pavements, Thessaloniki.* Thessaloniki: AUTh.

Asphalt Institute MS-2. *Mix Design Methods for Asphalt Concrete and Other Hot-Mix Types,* 6th Edition. Manual Series 2. Lexington, KY: Asphalt Institute.

Asphalt Institute MS-4. 2007. *The Asphalt Handbook,* 7th Edition. Manual Series 4. Lexington, KY: Asphalt Institute.

Asphalt Institute SP-2. 2001. *Superpave Mix Design,* 3rd Edition. Superpave Series No. 2. Lexington, KY: Asphalt Institute.

ASTM C 29/C 29M. 2009. *Standard test method for bulk density ('unit weight') and voids in aggregate.* West Conshohocken, PA: ASTM International.

ASTM C 88. 2013. *Standard test method for soundness of aggregates by use of sodium sulfate or magnesium sulphate*. West Conshohocken, PA: ASTM International.

ASTM C 127. 2012. *Standard test method for density, relative density (specific gravity), and absorption of coarse aggregate*. West Conshohocken, PA: ASTM International.

ASTM C 128. 2012. *Standard test method for density, relative density (specific gravity), and absorption of fine aggregate*. West Conshohocken, PA: ASTM International.

ASTM C 131. 2006. *Standard test method for resistance to degradation of small-size coarse aggregate by abrasion and impact in the Los Angeles machine*. West Conshohocken, PA: ASTM International.

ASTM C 142/C 142M. 2010. *Standard test method for clay lumps and friable particles in aggregates*. West Conshohocken, PA: ASTM International.

ASTM C 535. 2012. *Standard test method for resistance to degradation of large-size coarse aggregate by abrasion and impact in the Los Angeles machine*. West Conshohocken, PA: ASTM International.

ASTM C 1252. 2006. *Standard test methods for uncompacted void content of fine aggregate (as influenced by particle shape, surface texture, and grading)*. West Conshohocken, PA: ASTM International.

ASTM D 92-12b. 2012. *Standard test method for flash and fire points by Cleveland open cup tester*. West Conshohocken, PA: ASTM International.

ASTM D 113. 2007. *Standard test method for ductility of bituminous materials*. West Conshohocken, PA: ASTM International.

ASTM D 242/D 242M. 2009. *Standard specification for mineral filler for bituminous paving mixtures*. West Conshohocken, PA: ASTM International.

ASTM D 946M-09a. 2009. *Standard specification for penetration-graded asphalt cement for use in pavement construction*. West Conshohocken, PA: ASTM International.

ASTM D 1188-07e1. 2007. *Standard test method for bulk specific gravity and density of compacted bituminous mixtures using coated specimens*. West Conshohocken, PA: ASTM International.

ASTM D 2041/D 2041M. 2011. *Standard test method for theoretical maximum specific gravity and density of bituminous paving mixtures*. West Conshohocken, PA: ASTM International.

ASTM D 2170/D 2170M. 2010. *Standard test method for kinematic viscosity of asphalts (bitumens)*. West Conshohocken, PA: ASTM International.

ASTM D 2171/D 2171M. 2010. *Standard test method for viscosity of asphalts by vacuum capillary viscometer*. West Conshohocken, PA: ASTM International.

ASTM D 2419. 2009. *Standard test method for sand equivalent value of soils and fine aggregate*. West Conshohocken, PA: ASTM International.

ASTM D 2726. 2011. *Standard test method for bulk specific gravity and density of non-absorptive compacted bituminous mixtures*. West Conshohocken, PA: ASTM International.

ASTM D 3203/D 3203M. 2011. *Standard test method for percent air voids in compacted dense and open bituminous paving mixtures*. West Conshohocken, PA: ASTM International.

ASTM D 3381/D 3381M. 2012. *Standard specification for viscosity-graded asphalt cement for use in pavement construction*. West Conshohocken, PA: ASTM International.

ASTM D 3515-96. 2010. *Standard specification for hot-mixed, hot-laid bituminous paving mixtures*. West Conshohocken, PA: ASTM International. (Withdrawn, no replacement).

ASTM D 4013. 2009. *Standard practice for preparation of test specimens of bituminous mixtures by means of gyratory shear compactor*. West Conshohocken, PA: ASTM International.

ASTM D 4318. 2010. *Standard test methods for liquid limit, plastic limit, and plasticity index of soils*. West Conshohocken, PA: ASTM International.

ASTM D 4791. 2010. *Standard test method for flat particles, elongated particles, or flat and elongated particles in coarse aggregate*. West Conshohocken, PA: ASTM International.

ASTM D 4867/D 4867M. 2009. *Standard test method for effect of moisture on asphalt concrete paving mixtures*. West Conshohocken, PA: ASTM International.

ASTM D 5581-07ae1. 2007. *Standard test method for resistance to plastic flow of bituminous mixtures using Marshall apparatus (6 inch-diameter specimen)*. West Conshohocken, PA: ASTM International.

ASTM D 5801. 2012. *Standard test method for toughness and tenacity of bituminous materials*. West Conshohocken, PA: ASTM International.

ASTM D 5821. 2006. *Standard test method for determining the percentage of fractured particles in coarse aggregate*. West Conshohocken, PA: ASTM International.

ASTM D 6373-07e01. 2007. *Standard specification for performance graded asphalt binder.* West Conshohocken, PA: ASTM International.

ASTM D 6390. 2011. *Standard test method for determination of draindown characteristics in uncompacted asphalt mixtures.* West Conshohocken, PA: ASTM International.

ASTM D 6752/D 6752M. 2011. *Standard test method for bulk specific gravity and density of compacted bituminous mixtures using automatic vacuum sealing method.* West Conshohocken, PA: ASTM International.

ASTM D 6857/D 6857M. 2011. *Standard test method for maximum specific gravity and density of bituminous paving mixtures using automatic vacuum sealing method.* West Conshohocken, PA: ASTM International.

ASTM D 6925. 2009. *Standard test method for preparation and determination of the relative density of hot mix asphalt (HMA) specimens by means of the Superpave gyratory compactor.* West Conshohocken, PA: ASTM International.

ASTM D 6926. 2010. *Standard practice for preparation of bituminous specimens using Marshall apparatus.* West Conshohocken, PA: ASTM International.

ASTM D 6927. 2006. *Standard test method for Marshall stability and flow of bituminous mixtures.* West Conshohocken, PA: ASTM International.

ASTM D 6932/D 6932M. 2008. *Standard guide for materials and construction of open-graded friction course plant mixtures.* West Conshohocken, PA: ASTM International.

ASTM D 6995. 2005. *Standard test method for determining field VMA based on the maximum specific gravity of the mix (G_{mm}).* West Conshohocken, PA: ASTM International.

ASTM D 7064/D 7064M-08e1. 2008. *Standard practice for open-graded friction courses (OGFC) mix design.* West Conshohocken, PA: ASTM International.

Austerman A.J., W.S. Mogawer, and R. Bonaquist. 2009. Investigation of the influence of warm mix asphalt additive dose on the workability, cracking susceptibility, and moisture susceptibility of asphalt mixtures containing reclaimed asphalt pavement. *Proceedings of 54th Annual Conference of the Canadian Technical Asphalt Association.* Laval, Quebec, Canada: Polyscience Publications Inc.

Bańkowski W., M. Tušar, and L.G. Wiman. 2009. Laboratory and field implementation of high modulus asphalt concrete. Requirements for HMAC mix design and pavement design. *Sustainable Pavements for European New Member States (SPENS), Sixth Framework Programme: Sustainable Surface Transport.* European Commission, DG Research.

Baughan C.J., L. Chinn, G. Harris, R. Stait, and S. Phillips. 2002. *Resurfacing a Motorway with Porous Asphalt: Effects on Rural Noise Exposure and Community Response.* TRL536. Crowthorne, UK: Transport Research Laboratory.

Brown J.R. 1973. *Pervious Bitumen-Macadam Surfacings Laid to Reduce Splash and Spray at Stonebridge, Warwickshire.* TRL Laboratory Report LR563. Crowthorne, UK: Transport Research Laboratory.

BS PD 6691. 2007. *Guidance on the use of BS EN 13108 bituminous mixtures-Material specifications*: British Standards Institution.

Capitão S. and L. Picado-Santos. 2006. Applications, properties and design of high modulus bituminous mixtures. *Road Materials and Pavements,* Vol. 7, No. 1, pp. 103–117. Taylor & Francis.

Carswell J. 1987. *The Effect of EVA Modified Bitumen on Roiled Asphalt Containing Different Fine Aggregates.* TRRL RR122. Wokingham, UK: Transport Research Laboratory.

CEN EN 1097-6. 2013. *Tests for mechanical and physical properties of aggregates – Part 6: Determination of particle density and water absorption.* Brussels: CEN.

CEN EN 12591. 2009. *Bitumen and bituminous binders – Specifications for paving grade bitumens.* Brussels: CEN.

CEN EN 12697-5:2009/AC. 2012. *Bituminous mixtures – Test methods for hot mix asphalt – Part 5: Determination of the maximum density.* Brussels: CEN.

CEN EN 12697-6. 2012. *Bituminous mixtures – Test methods for hot mix asphalt – Part 6: Determination of bulk density of bituminous specimens.* Brussels: CEN.

CEN EN 12697-8. 2003. *Bituminous mixtures – Test methods for hot mix asphalt – Part 8: Determination of void characteristics of bituminous specimens.* Brussels: CEN.

CEN EN 12697-11. 2012. *Bituminous mixtures – Test methods for hot mix asphalt – Part 11: Determination of the affinity between aggregate and bitumen.* Brussels: CEN.

CEN EN 12697-12. 2008. *Bituminous mixtures – Test methods for hot mix asphalt – Part 12: Determination of the water sensitivity of bituminous specimens.* Brussels: CEN.

CEN EN 12697-13:2000/AC. 2001. *Bituminous mixtures – Test methods for hot mix asphalt – Part 13: Temperature measurement.* Brussels: CEN.

CEN EN 12697-16. 2004. *Bituminous mixtures – Test methods for hot mix asphalt – Part 16: Abrasion by studded tyres.* Brussels: CEN.

CEN EN 12697-17:2004+A1. 2007. *Bituminous mixtures – Test methods for hot mix asphalt – Part 17: Particle loss of porous asphalt specimen.* Brussels: CEN.

CEN EN 12697-18. 2004. *Bituminous mixtures – Test methods for hot mix asphalt – Part 18: Binder drainage.* Brussels: CEN.

CEN EN 12697-22:2003+A1. 2007. *Bituminous mixtures – Test methods for hot mix asphalt – Part 22: Wheel tracking.* Brussels: CEN.

CEN EN 12697-23. 2003. *Bituminous mixtures – Test methods for hot mix asphalt – Part 23: Determination of the indirect tensile strength of bituminous specimens.* Brussels: CEN.

CEN EN 12697-24. 2012. *Bituminous mixtures – Test methods for hot mix asphalt – Part 24: Resistance to fatigue.* Brussels: CEN.

CEN EN 12697-25. 2005. *Bituminous mixtures – Test methods for hot mix asphalt – Part 25: Cyclic compression test.* Brussels: CEN.

CEN EN 12697-26. 2012. *Bituminous mixtures – Test methods for hot mix asphalt – Part 26: Stiffness.* Brussels: CEN.

CEN EN 12697-30. 2012. *Bituminous mixtures – Test methods for hot mix asphalt – Part 30: Specimen preparation by impact compactor.* Brussels: CEN.

CEN EN 12697-31. 2007. *Bituminous mixtures – Test methods for hot mix asphalt – Part 31: Specimen preparation by gyratory compactor.* Brussels: CEN.

CEN EN 12697-32:2003+A1. 2007. *Bituminous mixtures – Test methods for hot mix asphalt – Part 32: Laboratory compaction of bituminous mixtures by vibratory compactor.* Brussels: CEN.

CEN EN 12697-33:2003+A1. 2007. *Bituminous mixtures – Test methods for hot mix asphalt – Part 33: Specimen prepared by roller compactor.* Brussels: CEN.

CEN EN 12697-34. 2012. *Bituminous mixtures – Test methods for hot mix asphalt – Part 34: Marshall test.* Brussels: CEN.

CEN EN 12697-41. 2013. *Bituminous mixtures – Test methods for hot mix asphalt – Part 41: Resistance to de-icing fluids.* Brussels: CEN.

CEN EN 12697-43. 2005. *Bituminous mixtures – Test methods for hot mix asphalt – Part 43: Resistance to fuel.* Brussels: CEN.

CEN EN 13043:2002/AC. 2004. *Aggregates for bituminous mixtures and surface treatments for roads, airfields and other trafficked areas.* Brussels: CEN.

CEN EN 13108-1:2006/AC. 2008. *Bituminous mixtures – Material specifications – Part 1: Asphalt concrete.* Brussels: CEN.

CEN EN 13108-2:2006/AC. 2008. *Bituminous mixtures – Material specifications – Part 2: Asphalt concrete for very thin layers.* Brussels: CEN.

CEN EN 13108-3:2006/AC. 2008. *Bituminous mixtures – Material specifications – Part 3: Soft asphalt.* Brussels: CEN.

CEN EN 13108-4:2006/AC. 2008. *Bituminous mixtures – Material specifications – Part 4: Hot rolled asphalt.* Brussels: CEN.

CEN EN 13108-5:2006/AC. 2008. *Bituminous mixtures – Material specifications – Part 5: Stone mastic asphalt.* Brussels: CEN.

CEN EN 13108-6:2006/AC. 2008. *Bituminous mixtures – Material specifications – Part 6: Mastic asphalt.* Brussels: CEN.

CEN EN 13108-7:2008/AC. 2008. *Bituminous mixtures – Material specifications – Part 7: Porous asphalt.* Brussels: CEN.

CEN EN 13108-8. 2005. *Bituminous mixtures – Material specifications – Part 8: Reclaimed asphalt.* Brussels: CEN.

CEN EN 13108-20:2006/AC. 2008. *Bituminous mixtures – Material specifications – Part 20: Type testing.* Brussels: CEN.

CEN EN 13108-21:2006/AC. 2008. *Bituminous mixtures – Material specifications – Part 21: Factory production control.* Brussels: CEN.

CEN EN 13501-1:2007+A1. 2009. *Fire classification of construction products and building elements – Part 1: Classification using test data from reaction to fire tests.* Brussels: CEN.

CEN EN 13924:2006/AC. 2006. *Bitumen and bituminous binders – Specifications for hard paving grade bitumens.* Brussels: CEN.

CEN EN 14023. 2010. *Bitumen and bituminous binders – Specification framework for polymer modified bitumens.* Brussels: CEN.

Controls Srl. 2014. Available at http://www.controls-group.com.

Corte J.-F. and H. Di Benedetto. 2004. *Matériaux Routiers Bitumineux 1: Description et propriétés des constituents,* 238 p. Paris: Lavoisier.

Croteau J.M. and B. Tessier. 2008. Warm mix asphalt paving technologies: A road builder's perspective. *Proceedings of 2008 Annual Conference of the Transportation Association of Canada.* Laval, Quebec, Canada: Polyscience Publications Inc.

D' Angelo J., E. Harm, J. Bartoszek, G. Baumgardner, M. Corrigan, J. Cowsert, T. Harman, M. Jamshidi, W. Jones, D. Newcomb, B. Prowell, R. Sines, and B. Yeaton. 2008. *Warm-Mix Asphalt: European Practice.* Technical Report FHWA-PL-08-007. Washington, DC: FHWA, Office of International Programs, Department of Transportation.

Daines M.E. 1992. *Trials on Porous Asphalts and Rolled Asphalt on the A38 at Burton.* TRRL Supplementary Report 323. Wokingham, UK: Transport Research Laboratory.

Decoene Y. 1989. Knowledge acquired after 10 years of research on porous asphalt in Belgium. *Proceedings of the 4th Eurobitume Symposium,* p. 762. Madrid, Spain.

Delorme J.-L., C. de la Roche, and L. Wendling. 2007. *LPC Bituminous Mixtures Design Guide.* Paris Cedex 15, France: Laboratoire Central des Ponts et Chaussées (LCPC).

Denning J.H. and J. Carswell. 1981a. *Improvement in Rolled Asphalt Surfacings by the Addition of Organic Polymers.* TRRL LR989. Wokingham, UK: Transport Research Laboratory.

Denning J.H. and J. Carswell. 1981b. *Improvement in Rolled Asphalt Surfacings by the Addition of Sulfur.* TRRL LR963. Wokingham, UK: Transport Research Laboratory.

Denning J.H. and J. Carswell. 1983. *Assessment of 'Novophalt' as a Binder of Rolled Asphalt Wearing Course.* TRRL LR1101. Wokingham, UK: Transport Research Laboratory.

Diefenderfer S.D. and A.J. Hearton. 2010. *Performance of Virginia's Warm-Mix Asphalt Trial Section.* Final Report VTC 10-R17. Charlottesville, VA: Transportation Research Council.

EAPA. 2010. *The Use of Warm Mix Asphalt.* Position paper. Brussels, Belgium: European Asphalt Pavement Association.

EAPA. 2011. *Asphalt Figures 2010.* Brussels, Belgium: European Asphalt Pavement Association.

FAA. 2011a. AC 150/5370-10F, Part 5, Item P-401. *Flexible Surface Courses, Item P-401, Plant Mix Bituminous Pavements.* Washington, DC: FAA.

FAA. 2011b. AC 150/5370-10F, Part 5, Item P-402. *Porous Friction Course.* Washington, DC: FAA.

FAA. 2011c. AC 150/5370-10F, Part 5, Item 403. *Plant Mix Bituminous Pavements (Base, Levelling or Surface Course).* Washington, DC: FAA.

Green E.H. and F.V. Montgomery. 1972. *Coated Chippings for Rolled Asphalt.* TRRL Report LR456. Wokingham, UK: Transport Research Laboratory.

Jacobs F.A. 1980. *A Study of Blends of Trinidad Lake Asphalt and Bitumen in Rolled Asphalt.* TRRL SR561. Wokingham, UK: Transport Research Laboratory.

Li Y., Y. Tan, and L. Meng. 2011. Application study on high modulus asphalt concrete in bridge pavement. *Advanced Materials Research,* pp. 243–249. Switzerland: Trans Tech Publications.

Mathews D.H. and B.W. Ferne. 1970. *Trials of the Manufacture and Machine Laying of Mastic Asphalt.* RRL Report RL298. Wokingham, UK: Transport Research Laboratory.

Maycock G. 1966. *The Problem of Water Thrown up by Vehicles on Wet Roads.* LR4. Wokingham, UK: Transport Research Laboratory.

Morgan P.A., R.E. Stait, S. Reeves, and M. Clifton. 2007. *The Feasibility of Using Twin-Layer Porous Asphalt Surfaces on England's Strategic Road Network.* PPR 433. Wokingham, UK: Transport Research Laboratory.

NAPA. 2011. *Warm-Mix Asphalt: Best Practices*, 3rd Edition. Lanham, MD: National Asphalt Pavement Association.

NCHRP Report 691. 2011. *Mix Design Practices for Warm Mix Asphalt*. Washington, DC: Transportation Research Board.

Nelson P.M. and N.F. Ross. 1981. *Noise from Vehicles Running on Open Textured Road Surfaces*. SR 696. Wokingham, UK: Transport Research laboratory.

Nicholls J.C. 1997. *Review of UK Porous Asphalt Trials*. TRL264. Crowthorne, UK: Transport Research Laboratory.

Nicholls J.C. 1998. *Specification Trials of High-Performance Hot Rolled Asphalt Wearing Courses*. TRL Report 315. Crowthorne, UK: Transport Research Laboratory.

Nicholls J.C. 2001. *Material Performance of Porous Asphalt, Including When Laid Over Concrete*. TRL499. Crowthorne, UK: Transport Research Laboratory.

Nicholls J.C. and I.G. Carswell. 2001a. *Effectiveness of Edge Drainage Details for Use with Porous Asphalt*. TRL376. Crowthorne, UK: Transport Research Laboratory.

Nicholls J.C. and I.G. Carswell. 2001b. *The Design of Porous Asphalt Mixtures to Performance-Related Criteria*. TRL497. Crowthorne, UK: Transport Research Laboratory.

Nicholls J.C. and D. James. 2013. Literature review of lower temperature asphalt systems. *Construction Materials, ICE Proceedings*. Vol. 166, pp. 276–285. ICE Publishing.

Nikolaides A. and E. Manthos. 2014. Performance of AC-12.5 designed by Marshall and Superpave procedure. *Proceedings of 3rd International Conference on Transportation Infrastructures—ICTI 2014*. Pisa, Italy.

Nkgapele M., E. Denneman and J.K. Anochie-Boateng. 2012. Construction of a high modulus asphalt (HiMA) trial section Ethekwini: South Africa's first practical experience with design, manufacturing and paving of HiMA. *Proceedings of 31st Southern African Transportation Conference (SATC 2102)*. Document Transformation Technologies.

Nunn M.E. and T. Smith. 1994. *Evaluation of Enrobé à Module Élevé: A French High Modulus Roadbase Material*. Project Report PR66. Crowthorne, UK: Transport Research Laboratory.

Nunn M.E. and T. Smith. 1997. *Road Trials of High Modulus Base for Heavily Trafficked Roads*. TRL Report 231. Crowthorne, UK: Transport Research Laboratory.

Perret J., A.-G. Dumont, and J.-C. Turtschy. 2004. Assessment of resistance to rutting of high modulus bituminous mixtures using full-scale accelerated loading tests. *Proceedings of 3rd Eurasphalt & Eurobitume Congress, Vienna*, Vol. II, p. 1520.

Phillips S.M., P.M. Nelson, and G. Abbott. 1995. Reducing the noise from motorways: The acoustic performance of porous asphalt on the M4 at Cardiff. Paper to Acoustics '95. Volume 17: Part 4, *Proceedings of the Institute of Acoustics*. St. Albans, Hertfordshire: Institute of Acoustics.

Prowell B.D. 2007. *Warm Mix Asphalt – The International Technology Scanning Program-Summary Report*. Washington, DC: FHWA, U.S. Department of Transportation.

Roberts F.L., P.S. Kandhal, E.R. Brown, D. Lee, and T.W. Kennedy. 1996. *Hot Mix Asphalt Materials, Mixture Design and Construction*, 2nd Edition. Lanham, MD: NAPA Education Foundation.

Sanders P.J. and M. Nunn. 2005. *The Application of Enrobé à Module Élevé in Flexible Pavements*. TRL Report TRL636. Crowthorne, UK: Transport Research Laboratory.

SETRA. 2008. *The Use of Standards for Hot Mixtures*. Technical Guide. Paris: SETRA.

Weston D., M. Nunn, A. Brown and D. Lawrence. 2001. *Development of a Performance-Based Surfacing Specification for High Performance Asphalt Pavements*. TRL Report 456. Crowthorne, UK: Transport Research Laboratory.

Wu C., B. Jing, and X. Li. 2011. Performance evaluation of high-modulus asphalt mixtures. *Advanced Materials Research*, Vols. 311–313, p. 2138. Switzerland: Trans Tech Publications.

Chapter 6

Cold asphalts

6.1 GENERAL

Cold asphalts are bituminous mixtures in which the binder material is bituminous emulsion. This chapter does not cover cold asphalts produced with cut-back or flushed bitumen.

The term *cold* derives from the fact that mixing, hence laying, is carried out at ambient temperatures in contrast to hot asphalts, where bitumen and aggregate are heated at elevated temperatures.

Due to the fact that cold asphalts are produced and laid at ambient temperatures, they are more easily to handle; they can be transported to longer distances compared to hot asphalts and they can successfully be laid/compacted at ambient temperature (greater than 7°C and rising).

Additionally, in comparison to hot asphalts, cold asphalts require less total energy to be produced and laid, the total greenhouse emissions is less and the working conditions and labour safety are improved.

The production of cold asphalts is carried out either 'in-place' or in stationary plant. When produced in a stationary plant, the plant is smaller in size and much simpler compared to a hot mix plant. In situ production has the additional advantage of eliminating the mixture transportation cost.

The limitations of the cold asphalts may be summarised in the following: (a) they cannot be laid if rain is expected after laying; (b) although they can be used for all pavement layers, as surface course, cold asphalts are typically suitable for medium and light traffic; (c) quality control needs special attention particularly when mixed in-place.

6.2 CHARACTERISTIC TYPES OF COLD ASPHALTS

Cold asphalts are divided into two general categories, those for surface, binder, base or sub-base courses, or for pothole filling known as cold asphalt mixtures (CAM), and those for slurry surfacing (SS), known as slurry seals (SSl) or micro-surfacings (MS).

The gradings of the CAM can be dense or open similar to asphalt concretes. The gradings of the SS are such as to produce finer dense mixture.

The difference between CAM and SS apart from their grading and usage is that SS contain much more water (extra water is added) and are more workable.

Despite the environmental, energy-saving and safety advantages of cold asphalts, they are not as popular as hot asphalts. In Europe, the total production of cold asphalts during 2011 was less than 1.5% of that of hot asphalts: 4×10^6 tonnes of cold asphalts versus 324×10^6 tonnes of hot asphalts produced (EAPA 2011). The countries with the highest production of cold asphalts, from those listed in EAPA's figures, are France and Turkey, with more than 1×10^6 annual tonnages. Mexico has also been reported to have an annual production of cold asphalts of 3.3×10^6 tonnes in 2010 (EAPA 2011).

Depending on the market conditions, size of project and distance of project from mixing plant, the cost for constructing a bituminous layer with CAM could be less than that constructed with hot asphalt.

6.3 DENSE-GRADED COLD ASPHALTS

Dense-graded cold asphalts (DGCAs) are made up of bituminous emulsion and aggregate mixed together so as to result in a dense-graded mixture. The particle size distribution is continuous. Mixing, laying and compaction stages do not require any input of heat. However, in some cases when high bitumen content emulsion is used, the bitumen emulsion may be heated up to approximately 45°C.

As with any dense-graded mixture, DGCAs possess greater stability and stiffness, and are far less permeable to water, when compared to open-graded cold asphalts (OGCAs).

6.3.1 Constituent materials

6.3.1.1 Binder

The bituminous emulsion, preferably the cationic type, is produced from paving grade bitumen and should comply with the national specifications (CEN EN 13808 2013, ASTM D 2397 2012 or AASHTO M 208 2009, ASTM D 977 2012 or AASHTO M 140 2013, or Asphalt Institute 2008, for the European or US market). Useful information for selecting the appropriate bituminous emulsion can be found in an asphalt cold mix manual (Asphalt Institute MS-19 2008).

6.3.1.2 Aggregates

Aggregates, as in hot asphalts, are preferably crushed materials of any appropriate rock or natural deposits. The use of uncrushed natural deposits should be avoided.

In all cases, the properties of the aggregates should satisfy the requirements for bituminous mixtures, for example, CEN EN 13043 (2004) or Superpave requirements – see Section 5.4.2 or other national specifications.

Particular attention should be given to the aggregate cleanness. All aggregates should be clean and free from disintegrated materials, clay and other materials affecting aggregate coating with bitumen and thus the stability and strength of the mixture.

It is also recommended that the aggregate absorption should not exceed the value of 2%, particularly when the cold mixture is going to be used for surface course. Aggregates with higher absorption can be used only for layers below the surface course provided the water absorption is taken into consideration for the determination of the effective binder content.

6.3.1.3 Filler

The filler used should be derived from a sound and clean parent material appropriate for the production of coarse and fine aggregates. In some cases where additional filler is required, cement may also be used to a maximum percentage of 2% per mass of total aggregate. In all cases, the filler material should meet the requirements of CEN EN 13043 (2004), AASHTO M 17 (2001) or ASTM D 242/242M (2009), or other national specifications.

Table 6.1 Aggregate grading limits for DGCAs

Sieve size (mm)	Grading of DGCAs					
	Per cent passing by weight (%)					
	37.5 mm	25.0 mm	19.0 mm	12.5 mm	9.5 mm	4.75 mm[a]
50.0 (2″)	100	—	—	—	—	
37.5 (1 1/2″)	90–100	100	—	—	—	—
25.0 (1″)	—	90–100	100	—	—	—
19.0 (3/4″)	60–80	—	90–100	100	—	—
12.5 (1/2″)	—	60–80	—	90–100	100	100
9.5 (3/8″)	—	—	60–80	—	90–100	—
4.75 (No. 4)	20–55	25–60	35–65	45–70	60–80	75–100
2.36 (No. 8)	10–40	15–45	20–50	25–55	35–65	—
0.600 (No. 30)	—	—	—	—	—	—
0.300 (No. 50)	2–16	3–18	3–20	5–20	6–25	15–30
0.075 (No. 200)	0–5	1–7	2–8	2–9	2–10	5–12
Sand equivalent			≥40%			≥40%
LA abrasion[b]			≤40			—
% crushed faces			≥65%			—

Source: Adapted from Asphalt Institute, *Basic Asphalt Emulsion Manual*, Manual Series No. 19 (MS-19), 4th Edition, Lexington, KY: Asphalt Institute, 2008. With permission.

[a] Sand bitumen emulsion mixture.
[b] At 500 revolutions.

6.3.1.4 Aggregate mixtures

The aggregate of the target mixture is proposed to fall within the limiting values stated in Table 6.1. Table 6.1 also gives the limiting values for sand equivalent, LA abrasion and per cent of crushed faces specified by Asphalt Institute (2008) for DGCAs.

Grading VI is for sand cold asphalt used for laying a waterproofing layer, and it should always be covered with a layer of hot asphalt or cold asphalt.

6.3.1.5 Added water

In DGCAs, particularly when aggregates are dry, the addition of water is almost inevitable, so as to ensure good coating of aggregates with bitumen. The water should be clean and potable and the required amount ranges from 1.5% to 4% by mass of aggregate. The required percentage is determined at the stage of mix composition (mix design), as described below.

6.3.1.6 Chemical additive

In some cases, the addition of a chemical decelerator, and in rare cases accelerator, is required to modify the breaking time of the bituminous emulsion. When required, the additive is dissolved in water and its type and concentration are declared in the composition (mix) design.

6.3.2 Composition of DGCAs

The composition of dense-graded cold asphalts is carried out by in-house methodologies developed by organisations or by laboratories of the public/private sector, such as those described in Asphalt Institute (2008), Hughes (1981), Chevron (1979) and so on. There has been no composition procedure for cold asphalts as widely accepted as Marshall, Superpave or EN procedures used for hot asphalts.

In this chapter, the design methodology called modified Marshall method for DGCAs, developed by the author, is going to be described in detail (Cabrera and Nikolaides 1988; Nikolaides 1983).

The methodology is based on volumetric and empirical properties of DGCAs. It has been presented in conferences (Nikolaides 1992, 1993) and has been accepted for use in Greece (MPW D14 1994) and Indonesia (Indonesian MPW 1990).

6.3.3 Modified Marshall method for DGCAs

The aim of the design method is to determine the target binder content for DGCAs to have a satisfactory and lasting performance. It utilises the Marshall compaction and testing apparatuses and uses volumetric, Marshall and permanent deformation properties, as well as bitumen film thickness.

In brief, the design criteria are the following:

- Soaked Marshall stability
- Retained stability
- Total void content
- Specimen water absorption
- Bitumen film thickness
- Degree of coating
- Creep coefficient

Given the suitability of the constituent materials, the methodology consists of two stages. In the first stage, the compatibility of the bituminous emulsion with respect to the selected aggregate is examined and the per cent of added water before mixing is determined. In the second stage, the properties of the cold asphalt are determined for optimum performance.

6.3.3.1 Compatibility of bituminous emulsion and determination of per cent of added water

The stage where the compatibility of the bituminous emulsion with the selected aggregate is examined is necessary since there are a variety of aggregates to be used for a given bitumen emulsion. The properties of the cold asphalt are directly related to the capability of the bitumen emulsion to coat the aggregate particles with bitumen. This is determined by executing the coating test.

As a result of the coating test, the required amount of added water for wetting the aggregates before mixing is determined. This should range from 1.0% to 4% by weight of dry aggregate; otherwise, the bituminous emulsion is rejected.

The coating test is described in detail in Annex 6.A, given at the end of this chapter.

6.3.3.2 Determination of mixture properties for optimum performance

The stage for determining the properties of the cold asphalt consists of determining the target mix in order to satisfy the requirements stated in Table 6.2.

The specimens, 100 mm in diameter by approximately 62.5 mm in height, are compacted by a Marshall compactor (50 blows) and tested at 21°C ± 1°C after curing for 48 h. A

Table 6.2 Design requirements of dense-graded cold bituminous mixtures (for optimum performance)

Property	Types of DGCAs					
	37.5 mm	25.0 mm	19.0 mm	12.5 mm	9.5 mm	4.75 mm
Soaked stability (kN)	≥2.5					2.0
Retained stability (%) (48 h in water bath, room temperature)	≥50					≥50
Total void content (%)ᵃ	7–13					6–12
Water absorption of specimen (%) (48 h in water, room temperature)	≤4					≤4
Bitumen film thickness (μm)	≥7					≥7
Degree of coating (%)	≥85					≥85
Behaviour in permanent deformation (when ESAL > 3000 per day)	The target binder content should be less than the maximum permissible determined (see Section 6.3.3.3)					—
Recommended layer thickness (mm)ᵇ	80 to 150	50 to 100	40 to 100	30 to 75	25 to 75	25 to 50

Note: ESAL, equivalent standard axle load.

ᵃ Total void content = air voids + voids filled with water.
ᵇ It may be changed after local experience and compaction tests in situ.

number of cured specimens are tested dry and an equal number of cured specimens are tested after soaking in water for 48 h.

Apart from the dry and soaked stability, the total void content after curing, the water absorption after soaking and the bitumen film thickness are determined. The water absorption, together with the thickness of the bitumen film, ensures that premature ageing of the bitumen will not occur.

In addition to the above, the maximum permissible bitumen content is determined by executing the triaxial cyclic compression test according to CEN EN 12697 (2012), Method B, or the unconfined static creep test (see Section 7.6.3). This ensures sufficient resistance of the CAM to permanent deformation.

A detailed description of specimen production and Marshall testing is provided in Annex 6.B, and a detailed description of the soaking test–capillary water absorption test is provided in Annex 6.C; both annexes are found at the end of this chapter.

Typical results obtained after carrying out the modified Marshall design method for DGCAs are shown in Figure 6.1.

The soaked Marshall stability usually retains its value or shows an increase as the bitumen content increases (Figure 6.1a). After immersing the specimens in water for 24 h, the retained stability increases as the bitumen content increases (Figure 6.1b). The total void content decreases as the bitumen content increases (Figure 6.1c). The absorbed water content decreases as the bitumen content increases and the bitumen film thickness constantly increases as the bitumen content increases (Figure 6.1d and e, respectively). The dry bulk density curve is used to determine the target density to be achieved on site, for the target bitumen content to be determined. The specimens for dry bulk density determination are extracted 48 to 96 h after laying.

The target bitumen content of the mixture is determined from the diagrams plotted and considering the requirements of Table 6.2. The aim is to determine the target bitumen content in order for the target mix to have as few total voids and as low a percentage of absorbed water as possible, as well as soaked stability, retained stability and bitumen film thickness values as high as possible.

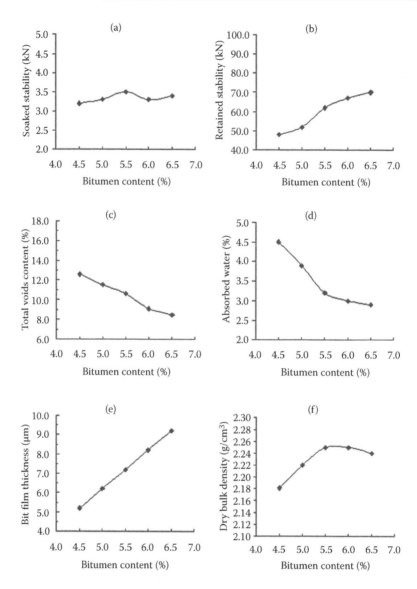

Figure 6.1 Typical diagrams derived from modified Marshall mix design for dense-graded cold mixtures. (a) Soaked stability vs. bitumen content. (b) Retained stability vs. bitumen content. (c) Total voids vs. bitumen content. (d) Absorbed water vs. bitumen content. (e) Bitumen film thickness vs. bitumen content. (f) Dry bulk density vs. bitumen content.

The target bitumen content should be lower than the maximum permissible bitumen determined by using the creep coefficient procedure, when the mixture is to be used in sections with more than 3000 ESAL (equivalent standard axle load) per day. For details on the procedure used to determine the creep coefficient, see Section 6.3.3.3.

6.3.3.3 Creep coefficient

The creep coefficient, B, is the slope from the least square linear fit on the log ε_n versus log n-values (n = load applications) determined from the triaxial cyclic compression test in

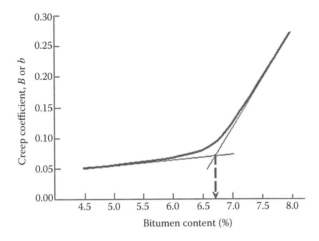

Figure 6.2 Determination of maximum permissible bitumen content.

accordance to CEN EN 12695-25, Method B. A description of the triaxial cyclic compression test is also given in Section 7.6.2.2.

The creep coefficient may also be determined by executing the static creep test. In this case, the creep coefficient, b, is the slope of the linear fit on the log S_{bit} versus log $S_{mix,creep}$. The stiffness of the bitumen, S_{bit}, is determined from the Van der Poel nomograph, and the static creep stiffness, $S_{mix,creep}$, is determined from the static creep test, at various stages of loading. More information regarding the static creep test is given in Section 7.6.3.

Two specimens per bitumen content (use five levels of bitumen content), after curing for Marshall testing, are subjected to either the triaxial cyclic compression test or the static creep test at 40°C ± 0.5°C.

The creep coefficient B or b is then plotted against bitumen content and a graph similar to Figure 6.2 is obtained. By drawing two tangent straight lines to the concave curve as shown in Figure 6.2, the maximum permissible bitumen content is determined at the point of intersection.

Cold dense-graded mixtures with bitumen content above the maximum permissible bitumen content are more susceptible to permanent deformation and should be rejected.

If all requirements stated in Table 6.2 are fulfilled at binder contents above the maximum permissible, the mix design is repeated by altering the aggregate gradation and possibly changing the bituminous emulsion used.

6.3.4 The ravelling test

The ravelling test, carried out in accordance to ASTM D 7196 (2012), is useful for classifying the curing and formulation of cold asphalt samples through ravel testing of compacted specimens. This performance test should be used to rank the mix conditions and approximate curing time for return to traffic and resistance to weather damage.

The test method measures the resistance to ravelling characteristics of cold asphalts with or without recycled asphalt pavement (reclaimed asphalt) by simulating an abrasion similar to early return to traffic.

A sample of cold asphalt at a quantity of approximately 2700 g of blended mixture (Method A), or laboratory-blended mixture, is compacted in a gyratory compactor (20 gyrations,

150 mm diameter mould, ASTM D 7229 2008). The specimen is then cured under specified conditions for a designated period, as prescribed by the following mix design method.

After the assigned period of curing, the sample is placed in the abrasion tester, similar to the one used in the wet truck abrasion test for slurry surfacing, and a rotating rubber hose exerts an abrasion force on the specimen for a certain period (typically 15 min).

The abraded loss of material is calculated as a percentage of mass loss using the following equation:

$$\% \text{ mass loss} = [((A - B)/A)] \times 100,$$

where A is the specimen mass before testing (g) and B is the specimen mass after abrasion (g).

The average value of two specimens is reported to the nearest 0.1%.

More details for the test can be found in ASTM D 7196 (2012).

6.4 OPEN-GRADED COLD ASPHALTS

OGCAs provide high air voids to drain water through the mixture. They have been used successfully for base courses and surface courses.

Other characteristics of these mixtures are that they show good resistance to reflection cracking and rutting, but their stiffness is much lower in comparison to dense-graded cold mixtures (approximately 50% to 60% of the dense-graded cold mixtures). Because of their high permeability, the underlying layer should always be of dense-graded mixture.

6.4.1 Constituent materials

6.4.1.1 Aggregates

The aggregate property requirements are similar to those for dense-graded cold mixtures.

The grading is such as to produce an open-graded mixture. Table 6.3 shows the gradations proposed by the Asphalt Institute (Asphalt Institute MS-19 2008).

Table 6.3 Aggregate gradation limits for OGCAs

| Sieve size (mm) | Percentage passing (%) | | | |
| | Base course | | | |
	25.0 mm	19.0 mm	9.5 mm	Wearing course
37.5 (1 1/2″)	100	—	—	—
25.0 (1″)	95–100	100	—	—
19.0 (3/4″)	—	90–100	—	—
12.5 (1/2″)	25–60	—	100	—
9.5 (3/8″)	—	20–55	85–100	100
4.75 (No. 4)	0–10	0–10	—	30–50
2.36 (No. 8)	0–5	0–5	0–10	5–15
1.18 (No. 16)	—	—	0–5	—
0.075 (No. 200)	0–2	0–2	0–2	0–2

Source: Adapted from Asphalt Institute, *Basic Asphalt Emulsion Manual*, Manual Series No. 19 (MS-19), 4th Edition, Lexington, KY: Asphalt Institute, 2008. With permission.

6.4.1.2 Bituminous emulsion

The bituminous emulsion, typically medium-setting cationic type with paving grade bitumen, should satisfy the requirements of the respective specification used.

6.4.1.3 Added water

In OGCAs, the addition of clean/potable water for wetting the aggregates before mixing is less than that required for dense cold mixtures. The required percentage is determined as in dense-graded mixtures but using steps of +0.5%. The added water for open-graded cold mixtures, if required, is normally not more than 3% by mass of aggregates.

6.4.1.4 Other additives

The addition of chemical additives or fillers in the mixture is not normally required in open-graded mixtures.

6.4.2 Composition of OGCAs

The composition of OGCAs is simpler than that of DGCAs. For a given aggregate grading, the determination of target bitumen content is based on the degree of coating of the aggregates, the sufficient workability of the mixture and drainage or runoff quantity.

The determination of the degree of coating is carried out in a manner similar to the coating test described in Annex 6.A. The only difference is that the nominal percentage of bitumen content in this case is less, that is, 2.5%, 3.0% and 3.5% for each base coarse mixture and 4.0% for the wearing course mixture.

The workability of the mixture is judged by the operator.

The target bitumen content is determined by the drainage or runoff test. This is similar to the one used for hot porous asphalt (Section 5.6.1.3.4), but it is advised to use the test method proposed by the Asphalt Institute (Asphalt Institute MS-19 2008).

To summarise the method, mixtures are made with varying bitumen emulsion content in 1% increments and subjected to a runoff test method.

On a 2000 g of dry aggregates mix, 40 g of water is added and is mixed thoroughly and left for 15 min covered with a clean cloth. Then, an appropriate amount of bitumen emulsion preheated to 60°C is added and hand mixed for 2 min. The mixture is then placed on a lightly dampened 2.36 mm sieve and is allowed to drain for 30 min at ambient temperature. The runoff is collected, dried and weighted and is expressed as grams of runoff.

The above is repeated with more bitumen emulsion, and increments of 1% are to be used; the results are plotted as mass of runoff against bitumen emulsion content by weight of aggregates. The target (optimum) emulsion content, hence bitumen content, is the one at 10 g of runoff. For more details, see Asphalt Institute (2008).

6.5 PRODUCTION OF COLD ASPHALT MIXTURES

The production of cold asphalt mixtures can be carried out either in a central or stationary mixing plant or in-place. The first allows the production of quality mixtures since more precise control of materials is possible.

Figure 6.3 Central mix plant for production of cold asphalts.

6.5.1 Production in a central plant

The central mixing plant for cold mixtures is simpler and smaller in size than its counterpart for hot mixtures, since the aggregate drier and the heating system of the binder at elevated temperatures are omitted.

A typical cold mix plant consists of the following: silos for aggregate storage, a feeding/weighing system for the aggregates, belt conveyors to transfer the aggregates to the mixer, emulsion storage tank(s), pumps to deliver the bitumen emulsion to the mixer, a measuring system for the bitumen emulsion, a water tank, a tank for chemical additives (if needed) and a pugmill mixer. A typical cold mix asphalt plant is shown in Figure 6.3.

The type of mixing plant is normally pugmill continuous. However, batch-type pugmills may also be used.

The cold bituminous mixture, in either case, is loaded directly to the transport vehicle (haul track).

When aggregates are wet, that is, containing more moisture than required, they should be aerated with a loader before placing them in the silos. The use of wet aggregates should be avoided to eliminate runoff effect.

6.5.2 Production in-place

Cold bituminous mixtures produced in-place (in situ) are mixed in a variety of mixing procedures. They can be mixed in a travel plant (travel plant mixing), by a motor grader (blade mixing) or by rotary/reclaimer mixers (rotary/reclaimer mixing).

Travel plant mixing and blade mixing are used when the cold bituminous mixture is to be produced from virgin aggregate material. Rotary/reclaimer mixing is normally used when the cold bituminous mixture consists of reclaimed material.

6.5.2.1 Travel plant mixing

The travel plants are self-propelled pugmill mixing plants. They are equipped with a hopper, a large bitumen emulsion tank, a small water tank, a belt conveyor system, an aggregate and bitumen emulsion proportioning system, a pugmill mixer and a screed. The aggregate is fed into the hopper from the haul track. Figure 6.4 shows a typical travel plant. Some

Figure 6.4 Travel plant for cold bituminous mixtures.

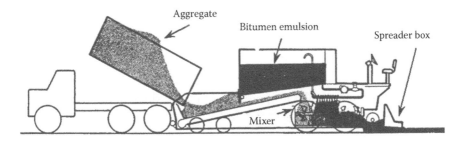

Figure 6.5 Schematic representation of a travel plant operation for the production of cold asphalt.

travel plants, not very common, do not have bituminous emulsion storage tanks; the bitumen emulsion is supplied by an emulsion tanker travelling in front or beside the travel plant.

A schematic representation of the travel plant operation is shown in Figure 6.5.

The proportioning system of aggregates, bitumen emulsion and water (if required) is independent on the travelling speed. The quantity of aggregate delivered to the mixer is determined by the opening height of the floodgates, while emulsion and water quantities are measured by flow meters.

The advantage of travel plant mixing compared to central plant mixing is flexibility and reduction of mixing/transporting cost.

To ensure good production quality, like in any mixing plant, the following are required: periodical and scrupulous calibration of the feed systems, periodic maintenance of all parts and generally, keeping the plant in good working order.

6.5.2.2 Blade mixing

In blade mixing, the bitumen emulsion is sprayed by a distributor on a flattened windrow of imported aggregate, or reclaimed aggregate, moving ahead of the motor grader. Mixing is carried out by the blade of the motor grader through a series of turning and rolling actions. The water needed to be added to the aggregate material is sprayed before the application of the bituminous emulsion.

Blade mixing is the least precise of the mixed-in-place methods since everything is dependent on the experience of the motor grader operator. In an attempt to achieve uniform and adequate coating, it is possible that bitumen coating is stripped from the aggregate.

For the above reasons, blade mixing is only used when short lengths or small areas are to be laid with cold asphalts or when stabilisation with bitumen emulsion is to be carried out.

6.5.2.3 Rotary/reclaimer mixing

Rotary/reclaimer mixing requires the use of mobile rotary mixers or reclaimers (more powerful machines equipped with strong tines for pulverisation of old bituminous layers or other road materials).

Both consist of a mixing chamber, with a spray bar, which is open at the bottom, and a transverse rotating shaft equipped with cutting blades or tines. The bitumen emulsion, or water, is supplied to the mixing chamber by a tanker moving in front of the mixers. For machines without a spray bar inside the mixing chamber, multiple passes are required, adding water and bitumen emulsion using water and bituminous emulsion sprayers.

This type of in-place mixing is quite common in cold in-place recycling.

6.6 LAYING AND COMPACTION OF COLD ASPHALT MIXTURES

6.6.1 Laying

Cold asphalts produced in a central mixing plant are laid with the usual asphalt pavers having the screed's heating and vibration system switched off.

The thickness of the mat should be appropriate, not too thick or not too thin, depending on the maximum dimension of the largest aggregate.

Preparation of the surface before laying is similar to the one followed for hot asphalt laying (clean surface, application of tack coating or prime coating, etc.).

Precautions: Laying should not start if rain is forecasted and laying should stop immediately when rain starts. It is always preferable to lay cold asphalts in several thin layers rather than in one thick layer.

6.6.2 Compaction

Compaction of the cold asphalts begins when the emulsion starts to break (the colour of the mixture changes from brown to dark-brown/black) provided the mixture can support the roller without excessive displacement.

The rollers most frequently used are pneumatic tyre rollers (8 to 12 tonnes) and static steel-wheel rollers (6–10 tonnes). Vibratory smooth steel-wheel rollers may also be used preferably without vibration.

The initial rolling is usually carried out with a static steel roller, while the intermediate and final rolling is usually carried out with a pneumatic tyre roller. The compaction with a vibratory smooth steel-wheel roller using the vibrations may be considered only along joints.

The number of passes is always determined from a trial section constructed near the project site using the job mix formula. It is recommended that the number of passes should be enough to achieve a degree of compaction greater than 95% of the target dry bulk density.

In some cases, in order to avoid aggregate particles sticking to the steel drum or tyre surface, a small quantity of sand is spread on the surface to be compacted. The technique of wetting the surface of the drum or of the tyres with a small amount water must be avoided. If wetting is inevitable, it should be carried out outside the area to be compacted and not during compaction.

Precautions: When ravelling occurs after opening to traffic, the loose material should be removed as quickly as possible to prevent further damage to the surface. If ravelling continues, necessary measures must be taken after determining the cause of ravelling.

6.7 QUALITY CONTROL OF COLD ASPHALT MIXTURES

Quality control consists of both mixture and construction inspection. Sampling, testing, thickness and degree of compaction procedures are similar to those applied for hot asphalts.

Mix samples are taken from either the delivery trucks or the paved layer before compaction.

The determination of bitumen content and aggregate gradation is carried out with the same test methods as for hot asphalts after the water has been completely evaporated from the mixture.

The degree of compaction determination may be carried out using the sand-cone method or by core extraction. Core extraction is preferred when layer thickness is also needed to be checked. However, coring takes place a few days after compaction, possibly even after 2–3 weeks, to be able to extract intact cores.

Sampling of constituent materials is carried out in the central mixing plant or upon delivery of the materials on site. Aggregate testing procedures are the same as those for hot mixtures. Bitumen emulsion testing, as well as tests on the recovered bitumen, follows the requirements and test methods specified in the standard used.

Finally, surface evenness is measured with the same equipment or devices used in layers constructed with hot asphalts.

6.8 COLD ASPHALTS FOR SLURRY SURFACING

Cold asphalts for slurry surfacing are commonly used worldwide for surface maintenance, for surface maintenance prevention and, above all, for the restoration of surface skid resistance of highways, city streets and airport pavements.

Slurry surfacing is surface treatment consisting of a mixture of aggregates, bituminous emulsion, water and additives, which are mixed and laid in-place. A slurry surfacing product may consist of one or more layers.

The slurry surfacing in which the bituminous emulsion is polymer modified is known as micro-surfacing. Micro-surfacing is normally made with larger-sized aggregates and is considered as high-performance surfacing material.

When slurry surfacing is made of small-sized aggregates and the bituminous emulsion is not modified, it is called slurry seal.

The thickness of the single layer of slurry surfacing ranges from 4 to 12 mm. The cold asphalt mixture for slurry surfacing is produced and laid in situ with a purposely built travel plant.

Slurry seals and in particular micro-surfacing, apart from being laid faster than any other bituminous mixture, have the following advantages and properties:

a. They provide an excellent anti-skidding surface.
b. They act as a sealing layer.
c. They do not substantially raise the pavement's height; as a consequence, adjustment of road manhole covers is not required.
d. They repair small surface irregularities.
e. They provide safer working conditions compared to hot asphalts.
f. They cause less atmospheric pollution compared to hot asphalts.
g. Under normal market conditions, they provide the most cost-effective solution for restoration of surface skid resistance.

Cold asphalts for slurry surfacing are specified by international standards, specifications and guidelines, such as CEN EN 12273 (2008) for slurry surfacing, ASTM D 3910 (2011) or International Slurry Seal Association (ISSA) A105 (2010) for slurry seals and ASTM D 6372 (2010) or ISSA A143 (2010) for micro-surfacing.

6.8.1 Constituent materials

Materials used for the production of cold asphalts for slurry surfacing are bituminous emulsion, graded aggregates (medium and fine size), water and additives. The additives may be mineral filler, added to the aggregate mix, or chemicals, diluted to water, to improve mixture consistency and to adjust mixture breaking and curing properties.

Sometimes, slurry surfacing may contain fibres (cellulose or others), which are added to gain strength and increase the durability of slurry surfacing. The latter, in some countries, is a proprietary technique.

6.8.1.1 Bituminous emulsion

The bituminous emulsion is preferably the cationic type, and for high performance, it should be polymer modified.

With the use of a cationic-type bituminous emulsion, better adhesion of aggregates is achieved, the surface opens quicker to the traffic and is less susceptible to rain after construction, early stage ravelling is minimised and life expectancy is increased.

Cationic bituminous emulsions should comply with CEN EN 13808 (2013), ASTM D 2397 (2012), AASHTO M 208 (2009) or any other national relevant standard.

With regard to micro-surfacing, the penetration grade bitumen, according to ISSA (ISSA A143 2010), should be within the range of 40 to 90 dmm (pen) (ASTM D 5 2013 or AASHTO T 49 2011) at 25°C, and the softening point should be ≥57°C (ASTM D 36 2012 or AASHTO T 53 2009). Additionally, the requirement for settlement and storage stability of bitumen emulsion, after 24 h (ASTM D 6930 2010 or AASHTO T 59 2013), should be ≤1%.

6.8.1.2 Aggregates

Aggregates should be derived from hard rocks, suitable for surfacing, with properties satisfying the requirements of CEN EN 13043 (2004) or any other national specification.

Among aggregate properties, particular attention should be given to aggregate cleanness (methylene blue test) (CEN EN 933-9+A1 2013) or sand equivalent (CEN EN 933-8 2012), resistance to fragmentation/crushing (Los Angeles) (CEN EN 1097-2 2010), resistance to wear micro-Deval (MDV) (CEN EN 1097-1 2011), resistance to polishing/abrasion (PSV or AAV) (CEN EN 1097-8 2009) and, when needed, resistance to abrasion by studded tyres (CEN EN 1097-9 2014).

Indicatively, French specification NF EN 12273 (2008), for slurry surfacing, requires PSV ≥ 50 or PSV ≥ 56 and MDV ≤ 20 or MDV ≤ 15, depending on the traffic category. With respect to methylene blue value, MB_F, the French specifications assume that it should be ≤10 g/kg as in all hot bituminous mixtures (Delorme et al. 2007).

The Asphalt Institute's and ISSA's aggregate requirements for micro-surfacing are as follows: Los Angeles abrasion loss (ASTM C 131 2006 or AASHTO T 96 2010), grading C or D, ≤30%, and soundness of aggregate (ASTM C 88 2013 or AASHTO T 104 2011) ≤ 15%, when sodium sulfate is used, or ≤ 25%, when magnesium sulfate is used (Asphalt Institute MS-19 2008; ISSA A143 2010). With regard to sand equivalent (ASTM D 2419 2009 or AASHTO T 176 2008), the Asphalt Institute's requirement is ≥60 (Asphalt Institute MS-19 2008) and ISSA's requirement is ≥ 65 (ISSA A143 2010).

With respect to slurry seals, the aggregate requirements by the Asphalt Institute and ISSA are as follows: Los Angeles abrasion loss, grading C or D, ≤35%, sand equivalent ≥45 (Asphalt Institute MS-19 2008; ISSA A105 2010). ISSA also has a requirement for soundness of aggregate, which is the same as for aggregates for micro-surfacing.

It must be stated that the requirement for Los Angeles abrasion loss alone does not ensure good durability against polishing and abrasion caused by the traffic.

With regard to CEN prEN 16333 (2014) for slurry surfacing on airfields, the percentage of crushed and broken surfaces should be 100% ($C_{100/0}$), the resistance to fragmentation, Los Angeles loss ≤20 or ≤15 (LA_{20} or LA_{15}) and the resistance to polishing of aggregate (PSV) ≥50 (PSV_{50}).

6.8.1.3 Additives

When mineral filler is used to improve mixture consistency and to adjust mixture breaking and curing properties, it can be Portland cement, hydrated lime, limestone dust, fly ash or other approved filler complying with the requirements of the appropriate standard. The percentage of filler added typically ranged from 0% to 3%, by weight of aggregate.

When a chemical additive is used to accelerate or retard the break/set of the slurry surfacing, it can be any suitable water-soluble chemical, complying with the requirements of the appropriate standard.

6.8.2 Aggregate gradation

The European standard CEN EN 12273 (2008) for slurry surfacing does not specify gradation limits. The target gradating and its tolerance range are left to be determined and declared by the producer/manufacturer.

However, the American standard ASTM D 3910 (2011) and the guidelines of the ISSA (ISSA 105 and ISSA 143) specify gradation limits for slurry surfacing. The gradation limits specified per type of slurry surfacing mix, which are the same for both organisations, are shown in Table 6.4.

Type 1 slurry surfacing is suitable to seal cracks, fill voids, repair moderate surface distresses (surface erosion) and provide protection.

Type 2 slurry surfacing is suitable to fill surface voids, correct severe surface erosion conditions and provide a minimum wearing surface.

Type 3 slurry surfacing is suitable to provide, in addition, maximum skid resistance and an improved wearing surface. This type of micro-surfacing is appropriate for heavily trafficked pavements and rut filling.

Type 4 slurry surfacing, which is proposed by the author, is a coarser micro-surfacing than Type 3. It has been used successfully for many years in Greece and elsewhere when higher surface macro-texture is required. By using this type of slurry surfacing, the resultant thickness of a single-layer micro-surfacing is in excess of 12 mm. Additionally, for improved durability, it is advised to be laid over a Type 2 slurry surfacing.

6.8.3 Mix design for slurry surfacing

Mix design procedure for slurry surfacing is provided by ASTM D 3910 (2011) for slurry seal and ASTM D 6372 (2010) for micro-surfacing.

The European standards (ENs) do not specify a mix design procedure for determining the target composition of the slurry surfacing mix. They specify test methods that are used for determining useful characteristic properties of the slurry surfacing mix such as consistency, cohesion and wearing under wet track abrasion conditions. Additionally, they provide a test method for determining the suitability of aggregates and cationic emulsions for slurry surfacings.

However, they specify performance-related technical requirements based on visual assessment of defects, which will be covered in subsequent sections. The determination of the target mix is left to the producer/manufacturer.

Nevertheless, the draft EN (CEN prEN 16333 2014) related to slurry surfacing specifications for airfields suggests a mix design procedure, very similar to ASTM D 6372 (2010).

Table 6.4 Types of slurry seals and indicative ranges of bitumen content and application rates of dry aggregates

Sieve	Type 1[a] (2.36 mm)	Type 2[a,c] (4.75 mm)	Type 3[a,c] (4.75 mm)	Type 4[e] (9.5 mm)	Permissible deviation[a]
12.5 mm	—	—	—	100	±5%
9.5 mm	—	100	100	85–100	±5%
4.75 mm	100	90–100	70–90	60–87	±5%
2.36 mm	90–100	65–90	45–70	40–60	±5%
1.18 mm	65–90	45–70	28–50	28–45	±5%
600 μm	40–65	30–50	19–34	19–34	±5%
300 μm	25–42	18–30	12–25	12–25	±4%
150 μm	15–30	10–21	7–18	8–17	±3%
75 μm	10–20	5–15	5–15	4–8	±2%
Indicative range of binder content % by wt. of dry aggregate	10–16[b]	7.5–13.5[b] (5.5–9.5)[c] (5.5–10.5)[d]	6.5–12.0[b]	6.0–8.5[e]	From the target % max ±1%[f]
Range of application rate of dry aggregate (kg/m²)	3.6–5.4[b]	5.4–9.1 (5.4–10.8)[d]	8.2–13.6 (8.1–16.3)[d]	9–14[e]	±1.1%[g]

[a] According to ASTM 3910 (2011), ASTM D 6372 (2010), ISSA A105 (2010), and ISSA A143 (2010) for slurry seal and micro-surfacing.
[b] For slurry seal.
[c] For micro-surfacing.
[d] For micro-surfacing, according to ISSA A143 (2010).
[e] Recommended by the author.
[f] For slurry seal according to ISSA A105 (2010).
[g] For slurry seal, when the pavement surface is in relative good condition and surface texture does not change significantly, according to ISSA A105 (2010).

6.8.3.1 Mix design procedure for slurry seal

The mix design procedure for slurry seal according to ASTM D 3910 (2011) requires the execution of the following tests: (a) consistency test, (b) set time test, (c) cure test and (d) wet track abrasion test, all described in ASTM D 3910 (2011).

A similar mix design procedure is also proposed by ISSA A105 (2010).

The *consistency test* is used to determine the optimum mix design (proper ratio of aggregate, filler, water and bitumen emulsion) as related to proper consistency for pavement surface placement.

Several mixtures are made using dried aggregate and various ratios of bitumen emulsion, water and additive (Portland cement or lime, and chemical additive). Mixing time shall be within the range of 1 to 3 min when mixed at 25°C ± 1°C. After mixing, the slurry mix is placed in a conical form, which is then removed and allows it to flow on a metal plate having engraved concentric circles.

A flow of 2 to 3 cm is considered to be the consistency normally required for a workable field mix.

The consistency test is also included in the mix design method recommended by ISSA A105 (2010), and the test is carried out using a similar procedure described in ISSA TB-106 (1990).

The *set time test* is used to determine the time required for the slurry mat to reach the initial set, indicated by the appearance of clean water, with the paper blob method.

The slurry mix or mixtures that satisfy the desired consistency is used to determine its setting characteristic. A mix passing the consistency test is poured onto a 152 by 152 mm asphalt felt pad and shaped to 6 mm thickness using a 6 mm template. At the end of 15 min, at 25° ± 1°C and 50% ± 5% relative humidity, a white paper towel or tissue is lightly pressed or blotted on the slurry surface. If no brown stain is transferred to the paper, the slurry is considered set. If a brown stain does appear, repeat the blot procedure at 15 min intervals. After 3 h of blotting, 30 min (or longer) blot intervals would be suitable.

Properly designed slurry should be set at the end of 12 h. For the quick-setting slurry seal, a much shorter set time is considered acceptable.

The *cure time* test is used to determine initial cohesion of slurry mat and resistance to traffic. Total cure of a slurry seal mat is obtained when complete cohesion between bitumen-coated aggregate particles occurs.

A slurry seal mix of optimum design obtained from use of the consistency test is spread onto a roofing felt pad to a thickness not exceeding the height of the largest aggregate fragment present in the mix. After the slurry mat has set (set time), the mat is placed in a cohesion testing device to measure cure time. A torque is applied by a rubber foot twisted on the surface every 15–30 min intervals.

The time required to reach a constant maximum torque, or until the rubber foot rides freely over the slurry mat without any aggregate particles being dislodged, is recorded as the cure time. A properly designed slurry mix should be completely cured at the end of 24 h after placement.

The *wet track abrasion test* covers measurement of the wearing qualities of slurry seal under wet abrasion conditions.

A slurry mixture of fine graded aggregate, asphalt emulsion and water that satisfies the desired consistency (consistency test) is prepared. The slurry seal mix is formed into a disk by the use of a circular template resting on roofing felt and dried to a constant weight at 60°C. The cured specimen is placed in a water bath for 1 h and then mechanically abraded underwater with a rubber hose for 5 min, when the Hobart Model C-100 equipment is used (slightly different times are required when other Hobart models are used). The loss in weight expressed as grams per square metre is reported as the wear value (WTAT loss).

A slurry seal mix should not show a loss of more than 807 g/m^2.

The wet track abrasion test is also included in the mix design method proposed by ISSA A105 (2010) and is carried out using a similar procedure described in ISSA TB-100 (1990).

The *target mix* is the one with the right consistency satisfying all the abovementioned requirements.

6.8.3.2 Mix design procedure for micro-surfacing

The mix design procedure for micro-surfacing according to ASTM D 6372 (2010) requires the execution of the following tests: (a) cohesion test, (b) wet track abrasion test, (c) loaded wheel test and (d) classification test, all described in ASTM D 6372 (2010).

A similar mix design procedure is also proposed by ISSA A143 (2010).

The aim of the mix design is to select a quick-setting target mix so that the micro-surfacing is opened to traffic as quickly as possible (within 1 h after placement at 24°C temperature and 50% or less humidity) and, after curing and initial traffic consolidation, resists compaction and maintains its high friction properties throughout the service life of the mixture.

The *cohesion test* is used to determine two set times of the micro-surfacing mixture, the 'set time' and the 'early rolling traffic time'.

Set time is defined as the elapsed time after casting a specimen of the micro-surfacing mixture wherein it cannot be remixed and no lateral displacement is possible when it is compacted. It is also defined as the time when there are no signs of free emulsion when pressed with an absorptive paper towel and there is no free emulsion diluted and washed away when rinsed with water.

Early rolling traffic time is defined as the time at which the micro-surfacing mixture will accept rolling traffic without picking or deformation.

Both set times are determined by employing the cohesion testing device, used to measure cure time of slurry seal, measuring the torque of a micro-surfacing mixture as it coalesces and develops cohesive strength. The amount of torque developed plotted over time shows how the mixture is developing resistance to movement.

After 30 min (representing the set time), the minimum torque developed should be ≥12 kg/cm, and after 60 min (representing the rolling traffic time), it should be ≥20 kg/cm (ISSA A143 2010).

The cohesion test is also included in the mix design method proposed by ISSA A143 (2010) and is carried out using a similar procedure described in ISSA TB-139 (1990).

The *wet track abrasion test* is used to determine the minimum bitumen content and the mixture's resistance to stripping. The test is the same as the one used in mix design for slurry seals, briefly described in the above paragraph.

A micro-surfacing mix should not show a loss of more than 538 g/m^2 after 1 h of soak period, and no more than 807 g/cm^2 after a 6-day soak period (ISSA A143 2010).

The wet track abrasion test is also included in the mix design method proposed by ISSA A143 (2010) and is carried out using a similar procedure described in ISSA TB-100 (1990).

The *loaded wheel test* measures the amount of compaction and displacement characteristics of a micro-surfacing mixture under simulated rolling traffic compaction.

A sample of normally 12.7 mm thick by 50.8 mm long and 38.1 cm wide consisting of the micro-surfacing mix is prepared and dried to a constant weight in a forced draft oven at 60°C for 18 to 20 h. After cooling the specimen and weighing its net mass, it is mounted in the loaded wheel track machine and subjected to 1000 cycles using a load of 56.7 kg, at a temperature of 22°C ± 2°C. After completion, the lateral displacement (increase) is measured and expressed as per cent increase of the specimen after being subjected to wheel compaction.

The allowable lateral increase is 5% maximum and the maximum specific gravity (ISSA A143 2010).

The loaded wheel test is also included in the mix design method proposed by ISSA A143 (2010) and is carried out using a similar procedure described in ISSA TB-147 (1990).

The *classification test* determines the relative compatibility between an aggregate filler of specific gradation and an emulsified bitumen residue.

A pill sample is prepared from an aggregate mix, filler, water and specified bitumen content. The pill sample is soaked for 5 days, and after drying, it is placed in a cylinder of an abrasion machine containing water, which runs for 3 h at 20 revolutions per minute. Upon completion, the abraded pill is placed in boiling water for 30 min. The remains of the boiled pill are surface dried and weighted. After air drying for 24 h, the percentage of aggregate filler particles fully coated with bitumen is estimated.

The test procedure provides a rating system or grading values for abrasion loss, in grams, adhesion after 30 min boil, in percentage coated, and integrity after 30 min boiling, in percentage of retained mass. The average value of quadruplet specimens is reported.

The requirement is to get more than 11 grade points. More details regarding the test can be found in ASTM D 6372 (2010) and ISSA TB-144 (1990).

The classification test is also included in the mix design method proposed by ISSA A143 (2010) and is carried out using a similar procedure described in ISSA TB-144 (1990).

6.8.3.3 Tests on slurry surfacing mix according to ENs

The tests proposed by ENs for slurry surfacing mix to determine useful characteristic properties and compatibility of the mix constituents are very similar to some proposed by American standards. Namely, the tests proposed are (a) consistency, (b) determination of cohesion of the mix, (c) determination of wearing and (d) shaking abrasion test.

The *consistency test* is carried out according to CEN EN 12274-3 (2002) and is used to determine the amount of water to form a stable workable mixture. The test is similar to the consistency test described in ASTM D 3910 (2011), a brief description of which is given in Section 6.8.3.1 (consistency test).

The *determination of cohesion of the mix* is carried out according to CEN EN 12274-4 (2003). The test method measures the development of cohesive strength and defines the time when the slurry seal or micro-surfacing can accept traffic. The test is similar to the cohesion test described in ASTM D 6372 (2010), a brief description of which is given in Section 6.8.3.2 (cohesion test).

The *determination of wearing* is carried out according to CEN EN 12274-5 (2003). The test method determines the minimum binder content of the mix under wet track abrasion conditions. It also covers the compatibility between aggregate fillers and cationic bitumen emulsion. The test is similar to wet track abrasion loss described in ASTM D 3910 (2011), a brief description of which is given in Section 6.8.3.1.

The *shaking abrasion test* is carried out according to CEN EN 12274-7. The test method determines the suitability of aggregates and cationic emulsions for slurry surfacings, and where appropriate, the effect of individual additives. The test is almost similar to the classification test described in ASTM D 6372 (2010), a brief description of which is given in Section 6.8.3.2.

The principle of the shaking abrasion test is used to determine the water sensitivity of mixes for slurry surfacings consisting of 0/2 mm aggregate and cationic emulsion for slurry surfacing.

The test measures the loss of material from standard specimens when cylindrical specimens, approximately 30 mm in diameter and 25 mm in height, of compacted material are placed in water-filled cylinders that are rotated end over end in a suitable device (a specially build mechanical shaker). The test uses mixtures for slurry surfacing using the materials used for producing slurry surfacing but made to a standard grading and binder content prepared at room temperature.

Four cylindrical specimens prepared using a standardised mix for slurry surfacing are tested in each set of tests. The specimens are statically compacted and then conditioned by storage in water in a vacuum prior to testing. The results after subjecting the samples in shaking abrasion for 180 min are expressed as the abrasion of a test specimen, in percentage by mass. The abrasion loss is determined by using the following equation:

$$AR' = (m_f - m_{ar})/m_f \times 100,$$

where AR' is the abrasion of a test specimen (% by mass), m_f is the mass of a wet test specimen prior to abrasion (g) and m_{ar} is the mass of a wet test specimen after abrasion (g).

The average abrasion of the four samples, AR, is expressed in percentage by mass, rounded to 0.1% to represent the average of the four individual values obtained.

The samples, after being abraded, do not follow a boiling stage, as required by the classification test as described in ASTM D 6372 (2010).

More details regarding the shaking abrasion test can be found in CEN EN 12274-7.

Table 6.5 Mix design requirements – recommendations for slurry surfacing on airfields

Properly	Requirement	Recommendation
Mix time (ISSA TB-113 1990)	—	≥90 s
Consistency (CEN EN 12274-3 2002)	—	2 to 3 cm
Determination of cohesion of the mix (CEN EN 12274-4 2003)	—	≥12 kg/cm after 30 min ≥20 kg/cm after 60 min
Determination of wearing (CEN EN 12274-5 2003)	≤500 g/m²	—

Source: Reproduced from CEN prEN 16333, *Slurry surfacing – Specification for airfields*, Brussels: CEN, 2014. With permission (© CEN).

6.8.3.4 *Mix design for slurry surfacing on airfields*

The draft European standard CEN prEN 16333 (2014) (WI approved dated 23/5/2014) specifies a mix design procedure when slurry surfacing is to be used on airfields. Until the development of an EN specifying a mix design method for slurry surfacing and in other areas, the author proposes this design method to be used in any micro-surfacing applications.

The procedure requires the following tests: (a) mix time, (b) consistency test, (c) determination of cohesion of the mix and (d) determination of wearing.

The *mix test* is carried out in accordance to the ISSA TB-113 (1990) method. The mix time test predicts how long the slurry can be mixed in the machine before it begins to stiffen and break. The mixing test should also be checked at the highest temperature expected during construction (CEN prEN 16333 2014).

The *consistency test* is carried out according to CEN EN 12274-3 (2002) and is used to determine the amount of water to form a stable workable mixture. A brief description is given in Section 6.8.3.1 (consistency test).

The *determination of cohesion of the mix* is carried out according to CEN EN 12274-4 (2003). It measures the development of cohesive strength and determines the time when the slurry seal or micro-surfacing can accept traffic. A brief description is given in Section 6.8.3.2 (cohesion test).

The *determination of wearing* is carried out according to CEN EN 12274-5 (2003). The test method determines the minimum binder content of the mix under wet track abrasion conditions. It also covers the compatibility between aggregate fillers and cationic bitumen emulsion. The test is similar to wet track abrasion loss briefly described in Section 6.8.3.1. The main differences between the wearing test and the wet track abrasion test is that slightly different abrasion machines are used and the run time for the wearing test is longer (300 ± 2 s). The requirements/recommendations for determining the target mix are as shown in Table 6.5.

6.8.4 Technical requirements of slurry surfacing based on visual assessment and surface characteristics

In accordance to CEN EN 12273 (2008), the slurry surfacing after a certain period should not have more surface defects than specified at the stage of determining the target mix, and its surface characteristics should satisfy the requirements.

The defects are assessed visually between 11 and 13 months after installation. During the same period, the surface characteristic, expressed in depth of macro-texture, is also determined.

The surface defects assessed together with the performance categories to be selected are shown in Table 6.6.

The high categories (4 and 5) are selected for sites with high traffic volume, whereas the low category (category 1) is selected for sites with low traffic.

Table 6.6 Required surface characteristics related to defects of slurry surfacing

Technical requirement visual assessment of defects	Reference	Unit	Performance categories					
			0	*1*	*2*	*3*	*4*	*5*
P_1 – bleeding, fatting up and tracking	CEN EN-12274-8 (2005)	%	NPD	≤8	≤2	≤0.5	≤0.2	
P_2 – delamination, loss of aggregate, wearing, lane joint gaps, rutting or slippage		%	NPD	≤8	≤2	≤0.5	≤0.2	
P_3 – corrugation, bumps and ridges		%	NPD	≤8	≤2	≤0.5	≤0.2	
$P_{4(n)}$ – groups of small and repetitive defects in not more than rectangles (n)		%	NPD	≤20 (20)	≤5 (6)	≤1 (2)	≤0.2 (1)	
L – longitudinal grooves (score marks)		m	NPD	<20	<10	<5	<1	
Surface characteristics								
Macrotexture	CEN EN 13036-1 (2010)	mm	NPD	≥0.2	≥0.4	≥0.6	≥0.8	≥1.0
Noise generation[a]		mm	Declared maximum value					
Macrotexture								

Source: Adapted from CEN EN 12273, *Slurry Surfacing – Requirements*, Brussels: CEN, 2008. With permission (© CEN).
Note: NPD, no performance determined.

[a] If the site configuration permits, then CEN EN ISO 11819-1 (2001) may be used.

The selection of categories for all technical requirements should be made to avoid technically incompatible combinations, for example, high macro texture category 4 and high bleeding defect category 1.

The visual assessment of slurry surfacing defects is carried out according to CEN EN 12274-8 (2005). The standard specifies a qualitative test method and a quantitative test method and is applicable to all slurry surfacing (roads, airfields and other areas).

The qualitative assessment method is an estimated 'drive-over' method, while the quantitative assessment is a measured method. Both methods have identical records and thus both may be used to check the specification for visual assessment of defects.

In addition, CEN EN 12274-8 (2005) also provides photographic records of sites and defects that may assist the assessment study.

More details regarding the visual assessment study can be found in CEN EN 12274-8 (2005).

When slurry surfacing is going to be used on airfields, the technical requirements regarding surface defects after 12 months, for Type Approval Installation Trial (TAIT), shall be in accordance with category 1 and those regarding construction trial (immediately after construction) shall be in accordance with category 2, as shown in Table 6.7.

For slurry surfacing on airfields, the visual assessment of defects is carried out according to CEN EN 12274-8 (2005).

6.8.5 Mixing and laying of slurry surfacing

Before laying the very thin layer of slurry surfacing, like in any other asphalt works, the pavement surface must be cleaned thoroughly so as to be free from dust, oils, mud and so on. Additionally, some repairs may be needed to be carried out before laying. In some cases, the same slurry surfacing mix may also be used for minor repairs.

The use of tack coat is not usually required unless the old surface is smooth or the construction is conducted on an uphill/downhill section. However, when used, a light tack coat is applied using a diluted bitumen emulsion at a rate of 60 to 100 g/m^2 of bitumen residue.

Table 6.7 Required surface characteristics related to defects of slurry surfacing on airfields

Technical requirement visual assessment of defects	Reference	Unit	Performance categories		
			Category 1		Category 2
P_1 – bleeding, fatting up and tracking	EN-12274-8 (2005)	%	≤0		≤0
P_2 – delamination, loss of aggregate, wearing, lane joint gaps, rutting or slippage		%	0		0
P_3 – corrugation, bumps and ridges		%	≤0.2		≤0.05
P_4 – groups of small and repetitive defects in not more than rectangles (n)		%	≤0.1		≤0.05
L – longitudinal grooves (score marks)		m	<1		0
Surface characteristics			Category 0	Category 1	Category 2
Macrotexture	CEN EN 13036-1 (2010)	mm	NPD	≥0.6	≥0.6
Skid resistance	ICAO		NPD	ICAO Annex 14	National requirements

Source: Reproduced from CEN prEN 16333, *Slurry surfacing – Specification for airfields*, Brussels: CEN, 2014. With permission (© CEN).

Note: NPD, no performance determined.

The existing road marking should always be sprayed with bituminous emulsion, similar to that used for tack coating.

Mixing and laying of the slurry surfacing mix are carried out by a self-contained, continuous-flow mixing-and-laying unit machine. It is a truck-mounted unit that carries an aggregate bin, a filler bin, a bitumen emulsion tank, a water tank and a small chemical additive tank (sometimes optional). All mix constituents are delivered to the mixer at a predetermined ratio using electronic or otherwise control systems.

The mixer in the slurry surfacing machines is either a double-shafted pugmill (more common) or the ribbon type, from which the slurry is discharged into a spreader box.

The spreader box is equipped with hydraulically powered augers (paddle type) to help the mixture uniformly spread across the spreader box and flexible squeegees. The width of the spreader box can be extended, usually by adding extension boxes. A typical slurry device is shown in Figure 6.6.

Figure 6.6 Typical slurry surfacing machine (application of micro-surfacing in airport runway).

An alternative type of slurry surfacing machine also exists in which the aggregate mix is fed to the unit by a delivery truck. This type of slurry machine carries larger-capacity bitumen emulsion and water tanks, which affects (increases) the daily paving output.

The thin slurry mat laid usually does not need *rolling*. However, in low-traffic sections, airports/taxiways or sections with high slopes, it is highly recommended to use pneumatic rolling after the emulsion breaks (usually within 30 min for cationic emulsions, indicated by the appearance of clean water). The aim of rolling is not to compact the layer but to assist the quick removal of the water.

It is recommended not to lay slurry surfacing when ambient temperature is 7°C and falling. Additionally, no slurry surfacing should be applied when there is a possibility that the ambient temperature falls below zero within the next 24 h after finishing laying.

The relative humidity and wind conditions also affect the breaking time of the bitumen emulsion; thus, necessary adjustments are needed to the amount of water added or the dosage of the chemical additive.

When laying a slurry surfacing mix, it is important to apply a homogeneous mix, without lumps, uncoated aggregates or aggregate segregation. Oversized aggregates should also be excluded or removed from the mix since they are causing streaks. If this is inevitable, streaks should be repaired immediately as they appear with a hand squeegee. Additionally, no excess build-up, uncovered areas or unsightly appearance should be permitted on the longitudinal or transverse joint.

When laying slurry surfacing, needless to say, all appropriate traffic control precautions must be taken to avoid accidents and traffic delays.

Finally, to ensure quality work, before mixing/laying, it is advisable to place one or two trial mixes. Additionally, frequent calibrations of the feeding and proportioning–metering devices of the slurry machine are also necessary.

6.8.6 Quality control during laying slurry surfacing

Quality control during laying consists of sampling and testing the slurry surfacing mix with respect to (a) residual bitumen content, (b) aggregate gradation, (c) application rate of dry aggregate and (d) properties of bitumen residue, so as to comply with the target mix and the declared properties of the bitumen residue and rate of aggregate spread.

For the determination of the residual bitumen content and aggregate gradation, the representative samples are taken from the slurry machine, at the discharge point, according to CEN EN 12274-1.

The determination of the residual bitumen content is carried out according to CEN EN 12274-2 (2003) followed by CEN EN 12697-1 (2012). Alternatively, after the mixture has been dried completely, the determination of the residual bitumen content may be carried out in accordance to ASTM D 2172 (2011) or AASHTO T 164 (2013). The acceptable tolerances from the target mix, according to ISSA A105 (2010), for the residual bitumen content should be ±1.0% by mass of dry aggregate.

The determination of the aggregate gradation after extraction is carried out mechanically as per hot asphalts, according to CEN EN 933-1 (2012), ASTM C 136 (2006) or AASHTO T 27 (2011) and ASTM C 117 (2013) or AASHTO T 11 (2009). The tolerance ranges from the target gradation should be as those given in Table 6.4.

The frequency of sampling for determination of residual bitumen content and aggregate gradation is recommended to be at least one every 5000 m².

The application rate of dry aggregate (rate of spread) is recommended to be determined on a daily basis by dividing the amount of aggregate used with the slurry microsurfacing area laid per day. Other methods may also be used, such as placing a roofing felt pad, or other material, of known area (say 1 m × 1 m), on the surface before laying the slurry surfacing. The amount of mix is carefully collected and dried in the laboratory. By knowing the amount of residual bitumen, the application rate of dry aggregates can be determined, an as average of three samples. The accepted tolerances, according to ISSA A105 (2010), should be ±1.1 kg/m^2, when the old texture surface does not change significantly.

The properties of the residual bitumen, particularly when polymer-modified bitumen is used, should be carried out as per residual bitumen extracted from hot mixtures, after full evaporation of water from the bitumen emulsion. The determination of the properties required, particularly of the elastic recovery, is carried out using the same relevant standards as in hot asphalts (CEN EN 13398 2010). Sampling of the bitumen emulsion is recommended to be carried out from the tank of the slurry machine, at a proposed frequency of every 5 tonnes of delivered bitumen emulsion. It is recommended that that the measured elastic recovery should be ≥50%.

6.8.7 Evaluation of conformity factory production control of slurry surfacing

The evaluation of conformity of slurry surfacing, according to CEN EN 12273 (2008), is demonstrated by Factory Production Control (FPC) and TAIT.

The producer shall establish, document and maintain an FPC system to ensure that the slurry surfacing placed on the market conforms to the stated performance characteristics. The FPC system shall consist of procedures, regular inspections, tests or assessments and the use of the results to control incoming materials, equipment, the production process and the product.

Where the producer purchases constituent materials or has the slurry surfacing designed, or parts of the production or testing carried out by sub-contracting, the FPC of the supplier or sub-contractor may be taken into account. However, where this occurs, the producer shall retain the overall control of the slurry surfacing and ensure that he receives all the information that is necessary to fulfil the requirements according to CEN EN 12273 (2008). The producer who sub-contracts all of his activities may, in no circumstances, discharge himself of his responsibilities to a sub-contractor.

General and specific requirements for FPC together with minimum test frequencies for FPC, such as equipment calibration requirements, inspection and test frequencies for aggregates, inspection and test frequencies for bituminous emulsions, control of the water, control of the additives (including cement, lime, fibres and chemicals), controls during the process and installation of the slurry surfacing and inspection and test frequencies measured after installation, are given in CEN EN 12273 (2008), Annexes 6.A and 6.B.

A TAIT consists of a defined section where slurry surfacing has been installed using FPC and which has been submitted to performance tests after a period of 1 year. Detailed information is recorded to clearly identify the product, its performance and the intended uses. The producer carries out one TAIT to cover each product family he wishes to place on the market. The TAIT is synonymous with the Initial Type Test, which demonstrates that the characteristics of the slurry surfacing comply with the requirements of the technical specification. More details regarding TAIT can be found in CEN EN 12273 (2008), Annex 6.C.

For slurry surfacing on airfields, the FPC and TAIT procedures are carried out in accordance with CEN prEN 16333 (2014), Annexes 6.A and 6.B, respectively.

REFERENCES

AASHTO M 17. 2011. *Mineral filler for bituminous paving mixtures.* Washington, DC: American Association of State Highway and Transportation Officials.

AASHTO M 140. 2013. *Emulsified asphalt.* Washington, DC: American Association of State Highway and Transportation Officials.

AASHTO M 208. 2009. *Cationic emulsified asphalt.* Washington, DC: American Association of State Highway and Transportation Officials.

AASHTO T 11. 2009. *Materials finer than 0.075 mm (no. 200) sieve in mineral aggregates by washing.* Washington, DC: American Association of State Highway and Transportation Officials.

AASHTO T 27. 2011. *Sieve analysis of fine and coarse aggregates.* Washington, DC: American Association of State Highway and Transportation Officials.

AASHTO T 49. 2011. *Penetration of bituminous materials.* Washington, DC: American Association of State Highway and Transportation Officials.

AASHTO T 53. 2009. *Softening point of bitumen (ring-and-ball apparatus).* Washington, DC: American Association of State Highway and Transportation Officials.

AASHTO T 59. 2013. *Emulsified asphalts.* Washington, DC: American Association of State Highway and Transportation Officials.

AASHTO T 96. 2010. *Resistance to degradation of small-size coarse aggregate by abrasion and impact in the Los Angeles Machine.* Washington, DC: American Association of State Highway and Transportation Officials.

AASHTO T 104. 2011. *Soundness of aggregate by use of sodium sulfate or magnesium sulphate.* Washington, DC: American Association of State Highway and Transportation Officials.

AASHTO T 164. 2013. *Quantitative extraction of asphalt binder from hot mix asphalt (HMA).* Washington, DC: American Association of State Highway and Transportation Officials.

AASHTO T 176. 2008. *Plastic fines in graded aggregates and soils by use of the sand equivalent test.* Washington, DC: American Association of State Highway and Transportation Officials.

Asphalt Institute MS-19. 2008. *Basic Asphalt Emulsion Manual,* 4th Edition. Manual Series No. 19. Lexington, KY: Asphalt Institute.

ASTM C 88. 2013. *Standard test method for soundness of aggregates by use of sodium sulfate or magnesium sulfate.* West Conshohocken, PA: ASTM International.

ASTM C 117. 2013. *Standard test method for materials finer than 0.075 mm (no. 200) sieve in mineral aggregates by washing.* West Conshohocken, PA: ASTM International.

ASTM C 131. 2006. *Standard test method for resistance to degradation of small-size coarse aggregate by abrasion and impact in the Los Angeles machine.* West Conshohocken, PA: ASTM International.

ASTM C 136. 2006. *Standard test method for sieve analysis of fine and coarse aggregates.* West Conshohocken, PA: ASTM International.

ASTM D 5/D 5M. 2013. *Standard test method for penetration of bituminous materials.* West Conshohocken, PA: ASTM International.

ASTM D 36/D 36M. 2012. *Standard test method for softening point of bitumen (ring-and-ball apparatus).* West Conshohocken, PA: ASTM International.

ASTM D 242/D 242M. 2009. *Standard specification for mineral filler for bituminous paving mixtures.* West Conshohocken, PA: ASTM International.

ASTM D 977-12b. 2012. *Standard specification for emulsified asphalt.* West Conshohocken, PA: ASTM International.

ASTM D 2172/D 2172M. 2011. *Standard test methods for quantitative extraction of bitumen from bituminous paving mixtures.* West Conshohocken, PA: ASTM International.

ASTM D 2397. 2012. *Standard specification for cationic emulsified asphalt.* West Conshohocken, PA: ASTM International.

ASTM D 2419. 2009. *Standard test method for sand equivalent value of soils and fine aggregate.* West Conshohocken, PA: ASTM International.

ASTM D 3910. 2011. *Standard practices for design, testing, and construction of slurry seal.* West Conshohocken, PA: ASTM International.

ASTM D 6372. 2010. *Standard practice for design, testing, and construction of micro-surfacing.* West Conshohocken, PA: ASTM International.

ASTM D 6930. 2010. *Standard test method for settlement and storage stability of emulsified asphalts.* West Conshohocken, PA: ASTM International.

ASTM D 7196. 2012. *Standard test method for raveling test of cold mixed emulsified asphalt samples.* West Conshohocken, PA: ASTM International.

ASTM D 7229. 2008. *Standard test method for preparation and determination of the bulk specific gravity of dense-graded cold mix asphalt (CMA) specimens by means of the Superpave gyratory compactor.* West Conshohocken, PA: ASTM International.

Cabrera J.C. and A.F. Nikolaides. 1988. Creep performance of cold dense bituminous mixtures. *Journal of the Institution of Highways and Transport,* Vol. 35, No. 10, p. 7.

CEN EN 933-1. 2012. *Tests for geometrical properties of aggregates – Part 1: Determination of particle size distribution – Sieving method.* Brussels: CEN.

CEN EN 933-8. 2012. *Tests for geometrical properties of aggregates – Part 8: Assessment of fines – Sand equivalent test.* Brussels: CEN.

CEN EN 933-9:2009+A1. 2013. *Tests for geometrical properties of aggregates – Part 9: Assessment of fines – Methylene blue test.* Brussels: CEN.

CEN EN 1097-1. 2011. *Tests for mechanical and physical properties of aggregates – Part 1: Determination of the resistance to wear (micro-Deval).* Brussels: CEN.

CEN EN 1097-2. 2010. *Tests for mechanical and physical properties of aggregates – Part 2: Methods for the determination of the resistance to fragmentation.* Brussels: CEN.

CEN EN 1097-8. 2009. *Tests for mechanical and physical properties of aggregates – Part 8: Determination of the polished stone value.* Brussels: CEN.

CEN EN 1097-9. 2014. *Tests for mechanical and physical properties of aggregate – Part 9: Determination of the resistance to wear by abrasion from studded tyres – Nordic test.* Brussels: CEN.

CEN EN 12273. 2008. *Slurry surfacing – Requirements.* Brussels: CEN.

CEN EN 12274-2. 2003. *Slurry surfacing – Test methods – Part 2: Determination of residual binder content.* Brussels: CEN.

CEN EN 12274-3. 2002. *Slurry surfacing – Test methods – Part 3: Consistency.* Brussels: CEN.

CEN EN 12274-4. 2003. *Slurry surfacing – Test methods – Part 4: Determination of cohesion of the mix.* Brussels: CEN.

CEN EN 12274-5. 2003. *Slurry surfacing – Test method – Part 5: Determination of wearing.* Brussels: CEN.

CEN EN 12274-8. 2005. *Slurry surfacing – Test methods – Part 8: Visual assessment of defects.* Brussels: CEN.

CEN EN 12697-1. 2012. *Bituminous mixtures – Test methods for hot mix asphalt – Part 1: Soluble binder content.* Brussels: CEN.

CEN EN 13036-1. 2010. *Road and airfield surface characteristics – Test methods – Part 1: Measurement of pavement macrotexture depth using a volumetric patch technique.* Brussels: CEN.

CEN EN 13043:2002/AC. 2004. *Aggregates for bituminous mixtures and surface treatments for roads, airfields and other trafficked areas.* Brussels: CEN.

CEN EN 13398. 2010. *Bitumen and bituminous binders – Determination of the elastic recovery of modified bitumen.* Brussels: CEN.

CEN EN 13808. 2013. *Bitumen and bituminous binders – Framework for specifying cationic bituminous emulsions.* Brussels: CEN.

CEN EN ISO 11819-1. 2001. *Acoustics – Measurement of the influence of road surfaces on traffic noise – Part 1: Statistical pass-by method.* Brussels: CEN.

CEN prEN 16333. 2014. *Slurry surfacing – Specification for airfields.* Brussels: CEN.

Chevron. 1979. *Bituminous Mix Design Manual.* Asphalt Division. El-Paso, TX: Chevron.

Delorme J.-L., C. de la Roche, and L. Wendling. 2007. *LPC Bituminous Mixtures Design Guide.* Paris: Laboratoire Central des Ponts et Chaussées.

EAPA. 2011. *Asphalts in Figures 2011.* Brussels, Belgium: European Asphalt Pavement Association.

Hughes C.S. 1981. *Illinois Method for Design of Dense-Graded Emulsion Base Mixes.* FHWA Report, FHWA/VA-81/R53, Charlottesville, VA: Virginia Highway and Transportation Research Council.

Indonesian MPW. 1990. *Paving Specifications Utilizing Bitumen Emulsions: Section 6.10 – Dense-Graded Emulsion Mixtures.* Directorate General of Highways. Jakarta: Ministry of Public Works.

ISSA A105. 2010. *Recommended performance guidelines, for emulsified asphalt slurry seal.* Annapolis, MD: International Slurry Seal Association.

ISSA A143. 2010. *Recommended performance guidelines, for micro-surfacing.* Annapolis, MD: International Slurry Seal Association.

ISSA TB-100. 1990. *Wet track abrasion of slurry surfaces.* Annapolis, MD: International Slurry Seal Association.

ISSA TB-106. 1990. *Measurement of slurry seal consistency.* Annapolis, MD: International Slurry Seal Association.

ISSA TB-113. 1990. *Trial mix procedure for slurry seal design.* Annapolis, MD: International Slurry Seal Association.

ISSA TB-139. 1990. *Classifying emulsified asphalt/aggregate mixture systems by modified cohesion tester measurement of set and cure characteristics.* Annapolis, MD: International Slurry Seal Association.

ISSA TB-144. 1990. *Test method for classification of aggregate filler-bitumen compatibility by Schultze-Breuer and Ruck procedures.* Annapolis, MD: International Slurry Seal Association.

ISSA TB-147. 1990. *Measurement of stability and resistance to compaction, vertical displacement of multi-layers fine aggregate cold mixes – Method A.* Annapolis, MD: International Slurry Seal Association.

MPW D14. 1994. *Technical terms for construction of pavement layers with cold bituminous mixtures produced in stationary or mobile mix plant.* Athens: Ministry of Public Works.

NF EN 12273. 2008. *Slurry Surfacing – Requirements.* French standard approved, Classification index, pp. 98–856. Paris: AFNOR.

Nikolaides A.F. 1983. Design of cold dense-graded bituminous emulsion mixtures and evaluation of their engineering properties. Ph.D. Thesis, Leeds: Leeds University.

Nikolaides A.F. 1992. Cold dense-graded bituminous mixtures: Proposed mix design method. *Proceedings of 1st National Conference 'Bituminous Mixtures and Flexible Pavements'.* Thessaloniki, Greece.

Nikolaides A.F. 1993. Proposed design method for cold dense-graded mixtures. *Proceedings of 5th Eurobitume Congress*, Session 3, p. 616. Stockholm.

ANNEX 6.A: COATING TEST

1. General
   a. This test covers the procedure for determining the suitability of the bituminous emulsion. Additionally, the percentage of added water is determined to improve the coating ability of the bitumen emulsion. The suitability of the bituminous emulsion is judged by its capability to coat as much aggregate surface as possible.
   b. The degree of coating is assessed visually by inspecting the aggregate particles individually. The degree of coating is expressed as percentage of coated surface from the total aggregate surface.
   c. Bituminous emulsion's coating ability depends directly on the water added to the aggregates before mixing. For this reason, the coating test is conducted at various percentages of added water. The test takes place at room temperature ($21°C \pm 3°C$).
2. Methodology
   a. A sufficient quantity of aggregate mixture is dried in an oven at $110°C \pm 2°C$ until constant mass is obtained.
   b. Approximately 500 g of dry aggregates is weighed, W_d, and then placed in a mixing bowl.

c. A certain quantity of clean water is added in the bowl and the contents are mixed well for 20–40 s. The mixing time should be as short as possible to avoid evaporation during mixing. However, it should be sufficient so that water is dispersed evenly. For DGCAs, start with 1% of added water, by weight of dry aggregates, and increase in steps of 1%. For open cold mixtures, start with 0% added water and increase in steps of 0.5%.

d. Immediately after mixing the aggregates with the added water, a required quantity of bituminous emulsion is added. The required quantity of bituminous emulsion (E_{add-c}) is determined from the nominal bitumen content (b_{nom}) required for coating test, given in Table 6.A.1.

   The required amount of bituminous emulsion is calculated using the following equation:

$$E_{add-c} = [b_{nom}/P_b] \times W_d \times \rho_w,$$

   where E_{add-c} is the mass of the bitumen emulsion for the coating test (g), b_{nom} is the nominal bitumen content from Table 6.A.1 (%), P_b is the binder content in the bituminous emulsion (%), W_d is the mass of the dry aggregate (g) and ρ_w is the density of water (g/cm^3, always taken at 1.0 g/cm^3).

e. The moist aggregates and the added bitumen emulsion are mixed well for approximately 30 s. Excessive mixing should be avoided since it may cause stripping or coagulation.

f. After mixing, the mixture is transferred to a metal tray and is allowed to dry. To reduce the drying period, an electric dryer may be used. It is noted that the purpose of drying is not to remove the water completely from the mix but to remove most of it to facilitate the estimation of the coated surface. The drying period and the method of drying should be kept the same for all mixtures.

g. Determine the degree of coating as a percentage of the coated area to the total aggregate surface.

h. Repeat steps (b) to (g) by adding more water to the dried aggregate mixture until a degree of coating greater than 75% is achieved. Very 'stiff' or very 'loose' mixtures during this test must also be rejected.

i. If the degree of coating is ≥75%, the bituminous emulsion is acceptable; proceed with further testing for the determination of target bitumen content.

j. If the degree of coating is <75%, increase the amount of bituminous emulsion by 10% and repeat steps (b) to (h).

k. When a successful mixture is achieved with the new quantity of bitumen emulsion, use the corresponding residual bitumen content as nominal bitumen content for further testing for determination of target bitumen content.

l. If an acceptable mixture cannot be produced after step (k), examine the use of using a decelerator diluted into the added water (recommended percentage of decelerator diluted in water is 0.5% to 2% by mass of water).

Table 6.A.1 Nominal bitumen content for coating test per type of DGCAs

Types of DGCAs					
37.5 mm	25.0 mm	19.0 mm	12.5 mm	9.5 mm	4.75 mm
Nominal bitumen content (%), by total mass					
4.0	4.5	5.0	5.5	6.0	7.0

3. Results

The degree of coating and the respective percentage of added water are reported. The bituminous emulsion is acceptable for further testing when the degree of coating is greater than 75%.

When no appreciable difference in the degree of coating greater than 75% is observed by increasing the amount of added water, always use the lowest percentage of added water.

ANNEX 6.B: MODIFIED MARSHALL MIX DESIGN FOR DENSE-GRADED COLD ASPHALT (DGCA)

6.B.1 Specimen production and Marshall testing

1. General
   a. This method is conducted to determine the volumetric and modified Marshall properties, and together with the creep coefficient procedure, it will help determine the target bitumen content of DGCA.
   b. The procedure involves two stages. At the first stage, the optimum water content for maximum compacted density is determined. At the second stage, after compacting all specimens at optimum water content, the target bitumen content is determined using volumetric and modified Marshall properties.
   c. The cylindrical specimens at all stages are compacted using 50 blows on each specimen side, regardless of traffic volume, and then tested using the Marshall apparatuses.
   d. Curing of the specimens before testing is required; 24 h within the moulds and 24 h of the extracted specimens in a ventilated oven at 40°C ± 1°C. Testing is carried out at 21°C ± 1°C.
   e. The criteria used to determine the target bitumen content are soaked stability, retained stability, total void content, absorbed water and bitumen film thickness.
   f. When DGCA is going to be used in pavements with medium to high traffic (ESAL > 3000 per day), the target binder content should satisfy the creep coefficient requirement.
2. Equipment
   a. The equipment required is the similar to that required for the Marshall hot mix design method.
   b. An air-controlled cabinet is needed either for curing the specimens at 40°C or for obtaining the required test temperature of 20°C ± 1°C.
   c. A larger number of moulds is required, than in hot mixing, since the specimens remain in the moulds for 24 h. The number required is equal to the laboratory's daily production of specimens (recommended number of moulds ≥20).
3. Methodology
   Stage 1: Determination of optimum water content at compaction
   a. A sufficient quantity of dry aggregate mixture is prepared for the production of 10 or more Marshall specimens (approximately 1.1 kg is required per specimen). The specimens are compacted in groups of two or more at five different water contents, using the procedure described below.
   b. The total mass of dry aggregate for the production of at least 10 specimens is added to the mixer together with a mass of water corresponding to the added water content determined from the coating test increased by 1%. The mixture is mixed thoroughly until all water distributes uniformly around the aggregate surface. Mixing may be stopped to examine if fine aggregates have stuck to the

walls of the mixer. If such were the case, they are detached with the aid of an appropriate metal medium (spatula or trowel) and the mixing is continued.

  c. The required bituminous emulsion quantity, which corresponds to the nominal bitumen content (Table 6.A.1), is added and all the contents are mixed very well, at room temperature, for approximately 45 s. Mixing may be interrupted to examine if aggregates remained uncoated to the sides of the bottom or the walls of the mixer. Use a spatula or trowel to remove them and continue mixing. Prolonged mixing should be avoided, since stripping or balling may occur.

  d. After mixing, the mixture is divided into five approximately equal parts and placed in five trays or pans. Each tray containing the mixture is immediately weighed and its weight is recorded (mass A_1 to A_5).

  e. All trays are left for a sufficient time (approximately 20 to 45 min, depending on the bitumen emulsion) until the surface colour changes from brown to black (breakdown of bitumen emulsion occurs).

  f. At this stage, the first tray is weighted (weight B_2) and then a sample of approximately 250 g of mixture is placed in an oven at 110°C for water content determination. This will be the water at compaction (P_{W-C}). The rest of the mixture is compacted immediately into two cylindrical specimens.

  g. Compaction of the two cylindrical specimens (100 mm in diameter by 60 mm height) is carried out at room temperature (21°C ± 3°C) using a Marshall compactor and applying 50 blows on each side of the specimen.

  h. After compaction, the moulds containing the specimens are placed over a perforated shelf and left for 24 h (or 48 h when anionic emulsion is used) at room temperature (21°C ± 3°C).

  i. Then, the specimens are extracted from the moulds, and their mass (in grams) and height (in centimetres) are measured.

  j. After that, the two specimens break into pieces with the use of a hammer. Two samples per specimen, approximately 250 g each, are taken and placed into bowls for water content determination. This will be the water content after 24 h of curing, wherein the height and weight of the specimen are measured for volume/density determinations.

  k. The bowls with the samples are weighed to an accuracy of 0.01 g and placed in a ventilated oven at 110°C ± 2°C, until a constant weight is achieved. The weight loss ratio by dry weight of the bituminous mixture multiplied by 100 represents the percentage of content contained in the mixture after 24 h of curing. The mean value of the two results, per specimen, is expressed to the nearest 0.01% and is used to determine the dry bulk density of the specimen.

  l. Repeat steps (f) to (k) but for lesser water content at compaction, allowing more water to evaporate from the second, third and so on prepared mixture. The water content at compaction should differ by approximately 0.5% to 1%. To achieve this, successive weighing of the mixtures in the pans is necessary. Record mass B_2, B_3 and so on and compare them with the initial mass A_2, A_3 and so on.

  m. From the five pairs of values, plot the water content during compaction and dry bulk density. The optimum water content at compaction is determined from this graph and is the one for maximum dry bulk density. A complete example of laboratory results and calculations is given in Table 6.B.1 and Figure 6.B.1.

All of the following specimens, which are to be produced for the determination of parameters required for the mixture composition, are to be compacted at water content for maximum density.

Table 6.B.1 Laboratory results for the determination of optimum water content at compaction

Sample	Water content at compaction (%)[a]	Mass of specimen in air (g)	Height of specimen (cm)	Volume of specimen (cm³)	Wet bulk density (g/cm³)	Water content after 24 h (%)[a]	Dry bulk density (g/cm³)
	A	B	C	D	E[b]	F	H[c]
1	4.15	1023.7	58.3	472.42	2.167	1.28	2.140
2	4.15	1023.6	57.8	468.37	2.185	1.12	2.161
Average	4.15				2.176	1.20	2.151
3	4.45	1014.6	57.3	464.31	2.185	1.45	2.154
4	4.45	1022.8	57.8	468.37	2.184	1.31	2.156
Average	4.45				2.185	1.38	2.155
5	4.8	1001.2	55.5	449.73	2.226	1.54	2.192
6	4.8	988.5	55.0	445.68	2.218	1.39	2.187
Average	4.8				2.222	1.47	2.190
7	5.12	1008.6	55.7	451.35	2.235	1.51	2.201
8	5.12	1014.8	56.0	453.78	2.236	1.51	2.203
Average	5.12				2.236	1.51	2.202
9	5.21	986.3	55.2	447.30	2.205	1.62	2.170
10	5.21	983.6	55.4	448.92	2.191	1.63	2.156
Average	5.21				2.198	1.63	2.163

[a] By mass of dry bituminous mixture.
[b] $E = B/D$.
[c] $[B - (F \times B)/100]/D$.

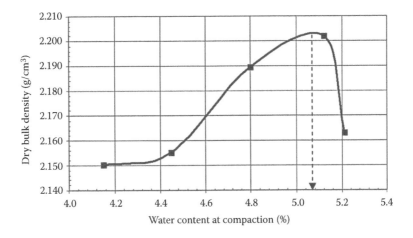

Figure 6.B.1 Graphical representation of results of dry bulk density versus water content at compaction.

Stage 2: Determination of parameters for target binder content
a. An appropriate total quantity of dry aggregate mixture (approximately 35 kg) is prepared for the production of six specimens per bitumen content (usually five different levels of binder content are used).
b. A batch of approximately 7 kg of aggregates, for the production of six specimens, is placed in a mixer together with the quantity of added water determined by the coating test (Annex 6.A). After thorough mixing, add a quantity of bituminous

emulsion and continue mixing. It is recommended to start with a quantity of emulsion corresponding to the nominal residual binder content given in Table 6.A.1.

c. After mixing, the mixture is transferred into a tray and it is weighed. The mixture is left at room temperature until it acquires a water content equal to the optimum water content for compaction, as determined in Stage 1(f). This is achieved with continuous weight measurement of the tray containing the mixture. The mass of water that should be evaporated can be calculated by the following equation:

$$W_w = W_{mix} \times P_A \times (P_{TW} - P_{W\text{-}C})/100,$$

where W_w is the mass of water to be evaporated (g), W_{mix} is the mass of cold mixture (g), P_A is the aggregate content in the mixture (%), P_{TW} is the total water content in the mixture at mixing (water added plus water contained in the emulsion) (%) and $P_{W\text{-}C}$ is the water content at compaction (%) (as determined in Stage 1[f], Annex 6.B).

d. When the mixture reaches the desired water content, all six specimens are compacted (as is Stage 1[g]).

e. After compaction, specimens are left in the moulds for 24 h and then extracted with the aid of a hydraulic jack. They are numbered and placed in a ventilated oven at 35°C $\pm$ 1°C for 24 h curing.

f. Repeat steps (b) to (f) using a different amount of bituminous emulsion to produce mixtures with higher and lower than the nominal bitumen content (use steps of $\pm 0.5\%$ bitumen content).

g. After completion of 24 h curing in the oven, the height and weight of all specimens are recorded.

h. Three of the specimens, per mixture of different bitumen content, are tested for Marshall stability at room temperature (21°C $\pm$ 3°C). The values determined are corrected owing to difference in volume using the same correction coefficients as in hot bituminous mixtures. The stability finally determined is designated as the modified Marshall stability (MS_{mod}). The results of the modified Marshall stability and soaked stability is measured in kilonewtons.

i. From the broken specimens, determine the water content at testing.

j. The remaining three specimens, per mixture of different bitumen content, are placed in a water bath for 48 h to soak (see Annex 6.C).

k. After 48 h of soaking, the specimens are weighted to determine the water absorption content and then tested for Marshall stability. The value determined, after volume correction is applied, is designated as the soaked Marshall stability (SS).

l. The percentage of retained stability (RS) is then calculated using the following equation:

$$RS = SS/MS_{mod} \times 100 \ (\%).$$

The result is rounded to the nearest 1%.

4. Results

The results of soaked Marshall stability, retained stability, dry bulk density, total void content, the absorbed water after immersing specimens into water for 48 h and bitumen film thickness are calculated and reported for each specimen. The average values per binder content are plotted as in diagrams shown in Figure 6.1. To facilitate the calculations, Table 6.B.2 and relations given below may be used.

Table 6.B.2 Laboratory form to be used in modified Marshall mix design for cold dense-graded bituminous mixtures

Specimen number	Binder content in the mix (%)	Maximum density of bituminous mix (g/cm³)	Mass of specimen in air (g)	Height of specimen (mm)	Volume of specimen (cm³)	Bulk density of specimen (g/cm³)	Water content at testing (%)	Dry bulk density (g/cm³)	Total Void content (%)
Column: A	B	C	D	E	F	G	H	I	J
1									
2									
3									
Average									
4									
etc.									

Specimen number	Weight of soaked specimen (48 h) in water (g)	% of Water absorbed (%)	Modified stability (kN)	Corrected stability (kN)	Soaked stability (kN)	Corrected soaked stability (kN)	Retained stability (%)	Bitumen film thickness (µm)
Column: A	K	L	M	N	O	P	Q	R
1								
2								
3								
Average								
4								
etc.								

5. Useful relations

For the calculation of the properties mentioned above, the following relations related to Table 6.B.2 are used:

Column C: $100/[(P_{agg}/G_{agg}) + (P_{bit}/G_{bit})]$ (g/cm^3),

where P_{agg}, P_{bit} are the percentages of aggregates and bitumen in the mix and G_{agg}, G_{bit} are the bulk density of the aggregate mix and the density of bitumen, respectively (g/cm^3).

Column F: $[(\pi \times d^2)/4] \times$ column E (cm^3),

where d is the diameter of the specimen (cm).

Column G: D/F (g/cm^3),

where D and H are the results from columns D and F, respectively.

Column I: $[G \times (100 + B)]/(100 + B + H)$ (g/cm^3),

where G, B and H are the results from columns G, B and H, respectively.

Column J: (C − I)/C (%),

where C and I are the results from columns C and I, respectively.

Column L: $[(K − D)/D] \times 100$ (%),

where K and D are the results from columns K and D, respectively.

Column N: The values of column N are determined from the values in column M, taken in the laboratory, multiplied by the correction factors given in Table 5.10.

Column P: The values of column P are determined from the values in column O, taken in the laboratory, multiplied by the correction factors given in Table 5.12.

Column Q: (P/N) × 100 (%),

where P and N are the results from columns P and N, respectively.

Column R: The thickness of the bitumen film is determined according to Section 5.4.2.8.

Note: The values in all other columns are taken from laboratory measurements.

ANNEX 6.C: SOAKING TEST – CAPILLARY WATER ABSORPTION TEST

1. General
   a. DGCAs, as all cold asphalts, do not develop their maximum stability soon after laying and compaction. The maximum mixture stability is achieved after some

months. It depends on the climatic conditions, as well as site conditions. During this period, DGCAs are likely to be exposed to the effect of water, and as a consequence, their stability will be reduced. For this prior testing for soaked stability, the specimens are subjected to soaking.

2. Purpose

a. This test covers the case where Marshall specimens are exposed to water. During this exposure, capillary water absorption takes place. The total duration of specimen exposure to water is 48 h. After this period, the water absorption percentage is determined. The test takes place at room temperatures and under normal atmospheric pressure conditions using a water bath capable of accommodating 20 or more specimens.

3. Equipment

The basic equipment and materials needed are as follows:

a. A metallic or plastic water bath with dimensions approximately 1200 × 600 mm, having a layer of pure sand or fine aggregates at the bottom of thickness 15–20 mm

b. A balance with a capacity to weigh 1500 g or more and with an accuracy of 0.1 g or more

c. Towels or cloths to clean the surface of the specimens after immersing in water

d. Pans or trays of sufficient dimensions for specimen placement and transfer

4. Methodology

a. Marshall specimens that have been prepared as mentioned in Annex 6.B are weighed in air (mass M_1) and then placed on the sand or aggregate layer of the empty water bath.

b. The water bath is filled with tap water so that specimens are covered with water to the middle.

c. Specimens are allowed to rest for 24 h and then they turned upside down and remained in the water for another 24 h. The water level must remain the same during immersion periods (add some amount of water if needed).

d. Once the immersion period of 48 h has expired, the surface of specimens is wiped with a towel and the specimens are weighed in air (M_2).

e. Specimens are now ready to be tested for soaked stability, as described in Annex 6.B.

5. Results

The water percentage absorbed during soaking is reported as water absorption content. The following formula is used to calculate the above percentage:

$$\text{Water absorption content} = (M_2 - M_1)/M_1 \times 100,$$

where M_1 is the mass of the specimen before soaking (g) and M_2 is the mass of the specimen after soaking for 48 h in water (g).

Chapter 7

Fundamental mechanical properties of asphalts and laboratory tests

7.1 GENERAL

Fundamental mechanical properties of materials constitute the necessary data for designing any structure. Elasticity modulus, compressive strength, tensile strength and shear strength can be designated as fundamental mechanical properties. Their resistance to fatigue and permanent deformation can be characterised as mechanical performance.

Something similar also applies to the analytical design approach of flexible pavements. The main difference with other structures composed of reinforced concrete, or steel, lies in the fact that asphalts (bituminous mixtures) do not behave as elastic material but as viscoelastic material owing to the presence of bitumen. The rest of the structural materials of pavements, such as compacted aggregates, stabilised aggregates or soil with binders other than bitumen, as well as untreated soil, may be characterised as materials with elastic behaviour.

The fundamental property of asphalts related to stress and strain is called stiffness modulus, which is distinguished into dynamic and static stiffness modulus.

The mechanical behaviour of asphalts is determined by resistance to tensile forces, known as resistance to fatigue cracking, and resistance to compressive forces, known as resistance to permanent deformation. The first is expressed by the relationship between tensile strain and number of loadings, known as the fatigue equation. The second is expressed by the relationship between compressive strain and number of loadings, known as the creep equation.

All the above parameters are determined in the laboratory by tests simulating, as much as possible, real on-site conditions, namely, loading time and conditions, temperature, degree of layer compaction and so on.

The execution of the tests for determining the mechanical properties and performance of asphalts presupposes the use of complex and expensive apparatuses, not available to every laboratory. For this reason, the early developed relevant nomographs/equations are still in use.

Determination of fundamental mechanical properties and behaviour of asphalts is not only used in analytical pavement design methodologies. Nowadays, their use has been extended to asphalt's mix design, known as performance-based mix design.

This chapter describes stiffness modulus tests under dynamic or static loading, wheel-tracking tests and fatigue tests, and also provides the relevant nomographs still in use.

7.2 STIFFNESS MODULUS OF ASPHALTS

Asphalts contain two materials with different mechanical behaviours: bitumen, with viscoelastic behaviour under normal loading conditions, and aggregates, with almost elastic behaviour. The bitumen's viscoelastic behaviour prevails and thus the asphalt behaves under normal loading conditions as a viscoelastic material; that is, it responds to loading in both an elastic and a viscous manner.

Considering the above, it is obvious that the asphalts cannot be characterised by the fundamental property of Young's modulus, or elastic modulus, used for linear elastic materials.

Asphalts and viscoelastic materials are characterised by their stiffness modulus or, for simplicity, stiffness, designated as S or E. The designation E has been adopted by the CEN EN 12697-26 (2012) standard.

Stiffness modulus was initially suggested by Van der Poel (1954). This most basic characteristic and fundamental property, in contrast to the elastic modulus where the deformation only depends on the applied load, depends on the temperature (T) and the loading rate (t). Hence, for a given asphalt, its stiffness modulus (E or S) is expressed as

$$E \text{ or } S = (\sigma/\varepsilon)_{T, t},$$

where σ is the applied stress, ε is the resulting strain, T is the test temperature and t is the loading rate.

When the applied load is dynamic (short loading rate), the stiffness measured is called elastic or dynamic stiffness modulus or, for simplicity, stiffness. When the applied load is static (long loading time), the stiffness measured is called static stiffness (modulus), S_{creep}.

The most commonly used stiffness modulus is the elastic or dynamic stiffness modulus. It is used for ranking bituminous mixtures (asphalts), for estimating their structural behaviour in the road, as data for compliance with specifications requirements and as a property of asphalt for pavement design analytical calculations or procedures.

The static stiffness modulus provides information relative to the permanent deformation performance of the asphalt, but since this information is also obtained from dynamic testing using higher loading rates, it is rarely used nowadays.

Over the last years, the terms *complex modulus, phase angle* and *secant modulus* emerged and are used.

According to CEN EN 12697-26 (2012), *complex modulus*, E^*, is defined as the relationship between stress and strain for a linear viscoelastic material submitted to a sinusoidal load wave form at time t, where applied stress $\sigma \times \sin(\omega \times t)$ results in a strain $\varepsilon \times \sin(\omega \times (t - \Phi))$ that has a phase angle Φ, with respect to stress (ω = angular speed, in radians per second). The amplitude of strain and the phase angle are functions of the loading frequency, f, and the test temperature, Θ.

The complex modulus is characterised by a pair of two components. This pair can be expressed in two ways:

a. The real component E_1 and the imaginary component E_2, which are determined as

$$E_1 = |E^*| \times \cos(\Phi) \text{ and } E_2 = |E^*| \times \sin(\Phi)$$

b. The absolute value of the complex modulus $|E^*|$ and the phase angle, Φ, which are determined as

$$|E^*| = \sqrt{(E_1^2 + E_2^2)} \text{ and } \Phi = \arctan(E_2/E_1)$$

The second characterisation (b) is more often used in practice. In linear elastic multi-layer calculations, for instance, the absolute value of the complex modulus, $|E^*|$, which is equivalent to the stiffness modulus, E, is used as input.

For purely elastic materials, the phase angle is zero; hence, the complex modulus is equal to Young's modulus. In bituminous materials, this happens at a very low temperature and at a very high loading frequency. Then, the complex modulus reaches its highest possible value, noted as E_∞.

The *secant modulus*, $E(t)$, is defined as the relationship between stress and strain at the loading time, t, for a material subjected to controlled strain rate loading; that is:

$$E(t) = \sigma(t)/\varepsilon(t),$$

where σ and ε are stress and strain at time t.

The strain law is

$$\varepsilon(t) = \alpha_i \times t^n,$$

where α_i and n are constants determined from tests.

For linear viscoelastic materials, the secant modulus obtained for different values of α_i at the same temperature depends on the loading time, t, only (CEN EN 12697-26 2012).

According to American practice, the stiffness modulus is called *dynamic modulus*, which again is the absolute value of the complex modulus (E^*) and is designated also as $|E^*|$.

Apart from the dynamic modulus, the term *resilient modulus*, M_R, is also used in the evaluation of the bituminous mixture's quality and as an input for pavement design, evaluation or analysis. Resilient modulus is quite similar to the stiffness value determined by the indirect tension test in CEN EN 12697-26 (2012), Annex C; the differences are outlined in Section 7.4.

Additionally, the American practice for specification testing or quality control of bituminous mixtures also uses the properties *complex shear modulus*, G^*, and *phase angle*, δ (see Section 7.4), using either the Superpave shear tester (SST) (see Section 7.4.9) or the dynamic shear rheometer (DSR) (see Section 7.4.10).

All EN and American standard testing procedures for the determination of the abovementioned properties are described in Section 7.4.

7.3 TYPES OF LOADING

The typical type of loading used in all stiffness modulus laboratory test determinations is either sinusoidal, pulse or controlled strain rate loading. The pulse or the sinusoidal type of loading is explained in Figure 7.1.

In a *pulse type* of loading, the load has a haversine wave form as shown in (if not similar to) Figure 7.1a, and is characterised by the rise time, t_{rs}, the maximum applied load known as amplitude or peak load and the frequency or pulse repetition period of the applied load.

The rise time is measured from when the load pulse commences and is the time taken for the applied load to increase from initial contact load to maximum value.

According to CEN EN 12697-26 (2012), the rise time is fixed to 124 ± 4 ms. The peak load is adjusted to achieve a target peak transient horizontal deformation of 0.005% of the specimen diameter. Experience dictates that the suitable values of peak horizontal deformation are 5 ± 2 microstrain (10^{-6} m/m) for a 100 mm diameter specimen and 7 ± 2 microstrain (10^{-6} m/m) for a 150 mm diameter specimen. The pulse repetition period is 3.0 ± 4 s.

In a *sinusoidal type* of loading, the load has a wave form shown in Figure 7.1b and is characterised by the load amplitude and loading frequency.

The amplitude of the load should be such that no damage can be generated during the time needed to perform the measurements. The load frequencies used are device dependent and range from 0.1 to 50 Hz.

(a)

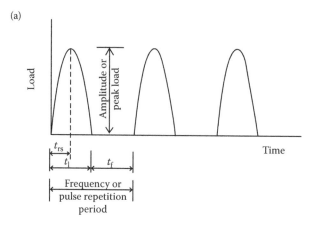

(b)

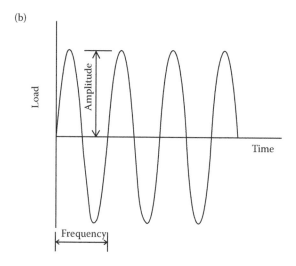

Figure 7.1 Typical forms of dynamic loading for laboratory test methods. (a) Haversine (pulse) loading. (b) Sinusoidal (harmonic) loading.

CEN EN 12697-26 (2012) suggests that, in order for most bituminous mixtures to determine the load amplitude, the strain should be kept at a level lower than 50 microstrain (50×10^{-6} m/m) to prevent fatigue damage. As for the loading frequencies and in order to allow a logarithmic presentation of isotherms, a typical set of frequencies is recommended: 0.1, 0.2, 0.5, 1, 2, 5, 20 and 50 Hz.

In a *controlled strain rate* type of loading, used in uniaxial direct tensile tests, a controlled rate displacement is applied to the specimen in direct tension to provide a constant strain rate.

According to CEN EN 12697-26 (2012), the maximum strain amplitudes during the test are related to the stiffness of the bituminous mixture and the test temperature, as shown in Table 7.1.

It is noted that the stiffness is determined by a preliminary direct tension test under the following conditions: strain amplitude, 50 microstrain; temperature, 10°C; loading force (F), >200 N; loading times, 3 and 300 s.

Table 7.1 Strain to be applied during a controlled strain rate loading test

Test temperature, Θ (°C)	Stiffness, 10°C, 3 s		Stiffness, 10°C, 300 s	
	<7.5 GPa	≥7.5 GPa	<1 GPa	≥1 GPa
	Strain amplitude (microstrain)			
≤10	100	50	—	—
10 ≤ Θ < 20	—	—	200	100
20 ≤ Θ ≤ 40	—	—	300	200

Source: Reproduced from CEN EN 12697-26, *Bituminous Mixtures – Test Methods for Hot Mix Asphalt – Part 26: Stiffness*, Brussels: CEN, 2012. With permission (© CEN).

7.4 DETERMINATION OF STIFFNESS MODULUS AND OTHER MODULI

The stiffness modulus determination, according to CEN EN 12697-26 (2012), can be carried out with one of the following tests:

Indirect tensile test
  a. Indirect tension test on cylindrical specimens (IT-CY)
  b. Cyclic indirect tension test on cylindrical specimens (CIT-CY)
Bending tests
  c. Two-point bending test on trapezoidal specimens (2PB-TR)
  d. Two-point bending test on prismatic specimens (2PB-PR)
  e. Three-point bending test on prismatic specimens (3PB-PR)
  f. Four-point bending test on prismatic specimens (4PB-PR)
Direct uniaxial tests
  g. Direct tension–compression test on cylindrical specimens (DTC-CY)
  h. Direct tension test on cylindrical specimens (DT-CY)
  i. Direct tension test on prismatic specimens (DT-PR)

All stiffness test methods are considered equivalent and are recommended to be used (CEN EN 13108-20 2008) in specifying asphalt concrete and hot rolled asphalt in terms of fundamental, performance-based properties.

The tests, apart from (b), are carried out under controlled strain conditions, and all of them are considered to be equally capable in determining stiffness modulus.

The measurements obtained during any test are applied force, F, in Newtons (N); displacement, z, in millimetres (mm); and their phase angle, Φ, in degrees (°).

The two components of the complex modulus, real component E_1 and imaginary component E_2, when required, are calculated by the following equations:

$$E_1 = \gamma \times (F/z \times \cos(\Phi) + 10^{-6} \times \mu \times \omega^2)$$

$$E_2 = \gamma \times (F/z) \times \sin(\Phi),$$

where γ is the form factor as a function of specimen size and form (1/mm) and μ is the mass factor, which is a function of the mass of the specimen, M (in grams), and the mass of movable parts, m (in grams), that influence the resultant force by their inertial effects.

The form and mass factor are determined by equations provided in CEN EN 12697-26 (2012) and depend on the test procedure and type of loading followed.

The stiffness modulus (the absolute value of complex modulus $|E^*|$) and the phase angle, Φ, an equivalent representation of the complex modulus, are derived using the equations given in Section 7.2.

In the American practice, the dynamic modulus, $|E^*|$, and the phase angle, θ, are determined over a range of temperatures and loading frequencies by the test procedure specified in AASHTO T 342 (2011), see also Section 7.4.7. According to AASHTO T 342 (2011), phase angle, θ, is the angle in degrees between sinusoidal peak stress and the resulting peak strain in a controlled stress test.

The resilient modulus, M_R, is determined by the indirect tension test according to ASTM D 7369 (2011).

Finally, the complex shear modulus, G^*, and phase angle, δ, are determined by using either the SST according to ASTM D 7312 (2010) (AASHTO T 320 2011), see also Section 7.4.9, or the DSR according to ASTM D 7552 (2009), see also Section 7.4.10. According to ASTM D 7312 (2010), phase angle, δ, is a measure of the time lag between the applied stress and the resulting strain, or applied strain and resulting stress, in a viscoelastic material.

7.4.1 Indirect tensile test on cylindrical specimens (IT-CY)

This method measures the elastic stiffness of bituminous mixtures by the indirect tensile test, using cylindrical specimens of various diameters and thickness, manufactured in the laboratory or cored from a road bituminous layer. The form of loading is pulse loading (see Figure 7.1a), and the test is carried out under controlled strain conditions. This test is perhaps the most popular of all other tests for stiffness determination.

The diameter of the specimen to be tested could be 80, 100, 120 150 or 200 mm and the height could be between 30 and 75 mm. Both dimensions are chosen relative to the nominal maximum aggregate size of the mixtures. The most common dimensions of specimens are 100 or 150 mm in diameter and 40, 50 or 60 mm in height.

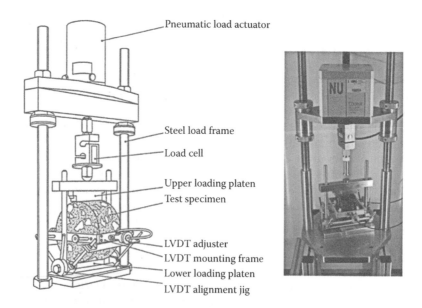

Figure 7.2 Schematic representation of the indirect tension test measuring device. (Adapted from CEN EN 12697-26, *Bituminous Mixtures – Test Methods for Hot Mix Asphalt – Part 26: Stiffness*, Brussels: CEN, 2012. With permission [© CEN] and NU ITT device [Cooper Research Technology Ltd.].)

The cylindrical specimen is placed into a steel load frame of the testing device shown in Figure 7.2. Pulse loading is applied to the specimen so that it results in a peak transient horizontal deformation of 0.005% of the specimen diameter (for a 100 mm diameter specimen, the resulting peak horizontal deformation is approximately 5 μm, and for a 150 mm diameter specimen, the corresponding value is approximately 7 μm). The time taken for the applied load to reach its maximum value, known as rise time, is set to 124 ± 4 ms. The loading cycle (pulse repetition period) is 3 ± 0.1 s.

The test is typically performed at a temperature of 20°C ± 0.5°C, but any other temperature, usually less than 20°C, may be used. The temperature during testing should be kept constant; thus, the testing device is placed in an appropriate climatic/temperature chamber.

After placing the specimen and before starting to record the results, at least 10 conditioning load pulses (pre-loading) are applied, in order to enable the equipment to adjust the load magnitude and duration to give the specified horizontal transient (diametral) deformation and time.

After the conditioning load pulses, five more load pulses are applied, during which the applied load and the resulting horizontal deformation are recorded. For each load pulse, the stiffness modulus is determined using the following equation:

$$E \text{ (or } S) = [F \times (\nu + 0.27)]/(z \times h),$$

where E or S is the stiffness modulus (MPa), F is the peak value of applied vertical load (maximum value) (N), ν is the Poisson ratio (if not determined, a value of 0.35 may be assumed for all temperatures), z is the amplitude of horizontal deformation obtained during the load cycle (mm) and h is the mean thickness of the specimen (mm).

After completing the above procedure, the specimen is rotated by 90° ± 10°, and the same procedure for determining the stiffness modulus is repeated in this new position.

The stiffness modulus of the sample tested is the average value of two measurements, provided that the second measurement is within +10% or –20% of the mean value recorded for the first measurement. Otherwise, the results are rejected.

It is noted that the test control and the recording and processing of the results are carried out with the aid of appropriate software.

The test, as mentioned, is usually conducted at 20°C ± 0.5°C; however, other favourable temperatures may be 2°C or 10°C. For a more accurate determination of specimen temperature, a dummy specimen is also placed in the thermostatic chamber, and with the help of a thermocouple, the specimen's temperature is monitored continuously.

More information about the test procedure is given in CEN EN 12697-26 (2012), Annex C.

7.4.1.1 Indicative stiffness modulus results

A variety of hot and cold bituminous mixtures have been tested in the Highway Engineering Laboratory of the Aristotle University of Thessaloniki (AUTh) for stiffness determination by indirect tensile method using the NU ITT device. Some of the results obtained are as shown in Table 7.2.

7.4.2 Cyclic indirect tension test on cylindrical specimens (CIT-CY)

This method measures the stiffness of bituminous mixtures by the cyclic indirect tensile test, using cylindrical specimens, manufactured in the laboratory or cored from a road layer.

Table 7.2 Indicative mean stiffness values for hot and cold asphalts

Type of asphalt	Stiffness, S, at 20°C (MPa)					
	40/50 grade bitumen	50/70 grade bitumen		70/100 grade bitumen	PmB: 25–55/75	
Dense AC-12.5 mm[a]						
4%	6200[a]	5350[a]	5200[b]	3200[a]	5800[a,e]	5800[b,e]
5%	6000[a]	4600[a]	4550[b]	3000[a]	4950[a,e]	4850[b,e]
6%	5300[a]	4000[a]	3850[b]	2700[a]	4100[a,e]	4050[b,e]
AC for very thin layers						
4.0%	—	3050[b]		—	3300[b,e]	
5.0%	—	2400[b]		—	3000[b,e]	
6.0%	—	2000[b]		—	2450[b,e]	
Porous asphalt						
4.0%	—	4200[c]		—	—	
4.5%	—	2700[d]		—	—	
5.0%	—	3200[c]		—	3200[c,e]	
Cold dense AC-12.5 mm	Cationic bituminous emulsion (with 70/100 grade)				—	
4%	1500[a]				—	
5%	1100[a]				—	
6%	900[a]				—	
7%	750[a]				—	

[a] With limestone aggregates.
[b] With diabase.
[c] With andesite.
[d] With limestone and andesite aggregates.
[e] SBS polymer-modified bitumen.

The form of loading is sinusoidal (see Figure 7.1b), and the test is carried out under force controlled conditions, without rest periods.

According to CEN EN 12697-26 (2012), Annex F, three sizes of cylindrical specimens are used depending on the maximum grain size. For maximum grain size ≤16 mm, the size of the specimen is 100 ± 3 mm diameter by 40 ± 2 mm height; for maximum grain size >16 mm and <32 mm, the size of the specimen is 150 ± 3 mm diameter by 60 ± 2 mm height; for maximum grain size ≥32 mm, the size of the specimen is 150 ± 3 mm diameter by 90 ± 2 mm height.

The cylindrical specimen is placed into the steel load frame of the testing device, similar to the one used in the indirect tensile test. The measurement of horizontal deformation can be carried out by load transducers (linear variable displacement transducers [LVDTs]), with an arrangement similar to Figure 7.2, or by strain gauges with extensometers (see Figure 7.3).

The test should be performed at a constant load and at a certain load frequency and temperature. The applied load should be within a lower and an upper load level. The lower load level is determined so that a resultant stress of 0.035 MPa is arrived at. The upper load level limits are determined in such a manner that the initial horizontal strains in the specimen centre are in a range from 0.05‰ to 0.10‰.

When a master curve is to be determined, multistage tests are conducted at different temperatures and loading frequencies. The recommended minimum test temperatures are –10°C, 0°C, 10°C and 20°C; if more are required, they should be chosen between –10°C

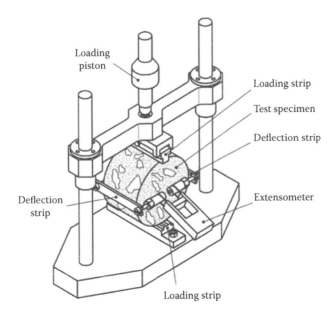

Figure 7.3 Schematic representation of the cyclic indirect tension test measuring device. (Adapted from CEN EN 12697-26, *Bituminous Mixtures – Test Methods for Hot Mix Asphalt – Part 26: Stiffness,* Brussels: CEN, 2012. With permission [© CEN].)

and 20°C. At each test temperature, four specimens are tested. An additional specimen is normally needed for estimating the required condition.

The recommended load frequencies are 10, 5, 1 and 0.1 Hz. For each load frequency, the number of load cycles is determined, which are 110, 100, 20 and 10, respectively.

Generally, the test conditions should be chosen to avoid any damage of the specimen during testing.

More details regarding the cyclic indirect tension test is given in CEN EN 12697-26 (2012), Annex F.

7.4.3 Two-point bending test on trapezoidal (2PB-TR) or prismatic (2PB-PR) specimens

The two-point bending test on trapezoidal specimens (and later on prismatic specimens) was initially developed in Shell laboratories in Holland and soon adopted by LPC in France for determining the stiffness modulus and the asphalt fatigue performance (Moutier 1990). Today, this test is the recommended type of stiffness test in France.

In this test method, the stiffness modulus of bituminous mixtures is measured using the cantilever bending test. A sinusoidal force, $F = F_o \times \sin(\omega \times t)$, or a sinusoidal deflection, $z = z_o \times \sin(\omega \times t)$, is applied to the head of a trapezoidal or prismatic specimen, which is glued at its base to a stand fixed to a rigid chassis of L shape (see Figures 7.4 and 7.5).

The force, F_o, or the deflection, z_o, is such that it causes a strain $\varepsilon \leq 50 \times 10^{-6}$ in the most heavily stressed part of the specimen (about in the middle), which is supposed to correspond to the linear range of bituminous mixture.

The specimen is subjected to a sinusoidal force, at the target frequency ±5%, applied at the head of the specimen from a time of 30 s (minimum) to a time of 2 min (maximum) to an imposed deflection corresponding with a strain, ε, less than 50 microstrain (μm/m).

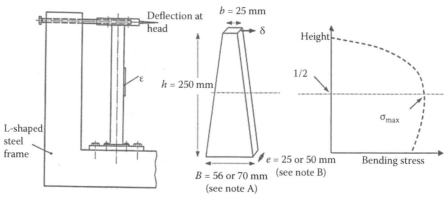

Note A: 56 mm only for asphalts where D ≤ 14 mm
Note B: 50 mm only for asphalts where 20 < D ≤ 40 mm

Figure 7.4 Schematic representation of the two-point bending test on trapezoidal specimens (application of force in the horizontal direction).

Figure 7.5 A two-point bending device on trapezoidal specimens. (Courtesy of Cooper Research Technology Ltd.)

The force, the deflection and the phase angle Φ are measured at the last 10 s of the test and complex modulus, E^*, at a required temperature and frequency is determined, using the equations given in Sections 7.2 and 7.4.

The conditions of temperature and loading frequency adopted for a single determination of complex modulus are 15°C and 10 Hz (0.02 s), respectively.

If the master curve is to be determined, the complex modulus should be determined at not less than four temperatures separated by not more than 10°C, and for each temperature at not less than three frequencies evenly spaced on a logarithmic scale with a minimum ratio of 10 between the extreme frequencies.

According to CEN EN 12697-26 (2012), Annex A, specimens are produced by sawing from slabs made in the laboratory according to CEN EN 12697-33 (2007) or from slabs extracted from road surfaces having a thickness ≥60 mm.

Trapezoidal or prismatic specimens have appropriate dimensions depending on the nominal maximum aggregate size, D. For trapezoidal specimens, the critical values of D are as follows: D ≤ 14 mm, D between 14 and ≤22 mm and D > 22 mm. Figure 7.4 gives an indication of the required dimensions.

For prismatic specimens, the critical values of D are as follows: D ≤ 22 mm or D > 22 mm. In the first case, the size of the specimen is 40 mm × 40 mm × 120 mm. In the second case, the size of the specimen is 80 mm × 80 mm × 240 mm.

More information about the test is given in CEN EN 12697-26 (2012), Annex A.

7.4.3.1 Limiting values

The French design guide (Delorme et al. 2007; SETRA 2008) has set limiting stiffness values for designing high stiffness asphalt concretes (AC-BBME) or high-modulus bituminous mixtures (AC-EME). The minimum values of stiffness (S or E) when the test is carried out at 15°C and 10 Hz at 0.02 s are as follows:

- 9000 MPa for high-stiffness asphalt concrete (AC-BBME) class 1
- 11,000 MPa for high stiffness asphalt concrete (AC-BBME) class 2 and 3
- 14,000 MPa for high-modulus bituminous mixtures (AC-EME 1 or AC-EME 2)

7.4.4 Three- or four-point bending test on prismatic specimens (3PB-PR or 4PB-PR)

With this test method, the stiffness modulus of bituminous mixtures is measured using the bending test. The prismatic specimen is subjected to three-point or to four-point periodic bending with free rotation and (horizontal) translation at all load and reaction points.

The bending is achieved by the movement of the centre load point(s) in vertical direction perpendicular to the longitudinal axis of the specimen. The two end points of the beam remain fixed (clamped). The applied periodical loading (force) is sinusoidal to obtain the required strain, ε, amplitude of 50 ± 3 microstrain. A schematic representation of the four-point bending test is given in Figure 7.6.

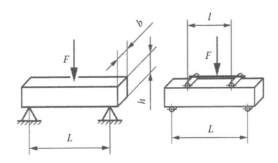

Figure 7.6 Schematic representation of the basic principles of the three-point and four-point bending test.

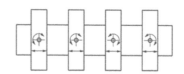

Free translation and rotation

Figure 7.7 A four-point bending main part for the four-point bending device on prismatic specimens. (Courtesy of Controls Srl.)

The main part of a four-point bending apparatus together with the free translation and rotation principle of the beam is shown in Figure 7.7.

According to CEN EN 12697-26 (2012), the specimen has the shape of a prismatic beam with a total length not exceeding the effective length by more than 10%. The effective length, L, should not be less than six times whatever the highest value is for the width, B, or the height, H. The width and the height are at least three times the maximum grain size, D, in the tested bituminous mixture.

The specimens are produced and prepared the same way as specimens for the two point bending test.

During the test, the force needed for the deformation of the specimen is measured as a function of time. The force, F_0, the deflection, z_0, and the phase angle, Φ, are recorded, together with the test temperature and frequency.

The initial stiffness modulus is determined as the modulus for a load cycle between the 45th and the 100th load repetition (typically determined at the 100th load repetition).

The complex modulus, E^*, at a required temperature and frequency is determined using the equations given in Sections 7.2 and 7.4.

The conditions of temperature and loading frequency typically used are 15°C and 10 Hz for the three-point bending test and 20°C and 8 Hz for the four-point bending test, respectively (CEN EN 13108-20/AC 2008).

More information about the test is given in CEN EN 12697-26 (2012), Annex B.

7.4.5 Direct tension–compression test on cylindrical specimens (DTC-CY)

In this test, the cylindrical specimen is glued on two steel plates and is subjected to a sinusoidal strain, $\varepsilon = \varepsilon_0 \times \sin(\omega \times t)$, of $\leq 25 \times 10^{-6}$ in the linear range of the bituminous mixtures. A schematic representation of the test is given in Figure 7.8.

Measuring the applied force, F_0, and the phase angle, Φ, the complex modulus, E^*, is calculated at the required temperature (usually 15°C) and for the required frequency (usually 10 Hz).

If the master curve is to be determined, the complex modulus is determined at not less than four different temperatures, separated by not more than 10°C and for each temperature at not less than three frequencies.

The typical four temperatures are 10°C, 20°C, 30°C and 40°C, and the typical six frequencies are 0.1, 0.3, 1.0, 3.0, 10 and 20 Hz.

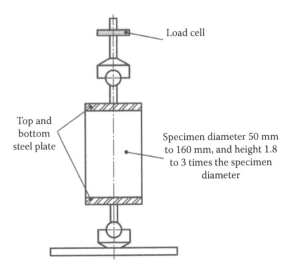

Figure 7.8 Schematic representation of the direct tension–compression test. (Adapted from CEN EN 12697-26, *Bituminous Mixtures – Test Methods for Hot Mix Asphalt – Part 26: Stiffness*, Brussels: CEN, 2012. With permission [© CEN].)

According to CEN EN 12697-26 (2012), Annex D, the cylindrical specimens are obtained either by core drilling and sawing slabs (made in the laboratory in accordance with CEN EN 12697-33 2007 or form the road) or by a gyratory compactor in accordance with CEN EN 12697-31 (2007). The sizes of the specimens are as indicated in Figure 7.5. In order to stabilise the specimens, they are stored between 2 weeks and 2 months before testing.

The conditions of temperature and loading frequency, which have been adopted for the determination of complex stiffness modulus, are 15°C and 10 Hz, respectively (CEN EN 13108-20/AC 2008).

More information about the test is given in CEN EN 12697-26 (2012), Annex D.

7.4.6 Direct tension test on cylindrical or prismatic specimens (DT-CY)

In this test, a uniaxial tensile load is applied to a cylindrical or prismatic specimen, at given temperatures and loading times, at a controlled strain rate, once both ends of the specimen are glued to two metal plates. During the test, the applied tensile load and the resulting strain are recorded, and the stiffness modulus for the required loading time and required temperature is determined from the master curve developed. The principle of the direct tension test is shown in Figure 7.9.

The cylindrical specimens have the same sizes and are produced as the specimens for the direct tensile-compression test. The prismatic specimens are sawed in the desired dimensions from slabs compacted in the laboratory or taken from the road.

According to CEN EN 12697-26 (2012), Annex E, the specimens are stabilised before and after each element test (chosen test temperature Θ_j, level of strain and loading time, t_i). Stabilisation consists of temperature stabilisation, preliminary mechanical stabilisation before testing and mechanical stabilisation between tests.

During temperature stabilisation, the specimens are kept to the test temperature for at least 4 h, if the diameter or the width is ≤100 mm, and for at least 8 h, in other cases.

Preliminary mechanical stabilisation, after being fitted to the test machine by applying a load of <100 N, consists of keeping the specimen without displacement of the machine for at least 60 min for test temperatures less than −5°C and for at least 30 min for other test temperatures.

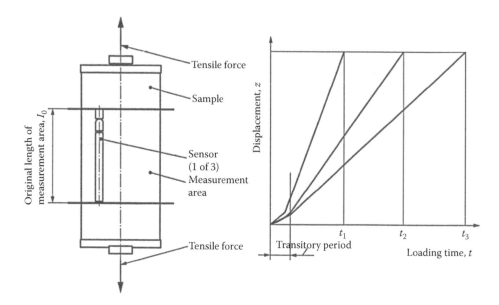

Figure 7.9 Schematic representation of the direct tension test. (Adapted from CEN EN 12697-26, *Bituminous Mixtures – Test Methods for Hot Mix Asphalt – Part 26: Stiffness*, Brussels: CEN, 2012. With permission [© CEN].)

Mechanical stabilisation between tests consists of cancelling the deformation by applying a compression load ≤300 N and keeping the specimen without uniaxial load for at least 100 s.

At each element test, the stress $\sigma\,(t_i, \Theta_j)$ is determined for the deformation ε_j.

7.4.6.1 Master curve

At least four loading times are necessary for at least one temperature, and at least two loading times are necessary for other test temperatures, to construct the isotherms of the stiffness in terms of loading times and from that to derive the master curve.

The stiffness modulus at any combination of temperature and loading time is derived from the master curve determined.

The derivation of master curve from isotherms is explained in detail in CEN EN 12697-26 (2012), Annex G.

7.4.6.2 Stiffness modulus

For the determination of stiffness modulus, the conditions of temperature and loading time typically used are 15°C and 0.02 s, respectively (CEN EN 13108-20/AC 2008).

More details about the direct tension test are given in CEN EN 12697-26 (2012), Annex E.

7.4.7 Dynamic modulus test

The dynamic modulus test is conducted in accordance to the AASHTO T 342 (2011) or AASHTO TP 79 (2013) procedures. Both determine the dynamic modulus and phase angle over a range of temperatures and loading frequencies.

According to AASHTO T 342 (2011), the dynamic modulus, $|E^*|$, is the absolute (normal) value of the complex modulus, E^*, calculated by dividing the maximum (peak-to-peak) stress by the recoverable (peak-to-peak) axial strain for a material subjected to haversine (pulse) vertical load. The complex modulus, E^*, is the relationship between stress and strain for a linear viscoelastic material.

The notification used for the dynamic modulus, $|E^*|$, is confusing since the same notification is used by CEN EN 12697-26 (2012) for the absolute value of the complex modulus, since the former uses the recoverable axial (vertical) strain and the latter uses the resulting axial (horizontal) strain.

The axial compressive stress is applied to the cylindrical specimen at a given temperature and loading frequency (time). The applied stress and the resulting recoverable axial strain response of the specimen is measured and used to calculate the dynamic modulus and phase angle, θ, at the given testing conditions (temperature and loading frequency). Figure 7.10 shows a schematic representation of the device used for the dynamic modulus test.

The dynamic modulus values measured over a range of temperatures and loading frequencies can be shifted into a master curve for characterising bituminous mixture for pavement thickness design and performance analysis.

The specimens are cored from gyratory-compacted specimens with an average diameter of 100 to 104 mm and a height of 170 mm.

Laboratory-prepared specimens, before testing, are temperature conditioned (aged) in accordance with the 4 h short-term oven conditioning, at a temperature of 135°C (the procedure was specified in AASHTO R 20). Specimens cored from bituminous layers on site do not need to be conditioned (aged) before testing.

The test conditions for the development of master curves are recommended to be conducted at –10°C, 4.4°C, 21.1°C, 37.8°C and 54°C at loading frequencies of 0.1, 0.5, 1.0, 5, 10 and 25 Hz at each temperature. Each test specimen, individually instrumented with LVDT brackets, should be tested for each of the 30 combinations of temperature and frequency of loading starting with the lowest temperature and proceeding to the highest. Testing at a

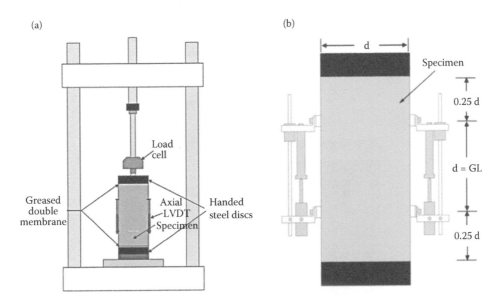

Figure 7.10 Schematic representation of the dynamic modulus test. (a) Sample in testing device. (b) Details on the position of the LVDT (GL, gauge length; d, specimen diameter). (From AASHTO T 342, *Determining Dynamic Modulus of Hot Mix Asphalt [HMA]*, Washington, DC: American Association of State Highway and Transportation Officials, 2011. With permission.)

given temperature should begin with the highest frequency of loading and proceed to the lowest.

In order to reach the equilibrium targeted testing temperature, the specimens are placed and left in the environmental chamber for a certain period specified in AASHTO T 342 (2011).

Before testing, a contact load, P_{min}, is applied equal to 5% of the dynamic load that will be applied to the specimen. The haversine dynamic load, $P_{dynamic}$, to be applied should be such to obtain axial strains between 50 and 150 microstrain.

The dynamic load to be applied depends on the specimen's stiffness, relative to the temperature, and generally ranges between 35 kPa (for 54°C) and 2800 kPa (for –10°C).

At the beginning of testing, the specimen is preconditioned with 200 cycles at 25 Hz and then the specimen is loaded for a number of cycles depending on the frequency used (see Table 7.3).

Table 7.3 Number of cycles for testing

Frequency (Hz)	Number of cycles
25	200
10	200
5	100
1	20
0.5	15
0.1	15

Source: AASHTO T 342, *Determining Dynamic Modulus of Hot Mix Asphalt (HMA)*, Washington, DC: American Association of State Highway and Transportation Officials, 2011. With permission.

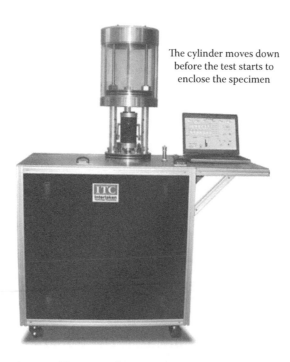

The cylinder moves down before the test starts to enclose the specimen

Figure 7.11 The AMPT equipment. (Courtesy of Interlaken Technology.)

The procedure for calculating the dynamic modulus, $|E^*|$, and the phase angle, $\theta(\omega)$, using the data from a specific loading frequency, ω, is presented in the relevant standard.

More details for the dynamic modulus test can be found in AASHTO T 342 (2011).

According to AASHTO TP 79 (2013), the test is conducted at three temperatures, 4°C, 20°C and at a high temperature depending on the performance grade of the binder used in the mixture. The frequencies used are 0.1 and 10 Hz for 4°C and 20°C and at four frequencies between 0.1 and 10 Hz for the high temperature. Three specimens are tested and the results are averaged. The test is conducted using a recently developed asphalt mixture performance tester (AMPT) as a simple fundamental performance tester for Superpave mix design (see Figure 7.11).

After test completion using either procedure, the results are analysed for consistency and the dynamic modulus data are collected at all temperatures and frequencies used to construct a master curve at a reference temperature according to AASHTO PP 62 (2010) and AASHTO PP 61 (2013), respectively.

7.4.8 Resilient modulus test by indirect tension

This test method is used for determining the resilient modulus by indirect tension test according to ASTM D 7369 (2011). The test is similar to but not exactly the same as the indirect tensile test specified in CEN EN 12697-26 (2012), Annex C.

Particularly, although the repetitive compressive load applied is a haversine and is applied along the vertical diametral plane of a specimen, the resilient Poisson ratio, μ, is first calculated using the recoverable vertical and horizontal deformations, and subsequently two separate resilient moduli, M_R, are determined: the instantaneous resilient modulus and the total resilient modulus.

The instantaneous resilient modulus is calculated using the instantaneous recoverable deformation that occurs during the unloading portion of one load–unload cycle.

The total resilient modulus is calculated using total recoverable deformation, which includes both the instantaneous recoverable and the time-dependent continuing recoverable deformation during the rest period portion of one cycle.

Additionally, in the resilient modulus test, the loading time is approximately 0.1 s and the resting period is approximately 0.9 s, compared to approximately 1 s loading time and 2 s resting period used in the indirect tension test.

In the EN indirect tensile test, the amplitude of the horizontal deformation obtained during load cycle is used for the determination of the modulus called stiffness, E.

In the resilient modulus test, the cylindrical specimens have a diameter of 101.6 ± 3.8 mm or 152.4 ± 9 mm and thickness ranging from 38.1 to 63.5 mm. The specimens are compacted in the laboratory with a gyratory compactor, or a Marshall compactor, or derived from field coring compacted asphalt layer(s).

The test is conducted at 25°C, and preconditioning is necessary before taking the final measurements. The number of preconditioning load cycles is 100 and the subsequent number of load cycles for taking final measurements is normally 5.

The selection of the applied load during preconditioning and loading is based on the indirect tensile strength, determined as specified in the ASTM D 6931 (2012) test method. Tensile stress levels from 10% to 20% of the tensile strength measured at 25°C are used in conducting the test at temperatures of 25°C ± 1°C. Specimen contact loads are maintained during testing, which is 4% of the maximum load ($0.04\ P_{max}$) and is not less than 22.2 N and not more than 89.0 N.

The test assembly and device used are similar to the ones used in the EN indirect tension test, except for the way deformations are measured. Both horizontal and vertical

(a)   (b)

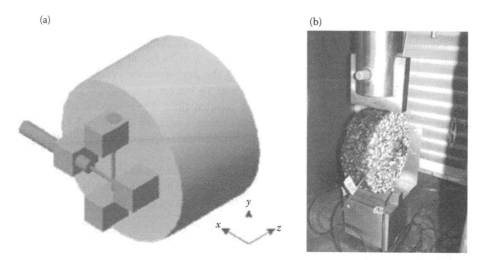

Figure 7.12 (a) LVDTs mounted on the specimen and (b) specimen during indirect testing. (Reprinted from ASTM D 7369, *Standard Test Method for Determining the Resilient Modulus of Bituminous Mixtures by Indirect Tension Test*, West Conshohocken, Pennsylvania, US: ASTM International, 2011. With permission [© ASTM International].)

deformations are measured on the surface of the specimen by mounting LVDTs between gauge points along the horizontal and vertical diameters (see Figure 7.12).

Once the Poisson ratio is determined, the resilient modulus is determined by the following formula:

$$M_R = [P_{cyclic}/(\delta_h \times t)] \times (I1 - I2 \times \mu),$$

where M_R is the instantaneous or total resilient modulus (MPa), $P_{cyclic} = (P_{max} - P_{contact})$ is the cycle load applied to specimen (N), P_{max} is the maximum applied load (N), $P_{contact}$ is the contact load (N), δ_h is the recoverable horizontal deformation (instantaneous or total) (mm), t is the specimen thickness (mm), $I1, I2$ are constant values depending on the gauge length provided by Table 2 of ASTM D 7369 (2011) and μ is the instantaneous or total Poisson ratio.

After defining the resilient modulus, the specimen is rotated by 90° and the same above-mentioned pre-loading/loading procedure is repeated.

The resilient modulus (instantaneous or total) is determined by the average values obtained by two loading positions and for three specimens, per type of bituminous mixture.

More information about the test is given in ASTM D 7369 (2011).

7.4.9 Complex shear modulus and permanent shear strain test

This test determines the stiffness complex shear modulus, as well as the permanent shear strain of bituminous mixtures, using the apparatus developed during the Superpave program, known as SST. The test is specified by ASTM D 7312 (2010) or AASHTO T 320 (2011).

The complex shear modulus determines the performance of the bituminous mixture in shear, whereas the permanent shear strain is related to permanent deformation and pavement rutting.

The *complex shear modulus, G**, defines the relationship between shear stress and strain for a linear viscoelastic material.

The *permanent shear strain*, γ_p, is the non-recoverable shear strain resulting from a shear load.

The test includes two procedures: procedure A is called frequency sweep test and procedure B is called repeated shear test at constant height (of specimen).

In the frequency sweep test, a repeated sinusoidal shear loading is applied at 10 frequencies and at a given temperature while a varying axial load is applied to prevent dilation of the specimen. The loads and deformations are used to calculate the complex shear modulus, G^*, and phase angle, δ, of the specimen at each frequency.

It is reminded that phase angle (δ) is defined as the time lag between the applied stress and the resulting strain, or the applied strain and the resulting stress, in a viscoelastic material.

In the repeated shear test, a repeated haversine shear stress is applied to the specimen while a varying axial load prevents dilation. From this test, the permanent shear strain (γ_p) is determined.

Both procedures use cylindrical specimens produced in the laboratory (preferably compacted by a gyratory compactor) or derived from field coring. The specimen's size is usually 150 mm in diameter and 50 mm in height. If the maximum nominal size of the aggregate particle (D) is 12.5, 9.5 or 4.75 mm, the height may be up to 38 mm. Generally, the diameter-to-height ratio should be 3:1 or greater. Mixtures with maximum nominal size of aggregates D > 19 mm may be tested, but it is not recommended by the specification.

The two parallel surfaces of the specimens are bonded to the loaded platens with adhesive (epoxy resin compound) before testing in both procedures.

After temperature conditioning, attach the shear and axial LVDTs to the specimen platens and place the specimen to the testing device. Allow the specimen temperature to stabilise for a minimum of 20 min and a maximum of 60 min and then apply a sinusoidal or haversine loading, according to the procedure followed. A schematic representation of the test is given in Figure 7.13.

In procedure B, a repeated haversine shear stress of 69 ± 5 kPa (approximately 1220 N for a 150 mm diameter specimen) is applied for 0.1 s followed by a 0.6 s rest period. During the test, the specimen height remains constant (within the range of ±0.0013 mm), by applying sufficient axial load. The test is performed for 5000 or 10,000 loadings. The 5000 loadings are applied within approximately 90 min. During the test, the shear and axial deformation, as well as the shear and axial load, are recorded.

The permanent shear strain (γ_p) is calculated from the following formula:

$$\gamma_p = (\delta_f - \delta_{ini})/h,$$

where γ_p is the permanent shear strain, δ_f is the final shear deformation (mm), δ_{ini} is the initial shear deformation (nominally zero) (mm) and h is the specimen height (mm).

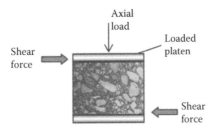

Figure 7.13 Schematic representation of testing for the determination of complex shear modulus and permanent shear strain.

More information about the test method is given in ASTM D 7369 (2011) or AASTHO T 320 (2011).

7.4.10 Complex shear modulus test using a DSR

This test method determines again the complex modulus (G^*) and the phase angle (δ), but using torsion rectangular geometry on a DSR to prismatic specimens. The test is specified by ASTM D 7552 (2009).

According to the specification, the test is applicable to bituminous mixtures having complex shear modulus values greater than 1×10^4 Pa when tested over a range of temperatures from 10°C to 76°C at frequencies of 0.01 to 25 Hz and strains of 0.001% to 0.1%.

Because of the geometry of the specimens tested, the test is not applicable to open-graded asphalts or stone mastic asphalts (SMAs). It is appropriate for dense-graded mixtures with 19 mm or smaller nominal maximum aggregate size.

Prismatic specimens nominally 49 ± 2 mm in length, 12 ± 2 mm in width and 9 ± 1.5 mm in thickness are cut from cylindrical specimens 150 mm in diameter, compacted in the laboratory (Marshall or gyratory compaction) or cored from the field.

During testing, one of the fixtures is rotated with respect to the other at a pre-selected% strain (usually 0.01%) and a range of 10 frequencies at the selected temperatures. The frequency ranges from 10 to 0.01 Hz.

From the data collected, the complex shear modulus and phase angle are determined at different temperatures and loading frequencies, as well as master curves at a certain frequency, if desired.

More information about the test is given in ASTM D 7552 (2009).

7.5 PREDICTION OF ASPHALT STIFFNESS

If the laboratory determination of the stiffness of bituminous mixtures is not feasible, the stiffness can be estimated by empirical methods.

One of the most widely accepted empirical method for predicting the asphalt stiffness at any temperature and loading time is the use of a nomograph developed by Shell (Bonnaure et al. 1977) (see Figure 7.10). The development of the nomograph is based on the fundamental research carried out by Van der Poel and other researchers (Heukelom and Klomp 1964; Van Draat and Sommer 1965).

The use of the nomograph is very simple and the data required are the stiffness of the bitumen, S_{bit}, in pascals, the percentage volume of bitumen and the percentage volume of mineral aggregates (see the example in Figure 7.14).

The stiffness modulus of the bitumen for a given temperature and loading time is determined from Van der Poel's nomograph (see Figure 4.8). The use of penetration and softening point values determined from extracted bitumen rather than bitumen before mixing is recommended.

The volume of bitumen and the volume of the mineral aggregates in the mixture are easily determined by volumetric calculations.

Apart from the nomograph in Figure 7.10, the stiffness value can be calculated from the equations used to develop the nomograph, which are given in Bonnaure et al. (1977).

The University of Nottingham has also developed a method for calculating asphalt stiffness (Brown 1980). The data required are the stiffness modulus of the bitumen, S_{bit}, in pascals, determined from Van der Poel's nomograph, and the voids in the mineral aggregate (VMA). The equation used to calculate the stiffness is as follows:

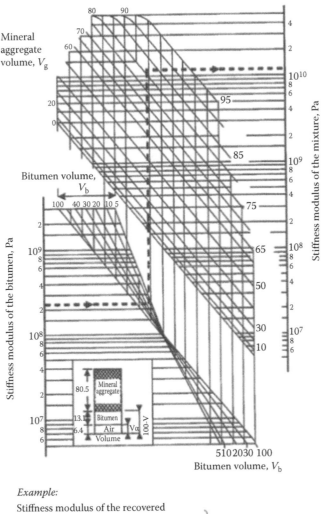

Figure 7.14 Nomograph for evaluating the stiffness modulus of asphalts. (From Bonnaure F. et al., A new method of predicting the stiffness of asphalt paving mixtures. *Proceedings of the Association of Asphalt Paving Technologists,* Vol. 46, p. 66, 1977; Taken from Shell Bitumen, *The Shell Bitumen Handbook,* Chertsey, Surry, UK: Shell Bitumen UK, 1990.)

$$S_{\mathrm{m}} = S_{\mathrm{bit}} \times [1 + (257.2 - 2.5 \times \mathrm{VMA})/(n \times (\mathrm{VMA} - 3))]^n,$$

where S_{m} is the asphalt stiffness (MPa), S_{bit} is the bitumen stiffness (MPa), VMA denotes the voids in the mineral aggregate (%) and $n = 0.83 \times \log(4 \times 10^6/S_{\mathrm{bit}})$.

Both methods/procedures can only be applied when the stiffness of bitumen exceeds 5 MPa, and for the Nottingham method, it can be applied only to asphalts with VMA values ranging from 12% to 30% (Bonnaure et al. 1977; Brown 1980).

7.5.1 Other prediction models of asphalt stiffness

Ever since the development of Bonnaure's and Brown's models, several other researchers, primarily in the United States, developed relationships for predicting asphalt stiffness, using the equivalent term dynamic modulus, $|E^*|$. A list of all predictive models until 2003 can be found in Bari and Witczak (2006).

Originally, the most wildly used predictive equation was Witczak's equation (Andrei et al. 1999; Witczak 2005), also known as the original Witczak equation. The dynamic modulus was determined with respect to percentage passing through certain sieves, binder viscosity, air voids, effective binder content and loading frequency.

Witczak's original predictive model was reformulated to include the dynamic shear modulus of the bitumen $|G^*|_b$ and the binder phase angle associated with $|G^*|_b$ (Bari 2005; Bari and Witczak 2006, 2007). The new equation developed is know as the modified Witczak equation (model) and is as follows:

$$|E^*| = -0.349 + 0.754\left(|G^*|_b^{|-0.0052}\right)\begin{pmatrix} 6.65 - 0.32p_{200} + 0.0027(p_{200})^2 + 0.011p_4 \\ -0.0001(p_4)^2 + 0.006p_{3/8} - 0.00014(p_{3/8})^2 \\ -0.08V_a - 1.06\left(\dfrac{V_{beff}}{V_{beff} + V_a}\right) \end{pmatrix}$$

$$+ \frac{2.558 + 0.032V_a + 0.713\left(\dfrac{V_{beff}}{V_{beff} + V_a}\right) + 0.0124p_{3/4} - 0.0001(p_{3/8})^2 - 0.0098p_{3/4}}{1 + \exp(-0.7814 - 0.5785\log|G^*|_b + 0.8834\log\delta_b)},$$

where $|E^*|$ is the asphalt dynamic modulus (psi), $|G^*|_b$ is the dynamic shear modulus of binder (psi), p_{200} is the percentage passing through the 0.075 mm sieve, p_4 is the cumulative percentage retained on the No. 4 (4.76 mm) sieve, $p_{3/8}$ is the cumulative percentage retained on the 3/8-in. (9.56 mm) sieve, $p_{3/4}$ is the cumulative percentage retained on the 3/4-in. (9.56 mm) sieve, V_a is the percentage of air voids (% by volume), V_{beff} is the percentage of effective bitumen content (% by volume) and δ_b is the binder phase angle associated with $|G^*|_b$ (degrees).

Another commonly used equation is the Hirsch predictive equation (Christensen et al. 2003). In this case, the dynamic modulus is determined with respect to voids in the mineral aggregate, voids filled with binder and dynamic shear modulus of bitumen. The equation developed is as follows:

$$|E^*| = P_c\left[4,200,000\left(1 - \frac{VMA}{100}\right) + 3|G^*|_b\left(\frac{VFA \times VMA}{10,000}\right)\right]$$

$$+ \frac{(1 - P_c)}{\dfrac{\left(1 - \dfrac{VMA}{100}\right)}{4,200,000} + \dfrac{VMA}{3|G^*|_b(VFA)}},$$

where $|E^*|$ is the dynamic modulus of asphalt (psi), VMA denotes the voids in the mineral aggregate (%), VFA indicates the voids filled with bitumen (%) and P_c is the aggregate con-

tact volume $= \dfrac{\left(20 + 3|G^*|_b(VFA)/(VMA)\right)^{0.58}}{650 + \left(3|G^*|_b(VFA)/(VMA)\right)^{0.58}}.$

The phase angle, θ, can be calculated from the following equation:

$$\theta = -21(\log P_c)^2 - 55 \log P_c.$$

Evaluation of the Hirsch model has shown a slight improvement over Witczak's predictive model (Christensen et al. 2003; Dongre et al. 2005).

A more simplified predictive model has been developed by Al-Khateeb et al. (2006), which, for the determination of dynamic modulus, uses only two parameters: the voids in the mineral aggregate and the dynamic shear modulus of the binder. As concluded, the model is capable of predicting the dynamic modulus of an asphalt concrete at a broader range of temperatures and loading frequencies than the Hirsch model. It also has the advantage of estimating the dynamic modulus of an asphalt concrete with modified bitumen (Al-Khateeb et al. 2006). The mathematical formulation of the model developed is as follows:

$$|E^*|_m = 3\left(\frac{100 - \text{VMA}}{100}\right)\left(\frac{(1+0.0326|G^*|_b)^{0.5}}{150 + 0.0120(|G^*|_b)^{0.5}}\right)|G^*|_g,$$

where $|E^*|$ is the dynamic modulus of asphalt (Pa), VMA denotes the voids in the mineral aggregate (%), $|G^*|_b$ is the dynamic shear modulus of binder (Pa) and $|G^*|_g$ is the dynamic shear modulus of binder at the glassy state (assumed to be 999,050 kPa).

Dynamic moduli on models have also been developed employing advanced computing tools such as artificial neutral networks (ANNs) and utilising the material properties (binder, volumetric and resilient properties) readily available from an extensive independent database, that is, not purposely produced laboratory specimens. Details regarding the development of the ANN models are given in Kim et al. (2011).

Evaluation studies on all predictive models carried out by various researchers have indicated that, in general, they provide reasonable predictions of dynamic modulus $|E^*|$. Their predictive power varies with one or more parameters such as nominal size aggregate, temperature, frequency or air voids.

Mohammad et al. (2007) evaluated the original Witczak and Hirsch models and found that the reliability of the Witczak model increases for higher nominal maximum aggregate size, whereas the reliability of the Hirsch model increases for lower nominal maximum aggregate size.

Other evaluation studies on Witczak, Hirsch and Al-Khateeb models showed that predictions were inaccurate at low frequencies/high temperatures (Kim et al. 2011; Sakhaeifar et al. 2009) or varied with temperature and air void levels of compacted specimens (Singh et al. 2011).

Independent evaluations of these models were performed in various studies such as Azari et al. (2007) and Robbins and Timm (2011). These studies consistently showed inaccuracies of statistical models at certain frequencies and temperatures.

Additionally, another study found that Witczak's models generally tend to overemphasise the influence of temperature and understate the influence of other mixture characteristics. Model accuracy also tends to fall off at the low and high temperature extremes (Ceylan et al. 2009).

The ANN-based dynamic modulus models using the same input variables were found to exhibit significantly better overall prediction accuracy, better local accuracy at high and low temperature extremes, less prediction bias and better balance between temperature and mixture influences than do their regression-based counterparts (Ceylan et al. 2009).

Because of the above and since the dynamic modulus |E*| has been identified as the main hot mix asphalt property in the *Mechanistic Empirical Pavement Design Guide* (MEPDG), it is necessary to calibrate the models to improve their accuracy. Michigan, in preparation for the implementation of the MEPDG, recommends that the designer should use either the locally calibrated modified Witczak model or the ANN model to predict |E*|. Both of these models are available in the DYNAMOD software developed. The ANN model, in general, is more accurate than the locally calibrated modified Witczak model (Kutay and Jamrah 2013).

7.6 PERMANENT DEFORMATION OF ASPHALTS

In order to determine the permanent deformation behaviour of asphalt, the low-stiffness response of the asphalt, that is, its response at high temperature and long loading time, must be analysed. Under these conditions, the stiffness of the mixture depends not only on the bitumen's hardness and the volume of the aggregate and the bitumen in the mix but also on the aggregate grading, aggregate shape and texture, degree of interlocking and degree of compaction.

The simplest test used to study the deformation behaviour of asphalts was the static unconfined uniaxial compression test, termed the creep test, developed in the 1970s by Shell Bitumen (Hill 1973). The specimen was subjected to static axial compressive load over a long period (1 h). The test procedure was very simple and required low-cost equipment. In addition, Shell Bitumen developed a rut prediction procedure based on results of the creep test but soon realised that it underestimated rut depths measured in trial pavements (Hill et al. 1974). This was attributed to the effects of dynamic loading producing higher deformation in the wheel-tracking test (Van de Loo 1974).

The inability of the static creep test to simulate the real loading conditions on site and to reflect better the improved performance of binder modifiers led to the establishment of the dynamic (cyclic) loading test at the beginning of the 1980s (Finn et al. 1983; Valkering et al. 1990).

Although the repeated load axial test is a potential performance test to assess the deformation resistance of asphalt (BSI 1996), various researchers (Monismith and Taybali 1988; Nunn et al. 1999) proved that the application of confining (lateral) pressure had a significantly positive effect on asphalt performance when it was tested under axial loading.

Further studies have demonstrated that although an unconfined axial loading test can discriminate between asphalt mixtures of the same composition utilising different binders and binder contents, it cannot discriminate between mixtures with different aggregate gradations and assess the effects of aggregate gradation and shape on the resistance to permanent deformation. Additionally, the confined axial loading test (VRLA-vacuum) measurements correlated well with those from the BS wheel-tracking test (Nunn et al. 2000).

As a consequence, nowadays, the creep test is carried out under dynamic (repetitive) loading using confining latter pressure. The latter pressure is applied directly by introducing a vacuum pressure or indirectly by using a load area smaller than the diameter of the specimen. The tests are known as cyclic compression tests with confinement. Needless to say, the loading/unloading periods are much longer than those used for defining the elastic stiffness modulus.

The use of the static creep test without confining lateral pressure, which was the first to be introduced, is currently almost obsolete.

Better simulation in the laboratory of the stress conditions that exist in a pavement related to rut formation provides the loading of the specimen by a moving wheel. Hence, the wheel-tracking test was developed to determine the performance of the asphalt in permanent deformation, particularly its susceptibility to deform under load.

The wheel-tracking test and the cyclic compression tests, apart from determining the resistance to permanent deformation of a bituminous mixture, also make it possible to rank various mixes or to check on the acceptability of a given mix.

Please be reminded that the permanent deformation behaviour of the bituminous mixtures may also be assessed by the permanent shear strain determined by the repeated shear test at constant height (see Section 7.4.9).

7.6.1 Behaviour of bituminous mixture under creep (static) and cyclic compressive loading

In the case of original creep (static) loading, one cycle of loading is typically 1 h. In the cyclic compression tests with pulse loading, the load cycle may be 1 or 2 s, depending on the test employed, while with haversinusoidal loading, the frequency typically ranges from 1 to 5 Hz. Please be reminded that one load cycle includes loading and rest period.

During loading time (t), the bituminous mixture deforms; the deformation (δl) and, thus, the strain (ε) increase rapidly at the beginning and then become quasi-constant. After removal of applied stress (rest period), part of the total deformation is recovered instantaneously (the elastic deformation), another part of deformation is recovered gradually and is time dependent (viscoelastic deformation) and another part of the deformation cannot be recovered (viscous or plastic or permanent deformation). Figure 7.15 explains the above, in terms of strain, which is known as creep behaviour. As it can be seen, the above behaviour is similar to the viscoelastic behaviour of the bitumen (see Section 4.21.1).

In a repetitive loading (cycle loading), the sum of the infinitesimal unrecoverable deformations is the cause of pavement deformation, which appears in the form of rutting.

A representation of cyclic loading (almost square pulse loading) is shown in Figure 7.16.

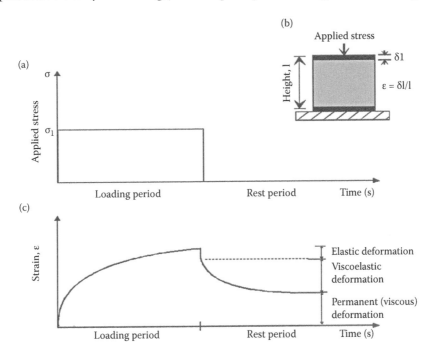

Figure 7.15 Schematic representation of bituminous mixture creep behaviour under single loading cycle. (a) Schematic representation of stress application. (b) Specimen. (c) Deformation curve during loading/unloading.

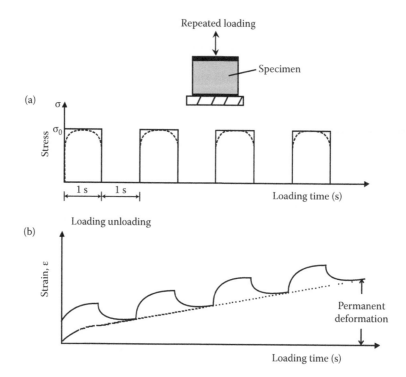

Figure 7.16 Schematic representation of the development of permanent deformation under cyclic block-pulse loading. (a) Schematic representation of cyclic block-pulse load. (b) Change in permanent deformation.

If the stiffness modulus ($S_{m,creep}$) is to be determined, the applied stress and the total permanent deformation are taken into account.

The assessment of the asphalt performance in resistance to permanent deformation is determined with reference to either creep rate and creep modulus or creep rate and cumulative strain after n load applications, or permanent shear strain, depending on the test employed.

According to EN test methods, the total number of pulses applied is 3600 (approximately 2 h), for the uniaxial cyclic compression test with confinement, or 10,000 pulses (approximately 6 h), for the triaxial cyclic compression test (vacuum induced).

According to the US test method, the total number of pulses is 5000 (approximately 90 min) or 10,000 (approximately 180 min) for the complex shear modulus and permanent shear strain test.

The total duration of loading in case the uniaxial static (creep) test is employed is 1 h.

7.6.2 Cyclic compression tests

According to CEN EN 12697-25 (2005), two test methods are described to determine the resistance to permanent deformation of a bituminous mixture by cyclic compression with confinement. The tests make it possible to rank various mixes or to check on the acceptability of a given mix. They do not allow raking a quantitative prediction of rutting in the field to be made.

Test *method A* determines the creep characteristics of bituminous mixtures by means of a uniaxial cyclic compression test with some confinement present. To achieve a certain confinement, the diameter of the loading platen is smaller than that of the sample. In this test, a cylindrical specimen is subjected to a cyclic axial stress. The type of confinement is suitable to predict realistic rutting behaviour of gap-graded mixtures, especially with a large stone fraction.

Test *method B* determines the creep characteristics of bituminous mixtures by means of the triaxial cyclic compression test. In this test, the confining stress is induced by vacuum and the cylindrical specimen is subjected to a cyclic axial stress. This test is most often used for the purpose of evaluation and development of new types of mixtures.

Both test methods area carried out at elevated temperatures of ≥40°C (typically 40°C, 45°C or 60°C).

According to the American practice, the resistance to permanent deformation, or resistance to rutting as it is called, of the bituminous mixtures is determined with the recently introduced flow number test. The test is a uniaxial repeated (dynamic or cyclic) compression test executed with a test device specially developed for Superpave mix design procedure. The test device is known as AMPT, which is also capable of determining the dynamic modulus of the mixture. The test is carried out at an elevated temperature determined from the average 7-day maximum pavement temperature 20 mm below the surface (see Section 7.6.2.3). The test is carried out in accordance to AASHTO TP 79 (2013).

Additionally, the performance of a bituminous mixture to rut can be evaluated by the indirect tensile strength test at high temperatures. The test, known as high-temperature indirect tensile test (HT-IDT), is performed as described in AASHTO T 283 (2011) for unconditioned (dry) specimens, but at test temperature 10°C below the average 7-day maximum pavement temperature 20 mm below the surface (see Section 7.6.2.4).

7.6.2.1 *Uniaxial cyclic compression test with confinement – method A*

In the uniaxial cyclic compression test, the cylindrical specimen is subjected to a cyclic axial block-pulse load (pressure) (see Figure 7.12a). There is no additional lateral confinement pressure applied. The confinement pressure is induced indirectly by using a smaller diameter loading plate (100 mm) than the diameter of the specimen (150 mm) (see Figure 7.17).

The specimens 60 ± 2 mm in height are produced in the laboratory by using a gyratory or an impact compactor (CEN EN 12697-31 2007 or CEN EN 12697-30 2012) or cored from the road or laboratory-compacted slab (CEN EN 12697-27 2000 or CEN EN 12697-33 2007).

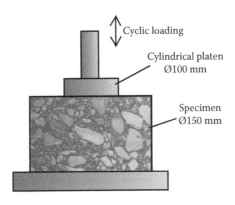

Figure 7.17 Schematic representation of uniaxial cyclic compression test with confinement.

Specimens are tested not before 2 days after compaction and all specimens are stored at a temperature between 5°C and 25°C. The minimum number of specimens for each bituminous mixture tested is five.

The typical test temperature is 40°C, and before testing, the specimens are left in a chamber for at least 4 h (up to 7 h maximum) to be stabilised at the test temperature.

Then, the specimens are positioned well centred coaxially with the test axis between the two platens. A constant pre-load of $(72 \pm 7) \times 10^{-3}$ kN is applied for 10 min, after which the test load of 100 kPa is applied for 3600 load applications. The loading time is 1 s and the rest time is also 1 s; hence, the duration of the test is 2 h.

During the test, regular measurements of the total permanent deformation are made. As a minimum requirement, readings are taken after the following loading applications: 2, 4, 6, 8, 10, 20, 40, 60, 80, 100, 200, 300, 400, ..., 3600.

The cumulative axial strain at the corresponding loading application is plotted in a diagram, to obtain the creep curve. A typical creep curve diagram is shown in Figure 7.18.

Permanent deformation in terms of cumulative axial strain (ε_n) and the characteristic creep properties creep rate (f_c) and creep stiffness $E_{n\text{-}c}$ are determined using the following equations.

Particularly, after n loading applications, the cumulative axial strain is determined from the following equation:

$$\varepsilon_n = 100 \times [(h_o - h_n)/h_o,$$

where ε_n is the cumulative axial strain after n load applications (%), h_o is the average height as measured by displacement transducers after pre-load of the specimen (mm), h_n is the average height as measured by displacement transducers after n load applications of the specimen (mm) and n is the number of load applications.

The cumulative axial strain is calculated at 3600 load applications.

The creep rate, f_c, and the creep modulus, $E_{n\text{-}c}$, are calculated from the following equations:

$$f_c = (\varepsilon_{n1} - \varepsilon_{n2})/(n_1 - n_2),$$

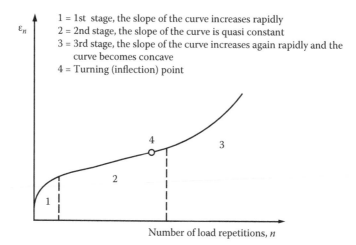

1 = 1st stage, the slope of the curve increases rapidly
2 = 2nd stage, the slope of the curve is quasi constant
3 = 3rd stage, the slope of the curve increases again rapidly and the curve becomes concave
4 = Turning (inflection) point

Number of load repetitions, n

Figure 7.18 Schematic representation of a typical creep curve.

where f_c is the creep rate (microstrain/load pulse); ε_{n1}, ε_{n2} are the cumulative axial strain after n_1, n_2 load applications; and

$$E_{n-c} = (\sigma/\varepsilon_n) \times 1000,$$

where E_{n-c} is the creep modulus after n load applications (MPa), σ is the applied stress (kPa) and ε_n is the cumulative strain after n load applications (%).

If the permanent deformation exceeds 40,000 microstrain and the drawn graph shows that the inflection point has been passed, then the extrapolated inclination line with the least slope gives the creep rate.

7.6.2.2 Triaxal cyclic compression test – method B

In the triaxial cyclic compression test, the cylindrical specimen is subjected to a confining pressure, σ_c, on which a cyclic axial pressure is superposed.

The cyclic axial pressure can be hanersinusoidal pressure, $\sigma_a(t)$, with amplitude σ_v and frequency between 1 and 5 Hz; typically, 3 Hz is used (recommended by CEN EN 13108-20 2008). It can also be block-pulse pressure, with height σ_B, at a cycle of 2 s (T_1 loading and T_0 resting) (see Figure 7.19).

The confining pressure may be either dynamic or static. However, the static type is typically used. In both cases, before starting the test, a small pre-loading axial stress, not exceeding $0.02[2\sigma_v + \sigma_c]$ in the case of haversinusoidal loading and $0.02[\sigma_v + \sigma_c]$ in the case of block-pulse loading, is applied.

Once the specimen is enclosed in an elastic membrane for protection, the confining pressure is applied using three alternative ways: (a) through a rubber socket (or foil) using water, oil or air; (b) through a pressure ring, which is placed around the specimen; or (c) with partial vacuum applied between the protective rubber membrane and the specimen. The confining pressure is in the 50 to 200 kPa range. The arrangement in which the confining stress is applied by partial vacuum is as shown in Figure 7.20.

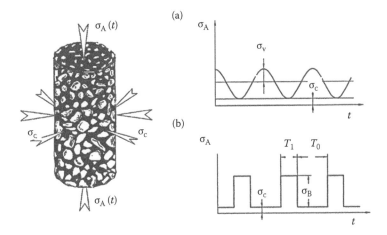

Figure 7.19 Schematic representation of pressures exerted on a cylindrical specimen in the case of (a) haversinusoidal cyclic loading, or (b) block-pulse cyclic loading. (Adapted from CEN EN 12697-25, *Bituminous Mixtures – Test Methods for Hot Mix Asphalt – Part 25: Cyclic Compression Test*, Brussels: CEN, 2005. With permission [© CEN].)

Figure 7.20 Triaxial cyclic compression test device making use of a partial vacuum as confining pressure.

The cylindrical specimens should have a minimum diameter of 50 mm and a minimum height of 50 mm, if the nominal maximum aggregate size is ≤16 mm. Alternatively, the minimum diameter should be 75 mm and the minimum height should be 75 mm, if the nominal maximum aggregate size is >16 mm. It is recommended to use a height-to-diameter ratio of 0.6, if the nominal aggregate size is ≤16 mm, and 0.8, if the nominal aggregate size is >16 mm. At least two specimens are required to be tested per type of bituminous mixture (CEN EN 12697-25 2005).

The specimens are produced in the laboratory by using a gyratory or an impact compactor (CEN EN 12697-31 2007 or CEN EN 12697-30 2012) or cored from the road or laboratory-compacted slab (CEN EN 12697-27 2000 or CEN EN 12697-33 2007). Specimens are tested not before 2 days after compaction and all specimens are stored at a temperature between 5°C and 25°C.

The test is conducted at 30°C to 50°C; typically, a temperature of 50°C is used for asphalts for surface courses and a temperature of 40°C is used for asphalts for binder and base courses.

After applying the confining pressure and the pre-load, the axial pressure is applied. The amplitude of pressure, σ_v, for haversinusoidal loading, ranges from 100 to 300 kPa. The specification CEN EN 12697-25 (2005) suggests that the amplitude of the haversinusoidal load should be taken two or three times larger than the confining stress.

For block-pulse loading, pressures σ_B of 100 kPa up to 700 kPa are used. The axial stress is usually 300 kPa for asphalts for surface courses and 200 kPa for asphalts for binder and base courses.

During the test, the permanent axial deformation is recorded on a regular basis, namely, every 10th (loading/unloading) cycle up to 1000 loading cycles and then every 500th loading cycle. The total number of loading/unloading cycles is 10,000; hence, the loading duration is approximately 5.6 h.

The test may stop earlier in case the deformation is too large (which possibly appears during the last stages of the test) and there is a risk of damaging the equipment. In this case, the strain should be at least 6%.

The cumulative strain, ε_n, is calculated from the measurements taken after n cycles, and in this particular case, $n = 10{,}000$. The cumulative strain is calculated using the same equation as in method A (see Section 7.6.2.1).

The bituminous mixture's resistance to permanent deformation is assessed in terms of creep rate, f_c (Method 1, Section 7.6.2.2.1), or in terms of parameters B and $\varepsilon_{1000,\text{calc}}$ (Method 2, Section 7.6.2.2.2).

7.6.2.2.1 Method 1

The creep rate, f_c, is determined by the creep curve. The slope of the creep curve is determined by the least-squares fit of the (quasi) linear part of the creep curve (stage 2 in Figure 7.14). The linear equation is of the following form:

$$\varepsilon_n = A_1 + B_1 \times n,$$

where ε_n is the cumulative axial strain after n cycles and n is the number of (loading/unloading) cycles.

Then, the creep rate, f_c, is calculated from the following equation:

$$f_c = B_1 \times 10^4 \ (\text{microstrain/loading cycle}),$$

where B_1 is the slope of the linear part of the curve.

This method is simple but its disadvantage is that it is only a poor representation of the deformation curve. Furthermore, the slope f_c depends highly on the selected interval used for curve fitting, because there is generally no part with real constant slope in the creep curve.

7.6.2.2.2 Method 2

The constant B and the cumulative axial strain, $\varepsilon_{1000,\text{calc}}$, are determined by the creep curve.

The constant B is determined from the least-squares power fit of the (quasi) linear part of the creep curve, using the equation

$$\varepsilon_n = A \times n^B,$$

or equivalent to the least-squares linear fit of the ($\log \varepsilon_n - \log n$) values, using the equation

$$\log \varepsilon_n = \log A + B \times \log n,$$

where ε_n is the cumulative axial strain after n load applications (%) and B is the power of the least-squares power fit or the slope from the least-squares linear fit of the $\log \varepsilon_n$ versus $\log n$ values.

The cumulative strain $\varepsilon_{1000,\text{calc}}$, in per cent (%), is calculated after 1000 loading/unloading cycles using the following equation:

$$\varepsilon_{1000,\text{calc}} = A \times 1000^B,$$

where A and B are constants from the above equations.

This method may seem more complicated but has the advantage that, in many cases, a clear linear part is observed in the creep curve.

The particular case where the creep curve goes from stage 1 to stage 3 (see Figure 7.18) indicates a bituminous mixture being very sensitive to permanent deformation for the given test conditions. As a consequence, the mixture is rejected and is considered unsuitable for use.

Figure 7.21 presents creep results from the triaxial cyclic compression test, using asphalt concrete mixtures with nominal maximum size aggregate 20 mm (AC 20 mm) with varied

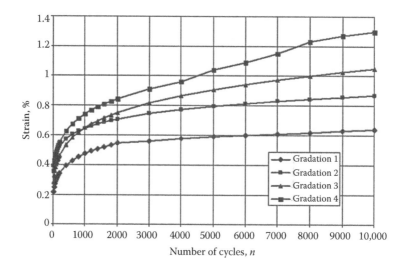

Figure 7.21 Typical creep curves under triaxial cyclic compression.

Table 7.4 Creep results under triaxial cyclic compression

	Method 1		Method 2	
Gradation	f_c (microstrain/cycle)	R^2	B (constant)	$\varepsilon_{1000,calc}$ (%)
1	0.28	0.99	0.107	0.51
2	0.45	0.96	0.129	0.65
3	0.82	0.97	0.209	0.65
4	1.22	0.99	0.254	0.71

Note: R^2 = correlation coefficient of linear regression.

aggregate gradations. The test was conducted at 40°C, with 50 kPa confining pressure and 200 kPa axial pressure.

The relevant creep parameters obtained to determine AC 20 mm resistance to permanent deformation using both calculations are given in Table 7.4. The results were obtained from tests carried out at the Highway Engineering Laboratory of the Aristotle University of Thessaloniki.

7.6.2.3 Flow number test using the AMPT

In the flow number test using the asphalt mixture performance analyser (AMPT) (see Figure 7.9), the cylindrical specimen is subjected to repeated axial compressive load cycles. The test procedure is carried out according to AASHTO TP 79 (2013).

The specimens, a minimum of three per type of mix, are 150 mm high and 100 mm in diameter and are obtained by coring gyratory-compacted samples.

Each load cycle consists of a pulse load applied for 0.1 s, followed by a rest period of 0.9 s.

The test is typically conducted at the average 7-day maximum pavement temperature 20 mm below the surface for a given location, at 50% reliability as determined by the LTPPBind (LTPP Products Online 2013) software.

The test is an unconfined test but there is continuing research to evaluate whether unconfined or confined testing simulates the field conditions better.

The permanent axial strain is measured during load cycles and the strain rate is calculated (microstrain/cycle). Initially, the sample experiences high strain, as the material is seated at the beginning of the test. Then, the strain rate settles down to a constant value. Later, as the material becomes unstable, the strain rate increases again. The flow number is defined as the number of load cycles corresponding to the minimum permanent strain rate. A higher flow number indicates a longer time for the material to become unstable and therefore a more rut-resistant bituminous mixture.

The average value of three tests is taken as the representative flow number for the mix tested.

The flow number is also used to determine the rut resistance performance of the pavement with the use of AASHTOWare Pavement ME Design Pavement Design and Analysis Software.

More information about the test procedure is given in AASHTO TP 79 (2013).

7.6.2.4 Indirect tensile strength at high temperatures

The indirect tensile strength test at high temperatures (HT-IDT) has been found to correlate well with rut depth (Transportation Research Circular C-C068 2004) and can be used as an alternative mix design to evaluate the rut resistance of hot mix bituminous mixture (NCHRP Report 673 2011).

The HT-IDT is carried out according to AASHTO T 283 (2011) (ASTM D 6931 2012) for unconditioned, that is, dry, specimens, but at a different test temperature. The test temperature is 10°C lower than the average 7-day maximum pavement temperature 20 mm below the surface for a given location, at 50% reliability as determined by the LTPPBind (LTPP Products Online 2013) software.

Unlike AASHTO T 283 (2011), the specimens in this procedure are compacted by a gyratory compactor to produce an air void content of approximately 4%.

The specimens are conditioned at the test temperature for at least 2 h, and if a water bath is used, they should be tightly sealed in plastic bags to prevent them from getting wet.

7.6.3 Static creep test

Although the static creep test is not included in ENs, some countries still use it.

The cylindrical specimen is subjected to a static constant coaxial stress for a certain period, at a certain testing temperature. Testing conditions have been established in an international conference (Colloquium 77 1977), after the basic research carried out by Van der Loo (1974). The established testing conditions are as follows: applied constant stress (σ_o), 0.1 MPa; test duration, 1 h; temperature, 40°C. Before applying the test stress to the specimen, a pre-loading stress is applied for 2 min (0.01 MPa) in order to ensure good contact between the loading plate and the specimen surface.

The specimen should have parallel and fairly smooth surfaces, so that the friction between the metal plates and the specimen is eliminated. The latter is ensured by coating the specimen surfaces with graphite grease. The deformation in the specimen is measured at regular intervals, typically after 5, 10, 20, 40, 80, 120, 300, 600, 2400 and 3600 s.

The apparatus used is much simpler than the apparatus used in the cyclic compression test (Cabrera and Nikolaides 1987; Jongeneel et al. 1985). It is mentioned that the CRT-UTM-NU (Cooper Research Technology Ltd. 2014) apparatus is also capable of performing the static creep test.

The creep curves obtained are similar to the ones shown in Figure 7.18.

The resistance to permanent deformation of the bituminous mixture is assessed in terms of creep stiffness modulus, S_{creep}, determined using the following equation:

$$S_{creep} = \sigma_o/\varepsilon_{3600},$$

where σ_o is the applied constant stress (0.1 MPa) and ε_{3600} is the cumulative strain after 3600 s.

Cabrera and Nikolaides (1988) proposed to also use the slope coefficient b, resulting from the linear logarithmic relationship of log S_{creep} with S_{bit} shown below:

$$\log S_{creep} = \log a + b \times \log S_{bit},$$

where S_{bit} is the bitumen stiffness modulus (N/m^2) from the Van der Poel (1954) nomograph, at loading times of 5, 10, 20, ..., 3200 s and at 40°C to SP (SP = bitumen's softening point); and S_{creep} is the creep stiffness modulus (N/m^2) of the bituminous mixture at loading times of 5, 10, 20, ..., 3600 s.

In mix design calculations, the mix with the lowest value of slope coefficient, b, and the highest creep stiffness modulus, S_{creep}, is more resistant to permanent deformation and should be selected as the target mix.

By conducting the static creep test, it is possible to estimate the oncoming permanent deformation (rutting) of the asphalt layers after n years in service.

The total permanent deformation, rutting, is the sum of the permanent deformation of each bituminous slayer. The permanent deformation of each layer can be determined by using the mathematical equation developed by Hill et al. (1974) shown below:

$$\delta h_i = C_m \times h_i \times \sigma_{av}/S_{creep},$$

where δh_i is the rut depth of bituminous layer i (mm); C_m is the static and dynamic loading correlation coefficient, varying from 1.0 to 2.0, depending on the bituminous mixture (for asphalt concrete, it varies from 1.2 to 1.6); h_i is the thickness of the bituminous layer (mm); σ_{av} is the average pavement stress (N/m^2) relative to the axial loading and stress distribution, for a typical standard axle, $\sigma_{av} = Z \times 6 \times 10^5$ (where Z is the correlation coefficient between strains developed in the pavement and during the creep test [obtained from tables]); and S_{creep} is the static stiffness modulus (N/m^2) at a certain bitumen viscosity, obtained from creep line by plotting S_{creep} and S_{bit} in a double logarithmic scale. The value of viscosity is relative to the traffic speed and the total number of equivalent standard axles.

For more information, see Shell's pavement design manual (Shell International Petroleum Co. 1978).

7.6.4 Permanent shear strain and complex shear modulus test using the SST

The permanent shear strain and complex shear modulus test, procedure B (at constant height), using the SST can be used to assess the performance of the bituminous mixture in wheel rutting, by using the shear strain parameter.

The test is carried out in accordance to ASTM D 7312 (2010) and AASHTO T 320 (2011) and is outlined in Section 7.4.9.

However, the test is a complicated, expensive piece of equipment and procedure B can be difficult to run. Therefore, it is not recommended that commercial laboratories, hot mix producers and similar organisations purchase the SST devices for use in routine mix design work. Either the asphalt mix pavement analyser (AMPT) or the device for indirect tensile strength is far better suited for routine use in hot mix asphalt mix design (NCHRP Report 673 2011).

7.6.5 Wheel-tracking tests by European norms

The wheel-tracking test was developed for the determination of asphalt susceptibility to deformation under moving load simulating conditions that exist on the road. The tests are all described in CEN EN 12697-22 (2007).

The test is applicable to asphalts with upper sieve size D ≤ 32 mm. Specimens are produced in the laboratory or cut from the pavement and held in a mould with their upper surface flush with the edge of the mould.

The susceptibility of the bituminous mixtures to deform is assessed by the rut formed by repeated passes of a loaded wheel at constant temperature.

According to CEN EN 12697-22 (2007), there are three alternative types of device to be used: the large-size device, the extra-large-size device and the small-size device.

With the large-size device and the extra-large-size device, the specimens are conditioned in air during testing, while with the small-size device, the specimens are conditioned in air or in water, during testing. It is noted that the large-size and the extra-large-size devices are not suitable for use with cylindrical cores.

Apart from the above alternative testing procedures, if wheel-tracking test is executed by a small-size device, two alternative procedures may be followed.

To summarise, there are five different testing methods used:

a. A large-size device, testing in air
b. An extra-large-size device, testing in air
c. A small-size device, procedure A, testing in air
d. A small-size device, procedure B, testing in air
e. A small-size device, procedure B, testing in water

The minimum number of specimens to be tested and their dimensions depend on the device used, which are shown in Table 7.5.

The specimens produced in the laboratory are compacted with a dynamic compaction apparatus, such as a roller compactor (perhaps the most commonly used) or a vibrating

Table 7.5 Minimum number and sizes of specimens

Device	Specimen		
	Number	*Length/width*	*Thickness*
Large	2	500 × 180 mm	50 mm for ≤50 mm[a,b]
			100 mm for >50 mm
Extra large	2	700 × 500 mm	60 mm[b]
Small, Proc. A, in air	6	300 × 300 mm or	25 mm for D < 8 mm
Small, Proc. B, in air	2	Ø200 mm	40 mm for D ≥ 8 to <16 mm
Small, Proc. B, in water	2		60 mm for D ≥ 16 to ≤22 mm
			80 mm for D > 22 to ≤32 mm

Note: D, upper sieve size.

[a] Layer thickness to be laid.
[b] If the thickness of the layer is to be specified, the thickness of the specimen shall be 30, 50, 60, 75 or 100 mm and generally 2.5 times the upper sieve size of the mix.

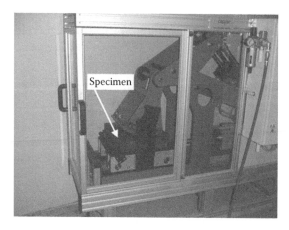

Figure 7.22 The roller compactor at the AUTh Highway Engineering Laboratory.

compactor, according to CEN EN 12697-33 (2007) or CEN EN 12697-32 (2007), respectively. Figure 7.22 shows a roller compactor device.

The density of specimens produced in the laboratory or derived from a compacted layer is determined by one of the methods described in CEN EN 12697-6 (2012) or even by gamma rays, in accordance with CEN EN 12697-7 (2002). Air voids are also determined and air voids in specimens to be tested should be almost the same.

In all cases, the specimens produced in the laboratory should be kept at a temperature not higher than 25°C for at least 2 days before testing.

7.6.5.1 Testing with the use of a large-size device, in air

This test procedure uses a pneumatic tyre having a track width of 80 mm and a tyre pressure of 600 kPa.

Before the test, the specimens are subjected to a conditioning run for 1000 loading cycles at a temperature between 15°C and 25°C. It is highlighted that one loading cycle is two passes (outward and return) of the loaded wheel.

After the conditioning run, the specimens are conditioned for 12–16 h at the selected test temperature applying 600 kPa wheel pressure, and afterwards, the 5 kN testing rolling load is applied for 30,000 cyclic loadings with 1 Hz (1 s) frequency, along the centre line of the specimen.

The rut depth is measured at 15 predetermined locations. The test is completed after the required number of cycles is reached (30,000) or if the mean rut depth measured at the 15 locations exceeds 18 mm.

The test temperature is usually 60°C or 50°C. The test temperature is measured both in the ventilated enclosure and within the specimen. The temperature within the specimen should be maintained at ±2°C from the selected test temperature.

The oncoming deformation during loading is measured in 15 predetermined locations after 1000, 2000, 3000 10,000 and 30,000 load cycles (N). From the 15 measurements obtained after n number of loadings, the proportional rut depth is determined (P_i). The proportional rut depth is determined by using the following equation:

$$P_i = 100 \times \sum_{j=1}^{15} \frac{(m_{ij} - m_{0j})}{(15 \times h)},$$

where P_i is the measured proportional rut depth (%), m_{ij} is the local deformation (mm), m_{0j} is the initial measurement at j location (mm) and h is the specimen thickness (mm).

Using the above equation, the proportional rut depth can be determined at all loadings. By plotting a graph $\ln(P_i)$ against $\ln(N)$, the best linear fit of the values can be drawn from which the proportional rut depth (P_i) at any number of loading cycle can be determined.

The mean proportional rut depth of two or more specimens tested for the same number of loading cycles expresses the bituminous mixture proportional rut depth.

7.6.5.2 Testing with the use of an extra-large-size device, in air

This test procedure is similar to the previous one but the specimen dimensions and the pneumatic tyre track width are different and no pre-loading of the specimen is required. Additionally, the conditional number of runs is greater.

The size of the specimens is as shown in Table 7.4. The pneumatic tyre track width is 110 mm; the tyre pressure is 600 kPa, same as in the large device procedure.

The specimens are conditioned for 14–18 h at the selected test temperature before testing. Then, the testing rolling load of 10 kN is applied for 30,000 load cycles, with a travel time (frequency) of 2.5 s.

The rut depth is measured by laser sensors at three cross sections (one at the central axis of the specimen perpendicular to the motion and two at a distance +150 mm and –150 mm from it). The test is completed after the required number of cycles is reached (30,000) or if the mean rut depth measured at the three cross sections exceeds 20 mm.

The test temperature is usually 60°C or 50°C and is recorded as in the previous test procedure.

The oncoming deformation during loading is measured after 100, 200, 500, 1000, 2000, 3000, 4000, 6000, 8000, 10,000, 12,000, 14,000 and 30,000 load cycles.

The proportional rut depth, P_i, is determined by the rut depth measurements received from the three cross sections using the following equation:

$$P_i = 100 \times \frac{(m_1 + ... + m_n)}{n \times h},$$

where P_i is the measured proportional rut depth (%), m_n is the measured rut depth in the measured cross sections (mm), n is the number of measured cross sections and h is the specimen thickness (mm).

The mean proportional rut depth of two or more specimens tested for the same number of loading cycles expresses the bituminous mixture proportional rut depth.

7.6.5.3 Testing with the use of a small-size device, procedures A and B, in air

In these test procedures, the size of the specimen is smaller (see Table 7.3) and loading conditions are different from those in the large device and extra large device test procedures. Additionally, in contrast to the previous two devices, the applied load is in a fixed position and the table carrying the specimen's mould is moving backwards and forwards. A typical small-size device used for procedure A and procedure B is shown in Figure 7.23.

The loading wheel carries a treadless tyre made of solid rubber, having a rectangular cross profile with a width of 50 mm.

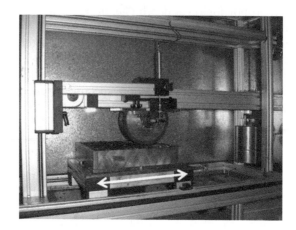

Figure 7.23 Wheel-tracking small-size device at the AUTh Highway Engineering Laboratory.

Before testing, the specimens are conditioned at the selected test temperature for at least 4 h, if their thickness is ≤60 mm, or at least 6 h, if their thickness is >60 mm, and not more than 24 h.

The magnitude of the applied load and the loading frequency is the same for both Procedures A (Section 7.6.5.3.1) and B (Section 7.6.5.3.2). In particular, the applied load is 700 N and the loading frequency is 26.5 load cycles per 60 s.

7.6.5.3.1 Procedure A

Before the start of recording of the deformations, five conditioning runs are applied to the specimen using the same test load, 700 N.

After conditioning runs, tracking is continued for 1000 load cycles or until a rut depth of 15 mm is reached, whichever is shorter.

The test temperature is usually 45°C or 60°C. The temperature is recorded, as in the previous test procedures, within the specimen and in the ventilated enclosure at a point close to the specimen. The temperature within the specimen should be maintained at ±1°C from the selected test temperature.

The oncoming vertical displacement during loading is measured in reference to the initial position of the loaded wheel by an automatic displacement measuring device, or dial gauge, set within 10 mm from the centre point of the loaded area at the midpoint of traverse. Loading is applied by the use of a weighted cantilever arm or by a dead load. Readings of vertical displacement are taken every 25 load cycles.

The parameters used to assess the resistance to permanent deformation of the bituminous mixture, using procedure A, are the rut depth and the wheel-tracking rate.

The *rut depth*, RD, is the mean value of the vertical displacement after 1000 load cycles determined from the number of samples tested (at least six), expressed in millimetres.

The *wheel-tracking rate*, WTR, is calculated from the following equation:

$$\text{WTR} = 10.4 \times \text{TR}_{\text{m}} \times (w/L),$$

where WTR is the wheel-tracking rate (μm/cycle); TR_{m} is the mean value of tracking rate, TR, of specimens tested (at least six) (μm/cycle) ($\text{TR} = 3 \times r_n + r_{n-1} + r_{n-2} - 3 \times r_{n-3}$, where n is the total number of readings taken at 100 load cycle intervals up to 1000 load cycles, excluding the initial reading, and r_i is the change in vertical displacement from the initial value, r_0, to the relevant reading [mm]. If less than eight readings are taken, the relevant equations to calculate TR are given in CEN EN 12697-22 2007.); w is the width of the tyre applying the load (mm) and L is the applied load (N).

7.6.5.3.2 Procedure B

Before the start of the recording of the deformations, as in procedure A, five conditioning runs are applied to the specimen using the same test load, 700 N.

After conditioning runs, tracking is continued for 10,000 load cycles or until a rut depth of 20 mm is reached, whichever is shorter.

The test temperature is usually 45°C, 50°C or 60°C. The temperature is recorded, as in the previous test procedures, within the specimen and in the ventilated enclosure at a point close to the specimen. The temperature within the specimen should be maintained at ±1°C from the selected test temperature.

The oncoming vertical displacement during loading is measured initially, six or seven times in the first hour, and at least one reading every 500 cycles thereafter. The vertical position of the wheel is defined as the mean value of the profile of the specimen on a length of ±50 mm about the centre of the loading area at the midpoint of traverse, measured in at least 25 points approximately equally spaced. The loading is applied, as in procedure A, by the use of a weighted cantilever arm or by a dead load.

The parameters used to assess the resistance to permanent deformation of the bituminous mixture, using procedure B, are the wheel-tracking slope, the proportional rut depth and the rut depth.

The *wheel-tracking slope*, WTS_{AIR}, is the average wheel-tracing slope of the specimens tested (at least two). The wheel-tracking slope of each specimen is calculated using the following equation:

$$WTS_{AIR} = (d_{10,000} - d_{5000})/5,$$

where WTS_{AIR} is the wheel-tracking slope (mm per 10^3 load cycles), $d_{10,000}$ is the rut depth after 10,000 load cycles (mm) and d_{5000} is the rut depth after 5000 load cycles (mm).

The *proportional rut depth*, PRD_{AIR}, is the mean value of the proportional rut depth of the specimens tested (at least two). The proportional rut depth of each specimen is determined by dividing the rut depth (vertical displacement), in millimetres, after n cycles (= 10,000) by the initial specimen height, in millimetres. The result is expressed in percentage (%) to one decimal point.

The *rut depth*, RD_{AIR}, is the mean value of the rut depth (vertical displacement) of the specimens tested (at least two) measured after 10,000 load cycles, in millimetres, to one decimal point.

The test devices used are operating on specially written software, and all the above wheel-tracking parameters are calculated automatically by the same software.

7.6.5.4 Testing with the use of a small-size device, procedure B, in water

This test procedure is exactly the same as procedure B in air, apart from the fact that the specimens are tested in water.

The specimens are placed in water at the specified temperature until the temperature equilibrium is reached within the specimen; this takes place in not less than 1 h.

In order to distinguish the wheel-tracking parameters wheel-tracking slope, proportional rut depth and rut depth from those obtained in air, they are designated as WTS_W, PRD_W and RD_W, respectively.

7.6.5.5 Wheel-tracking test by a small device, BS 598-110 (1996), in air

Before the development of CEN EN 12697-22 (2007), the wheel-tracking test was carried out, in some countries, in accordance with BS 598-110 (1996). The procedure specified was quite similar to procedure A, using a small-size device, but with some differences.

The differences between BS and EN procedures are identified to be the following: (a) magnitude of the applied load in the BS procedure, 520 N, against that in the EN procedure, 700 N; (b) loading frequency, 21 cycles/min, against 26.5 cycles/min; (c) loading time, 45 min, against 1000 loadings, (d) no pre-loading is required, 0 cycles against 5 cycles; and (e) the wheel-tracking rate is expressed in millimetres per hour, against microns per loading cycle.

The wheel-tracking test in the BS procedure had gained popularity because of the size and simplicity of the device required to be used and a lot of research has been carried out using this procedure.

7.6.6 Evaluation of asphalts based on rut resistance employing European wheel-tracking test procedures

The wheel-tracking parameters are used for the evaluation of conformity of the bituminous mixtures and for the selection of the target mix, during mix design, with respect to their rut resistance. Some countries have determined limiting requirements.

Tables 7.6 through 7.8 give the limiting wheel-tracking values recommended to be used in the United Kingdom to determine the target composition of asphalt concrete, hot rolled asphalt and SMA mixtures.

Table 7.9 gives the limiting values of wheel-tracking values used in France for the determination of the target composition of various bituminous mixtures.

A study was carried out to examine the effect of volumetric property changes, owing to bitumen content variation, and the degree of compaction on wheel-tracking performance, using CEN EN 12697-2, procedure B in air, and BS 598-110 (1996), testing procedures at two testing temperatures (45°C and 60°C); asphalt concrete 19 mm with 50/70 grade bitumen was used (Nikolaides and Manthos 2008). The results of the study showed that the BS procedure was not sensitive to volumetric variations based on statistical significance of

Table 7.6 Limiting wheel-tracking test requirements for dense asphalt concrete for base and binder courses

			Category WTS_{AIR}	Category PRD_{AIR}		
No.	Site description	Test temperature (°C)	Maximum wheel-tracking slope[a] (mm/1000 cycles)	Maximum proportional rut depth[a] (%)	Maximum rut rate[b] (mm/h)	Maximum rut depth[b] (mm)
1	Moderately to heavily stressed sites requiring high rut resistance	45	$WTS_{AIR\ 1.0}$	$PRD_{AIR\ 9.0}$	2.0	4.0
2	Very highly stressed sites requiring very high rut resistance	60	$WTS_{AIR\ 1.0}$	$PRD_{AIR\ NR}$	5.0	7.0
3	Other sites	N/A	$WTS_{AIR\ NR}$	$PRD_{AIR\ NR}$	—	—

Source: Extracts from BSI PD 6691, *Guidance on the use of BS EN 13108 Bituminous mixtures – Materials specifications*, London: BSI, 2010. With permission.

Note: NR, not required.

[a] Determined in accordance to CEN EN 12697-22 (2007), small device, procedure B.
[b] Determined in accordance to BS 598-110 (1996).

Table 7.7 Limiting wheel-tracking test requirements for hot rolled asphalt

| | | | Category WTR_{AIR} | | Category RD_{AIR} | |
| | | | | | | |
No.	Site description	Test temperature (°C)	Maximum wheel-tracking rate[a] (μm/cycle)	Maximum rut depth[a] (mm)	Maximum rut rate[b] (mm/h)	Maximum rut depth[b] (mm)
1	Moderately to heavily stressed sites requiring high rut resistance	45	$WTR_{AIR\,7.5}$	$RD_{AIR\,5.0}$	2.0	4.0
2	Very highly stressed sites requiring very high rut resistance	60	$WTR_{AIR\,15.0}$	$RD_{AIR\,7.0}$	5.0	7.0
3	Other sites	N/A	$WTR_{AIR\,NR}$	$RD_{AIR\,NR}$	—	—

Source: Extracts from BSI PD 6691, *Guidance on the use of BS EN 13108 Bituminous mixtures – Materials specifications*, London: BSI, 2010. With permission.

Note: NR, not required.

[a] Determined in accordance to CEN EN 12697-22 (2007), small device, procedure A.
[b] Determined in accordance to BS 598-110 (1996).

Table 7.8 Limiting wheel-tracking test requirements for SMA

No.	Site description	Test temperature (°C)	Maximum wheel-tracking slope[a] (mm/1000 cycles)	Maximum rut rate[b] (mm/h)	Maximum rut depth[b] (mm)
1	Moderately to heavily stressed sites requiring high rut resistance	45	1.0 ($WTS_{AIR\,1.0}$)	2.0	4.0
2	Very highly stressed sites requiring very high rut resistance	60	1.0 ($WTS_{AIR\,1.0}$)	5.0	7.0
3	Other sites	N/A	$WTS_{AIR\,NR}$	—	—

Source: Extracts from BSI PD 6691, *Guidance on the use of BS EN 13108 Bituminous mixtures – Materials specifications*, London: BSI, 2010. With permission.

Note: NR, not required.

[a] Determined in accordance to CEN EN 12697-22 (2007), small device, procedure B.
[b] Determined in accordance to BS 598-110 (1996).

differences in wheel-tracking rate and depth. Additionally, all mixtures tested were found to satisfy the requirements based on the BS testing procedure (see Table 7.5). On the contrary, the CEN EN procedure was sensitive to the parameters examined. Some mixtures possessed proportional rut depth above the highest category ($PRD_{9.0}$) set by CEN EN 13108-1 (2008), and in some occasions, the wheel-tracking slope was close to the highest wheel-tracking slope ($WTS_{1.0}$), also set by CEN EN 13108-1 (2008).

Results similar to the ones above were obtained when the effect of type of bitumen (50/70 grade and SBS [styrene–butadiene–styrene]-modified bitumen) was examined on wheel-tracking performance employing the abovementioned two testing procedures (Nikolaides and Manthos 2009).

In a study to compare the permanent deformation behaviour of asphalt concrete (AC 20 mm) when gradation and binder content were varied but within the tolerance limits using different testing procedures, it was found that the ranking of the mixtures was the same regardless of the procedure used, with the triaxial cyclic compression test included

Table 7.9 Limiting wheel-tracking test requirements by French design guide

Type of mix	Class	Number of cycles	Maximum proportional rut depth (%) (category PRD)[a]
AC	1	30,000	≤10% (P_{10})
	2		≤7.5% ($P_{7.5}$)
	3		≤5% (P_5)
AC-thin	1	3000	≤15% (P_{15})
	2	10,000	≤15% (P_{15})
	3	30,000	≤10% (P_{10})
AC-Airf	1	10,000	≤10% (P_{10})
	2		≤7.5% ($P_{7.5}$)
	3		≤5% (P_5)
AC-VTL10	1 and 2	3000	≤15% (P_{15})
AC-VTL6	1 and 2		≤20% (P_{20})
AC-GB	2 and 3	10,000	≤10% (P_{10})
	4	30,000	≤10% (P_{10})
EME	1 and 2	30,000	≤7.5% ($P_{7.5}$)

Source: Adapted from Delorme, J.-L. et al., *LPC Bituminous Mixtures Design Guide*, Laboratoire Central des Ponts et Chaussées, Paris: LCPC, 2007; SETRA, *Technical Guide: The Use of Standards for Hot Mixes*, Paris: SETRA, 2008.

Note: AC, asphalt concrete and high stiffness asphalt concrete for surface or binder course, AC10 and AC14, with thickness between 5 and 9 cm.; AC-Airf, asphalt concrete for airfields for surface and binder course, AC10 or AC14; AC-GB, asphalt for road base, AC20, with thickness between 8 and 16 cm.; AC-thin, asphalt concrete for surface or binder course, AC10 and AC14, with thickness between 3 and 5 cm.; AC-VTL6, asphalt concrete for very thin layers with D = 6 mm.; AC-VTL10, asphalt concrete for very thin layers with D = 10 mm.; EME, high stiffness modulus mixtures, AC10, with thickness between 7 and 13 cm, or AC14, with thickness between 9 and 15 cm.

[a] Determined in accordance to CEN EN 12697-22 (2007), large device.

(Nikolaides and Manthos 2014). Additionally, the superiority of the mixtures with aggregate gradation not passing through the restricted zone proposed by the Asphalt Institute (2001) was clearly shown. Furthermore, concerns have been expressed regarding the proposed limiting values by the BS PD 6691 (2010) design guide for asphalt concrete.

On the basis of all the above findings, the author proposes the abundance of BS wheel-tracking procedure and its replacement with CEN EN 12697-22 (2007), procedure B in air, for assessing the performance of asphalt concrete in permanent deformation. He also considers that the limits set by BSI PD 6691 (2010) for asphalt concrete are high and need to be revised. Additionally, he proposes that future European limiting wheel-tracking values for asphalts should differentiate not only among traffic volumes but also among site ambient temperatures (average maximum).

7.6.7 Wheel-tracking tests by American standards

The development and use of the wheel-tracking test are relatively recent in the United States. The wheel-tracking test procedures used in the United States for the determination of the resistance to permanent deformation, or susceptibility to rutting, of the bituminous mixtures are (a) the asphalt pavement analyser (APA) test and (b) the Hamburg wheel track (HWT) test.

The APA test is a relatively new test growing in popularity among US pavement agencies for evaluating rut resistance. The test procedure is specified by AASHTO T 340 (2010).

The HWT test is one of the first wheel-tracking tests, developed in Germany, and is used in the United States by various agencies. Today, it is not as widely used as the APA tester. The test procedure is specified by AASHTO T 324 (2011).

Figure 7.24 Specimens during the wheel-tracking test with the APA device. (Courtesy of Pavement Technology Inc.)

7.6.7.1 Wheel-tracking test with the use of the APA device

The wheel-tracking test using the APA for evaluating the rut resistance of bituminous mixtures is carried out in accordance to AASHTO T 340 (2010). The test is used as part of fundamental performance tests for Superpave mix design.

The device (APA), a variation of the Hamburg wheel-tracking device, provides the ability of testing simultaneously two or six cylindrical specimens 150 mm in diameter (see Figure 7.24). Alternatively, the device may test simultaneously three prismatic specimens (beams) 125 mm in width by 300 mm in length (see also Figure 7.24). The specimens' thickness in both cases are typically 75 mm.

The cylindrical specimens are compacted with a gyratory compactor for 4% ± 1% air voids (typical value). The prismatic specimens are compacted with a vibratory compactor, again for 4% ± 1%.

On the specimens, a load of 445 N (100 lb) is applied by a solid (or concave) steel wheel run atop a rubber pneumatic hose, pressurised with a 690 kPa (100 psi) pressure, positioned at the middle of the specimen surface. The above simulates the vehicle's pneumatic tyre.

Prior to testing, 50 seating cycles are applied to the specimen using the test load.

The test duration is 8000 load cycles (test cycles), approximately 2 h and 15 min, and the test temperature is typically 64°C. The test is typically run in air.

During loading, the rut depth is measured (typically at two points in the case of the cylindrical specimen or in five points in the case of the prismatic beam).

At the end of the test, after 8000 test cycles, the rut depth is measured. The average value of six specimens is the representative rut depth value of the bituminous mixture tested. This value should not be higher the limiting value specified.

More information about the test is given in AASTHO T 340 (2010).

7.6.7.2 Wheel-tracking test with the use of the Hamburg device

This wheel-tracking test with the use of Hamburg device, developed in Germany during the 1970s, is carried out in accordance to AASHTO T 324 (2011).

The test can determine not only the bituminous mixture's resistance to permanent deformation but also its moisture susceptibility.

The device can simultaneously test four cylindrical specimens 150 mm in diameter and 62 mm thick. Alternatively, prismatic specimens 320 mm long by 240 mm wide may be

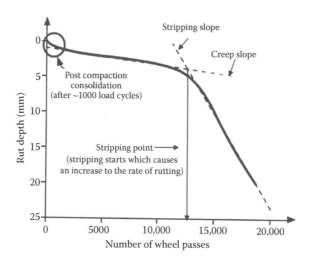

Figure 7.25 Typical creep curve obtained using the laboratory wheel-tracking device. (Courtesy of Pavement Interactive, Laboratory wheel-tracking devices, July 1, 2011.)

used. The thickness can be 40, 80 or 120 mm, depending on the maximum nominal size of aggregates. As a guide, the thickness of the specimen should not be less than three times the nominal maximum aggregate size.

Dense asphalt concrete specimens are compacted at 7% ± 1% air voids. If SMA specimens are tested, compaction is such to achieve 5.5% ± 0.5% air voids.

In contrast to the APA device, the solid metal wheel load, 204 mm in diameter and 47 mm wide, runs directly on the specimen's surface without the interference of the rubber pneumatic hose. The applied load is 705 ± 2 N (158 ± 0.5 lb).

The specimens are immersed in water, and the test temperature is typically 50°C.

The test duration is 10,000 load cycles (20,000 wheel passes), approximately 6 h and 20 min (if a loading frequency of 26.5 cycles/min is used).

Rut depth measurements are taken at intervals of 10, 25, 50 or 100 loading cycles by LVDTs.

The rut depth after a specified number of wheel passes, 10,000, 15,000 or 20,000, is measured and the average of at least four values is the representative rut depth of the bituminous mixture tested. This value, after a certain number of passes, should not be higher than the limiting value specified, or alternatively, at a specified rut depth, the number of passes should be higher than the limiting number specified.

As mentioned above, the test may also assess the negative effect of water on the bituminous mixture (moisture susceptibility of aggregates). This is achieved by determining the 'stripping' inflection point, which is the point of intersection of two straight lines (see Figure 7.25).

When the stripping inflection point appears in ≤5000 wheel passes, the bituminous mixture is considered susceptible to moisture.

More information about the test can be found in AASTHO T 324 (2011).

7.6.8 Evaluation of bituminous mixtures on rut resistance employing US testing procedures

The evaluation of the bituminous mixtures with respect to rut resistance by employing US testing procedures can be carried out by using one of the following parameters/criteria:

Table 7.10 Recommended limiting values for evaluation of rut resistance performance

	AMPT	APA	HWTT	SST/RSCH	IDT-HT
Traffic level (10⁶ ESALs)	Minimum FN (cycles)	Maximum rut depth (mm)	Minimum number of passes to 0.5-inch rut depth	Maximum MPSS (%)	Minimum HT/IDT strength (kPa)
<3	—	—	10,000[a]	—	—
3 to <10	50	5	15,000[b]	3.4	270
10 to <30	190	4	20,000[c]	2.1	380
≥30	740	3		0.8	500

Source: Adapted from NCHRP Report 673, *A Manual for Design of Hot Mix Asphalt with Commentary*, Washington, DC: TRB, 2011.

Note: AMPT, asphalt mixture performance tester; APA, asphalt pavement analyser; FN, flow number; HWTT, Hamburg wheel track tester; IDT-HT, indirect tensile strength at high-temperature test; MPSS, maximum permanent shear strain; SST/RSCH, superpave shear tester/repeated shear constant height.

[a] For binder grade PG 64 or lower.
[b] For binder grade PG 70 or lower.
[c] For binder grade PG 76 or higher.

a. Flow number (FN) measured by the AMPT procedure
b. Rut depth (RD) measured by the APA procedure
c. Minimum number of passes to 0.5-inch rut depth ($N_{passes-0.5inRD}$) measured by the Hamburg wheel-tracking test
d. Maximum permanent shear strain (MPSS) determined by the SST, procedure B (at constant head) (SST/RSCH)
e. High-temperature indirect tensile strength (HT/IDT) determined by the indirect tensile strength at high temperature test (IDT-HT).

The recommended limiting values set for all the above parameters are as shown in Table 7.10.

The above parameters/criteria are used at the mix design stage for the determination of the target mix or for the evaluation of its conformity to the declared value by the producer.

7.7 FATIGUE OF ASPHALTS

Fatigue, in general, is the progressive and localised structural damage that occurs when a material is subjected to loading and unloading (cyclic loading). The applied tensile stress is always less than the tensile stress required to fracture the material after only one loading, that is, the material's tensile strength. CEN EN 12697-24 (2012) defines fatigue as the reduction of strength of a material under repeated loading when compared to the strength under a single load.

Layers of pavements under the effect of axial loads are subjected to repeated tension. The magnitude of the tensile strain developed in pavements under a certain load mainly depends on the stiffness of the layers. For flexible pavements and after detailed measurements in situ, it was found (Pell 1967) that the magnitude of the tensile strain ranges between 30 and 200 microstrain, for a standard axle load (8 tonnes). This tensile strain generates conditions for fatigue of asphalt layers to occur, which appears as cracking.

Pavement cracking, owing to the fatigue of asphalt layers, is one of the main modes of failure.

Asphalt resistance to fatigue depends on its composition and properties of constituent materials. Thus, each type of asphalt possesses different fatigue behaviour. This behaviour should be determined in order to design the pavement so that it does not crack prematurely.

7.7.1 Determination of fatigue characteristics from laboratory testing

The fatigue, or better, the behaviour of asphalts to fatigue, can be determined in the laboratory under controlled testing conditions. Tests and devices used are similar to those abovementioned for the determination of the elastic stiffness modulus, except for the uniaxial test under repetitive load. All devices have the capability of applying tensile stress, which, after n number of repeated loadings, will cause cracking and failure of the specimen. Thus, the only essential difference between the stiffness modulus measurement procedure and the determination of the asphalt's behaviour to fatigue is the duration of loading.

During loading of the asphalt specimen, the start of cracking is impossible to be seen by the naked eye. It can only be detected by plotting the number of load applications and resulting strain. The point where an abrupt and significant change occurred in the slope of the curve is considered, by some researchers, as the point of fatigue failure (see Figure 7.26).

CEN EN 12697-24 (2012) considers the number of the load applications when the complex stiffness modulus of the asphalt has decreased to half its initial value as a conventional criterion of fatigue failure.

Alternatively, ASTM D 7460 (2010) considers as failure point the number of cycles to failure that corresponds to the maximum (or peak) normalised complex modulus × cycles when plotted versus number of cycles (see Section 7.7.2).

The fatigue point depends not only on the asphalt composition and its properties but also on the magnitude of the applied stress, the test temperature and the loading time.

Regardless of asphalt type, magnitude of stress, test temperature and method of measurement, the asphalt will reach fatigue after a number of load applications, N_f. This number is also known as the service life (for fatigue) of the asphalt and, further, of the pavement. By changing only the magnitude of the applied stress, a new number of load applications is obtained. Thus, for a given test temperature, if the results of the applied stress and the number of load applications (loading cycles) are plotted on coordinate axes of a logarithmic

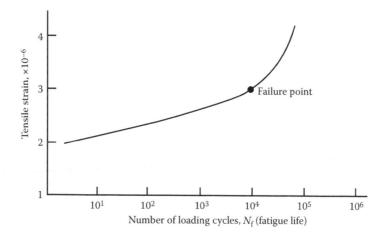

Figure 7.26 Determination of asphalt fatigue point.

scale, a straight line is obtained, known as fatigue line. Figure 7.27 shows typical fatigue lines at various test temperatures, at a given load application frequency. Needless to say, the higher the applied stress, the smaller the number of loading cycles until the asphalt reaches the fatigue stage.

Figure 7.27 also shows the impact of temperature on the asphalt fatigue life. The higher the temperature, the lower the fatigue life is, at a given applied stress. Similar results are obtained if the frequency of loading application varies. The smaller the loading frequency, the higher the number of loadings is, for the same magnitude of applied stress.

If fatigue lives (number of loading cycles) are plotted in terms of constant strain, results from different stiffness of asphalt, owing to temperature and time of loading, coincide in one straight line as shown in Figure 7.28. Thus, the criterion for failure is strain (temperature

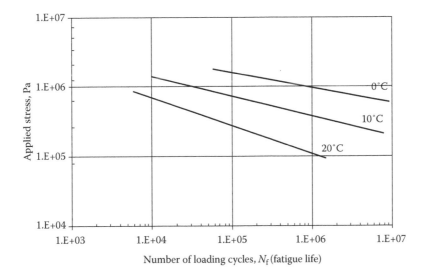

Figure 7.27 Typical fatigue lines of asphalt at different temperatures.

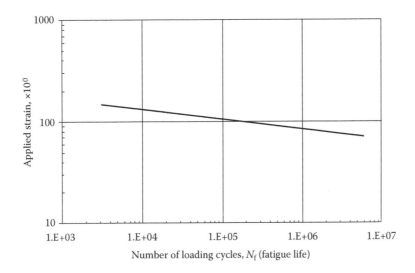

Figure 7.28 Typical fatigue line in terms of strain criterion.

and time of loading affect stiffness), and this effect is known as 'the strain criterion' (Pell and Taylor 1969).

The general equation defining the fatigue life in terms of tensile strain, as a criterion, is as follows:

$$N_f = k \times \left(\frac{1}{\varepsilon_o}\right)^n,$$

where N_f is the number of load applications (loading cycles) to conventional failure, ε_o is the tensile strain at the centre of the specimen (microstrain, $\mu\varepsilon$), k and n are constants (factors) depending on the composition and properties of the asphalt; n is the slope of the fatigue life line.

However, if applied stress is taken as a criterion, a different fatigue line is obtained. Van Dijk (1975) suggested that the differences in fatigue life determined under controlled conditions of strain or stress can be explained by the dissipated energy concept. This is the energy lost from the system owing to fatigue damage per cycle summed for the entire time of loading (life) of the specimen. He also stated that, for a given bituminous mix, the relationship between dissipated energy and the number of load repetitions to failure is valid and independent of testing method (controlled strain or stress) and temperature.

Himeno et al. (1987) applied the dissipated energy concept to the failure of an asphalt layer in a pavement. Rowe (1993, 1996) showed that the dissipated energy can be used to predict life to crack initiation with good accuracy.

Ghuzlan and Carpenter (2006) found that the number of load cycles to 50% reduction in initial stiffness correlated highly with the new failure point introduced (plateau value [PV] and number of load cycles to true failure [N_{tf}]). Maggiore et al. (2012) evaluated fatigue of asphalt mixtures paying attention to the dissipated energy criteria developed, using two methods to measure dissipated energy in tension–compression fatigue tests.

Carpenter and Shen (2006) and Shen and Carpenter (2007) studied the healing phenomenon observed in relation to fatigue behaviour of hot asphalts considering the dissipated energy concept.

It is noted that the updated European standard for resistance of hot asphalts to fatigue (CEN EN 12697-24 2012) has implemented the dissipating energy concept in the recommended testing procedures.

The fatigue characteristics of an asphalt material are influenced significantly by its composition. The stiffness of the mixture is also influenced by the composition, and distinction between the two is important.

The fatigue of the asphalt material is affected by various parameters such as void content, grade and type of bitumen, filler content, shape of aggregate, gradation of aggregate and so on – its fatigue characteristics. One of the purposes of fatigue testing, apart from ranking the material, is to determine the resulting improvement (or the lack of it) in the fatigue performance of the asphalt material.

Figure 7.29 gives a general example of two asphalt materials where the fatigue line of an asphalt material (line I) has been shifted to the right because of the effect of one of the abovementioned parameters (line II). For the same tensile strain induced in the specimen or in the asphalt layer, ε_{t1} or ε_{t2}, the fatigue life increases. Hence, the asphalt material represented with fatigue line II has better fatigue performance and is preferable to the asphalt material represented by fatigue line I.

The fatigue line could also be used to determine the permissible maximum tensile strain to be induced at the asphalt layer to achieve the desired service life of the pavement. This

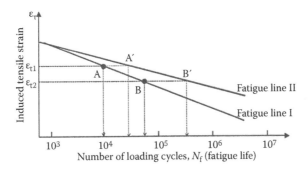

Figure 7.29 Fatigue life change owing to mix composition improvement.

constitutes a useful parameter in pavement design analysis for the determination of the asphalt layer thickness.

The following paragraphs outline and briefly describe the current test methods specified by the European and American standards for characterising the fatigue of the bituminous mixtures.

7.7.2 Resistance to fatigue tests

The European standard related to resistance to fatigue of hot mix asphalt, CEN EN 12697-24 (2012), specifies five alternative test methods for characterising the fatigue of bituminous mixtures. Four of the tests are bending direct tensile tests and one is an indirect tensile test.

In detail, the five tests methods specified by CEN EN 12697-24 (2012) are as follows:

Indirect shear test
  a. Indirect tensile test of cylindrical specimens, IT-CY
Bending tests
  b. Two-point bending test on trapezoidal specimens, 2PB-TR
  c. Two-point bending test on prismatic specimens, 2PB-PR
  d. Three-point bending test on prismatic specimens, 3PB-PR
  e. Four-point bending test on prismatic specimens, 4PB-PR

According to CEN EN 12697-24 (2012), the resistance to fatigue test procedures are used

  a. To rank the bituminous mixtures on the basis of resistance to fatigue
  b. As a guide to the relative performance of the pavement
  c. To obtain data for estimating the structural behaviour of the road
  d. To judge test data according to the specifications for bituminous mixtures

The tests are performed on laboratory-prepared samples or cored from road layer samples, under sinusoidal loading, or haversine loading in the case of indirect tensile test, using different types of specimens and supports.

7.7.2.1 Failure criteria

The criterion for failure (conventional failure criterion) in all test methods is defined as the number of load application, $N_{f/50}$, when the complex stiffness modulus, S_{mix}, has decreased to half its initial value; the initial stiffness modulus, $S_{mix,0}$, is the complex stiffness modulus after 100 load applications.

7.7.2.2 Age of the specimens

Prior to testing, the specimens are aged by placing them on a flat surface at a temperature of not more than 20°C for 14 to 42 days from the time of their manufacture. In the case of samples requiring cutting or gluing, the cutting is performed no more than 8 days after compaction of the asphalt and the gluing is carried out at least 2 weeks after cutting.

The results obtained from the different test methods are not assured to be comparable.

With regard to American practice, fatigue test is performed according to ASTM D 7460 (2010) or AASHTO T 321 (2011). The standard specifies one procedure/test method, the four-point flexural bending test on prismatic (beam) specimens.

The test is performed on specimens sawed from laboratory or field-compacted asphalt material (asphalt concrete), which are subjected to repeated haversine loading/displacement.

The criterion for failure in this test procedure is defined as the number of cycles to failure, N_f, which corresponds to the maximum (or peak) normalised modulus × loading cycles when plotted against number of loading cycles. The normalised modulus × cycles is the ratio of (maximum beam stiffness modulus × loading cycles) to (initial beam stiffness × loading cycles of initial beam stiffness). The initial beam stiffness modulus is the beam stiffness modulus determined after 50 load applications (loading cycles).

Before compacting the specimens, the prepared mixtures are conditioned with a short-term ageing process, as defined in AASHTO R 30 (2010) (mixtures left in the oven at 135°C ± 3.0°C for 4 h ± 5 min).

7.7.2.3 Indirect tensile test on cylindrical-shaped specimens

This test method determines the asphalt behaviour under repeated load fatigue testing with a constant load mode using an indirect tensile load.

The specimens are of cylindrical shape and prepared in the laboratory, drilled from laboratory-prepared slabs or prepared from drilled core taken from the road (pavement). The compaction of specimens in the laboratory is carried out using a gyratory compactor according to CEN EN 12697-31 (2007). If the specimens are drilled from laboratory slabs or prepared from cores taken from the road, their preparation is conducted according to CEN EN 12697-33 (2007) and CEN EN 12697-27 (2000), respectively.

The cylindrical specimens for asphalts with maximum aggregate size up to 25 mm have a diameter of 100 mm and a thickness of at least 40 mm. If the maximum aggregate size is higher, up to 38 mm, the specimens have a diameter of 150 mm and a thickness of at least 60 mm.

The test is carried out with at least three specimens per stress level, for laboratory-manufactured specimens, or at least five specimens, for cores from the road. The specimens after ageing are conditioned to the specified test temperature for at least 4 h prior to testing. The test temperature is usually 20°C, 15°C or 10°C.

Three different constant stress levels of pulse form are applied to the specimens at 0.1 s loading time and 0.4 s unloading time such that the resulting strain is within the 100 to 400 μm/m range. The test is recommended to start at a loading amplitude of 250 kPa.

The specimen is positioned in the loading device so that the axis of the deformation strips is 90° ± 5° to the axis of loading strips. During the test, the load and the resulting horizontal deformation are recorded at pre-selected intervals (number of loadings). A schematic representation of the test device is given in Figure 7.30.

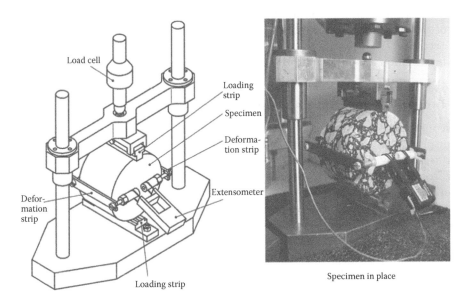

Figure 7.30 Schematic representation and photo of an indirect tensile device for fatigue test.

The test continues until complete fracture of the specimen. During the test, the tensile strain, at an assumed Poisson's ratio of 0.35, and stress at the centre of the specimen are calculated using the following equations:

$$\varepsilon_o = 2.1 \times \left(\frac{\Delta H}{\Omega} \right)$$

and

$$\sigma = \frac{2 \times F}{\pi \times t \times \Omega},$$

where ΔH is the horizontal deformation (mm), Ω is the specimen diameter (mm), F is the maximum applied load (N) and t is the specimen height (mm).

The number of specimens tested at each level of stress is recommended to be at least three, if specimens are laboratory manufactured, or at least five, if cores are taken from the road. To construct the fatigue line, at least three levels of stresses are recommended to be used.

At each level of stress and at the conventional failure point, the number of load applications and the corresponding tensile strain at the centre of the specimen are used as data to construct the fatigue line of the material tested.

The fatigue line is the least-squares regression relationship of the above data plotted in a double logarithmic scale with the data of the logarithm of load applications (fatigue life) as a dependent variable according to the following formula:

$$\log(N_f) = k + n \times \log(\varepsilon_o),$$

where N_f is the number of load applications, k and n are material constants and ε_o is the tensile strain at the centre of the specimen.

If the correlation R^2 of the best-fit line is less than 0.9, it is recommended to increase the number of specimens. It is noted that the above equation is the logarithmic fatigue equation given in Section 7.7.1.

More information about the indirect fatigue test is given in CEN EN 12607-24, Annex E.

7.7.2.4 Two-point bending test on trapezoidal-shaped specimens

This test method determines the asphalt's behaviour under fatigue loading with controlled displacement by two-point bending using trapezoidal specimens. The method is used for asphalts with maximum aggregate size up to 20 mm. For asphalts with maximum aggregate size between 20 and 40 mm, the test can be performed with adapted specimen sizes.

The specimens have an isosceles trapezoidal shape; their sizes are the same as those for the stiffness test using trapezoidal specimens (see Figure 7.4). They are produced from asphalt slabs compacted in the laboratory according to CEN EN 12697-33 (2007) or from slabs taken from road layers having a thickness at least 40 mm, if asphalts have a maximum aggregate size of up to 20 mm; in other cases, the layer thickness should be at least 60 mm. In any case, the specimens are sawed in the desired dimensions and are aged/stored before testing.

At least six specimens per applied deformation level are required to conduct the test. At least 18 specimens are required to complete the fatigue test.

Before fitting the specimen to the test machine, each specimen is glued by its large base into the groove, approximately 2 mm deep, of a metal base.

After the test temperature has been achieved in the specimen that is placed in a thermostatic chamber, a horizontal sinusoidal loading at 25 Hz frequency is applied to the head of the specimen, resulting in a constant displacement amplitude.

The test temperature is typically 10°C and the test starts with a displacement amplitude of ±5 μm.

During the test, the displacement, z, is measured at a predetermined number of load cycles, until complete fracture of the specimen. From the applied displacement, the maximum strain is calculated (see relevant equations in CEN EN 12697-24 2012).

The strain, ε, and the number of load cycles, N_f, at the conventional failure point are recorded so as to be used for the construction of the fatigue line.

At least three different levels of deformation (strain) are applied to the specimens, for the construction of the fatigue line. The deformations shall be such that at least 1/3 of the specimens provide results with $N_f \geq 10^6$ and at least 1/3 of the specimens provide results with $N_f \leq 10^6$.

The asphalt's fatigue line is drawn in the same way as described in Section 7.7.2.3.

On completion of the test, it is recommended that the strain corresponding to 10^6 cycles, ε_6, the quality index, $\Delta\varepsilon_6$, and the slope of the fatigue line are reported.

More information is given in CEN EN 12697-24 (2012), Annex A.

7.7.2.4.1 Limiting values

The fatigue test is required by CEN EN 13108-1 (2008) to be used for the mix design of asphalt concrete using the fundamental approach.

French design guidelines (Delorme et al. 2007; SETRA 2008) have set limiting values for bituminous mixtures related to the fatigue characteristic value of ε_6. The limiting values of strain, ε_6, for all bituminous mixtures where the fatigue strain is a design parameter, when the fatigue test is performed at 10°C and 25 Hz, are as shown in Table 7.11.

Table 7.11 Limiting fatigue test requirements by French design guide

Type of mix	Class	Strain at 10^6 load cycles, $\varepsilon_6{}^a$ (microstrain)
AC	1, 2 and 3	—
AC-HS	1, 2 and 3	≥ 100 (ε_{6-100})
AC-thin	1, 2 and 3	Not required
AC-Airf	1, 2 and 3	Not required
AC-VTL10	1 and 2	Not required
AC-VTL6	1 and 2	Not required
AC-GB	2	≥ 80 (ε_{6-80})
	3	≥ 90 (ε_{6-90})
	4	≥ 100 (ε_{6-100})
EME	1	≥ 100 (ε_{6-100})
	2	≥ 130 (ε_{6-130})

Source: Adapted from Delorme, J.-L. et al., *LPC Bituminous Mixtures Design Guide*, Laboratoire Central des Ponts et Chaussées, Paris: LCPC, 2007; SETRA, *Technical Guide: The Use of Standards for Hot Mixes*, Paris: SETRA, 2008.

Note: AC, asphalt concrete and high stiffness asphalt concrete for surface or binder course, AC10 and AC14, with thickness between 5 and 9 cm; AC-HS, asphalt concrete with high stiffness for surface or binder course, AC10 and AC14, with thickness between 5 and 9 cm; AC-thin, asphalt concrete for surface or binder course, AC10 and AC14, with thickness between 3 and 5 cm; AC-Airf, asphalt concrete for airfields for surface and binder course, AC10 or AC14; AC-VTL10, asphalt concrete for very thin layers with D = 10 mm; AC-VTL6, asphalt concrete for very thin layers with D = 6 mm; AC-GB, asphalt for road base, AC20, with thickness between 8 and 16 cm; EME, high stiffness modulus mixtures, AC10, with thickness between 7 and 13 cm, or AC14, with thickness between 9 and 15 cm.

[a] Determined in accordance to CEN EN 12697-24 (2012), Annex A.

7.7.2.5 Two-point bending test on prismatic-shaped specimens

This test method is essentially the same as the two-point bending test on trapezoidal specimens, except for the shape of the specimens. The test is carried out under controlled strength so as to achieve the intended displacement amplitude.

In this test, specimens have a square prismatic shape with 40 mm × 40 mm × 160 mm dimensions, for mixtures with maximum aggregate size D ≤ 22 mm, or 80 mm × 80 mm × 320 mm, for mixtures with maximum aggregate size D > 22 mm. The height of the specimen when D ≤ 22 mm is 160 mm, and when D > 22 mm, the height is 320 mm.

The prismatic shape specimens are obtained with the same procedures as in trapezoidal shape specimens. The frequency of the applied displacement to the specimen is also the same, 25 Hz.

The intended displacement amplitude is related to the intended tension by the following equation:

$$P_{ij} = \frac{\sigma_{j\max}}{K_{\sigma,i}},$$

where P_{ij} is the amplitude of the strength applied to the head (N); $\sigma_{j\max}$ is the greatest relative tension of the specimen, corresponding to the strength applied to the head; and $K_{\sigma,i}$ is the constant for consideration of the geometry at constant strength.

The test is carried out at not less than three levels of tension with a minimum of six specimens per level. The levels of tension are such that the average fatigue life of the series lies between 10^4 and 10^6 cycles for a minimum of two levels and between 10^6 and 10^7 cycles for at least one level.

The fatigue line is drawn by making a linear regression between the natural logarithm of $\sigma_{j\,max}$ and number of cycles.

The result is expressed in tension, σ_6, at 10^6 load cycles.

More information is given in CEN EN 12697-25 (2005), Annex B.

7.7.2.6 Three-point bending test on prismatic-shaped specimens

This test method determines the behaviour of the asphalts under repetitive fatigue loading with controlled displacement by three-point bending, using prismatic beam-shaped specimens of equal width and height. The test can be used for asphalts with maximum aggregate size D up to 22 mm.

The prismatic beam-shaped specimens are obtained from samples manufactured in the laboratory using a roller compactor according to CEN EN 12697-33 (2007) or from samples cored from road layers with thickness ≥50 mm. In any case, the specimens are cut in the desired dimensions, 50 mm × 50 mm × 300 mm, and are tested after ageing (stored at 20°C for 14 to 42 days).

At least 10 element tests are required in order to characterise the asphalt's behaviour through the determination of the fatigue law in terms of strain (relation between strain and number of load cycles at failure) and the associated energy law. Element tests are tests carried out at different displacement amplitudes.

The prismatic beam is clamped in both edges and a constant amplitude sinusoidal displacement is applied to the mid-span point of the beam by a piston rod, with a wave frequency of 10 Hz.

The criterion for failure in this test is when the amplitude of the cyclic load calculated at cycle N is half of the amplitude of the cyclic load calculated at cycle 200.

The stress, σ, and the tensile strain of the mixture at the mid-span point are first calculated per cycle, followed by the dynamic modulus, phase angle and density of anticipated energy per cycle.

Then, the controlled displacement fatigue law and the energy law are determined from the results of not less than 10 element tests.

The test report includes, among others, the fatigue law constants, the energy law constants and the strain for 10^6 cycles.

More information for the determination of the above parameters and other details are given in CEN EN 12697-24 (2012), Annex C.

7.7.2.7 Four-point bending test on prismatic-shaped specimens

This test method determines the asphalt's behaviour under repetitive fatigue loading in a four-point bending test equipment in which the inner and outer clamps are symmetrically placed and a slender rectangular-shaped specimen (prismatic beam) is used. The prismatic beam is subjected to four-point periodic sinusoidal bending with free rotation and translation at all load and reaction points.

The height, H, and width, B, of the specimens depend on the maximum aggregate size, D, and should be at least 3 × D (or 2.5 × D in the case of large values of D). The difference between the maximum and minimum measured value of the width and height is not greater than 1.0 mm.

The effective length, L (from clamp to clamp), should be at least six times the highest value of width, B, or height, H. The total specimen length (L_{tot}) should not be higher than 1.1 × L. Hence, for asphalts with D = 20, the specimen dimensions could be 60 mm × 60 mm × 380 mm.

Bending is achieved by applying load to two points of the prismatic beam at $L/3$ and $L/3$ distance from the clamped ends of the specimen. The layout of the test device is similar to the one shown in Section 7.4.4 (Figure 7.6) for four-point bending. The periodic loading is sinusoidal and it can cause a constant moment and hence a constant strain between the inner clamps (in the middle of the specimen).

The specimens are produced and prepared as in the three-point bending test and the minimum number is 18, 6 specimens for each of the three levels of chosen loading mode (constant deflection or constant force).

The force applied, the oncoming central displacement and the lag between the load and the displacement are recorded (phase angle) during the test and on a regular basis after the first 100 loadings.

The test is conducted at 20°C, 10°C or even at 0°C using three levels of the chosen loading mode in such a way that the fatigue lives are within 10^4 to 2×10^6 cycles. Failure is defined as the point (number of cycles) where the asphalt stiffness modulus (S_{mix}) drops to half the initial stiffness modulus S_{mix} value (at the 100th loading cycle).

Using the obtained data of force, deflection and phase lag between these two signals measured at load cycles n, the complex stiffness modulus (dynamic modulus) is calculated. Additionally, the following may also be calculated: material phase lag, dissipated energy per cycle and cumulative dissipated energy up to cycle n.

The shape of the fatigue line is expressed in the following equation:

$$\ln(N_{i,j,k}) = A_0 + A_1 \times \ln(\varepsilon_i),$$

where i is the specimen number, j is the chosen failure criteria, k is the set of test conditions and ε_i is the initial strain amplitude measured at the 100th load cycle.

The initial strain corresponding with a fatigue life of 10^6 cycles, ε_6, is determined from the fatigue equation.

More information is given in CEN EN 12697-24 (2012), Annex D.

7.7.2.8 Repeated flexural bending test according to ASTM D 7460 (2010)

This test method determines a unique failure point for estimating the fatigue life of a prismatic beam-shaped specimen, 380 mm long by 50 mm thick by 63 mm wide asphalt concrete, and is described by ASTM D 7460 (2010).

The test is a four-point flexural bending where a haversine loading (displacement) is applied at the central H-frame third points of a beam specimen, while the outer points are held in an articulating fixed position. The frequencies used range from 5 to 50 Hz. This produces a constant bending moment over the centre third ($L/3$) span, having a length of 118.5 to 119 mm between the H-frame contact points on the beam specimen. The level of desired strain is pre-calculated and is used as an input for the displacement control. The device used (Figure 7.31) is similar to the device used in the four-point bending test on the prismatic specimen according to CEN EN 12697-24 (2012), Annex D (see Section 7.7.2.7).

The specimens are produced from slab(s) or beam(s) compacted in accordance with AASHTO PP 3 (1994) or ASTM WK34713 (2014). The number of replicate specimens to be prepared is nine; six specimens are tested at different strain levels in order to develop the fatigue line. The extra specimens may also be tested as desired, if the data appear to include an outlier or if a beam failure occurs directly at a clamp.

Before compacting for the laboratory-produced specimens, the prepared mixtures are conditioned with a short-term ageing process, as defined in AASHTO R 30 (2010) (mixtures were left in the oven at 135°C ± 3.0°C for 4 h ± 5 min).

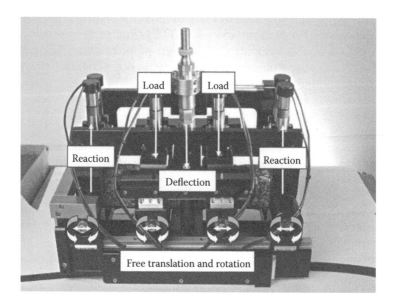

Figure 7.31 Main part of a fatigue test apparatus (COX) (load and freedom characteristics). (Reprinted from ASTM D 7460, *Standard Test Method for Determining Fatigue Failure of Compacted Asphalt Concrete Subjected to Repeated Flexural Bending*, West Conshohocken, Pennsylvania, US: ASTM International, 2010. With permission [© ASTM International].)

The initial desired strain for conventional asphalt concretes is recommended to range between 200 and 800 microstrain. The displacement level is selected so that the specimen undergoes at least 10,000 loadings prior to failure.

During the test and after the first 50 loadings, the force applied and the oncoming central displacement are recorded at regular loading intervals.

The test is usually conducted at 20°C. However, the test may be conducted at the effective test temperature for equivalent pavement fatigue damage using the equation developed during the Strategic Highway Research Program:

$$T_{eff,Fatigue} = 0.8 \times (MAPT) - 2.7,$$

where MAPT is the mean annual pavement temperature ($= T_{20mm}$) (°C), determined by the equation for T_{20mm} given in Section 3.4.2.

Failure, in this test, is defined as the number of loadings where the normalised complex modulus (NCM or NM) (see Section 7.7.2) obtains the highest value. The number of loadings at failure (N_f) expresses the asphalt's resistance to fatigue.

From the data collected, N_f at different strain levels on the fatigue line (equation) is determined. More information is given in ASTM D 7460 (2010).

Similar is the test described in AASHTO T 321 (2011).

7.7.3 Prediction of fatigue performance

The laboratory determination of the fatigue equation presupposes the existence of the appropriate device.

Some laboratories and organisations have developed various methods/techniques (equations or nomograms) from which the fatigue performance of asphalt can be predicted more easily with reasonable accuracy. The predictive methods are based on asphalt stiffness, bitumen properties (penetration index, volume of bitumen or softening point) or air voids.

A method for predicting the fatigue life of asphalts by the use of a nomograph has been developed by Bonnaure et al. (1980). The required data are percentage volume of bitumen, V_b, penetration index of the bitumen (PI), stiffness modulus of the mixture, S_{mix}, and initial strain level. From the same nomograph, if the number of load cycles is known, the initial strain (permissible strain), under constant strain or constant stress condition, can be predicted.

From fatigue studies carried out with a wheel-tracking machine on asphalt slabs under a rolling wheel, it was concluded that the crack patterns observed were very similar to those defined as alligator (fatigue) cracking in practice and that the results of controlled-strain bending tests could be used as a criterion of fatigue (Shell International Petroleum Co. 1978). Fatigue measurements made on several mix types and two alternative approaches to fatigue prediction have been developed.

In the first approach, the permissible asphalt strain to fatigue, ε_{fat}, can be predicted when two parameters are known: the stiffness modulus of the asphalt and the volume of the bitumen in the mix. The equation developed is as follows (Shell International Petroleum Co. 1978; Van Dijk 1975):

$$\varepsilon_{fat} = (0.856 \times V_b + 1.8) \times S_{mix}^{-0.36} \times N_{fat}^{-0.2},$$

where ε_{fat} is the tensile strain to fatigue (microstrain), S_{mix} is the stiffness modulus of asphalt (Pa), V_b is the bitumen volume in the mix (%) and N_{fat} is the number of loadings for fatigue failure.

In the second approach, the dissipated energy concept was used. The permissible asphalt strain to fatigue, ε_{fat}, can be predicted when the following parameters are known: the stiffness modulus of the asphalt, S_{mix} (in pascals); the initial phase angle between stress and strain, ϕ_o; and a mix parameter, C (in joules per cubic metre). For more details, see Shell International Petroleum Co. (1978) and Van Dijk and Visser (1977).

It has also been stated (Shell International Petroleum Co. 1978) that fatigue data obtained in the laboratory cannot be applied directly to thickness design since, in practice, the mode of loading and the spectrum of strain values are different from the laboratory conditions. There is also an indication that some healing occurs in practice and that intermittent loading has a less damaging effect than continuous loading. For these and other reasons such as transverse distribution of wheel loads, as well as temperature and asphalt thickness, the ultimate correction factor is of the order of 10 to 20 towards a higher design life (Shell International Petroleum Co. 1978).

Another nomograph was developed by Cooper and Pell (1974) referring to dense-graded macadams and hot rolled asphalts. The nomograph is expressed by the following mathematical equation:

$$\log N_f = 15.8 \times \log \varepsilon_t - 40.7 - (5.13 \times \log \varepsilon_t - 14.39) \times \log V_b$$

$$- (8.63 \times \log \varepsilon_t - 24.2) \times \log SP,$$

where N_f is the number of load repetitions to failure, ε_t is the tensile strain (microstrain), V_b is the bitumen volume in the mixture (%) and SP is the softening point (°C).

Furthermore, a similar fatigue prediction equation has been developed and used by the Asphalt Institute to develop the design charts for the thickness design manual (Asphalt Institute 1990, 1999). The equation developed, where failure is defined as 20% or greater fatigue cracking on the pavement surface, is as follows:

$$N_f = 18.4 \times 10^M \times 0.004325 \times \varepsilon_t^{-3.291} \times E_{ac}^{0.854},$$

where N_f is the number of load repetitions to failure, ε_t is the magnitude of tensile strain at the bottom of the asphalt concrete layer caused by the wheel load (microstrain), E_{ac} is the modulus of asphalt concrete (psi) and $M = 4.84 \times [(V_b/(V_b + V_v)) - 0.69]$, where V_b is the bitumen volume in the mixture (%) and V_v is the volume of air voids (%).

7.8 THERMAL CRACKING OF ASPHALTS

Thermal cracking, also called low-temperature cracking, of flexible pavements is a critical issue for many highway authorities of countries with predominately low to very low ambient temperatures.

It was as early as 1965 that Monismith et al. (1965) suggested that, in the very cold winters of Northern America, tensile stresses induced in the wearing course could exceed the breaking strength of the material and lead directly to transverse cracking of the pavement.

Later on, Dauzats and Rampal (1987) published results that surface-initiated cracks in France were initially caused by thermal stresses and then further propagated by traffic loads. Gerritsen et al. (1987) reported that pavements in the Netherlands were experiencing premature cracking in the wearing courses and those observed outside the wheel paths were associated with asphalts having low strength characteristics at low temperatures.

In a recent study (Romero et al. 2011) on the premature surface cracking in flexible pavements in Utah, it was reported that the cracks were not attributed to pavement structural deficiency but to the low-temperature performance of the bituminous mixtures. Numerous other studies have found that the main form of deterioration in asphalt pavements within the freezing areas of the United States and Canada is thermal cracking (Marasteanu et al. 2007).

The US practice uses the creep compliance and indirect tensile strength test, according to AASHTO T 322 (2011), as a suitable performance test for low-temperature cracking in hot mix asphalt mix design. The same procedure was selected as the material characterisation test method for the prediction of low-temperature cracking of flexible pavements in the AASHTO (2008) *Mechanistic-Empirical Pavement Design Guide*.

A recent study concluded that the bending beam rheometer test on bituminous binders is a vital tool that can be used to control pavement performance at low temperatures (Romero et al. 2011). In the same report, a draft specification was presented along with examples to demonstrate how the BBR test procedure can be adopted to facilitate quality control/quality acceptance operations in asphalt construction.

Finally, the fracture energy, G_f, of asphalt–aggregate mixtures using the disc-shaped compact tension geometry may also be used to describe the fracture resistance of asphalt concrete. The test is performed in accordance to ASTM D 7313 (2013) specifications.

In the European specifications, the thermal cracking issue of the bituminous mixtures has not yet been addressed.

7.8.1 Creep compliance and strength of hot mix asphalts using the indirect tensile test device

The creep compliance and strength test using the indirect tensile test device is carried out in accordance to AASHTO T 322 (2011) specifications.

The loading device used is similar to the one used for the determination of indirect tensile strength at 25°C (ASTM D 6931 2012 or AASHTO T 283 2011), but the apparatus is equipped with an environmental chamber with a range of temperatures from −30°C to +30°C, and the rate of applying the load for strength determination is slower (12.5 mm/min instead of 50 mm/min).

The test method determines the creep compliance, $D(t)$, the tensile strength, $S_{t,n}$, and Poisson's ratio, v, of hot mix asphalts. The procedure is applied to test specimens having a maximum aggregate size of 38 mm or less.

According to AAHTO T 322 (2011), creep compliance, $D(t)$, is defined as the time-dependent strain divided by the applied stress.

The tensile strength is determined at the maximum load achieved for each specimen.

The tensile creep and tensile strength test data are required for Superpave mixtures to determine the master relaxation modulus curve and fracture parameters. The master relaxation modulus curve controls thermal crack development, while the fracture parameter defines a mixture's resistance to fracture. The tensile creep data may be used to evaluate the relative quality of materials.

The specimens, typically three replicas per type of mix tested, can be gyratory compacted, in accordance with AASHTO T 312 (2012), or cored from laboratory-compacted slabs, in accordance with AASHTO PP 3 (1994), The specimens can also be obtained by coring from pavements in accordance with ASTM D 5361.

The size of the specimens is 150 mm in diameter and 38 to 50 mm in height. In all cases, the specimens should have smooth and parallel surfaces; this is achieved by sawing both sides of the specimen.

Bulk specific gravity and void determination is carried out using the saturated surface-dried method (AASTHO T 166 2013). However, for high-absorption specimens (more than 2% water absorption), it is recommended to use an impermeable plastic film rather than a paraffin-coated coating, as specified in AASHTO T 166 (2013). Christensen and Bonaquist (2004) have used the vacuum sealing method (ASTM D 6752 2011) for all specimens tested.

The horizontal and vertical deformation necessary to determine, primarily, the creep compliance and Poisson's ratio are measured by LVDTs, or other types of transducers/gauges, on both specimen faces (see Figure 7.32).

After gluing the LVDTs, the specimen is placed in the loading device and conditioned to reach the test temperature. The use of a dummy IDT specimen for monitoring the temperature is recommended.

The test temperatures to be used depend on the binder grade. These are recommended by AASHTO T 322 (2011) to be (a) –30°C, –20°C and 10°C, when binder grade PG XX-34 or softer is used; (b) –20°C, –10°C and 0°C, when binder graded PG XX-28 and PG XX-22 are used; and (c) –10°C, 0°C and +10°C, when binder grade PG XX-16 is used. For mixtures subjected to severe age hardening, the test temperatures are recommended to be increased by 10°C.

After the test temperature has been reached, the specimen is subjected to a static load, without impact, that produces a horizontal deformation of 0.00125 to 0.0190 mm, for a period of 100 s.

To achieve the above deformations, a load protocol shown in Table 7.12 has been suggested.

Table 7.12 Guidelines for applied load in the IDT creep test

Test temperatures	Initial applied load (kN)	Other possible applied loads (kN)
Lowest	40	Deformation < 0.01 mm: 80 Deformation > 0.02 mm: 20 or 10
Intermediate	10	Deformation < 0.01 mm: 20 or 40 Deformation > 0.02 mm: 5 or 2
Highest	5	Deformation < 0.01 mm: 10 or 20 Deformation > 0.02 mm: 2 or 1

Source: Christensen D.W. and R.F. Bonaquist, *Evaluation of Indirect Test (IDT) Procedures for Low-Temperature Performance of Hot Mix Asphalt*. NCHRP Report 530. Washington, DC: Transportation Research Board, 2004.

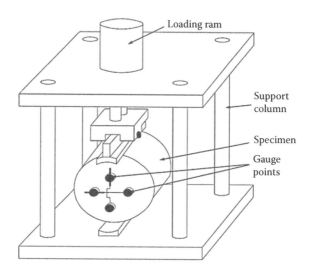

Figure 7.32 Specimen loading frame with four support columns and specimen in place. (Adapted from AASHTO T 322, *Determining the Creep Compliance and Strength of Hot Mix Asphalt [HMA] Using the Indirect Tensile Test Device*, Washington, DC: American Association of State Highway and Transportation Officials, 2011. With permission.)

After the 100 s period, the creep compliance is computed from the equation given in AASHTO T 322 (2011), using the average values of thickness (mm), diameter (mm) and applied creep load (kPa) for the three replicates.

Poisson's ratio is computed from the equation also given in AASHTO T 322 (2011), using the mean horizontal and vertical deformations at time corresponding to half of the total creep time, that is, 50 s.

After the creep test has been completed at each temperature, the tensile strength is determined by applying a load to the specimen at a rate 12.5 mm/min of ram (vertical) movement. The tensile strength is determined at the maximum load achieved, using the following equation:

$$S_{t,n} = \frac{2 \times P_{f,n}}{\pi \times b_n \times D_n},$$

where $S_{t,n}$ is the tensile strength of specimen n (Pa), b_n is the thickness of specimen n (mm), D_n is the diameter of specimen n (mm) and $P_{f,n}$ is the maximum load observed for the specimen n (N).

The average of the three tensile strength values is reported as the representative tensile strength of the tested material.

The tensile strength is normally determined at the middle temperature used for the creep test.

The IDT test for the determination of the tensile strength may be performed without the LVDTs to avoid damaging the transducers. In this case, the following empirical relationship developed is proposed to be used (Christensen and Bonaquist 2004):

Tensile strength = (0.78 × IDT strength) + 38,

where Tensile strength is the strength corrected to the AASHTO T 322 (2011) procedure and IDT strength is the strength calculated as a function of maximum load.

More details on the creep compliance and strength test can be found in AASHTO T 322 (2011).

To minimise or eliminate thermal cracking, reduce the creep compliance or increase the indirect tensile strength of the hot mix asphalt surface mixture. Thermal cracking is also eliminated by using softer bitumen in the surface layer, increasing the bitumen content of the surface mixture or increasing the thickness of the hot mix layers (AASHTO 2008).

The increase of air voids has been found to increase the creep compliance and to decrease the indirect tensile strength, while the presence of recycled asphalt pavement increases the indirect tensile strength and decreases the creep compliance (Richardson and Lusher 2008).

7.8.2 Determination of fracture energy of asphalt–aggregate mixtures using the disc-shaped compact tension geometry

This test method for the determination of the fracture energy, G_f, of asphalts using disc-shaped compact tension geometry is carried out in accordance with ASTM D 7313 (2013).

The fracture energy can be utilised as a parameter to describe the fracture resistance of asphalt concrete, and the fracture energy parameter is particularly useful in the evaluation of mixtures with ductile binders, such as polymer-modified bitumen. It has been shown to discriminate between these materials more broadly than the indirect tensile strength parameter determined by AASHTO T 322 (2011) (Wagoner et al. 2006).

The test is generally valid at temperatures of 10°C and below, or for material and temperature combinations that produce valid material fracture.

More details and description of the test are given in ASTM D 7313 (2013).

REFERENCES

AASHTO. 2008. *Mechanistic-empirical pavement design guide – A manual of practice*, 3rd Edition. Washington, DC: American Association of State Highway and Transportation Officials.

AASHTO PP 3. 1994. *Provisional standard for hot mix asphalt (HMA) specimens by means of rolling wheel compactor*. Washington, DC: American Association of State Highway and Transportation Officials. (1996), last printed in 2002.

AASHTO PP 61. 2013. *Practice for developing dynamic modulus master curves for hot mix asphalt (HMA) using the asphalt mixture performance tester (AMPT)*. Washington, DC: American Association of State Highway and Transportation Officials.

AASHTO PP 62. 2010. *Developing dynamic modulus master curves for hot mix asphalt (HMA)*. Washington, DC: American Association of State Highway and Transportation Officials.

AASHTO R 30. 2010. *Mixture conditioning of hot mix asphalt (HMA)*. Washington, DC: American Association of State Highway and Transportation Officials.

AASHTO T 166. 2013. *Bulk specific gravity (Gmb) of compacted hot mix asphalt (HMA) using saturated surface-dry specimens*. Washington, DC: American Association of State Highway and Transportation Officials.

AASHTO T 283. 2011. *Resistance of compacted hot mix asphalt (HMA) to moisture-induced damage*. Washington, DC: American Association of State Highway and Transportation Officials.

AASHTO T 312. 2012. *Preparing and determining the density of hot mix asphalt (HMA) specimens by means of the Superpave gyratory compactor*. Washington, DC: American Association of State Highway and Transportation Officials.

AASHTO T 320. 2011. *Determining the permanent shear strain and stiffness of asphalt mixtures using the Superpave shear tester (SST)*. Washington, DC: American Association of State Highway and Transportation Officials.

AASHTO T 321. 2011. *Determining the fatigue life of compacted hot mix asphalt (HMA) subjected to repeated flexural bending.* Washington, DC: American Association of State Highway and Transportation Officials.

AASHTO T 322-7. 2011. *Determining the creep compliance and strength of hot mix asphalt (HMA) using the indirect tensile test device.* Washington, DC: American Association of State Highway and Transportation Officials.

AASHTO T 324. 2011. *Hamburg wheel-track testing of compacted hot mix asphalt (HMA).* Washington, DC: American Association of State Highway and Transportation Officials.

AASHTO T 340. 2010. *Determining rutting susceptibility of hot mix asphalt (HMA) using the asphalt pavement analyzer (APA).* Washington, DC: American Association of State Highway and Transportation Officials.

AASHTO T 342. 2011. *Determining dynamic modulus of hot mix asphalt (HMA).* Washington, DC: American Association of State Highway and Transportation Officials.

AASHTO TP 79. 2013. *Provisional standard method of test for determining the dynamic modulus and flow number for hot mix asphalt (HMA) using the asphalt mixture performance tester (AMPT).* Washington, DC: American Association of State Highway and Transportation Officials.

Al-Khateeb G., A. Shenoy, N. Gibson, and T. Harman. 2006. A new simplistic model for dynamic modulus predictions of asphalt paving mixtures. *Journal of the Association of Asphalt Paving Technologists,* Vol. 75E, pp. 1–40.

Andrei D., M.W. Witczak, and W. Mirza. 1999. *Development of Revised Predictive Model for the Dynamic (Complex) Modulus of Asphalt Mixtures.* Interteam Technical Report, NCHRP Project 1-37A. Maryland, MD: University of Maryland.

Asphalt Institute. 1990. *Computer Program CAMA, CP-6, Version 2.0.* Lexington, KY: Asphalt Institute.

Asphalt Institute MS-1. 1999. *Thickness Design, Asphalt Pavements for Highways & Streets.* Manual Series No. 1. Lexington, KY: Asphalt Institute.

Asphalt Institute SP-2. 2001. *Superpave Mix Design,* 3rd Edition. Superpave Series No. 2. Lexington, KY: Asphalt Institute.

ASTM D 6752/D 6752M. 2011. *Bulk specific gravity and density of compacted bituminous mixtures using automatic vacuum sealing method.* West Conshohocken, PA: ASTM International.

ASTM D 6931. 2012. *Standard test method for indirect tensile (IDT) strength of bituminous mixtures.* West Conshohocken, PA: ASTM International.

ASTM D 7312. 2010. *Standard test method for determining the permanent shear strain and complex shear modulus of asphalt mixtures using the Superpave shear tester (SST).* West Conshohocken, PA: ASTM International.

ASTM D 7313. 2013. *Standard test method for determining fracture energy of asphalt–aggregate mixtures using the disc-shaped compact tension geometry.* West Conshohocken, PA: ASTM International.

ASTM D 7369. 2011. *Standard test method for determining the resilient modulus of bituminous mixtures by indirect tension test.* West Conshohocken, PA: ASTM International.

ASTM D 7460. 2010. *Standard test method for determining fatigue failure of compacted asphalt concrete subjected to repeated flexural bending.* West Conshohocken, PA: ASTM International.

ASTM D 7552. 2009. *Standard test method for determining the complex shear modulus (G*) of bituminous mixtures using dynamic shear rheometer.* West Conshohocken, PA: ASTM International.

ASTM WK34713. 2014. Under development within ASTM Committee D.04. *New practice for preparation of slab samples using a rolling wheel compactor for the repeated flexural bending.* West Conshohocken, PA: ASTM International.

Azari H., G. Al-Khateeb, A. Shenoy, and N.H. Gibson. 2007. Comparison of simple performance test |E*| of accelerated loading facility mixtures and prediction |E*|: Use of NCHRP 1-37A and Witczak's new equations. *Transportation Research Record: Journal of the Transportation Research Board,* Vol. 1998, pp. 1–9.

Bari J. 2005. Development of a new revised version of the Witczak E* predictive models for hot mix asphalt mixtures. Arizona, AZ: PhD Dissertation, Arizona State University.

Bari J. and M.W. Witczak. 2006. Development of a new revised version of the Witczak E* predictive model for hot mix asphalt mixtures. *Journal of the Association of Asphalt Paving Technologists from the Proceedings of the Technical Sessions,* Vol. 75, pp. 381–423. Savannah, GA.

Bari J. and M.W. Witczak. 2007. New predictive models for the viscosity and complex shear modulus of asphalt binders for use with the Mechanistic-Empirical Pavement Design Guide. *Transportation Research Record: Journal of the Transportation Research Board*, No. 2001, pp. 9–19.

Bonnaure F., G. Gest, A. Gravois, and P. Uge. 1977. A new method of predicting the stiffness of asphalt paving mixtures. *Proceedings of the Association of Asphalt Paving Technologists*, Vol. 46, p. 66.

Bonnaure F., A. Gravois, and J. Udron. 1980. A new method for predicting the fatigue life of bituminous mixes. *Proceedings of the Association of Asphalt Paving Technologists*, Vol. 49, p. 499.

Brown S.F. 1980. *An Introduction to Analytical Design of Bituminous Pavements.* Nottingham, UK: University of Nottingham.

BS 598-110. 1996. *Sampling and examination of bituminous mixtures for roads and other paved areas, Part 110: Methods of test for the determination of wheel-tracking rate.* London: British Standards Institution.

BSI PD 6691. 2010. *Guidance on the use of BS EN 13108 Bituminous mixtures – Materials specifications.* London: BSI.

Cabrera J.C. and A.F. Nikolaides. 1987. CANIK UL – A new creep testing machine. *Journal of the Institution of Highways and Transportation*, Vol. 34, No. 11, p. 33.

Cabrera J.C. and A.F. Nikolaides. 1988. Creep performance of cold dense bituminous mixtures. *Journal of the Institution of Highways and Transport*, Vol. 35, No. 10, pp. 7–15.

Carpenter S.H. and S. Shen. 2006. Dissipated energy approach to study hot-mix asphalt healing in fatigue. *Journal of the Transportation Research Record*, Vol. 1970, pp. 178–185. Washington, DC: Transportation Research Board.

CEN EN 12697-6. 2012. *Bituminous mixtures – Test methods for hot mix asphalt – Part 6: Determination of bulk density of bituminous specimens.* Brussels: CEN.

CEN EN 12697-7. 2002. *Bituminous mixtures – Test methods for hot mix asphalt – Part 7: Determination of bulk density of bituminous specimens by gamma rays.* Brussels: CEN.

CEN EN 12697-22:2003+A1. 2007. *Bituminous mixtures – Test methods for hot mix asphalt – Part 22: Wheel tracking.* Brussels: CEN.

CEN EN 12697-24. 2012. *Bituminous mixtures – Test methods for hot mix asphalt – Part 24: Resistance to fatigue.* Brussels: CEN.

CEN EN 12697-25. 2005. *Bituminous mixtures – Test methods for hot mix asphalt – Part 25: Cyclic compression test.* Brussels: CEN.

CEN EN 12697-26. 2012. *Bituminous mixtures – Test methods for hot mix asphalt – Part 26: Stiffness.* Brussels: CEN.

CEN EN 12697-27. 2000. *Bituminous mixtures – Test methods for hot mix asphalt – Part 27: Sampling.* Brussels: CEN.

CEN EN 12697-30. 2012. *Bituminous mixtures – Test methods for hot mix asphalt – Part 30: Specimen preparation by impact compactor.* Brussels: CEN.

CEN EN 12697-31. 2007. *Bituminous mixtures – Test methods for hot mix asphalt – Part 31: Specimen preparation by gyratory compactor.* Brussels: CEN.

CEN EN 12697-32 +A1. 2007. *Bituminous mixtures – Test methods for hot mix asphalt – Part 32: Laboratory compaction of bituminous mixtures by vibratory compactor.* Brussels: CEN.

CEN EN 12697-33:2003+A1. 2007. *Bituminous mixtures – Test methods for hot mix asphalt – Part 33: Specimen preparation by roller compactor.* Brussels: CEN.

CEN EN 13108-1:2006/AC. 2008. *Bituminous mixtures – Material specifications – Part 1: Asphalt concrete.* Brussels: CEN.

CEN EN 13108-20:2006/AC. 2008. *Bituminous mixtures – Material specification – Part 20: Type testing.* Brussels: CEN.

Ceylan H., C. Schwartz, S. Kim, and K. Gopalakrishnan. 2009. Accuracy of predictive models for dynamic modulus of hot-mix asphalt. *Journal of Materials in Civil Engineering*, Vol. 21, No. 6, pp. 286–293.

Christensen D.W. and R.F. Bonaquist. 2004. *Evaluation of Indirect Test (IDT) Procedures for Low-Temperature Performance of Hot Mix Asphalt.* NCHRP Report 530. Washington, DC: Transportation Research Board.

Christensen D.W., T.K. Pellinen, and R.F. Bonaquist. 2003. Hirsch model for estimating the modulus of asphalt concrete. *Journal of the Association of Asphalt Paving Technologists*, Vol. 72, pp. 97–121.

Colloquium 77. 1977. *Plastic Deformability of Bituminous Mixtures (Plastitische Verformbarkeit von Asphaltmischungen)*. Zurich: Eidgenössisch Technische Hochnische Zürich, Institut für Strassen-, Eisenbahn- und Felsbau, Mitteilung Nr. 37.

Controls Srl. 2014. Available at http://www.controls-group.com.

Cooper K.E. and P.S. Pell. 1974. *The Effect of Mix Variables on the Fatigue Strength of Bituminous Materials*. TRRL Report LR 633. Crowthorne, UK: Transport Research Laboratory.

Cooper Research Technology Ltd. 2014. Available at http://www.cooper.co.uk.

Dauzats M. and A. Rampal. 1987. Mechanism of surface cracking in wearing courses. *Proceedings, 6th International Conference Structural Design of Asphalt Pavements*, pp. 232–247. Ann Arbor, MI: The University of Michigan.

Delorme J.-L., C. de la Roche, and L. Wendling. 2007. *LPC Bituminous Mixtures Design Guide*. Laboratoire Central des Ponts et Chaussées. Paris: LCPC.

Dongre R.L., J. Myers, C. D'Angelo, C. Paugh, and J. Gudimettla. 2005. Field evaluation of Witczak and Hirsch models for predicting dynamic modulus of hot-mix asphalt. *Journal of the Association of Asphalt Paving Technologists from the Proceedings of the Technical Sessions*, Vol. 74, pp. 381–442.

Finn F.N., C.L. Monismith, and N.J. Markevich. 1983. Pavement performance and asphalt concrete mix design. *Proceedings of the Association of Asphalt Paving technologists*, Vol. 52, p. 121.

Gerritsen A.H., C.A.P.M. van Gurp, J.P.J. van der Heide, A.A.A. Molenaar, and A.C. Pronk. 1987. Prediction and prevention of surface cracking in asphaltic pavements. *Proceedings, 6th International Conference Structural Design of Asphalt Pavements*, pp. 378–391. Ann Arbor, MI: The University of Michigan.

Ghuzlan K.A. and S.H. Carpenter. 2006. Fatigue damage analysis in asphalt concrete mixtures using the dissipated energy approach. *Canadian Journal of Civil Engineering*, Vol. 33, No. 7, pp. 890–901.

Heukelom W. and A.J.G. Klomp. 1964. Road design and dynamic loading. *Proceedings of the Association of Asphalt Paving technologists*, Vol. 33, p. 92.

Hill J.F. 1973. The creep of asphalt mixes. *Journal of the Institution of Petroleum*, Vol. 59, No. 570, p. 247. The Institute of Petroleum.

Hill J.F., D. Brien, and P.J. Van de Loo. 1974. *The Correlation of Rutting and Creep Tests on Asphalt Mixes*. London: Institution of Petroleum, IP 74-001, p. 1.

Himeno K., T. Watanabe, and T. Maruyama. 1987. Estimation of fatigue life of pavement. *6th International Conference on Structural Design of Asphalt Pavements*. Ann Arbor, MI.

Interlaken Technology. 2014. Available at http://www.interlaken.com.

Jongeneel D.J., L.D. Haugh, and Gerritsen. 1985. Creep testing, results of European interlaboratory study of laboratory apparatuses and test procedures. *Proceedings of the 3rd Eurobitume Symposium*, Vol. 1, p. 295. The Hague, Netherlands.

Kim Y.R., B. Underwood, M. Sakhaei Far, N. Jackson, and J. Puccinelli. 2011. *LTPP Computer Parameter: Dynamic Modulus*. Federal Highway Administration Technical Report No FHWA-HRT-10-035. Springfield, VA: National Technical Information Service.

Kutay M.E. and A. Jamrah. 2013. *Preparation for Implementation of the Mechanistic-Empirical Pavement Design Guide in Michigan – Part 1: HMA Mixture Characterization*. Final Report RC-1593. Lansing, MI: Michigan Department of Transportation.

LTPP Products Online. 2013. Available at http://ltpp-products.comsoftware.

Maggiore C., J. Grenfell, G. Airey, and A.C. Collop. 2012. Evaluation of fatigue life using dissipated energy methods. *7th RILEM International Conference on Cracking in pavements*. RILEM Bookseries, Vol. 4. Netherlands: Springer.

Marasteanu M., A. Zofka, M. Turos, X. Li, R. Velasques, W. Buttlar, G. Paulino, A. Braham, E. Dave, J. Ojo, H. Bahia, R.C. Williams, J. Bausano, A. Gallistell, and J. McGraw. 2007. *Investigation of Low Temperature Cracking in Asphalt Pavements*. National Pooled Fund Study 776. St. Paul, MN: Minnesota Department of Transportation Report MN/RC 2007-43.

Mohammad L., S. Saadeh, S. Obularedd, and S. Cooper. 2007. Characterization of Louisiana asphalt mixtures using simple performance tests. *Proceedings of the 86th Annual Meeting of the Transportation Research Board (TRB)*. Washington, DC: National Research Council.

Monismith C.L., G.A. Secor, and K.E. Secor. 1965. Temperature induced stresses and deformation in asphalt concrete. *Proceedings, Association of Asphalt Paving Technologies*, Vol. 34, pp. 246–285.

Monismith C.L. and A.A. Taybali. 1988. Permanent deformation (rutting) considerations in asphalt concrete pavement sections. *Proceedings of the Association of Asphalt Paving Technologists*, Vol. 57, p. 414. Williamsburg, VA.

Moutier F. 1990. L'essai de fatigue LPC: Un essai vulgarisable? *Proceedings of the International RILEM Symposium, Mechanical Tests for Bituminous Mixes*, p. 540. Budapest.

NCHRP Report 673. 2011. *A Manual for Design of Hot Mix Asphalt with Commentary*. Washington, DC: TRB.

Nikolaides A. and E. Manthos. 2014. Laboratory evaluation of asphalt concrete mixtures to permanent deformation. *Proceedings of the Institution of Civil Engineers – Construction Materials*, Vol. 167, Issue CM4, pp. 201–213, London, UK: ICE.

Nikolaides A.F. and E. Manthos. 2008. The effect of volumetric properties of asphalt concrete mixture to wheel track rutting with respect to EN and BS rutting test methods. *Proceedings of the 7th International RILEM Symposium on Advanced Testing and Characterization of Bituminous Materials*. Rhodes, Greece.

Nikolaides A.F. and E. Manthos. 2009. Wheel tracking performance of asphalt concrete mixture with conventional and modified bitumen. *Proceedings of MAIREPAV 6, 6th International Conference on Maintenance and Rehabilitation of Pavements & Technological Control*, Vol. I, p. 213. Torino, Italy.

Nunn M.E., A. Brown, and D. Lawrence. 1999. An evaluation of practical tests to assess the deformation resistance of asphalt. *3rd European Conference on the Durability of Asphalt and Hydraulically Bound Materials*. University of Leeds, April.

Nunn M.E., D. Lawrence, and A. Brown. 2000. Development of a practical test to assess the deformation resistance of asphalt. *Proceedings of 2nd Eurasphalt & Eurobitume Congress*. Barcelona.

Pavement Interactive. 2011. Available at http://www.pavementinteractive.org/article/laboratory-wheel -tracking-devices.

Pavement Technology Inc. 2014. Available at http://www.pavementtechnology.com.

Pell P.S. 1967. Fatigue of asphalt pavement mixes. *Proceedings of 2nd International Conference on the Structural Design of Asphalt Pavement*. Ann Arbor, MI: University of Michigan.

Pell P.S. and I.F. Taylor. 1969. Asphaltic road materials in fatigue. *Proceedings of the Association of Asphalt Paving Technologists*, Vol. 38, p. 371.

Richardson D.N. and S.M. Lusher. 2008. *Determination of Creep Compliance and Tensile Strength of Hot-Mix Asphalt for Wearing Courses in Missouri*. Final report R105-052. Missouri Department of Transportation.

Robbins M.M. and D. Timm. 2011. Evaluation of dynamic modulus predictive equations for NCAT test track asphalt mixtures. *Proceedings of the Transportation Research Board 90th Annual Conference*, January 23–27.

Romero R., C.H. Ho, and K. VanFrank. 2011. *Development of Methods to Control Cold Temperature and Fatigue Cracking for Asphalt Mixtures*. Report No. UT-10.08. State of Utah: Utah Department of Transportation, Research Division.

Rowe G.M. 1993. Performance of asphalt mixtures in the trapezoidal fatigue test. *Proceedings of Association of Asphalt Paving Technologist*, Vol. 62. Seattle, WA: Association of Asphalt Paving Technologists.

Rowe G.M. 1996. Application of the dissipated energy concept to fatigue cracking in asphalt pavements. Nottingham, UK: Ph.D. Thesis, University of Nottingham.

Sakhaeifar M.S., S. Underwood, R. Ranjithan, and Y.R. Kim. 2009. The application of artificial neural networks for estimating the dynamic modulus of asphalt concrete. *Journal of the Transportation Research Board, Transportation Research Record*, Vol. 2127, pp. 173–186.

SETRA. 2008. *Technical Guide: The Use of Standards for Hot Mixes*. Paris: SETRA.

Shell Bitumen. 1990. *The Shell Bitumen Handbook*. Chertsey, Surry, UK: Shell Bitumen UK.

Shell International Petroleum Co. 1978. *Shell Pavement Design Manual*. London: Shell International Petroleum Company Ltd.

Shen S. and S.H. Carpenter. 2007. *Dissipated Energy Concepts for HMA Performance: Fatigue and Healing*. COE Report No. 29. Urbana, IL: University of Illinois.

Singh D., M. Zaman, and S. Commuri. 2011. Evaluation of predictive models for estimating dynamic modulus of hot-mix asphalt in Oklahoma. *Transportation Research Record, Journal of the Transportation Research Board*, Vol. 2210, pp. 57–72.

Transportation Research Circular C-C068. 2004. *New simple performance tests for asphalt mixes*. Washington, DC: TRB.

Valkering C.P., D.J. Lancon, E. De Hilster, and D.A. Stoker. 1990. Rutting resistance of asphalt mixes containing non-conventional and polymer modified binders. *Proceedings of the Association of Asphalt Paving Technologists*, Vol. 59, p. 590.

Van de Loo P.J. 1974. Creep testing, a simple tool to judge asphalt stability. *Proceedings of the Association of Asphalt Paving Technologists*, Vol. 43, p. 253.

Van der Poel C. 1954. A general system describing the visco-elastic properties of bitumen and its relation to routine test data. *Journal of Applied Chemistry*, Vol. 4, p. 221.

Van Dijk W. 1975. Practical fatigue characterization of bituminous mixes. *Proceedings of Association of Asphalt Paving Technologists*, Vol. 44, pp. 38–74.

Van Dijk W. and W. Visser. 1977. The energy approach to fatigue for pavement design. *Proceedings of Association of Asphalt Paving Technologists*, Vol. 46, pp. 1–40.

Van Draat W.E.F. and P. Sommer. 1965. An apparatus for determining the dynamic elastic modulus of asphalt. *Strasse und Autobahn*, Vol. 6, p. 201.

Wagoner, M.P., Buttlar, W.G. , Paulino, G.H., and Blankenship, P.I. 2006. Laboratory testing suite for characterization of asphalt concrete mixtures obtained from field cores. *Journal of the Association of Asphalt Paving Technologists from the Proceedings of the Technical Sessions*, Vol. 75, pp. 815–852. Savannah, GA.

Witczak M. 2005. *Simple Performance Tests: Summary of Recommended Methods and Database*. NCHRP Report 547. Transportation Research Board. Washington, DC: TRB.

Chapter 8

Production, transportation, laying and compaction of hot mix asphalts

8.1 GENERAL

The plant production of hot asphalts consists of hot mixing aggregates with bitumen in special plants, in order for the produced asphalt to be homogeneous and in accordance with the target mix composition. Before adding the bitumen, the aggregates are blended at predetermined portions so that the required (target) grading is achieved. Before or after blending the aggregates, they are heated to dry and to raise them to a temperature suitable for bitumen coating.

Drying/heating of aggregates, as well as mixing, was initially carried out manually. The mechanical drying/heating and mixing started in the beginning of the 20th century. The first machines used were based on cylindrical dryers and modified concrete mixers, heated by coal burning. Soon, paddle or pugmill mixers were developed, initially with a single shaft and later with two shafts. The energy source for heating was replaced with heavy oils (fuel oil), diesel fuel or liquid gas. The introduction of new fuel led to the development of the drum dryer, which was soon modified to serve as a dryer and mixer drum, if desired.

Hot mix asphalt plants are installed close to the aggregate production source (quarries), in suitable rural locations nearby expected projects or at the project site. The mixing plants can be stationary or mobile. The mobile plants provide flexibility and are less costly, but daily production outputs as high as those of stationary plants cannot be achieved.

8.2 TYPES OF ASPHALT PRODUCTION PLANTS

All types of asphalt production plants consist of five distinct parts: the cold aggregate feeder system, the dryer/heater, the dust collection system, the mixing system and the storage bins. The mixing system is basically the factor that determines their categorisation.

There are two basic categories of hot mix asphalt plants:

a. Batch plants
b. Drum-mix plants

In batch plants, the hot aggregates are stored in hot bins before mixing with bitumen in discrete batches in a pugmill (typically) mixer and then loaded onto trucks or transferred to a silo (or silos) for temporary storage.

In drum-mix plants, the mixing of the aggregate and the bitumen takes place in the dryer/heater drum, after which the asphalt is stored in a silo (or silos) before being loaded into delivery trucks.

Each of the above two categories split into two sub-categories. The first category splits into

a₁. Conventional batch plants (mostly contra flow heating)
a₂. Batch heater plants (contra flow)

The second category splits into

b₁. Parallel-flow drum-mix plants
b₂. Counterflow drum-mix plants

The decision to choose the most suitable asphalt plant is quite complicated. It is influenced by the market conditions related to demand and selling price of the asphalt, the typical hourly output capacity, the cost of purchasing the plant, the types of asphalt usually required to be produced, the capability of the plant to produce recycled asphalt, the amount of reclaimed asphalt (or reclaimed asphalt pavement [RAP]) to be used, the land space availability and the environmental restrictions (mainly emissions and noise).

The analytical description, as well as the reference to the pros and cons of each type, could help in the decision making of choosing the most appropriate plant.

Today, the predominant plant in the United States and New Zealand is the drum-mix plant. Batch plants prevail in Europe, South Africa and Australia (EAPA and NAPA 2011).

8.3 BATCH PLANTS

8.3.1 Conventional batch plant

In this type of mixing plant, the asphalt is produced in batches of predetermined weight and the heating of the aggregate precedes both the determination of the batch weight and the asphalt mixing. The batch weight may vary from 1000 to 5000 kg, depending on the plant model.

Cold and damp aggregates outflow from feed bins and a conveyor belt are fed into a rotating heater drum, where they dry and obtain the desired temperature. Feeder units are located under each bin to control the flow of the aggregates onto the conveyor belt.

To eliminate dust from the dried exhaust, dust collectors (bughouses) are fitted.

The hot aggregates are transported by a bucket elevator into the screening/separation unit, where they are stored temporarily in hot bins. The separation is carried out by a specified series of sieves.

From each hot bin, the aggregate is charged into a weigh hopper that is weighted at a predetermined amount and then discharged into the mixer. Fines or filler together with hot bitumen are also added to the mixer at predetermined amounts. If reclaimed asphalt is used, it is added either directly to the mixer, to the superheated aggregate before screening or in the heater drum.

Hot aggregates, reclaimed asphalt (if used), filler and hot bitumen are mixed for a certain period and the produced asphalt is directly unloaded onto the pending truck or transported by skip to a temporary storage silo. A schematic representation of a batch plant is given in Figure 8.1.

In Figure 8.1, the reclaimed asphalt is fed into the mixer. This restricts the amount of reclaimed asphalt to be added for the production of recycled mixture to approximately 15%. Other types of batch plants are also available in which the reclaimed asphalt is either

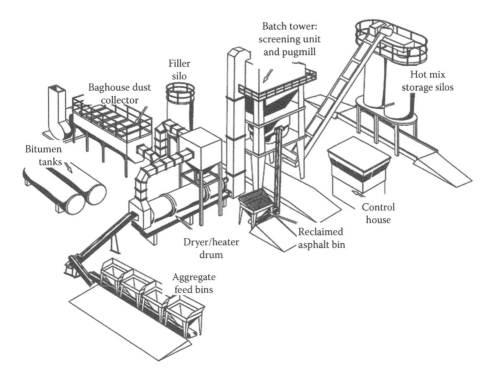

Figure 8.1 Typical layout of batch plant for hot mix asphalt. (From EAPA and NAPA, *The Asphalt Paving Industry: A Global Prospective*, 2nd Edition, GL 101. Brussels: EAPA, 2011; Courtesy of National Asphalt Pavement Association, http://www.asphaltpavement.org.)

blended with the superheated virgin aggregate in the bucket elevator or added in the rotating dryer/heater drum. In the first alternative technique, the percentage of reclaimed asphalt can increase up to 40%. In the second alternative technique, the percentage of reclaimed asphalt can be increased further and up to 50% of the resulted recycled mixture.

The required mixing time is affected by the type of asphalt produced, as well as the batch quantity. As a consequence, the mixing time and the batch quantity determine the hourly production output. Thus, in a 2000 kg asphalt concrete batch plant, where the mixing time is approximately 45 s, the maximum hourly output can reach 160 tonnes.

The basic advantages of a conventional batch plant, which is usually stationary, are the following: flexibility to produce all types of asphalts, small tonnages are possible to be produced, high percentage of reclaimed asphalt can be added, storage of mixed asphalt is not essential and emissions are within acceptable limits.

The basic disadvantages of a conventional batch plant are as follows: high capital cost, relative high maintenance cost, overall higher production cost over drum-mix plants, heat and aggregate material wasted on overflow material and capacity being restricted to mixing cycle and mixer size.

8.3.1.1 Basic components of a batch plant unit

8.3.1.1.1 Storage feeding of aggregates and reclaimed asphalt

Aggregates and reclaimed asphalt, if used, stocked in piles, are placed in cold feed holding bins using a loader or a truck. The number of cold feed bins is usually from three up to nine, depending on the number of different size aggregate used and hourly production output.

At the bottom of each bin, there is an opening gate that regulates the quantity flowing to the feeding system or to the weighing system. The width of the opening is checked frequently and the weighing system is calibrated frequently. The latter is significantly important for continuous-flow plants.

In certain models, the bins, particularly those of fine aggregates, are equipped with a vibrating system. This eliminates the development of 'caves' within the bins when aggregates are damp during discharge of aggregates to the conveyor belt.

The feeding unit consists of a system of conveyor belts, or elevators, which, in modern plants, are able to change speed and thus vary the supply of aggregate to the heater drum.

The proportion of coarse, medium and fine aggregate flown onto the feeding system is similar to that determined in the mix design, preventing excessive overflow of a certain hot bin.

8.3.1.1.2 Dryer/heater drum

The role of the drying and heating system is significantly important, given that the amount of water (moisture) usually contained in the cold aggregates should be removed within a short period and the aggregate should reach the appropriate temperature to be coated with bitumen.

The drying/heating unit consists of a rotating drum fitted with flights or steel angles and a gas or oil burner. In a batch mixing plant, the burner is placed at the lowest end of the drum (exit end). The drum is inclined with a small angle varying from 2° to 10°. By changing the inclination angle, the retention time (length of time during which the aggregates remain in the drum) varies; hence, drying and heating of the aggregates are effective. For a give length of drum, the retention time decreases as the slope angle increases. The retention time also decreases as the drum rotations, the drum diameter and the feed quantity increase. Correspondingly, the retention time increases as the drum slope decreases, the number of rotations decreases and the speed of the exhaust fan decreases, and, indirectly, as the aggregate size decreases. It should be noted that the stage of drying and heating the aggregates requires large quantities of fuel, which affects the production cost.

At the entering end of the drum, at the higher end, the dust collection system is fitted. Nowadays, there are basically two types of dust collection systems – the cyclone type and the backhouse type (use of fabric filters); the wet collection system used in the past is almost obsolete.

The cyclone type is used as the primary collector and the backhouse type is used as the secondary collector. The fines collected in the primary collector are coarse dust, which returns to the hot aggregate. The fines collected in the secondary collector are fine dust, which, if suitable, is collected in the filler silo.

8.3.1.1.3 Separation of hot aggregate unit

The separation or screening unit includes a series of sieves that separate hot aggregates (inclusive of reclaimed asphalt if used and fed before screening) into fractions (usually between four and six) of specified size and reject the oversized aggregates. The fractions of aggregates are stored temporarily in hot bins. Each bin is equipped with an overflow tube of appropriate diameter to prevent the material from overflowing into other bins.

8.3.1.1.4 Weighing system of aggregate material

The weigh hopper is positioned exactly under the hot bins. Each aggregate fraction is individually weighed in a specified proportion with the added filler to obtain the target gradation. Weighing usually begins from coarser to finer aggregates.

The weighing system should periodically be calibrated to attain gradations within the specified variation limits.

8.3.1.1.5 Supply of bituminous binder

The bituminous binder is kept in insulated and heated storage tanks of appropriate capacity. The tanks should be equipped with either agitators (stirrers) or circulation pumps, or both. Agitators are preferred when modified binders are to be used.

The tanks are usually heated with hot oil using a heat transfer oil system. With this system, in comparison to burners, local overheating, oxidation or carbonisation is prevented.

The bituminous binder is supplied to the mixer by an appropriate pump. The amount of binder needed per batch is measured volumetrically by a flow meter or by a weigh bucket. If bituminous binder quantity is measured volumetrically, the changes of the specific gravity of the binder with respect to operating temperatures should be taken into account. Table 8.A.1, in Annex A of this chapter, gives the typical specific gravity values of bituminous binders at various temperatures. Additionally, Table 8.A.2 gives the number of litres per tonne of bitumen at different temperatures for typical specific gravities of bituminous binders.

8.3.1.1.6 Mixing

Mixing of the pre-weighted aggregate, filler and bitumen is carried out in a pugmill mixer. This is a large semi-circular bowl with two shafts with mixing blades carrying adjustable hardened steel tips. At the beginning and for a short period, aggregates, filler and the additives, if used, are mixed alone (dry mixing). Dry mixing should be kept as short as possible in order to avoid early wear of the mixing bowl. Then, the bitumen is added and all are mixed further until the aggregates are completely covered with bitumen.

The total mixing time should be as long as necessary for uniform coating of aggregates with bitumen to be achieved. Longer-than-necessary mixing should be avoided not only because it decreases the hourly plant production but mainly because the bitumen is further oxidised. Furthermore, mixing for shorter than the appropriate period could lead to a mixture with uncoated aggregates. The latter quite often leads to an arbitrary increase of the bitumen quantity in the mixture and thus help achieve complete coating of aggregates. This results in production of asphalt with a surplus of bitumen.

Mixing time for a particular mix can be established by trial and error, and the procedure described in ASTM D 2489 (2008) or AASHTO T 195 (2011) may be used; it is based on estimating the percentage of coated aggregates, which should always be 100%.

The mixing blades of the shafts are placed in a certain angle, which can be altered to achieve better mixing. The mixer should be supplied with sufficient and appropriate mixture quantity so that good mixing is achieved. The quantity is appropriate when the tips of the blades are visible during mixing.

Proper mixing temperature is vital and is determined by the type of bituminous binder used. To achieve the required mixing temperature, aggregates and bitumen should have the appropriate temperature when entering the mixer.

After mixing, the batch is discharged onto waiting trucks or temporary stored in insulated and heated silos.

8.3.1.1.7 Measuring instruments and automations

The mixing plant is equipped with all necessary and appropriate instruments to measure the temperatures and quantities of all asphalt components and of the produced material at any critical stage of production. The feeding, heating, mixing and unloading procedures are completely automated. The system operator has all the information available in the panel of the control tower and can quickly and easily carry out any adjustments that may be needed.

8.3.2 Batch heater plants

The process in this type of plant is different from that in a conventional batch plant. The plant is called batch heater because aggregates are dried and heated in batches prior to be mixed with the bitumen and filler. The dried and hot batch of aggregate is then dropped directly into the pugmill mixer, without being temporarily stored.

Each different size aggregate is weighted before drying/heating and then fed to a shorter dryer/heater than that in a batch plant. Allowance is made for aggregates having a moisture content of less than 2%. Some plants have an additional weighing system to determine the dry aggregate weight. At moisture contents exceeding 2%, a separate dryer would normally be needed since the aggregate cannot dry efficiently.

Since there is no separation of the hot aggregate before mixing, the number of cold aggregate fractions used should be as many as possible (single-sized aggregates are used) to ensure consistency of the produced mixture. This increases the number of cold bins required.

For all the above, batch heater plants are better suited for mixtures with low sand content, such as open-graded mixtures, which typically require low to medium mixing temperature.

The total mixing time of a batch is almost 2 min and the typical hourly output ranges from 50 to 200 tonnes, depending on the model of the plant.

The percentage of reclaimed asphalt incorporated in the mix, when using this type of mixing plant, cannot be more than 10% to 15%.

The use of this type of batch plant is limited nowadays because of its high capital cost, its relatively high maintenance cost, the requirement of almost single-sized aggregate and its low output.

8.4 DRUM-MIX PLANTS

In a drum-mix plant, the aggregate material is heated and mixed with the bitumen in the same rotating drum, known as the drum mixer. Drum-mix plants are also known as continuous-flow plants; hence, the provision of storage bins is essential.

8.4.1 Parallel-flow drum-mix plants

The parallel-flow drum-mix plants use drum mixers in which the burner discharges super-heated air parallel to the direction of aggregate flow. The wet ambient aggregate is fed to the rotating drum where it is mixed with the bitumen and the asphalt is conveyed to the thermal insulated storage bins (silos) before being loaded to the trucks. The reclaimed asphalt, if used, is added to the drum-mixer after the first half of its length. The typical layout of a parallel-flow drum-mix plant is shown in Figure 8.2. The parallel-flow drum-mix plants first emerged in the early 1970s; thus, they are considered conventional and are often called just drum-mix plants.

The feeding system, particularly the proportioning part, is the key part of the plant. The asphalt's quality directly depends on its precise function. The cold feed bins discharge the aggregate onto the collecting conveyor, while the amount of each aggregate size is controlled by the size of bin gate opening. The collecting conveyor feeds the aggregates into a charging conveyor equipped with a belt weigher, enabling the aggregate flow rate to be maintained

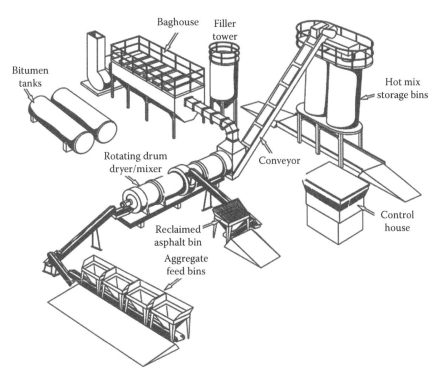

Figure 8.2 Typical layout of parallel-flow drum-mix plant for hot mix asphalt. (From EAPA and NAPA, *The Asphalt Paving Industry: A Global Prospective*, 2nd Edition, GL 101. Brussels: EAPA, 2011; Courtesy of National Asphalt Pavement Association, http://www.asphaltpavement.org.)

or altered taking into account the amount of moisture in the aggregate. This allows proper determination of the amount of filler and bitumen to be added.

The dryer/heater of aggregates differs from the one used in batch plants in the following aspects: (a) the burner is placed at the top end of the drum, (b) the configuration of the flights inside the drum is more complex and (c) the drum is longer.

The drum interior essentially has two zones. In the first zone, the aggregate mixture is heated and dried (drying/heating zone), whereas in the second zone, the hot aggregates are mixed with the hot bitumen (mixing zone) (see Figure 8.3).

The mixing zone is protected from the radiant heat and the burner flames by a special flame shield or another appropriate device. The hot bitumen is usually added under pressure from the discharge end. The filler and the reclaimed asphalt, if used, are also added in the mixing zone.

One of the advantages of the drum mixer is the fact that exhaust gases contain a very low percentage of dust. The reduction is due to the fact that the dust produced in the drying/heating zone is eliminated by the bitumen added in the mixer. However, environmental legislation imposes the use of dust collectors. The small dust (filler) quantity collected is usually re-supplied to the drum.

In general, a drum-mix plant is relatively simple and smaller in size than a batch plant. Regarding the hourly output, it may reach up to 500–550 tonnes per hour. Thus, this type of plant is considered as the most appropriate in cases of high demand.

The other advantages of the parallel-flow drum-mix plant are as follows: lower capital cost in comparison to batch plant, lower space requirement, higher outputs can be achieved,

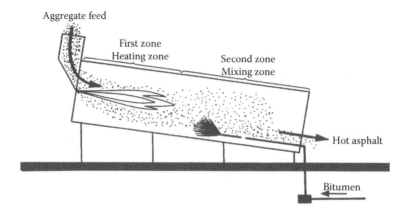

Figure 8.3 Distinct zones in a parallel-flow drum mixer. (From Asphalt Institute MS-3, *Asphalt Plant Manual*. 5th Edition Maryland: Asphalt Institute.)

adaptability to mobile units, relative lower maintenance cost and economical for long production runs.

The basic disadvantages of the parallel-flow drum-mix plant are the following: fume emission increases if high mixing temperatures are used, low percentage of reclaimed asphalt used (10% to 15%), uneconomical to produce small batches, difficulty to produce all types of asphalts, single-sized aggregate feed and wastage of material at the beginning and end of production.

8.4.2 Counterflow drum-mix plant

The counterflow drum-mix plant differs from the parallel-flow drum-mix plants only in terms of the mixing drum. In a counterflow drum, an extended burner is used, which is mounted at the discharge end of the drum and the mixing zone is completely separated from the drying and heating zone. Because of the position of the burner, the superheated air (flames) moves against the flow of the aggregate, hence the name *counterflow drum*. A typical layout of a counterflow drum is shown in Figure 8.4.

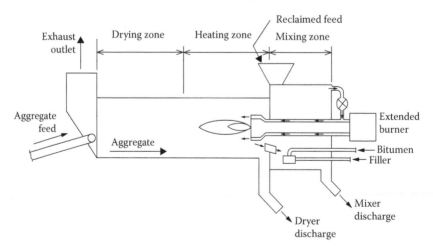

Figure 8.4 Counterflow drum mixer. (From Hunter R.N., *Asphalts in Road Construction*. London: Thomas Telford Publishing, 2000.)

Because of the separation of the mixing zone, fume emission is minimised. The bitumen, the filler and the reclaimed asphalt (if added) are all fed into the mixing zone. As a consequence, the drying zone can be used only as a dryer/heater of the aggregate if needed.

The efficient drying/heating of the aggregates, as well as the complete separation of the mixing zone, has the advantage of using relatively higher percentages of reclaimed asphalt.

All the other production stages are exactly the same as those in the parallel-flow drum-mix plant.

The basic advantages of a counterflow plant over a parallel-flow drum-mix plant are that it is more environment friendly, fuel efficiency increases and higher percentages of reclaimed asphalt can be used.

8.4.3 Newer continuous processing mixers/plants

The desire to improve production efficiency, reduce hydrocarbon emissions, increase fuel efficiency, reduce operating cost and increase the percentage of reclaimed asphalt blend with virgin material led to the development of alternative counterflow drum dryer/mixer techniques.

One such technique is the use of a separate dryer and a separate mixer, where the mixer can be either a mixing drum or a continuous pugmill. Figure 8.5 shows a counterflow dryer with a separate continuous pugmill mixer.

Another technique is the development of a double-barrel dryer/mixer (see Figure 8.6). In this technique, the aggregate is dried and heated in an inner rotating drum, while mixing of the constituent materials of the asphalt (aggregates, bitumen, reclaimed asphalt, etc.) takes place in the outer shell (barrel). This provides more efficient drying and mixing at a lower operation cost.

When recycling is predominant, the hot aggregate, fines and reclaimed asphalt pass through the double-barrel and mixing takes place in a twin-shaft pugmill or rotary mixer. Hence, the double-barrel dryer/mixer is called double RAP dryer (ASTEC Industries Inc. 2014). With the addition of the Green Pac, an option offered by the same company, the double RAP dryer produces warm mix asphalt.

The latest development in the continuous-flow mixing industry for hot mix asphalt is claimed to be the use of a drum for drying/heating the aggregate and a twin pugmill for

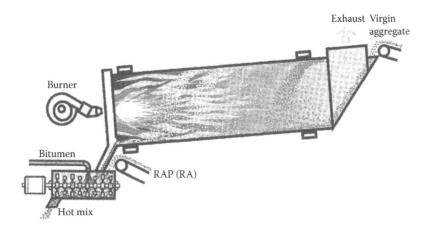

Figure 8.5 Separate dryer with a continuous pugmill mixer. (From Asphalt Institute, *The Asphalt Handbook*, MS-4, 7th Edition, Lexington, KY: Asphalt Institute, 2007. With permission.)

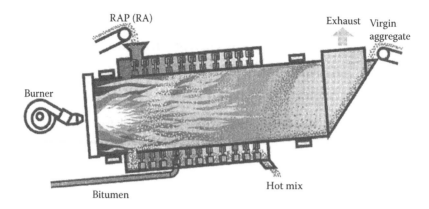

Figure 8.6 Double-barrel dryer/mixer. (From Asphalt Institute, *The Asphalt Handbook*, MS-4, 7th Edition, Lexington, KY: Asphalt Institute, 2007. With permission.)

mixing (AMMANN Group 2014). This combines the advantages of both techniques, batch and continuous.

The reclaimed asphalt, when used, is added to the drum; the filler and fibres, if required, are added to the hot aggregate/reclaimed asphalt after they are discharged from the drum and the binder is added to the mixer. The clear separation of the material heating and mixing processes reduces emissions and allows all the ingredients to be fed easily and accurately directly into the mixer. The other advantages are as follows: ease in recipe changes, reduction of stop/start losses, lower capital cost compared to batch plant, ability to incorporate high percentage of reclaimed asphalt (up to 50%) and ability to produce all types of warm or half warm mixed asphalts. A typical arrangement of a modern continuous-flow combination plant is shown in Figure 8.7.

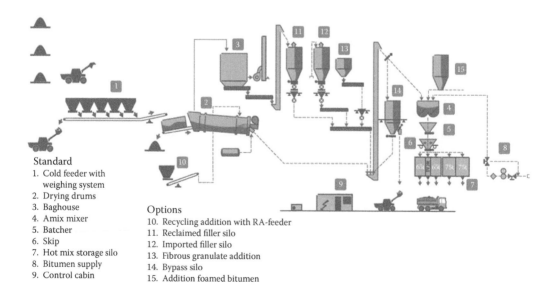

Standard
1. Cold feeder with weighing system
2. Drying drums
3. Baghouse
4. Amix mixer
5. Batcher
6. Skip
7. Hot mix storage silo
8. Bitumen supply
9. Control cabin

Options
10. Recycling addition with RA-feeder
11. Reclaimed filler silo
12. Imported filler silo
13. Fibrous granulate addition
14. Bypass silo
15. Addition foamed bitumen

Figure 8.7 Typical layout of a new generation continuous flow asphalt plant. (Courtesy of AMMANN Group.)

8.5 TRANSPORTATION OF HOT MIX ASPHALTS

The transportation and delivery of the hot mix asphalts from the plant to the paver are carried out with haul trucks, which should be insulated and sheeted (use of tarps) to prevent heat loss and to protect the hot asphalt from bad weather conditions (mainly rain, winds and low air temperatures). Additionally, the floor and the walls of the haul trucks must always be clean and free of foreign materials to avoid intermingling and contamination, and flat and free of dimples.

The haul trucks can be standard or semitrailer end-dump trucks, horizontal-discharge trucks or bottom-dump trailers. The end-dump trucks are the ones most commonly used throughout the world. The capacity of the standard end-dump truck is usually up to 20 tonnes, while the capacity of the semitrailer truck can go up to usually 25 tonnes. They discharge the hot asphalt directly to the paver hopper through an automatic tailgate opening by raising the truck bed.

The horizontal-discharge trucks use a conveyor system to deliver the hot asphalt to the paver hopper, without the need to raise the truck bed. Their capacity is higher than end-dump trucks and normally used in very large projects. The main advantages of the horizontal-discharge trucks are that segregation of the mix is minimised and that the problem with overhead obstructions (trees, bridges, cables, etc.) is eliminated. However, cleaning and frequent checking of the conveyor belt system are essential.

The transportation of the asphalt consists of three distinct stages: the loading, the transportation to the site and the tipping into the hopper of the paver.

During loading, the risk of asphalt segregation and the decrease of asphalt temperature should be minimised during loading. The first is ensured by loading as evenly as possible, that is, no asphalt piles with steep sides. The second is ensured by the quickest possible and uninterrupted loading.

During transportation, the smallest possible asphalt temperature loss and asphalt protection from rain should be ensured. For this, as it was mentioned, the trailer should be covered with an appropriate moist insulator, especially during the days where ambient temperatures are low and there is the possibility of rain during the transportation.

The driver's experience will prove to be vital during tipping into the hopper of the paver to ensure regular feeding and to prevent asphalt segregation. To facilitate asphalt unloading, the inside of the trailer may be coated with an approved release agent or a small filler amount. The use of diesel or kerosene is not recommended, since it may affect the properties of the hot asphalt.

8.6 LAYING/PAVING

Laying of the hot asphalt is carried out with special motorised units called pavers. Pavers are available in a wide variety of sizes capable of laying mats from as narrow as 1 m to up to 16 m wide. The minimum and the maximum range of paving width vary from one manufacturer's model to another.

The pavers are distinguished from the type of their traction, and there are two types: the wheeled pavers and the tracked pavers. In the first, the traction is provided via driving pneumatic wheels (wheeled paver), and in the second, the traction is provided via driving crawler tracks (tracked paver). The two different types of pavers are shown in Figure 8.8.

The wheeled pavers provide excellent manoeuvrability, have the advantage of being driven easily for short distances and require less purchase and operation cost. They can pave mats, normally up to 9 m wide. Wheeled pavers dominate the market in most countries.

(a)

(b)

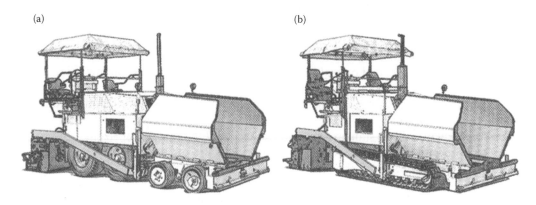

Figure 8.8 Types of pavers. (a) Wheeled pager (with pneumatic wheels). (b) Tracked paver (with crawler tracks).

The tracked pavers, owing to long and stable tracks, assure very good grip, stability and maximum traction even under arduous conditions and exert less pressure on the underlying layers. They can pave wider mats compared to wheel pavers, which can be as wide as 14 m or even 16 m.

Both asphalt pavers consist of two basic parts, the tractor unit and the screed (see Figure 8.9).

Some manufacturers offer pavers that are also equipped with a sprayer in front of the screed for spraying the tack coat. These pavers spray tack coat and apply the hot mix asphalt seconds later. They are known as spray pavers or integral pavers. An example is shown in Figure 8.10.

Finally, pavers that can simultaneously lay two asphalt layers (normally wearing and binder course) in a single pass have been developed. This results in better interlocking of the courses, elimination of tack coating and a reduction of the paving time. The method is known as compact asphalt laying method. An example of a paver is shown in Figure 8.11.

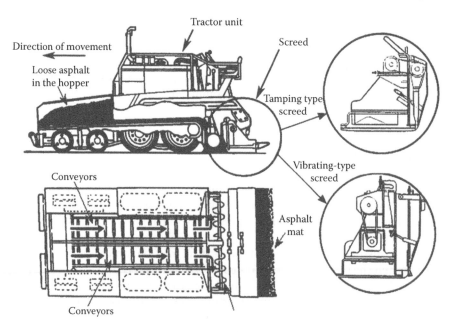

Figure 8.9 Typical representation of wheeled paver. (From Shell Bitumen, *The Shell Bitumen Handbook*. Surrey, UK: Shell Bitumen UK, 1990.)

Figure 8.10 Roadtec spray paver, SP-200e/ex. (Courtesy of ASTEC Industries Inc.)

Figure 8.11 Compact asphalt system. (Courtesy of Atlas Copco.)

8.6.1 Tractor unit

The tractor unit, as mentioned, moves on pneumatic wheels or crawler tracks and its aim is to (a) pull the screed, (b) receive the mixture from the truck and transfer it to the paver and (c) provide mechanical, electronic and hydraulic energy for the function of all systems of the device.

The basic components of a tractor unit are the following: hopper, push rollers, conveyor, flow control gates, side arms, augers and automations and control panel.

8.6.1.1 Hopper

The hopper is the area where the hot asphalt is discharged from the trucks (lorries). The sides of the hopper (wings) can fold hydraulically towards the centre to ensure that all the materials are 'pushed' onto the conveyor.

The capacity of the hopper depends on the model of the paver. In most models, its capacity varies from 5.5 to 6.5 m³, which in terms of mass is approximately 10 to 13 tonnes. The width of the hopper is normally a bit wider that the width of the delivery truck, that is, approximately 2.5 m, but in some models, it can be up to 3.5 m.

8.6.1.2 Push rollers

The rotating rollers are mounted at the front of the paver, and they are in contact with the tyres of the delivery truck and used to push the delivery truck.

8.6.1.3 Conveyor

The conveyor is the mechanism that moves the hot asphalt from the hopper through the paver to the augers in front of the screed.

8.6.1.4 Flow control gates

The flow control gates are adjustable metal plates that control the amount of material flowing out of the hopper by changing the height of the opening at the hopper.

8.6.1.5 Side arms

The side arms connect the screed to the tractor at the 'tow' points and tow the screed. The tow points are height adjustable and dictate the thickness of the mat.

8.6.1.6 Augers

The augers receive the material from the conveyor and distribute it transversely and uniformly to the full width of the screed. The augers' speed and height can be adjusted. Changing the height of the augers is required when the layer thickness or the mix coarseness is changed. Thus, the performance of the device is maximised.

The start and stop of the augers' operation, as well as the rotating velocity, are determined by the operator.

8.6.1.7 Automations and automatic level and slope control device

The tractor unit also carries all the automations for material handling and the automatic level (grade) and slope control device.

The automatic level control device caters to the layer's thickness and its evenness. Today, there are various types of sophisticated automatic control levelling systems, but all of them work on the same principle, to maintain the tow points at a constant height from a reference datum level.

The reference datum can be fixed or mobile with different attachments to suit each form.

Fixed datum can be a tensioned reference wire placed on one side and near the paver, the adjacent mat or even the kerb, if it is not misplaced. The tensioned reference wire is typically used in new roads when paving the first asphalt layer.

The mobile reference datum can be provided by a levelling (floating) averaging beam or by sonic sensors.

The levelling averaging beam is towed by the tractor alongside the paver, while the sonic sensors (usually four in number) are mounted on a rigid beam on the side of the paver.

Apart from the levelling control system, there is also the slope control system that is integrated with the levelling control system in modern pavers. The slope control system is particularly useful when laying base asphalt layers, thick asphalt layers or asphalt layers in general, on curved sections.

8.6.2 The screed

The screed is perhaps the most important part of the asphalt paver. It is essentially a wide flat piece of heated steel that levels and pre-compacts the asphalt placed in front of it by the augers to a specific width, grade and cross slope or crown profile.

All asphalt screeds are of the self-levelling floating type. In other words, the screed 'floats' (rides up or down) on the asphalt seeking the level where the path of its flat screed plate is parallel to the direction of pull. A floating-type screed is shown in Figure 8.12.

The forward motion (paver's stop and start and paver's speed change), the amount of head material (F3) and the angle of attack (F2) affect the performance of the screed in laying a uniform mat thickness. The weight of the screed practically remains constant during the laying process and can therefore be assumed to have no influence on the performance of the screed.

Frequent starts/stops of the paver should be avoided and uniform travel speed should be aimed during the laying process.

Changes of the angle of attack affect the horizontal and vertical forces acting on the screed and should be kept to as few as necessary, when altered manually (manual adjustment of the depth screws by screed operator).

Controlling the head of the material is perhaps the most important factor of the three affecting the performance of the screed. Once the head (amount) of the material has been established, it should be kept as constant as possible. Excess amount (volume) of material in front of the screed will cause the screed to rise and an increase in the thickness of the laying mat. Starving the screed of the material will result in settling of the screed and decrease in the thickness of the mat.

Ideally, the head of the material should be maintained at a height that is exposed to half of the auger and should not vary by more than ±25 mm (Asphalt Institute 2007). Today, many pavers have automatic controls that continuously regulate the flow of the asphalt from the hopper through the conveyors to the augers.

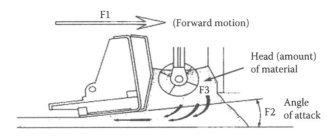

Figure 8.12 Floating screed details of an asphalt paver. (From Hunter R.N., *Bituminous Mixtures in Road Construction*. London: ICE Publishing, 1994.)

8.6.2.1 Screed components

The main components and adjustments of the screed are (a) the pre-compaction system, (b) the screed heating system and (c) the crown point.

The *pre-compaction system* characterises the screed and can be the vibratory type, tamper type, or pressure bar(s) type. Some manufactures offer a combination of the vibratory and tamper type. High-compaction screeds were introduced in the late 1990s.

In the vibrating screed (V-screed), the vibration is provided to the screed plate by hydraulic motors driving shafts that have eccentrically loaded weights positioned on them. Excess vibration should be avoided since it may cause bitumen to rise to the surface. Additionally, it is recommended to not use the vibrators when paving thin lifts in order to avoid possible fracturing of the aggregates. The frequency of the vibration in all modern vibrating screeds is adjustable. With the use of nuclear density gauge, it is possible to determine the proper frequency of the screed plate to achieve the required initial density for a given type of asphalt.

A tamping screed (T-screed) consists of a blade edge oscillating by tampers in a vertical or slightly inclined plane ahead of the screed plate. The tamping action strikes the hot asphalt laid in front of the screed, resulting in a degree of initial compaction. Tampers normally operate with high-amplitude oscillation at a comparatively low frequency. Some screeds may have double tamper blade edges (one behind the other) giving additional initial density to the hot asphalt being laid.

Most screeds nowadays are equipped with tamper and vibration action and are known as combined screeds (vibratory/tamper screeds, VT-screeds). The tampers tuck the asphalt laid and the vibrators compact to higher density. Hence, the use of combined screeds provides a high degree of pre-compaction, which means that less effort is required from the rollers during the rolling (compaction) procedure.

The pressure bar(s) pre-compaction system is offered by some manufactures as an additional system to achieve a higher screed compaction. Hence, the screeds are distinguished as 'high-compaction screeds'. The pressure bar(s) system, in contrast to the tamper, is in constant contact with the asphalt and pressed down by high-frequency hydraulic, variable impulses.

The pressure bar(s) system is equipped with one pressure bar (P1) or two pressure bars (P2) and always comes with a tamper system (T), such that the screed is distinguished as TP1 or TP2. Additionally, the pressure bars system may come with a tamper (T) and vibratory system (V), to achieve the highest possible screed compaction (pre-compaction). Thus, the screed is distinguished as a TVP2 screed.

The advantages of 'high-compaction screeds' are as follows: (a) they minimise the compaction effort on subsequent compaction equipment to achieve the required densities, (b) they open up the possibility of single-layer construction of thicker layer and (c) they minimise the effect of environmental factors on the cooling of asphalt mat.

With regard to the degree of compaction, more than 90% of the target density can be achieved with 'high-compaction screeds' compared to conventional screeds (T, V or TV) where only 75% to 85% can be achieved.

The *heating screed system* uses propane gas, diesel or electrically powered elements. The latter minimises fumes and provides a more even heat across the entire screed plate. Heating of the screed bottom plate at the start of the paving shift prevents hot asphalt from sticking to the screed, and eliminates temperature loss of the asphalt.

The *crown point* is positioned at the middle of the screed and grants the screed the ability to be crowned at the centre, yielding a positive (up to +5%) or a negative (up to –2.5%) crown to the profile of the asphalt being laid.

Modern extending screeds are capable of even handling positive or negative gull wing profiles, known as M or W profiles (Joseph Vögele AG 2014).

8.6.2.2 Type of screeds

There are two types of screeds, the fixed type and the variable type. In the fixed type, the width of the screed is fixed and, when needed, is changed manually. In the variable type, the width of the screed is changed telescopically by the push of a button. The latter is the more common type of screed, although in some applications, fixed screeds have been found to be the most economical option.

The operating width for variable-type screeds typically ranges from approximately 2.5 to 9.5 m, while for fixed/manually extended-type screeds, the operation width may even reach 16.0 m.

8.6.3 Critical points during paving

Given the fact that the asphalt arrives on site at the appropriate temperature and that the rate of material supply is such to ensure uninterrupted paving, the following additional critical points must be considered for successful paving:

a. *Head of material.* The head (volume or quantity) of the asphalt in front of the screed should be correct for the screed type and type of asphalt used. It should also be maintained constant throughout the paving operation. A higher (or a lower) amount will certainly affect the thickness and evenness of the layer. Figure 8.13 explains the effect of the head of material in front of the screed.

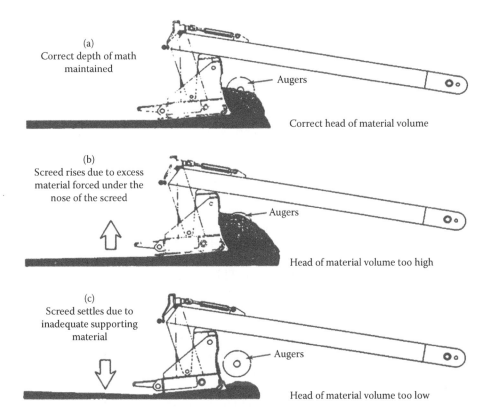

Figure 8.13 The effect of head of material. (From Blaw-Knox Construction Equipment Co. Ltd., *Paving Manual.* Rochester: Blaw-Knox Construction Ltd., 1986; Hunter R.N., *Asphalts in Road Construction.* London: Thomas Telford Publishing, 2000.)

b. *Paver speed motion.* The paver should travel at constant speed (typical travel speed, approximately 3 m/min) and frequent stoppages should be avoided. Both can result in uneven surface.

c. *Proper use of automations related to level and slope control.* Appropriate staff training is required, based on the understanding of basic system function principles.

d. *Heating and cleanness of the screed.* The screed should always be heated before paving (approximately 100°C to 150°C depending on the type of screed and asphalt), and its cleanness should be checked daily before paving operation begins. This will ensure smooth paving.

e. *Use of appropriate screed frequency or amplitude.* The appropriate vibrating frequency or tamping amplitude should be determined on the type of asphalt and thickness laid. This will ensure effective pre-compaction and will eliminate the possible rise of bitumen to the surface.

f. *Augers' height and control.* Position the augers to the appropriate height relative to the type of asphalt and screed used. Control also their turning speed to ensure an even distribution of the asphalt across the paving width and to avoid possible segregation of the bituminous material.

g. *Monitoring layer thickness.* The layer thickness shall be continuously monitored using a specifically modified metal ruler. Make fine adjustments, if needed.

h. *Inspect and repair irregularities or potholes before paving.* It is advisable, before paving, to inspect the surface to be paved for surface irregularities. They should be corrected before paving begins.

8.7 COMPACTION OF ASPHALTS

The compaction of asphalt layers is possibly the most critical stage of asphalt works. It is needed to achieve proper and uniform compaction, which in turn ensures a better long-lasting performance.

During compaction, the coated aggregates are compressed, are re-oriented and take such positions that the distance between them becomes the smallest possible. As a consequence, the air voids decrease and the mixture density increases.

Because of aggregate re-orientation, the stability of the mix and the strength of the asphalt and of the pavement increase.

The reduction of air voids results in a mixture that is less pervious to air and water and increases the stiffness of the asphalt. The reduction of permeability is mainly due to the fact that the number of interconnected voids decreases. In case of under-compaction, the number of interconnected voids is relatively large. However, the number of interconnected voids may also increase in case of over-compaction. This is more predominant in dense-graded asphalts owing to the development of internal 'hydraulic' pressures related to the excessive reduction of available space (voids). Considering the above, it is obvious that an optimum void content, not the minimum possible, is required to be achieved during compaction.

The reduction of permeability attributed to proper compaction has the benefit of not only permitting smaller water quantity to pass through the layer to the underlying layers, bearing the well-known negative results, but also allowing less air to move through the asphalt. As a consequence, the rate of bitumen oxidation decreases.

Powell and Leech (1983) found that the reduction of voids to the optimum content resulted in an asphalt stiffness increase by approximately 30% in macadam mixtures. This stiffness increase led to the decrease of the unbound base thickness by 8%. The positive effect of the

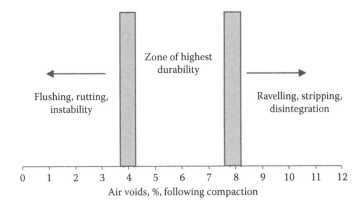

Figure 8.14 The effect of air voids obtained during compaction on the durability of asphalt concrete layer. (From Asphalt Institute, *The Asphalt Handbook*, MS-4, 7th Edition, Lexington, KY: Asphalt Institute, 2007. With permission.)

void reduction on the asphalt stiffness, as well as on the pavement service life, has also been documented by other researchers (Goddard et al. 1978; Lister and Powell 1977).

Improper compaction (over- or under-compaction) will certainly cause premature rutting.

Thus, the aim during compaction is to achieve an optimum void content and at the same time to ensure a smooth surface. An asphalt concrete immediately after laying has a void content ranging from 15% to 20%, using conventional screeds. The task of the rollers is to reduce this content to approximately 8% or less. Air voids of less than 3.5% after compaction should be avoided, since rutting, flushing and instability of the mix will most certainly occur. The effect of air voids on the pavement's performance is graphically illustrated in the Figure 8.14.

8.7.1 Factors affecting compaction

The factors affecting compaction are (a) aggregate material, (b) bitumen grade and compaction temperature, (c) environmental conditions, (d) layer thickness, (e) compaction equipment and (f) compaction procedure.

8.7.1.1 Aggregate material

Aggregates, with respect to their particle size distribution, shape and surface texture, directly affect the asphalt workability/compatibility. Open-graded mixtures have better workability and require a smaller compaction effort than dense-graded mixtures. Additionally, an increase of coarse aggregate content reduces the workability and increases the compaction effort.

When rounded and smooth-surfaced aggregates are contained in the mixture, its workability increases. On the contrary, the workability decreases in mixtures containing angular aggregates or aggregates with high surface texture. It is known that workability/compatibility is improved by the addition of natural sand or by the use of uncrushed aggregates instead of crushed aggregates. It is noted that improving workability/compatibility is not as demanding in counties with low ambient temperatures.

Finally, high percentage of filler can have a negative impact on the asphalt workability/compatibility.

8.7.1.2 Bitumen grade and compaction temperature

The grade or type of bitumen and its quantity in the mixture are the major factors affecting asphalt compaction. Each grade bitumen, at a specific temperature, has its respective 'hardness' and viscosity. As a consequence, when 'hard' bitumen is incorporated into the asphalt, the compaction at a given temperature is more 'difficult' than the one of asphalt containing 'soft' bitumen.

To phase this problem, the asphalt produced with hard bitumen is compacted, and thus produced, at higher temperatures. The appropriate compaction temperature is determined by the viscosity/temperature relationship (see Section 3.9.3).

As a general rule, compaction of asphalt with grade bitumen should never start when the mix temperature is less than 85°C to 90°C. In most cases, insufficient compaction is attributed to the low mix temperature starting the compaction. However, there may be the case that the temperature of the asphalt before compaction is high and the mix cannot sustain the weight of the roller. This situation is solved by simply waiting the temperature of the mix to drop, which only delays compaction procedure.

A point that should also be noted is the uniformity of the temperature throughout the mass of the mixture. As the mixture is laid, the mass coming in contact with the underlying layer immediately loses some heat because of thermal conductivity. The heat loss rate depends on the temperature difference between the underlying layer and the asphalt laid. The higher the difference, the higher and quicker the heat loss is. Similarly, the asphalt quantity on the surface loses heat quickly because of heat emission into the atmosphere. Thus, the temperature distribution curve in reference to the layer thickness has the form of the curve shown in Figure 8.15. The shape of the curve changes slightly when temperature is measured at different time lapsed from laying and for different mat thicknesses. More details can be found in Shell Bitumen (2003).

The above non-uniformity of temperature affects the density of layer with respect to depth. Research showed that the achieved densities follow the same shape as the curve shown in Figure 8.15 (Lister and Powell 1977).

Finally, the increase of bitumen content has, up to a point, a beneficial effect on the asphalt workability/compaction. This derives from the fact that the bitumen operates as a 'lubricant' medium between aggregates and decreases aggregate-to-aggregate friction during compaction. However, asphalts rich in bitumen, if they also contain harder bitumen, are more difficult to compact.

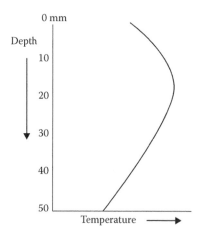

Figure 8.15 Indicative temperature distribution curve in the asphalt layer.

8.7.1.3 Environmental conditions

Environmental conditions, mainly ambient temperature and wind speed, are the main parameters affecting compaction and the duration during which compaction should be completed. Low ambient temperatures and high wind speeds demand a shorter duration of compaction.

8.7.1.4 Layer thickness

The layer thickness affects the ease in achieving the desired degree of compaction. In general, the thicker the layer, the easier it is to achieve the desired compaction, since it retains its heat for a longer period (lower rate of heat loss).

Layers with a thickness between 25 and 40 mm, if possible, should not be laid during cold winter months, or greater attention should be given to the duration of compaction, which should be as short as possible.

More information about this subject can be found in Brown (1980), Daines (1985) and Nicholls et al. (2008).

8.7.1.5 Compaction equipment

Effective compaction is related to the type of compaction equipment used. Asphalt layer compaction equipment consists of self-propelled vehicles, known as rollers, which compact the asphalt layer by the effect of their self-weight or by additionally imposing dynamic loading. The desired compaction is achieved by applying a certain number of passes of the rollers over the asphalt layer, known as compaction effort.

The number of passes is always determined in situ and it depends on the asphalt type, thickness of layer, weather conditions and type and weight of roller.

The surface of the cylinders or tyres that come in contact with the asphalt layer should be clean and in good condition. Dirt or oil on the surface of the cylinders or of the tyres will cause premature failures, while rough and worn surfaces may leave roller marks on the surface.

There are four types of rollers: (a) static steel-wheel rollers, (b) vibrating steel-wheel rollers, (c) pneumatic-tyre rollers and (d) combination rollers.

8.7.1.5.1 Static steel-wheel rollers

Static steel-wheel rollers were developed first and are used in all rolling (compaction) phases (initial or breakdown rolling, intermediate rolling and finish rolling). Static steel-wheel rollers are available in two basic types, the tandem steel-wheel rollers (also called double-drum rollers) and the three-steel-wheel rollers (also called steel drum rollers). The tandem steel-wheel roller bears a steel drum at the front and a steel drum at the back, while the three-steel-wheel roller, shown in Figure 8.16, bears two steel wheels at the front and a steel drum at the back.

Research showed that the three-steel-wheel roller may result in a non-uniform compaction compared to the tandem steel-wheel roller with which more uniform compaction can be ensured (Lister 1974). However, three-steel-wheel rollers provide better manoeuvrability.

The static weight of steel-wheel rollers varies from 3 to 13 tonnes depending on the model and ballast used. The typical static weight for compacting asphalt layers of thickness 4.0 to 6 cm is 8–12 tonnes. The diameter of drums or steel wheels and their width vary depending on the model used. The larger the drum diameter, the smaller the stress applied to the compacted surface, for the same static weight.

Figure 8.16 Three-wheel static roller. (Courtesy of Atlas Copco.)

8.7.1.5.2 *Vibrating rollers*

Vibrating rollers are widely used in all asphalt compaction works because they are very versatile and effective, especially when thick mats are to be compacted. They consist of two smooth-surface steel drums equipped with a vibratory system. They compact through a combination of static and dynamic loading. Modern vibrating rollers are capable of varying both frequency and amplitude. Figure 8.17 shows a double-drum vibrating roller.

The advantage of vibrating rollers is that the desired compaction is achieved in less number of passes in contrast to all the other rollers. Furthermore, vibrating rollers can achieve better compaction or compact more effectively thicker layers (Nunn 1985; Tunnicliff 1977).

The frequency and the amplitude of the drums can be altered by the operator. Normally, high frequency/high amplitude is used for thick layers and high frequency/low amplitude is used for thin layers.

They come in a range of width and dead weights. The static operating mass may vary from 2000 to 18,000 kg and the drum width varies from 0.8 to 2.13 m for typical asphalt works. In general, the static linear load for vibrating rollers is less than that for steel-wheel rollers.

Figure 8.17 Double-drum vibrating roller. (Courtesy of Atlas Copco.)

Vibrating rollers can operate as static rollers too. As a consequence, they can be used in all phases of compaction. It is noted that no vibration should be used during the final compaction phase, since there is a risk of damaging the structure of the asphalt and force the bitumen to the surface.

8.7.1.5.3 Pneumatic-tyre rollers

Pneumatic-tyre rollers consist of a number of tyres, usually seven or nine in total. The number between front and back tyres always differs by one tyre. An example of pneumatic-tyre roller with four tyres at the front and three at the back is shown in Figure 8.18.

Pneumatic-tyre rollers differ from the steel rollers in respect to the way they compact; compaction with pneumatic-tyre rollers takes place under a kneading action.

The advantage of pneumatic-tyre rollers is that the contact pressure can easily change by changing the tyre pressure. Modern pneumatic-tyre rollers have the ability to change the tyre pressure during operation.

As a result, the same roller can be used for all types of asphalts ranging from very flexible to very stiff, as well as for very thick to very thin layers; the roller's static load can be fluctuated by ballast placed in a space between front and rear tyres. The typical mass of pneumatic-tyre rollers for asphalt works ranges from 6 tonnes up to approximately 27 tonnes (with ballast) depending on the model.

The pneumatic-tyre roller is usually used at the intermediate and final phase of compaction. Using the pneumatic-tyre roller at the final compaction phase has the advantages of obtaining a better surface texture and eliminating hair cracks, which may be created when compacted with steel-wheel rollers. The latter is common when final compaction is carried out at a lower than appropriate temperature.

8.7.1.5.4 Combination rollers

This type of rollers is equipped with a vibrating drum at the front and pneumatic tyres at the rear (usually four). The vibrating drum ensures quicker compaction and the rubber wheels ensure denser and smoother surface. The operating maximum mass varies from approximately 2500 to 11,000 kg and the drum width varies from approximately 1030 to 1950 mm, depending on the model. An example of a vibrating/pneumatic roller is shown in Figure 8.19.

Figure 8.18 Pneumatic-tyre roller (four front and three back tyres). (Courtesy of AMMANN Group.)

Figure 8.19 Combination roller. (Courtesy of Atlas Copco.)

8.7.1.5.5 General remarks about rollers

Modern rollers are equipped with GPS and devices measuring the density and the temperature of the mat during rolling. Colour displays provide additional information of areas that need further compaction (coverage) or those where over-compaction should be prevented. The implementation of such devices is known as intelligent compaction (IC) technology.

With hot mix asphalt IC, tracking roller passes and surface temperatures provide the necessary means to maintain a consistent rolling pattern within optimal ranges of temperatures for 100% coverage of a construction area. Additionally, IC technologies can be especially beneficial to maintain consistent rolling patterns under lower visibility conditions such as night paving operations (Chang et al. 2011). Additional information for IC technology can be found in Gallivan (2011a and 2011b) and Chang et al. (2012).

Any type of roller can be used for the compaction of a hot asphalt layer. In general, lighter steel drum rollers or pneumatic rollers are used in mats of thickness less than 4 cm and heavier rollers are used in thicker mats.

Vibrating rollers should only be used to compact asphalt layers of thickness greater than 5 cm, unless the vibrating roller is capable of altering the frequency and the amplitude, and experience has shown that satisfactory compaction can be obtained for the type of asphalt used.

Pneumatic-tyre rollers are very useful for compacting thin and very thin asphalt layers or for primary compaction of overlays in general, or levelling courses over uneven surfaces. Pneumatic rollers have been found to eliminate the hairline cracks that appear on the finished surface when compaction is carried out at lower than permitted temperatures.

On thick mats, a typical rolling sequence consists of a vibrating roller for breaking down compaction, followed by a vibrating roller or a pneumatic roller for intermediate compaction and finishing with a static steel-wheel roller or vibrating roller operating in the static mode.

8.7.1.6 Rolling procedure

To achieve proper and effective compaction of asphalt layers, the following points are recommended:

1. Rolling should start as quickly as possible after asphalt has been laid, provided the temperature of the mix is not too high and has developed sufficient stability to withstand the weight of the roller.

2. Rolling consists of three consecutive phases: the initial or breakdown rolling, the intermediate rolling and the finish rolling.

   Most of the compaction is achieved during breakdown rolling. Intermediate rolling increases the density of the mix further and minimises all surface pores. During finish rolling, all roller traces and other surface deficiencies are removed. Between the three phases, there should be no time delay.

3. The number of rollers required to be used is determined by the width of paving lane; for a typical paving lane width of 3.5 to 3.75 m, usually two or more rollers are required. As for the width of the roller, it is usually chosen to be approximately equal to one-third of the width of the paving lane.

4. Rolling always starts from the lowest point of the mat, in case of transverse slope.

5. The roller moves twice over the same rolling path, by moving forwards and backwards; then, the roller changes rolling path (lane). The same applies to all subsequent rolling.

6. When longitudinal joint is formed, rolling starts from the joint. The roller moves over the hot mat with approximately 200 mm of its drum overlapping the already compacted mat. This is known as hot-side rolling.

   Figure 8.20 shows a typical hot-side rolling pattern of a steel roller during breakdown rolling when the longitudinal joint has been formed. Cold-side rolling, that is, placing the roller over the cold mat and gradually moving it towards the hot mat, should be avoided because the hot mat is cooling while the roller operates on the cold mat.

7. The length of the first and second rolling pass, and all subsequent passes, is mainly dependent on the thickness of the mat. Longer rolling lengths can be used on thick mats (more than 60 mm in thickness) in comparison to thin mats. Usually, an ideal length is between 30 and 40 m for a mat of asphalt concrete of 100 mm in thickness.

8. In case paving starts from a transverse cold joint, with or without the formation of longitudinal joint, a number of different techniques are used; in all cases, the transverse joint is cross rolled first.

9. During rolling, particularly at the start, the surface of cylinder or of the tyres is sprayed with a small quantity of water to avoid mixture adhering to the surface of the cylinder or the tyres.

10. Rollers should move at low speeds, not higher than 5 km/h for static or vibrating rollers and not higher than 8 km/h for rubber ones. The selected speed should be retained constant throughout rolling.

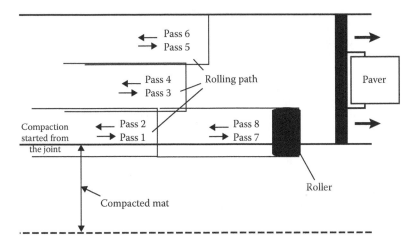

Figure 8.20 Typical rolling pattern after the formation of the longitudinal joint.

11. Abrupt changes of rolling path as well as the back-and-forth movement over short lengths should be avoided. Changing rolling lanes (paths) should be done outside the freshly laid material to avoid damaging it. Additionally, the roller is not permitted to stop and stand over a freshly laid mat.

12. With static steel-wheel rollers, rolling should progress with the drive wheel forward in the direction of paving. The greater weight of the front drum and the higher turning force tuck the material more efficiently. Otherwise, there is a danger of 'pushing' rather than tucking the asphalt. An exception to the above rolling rule occurs when compacting on high gradient (longitudinal slope more than 5%).

13. Breakdown rolling is recommended to be carried out with static or vibratory steel drum rollers, and finish rolling, with pneumatic rollers or static steel drum rollers. For levelling courses or thin layers (less than 40 mm), breakdown rolling is preferred to be carried out by pneumatic rollers.

Additional useful information about laying, compaction and the construction procedure of asphalts in general can be found in Asphalt Institute (2007), Nicholls et al. (2008) and Milster et al. (2011).

REFERENCES

AASHTO T 195. 2011. *Determining degree of particle coating of asphalt mixtures.* Washington, DC: American Association of State Highway and Transportation Officials.

AMMANN Group. 2014. Available at http://www.ammann-group.com.

Asphalt Institute MS-3. *Asphalt Plant Manual,* 5th Edition. MD: Asphalt Institute.

Asphalt Institute MS-4. 2007. *The Asphalt Handbook,* 7th Edition. Lexington, KY: Asphalt Institute.

ASTEC Industries Inc. 2014. Available at http://www.astecinc.com.

ASTM D 2489/D 2489M. 2008. *Standard practice for estimating degree of particle coating of bituminous-aggregate mixtures.* West Conshohocken, PA: ASTM International.

Atlas Copco. 2014. Available at http://www.atlascopco.com.

Blaw-Knox Construction Equipment Co. Ltd. 1986. *Paving Manual.* Rochester: Blaw-Knox Construction Ltd.

Brown J.R. 1980. *The Cooling Effects of Temperature and Wind on Rolled Asphalt Surfacings.* TRRL Supplementary Report SR 624. Crowthorne, UK: Transport Research Laboratory.

Chang G., Q. Xu, and J. Rutledge. 2012. *Intelligent Compaction: Quality Assurance for In-Place Density Acceptance Experiment Plan for the Asphalt IC Demonstration in Utah.* FHWA-RD-12. Springfield, VA: National Technical Information Service.

Chang G., Q. Xu, J. Rutledge, B. Horan, L. Michael, D. White, and P. Vennapusa. 2011. *Accelerated Implementation of Intelligent Compaction Technology for Embankment Subgrade Soils, Aggregate Base, and Asphalt Pavement Materials-Final Report.* FHWA-IF-12-002. Springfield, VA: National Technical Information Service.

Daines M.E. 1985. Cooling of bituminous layers and time available for their compaction. *Proceedings of 3rd Eurobitume Symposium,* p. 237. Brussels: Eurobitume.

EAPA and NAPA. 2011. *The Asphalt Paving Industry: A Global Prospective,* 2nd Edition, GL 101. Brussels: EAPA.

Gallivan V.L., G.K. Chang, and D.R. Horan. 2011a. Practical implementation of intelligent compaction technology in hot mix asphalt pavements. *Journal of the Association of Asphalt Paving Technologists,* Vol. 80, p. 1.

Gallivan V., G.K. Chang, and D.R. Horan. 2011b. Intelligent compaction for improving roadway construction. *Proceedings of GeoHunan International Conference: Immerging Technologies for Materials, Design, Rehabilitation and Inspection of Roadway Pavements,* pp. 117–124. American Society for Civil Engineers (ASCE).

Goddard R.T., W.D. Powell, and M.W. Applegate. 1978. *Fatigue Resistance of Dense Bitumen Macadams: The Effect of Mixture Variables and Temperature*. TRRL Supplementary Report SR 410. Crowthorne, UK: Transport Research Laboratory.

Hunter R.N. 1994. *Bituminous Mixtures in Road Construction*. London: ICE Publishing.

Hunter R.N. 2000. *Asphalts in Road Construction*. London: Thomas Telford Publishing.

Joseph Vögele AG. 2014. Available at http://www.voegele-ag.de.

Lister N.W. 1974. *Levels of Compaction of Dense Coated Macadam Achieved during Pavement Construction*. TRRL Laboratory Report LR 619. Crowthorne, UK: Transport Research Laboratory.

Lister N.W. and W.D. Powell. 1977. *The Compaction of Bituminous Base and Basecourse Materials and Its Relation to Pavement Performance*. TRRL Supplementary Report SR 260. Crowthorne, UK: Transport Research Laboratory.

Milster R., W. Emperhoff, K. Leipheim, C. Lips, and R. Mansfeld. 2011. *Guidance for Asphalt Paving Operations*, 2nd Edition, English translation January 2011. Deutscher Asphaltverband e.V. (German Asphalt Pavement Association). Bonn: DAV.

Nicholls J.C., M.J. McHale, and R.D. Griffiths. 2008. *Best Practice Guide for Durability of Asphalt Pavements*. TRL Report Note 42. Crowthorne, UK: Transport Research Laboratory.

Nunn M.E. 1985. *Compaction of coated macadam by Bomag BW 75ADL vibratory roller*. TRRL Research Report 56. Crowthorne, UK: Transport Research Laboratory.

Powell W.D. and D. Leech. 1983. *Compaction of Bituminous Road Materials Using Vibratory Rollers*. TRLL Laboratory Report LR 1102. Crowthorne, UK: Transport Research Laboratory.

Shell Bitumen. 1990. *The Shell Bitumen Handbook*. Surrey, UK: Shell Bitumen UK.

Shell Bitumen. 2003. *The Shell Bitumen Handbook*, 5th Edition. London: Thomas Telford Publishing.

Tunnicliff D.G. 1977. Symposium on vibratory compaction of asphalt pavements. *Journal of the Association of Asphalt Paving Technologists*, Vol. 46, p. 259.

ANNEX 8.A

Table 8.A.1 Typical specific gravities of bituminous binders at various temperatures

Temperature (°C)	Specific gravity at 25°C					
	1.00	*1.01*	*1.02*	*1.03*	*1.04*	*1.05*
15.5	1.006	1.016	1.026	1.036	1.046	1.056
25	1.000	1.010	1.020	1.030	1.040	1.050
45	0.988	0.998	1.008	1.018	1.028	1.038
60	0.979	0.989	0.999	1.009	1.019	1.029
90	0.961	0.971	0.981	0.991	1.001	1.011
100	0.955	0.965	0.975	0.985	0.995	1.005
110	0.949	0.959	0.969	0.979	0.989	0.999
120	0.943	0.953	0.963	0.973	0.983	0.993
130	0.937	0.947	0.957	0.967	0.977	0.987
140	0.931	0.941	0.951	0.961	0.971	0.981
150	0.925	0.935	0.945	0.955	0.965	0.975
160	0.919	0.929	0.939	0.949	0.959	0.969
170	0.913	0.923	0.933	0.943	0.953	0.963
180	0.907	0.917	0.927	0.937	0.947	0.957
190	0.901	0.911	0.921	0.931	0.941	0.951
200	0.895	0.905	0.915	0.925	0.936	0.945

Source: Shell Bitumen, *The Shell Bitumen Handbook*. Surrey, UK: Shell Bitumen UK, 1990.

Table 8.A.2 Conversion factors linking volume to mass for bitumen at various temperatures and specific gravities

| Temperature (°C) | Specific gravity at 25°C | | | | | |
	1.00	1.01	1.02	1.03	1.04	1.05
	l/tonne					
25	995	984	973	963	953	943
45	1010	999	988	978	968	958
60	1020	1009	998	988	978	968
90	1041	100	1019	1009	999	989
100	1047	1036	1026	1015	1005	995
110	1054	1043	1032	1022	1011	1001
120	1060	1049	1038	1028	1017	1007
130	1067	1056	1045	1034	1024	1013
140	1074	1063	1052	1041	1030	1019
150	1081	1070	1058	1047	1036	1026
160	1088	1076	1065	1054	1043	1032
170	1095	1083	1072	1060	1049	1038
180	1103	1091	1079	1067	1056	1045
190	1110	1098	1086	1074	1063	1052
200	1117	1105	1093	1082	1070	1058

Source: Shell Bitumen, *The Shell Bitumen Handbook*. Surrey, UK: Shell Bitumen UK, 1990.

Chapter 9

Quality control of production and acceptance of asphalts

9.1 GENERAL

Asphalt production control is necessary in order to ascertain that the produced asphalt complies with the mix formulation and to verify a good and stable mix plant operation.

Applying procedures and conducting quality control tests at the mixing plant during production will help ensure that the above is achieved. According to the European standard CEN EN 13108-21 (2008), this stage is called Evaluation of Conformity (EC) of all bituminous mixtures produced under specifications. The EC is composed of the Initial Type Testing (ITT) and the Factory Production Control (FPC).

The ITT consists of inspection and tests on the constituent materials so as to document their conformity with the requirements specified.

The FPC consists of inspection and tests on finished asphalt and is the permanent internal control of the production process.

In addition to the above, in each asphalt production plant, procedures are established and implemented to ensure good management, storage and delivery of the materials, as well as calibration and maintenance of the asphalt plant.

To ensure all the abovementioned, apart from the laboratory equipment, the asphalt producer should also have the following: (a) quality plan; (b) organisation plan with assigned responsibilities; (c) control procedures for material testing, handling, storage and delivery as well as for plant calibration and maintenance and (d) inspection and testing program.

In this chapter, an extensive reference is made only to the inspection and testing program on the constituent materials delivered in the plant (aggregates, bitumen, etc.) and on the final product, namely, the asphalt. Information about items (a) to (c), as well as more information about (d) and additional quality assurance procedures can be found in CEN EN 13108-21 (2008).

This chapter also covers the testing for acceptance of delivered and laid asphalt.

9.2 SAMPLING

The material control is conducted on representative samples received on a regular basis, given that it is impractical to test all deliveries and all daily production. Taking representative samples is of great importance, since otherwise results may lead to erroneous conclusions.

The representativeness of samples is directly related to the quantity taken to conduct the particular test. The sampling location does not affect the sample representativeness, since all sampling locations are theoretically equivalent.

Table 9.1 Minimum sampling quantities of asphalt and binder material

Nominal maximum aggregate size	Minimum weight of sample		
	Aggregates[a] (kg)	Asphalt[b] (kg)	Binder material[c]
2.36 mm	5	2	Bitumen:
4.75 mm	5	2	1 lt
9.5 mm	5	4	Bituminous
12.5 mm	10	5	emulsion:
19.0 mm	20	7	4 lt
25.0 mm	30	9	
37.5 mm	50	11	

[a] According to ASTM D 75 (2009) or AASHTO T 2 (2010).
[b] According to ASTM D 979 (2012) or AASHTO T 168 (2011).
[c] According to ASTM D 140 (2009) or AASHTO T 40 (2012).

The minimum representative sampling quantity for aggregate and asphalt is determined by the nominal maximum aggregate size (D) in combination with the type of the test to be conducted.

According to American standards, minimum aggregate and asphalt sampling quantities for grading and binder content determination are as shown in Table 9.1. The minimum bituminous binder sampling quantity for routine testing is also given in the same table.

According to European standards, the minimum sampling quantities are determined by the appropriate EN sampling method, as well as the test to be conducted.

Sampling should also be random regarding the time, the quantity and the length when in-place sampling takes place.

More information about random sampling is given in ASTM 3665 (2012) and AASHTO R 9 (2009).

9.3 METHODS OF MATERIAL SAMPLING

There are various sampling methods depending on the type of material and location of sampling. The usual appropriate sampling methods, which can be applied at the mixing plant or on site, are outlined below.

9.3.1 Methods of sampling aggregate materials

Aggregate sampling for conducting the required tests may be carried out from (a) a stockpile, (b) a conveyor belt and (c) hot bins.

9.3.1.1 Sampling from a stockpile

When sampling from a stockpile, great attention should be given since aggregates tend to segregate when stored in a stockpile. A good technique is to take a quantity of aggregates with a loader first and, from this quantity, to take the quantity needed for laboratory testing.

9.3.1.2 Sampling from a conveyor belt

Sampling from a conveyor belt is used for sampling both individual aggregate fractions and aggregate mixture. In most production plants, conveyor belt sampling presupposes stoppage of the conveyor belt.

9.3.1.3 Sampling from hot bins

Sampling from hot bins is applied only to batch mixing plants and is used exclusively for carrying out fine adjustment on the proportion of aggregate fractions to obtain the target gradation.

More information regarding the methods of aggregate sampling is given in CEN EN 932-1 (1996) and ASTM D 75 (2009) or AASHTO T 2 (2010).

9.3.2 Methods of sampling bituminous binders

Sampling of the bituminous binder material, namely, bitumen, modified bitumen or bituminous emulsion, can be carried out from a delivery vehicle tank, from a plant storage tank, from pipelines during unloading or from feed pipes to the mixer.

Sampling from a delivery vehicle tank or from pipelines during unloading is more representative for the bituminous binder delivered, while sampling from a storage tank or feed pipes to the mixer is more representative for the bituminous binder to be incorporated into the mixture.

The sampling procedure is conducted in accordance to CEN EN 58 (2012), ASTM D 140 (2009) or AASHTO T 40 (2012).

9.3.3 Methods of sampling hot mix asphalt

In general, asphalt sampling for conducting the required tests may be carried out (a) from a lorry load (Section 9.3.3.1), (b) from the augers of the paver (Section 9.3.3.2), (c) from a workable mixture in heaps (Section 9.3.3.3), (d) from laid but not rolled material (Section 9.3.3.4), (e) from laid and compacted material by coring (Section 9.3.3.5), (f) from laid and compacted material by sawing out slabs (Section 9.3.3.6), (g) from the slat conveyor of a continuous process plant (Section 9.3.3.7) and (h) from a stockpile, for coated chippings only (Section 9.3.3.8). In the case of mastic asphalt, sampling is carried out during discharge from a mixer transporter.

9.3.3.1 Sampling from a lorry load

Sampling may be conducted directly after loading the asphalt to the lorry or after haulage. The sampling method is simple and it does not require any particular equipment, just a sampling shovel. However, there is a risk of unrepresentative sample to be taken owing to possible asphalt segregation during loading or haulage. Additionally, a disadvantage of this sampling method is that there is uncertainty about the precise location of the material when laid in the pavement.

The minimum asphalt quantity required, according to CEN EN 12697-27 (2000), is 12 kg for material containing aggregate of nominal size, D, less than 16 mm or 28 kg when D ≥ 16 mm. It is stated that the above quantities are taken from four different positions, 100 mm below the surface of the material, as widely spaced as possible but not less than 300 mm from the side of the lorry.

9.3.3.2 Sampling from the augers of the paver

Sampling from the augers of the paver has the following advantages: (a) there is certainty of the location of the material in the pavement; (b) there is no interruption to paving operations; (c) no special equipment is needed, just a sampling shovel and (d) sampling is easy to perform. The disadvantage of this sampling method is that the sample may not be representative owing to possible asphalt segregation at the ends of the paver augers. Additionally, attention should be given to the sampler's personal safety.

The required quantity of sampling material, minimum of 28 kg, is taken in two increments of 7 kg from each side of the paver.

9.3.3.3 Sampling of workable material in heaps

This sampling method is similar to the sampling from a truck. The same sampling quantities are required and it has the same advantages and disadvantages as the sampling from a truck. One additional disadvantage is that there is the possibility of heap contamination by other materials and hence contamination of the sample received.

9.3.3.4 Sampling from laid but not rolled material

Sampling from laid but not rolled material is conducted in two ways, by using sampling trays or from a cut trench. In both cases, this sampling method is not recommended by CEN EN 12697-27 (2000) to be used in layers where the difference between the layer thickness and the nominal aggregate size, D, is less than 20 mm.

In the first case, the sampling steel trays are placed at a certain distance apart and in a position not to be damaged by the paver. The sampling tray with its content is removed quickly usually by pulling the wire attached to one corner for this purpose.

In the second case, sampling is conducted from a trench cut transversely across to the strip laid by a sampling shovel or scoop. The trench has a depth equal to the full depth of the layer of the material. The recommended sampling quantity is the same as the one required in the sampling from a lorry load.

Both alternative sampling methods have the advantages that there is certainty of the location of the material in the pavement, there is minimal risk of segregation and there is no interruption to paving operations. The disadvantages are that there is a possibility of affecting the finished surface and it requires more labour.

9.3.3.5 Sampling from laid and compacted material by coring

Sampling by coring has the advantage that there is certainty of the location of the material been sampled. However, it requires a core-cutting machine, it affects the pavement surface in case of insufficient restoration and cutting operation affects to some extent the aggregate grading (particularly of coarse mixtures).

For the determination of the aggregate gradation, the diameter of the cores is usually 150 mm or greater, in contrast to the diameter of the cores for determination of layer thickness, air voids and density, which can be 100 mm. The number of cores required in both cases is at least two, spaced usually approximately 10 cm apart.

9.3.3.6 Sampling from laid and compacted material by sawing out slabs

Slabs, approximately 300–400 mm wide by 300–400 mm long, are cut with a wheel cutting machine at the full depth of the course to be sampled. The advantages and disadvantages of this sampling method are similar to those in sampling by coring. The only difference is that the possibility to affect the aggregate gradation by using this sampling method is eliminated. For physical properties determinations, extract and transport the slab so that damage or distortion does not occur. One slab of the abovementioned size is sufficient for aggregate grading and binder content determination.

9.3.3.7 Sampling from the slat conveyor of a continuous process plant

In case of using a continuous process plant, asphalt sampling can be carried out from the slat conveyor, once a specially constructed sampling flap is fitted on the underside of the slat conveyor.

This method of sampling is easy and the material is immediately available for testing at the plant and observations of the material can detect errors in the product at an early stage. Additionally, there is no risk of an unrepresentative sample being taken owing to aggregate segregation and there is minimal risk to the sampler's personal safety. A disadvantage of this sampling method is that there is uncertainty of the precise location of the material in pavement.

9.3.3.8 Sampling coated chipping from stockpiles

A total quantity of coated chippings of approximately 25 kg is obtained from 10 different points of the heap using a shovel. Sampling is always made at least 10 cm from the outer surface of the heap.

More information about asphalt sampling according to European standards is given in CEN EN 12697-27 (2000), whereas information about the sampling according to American standards is given in ASTM D 979 (2012) and ASTM D 5361 (2011) or AASHTO T 168 (2011).

9.4 INSPECTION AND TESTING OF INCOMING CONSTITUENT MATERIALS AND DELIVERED PRODUCT

Part of the FPC system requirements specified by CEN EN 13108-21 (2008) is inspection and testing on incoming constituent materials and on finished bituminous mixture (delivered product). As constituent materials, the following are distinguished: aggregates, filler, binders, additives and reclaimed asphalt.

The aim of the inspection and testing of the above is to ensure product quality and compatibility with the predetermined requirements.

9.4.1 Inspection and testing of incoming constituent materials

The constituent materials, such as aggregates, filler, binders, additives and reclaimed asphalt, should be inspected and tested on a regular basis.

The inspection/test frequency and the types of tests per incoming constituent material are all covered in CEN EN 13108-21 (2008). Table 9.2 gives an example of minimum inspection/test and the frequency recommended for aggregate and binder.

9.4.2 Inspection and testing of finished bituminous mixture

The produced asphalt should also be inspected and tested on a regular basis.

The inspection frequency and the types of tests as recommended by CEN EN 13108-21 (2008) are given in Table 9.3.

Other mix characteristics to be tested regardless of asphalt type mixture, according to CEN EN 13108-21 (2008), are void content and, if reclaimed asphalt is used, the penetration and the softening point (R&B) of the reclaimed binder. In the case of mastic asphalt, only the indentation on cubes (CEN EN 12697-20 2012) is required to be determined.

The frequency for the above tests depends on the operating compliance level (OCL) of the plant. For OCL A, B or C, the frequency is every 10,000 t, 5000 t or 3000 t (CEN EN 13108-21 2008).

Table 9.2 Minimum inspection and test frequencies for aggregates

No.	Inspection/test	Purpose	Frequency
1	Tests for intrinsic properties of aggregate (strength, etc.)	To check suitability for intended use	Source approval before initial use[a]
2	Inspection of delivery ticket[b]	To check consignment is as ordered and from the correct source	Each delivery
3	Organoleptic check of stockpile[b,c]	For comparison with normal appearance with respect to source, grading, shape and impurities	Daily
4	Sieve analysis	To assess compliance with standard or other agreed grading	a. First delivery from new source b. In case of doubt following organoleptic check c. 1 per 2000 t
5	Shape, crushed particle index etc.	To assess compliance with standard or other agreed specification	a. First delivery from new source b. In case of doubt c. As indicated in the quality plan
6	Moisture content	Process control	As indicated in the quality plan

Source: Reproduced from CEN EN 13108-21, *Bituminous mixtures – Material specifications – Part 21: Factory control*, Brussels: CEN, 2008. With permission (© CEN).

Note: This table may include the results of tests and inspections by the supplier as part of his FPC.

[a] Updated in accordance with CEN EN 13043 (2004).
[b] Organoleptic check is defined as the assessment made with the senses (e.g. sight, touch, smell etc.).
[c] These requirements will not apply in the case of direct supplies from an aggregate production unit to an asphalt plant on the same site.

Table 9.3 Minimum inspection/test frequencies for delivered product

No.	Inspection/test	Purpose	Frequency
1	Organoleptic check on mixed asphalt[a]	For comparison with normal appearance with respect to grading, evenness of mixing and adequacy of coating	Every load
2	Temperature	To ensure material conforms with specification or other requirements	a. As scheduled b. Whenever samples are taken
3	Grading and binder content	To ensure material conforms to mix design	See Table 9.4
4	Other characteristics included in technical specifications, see note below	To assess conformity	See note below
5	Suitability of delivery vehicles by visual assessment	To check adequacy of insulation	Prior to first use and in case of doubt
6	Cleanliness of delivery vehicles by visual assessment	To avoid contamination	Every load prior to loading

Source: Reproduced from CEN EN 13108-21, *Bituminous mixtures – Material specifications – Part 21: Factory control*, Brussels: CEN, 2008. With permission (© CEN).

Note: Other mix characteristics to be tested regardless of asphalt type mixture, according to CEN EN 13108-21 (2008), are void content and, if reclaimed asphalt is used, the penetration and the softening point (R&B) of the reclaimed binder. In the case of mastic asphalt, only the indentation on cubes (CEN EN 12697-20 2012) is required to be determined. The frequency for the above tests depends on the operating compliance level (OCL) of the plant. For OCL A, B or C, the frequency is every 10,000 t, 5000 t or 3000 t (CEN EN 13108-21 2008).

[a] Organoleptic check is defined as the assessment with the senses (sight, touch, smell etc.).

Table 9.4 Minimum frequency for analysis of finished product (tonnes/test)

| No. | Level | Operating compliance level[a] | | |
		A	B	C
1	X[b]	600	300	150
2	Y[b]	1000	500	250
3	Z[c]	2000	1000	500

Source: Reproduced from CEN EN 13108-21, *Bituminous mixtures – Material specifications – Part 21: Factory control*, Brussels: CEN, 2008. With permission (© CEN).

[a] The operating compliance level (OCL) is determined according to experience, the plant's operating time and the invariability of the product produced. A new plant starts from OCL C.
[b] Usually, levels X and Y are used in contracts.
[c] Level Z shall be the minimum test frequency applicable for all purposes.

More information about the inspection and testing on constituent materials and finished bituminous mixture as well as information about FPC requirements so that the quality of the product is ensured, according to European standards, are given in CEN EN 13108-21 (2008).

Quality control requirements according to American practice are somehow similar to those according to the European practice. The only difference is that the required procedures are not as predetermined as in CEN EN 13108-21 (2008). Relevant information can be found in Asphalt Institute (2007).

9.5 ACCEPTANCE OF DELIVERED AND LAID ASPHALT

The acceptance of delivered and laid asphalt is usually based on the results obtained for the determination of

a. Binder content
b. Aggregate gradation
c. Mixture volumetric properties (voids, VMA or VFA)
d. Asphalt temperature
e. Degree of compaction
f. Compacted layer evenness
g. Layer thickness

The frequency of sampling/testing is always determined in contract documents. Sampling/testing frequencies that are usually used are given in Table 9.5.

Table 9.5 Frequency of sampling and testing of laid and compacted asphalts

Test/property	Frequency sampling/testing
Binder content	Every 1000 t
Gradation	
Bituminous mixture's volumetric properties (voids, etc.)	
Temperature of the bituminous mixture	Each delivery
Compaction achieved	Every 250–300 m (positions to be specified)
Layer thickness	
Roughness (evenness):	
– All measuring devices	As specified, usually upon completion of asphalt works
– With a 3 m straightedge	When required

9.5.1 Binder content

The binder content of hot mix asphalt for acceptance of the delivered product is determined from samples taken in situ. More information for sampling in situ is given in Section 9.3.3.

Binder content is determined using one of the following methods: (a) binder extraction method, (b) ignition method or (c) nuclear method.

In the first two methods, the remaining/recovered 'clean' aggregate is used for determining aggregate gradation and density; in the third method, only binder content determination can be carried out.

The binder extraction method is the most common and widely used method for determining the binder content in the mixture. It uses hydrocarbon solvents capable of dissolving bitumen and a binder extraction apparatus. Because of the high purchase and disposal costs of solvents, the risk to the operator's personal health and safety and the environmentally unfriendly nature of the hydrocarbon solvents, this test method was started to be replaced with the ignition method. The binder extraction method by European standards is conducted in accordance with CEN EN 12697-1 (2012), whereas that by American standards is in accordance with ASTM D 2172 (2011) or AASHTO T 164 (2013) and AASHTO T 319 (2008). A detailed description of the binder extraction test method is given in Section 9.6.1.

The ignition method uses a well-insulated furnace capable of raising the temperature at a very high level (above 580°C) and burning all the binder. The binder content is determined quickly (within approximately 40 min) by weight loss calculation that includes an aggregate calibration term. The ignition method, in many countries, has replaced the extraction method for reasons mentioned above and for the fact that the result is derived in a much shorter period. The ignition method according to European standards is conducted in accordance with CEN EN 12697-39 (2012), whereas the ignition method according to American standards is conducted in accordance with ASTM D 6307 (2010) or AASHTO T 308 (2010). A detailed description of the test method is given in Section 9.6.5.

The nuclear method is a non-destructive method and determines the binder content on both uncompacted asphalt and laboratory-compacted asphalt specimens. With this method, it is not possible to determine the aggregate gradation afterwards. The method uses an apparatus that utilises neutron thermalisation techniques. The nuclear method is quicker than the ignition method but it utilised radioactive material, which may be hazardous to the health of the users unless proper precautions are taken. The nuclear method for determination of the asphalt binder content is conducted according to ASTM D 4125 (2010) or AASHTO T 287 (2010). A detailed description of the test method is given in Section 9.6.6. There was no relevant European standard at the time of writing the book.

Regardless of the test method used, the determined binder content should be within the tolerance limits declared by the supplier or set by the relevant specification.

9.5.2 Aggregate gradation

The determination of the aggregate gradation of the asphalt sampled from the site is carried out by sieving after extracting or burning the binder from the bituminous mixture.

The sieving procedure is conducted according to CEN EN 933-1 (2012), ASTM C 117 (2013) and ASTM C 136 (2006) or AASHTO T 27 (2011) and AASHTO T 11 (2009).

More information about the aggregate sieve analysis is given in Section 2.11.1.

The aggregate gradation determined should be within the tolerance limits declared by the supplier or set by the relevant specification.

9.5.3 Volumetric properties of the asphalt

The volumetric properties of the asphalt such as air voids, voids in the mineral aggregate and voids filled with bitumen are calculated from the compacted asphalt specimens obtained from the site. The calculations for the determination of the volumetric properties are carried out according to CEN EN 12697-8 (2003), ASTM D 3203 (2011) or AASHTO T 269 (2011).

The volumetric properties determined should be within the tolerance limits declared by the supplier or set by the relevant specification.

9.5.4 Asphalt temperature

The temperature of the asphalt arriving on site is a critical parameter for effective paving and compaction operations.

For the acceptance of delivered product, the asphalt temperature is measured while the material is still in the arrived lorry. The temperature of the asphalt may also be measured while the asphalt is in the hopper or in augers area, as well as after it has been laid before rolling. The latter is useful information for laying and compaction procedures.

The temperature measuring devices are fitted with a probe and are conventional contact thermometers or digital contact thermometers with an accuracy of ±2°C.

When measurements are made while the asphalt is in the lorry, the probe of the measuring device is inserted into the material to a depth of at least 100 mm. At least four readings are taken at equally spaced intervals along each side of the lorry and at least 500 mm away from the edges of the load, and the average is determined.

When the temperature of the asphalt is measured in the mat before rolling, the temperature-sensitive element is positioned as close as possible to the mid-depth of the layer.

More information about the asphalt temperature measurement is given in CEN EN 12697-13 (2001).

Infrared thermometers are not advised to be used since readings are very sensitive to wind and moisture conditions and will certainly give erroneous results.

9.5.5 Compaction achieved

The compaction achieved (degree of compaction) after completion of rolling should always be within the pre-determined tolerance range. The degree of compaction is the ratio of bulk density obtained on site over the bulk density obtained in the laboratory for the target mix, expressed in percentage. The degree of compaction achieved by no means should be equal to 100%. For dense asphalt concrete, the targeted minimum degree of compaction on site is usually 95% and the maximum is 98%.

The bulk density achieved after completion of rolling is usually determined from extracted cores, following the procedure specified in CEN EN 12697-6 (2012), ASTM 2726 (2011), ASTM D 6752 (2011), AASHTO T 166 (2013) or AASHTO T 331 (2013).

The bulk density achieved alternative can be determined in place by using nuclear devices. This is a non-destructive method and the procedure is specified in CEN EN 12697-7 (2002) or ASTM D 2950 (2011).

Devices that use electromagnetic waves have been developed and used over the last years to also measure the density of asphalts in place. More information about the devices and the test procedure is given in ASTM D 7113 (2010) or AASHTO T 343 (2012).

The method to be employed for determining the degree of compaction for acceptance of the compacted asphalt layer should always be stated in the contract document.

9.5.6 Layer thickness

The thickness of the compacted layer is determined from cores, taken at specified locations, using a metal tape or rule, set of callipers, measurement jig or other device, capable of measuring specimen thicknesses. The measurement procedure is carried out in accordance to CEN EN 12697-36 or ASTM D 3549.

The thickness of the asphalt layer may also be determined by a non-destructive method using short-pulse radar, according to ASTM D 4748.

CEN EN 12697-36 also describes a non-destructive method for measuring the thickness of the asphalt layer by using an electromagnetic apparatus (eddy current principle) and an antipole fixed on the road prior to laying the asphalt.

The method employed for determining the thickness of the compacted asphalt layer, as well as the tolerance limits, should always be stated in the contract document.

Some contractors or supervisors determine the average thickness of the compacted mat from the tonnage of material used and the surface paved, usually per day. This is a rough determination of the layer thickness and serves only the purpose of avoiding systematic errors during paving. By no means can it be used as a method for accepting the asphalt that has been laid and compacted.

The final thickness of the asphalt layer should not deviate more than the permissible tolerance, specified in the contract document.

9.5.7 Surface irregularities and evenness (roughness)

The irregularities and evenness (or roughness) of the surface(s) or of the surface course are measured for compliance within the specified limits, which is a prime determinant of quality in new construction of asphalt works. Measurements are taken normally after completion of asphalt works, although daily measurements are not uncommon.

The surface irregularities and evenness (roughness) are measured by one of the following devices/methods: (a) straightedge beam (static device), (b) rolling profile devices (low-speed profile devices), (c) relative displacement or response-type devices (high-speed evenness/roughness meters) and (d) dynamically measured surface profile devices (high-speed profilometers).

The devices under (a) and (b) are usually employed after completion of short-duration asphalt works or during construction. The devices under (c) and (d) are employed after completion of long-duration asphalt works or for monitoring in-service surfaces.

The *straightedge beam* (usually 3 m long) is an apparatus measuring single irregularities attributed to quality defects in new surface(s) or surface course. It can also be used transversely to measure rut depth of in-service roads. This test method is not applicable to providing information on profile or general unevenness.

With the straightedge, the distance (the deviation) is measured between the surface and the plane of its measurement edge. This distance should be less than the maximum limit specified. Usually, the limit value for a newly constructed surface (surface of an individual layer) is 6 mm, and for a newly constructed surface course, it is 3 mm. The test method is carried out according to CEN EN 13036-7 (2003) or ASTM E 1703 (2010).

The *rolling devices*, such as rolling straightedge, travelling beam, profilographs (with uniformly spaced wheels or non-uniformly spaced wheels) and rolling inclinometer, are low-speed devices (walking speed) with moving reference datum measuring longitudinal or transverse profile.

When the rolling straightedge beam and the travelling beam (a similar device) are used, the number of surface irregularities is determined over a specified distance in the longitudinal direction. The number of surface irregularities should be less than the permitted number (see Section 15.3.12.2).

When the profilographs are used, the surface record is analysed to determine the rate of roughness and to identify bumps that exceed a specified threshold. For more details, see ASTM E 1274 (2012).

When the rolling inclinometer is used, the longitudinal and the transverse profiles of the travelled surface is determined, from which the roughness of the new pavement is quantified in terms of selective roughness index. The results are displayed by the on-board computer screen and also 'road roughness/smoothness' and 'must grind areas' can be reported in printed graphical and tabular report form. For more details, see ASTM E 2133 (2009).

The *relative displacement devices (high-speed evenness meters)* can be response-type device accelerometers mounted to a vehicle or wheel trailer. Such devices are as follows: TRRL Bump Integrator (BI), Mays ride meter, NAASRA roughness meter, Road meter PCA, ROMDAS bump integrator, all response-type devices, Canadian ARAN, Road Surface Tester (RST-SAAB) and accelerometer-type devices. When these devices are used, proper calibration of the equipment is vital and the output is best expressed in IRI (International Roughness Index).

The *dynamically measured surface profile devices* (high-speed profilometers) such as laser beams and precision accelerometers and gyros are mounted to the vehicle or inertia type mounted on a single or double trailer (APL profilometer-IFSTTAR). There are many high-speed profilometers available, some of which are as follows: ARAN (Furgo Roadware), RSP-DYNATEST, FRMS (Technical Research Centre – Finland), Road Assessment Vehicle (WDM – United Kingdom), Greenwood profilograph (Greenwood – Denmark), HARRIS (TRL – United Kingdom), TRACKS (Highways Agency – England-replaced HRM), ARAN (NCAT – Auburn University) and KJLAW profilographs (KJLAW – United States). The output of the result for surface evenness in all the above devices is also expressed in IRI.

Almost all of the abovementioned high-speed profilometers are capable of determining the transverse profile as well; some can also measure other surface characteristics such as texture, cracking, road geometry and so on.

The device to be employed for quality acceptance of the asphalt works with regard to surface roughness should always be specified in the contract document together with the respective limited value.

More details regarding surface roughness/evenness measurements and equipment used are given in Section 16.3.12.

Of course, the longitudinal profile can be measured by conventional survey *optical level and a graduated rod*. This method is labour-intensive with respect to other means for measuring the longitudinal profile. It is rarely used, and when used, it is mainly for validating other profile-measuring methods or for calibrating response-type roughness-measuring systems. Details can be found in ASTM E 1364 (2012).

9.6 METHODS FOR DETERMINATION OF BINDER CONTENT

9.6.1 Binder extraction methods

The determination of binder content of asphalts and reclaimed asphalts by the binder extraction method consists of the following steps:

    a. Binder extraction by dissolving in a hot or cold solvent
    b. Separation of mineral matter from the binder solution
    c. Determination of the binder quantity (by difference or binder recovery)
    d. Calculation of (soluble) binder content

Binder extraction with hot or cold solvent can be conducted with various devices, which characterise the binder extraction method used (hot extraction or cold extraction).

The separation of the mineral matter from the binder solution is conducted with various ways, depending on the selected extraction method.

The determination of the binder quantity is conducted either by mass or volume difference or binder recovery from a portion of the binder solution, depending on the selected extraction method.

Finally, the calculation of the binder content is conducted applying appropriate mathematical equations.

Alternative methods and procedures for the determination of binder content are given in Figure 9.1.

The solvent used is usually trichloroethylene, tetrachloroethylene, normal propyl bromide or methylene chloride. Any other solvent can be used, provided that the binder solubility is tested according to CEN EN 12592 (2007).

The quantity of asphalt sample required regardless of the method employed depends on the nominal maximum aggregate size. Table 9.6 gives the minimum and maximum mass of test portion of bituminous materials required for each determination of binder content.

The required mass of test portion is obtained from a larger mass of material sampled by quartering process after heating the bituminous material, following the procedure described in CEN EN 12697-28 (2000).

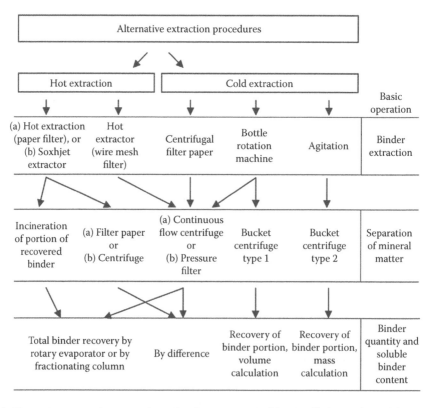

Figure 9.1 Alternative extraction procedures for determination of binder. (Adapted from CEN EN 12697-1, *Bituminous mixtures – Test methods for hot mix asphalt – Part 1: Soluble binder content*, Brussels: CEN, 2012. With permission [© CEN].)

Table 9.6 Mass of material for extraction test

Type of material	Largest size of aggregate (mm)	Minimum and maximum mass of test portion for each determination (g)
Bituminous mixtures	63 or 45	3,000–5,000
	40	2,500–4,000
	31.5	1,500–2,800
	22.4 or 20	1,000–2,000
	16, 14 or 12.5	800–1,400
	11.2, 10 or 8	300–1,000
	6.3, 5.6, 4 or 2	100–500
Coated chippings	All sizes	2,000–3,000

Source: Reproduced from CEN EN 12697-28, *Bituminous mixtures – Test methods for hot mix asphalt – Part 28: Preparation of samples for determining binder content, water content and grading*, Brussels: CEN, 2000. With permission (© CEN).

The maximum temperatures to be used in order to divide the bituminous material in smaller quantities and generally to heat it before extraction are as follows: 135°C for 25–60 dmm grade bitumen and 120°C for 60–330 dmm bitumen (CEN EN 12697-28 2000).

As mentioned above, after the extraction of the binder, the aggregates are dried and sieved for aggregate gradation determination. With some, the binder extraction method is also capable of determining the possible water contained in the mix. Of course, the determination of water content (if any) can be carried out before the binder extraction procedure by weight loss, after heating the mix in an oven for 2 to 3 h at 150–160°C (until a constant mass is obtained). The above presupposes the fact that no further tests are going to be carried out on the extracted bitumen.

9.6.1.1 Cold extraction using a centrifuge extractor method

The cold extraction method using a centrifuge extractor (centrifugal filter paper method) is perhaps the most common method worldwide for the determination of the soluble binder content.

The extraction is conducted using an extraction apparatus consisting of a bowl with a capacity of 1500 or 3000 g, with a safety cover, and an apparatus in which the bowl may be revolved at controlled variable speeds up to 3600 revolutions per minute (rpm). Additionally, the apparatus is provided with a container for catching the solvent thrown from the bowl, a drain for removing the solvent and an appropriate mechanical brake to stop the device turning.

A typical centrifugal extractor is presented in Figure 9.2 (Controls Srl 2014).

A sufficient asphalt test portion (Table 9.6) is weighted and placed in the bowl. The test portion is then fully covered with an appropriate solvent and sufficient time is allowed for the solvent to disintegrate the mixture (not longer than 1 h). A dry filter ring, after its mass has been determined, is placed around the edge of the bowl and then the cover is clamped tightly on the bowl.

The mass of the test portion, M, can alternatively be determined by weighing first the bowl and the filter ring together, mass M_A, and then by weighing the bowl, the filter paper and the added test portion altogether, mass M_B; thus, the mass of the test portion, M, is equal to $M_B - M_A$.

After placing a beaker or a flask under the drain to collect the extract, the centrifuge starts revolving until it reaches 3600 rpm. When the solvent ceased to flow from the drain, the machine stops and a new quantity of clean solvent is added (approximately 200 to 500 ml depending on the test portion), and the mixture is centrifuged again. The above procedure is repeated as many times as necessary until the extract is virtually colourless.

Figure 9.2 Centrifuge binder extractor. (Courtesy of Controls Srl.)

Once the process is completed, the filter ring is carefully removed and together with the extracted 'clean' aggregate is placed in a metal tray to dry to a constant weight in an oven at 110°C ± 5°C. After drying, the mass of the aggregates and the filter paper is determined.

When low-ash filter ring is used, an alternative procedure to determine the amount of filler stuck to it can be used. This is to burn the dried filter ring folded over the aggregate.

9.6.1.1.1 Separation of mineral matter from the binder solution

Since most probably the extract (binder solution) contains a small amount of fine mineral filler (certainly when the filler content in the initial test portion is more than 3%), the total quantity of extract collected is centrifuged further using the continuous-flow (filterless) binder extractor shown in Figure 9.3 (Controls Srl 2014) or a pressure filter apparatus.

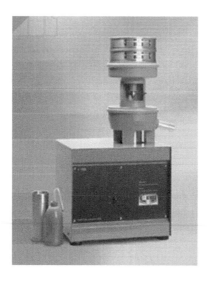

Figure 9.3 Centrifuge filler extractor. (Courtesy of Controls Srl.)

The fine mineral filler is determined after drying the cup containing the filler to a constant mass, in an oven at 110°C. The determination of the mineral filler is carried out by mass difference of the empty cup and cup containing the filler.

In the case of using a pressure filter apparatus, the fine mineral filler is determined by dry mass difference of the filter paper before and after filtering. More information on the continuous-flow centrifuge or pressure filter apparatus is given in CEN EN 12697-1 (2012), Annex B.2.

9.6.1.1.2 Binder quantity and determination of soluble binder content

The soluble binder content may be determined by the difference method or by the recovery method.

When the *difference method* is used, the soluble binder content, S, is determined by using the following equation:

$$S = \frac{100 \times [M - (M_1 + M_w)]}{(M - M_w)},$$

where S is the soluble binder content (%), M is the mass of undried test portion (g), M_1 is the mass of total recovered mineral matter (aggregates and filler) (g) and M_w is the mass of water in the undried test portion (if any) (g).

Note: The test sample is usually free of water/moisture. In case of uncertainty, dry the test sample in an oven at a temperature of 80°C ± 5°C to a constant mass, to avoid binder drainage. When the test portion mass is going to be determined by placing it in the bowl, the oven temperature can be 110°C ± 5°C.

When *the recovery method* is used, the total binder is first recovered from the binder solution collected from the continuous-flow extractor or pressure filter apparatus, by using either the rotary evaporation method, according to CEN EN 12697-3 (2013), or the fractionating column method, according to CEN EN 12697-4 (2005) (see Section 9.6.2).

It is reminded that the two binder recovery methods mentioned above are used when it is also necessary to determine the characteristic binder properties (penetration, softening, etc.).

The soluble binder content, S, when the binder is totally recovered is determined by the following equation:

$$S = \frac{100 \times M_b}{(M - M_w)},$$

where S is the soluble binder content (%), M is the mass of undried test portion (g), M_b is the mass of recovered binder (g) and M_w is the mass of water in the undried test portion (if any) (g).

More information about the centrifuge extraction method is given in CEN EN 12697-1 (2012), Annex B.1.5.

If the test is to be conducted according to American standards, ASTM 2172 (2011) or AASHTO T 164 (2013) is used. The procedure is very similar to that according to European standards.

9.6.1.2 Hot extraction paper filter method

The apparatus for the hot extraction paper filter method consists of a hot extractor, a graduated receiver, a condenser, a heater and a filter paper.

The hot extractor is a brass or steel pot that contains a cylindrical container from non-corrodible or brass gauze of approximately 1 to 2 mm aperture size or, alternatively, a spun copper tube, with a ledge at the bottom on which a removable gauze disc rests. Two types of graduated receivers are used depending on the density of the solvent used. An assembly of the apparatus used for hot extraction by using the paper filter method is shown in Figure 9.4.

The filter paper is fitted into the cylindrical container to form a complete lining and weigh them. Then, the test portion is carefully placed and all are weighted to an accuracy of 0.05 g. Afterwards, the container is placed in the pot and sufficient solvent is poured over the sample to permit refluxing. The pot is heated to ensure a steady reflux rate of two to five drops per second from the end of the condenser.

Any water present in the sample will be collected in the receiving tube, while the solvent will flow back over the sample and drain through the filter paper into the bottom of the pot. If the amount of water exceeds the capacity of the receiver, the distillation is discontinued and the measured portion of water is removed. The distillation starts again after re-assembling the apparatus and continues until the extraction is complete and the receiver ceases collecting water. The completion of extraction can only be determined reliably by dismantling the apparatus and examining the aggregate.

On completion of extraction, the mineral aggregate is removed with its container and dried to a constant mass.

The fine mineral matter present in the solution is determined by filtering the whole solution through a suitable-grade filter paper or centrifuging it (continuous-flow centrifuge).

The determination of the soluble binder content is carried out by difference as described in Section 9.6.1.1.

It is noted that the residual mineral matter in the binder extract may also be determined by incineration (see Section 9.6.3).

More information about the hot extraction paper filter method is given in CEN EN 12697-1 (2012), Annex B.1.1.

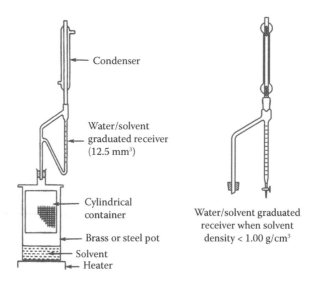

Condenser

Water/solvent graduated receiver (12.5 mm³)

Cylindrical container

Brass or steel pot

Solvent

Heater

Water/solvent graduated receiver when solvent density < 1.00 g/cm³

Figure 9.4 Assembled apparatus for hot extraction paper filter method.

9.6.1.3 Hot extraction wire mesh filter method

The hot extraction wire mesh filter method is similar to the hot extraction paper filter method. The only differences between the two methods are (a) the extraction pot (cylinder) can be made of glass or metal; (b) the extraction container is a wire basket made of 63 μm wire cloth or a metal cylinder with 63 μm wire cloth sieving medium, to which the extraction thimble of the fibre material is placed; and (c) a protective sieve is used.

The extraction procedure is similar to the hot extraction paper filter method and the soluble binder content is determined as in cold extraction using the centrifuge extractor method (see Section 9.6.1.1).

More information about the hot extraction wire mesh filter method is given in CEN EN 12697-1 (2012), Annex B.1.2

9.6.1.4 Hot extraction soxhlet extractor method

The Soxhlet extractor consists of glass extraction equipment known as modified Soxhlet extractor, extraction case/thimble (test sample receiver) or filtering cartridges, a 5000 ml flask, heating equipment and a desiccator or heated storage box to store the dried extraction case. An assembled Soxhlet extractor apparatus is shown in Figure 9.5.

The extraction case and the flask (if binder recovery is required) are weighted first. Then, the test portion is placed in the extraction case and is weighted again to an accuracy of 0.05 g.

The extraction case is placed in the extractor that has been filled with solvent so that most of the material in the extraction case is covered by solvent, and by switching on the heater, the extraction starts. The extraction stops when the solvent collected in the extractor becomes colourless.

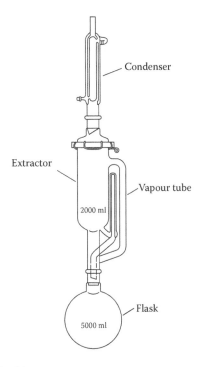

Figure 9.5 Assembled modified Soxhlet extractor apparatus.

The extraction case is then removed from the extractor and placed in the desiccator until all solvent is evaporated. After that, the mineral aggregate is removed from its container and dried to a constant mass.

In order to remove any fine material present in the solution at the end of the test, the whole solution is filtered through a suitable filter paper or centrifuged.

The soluble binder content is determined by difference, as described in Section 9.6.1.1.

More information about the Soxhlet extraction method is given in CEN EN 12697-1 (2012), Annex B.1.3.

9.6.1.5 Cold extraction bottle rotation machine method/ rotary bottle extraction method

The apparatus for the cold extraction bottle rotation machine method, also known as the rotary bottle extraction method, consists of metal bottles and a machine capable of rotating the bottles.

The capacity of the metal bottles is appropriate to the size of sample being analysed, for example, 600×10^3 mm^3, 2500×10^3 mm^3, 7000×10^3 mm^3 and $12,000 \times 10^3$ mm^3.

When binder determination is to be carried out by using the binder portion recovery method, volumetric flasks of appropriate capacity from 250×10^3 mm^3 (250 ml) to 2000×10^3 mm^3 (2000 ml) are also needed.

The test specimen is placed in the metal bottle and then the sufficient volume of solvent is added and the bottle with the contents rotates (less than 20 rpm) for a certain period.

The minimum rolling time depends on the type of material. The minimum rolling time for asphalt concretes and porous asphalts is 20 min, whereas the minimum rolling time for stone mastic asphalt, hot rolled asphalt and mastic asphalt is 30 min. In the case of coated chippings, the minimum rolling time is 10 min.

After rolling time, the bottle content is allowed to stand for at least 2 min; all aggregates are then removed and dried to a constant mass. The filler separation is carried out by a continuous-flow centrifuge, a pressure filter apparatus, or a bucket-type centrifuge-Type 1.

When a continuous-flow centrifuge or pressure filter apparatus is used, the determination of the soluble binder content is conducted by the weight difference method, as in the case of the extraction method with a centrifugal extractor (see Section 9.6.1.1).

When the bucket-type centrifuge-Type 1 is used, the soluble binder content is determined by volume calculation using the equation given in paragraph 5.5.4 of CEN EN 12697-1 (2012).

The bucket-type centrifuge-Type 1 apparatus is capable of taking four buckets fitted with centrifuge tubes of at least 50×10^3 mm capacity and of an acceleration of between 1.5×10^4 and 3×10^4 m/s^2. The time of centrifuging is determined from a graph provided by CEN EN 12697-1 (2012), Annex B.2.3.1.1. The procedure to recover a portion of binder quantity is described in CEN EN 12697-1 (2012), Annex B.3.1.

More information about the bottle rotation machine extraction method, in general, is given in CEN EN 12697-1 (2012), Annex B.1.4.

9.6.1.6 Cold extraction by agitation

The cold mix extraction (dissolution of bitumen) by agitation is conducted by using 2000×10^3 mm^3 (2000 ml) or 3000×10^3 mm^3 (3000 ml) sealed containers and a sufficient mass of solvent.

The test portion is placed in the container and is shaken for at least 30 min, after which the container is left standing up for at least 5 min to settle.

The mass of the solvent (tetrachloroethylene), M_p, is 1.6 times the mass of the sample, if the asphalt contains more than 5% binder, or 0.8 times the latter, if it contains less than 5% binder. For mastic asphalt, the mass of the solvent is 4 to 5 times the mass of the sample.

The temperature of the asphalt in all cases should be less than 90°C.

The separation of mineral filler from the solution is carried out by bucket-type centrifuge-Type 2. The bucket-type centrifuge-Type 2 is capable of producing a minimum acceleration of 40×10^3 m/s^2 fitted with buckets having a capacity greater than 20×10^3 mm. The procedure to recover a portion of binder quantity is described in CEN EN 12697-1 (2012), Annex B.3.2.

The soluble binder content is determined by mass calculation, using the equation given in paragraph 5.5.5 of CEN EN 12697-1 (2012).

More information about the extraction method by agitation is given in CEN EN 12697-1 (2012), Annex B.1.6.

9.6.2 Total binder recovery methods

The total binder recovery is carried out by using the rotary evaporator or the fractionating column method. Both methods are used mainly for the recovery of soluble binder from bituminous pavement materials in a form for further testing (penetration, softening point, viscosity, etc.).

The rotary evaporator method is suitable only for the recovery of paving grade bitumen. The fractionate column method is suitable for the recovery of paving grade bitumen and for mixtures containing volatile matter such as cut-back bitumen. However, the reference method for recovering paving grade bitumen is the rotary evaporator method. The rotary evaporator procedure is also recommended for the recovery of polymer-modified bitumens.

In both procedures, the bitumen is separated from the sample by dissolving in dichloromethane (or other suitable solvent). The separation is carried out by centrifuging or filtration. After removal of undissolved solids from the bitumen solution, the bitumen is recovered in the first case by vacuum distillation using a rotary evaporator. The temperature at which the distillation procedure is carried out is determined by the type of solvent used (for dichloromethane, the temperature during the first phase of distillation is 45°C and that during the second phase is 150°C; the maximum temperature is 175°C).

In the second case, the bitumen is concentrated by atmospheric distillation in a fractionating column. The last traces of solvent are removed from the concentrate by distillation at 100°C above the expected softening point or 175°C, whichever is higher, with the pressure reduced from atmospheric pressure, 100 to 20 kPa, and with the aid of a stream of carbon dioxide gas (when cut-backs contain white spirit or other high volatile fluxes, the use of carbon dioxide is omitted).

A typical rotary evaporator apparatus is shown in Figure 9.6 and a fractionating distillation apparatus is shown in Figure 9.7.

In both bitumen recovery methods, the temperatures required for the distillation procedure are obtained with the use of heated oil bath.

The introduction of the bitumen solution into the distillation flask is carried out at a slow rate to avoid the creation of a high volume of steam or excessive bubbling.

It was mentioned that binder recovery by rotary evaporator is usually shorter when the rotary evaporator procedure is used.

More information about binder recovery by rotary evaporator is given in CEN EN 12697-3 (2013). The respective American standard is ASTM D 5404 (2012).

More information about binder recovery by fractionating column is given in CEN EN 12697-4 (2005).

Similar to the fractionating column method for binder recovery is the Abson method. For more information, see ASTM D 1856 (2009).

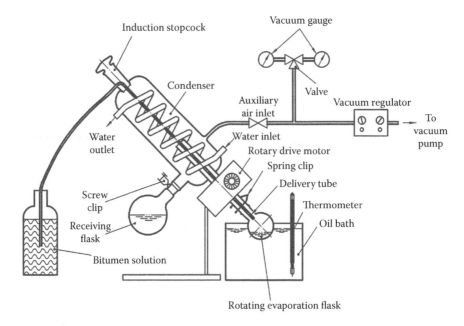

Figure 9.6 Typical rotary evaporator apparatus.

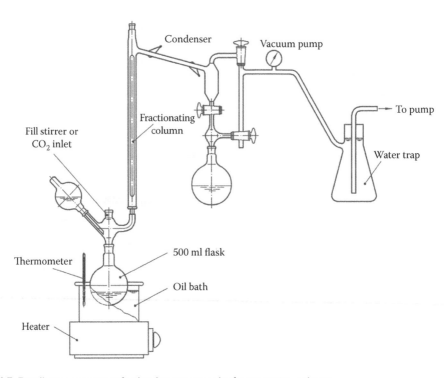

Figure 9.7 Distillation apparatus for binder recovery by fractionating column.

9.6.3 Determination of residual mineral matter in the binder extract by incineration

The incineration method for determining the residual matter in the binder extract is used by some extraction methods for residual binder content as an alternative method (see Figure 9.1).

The main apparatus used is a furnace capable of heating to 600°C. After determining the volume of the total binder solution collected free of mineral matter, a portion of the binder solution (approximately 100×10^3 mm^3) is placed into the ignition disk. The amount of solvent is evaporated until the sample is dried to a constant mass. The residue is placed in the furnace at 500°C to 600°C until ash residue is obtained.

After cooling, 5×10^3 mm^3 of saturated ammonium carbonate solution is added per gram of ash. The content is left to digest at room temperature for 1 h and dried to a constant mass, and after cooling, its mass is determined to 1 mg.

The mass of the mineral matter in the total volume of extract, M_2, is determined by using the following equation:

$$M_2 = G \times \frac{V_1}{V_1 - V_2},$$

where M_2 is the mass of the mineral matter in the total volume of extract (g), G is the mass of ash in aliquot (g), V_1 is the total volume of extract (mm^3) and V_2 is the volume of extract after removing the aliquot (mm^3).

The soluble binder content, S, is then calculated by using the following equation:

$$S = \frac{[(M - M_W) - (M_1 + M_2)] \times 100}{M - M_W},$$

where S is the soluble binder content (%), M is the mass of undried test portion (g), M_W is the mass of water in the undried test portion (g), M_1 is the mass of recovered mineral matter (g) and M_2 is the mass of the mineral matter in the total volume of the extract (g).

More information about the determination of residual matter in the binder extract by incineration is given in CEN EN 12697-1 (2012), Annex C.

9.6.4 Guidance on determination of soluble binder content of asphalts with polymer-modified binders

The same methods and procedures as those mentioned above, which are based on CEN EN 12697-1 (2012), can also be used for the extraction of asphalts containing modified bitumen.

However, according to CEN EN 12697-1 (2012), Annex D, great attention should be given on the selection of the appropriate solvent, the duration of extraction and the separation of the fine mineral matter.

With regard to the *solvents*, toluene, trichloroethylene, dichloromethane and tetrachloroethylene are generally able to satisfactorily dissolve the styrene butadiene styrene (SBS) modified binders. Toluene, trichloroethylene and tetrachloroethylene are sometimes suitable for ethylene vinyl acetate (EVA)-modified binders. Dichloromethane is generally not suitable for EVA-modified binders. Few solvents have been found to be suitable to dissolve atactic polypropylene (APP)-modified binders completely; the best solubility with such a modifier is obtained with hot xylene.

The solubility of the polymer-modified binders depends largely on the temperature of the solvent. This can lead to considerable differences between the results of cold and hot extraction methods.

With regard to the *duration of the extraction*, the extraction does not terminate when the solvent becomes colourless.

The duration for the extraction of the binder when the hot extraction (paper filter and wire mesh) method, a Soxhlet extractor or the centrifuge extractor method is used increases by approximately +10% of the time required for asphalts containing unmodified bitumen.

In the case of using the bottle rotation machine method, the time required for extraction increases correspondingly by 10 min from the minimum extraction time values given in Section 9.6.1.5.

When cold mix dissolution of bitumen by agitation is used, provided the solubility of the modified bitumen is sufficient, the container should be shaken for at least 45 min instead of 30 min.

With regard to *separation of fine mineral matter (filler)*, only the continuous-flow centrifuge and the bucket-type centrifuge, Type 1 and Type 2, can be used. In the continuous centrifuge flow method in particular, re-washing of the filler collected is recommended to be repeated, so the decanted solvent has become virtually colourless.

The pressure filter method is not recommended for polymer-modified binders because of risk of clogging.

For the determination of the binder quantity, if the soluble modified binder is not needed for further purposes, the difference method is the reference method.

More information about the determination of soluble binder content of mixtures with polymer-modified binders is given in CEN EN 12697-1 (2012), Annex D.

9.6.5 Quantitative extraction and recovery of the binder method

The quantitative extraction and recovery of the binder method have been developed with the primary intention for use when the physical properties of the recovered binder are to be determined.

It can also be used to determine the quantity of the soluble binder in the hot mix asphalt or reclaimed asphalt. The recovered aggregate may be used for sieve analysis or other aggregate testing.

In fact, this method is a combined one-step procedure for extracting and recovering the binder from the bituminous mix.

The binder extraction is carried out by washing the asphalt with solvent and filtering the solution, repeatedly in an extraction/filtration vessel. The extraction is carried out by rotating the vessel for 5 min at 30 rpm, while the vessel is positioned vertically (unlike the rotating bottle extraction method). The extraction vessel is equipped with three cloth sieves (2.00 cm, 300 mm and 75 mm).

Each filtrate is distilled under vacuum in a rotary evaporator. After recovery of the final filtrate, the solution is concentrated to approximately 300 ml and centrifuged to remove fine mineral filler. The decanted solution is distilled under vacuum to remove the solvent. Nitrogen gas is introduced during the final phase of distillation to drive off any remaining traces of solvents.

The recovered asphalt (distillation residue) is subjected to further physical testing as required and the quantity of soluble binder in the asphalt mixture can be calculated, if required. The recovered aggregate can also be used for sieve analysis, if desired.

More details on the quantitative extraction and recovery of the binder method are given in AASHTO T 319 (2008) or ASTM D 6847. However, ASTM D 6847 has been withdrawn in 2010. There is no replacement ever since probably because of the unpopularity of this test method.

9.6.6 Binder recovery by the ignition method

The ignition method for the determination of the binder content is an alternative to the more traditional methods of extracting the bitumen using solvents. Apart from obtaining the results much quicker than any traditional method, the ignition method is also environmentally friendlier and safer to the user.

The ignition method is appropriate for hot mix asphalts or reclaimed asphalts, containing unmodified or modified binders. After binder content determination, the remaining aggregate can be used for determining aggregate gradation and density, provided excessive breakdown of the aggregate particles does not occur at the temperature used.

The results obtained can be used for process control or to check the compliance of mixtures.

Prior testing calibration is needed, either on the complete mixture or on its components, aggregate and reclaimed asphalt pavement. The calibration procedure is described in the relevant standard.

The main apparatus is a furnace capable of reaching temperatures above 600°C that has sufficient capacity to receive the sample, with an internal balance, a reducing furnace emission system, a programming and data collection system and perforated sample metal baskets with a catch pan.

A sufficient mass of the sample is preheated in an oven at 110°C, weighted and placed in the furnace set at the target temperature, usually 540°C, also determined during the calibration.

The size of the test sample should be within a range relative to the maximum aggregate size of the asphalt. The range specified by CEN EN 12697-39 (2012) is as shown in Table 9.7.

The test continues to run until the sample mass has stabilised and the range of three consecutive weighings, taken at 1 min intervals, is within the constant mass limit shown in Table 9.7.

All weight measurement and calculation of the binder content are carried out by the automations of the furnace (method A). The test may also be carried out using a furnace without an internal weighing system (method B).

The binder content is determined by mass difference after binder ignition by a calculation that includes the calibration factor.

As mentioned previously, after binder content determination, the aggregate is allowed to cool at room temperature and then sieved for aggregate gradation determination or density determination.

More information about the test is given in CEN EN 12697-39 (2012), ASTM D 6307 (2010) or AASHTO T 308 (2010).

Table 9.7 Size of sample required for binder recovery by the ignition method

Nominal maximum aggregate size (mm)	Mass of test sample (g)	Maximum constant mass limit (if water is contained) (g)
4	1000–1400	0.15
5.6 or 6.3 or 8.0 or 10	1000–1600	0.15
11.2 or 12.5 or 14 or 16	1000–1700	0.20
20 or 22.4	1000–2400	0.25
31.5	1000–3000	0.30
40 or 45	1000–4000	0.40

Source: CEN EN 12697-39, *Bituminous mixtures – Test methods for hot mix asphalt – Part 39: Binder content by ignition*, Brussels: CEN, 2012. With permission (© CEN).

9.6.7 Binder content determination by the nuclear method

The nuclear method is developed for rapid determination of the binder content of uncompacted mixtures (test method A) and laboratory-compacted specimens (test method B). Test method A is suitable for rapid quality control and acceptance testing with regard to binder content. According to ASTM D 4125 (2010), the precision for method B has not been established yet. Therefore, method B cannot be used for acceptance or rejection of a material for purchasing purposes.

The binder content determination using the nuclear method is specified by ASTM D 4125 (2010) or AASHTO T 287 (2010). European standards have not yet included it as an alternative binder content determination test.

The nuclear method utilises neutron thermalisation techniques and is a non-destructive method. However, it must be stated that the nuclear method can only be used for binder content determination since it does not provide extracted aggregates for gradation analysis.

The apparatus consists of encapsulated and sealed nuclear (radioactive source), thermal neutron detectors and a read-out instrument. The detectors are sensitive to outside influences; therefore, any other source of neutron radiation is kept at least 10 m from the apparatus during use. The area around the apparatus should also be kept free of large amounts of hydrogenous material, such as water, plastics or asphalt during use.

The test results obtained are influenced by the types of aggregate, the source, the grade of the bitumen and the aggregate gradation. Hence, a calibration curve must be developed for each mix type and aggregate blend to be tested. A new calibration curve should be developed whenever there is a change in the source of asphalt or aggregate or a significant change in aggregate gradation. A new calibration curve should also be established for new or repaired apparatus. The calibration procedures for both methods A and B are described in the standards. Once the device has been calibrated, the binder content, expressed in percentage by mass of mixture, is determined in a very short period.

More information about binder content determination using the nuclear method is given in ASTM D 4125 (2010) or AASHTO T 287 (2010).

REFERENCES

AASHTO R 9. 2009. *Acceptance sampling plans for highway construction*. Washington, DC: American Association of State Highway and Transportation Officials.

AASHTO T 2. 2010. *Sampling of aggregates*. Washington, DC: American Association of State Highway and Transportation Officials.

AASHTO T 11. 2009. *Materials finer than 75 μm (no. 200) sieve in mineral aggregates by washing*. Washington, DC: American Association of State Highway and Transportation Officials.

AASHTO T 27. 2011. *Sieve analysis of fine and coarse aggregates*. Washington, DC: American Association of State Highway and Transportation Officials.

AASHTO T 40. 2012. *Sampling bituminous materials*. Washington, DC: American Association of State Highway and Transportation Officials.

AASHTO T 164. 2013. *Quantitative extraction of asphalt binder from hot mix asphalt (HMA)*. Washington, DC: American Association of State Highway and Transportation Officials.

AASHTO T 166. 2013. *Bulk specific gravity of compacted (Gmb) hot mix asphalt (HMA) using saturated surface-dry specimens*. Washington, DC: American Association of State Highway and Transportation Officials.

AASHTO T 168. 2011. *Sampling bituminous paving mixtures*. Washington, DC: American Association of State Highway and Transportation Officials.

AASHTO T 269. 2011. *Percent air voids in compacted dense and open asphalt mixtures.* Washington, DC: American Association of State Highway and Transportation Officials.

AASHTO T 287. 2010. *Asphalt binder content of asphalt mixtures by the nuclear method.* Washington, DC: American Association of State Highway and Transportation Officials.

AASHTO T 308. 2010. *Determining the asphalt binder content of hot mix asphalt (HMA) by the ignition method.* Washington, DC: American Association of State Highway and Transportation Officials.

AASHTO T 319. 2008. *Quantitative extraction and recovery of asphalt binder from asphalt mixtures.* Washington, DC: American Association of State Highway and Transportation Officials.

AASHTO T 331. 2013. *Bulk specific gravity (Gmb) and density of compacted hot mix asphalt (HMA) using automatic vacuum sealing method.* Washington, DC: American Association of State Highway and Transportation Officials.

AASHTO T 343. 2012. *Density of in-place hot mix asphalt (HMA) pavement by electronic surface contact devices.* Washington, DC: American Association of State Highway and Transportation Officials.

Asphalt Institute MS-4. 2007. *The Asphalt Handbook,* 7th Edition. Lexington, KY: Asphalt Institute.

ASTM C 117. 2013. *Standard test method for materials finer than 75 μm (no. 200) sieve in mineral aggregates by washing.* West Conshohocken, PA: ASTM International.

ASTM C 136. 2006. *Standard test method for sieve analysis of fine and coarse aggregates.* West Conshohocken, PA: ASTM International.

ASTM D 75/D 75M. 2009. *Standard practice for sampling aggregates.* West Conshohocken, PA: ASTM International.

ASTM D 140/D 140M. 2009. *Standard practice for sampling bituminous materials.* West Conshohocken, PA: ASTM International.

ASTM D 979/D 979M. 2012. *Standard practice for sampling bituminous paving mixtures.* West Conshohocken, PA: ASTM International.

ASTM D 1856. 2009. *Standard test method for recovery of asphalt from solution by Abson method.* West Conshohocken, PA: ASTM International.

ASTM D 2172/D 2172M. 2011. *Standard test method for quantitative extraction of bitumen from bituminous paving mixtures.* West Conshohocken, PA: ASTM International.

ASTM D 2726. 2011. *Standard test method for bulk specific gravity and density of non-absorptive compacted bituminous mixtures.* West Conshohocken, PA: ASTM International.

ASTM D 2950/D 2950M. 2011. *Standard test method for density of bituminous concrete in place by nuclear methods.* West Conshohocken, PA: ASTM International.

ASTM D 3203/D 3203M. 2011. *Standard test method for percent air voids in compacted dense and open bituminous paving mixtures.* West Conshohocken, PA: ASTM International.

ASTM D 3665. 2012. *Standard practice for random sampling of construction materials.* West Conshohocken, PA: ASTM International.

ASTM D 4125/D 4125M. 2010. *Standard test methods for asphalt content of bituminous mixtures by the nuclear method.* West Conshohocken, PA: ASTM International.

ASTM D 5361/D 5361M-11a. 2011. *Standard practice for sampling compacted bituminous mixtures for laboratory testing.* West Conshohocken, PA: ASTM International.

ASTM D 5404/D 5404M. 2012. *Standard practice for recovery of asphalt from solution using the rotary evaporator.* West Conshohocken, PA: ASTM International.

ASTM D 6307. 2010. *Standard test method for asphalt content of hot-mix asphalt by ignition method.* West Conshohocken, PA: ASTM International.

ASTM D 6752/D 6752M. 2011. *Standard test method for bulk specific gravity and density of compacted bituminous mixtures using automatic vacuum sealing method.* West Conshohocken, PA: ASTM International.

ASTM D 7113/D 7113M. 2010. *Standard test method for density of bituminous paving mixtures in place by the electromagnetic surface contact methods.* West Conshohocken, PA: ASTM International.

ASTM E 1274. 2012. *Standard test method for measuring pavement roughness using a profilograph.* West Conshohocken, PA: ASTM International.

ASTM E 1364. 2012. *Standard test method for measuring road roughness by static level method.* West Conshohocken, PA: ASTM International.

ASTM E 1703/1703M. 2010. *Standard test method for measuring rut-depth of pavement surfaces using a straightedge.* West Conshohocken, PA: ASTM International.

ASTM E 2133. 2009. *Standard test method for using a rolling inclinometer to measure longitudinal and transverse profiles of a traveled surface.* West Conshohocken, PA: ASTM International.

CEN EN 58. 2012. *Bitumen and bituminous binders-Sampling bituminous binders.* Brussels: CEN.

CEN EN 932-1. 1996. *Tests for general properties of aggregates – Part 1: Methods for sampling.* Brussels: CEN.

CEN EN 933-1. 2012. *Tests for geometrical properties of aggregates – Part 1: Determination of particle size distribution – Sieving method.* Brussels: CEN.

CEN EN 12592. 2007. *Bitumen and bituminous binders – Determination of solubility.* Brussels: CEN.

CEN EN 12697-1. 2012. *Bituminous mixtures – Test methods for hot mix asphalt – Part 1: Soluble binder content.* Brussels: CEN.

CEN EN 12697-3. 2013. *Bituminous mixtures – Test methods for hot mix asphalt – Part 3: Bitumen recovery: Rotary evaporator.* Brussels: CEN.

CEN EN 12697-4. 2005. *Bituminous mixtures – Test methods for hot mix asphalt – Part 4: Bitumen recovery: Fractionating column.* Brussels: CEN.

CEN EN 12697-6. 2012. *Bituminous mixtures – Test methods for hot mix asphalt – Part 6: Determination of bulk density of bituminous specimens.* Brussels: CEN.

CEN EN 12697-7. 2002. *Bituminous mixtures – Test methods for hot mix asphalt – Part 7: Determination of bulk density of bituminous specimens by gamma rays.* Brussels: CEN.

CEN EN 12697-8. 2003. *Bituminous mixtures – Test methods for hot mix asphalt – Part 8: Determination of void characteristics of bituminous specimens.* Brussels: CEN.

CEN EN 12697-13:2000/AC. 2001. *Bituminous mixtures – Test methods for hot mix asphalt – Part 13: Temperature measurement.* Brussels: CEN.

CEN EN 12697-20. 2012. *Bituminous mixtures – Test methods for hot mix asphalt – Part 20: Indentation using cube or cylindrical specimens.* Brussels: CEN.

CEN EN 12697-27. 2000. *Bituminous mixtures – Test methods for hot mix asphalt – Part 27: Sampling.* Brussels: CEN.

CEN EN 12697-28. 2000. *Bituminous mixtures – Test methods for hot mix asphalt – Part 28: Preparation of samples for determining binder content, water content and grading.* Brussels: CEN.

CEN EN 12697-39. 2012. *Bituminous mixtures – Test methods for hot mix asphalt – Part 39: Binder content by ignition.* Brussels: CEN.

CEN EN 13036-7. 2003. *Road and airfield surface characteristics – Test methods – Part 7: Irregularity measurement of pavement courses: The straightedge test.* Brussels: CEN.

CEN EN 13043:2002/AC. 2004. *Aggregates for bituminous mixtures and surface treatments for roads, airfields and other trafficked areas.* Brussels: CEN.

CEN EN 13108-21:2006/AC. 2008. *Bituminous mixtures – Material specifications – Part 21: Factory control.* Brussels: CEN.

Controls Srl. 2014. Available at http://www.controls-group.com.

Chapter 10

Layers of flexible pavement

10.1 GENERAL

A pavement is a set of superimposed layers of imported materials (selected, processed unbound and bound materials) that are placed on the natural soil for the creation of a road. A pavement is a complex structure that has to accomplish various functions different from each other.

The main structural function of a pavement is to sustain traffic loads and distribute them to the subgrade. The stresses transferred to the surface of the subgrade should be such as to cause minimal deformation of the subgrade soil layer. Additionally, part of the upper layers of the pavement structure should be almost impervious to water, so that the subgrade, as well as the unbound layers, is protected from the detrimental effect of surface water. Finally, the pavement surface should be skid resistant, resistant to the polishing action of tyres and even.

In general, the flexible pavement structure consists of two characteristic sets of layers with different mechanical properties and performance: the unbound or hydraulically bound aggregate layers, seated on the subgrade, and the bound asphalt (bituminous) layers, seated on the previous set of layers.

The above separation of the flexible pavement structure is based on the different mechanical performance of the layers and constitutes the base for the development of any flexible pavement design methodology.

The *asphalt layers* are distinguished into surface or wearing course, binder course and asphalt base course, while the *unbound* or *hydraulically bound aggregates* are distinguished into base course and sub-base course.

Quite often, because of the existence of weak subgrade, an extra layer is constructed or formed over which the pavement is constructed. This layer is known as *capping layer*.

A typical cross section of a flexible pavement is given in Figure 10.1.

In some countries, particularly in the United Kingdom, the base, the sub-base and the capping layer (if it exists) are characterised as foundation layers or as pavement foundation.

All the other layers are characterised as pavement upper layers or upper structure. The pavement upper layers can be one of the following: (a) all bound in bitumen (asphalt layers) or (b) some bound in bitumen and some bound in hydraulic binders.

The first case, together with the foundation layers, represents the flexible type of pavement, whereas the second case represents the flexible composite pavement.

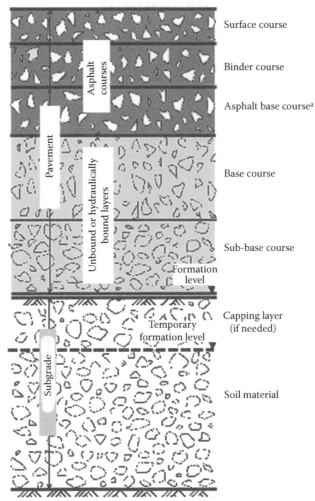

Figure 10.1 Typical cross section of a flexible pavement.

10.2 SUBGRADE

The subgrade is the formed and compacted soil, on which the pavement is constructed, extended to a depth that affects pavement design. This depth for conventional road construction projects extends from 600 to 700 mm below the surface of the subgrade.

In rare cases, when no soil formation or preparation (except topsoil-vegetation removal) is needed, the subgrade surface coincides with the surface of the natural soil. In more frequent cases, where an embankment or a capping layer is constructed, the surface of the subgrade is considered to be the surface of the final layer of the embankment or of the capping layer.

Generally, the subgrade surface determines the level on which the pavement is seated. This is known as formation level (see Figure 10.1).

10.2.1 Bearing capacity of subgrade and influencing factors

The bearing capacity or strength of the subgrade is crucial to pavement design, since the thicknesses of the overlying layers are directly related to it.

Subgrade bearing capacity depends on many factors, but mainly on the size of soil particles and soil cohesion, moisture content and degree of the compaction of soil or soil material used.

The suitability of the subgrade to provide the platform on which the pavement is to be constructed can be determined by classifying the material using any authorised soil classification system (ASTM D 2487 2011, AASHTO M 145 2012 or others). However, the determination of the bearing capacity or strength of the subgrade is obtained only after conducting laboratory or in situ tests.

The subgrade bearing capacity is expressed in terms of one of the following parameters: California bearing ratio (CBR), modulus of reaction (K), resistance R value, resilient modulus (M_r) or stiffness modulus (E).

The most common parameter used in pavement design methodologies for flexible pavements is CBR. Also, commonly used parameters are the resilient modulus (M_r) and the stiffness modulus (E). The modulus of reaction (K) is mainly used in some design methodologies for rigid pavements, whereas the resistance value (R) has a limited use, mainly in some US states. All the above test methods that measure the subgrade bearing capacity or strength are described in Chapter 1.

While there are various ways for measuring soil strength, the major problem that a highway engineer faces is determining a representative value throughout the service life of the pavement. A decisive parameter in this hard task is the estimation of the moisture content at which the subgrade will be exposed throughout the life of the pavement. This will be the moisture content at which the strength of the subgrade material should be measured. The moisture content of the subgrade is related to both the subgrade moisture content during compaction and to the subgrade moisture content during the pavement service life. The latter should be kept constant or should not exceed a certain value (equilibrium value).

The determination of the moisture content during compaction is an easy task given that it is determined in the laboratory for maximum compacted density, employing the Proctor test (modified method), according to CEN EN 13286-2 (2012) or ASTM D 1557-09 (2012), or other suitable test methods (see Chapter 1). The only thing the engineer has to do is to ensure that the same moisture content was used during construction/compaction.

The above is achieved with relative easiness when works are carried out during summer months or dry periods. If construction works are conducted during the winter months or wet periods, the subgrade moisture during compaction will surely be higher than the desired one, the compacted density will be lower and thus the bearing subgrade capacity will be lower.

Black and Lister (1979) proved that when subgrades, particularly clayey material, are in saturated condition during construction, the subsequent subgrade bearing capacity at equilibrium, after the end of construction works, is lower. As a result, the pavement service life decreases. It has been estimated that the service life in these cases is approximately half if construction works were carried out under unsaturated conditions. Even in cases where a capping layer was also laid at the same time with the subgrade works, the subgrade moisture content increased significantly owing to the rainfall during the period of construction, which had an adverse effect on the pavement's service life.

From the above, it is recommended that the subgrade works be conducted during periods where no rainfalls are expected and, if possible, that construction of unbound layers (base and sub-base) and the first asphalt layer be completed during the dry periods.

It is noted that when the base course is left uncovered during the winter months to continue works during the spring months, surface erosion is most likely to occur, which affects the construction cost since corrective works will be needed before laying the asphalt layers.

The precise determination of the representative moisture content of the subgrade during the pavement's service life is a difficult, if not impossible, task. This is because the subgrade moisture content during the pavement's service life is affected by various factors, the most important of which are the seasonal variation of the depth of the water table level, the type of soil material (soil classification), the construction conditions in terms of provision of drainage system and the thickness of superimposed layers (thickness of pavement).

For an engineer, the most appropriate technique to solve the above problem and design the pavement properly is to precisely follow the instructions given in the pavement design methodology decided to be used.

Some pavement design methodologies require the determination of the subgrade strength and the corresponding moisture content of the soil at representative months of the year and, if possible, every month. This implies sampling and laboratory testing of the soil material at representative periods (say spring–summer–autumn–winter months), or every month, for the determination of strength/moisture content relationship. Having established this relationship and following the instructions of the methodology, the design strength value of the subgrade can be determined.

Some other pavement design methodologies determine the design strength of the subgrade at fully saturated conditions. This may lead to over-design of the pavement, but it certainly safeguards the desired design life of the pavement.

Regardless of the procedure used to determine the design strength of the subgrade, special attention should be given to cuts when clay, silt, silty clay or sandy clay material are encountered. In this case, the natural moisture content of the subgrade should be kept at equilibrium with the provision of waterproofing membrane and roadside drainage system (see Figure 10.2) or drainage layer below formation level and roadside drainage system.

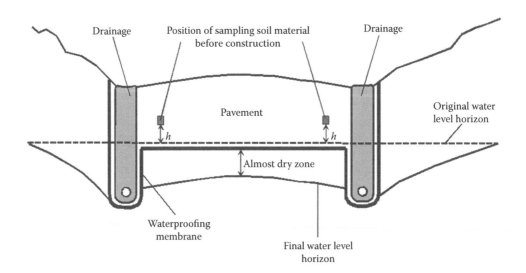

Figure 10.2 A typical cross-section of pavement protection from moisture fluctuation and lowering the water table in a cut.

The drainage layer consists of appropriate, almost single-size, granular materials, and it is usually 150 or 200 mm in thickness. The depth of the roadside trench below formation level depends on the capillary suction potential of the soil material.

Another parameter that should also be considered for the determination of the design subgrade strength is the uniformity of soil material along the entire length of the road. The soil may change several times, and if the road extends for many kilometres, pavement thickness may differ. If the road is short, the designer may select to use the lowest design subgrade strength. Soil stabilisation, locally (or in full length, if the soil material is very weak), may also be considered as another alternative.

10.2.2 Estimation of subgrade CBR

The CBR of the subgrade material can be estimated from tables derived after extensive laboratory work. Determining the design subgrade bearing capacity in terms of CBR constitutes a tedious and costly but necessary work. To reduce both effort and cost, some organisations have carried out extensive research from which tables and equations have been developed to estimate or predict the CBR values of various subgrade soil materials.

One of the most detailed studies for the CBR subgrade evaluation was conducted by Black and Lister (1979) and Powell et al. (1984). The results of Powell et al. (1984) have been incorporated and are used in the British pavement design methodology (Highways Agency 2006, 2009c), where it is not possible to collect material samples for executing CBR laboratory test. The results for estimating long-term CBR depending on soil type particularly for clay subgrades where moisture and plasticity index are significant issues, for high and low water table under different construction conditions, are shown in Table 10.1.

Table 10.1 Estimation of equilibrium subgrade CBR Value

Soil	PI	CBR high water table						CBR low water table					
		Construction conditions						Construction conditions					
		Poor		Average		Good		Poor		Average		Good	
		Thn	Thk	Thn	Thk	Thn	Thk	Thn	Thk	Thn	Thk	Thn	Thk
Clay	70	1.5	2	2	2	2	2	1.5	2	2	2	2	2.5
	60	1.5	2	2	2	2	2.5	1.5	2	2	2	2	2.5
	50	1.5	2	2	2.5	2	2.5	2	2	2	2.5	2	2.5
	40	2	2.5	2.5	3	2.5	3	2.5	2.5	3	3	3	3.5
Silty clay	30	2.5	3.5	3	4	3.5	5	3	3.5	4	4	4	6
Sandy clay	20	2.5	4	4	5	4.5	7	3	4	5	6	6	8
	10	1.5	3.5	3	6	3.5	7	2.5	4	4.5	7	6	>8
Silt		1	1	1	1	2	2	1	1	2	2	2	2
Sand													
– Poorly graded						20							
– Well graded						40							
Sandy gravel													
– Well graded						60							

Source: Powell W.D. et al., *The Structural Design of Bituminous Roads*. TRRL Laboratory Report LR1132. Crowthorne, UK: Transport Research Laboratory, 1984.

Note: PI, plasticity index; Thn, thin pavement; Thk, thick pavement.

A high water level is defined as when the water level is 300 mm beneath the formation level and is consistent with effective sub-soil drainage; a low water level is when the water level is 1 m below the formation level. Good construction conditions result in subgrades never getting wetter than their equilibrium moisture contents beneath the finished road. A thick pavement is 1200 mm deep, including a 650 mm capping layer, and is typical of motorway construction; a thin pavement is 300 mm deep. For pavements of intermediate thickness founded on plastic soils, an equilibrium value of CBR may be interpolated (Powell et al. 1984).

The CBR of soil materials may also be estimated from index properties or soil classification groups. For more information, see Sections 1.6.4 and 1.9.5, respectively.

10.3 CAPPING LAYER

The capping layer is the layer that is constructed between the subgrade and sub-base when the subgrade soil material is too weak. Its purpose is to improve the subgrade's bearing capacity and prepare an acceptable platform on which the pavement is to be constructed. The capping layer is usually needed when the subgrade CBR is lower than 5% and it is absolutely necessary when the CBR is lower than 2.5%.

The capping layer, provided pavement elevation permits, can be an additional layer after removing the vegetal top soil. Otherwise, the soil material is removed to a certain depth and is replaced with a suitable material. The thickness of the capping layer usually ranges from 300 to 600 mm, determined by the pavement design methodology followed.

The materials of the capping layer, in general, can be unbound granular material, including recycled aggregate of recycled bituminous planings, or stabilised with cement or lime subgrade material. Stabilisation is recommended when subgrade material has a CBR of less than 2%.

10.3.1 Materials for capping layer

The materials used for the construction of the capping layer or the formation of subgrade may be (a) collective granular soil material; (b) unbound mixtures of natural, artificial and recycled aggregates; and (c) soil material treated (stabilised) with hydraulically bound binders.

The collective granular material could be type GW, GP or GM according to the Unified Soil Classification System; type A-1-a or A-1-b according to the ASTM classification system; or type 6F1, 6F2 or 6F3 according to Highways Agency classification.

The unbound mixtures of natural, artificial and recycled aggregates should conform to CEN EN 13285 (2010) or any other relevant specification. The amount of recycled aggregates, originated from reclaimed granulated asphalt or bituminous planings, is recommended to be less than 50% except for type 6F (Highways Agency 2009a).

The hydraulic binders for soil stabilisation may be cement, lime, lime/cement or fly ash. Relevant European and American specifications for soil stabilisation are as follows: (a) for hydraulic binders, CEN EN 14227-13 (2006), and in particular for cement stabilisation, CEN EN 14227-10 (2006) or ASTM D 806 (2011); (b) for lime stabilisation, CEN EN 14227-11 (2006), ASTM C 977 (2010) or AASHTO M 216 (2013); and (c) for stabilisation with fly ash, CEN EN 14227-14 (2006) or ASTM C 593 (2011).

As a rule of thumb, cement binder is preferred to be used on soils with low plasticity index (PI lower than 10% to 12%). Higher plasticity soils are normally stabilised with lime.

In general, the capping material should conform to the national specification followed. Table 10.2 gives the characteristic property requirements of the capping materials recommended to be used in the United Kingdom.

Table 10.2 Characteristic properties of capping layer materials

Sieve size (mm) and other properties	Selected granular material			Granular material complying with CEN EN 13285 (2010)		Cement-stabilised material		Lime- or lime and cement-stabilised material	
	Fine 6F1	Coarse 6F2[a]	Coarse 6F3[b]	Fine 6F4	Coarse 6F5	Well graded 9A	Silty cohesive 9B	Well graded 9F[c]	Cohesive 9D, 9E[c], 9C[d]
Percentage passing (%), by mass									
125	—	100		0/31.5 mm and G_E, UF15 and OC75	0/80 mm and G_E, UF12 and OC75	100	100	Same as 9A	—
90	—	80–100				85–100	—		100
75	100	65–100				—	—		—
37.5	75–100	45–100				—	—		—
28	—	—				—	—		95–100
10	40–95	15–60				25–100	—		—
5.0	30–85	10–45				—	—		—
0.600	10–50	0–25				10–100	—		—
0.063	<15	0–12				<15	15–100		15–100
Moisture content at compaction	Optimum to (optimum − 2%)			Optimum to (optimum − 2%)		f	f	f	f
Los Angeles coefficient (LA)	≤60%	≤50%	—	≤60	≤50	—	—	—	—
% of reclaimed asphalt (R_a)	≤50%	≤50%	≥50%	≤50%	≤50%	—	—	—	—
Bitumen content of the R_a[e]	≤2%	≤2%	≤10%	≤2%	≤2%	—	—	—	—
Pulverisation	—			—	—	≥60%	g	≥60%	≥30%[h]
CBR	—			—	—	g	g	g	g
MCV (BS 1377-4 1990)	—			—	—	—	Max 12%[g]	—	f,i
Volume stability, if slag is used	—			Refer to CEN EN 13242 (2007)					—

Source: Adapted from Highways Agency, *The Manual of Contract Documents for Highway Works (MCDHW), Volume 1: Specification for Highway Works, Series 600: Earthworks*, London: Department for Transport, Highways Agency, 2009a (© Highways Agency).

Note: MCV = moisture condition value, limit of natural moisture for compaction.

a May include recycled aggregates with not more than 50% by mass of recycled bituminous planings and granulated asphalt, but excluding materials that contain tar.
b May include recycled aggregates with more than 50% by mass of recycled bituminous planings and granulated asphalt, but excluding materials that contain tar.
c Lime and cement stabilised.
d Conditioned pulverised fuel ash cohesive material.
e Limits shall not apply if R_a content is less than 20%.
f Min–max shall be specified.
g Min value shall be specified.
h For type 9C, the value is ≥60%.
i Not required for type 9C.

10.3.2 Soil stabilisation

Soil stabilisation is the process of improving the engineering properties of the soil to increase its bearing capacity and hence its strength. The process involves the use of hydraulic binders such as cement, lime, lime/cement mix and pulverised fuel ash often with lime or blast furnace slag by itself or mixed with cement. This is known as additive stabilisation or chemical stabilisation. In all cases, the hydraulic additives intermingle with the soil material at the project site. Additive stabilisation rarely takes place away from the project site.

Another type of stabilisation is mechanical stabilisation. Mechanical stabilisation is accomplished by mixing or blending soils of different gradations to obtain a material meeting the requirements. Soil blending may take place at the project site, in a central plant or in a borrowed site.

The decision to choose between additive stabilisation or mechanical stabilisation depends on the type of soil, the cost of the resulting material including transportation cost, availability and suitability of borrowed material and the availability of the required equipment.

It must be noted that, for the construction of a capping layer, both techniques are considered equally effective.

In the case of the additive stabilisation process, the percentage of the hydraulic binder to be added as well as the amount of water should always be determined in the laboratory by executing CBR, M_r (resilient modulus) or unconfined compressive strength tests, whichever is appropriate.

The minimum values of CBR or M_r to be achieved after chemical stabilisation should be at least equal to the value that an untreated granular soil material possesses, suitable for subgrade material. An addition of 2.0% cement or 2.5% of lime of total mass in a granular soil material normally gives satisfactory strength and performance of the stabilised material. However, higher than 2.5% of lime may be required to be added when the soil material has a lot of fines and has a high plasticity.

The choice between the types of additive to use for stabilisation depends almost exclusively on the plasticity of soil material. For a soil material with low to medium plasticity, a cement additive may be used, while lime additive is used for medium- to high-plasticity soils. Pulverised fuel ash, normally with lime, is used in soils with little or low plastic fines.

Chemical stabilisation is effective when, prior to the addition of the additives, soil has been pulverised such as all materials pass through a sieve of 31.5 mm. Another factor that also influences stabilisation is the moisture at compaction, since a certain amount of water will be taken for the chemical reaction at the early stage of mixing–curing.

Finally, for effective compaction, the thickness of the stabilised layer should be no more than 250 mm.

10.3.3 Compaction of capping layer

The capping layer materials are compacted using a suitable compaction plant. The capping layer may be compacted in one or more layers. The thickness of a single compacted layer is typically not less than 100 mm and not more than 250 mm. A lot of details regarding the type of compaction plant (equipment) may be used, and the corresponding number of passes to achieve a satisfactory compaction of the capping layer materials can be found in Highways Agency (2009a).

Table 10.3 gives summarised information for compacting capping layer materials with different compaction plants. In addition to the compaction plants given in Table 10.3, vibratory tamping rollers may also be used; for more information, see Highways Agency (2009a).

If the thickness of a single compacted capping layer is decided to be more than 250 mm, the number of passes for effective compaction should be determined on site in a demonstration area.

In all cases in Table 10.3, the number of passes refers to the moisture content at compaction or, in the case of stabilised material, the moisture condition value (MCV) at compaction, determined in the laboratory.

Prior to the compaction of the capping layer, the surface on which the capping layer is going to be constructed (cut or fill) should be rolled after formation and before laying the capping material. This consists of at least one pass of a static smoothed wheel roller of >2100 kg per metre width of roll, or at least one pass of vibrating roller of >700 kg per metre width of vibrating roll (Highways Agency 2009a).

10.3.4 Use of geotextiles and geotextile-related products

Geotextiles and geotextile-related products belong to the wider category of geosynthetic materials. Their role consists of providing at least one of the following three functions: filtration, separation or reinforcement. It is noted that in all cases, the function of separation is always in conjunction with filtration or reinforcement. Thus, to prescribe geotextiles for separation purpose alone is uncommon.

The use of geotextiles is recommended when the subgrade has a very low or low carrying capacity, CBR <2% or 2%–5%, respectively.

When using a geotextile, (a) the effectiveness of the capping layer increases, since no loss of coarse aggregate, forced in the soft soil material beneath during compaction, occurs; (b) more effective compaction of the overlaying unbound layer is achieved; and (c) reduction of shear strength of the capping layer material owing to clay particle intrusion is avoided.

The positive effect of the use of geotextile is documented by many field studies as well as comprehensive researches, namely, Potter and Currer (1981) and Robnett and Lai (1982). The measured surface deformation on the unbound layer laid over a subgrade with CBR 2% was much lower when a 450 g/m² polypropylene geotextile with a tensile strength of 10.5 kN/m was used (Potter and Currer 1981).

Geotextiles are constructed from synthetic fibres (polyamide, polyethylene, polyester or polypropylene) or other fibres, creating a textile of minimum water permeability, typically 10 l/m²/s (the opening of the pores usually ranges from 0.1 to 0.3 mm). Additionally, geotextiles should have sufficient tensile strength in order for the overlying layer to not deform and to have sufficient resistance to perforation and so that the geotextiles are not harmed or destroyed during discharge and compaction of granular aggregates.

Conclusively, the use of geotextiles is primarily to be able to achieve more effective compaction of the overlying layer and to protect the overlying layer from clay particle intrusion. The high tensile strength of the geotextiles contributes to the reduction of stresses imposed on the subgrade material and indirectly affects the bearing capacity of the subgrade.

A more effective result of the increase in subgrade bearing capacity is achieved with the use of geogrids, since they act as a kind of reinforcement. A geogrid can be used either on its own or in combination with a geotextile.

Geogrids usually consist of polyesters or woven fibres coated with PVC. The square mesh created has various dimensions that typically range from 10 mm × 10 mm to 50 mm × 50 mm. The mesh size to be used is determined by the size of aggregate particles of the overlying layer or of the layer in which the geogrid is to be embedded.

Geogrids have a much higher tensile strength than geotextiles; values usually range from 35 to 100 kN/m.

Table 10.3 Compaction of capping cayer materials with different compaction plants

Type of compaction plant	Category (unit mass in kg)	Recommended number of passes (N) for classes 6F1, 6F2, 6F3, 6F4, 6F5, 9A and 9F			Recommended number of passes (N) for classes 9B, 9D and 9E	
		Thickness of compacted layer (mm)				
		110 mm	150 mm	250 mm	150 mm	250 mm
Smoothed wheel roller (or vibratory roller operating without vibration)	Mass per metre width of roll:					
	>2100 to 2700	Unsuitable	Unsuitable	Unsuitable	Unsuitable	Unsuitable
	>2700 to 5400	16	Unsuitable	Unsuitable	Unsuitable	Unsuitable
	>5400	8	16	Unsuitable	12	Unsuitable
Grid roller	>2700 to 5400	Unsuitable	Unsuitable	Unsuitable	Unsuitable	Unsuitable
	>5400 to 8000	20	Unsuitable	Unsuitable	16	Unsuitable
	>8000	12	20	Unsuitable	8	Unsuitable
Deadweight tamping roller	4000 to 6000	12	20	Unsuitable	4	8
	>6000	8	12	20	3	6
Pneumatic-tyred roller	Mass per wheel:					
	<4000	Unsuitable	Unsuitable	Unsuitable	—	Unsuitable
	>4000 to 6000	12	Unsuitable	Unsuitable	a	16
	>6000 to 8000	12	Unsuitable	Unsuitable	b	8
	>8000 to 12,000	10	16	Unsuitable	c	4
	>12,000	8	12	Unsuitable	d	4
Vibratory roller (compaction speed 1.5 to 2.5 km/h)	Mass/metre width of vibrating roll:					
	<700	Unsuitable	Unsuitable	Unsuitable	Unsuitable	Unsuitable
	>700 to 1300	16	Unsuitable	Unsuitable	Unsuitable	Unsuitable
	>1300 to 1800	6	16	Unsuitable	Unsuitable	Unsuitable
	>1800 to 2300	4	6	12	12	Unsuitable
	>2300 to 2900	3	5	11	10	Unsuitable
	>2900 to 3600	3	5	10	10	Unsuitable
	>3600 to 4300	2	4	8	8	Unsuitable
	>4300 to 5000	2	4	7	8	Unsuitable
	>5000	2	3	6	6	12
Vibrating plate compactor	Mass/m^2 of base plate:					
	<1400	Unsuitable	Unsuitable	Unsuitable	Unsuitable	Unsuitable
	>1400 to 1800	8	Unsuitable	Unsuitable	10	Unsuitable
	>1800 to 2100	5	8	Unsuitable	8	Unsuitable
	>2100	3	6	12	6	Unsuitable
Vibro-tamper	Mass:					
	>50 to 65	4	8	Unsuitable	Unsuitable	Unsuitable
	>65 to 75	3	6	12	Unsuitable	Unsuitable
	>75 to 100	2	4	10	Unsuitable	Unsuitable
	>100	2	4	10	8	Unsuitable

(continued)

Table 10.3 Compaction of capping layer materials with different compaction plants (Continued)

Type of compaction plant	Category (unit mass in kg)	Recommended number of passes (N) for classes 6F1, 6F2, 6F3, 6F4, 6F5, 9A and 9F			Recommended number of passes (N) for classes 9B, 9D and 9E	
		Thickness of compacted layer (mm)				
		110 mm	150 mm	250 mm	150 mm	250 mm
Power rammer	Mass:					
	>100 to 500	5	8	Unsuitable	8	Unsuitable
	>500	5	8	14	6	10
Dropping-weight compactor	—	All categories are unsuitable				

Source: Adapted from Highways Agency, *The Manual of Contract Documents for Highway Works (MCDHW), Volume 1: Specification for Highway Works, Series 600: Earthworks*, London: Department for Transport, Highways Agency, 2009a (© Highways Agency).

a >1500 to 2000 : 12 (mass/wheel : N).
b >2000 to 2500 : 6 (mass/wheel : N).
c >2500 to 4000 : 5 (mass/wheel : N).
d >4000 to 6000 : 4 (mass/wheel : N).

The properties of the geotextiles and geotextile-related materials required for use in the road construction or for use in earthworks should conform to CEN EN 13249 (2014) and CEN EN 13251 (2014) or ASTM D 4759 (2011).

Although with the use of geotextiles, and even more with the use of geogrids, a reduction in the thickness of the unbound layers is possible, there is no pavement design methodology that caters for that. Determination of the resulting thickness reduction with the use of geotextiles and related materials is conducted, when needed, separately.

10.4 SUB-BASE COURSE

The *sub-base* is the first layer constructed, if necessary, over the subgrade (or the capping layer). This layer, certainly when the material used for its construction is the same as the base course material, is not a distinct layer. In a few countries, such as the United Kingdom, the sub-base indicates all the layers that are below the upper bound in bitumen or cement layers of the pavement.

The sub-base performs the following basic functions:

a. It reduces the loads and transfers them to the subgrade.
b. It eases the traffic of the worksite vehicles during construction.
c. It protects the base course materials from contamination from soil material (clay, silt, organic materials, etc.).
d. It acts as an anti-frost protective layer in cases where soil material is frost susceptible.
e. It may function as a drainage layer for the protection of subgrade from surface water passing through the pavement. This is possible when the pavement is thin, semi-dense asphalts have been used or when asphalt layers begin to crack.

In the case where the sub-base functions as a drainage layer, the use of waterproofing membrane is necessary on the interface between a subgrade and a sub-base course.

The thickness of the sub-base should be such that no high compressive and shear stresses develop in the subgrade under the influence of loads of worksite vehicles. The required thickness is determined according to the design methodology followed.

In the case where the sub-base course is considered as an anti-frost layer, its thickness is determined (increased) from the depth of frost penetration and the overall resulting pavement thickness.

10.4.1 Sub-base course material

Materials used for the construction of the sub-base course can be unbound or hydraulically bound aggregates obtained by processing natural or manufactured or recycled materials. In the case of hydraulically bound sub-base, the binder is cement, lime or fly ash.

In fact, sub-base materials are the same as base course materials. In some countries, sand-gravels are also permitted to be used for the construction of the sub-base course.

More information about sub-base materials is given in Sections 10.5.1 and 10.5.2.

10.5 BASE COURSE

The base course layer is positioned between the sub-base course (or subgrade, if there is no sub-base constructed) and the asphalt layers. Together with the asphalt layers, in a typical flexible pavement, it constitutes one of the main two structural elements of the pavement.

The base course layer performs the following functions:

a. It receives the loads exerted from the asphalt layers above, distributes them and passes them to the next layer below sub-base or subgrade.
b. It contributes to pavement strength.
c. It provides a good and even surface for laying the subsequent layers.
d. It contributes to the effective compaction of the overlying asphalt layers.

The base course consisted of a number of layers whose materials can be unbound aggregates or hydraulically bound aggregates.

In the case of full-depth asphalt pavement construction, the base course material obviously consists of bitumen-bound aggregates, that is, asphalts.

10.5.1 Materials for base and sub-base courses

The materials for the base and sub-base course are unbound aggregate mixtures or hydraulically bound aggregate mixtures. The aggregates may be crushed granular materials, manufactured materials from rock deposits or industrial by-products (slags) or recycled materials. The hydraulic binders, in the case of bound materials, are cement, fly ash, slag, lime, a mixture of some of them or factory-blended hydraulic binders for road use.

In the case of the sub-base course only, the granular material may also be uncrushed material conforming to the requirements.

The geometric, physical and chemical properties of the aggregates and the aggregate mix should comply with national specifications. In Europe, the relevant specifications are CEN EN 13242 (2007) and CEN EN 13285 (2010). In the United States, the relevant specification is ASTM D 2940 (2009).

As for the recycled materials, according to CEN EN 13242 (2007) and CEN EN 933-11 (2009), they may be recycled materials such as concrete products and concrete masonry units (Class R_c), unbound aggregate, natural stone or hydraulically bound aggregate (Class R_U), bituminous materials (Class R_a), clay masonry units (i.e. bricks and tiles), calcium silicate masonry units or aerated non-floating concrete (Class R_b), glass (Class R_g), floating material in volume (Class FL) and other materials in small quantities (less than 1% by mass) such as cohesive (i.e. clay and soil) and miscellaneous metals (ferrous and non-ferrous), non-floating wood, plastic and rubber or even gypsum plaster (classified as Class X).

10.5.2 Requirements of unbound aggregate mixtures according to European specifications

The requirements set by CEN EN 13285 (2010) for unbound mixtures to be used in bases and sub-bases for the construction or maintenance of pavements refer to (a) mixture requirements, (b) grading requirements, (c) aggregate requirements and (d) other requirements.

The *mixture requirements* consist of selecting and determining (a) the designation of the mixture, (b) the maximum or minimum (when required) fine content (passing through a 0.063 mm sieve) and (c) the percentage of oversize particles. All the above prescribe broadly the aggregate mixture.

The mixture designation is selected out of 15 mixtures, each one denoted by the ratio '0/D', where D is the nominal upper sieve size (in millimetres). Hence, a mixture designation of 0/31.5 means that the upper (D) sieve size or the nominal maximum size of aggregate is 31.5 mm and the mixture contains all fines.

The maximum fine content is selected out of five categories denoted by the letters 'UF' and a lower indexed number referring to a percentage value. Thus, a designation of UF_{12} means that the maximum fines content in the aggregate mixture is no more than 12%. When required, the lower percentage of fines is selected from three categories denoted by the letters 'LF' and a lower indexed number. Hence, a designation of LF_4 means than the lower fines content in the aggregate mixture is not less than 4%.

Similarly, the percentage of oversize particles is selected out of four categories denoted by the letters 'OC' and a lower indexed number referring to a percentage. Thus, a designation of OC_{85} means that the percentage passing through the upper (D) sieve lies within the range from 85% to 99%.

The *grading requirements* consist of determining (a) the overall grading range (limits), (b) the tolerance at each sieve from the supplier-declared value (SDV) to ensure production consistency and (c) the variation of the gradation of individual batch/sampling between successive sieves to ensure continuity of the gradation. In addition, the mean value calculated from all gradations should be within the SDV grading range.

The CEN EN 13285 (2010) specification, unlike other specifications, does not simply specify grading range per sieve for a given type of aggregate mixture. The grading range is determined after selecting the designation of the mixture (0/D) and its general category. From the designation of the aggregate mixture, the size of control sieves is determined, and from the general category chosen, the overall grading range is defined. The grading range is taken from a table provided by the specification.

The general categories to choose from are as follows: three for 'normal' (dense) mixtures (G_A, G_B and G_C), two for 'open' graded (G_o and G_p) and three for 'other' mixtures (G_E, G_U and G_V). The general category G_A has a slightly narrower grading range than G_B, and G_B has a slightly narrower grading range than G_C. The same applies for G_o and G_p. The categories for 'other mixtures' have a much broader grading range than all the above and do not require a supplier-declared grading range.

The supplier-declared grading range is also determined by the specification having selected the general category of the aggregate mixture.

The *aggregate requirements* consist of aggregate properties such as shape of coarse aggregate, percentage of crushed or broken particles and of totally rounded particles in coarse aggregates, fines quality, resistance to fragmentation of coarse aggregate, particle density, water absorption, resistance to wear of coarse aggregate, chemical requirements and durability requirements. These requirements shall be in accordance to CEN EN 13242 (2007).

The *other requirements* consist of additional aggregate requirements such as frost susceptibility, permeability, leaching or mechanical behaviour of unbound mixtures. These requirements must be considered and stated when conditions arise.

It must be noted that the water content of the mixture and the density of the installed layer are related to the control of the construction and they are not specified mixture requirements. However, the supplier is obliged to declare both values.

Table 10.4 gives mixture and grading requirements, proposed by the author, of two unbound types of mixtures recommended to be used. Type I is recommended to be used for both base and sub-base courses, while Type II is exclusively for sub-base courses.

Table 10.4 Recommended mixture and grading requirements for normal graded and other unbound mixtures for base and sub-base

Unbound mix	Type I (dense graded–all crushed)			Type II (selective granular material)		
Designation	0/31.5			0/31		
Max. fines	UF_9			UF_{12}		
Oversize	OC_{85}			OC_{80}		
General category	G_A			G_E		
	Range	SDV	Tolerance	Range	SDV	Tolerance
Sieve (mm)	% Passing, by mass			% Passing, by mass		
40	100	—	—	100	NR	NR
31.5	85–99	—	—	80–99		
20	—	—	—	—		
16	55–85	63–77	±8	50–90		
10	—	—	—	—		
8	35–65	43–57	±8	30–75		
4	22–50	30–42	±8	15–60		
2	15–40	22–33	±7	—		
1	10–35	15–30	±5	0–35		
0.5	0–20	5–15	±5	—		
0.063	0–9	—	—	0–12		

Difference between successive sieves					
Sieve (between sieves)		Permissible range (%)	Sieve (between sieves)		Permissible range (%)
—	—	—	20 mm	10 mm	7–30
16 mm	8 mm	10–25	—	—	—
—	—	—	10 mm	4 mm	7–30
8 mm	4 mm	10–25	NR	NR	NR
4 mm	2 mm	7–20	NR	NR	NR
2 mm	1 mm	4–15	NR	NR	NR

Note: SDV = supplier-declared value grading range (S); Tolerance = permitted tolerance, from supplier-declared value.

The aggregates in the Type I mixture should be all crushed material from rock deposits, crushed gravel, slugs or recycled material.

The aggregates in the Type II mixture should be natural sand and gravels with or without crushed aggregate materials as per Type I.

When reclaimed materials are to be used in the Type I mixture, such as asphalt (Class R_a) or other materials (Class X), the maximum permitted content recommended by the British specifications is less than 10% and 1%, respectively (% by mass) (Highways Agency 2009b).

Table 10.4 also gives the SDV range and the tolerance range per control sieve, for the Type I mixture. For Type II, there is no such requirement since it refers to basically natural material.

The meaning of SDVs and tolerances is that the supplier should specify the aggregate mixture gradation within the SDV range and all individual gradations from patches/tests should comply with the tolerances stated. By applying the tolerance range, the gradation is permitted to fall outside the SDV range.

At the bottom of Table 10.4, the permitted range of variation between successive sieves is also given, which, as mentioned earlier, ensures continuity of the gradation.

When open-graded unbound mixture is going to be used, Type III, specified by British specification (Highways Agency 2009b), may be used; the mixture and grading requirements are given in Table 10.5. The British specifications (Highways Agency 2009b) also allow the use of open-graded unbound mixtures with asphalt arisings (asphalt road planings or granulated asphalt); the mixture and grading requirements are also given in Table 10.5 (see Type IV).

The unbound mixture of Type III is made of materials as per the Type I mixture.

The unbound mixture of Type IV contains asphalt arisings and also contains crushed rock, crushed slag, crushed concrete or well-burnt non-plastic shale and up to 10% by mass of natural sand that passes through the 4 mm size test sieve. The permissible percentage of asphalt arisings should range between 50% and 100% provided the recovered bitumen content of the asphalt is not more than 10% (Highways Agency 2009b).

With regard to aggregate property requirements for all the above types of unbound mixture, the author proposes the use of Table 10.6.

10.5.2.1 Factory production quality control

The aggregate material produced for the construction of unbound layers should be periodically tested in order to ensure constant quality and compliance throughout the factory's production period.

The tests required for determining the aggregates' general properties and the minimum required test frequencies can be found in Table C.1 of CEN EN 13242 (2007).

10.5.3 Requirements of aggregates for unbound mixtures according to American standards

The requirements for graded aggregate material for unbound bases or sub-bases according to American standards are determined by ASTM D 2940 (2009) or AASHTO M 147 (2012).

According to ASTM D 2940 (2009), the coarse aggregate, retained on the 4.75 mm (No. 4) sieve, should be crushed stone, gravel or slag capable of withstanding the effects of handling, spreading and compacting without degradation and production of deleterious fines.

Table 10.5 Mixture and grading requirements for open-graded unbound mixture and unbound mixtures with asphalt arisings (for base and sub-base)

Unbound mix	Type III (open graded)			Type IV (with asphalt arisings)		
Designation	0/40			0/31.5		
Max. fines	UF$_5$			UF$_9$		
Oversize	OC$_{85}$			OC$_{75}$		
General category	G$_o$			G$_p$		
	Range	SDV	Tolerance	Range	SDV	Tolerance
Sieve (mm)	% Passing, by mass			% Passing, by mass		
	100	—	—	—	—	—
63	—	—	—	100	—	—
40	80–99	—	—	—	—	—
31.5	—	—	—	75–99	—	—
20	50–78	58–70	±8	—	—	—
16	—	—	—	43–81	54–72	±15
10	31–60	39–51	±8	—	—	—
8	—	—	—	23–66	33–52	±15
4	18–46	26–38	±8	12–53	21–38	±15
2	10–35	17–28	±7	6–42	14–27	±13
1	6–26	11–21	±5	3–32	9–20	±10
0.5	0–20	5–15	±5	—	—	—
0.063	0–5	—	—	0–9	—	—

Difference between successive sieves					
Sieve (between sieves)		Permissible range (%)	Sieve (between sieves)		Permissible range (%)
—	—	—	20 mm	10 mm	7–30
20 mm	10 mm	10–25	—	—	—
—	—	—	16 mm	10 mm	7–30
10 mm	4 mm	10–25	NR	NR	NR
8 mm	4 mm	—	NR	NR	NR
4 mm	2 mm	7–20	NR	NR	NR
2 mm	1 mm	4–15	NR	NR	NR

Source: Adapted from Highways Agency, The Manual of Contract Documents for Highway Works (MCDHW), Volume 1: Specification for Highway Works, Series 800: Road pavements – Unbound, cement and other hydraulically bound mixtures, London: Department for Transport, Highways Agency, 2009b (© Highways Agency).

Note: SDV = supplier-declared value grading range (S); Tolerance = permitted tolerance, from supplier-declared value.

The fine aggregate passing through the 4.75 mm (No. 4) sieve should normally consist of fines from the operation of crushing the coarse aggregate. The addition of natural sand or finer mineral matter, or both, is not prohibited, when they are suitable.

The aggregate grading and property requirements have been tabulated and are as shown in Table 10.7.

In addition to the requirements in Table 10.7, the fraction of the final mixture that passes through the 75 μm (No. 200) sieve should not exceed 60% of the fraction passing through the 0.600 mm (No. 30) sieve.

More details on sampling and testing frequencies can be found in ASTM D 2940 (2009).

Table 10.6 Recommended aggregate property requirements for unbound mixtures

Property	Type I	Type II	Type III	Type IV
$C_r{}^a$ and $C_{tr}{}^b$ for crushed rock and recycled aggregate materials (CEN EN 933-5 2004)				
$C_r{}^a$		≥90% (C_{90})		
$C_{tr}{}^b$		≤3% (C_3)		
$C_r{}^a$ and $C_{tr}{}^b$ for natural aggregate materials (CEN EN 933-5 2004)				
$C_r{}^a$	≥50% (C_{90})	NR[c]	NP[d]	NP[d]
$C_{tr}{}^b$	≤10% (C_3)			
Resistance to fragmentation (CEN EN 1097-2 2010)				
LA	≤40%	≤50%[e]	≤40%	≤50%[e]
Flakiness index (CEN EN 933-3 2012)				
FI		≤35%		
Magnesium sulfate soundness (CEN EN 1367-2 2009)				
MS		NR for frost-free or dry conditions in all climates. For saturated conditions: NR for Mediterranean climate, ≤25% for Atlantic and ≤18% for Continental climate		
Sand equivalent (CEN EN 933-8 2012)				
SE		≤40		NR[c]
Methylene blue (in fraction 0/2 mm, if fines >3%) (CEN EN 933-9 2013)				
MB		≤3		NR[c]
Water absorption				
WA_{24h}		NR. The supplier shall state the value for the aggregate used		
Volume stability of steel slags (BOF and EAF) (CEN EN 1744-1 2012, paragraph 19.3)				
V		≤5%	NP	≤5%
Dicalcium silicate and iron disintegration of air-cooled blast-furnace slag (CEN EN 1744-1 2012, paragraphs 19.1 and 19.2)				
'Bright' spots		Free from bright spots of yellow, bronze or a cinnamon colour on a violet background		

[a] C_r = crushed or broken particles.
[b] C_{tr} = totally rounded particles.
[c] NR = no requirement.
[d] NP = not permitted.
[e] In all pavements except motorways (if motorways, LA ≤ 40%).

10.5.4 Determination of water content for optimal density and wetting/laying procedures of unbound material

The determination of the water content for optimal dry density, required to be declared by the supplier, is carried out in the laboratory by compacting the unbound material at different water contents. Compaction of the unbound material can be carried out using proctor compaction in accordance with CEN EN 13286-2 (2012), ASTM D 1557 (2012) or AASHTO T 180 (2010); vibrocompression with controlled parameters in accordance with CEN EN 13286-3 (2003); vibrating hammer in accordance with CEN EN 13286-4 (2003); or vibrating table in accordance with CEN EN 13286-5 (2003).

Table 10.7 Grading and property requirements of aggregates for bases and sub-bases

| Sieve size (square openings) | Grading range | | Job mix tolerances | |
	Base	Sub-base	Base	Sub-base
	% Passing, by mass			
50 mm (2")	100	100	−2	−3
37.5 mm (1 1/2")	95–100	90–100	±5	±5
19 mm (3/4")	70–92	—	±8	—
9.5 mm (3/8")	50–70	—	±8	—
4.75 mm (No. 4)	35–55	30–60	±8	±10
0.600 mm (No. 30)	12–25	—	±5	—
0.075 mm (No. 200)	0–8[a]	0–12[a]	±3	±5
Two or more fractured faces[b]	≥75%	≥75%	—	—
Plasticity index (PI)[c]	≤4	≤4[d]		
Liquid limit[c]	≤25	≤25		
Sand equivalent (SE)	≥35	≥35[d]		
For steel slags:				
Expansion at 7 days[e]	≤0.5%	≤0.5%	—	—

Source: Adapted from ASTM D 2940, *Standard specification for graded aggregate material for bases or subbases for highways or airports,* West Conshohocken, Pennsylvania, US: ASTM International, 2009. With permission (© ASTM International).

[a] Determined by wet sieving and lower percentage shall be specified to prevent damage by frost action.
[b] For aggregates retained on the 9.5 mm sieve, according to ASTM D 5821 (2013).
[c] For the fraction passing through the 0.425 mm sieve, according to ASTM D 4318 (2010).
[d] For material at a greater depth than probable frost penetration, PI ≤ 6 and SE ≥ 30.
[e] Tested according to ASTM D 4792 (2013).

Wetting the base/sub-base unbound material to achieve the laboratory-determined optimum water content at compaction should be carried out in a suitable stationary or mobile mixing plant with appropriate water dosing system.

Laying the wet unbound material should be carried out by suitable pavers, similar to the ones used for laying the bituminous mixtures.

The quite common practice of wetting the unbound material by spraying water over the uncompacted layer laid by a motor grader, when permitted and employed, should ensure the use of the correct quantity of water and its uniform dispersion.

In all cases, great deviations of the water content from the laboratory-determined optimum value are not allowed. Generally, deviations of water content at compaction ranging from 2% lower to 1% higher than the optimum value determined are usually considered acceptable.

When wetted unbound material has to be transported, care must be taken to avoid evaporation of water during transportation.

10.5.5 Compaction of unbound base and sub-base layers

A fundamental prerequisite for a well-constructed base/sub-base is its sufficient compaction. Compaction is affected not only by the moisture content of the unbound mixture but also by the thickness of the layer to be compacted and hence the thickness of the compacted layer as well as the applied compaction energy.

The maximum permissible thickness of a compacted single layer, according to British specifications (Highways Agency 2009b), is 225 mm. If the thickness of the base/sub-base is more than 225 mm, compaction should be conducted in more than one layer. In the same specification, the minimum thickness of compacted layer is required to be 110 mm.

The compaction energy applied is related to the type and mass of compaction equipment used as well as the number of passes.

According to British specifications (Highways Agency 2009b), the compaction requirements for sufficient compaction of base/sub-base unbound mixtures using various types of compaction equipment (plant), when the compacted layer thickness is 110, 150 or 225 mm, are given in Table 10.8.

It should be mentioned that the most common type of compaction plant used to compact base/sub-base unbound mixtures is the vibratory roller.

Table 10.8 Compaction requirements for unbound mixtures

Type of compaction plant	Category	Number of passes for layers not exceeding the following compacted thicknesses		
		110 mm	150 mm	225 mm
Smooth-wheeled roller (or vibratory roller operating without vibration)	Mass per metre width of roll:			
	>2700–5400 kg	16	Unsuitable	Unsuitable
	>5400	8	16	Unsuitable
Pneumatic-tyred roller	Mass per wheel:			
	>4000 kg up to 6000 kg	12	Unsuitable	Unsuitable
	>6000 kg up to 8000 kg	12	Unsuitable	Unsuitable
	>8000 kg up to 12,000 kg	10	16	Unsuitable
	>12,000 kg	8	12	Unsuitable
Vibratory roller	Mass per metre width of vibrating roll:			
	>700 kg up to 1300 kg	16	Unsuitable	Unsuitable
	>1300 kg up to 1800 kg	6	16	Unsuitable
	>1800 kg up to 2300 kg	4	6	10
	>2300 kg up to 2900 kg	3	5	9
	>2900 kg up to 3600 kg	3	5	8
	>3600 kg up to 4300 kg	2	4	7
	>4300 kg up to 5000 kg	2	4	6
	>5000 kg	2	3	5
Vibrating-plate compactor	Mass per square metre of base plate:			
	>1400 kg up to 1800 kg	8	Unsuitable	Unsuitable
	>1800 kg up to 2100 kg	5	8	Unsuitable
	>2100 kg	3	6	10
Vibro-tamper	Mass:			
	>50 kg up to 65 kg	4	8	Unsuitable
	>65 kg up to 75 kg	3	6	10
	>75 kg	2	4	8
Power rammer	Mass:			
	>100 kg up to 500 kg	5	8	Unsuitable
	>500 kg	5	8	12

Source: Highways Agency, The Manual of Contract Documents for Highway Works (MCDHW), Volume 1: Specification for Highway Works, Series 800: Road pavements – Unbound, cement and other hydraulically bound mixtures, London: Department for Transport, Highways Agency, 2009b (© Highways Agency).

10.5.6 Strength and stiffness of unbound materials and unbound layers

The strength and the stiffness of the unbound materials to be used, or alternatively the stiffness of the compacted unbound layer, should be determined so as to ascertain the compliance with the requirements, dictated from the pavement design methodology followed.

The laboratory determination of the *strength* of the unbound materials is carried out using the CBR test (CEN EN 13286-47 2012, ASTM D 1883 2007 or AASHTO T 193 2013) or the *R* value test (ASTM D 2844 2013 or AASHTO T 190 2013).

The minimum laboratory CBR value usually required for natural selective material is 20% or 30%. As for the crushed unbound material, the minimum laboratory CBR value, when required, is usually 80%. When the strength is determined by the *R* value test, the minimum required values are 55 and 78 for the natural selective material and crushed unbound material, respectively (Asphalt Institute 1999).

The in situ determination of the strength of the compacted unbound layer is carried out by the Dynamic Cone Penetrometer (DCP) or by the plate bearing test. The DCP test is gaining grounds because of its simplicity in use. It must be reminded that the penetration index, determined by DCP, or *k* value, determined by the plate bearing test, can be correlated to the CBR value (see Sections 1.6.3 and 1.7).

The laboratory determination of the *stiffness* of the unbound material is carried out by the repeated load triaxial test (see Section 1.9.1) or by the springbox test (see Section 1.9.4).

The in situ determination of the stiffness of compacted unbound layer is carried out by the dynamic plate test using the light weight deflectometer or, in some countries, using the German dynamic plate equipment (see Section 1.9.3).

The falling weight deflectometer, usually performed to assess the in situ pavement layer stiffness, may also be used for the sub-base or base layers, but since the whole procedure is costly, it is rarely used for routine measurements.

The in situ stiffness determined by the dynamic plate test is in fact a determination of the surface elastic modulus affected by site conditions (moisture, temperature, etc.), the confinement of the loaded area, the thickness of the layer and the strength of the subgrade. The stiffness determined by the dynamic plate is also called surface modulus (E). Minimum requirements of surface modulus are set by some pavement design methodologies. The UK pavement design (Highways Agency 2006, 2009c), for example, requires a minimum value of surface modulus of 50 MPa, when unbound selective natural granular materials are used as foundation layer and a minimum of 100 MPa when unbound crushed materials are used as a foundation layer. It is reminded that the foundation layer is the layer on top of which the flexible or rigid layers are going to be constructed.

The above surface modulus is by no means the same as the layer's modulus, since stiffness modulus, among other things, is related to stress state and the non-linear (elastic) behaviour of the unbound material. As for stress state, high stresses result in higher stiffness modulus. This has been shown by various researchers such as Brown and Selig (1991). Unbound layers, when the load is applied at their surface, were found to have stiffness modulus even up to four times greater in comparison to the stiffness obtained when the same load was applied at the surface of the pavement (Brown and Dawson 1992). Highways Agency (2009c) states that an unbound foundation surface modulus of 100 MPa may comprise an unbound layer modulus of 150 MPa over a subgrade with a subgrade surface modulus of 60 MPa.

The layer modulus may also be determined in the laboratory by using the triaxial test with repeated axial cyclic stress, if in situ moisture and stress state conditions determined are simulated and applied in the laboratory. However, AASHTO (1993) provides a method

Table 10.9 Typical values of k_1 and k_2 for unbound base and sub-base materials

Moisture condition	For bases		For sub-bases	
	k_1	k_2	k_1	k_2
Dry	6000–10,000	0.5–0.7	6000–8000	0.4–0.6
Damp	4000–6000		4000–6000	
Wet	2000–4000		1500–4000	

Source: AASHTO, *Guide for design of pavement structures*, Washington, DC: American Association of State Highway and Transportation Officials, 1993. With permission.

to calculate the layer stiffness of the unbound base and sub-base courses. This is obtained by using the following equation:

$$E = k_1 \times \theta \times k_2,$$

where E is the elastic (resilient) layer modulus (psi) (1 psi = 6.8948 MPa); k_1, k_2 are regression constants, which are a function of material types; and θ is the stress state or sum of principal stresses $\sigma_1 + \sigma_2 + \sigma_3$ (psi).

The constants k_1 and k_2 are encouraged to be determined for each unbound material from the repeated triaxial test (e.g. $M_r = k_1 \times \theta \times k_2$); however, in the absence of these data, values given in Table 10.9 may be used.

A more simplified approach for estimating the layer's elastic modulus is proposed by the Shell pavement design methodology (Shell International 1985). In this case, the layer modulus can be estimated using the following equation:

$$E = k \times E_{sg},$$

where E is the modulus of unbound layer (Pa), k is the coefficient related to the thickness of the unbound layer and E_{sg} is the subgrade modulus (Pa).

The coefficient k is determined from the following relationship: $k = 0.2 \times h_2^{0.45}$ and $2 < k < 4$, where h_2 is the thickness of the unbound layer (mm).

The above equation, as it is stated, does not seem to be valid for high values of subgrade modulus (values above 2×10^8 Pa). Additionally, since the equation was developed using granular uncrushed and crushed natural or rock material in the unbound courses, other materials such as blast furnace slag may develop a much higher modulus. The opposite was found when rubble is used (Shell International 1985).

10.5.7 Hydraulically bound materials for bases and sub-bases

The base/sub-base course may be constructed using hydraulically bound mixtures (HBMs).

The hydraulic binder constituents are cement, lime, fly ash, slags, a mixture of some of them or factory-blended hydraulic binders for road use. The most commonly used hydraulic binder is the cement. The lime, when used, is almost exclusively used in a mix with other hydraulic binders.

The aggregates may be natural or crushed material having the same properties as those used for unbound courses.

The hydraulic bound material may, sometimes, consist of soil treated with one of the abovementioned hydraulic binders.

With the addition of hydraulic binders, the strength and the stiffness of the material and of the layer are increased to a certain extent and, when required, noticeable with the use of cement. However, the use of cement in the base/sub-base course by no means provides a mixture with strength equal to the strength required in the upper cement-bound layers of the pavement.

The use of hydraulic bound mixtures is encouraged and is effective, in terms of reduction of pavement's total thickness, when the subgrade material has low to medium strength (CBR values less than 15%). Additionally, the use of hydraulically bound material may be decided when the available unbound material do not comply with all requirements to be used on its own for the construction of the unbound layers.

Finally, the use of HBMs is recommended by the British pavement design methodology (Highways Agency 2006, 2009c) when the traffic volume is high (higher than 60 million of standard axles), that is, for long-term pavement design life.

In the case of using HBM, it is absolutely important to ensure good performance of the layer in shrinkage and thermal cracking and that the overlying asphalt layers have adequate thickness to reduce the risk of reflection cracking to an acceptable level. Pavements have suffered early-stage development of transverse reflective cracks, when no pre-cracking was induced and the thickness of the asphalt overlying layers was even 20 cm. Transverse crack was more frequent in areas where the compression strength of the cement-bound material was found to be high (C12/15 and C16/20) (Nikolaides 2009). Although these cracks do not present a structural problem, they often accelerate deterioration of the pavement by allowing water to enter lower pavement layers.

Overall, the decision to use hydraulically bound material should be based on cost criteria, design traffic demand, availability of plant and machinery for skilful construction, local experience and absence of suitable unbound materials.

10.5.8 European practice for HBMs

The European practice for the construction of the sub-base or base course with HBMs includes the use of the granular material bound with various hydraulic binders, as well as the use of soil treated with the same hydraulic binders.

10.5.8.1 Hydraulic binders or binder constituents

The hydraulic binders or constituents such as cement, fly ash, slags and lime should comply with CEN EN 197-1 (2011), CEN EN 14227-4 (2013), CEN EN 14227-2 (2013) and CEN EN 14227-11 (2006), respectively. Additionally, any other hydraulic road binder, that is, factory-blended hydraulic binder for road use, may also be used, provided it complies to CEN EN 12447-5.

The percentage of binder added to the mix is determined by mix design procedures. Mix design procedures shall conform to national regulations or to provisions valid at the place of use. The minimum binder content, when cement binder is used, is specified by CEN EN 14227-1 (2013) with respect to maximum nominal aggregate size of the mix.

Table 10.10 provides the minimum percentages required when cement is used as a hydraulic binder, together with the minimum values for all binders or binder constituents used today, as specified by British specifications (Highways Agency 2006).

It must be noted that the lime is normally used for lowering the plasticity index of the material and to temporarily modify the materials to ease construction. Permanent increase of the strength of the material can be achieved only when used in a mixture with other hydraulic binders. Hence, the use on its own is very limited.

Table 10.10 Minimum binder additions for HBM

Binder or binder constituent	Application	Minimum addition by dry mass of mixture	
		Mix-in-plant by mass batching (% by mass)	Mix-in-place or mix-in-plant by vol. batching (% by mass)
Cement	When used as the only binder	3%, 4%, 5%[a]	4%, 5%, 6%[a]
	When used with another binder	2%	3%
	When used as the only binder in soil treated by cement (SC)	3%	4%
Dry fly ash	When used with cement	4%	5%
	When used with lime	5%	6%
Wet (conditioned) fly ash	All applications	6%	8%
Fine granulated blast furnace slag	When used with cement	2%	3%
	When used with lime	3%	4%
Granulated blast furnace slag (GBS)	When used with lime	6%	8%
	When used with ASS (GBS + AAS ≥ 11%)	2.5%	3%
Air-cooled slag (ASS)	When used with GBS (ASS + GBS ≥ 11%)	2.5%	3%
Lime (quicklime or hydrated lime)	When used as the only binder in FABM 5	3%	4%
	When used with another binder	1.5%	2%
Hydrated road binder	All applications	3%	4%

Source: Highways Agency, *The Manual of Contract Documents for Highway Works (MCDHW), Volume 1: Specification for Highway Works, Series 800: Road pavements – Unbound, cement and other hydraulically bound mixtures*, London: Department for Transport, Highways Agency, 2009b (© Highways Agency).

[a] The percentages refer to mixtures with maximum nominal size aggregate >8 to 31.5 mm, 2 to 8 mm and <2 mm, respectively.

10.5.8.2 Aggregate requirements for HBMs

The aggregates for HBMs may be either crushed or uncrushed, or a combination of both, naturally occurring or artificial and recycled construction aggregate. They should all comply with CEN EN 13242 (2007).

The aggregate property requirements, mainly percentage of crushed and broken particles, resistance to fragmentations and fines quality (plasticity) and their requirements, depend on the HBM that will be used.

Table 10.11 gives the aggregate property requirements of all the HBMs proposed by British specifications (Highways Agency 2009b).

The aggregate grading requirements also depend on the specific HBM that will be used. With regard to cement-bound granular mixtures (CBGMs), which are the most common type of HBMs, aggregate grading should comply with CEN EN 14227-1 (2013). In particular, the aggregate grading of the job mix may fall within any of the three different types of grading envelopes (A, B and C) specified (see Table 10.12). Grading envelope A is very wide and covers all gradings with which practical experience in cement-bound granular mixture exists, including sands. Grading envelope B is narrower and grading classified by this envelope includes well-graded coarse aggregates with limited contents of fines (<0.063 mm). Grading envelope C is sub-divided into four even narrower envelopes referring to 0/31 mm, 0/20 mm, 0/14 mm and 0/10 mm size mixture. Apart from the 0/31 mm envelope, the other

Table 10.11 Aggregate property requirements for HBM

	Properties						
	CBGM			FABM 1	SBM B2 FABM 2	SBM B3 FABM 3	SBM B1-1, B1-2, B1-3
HBM designation	A	B	C	HRBBM 1	HRBBM 2	HRBBM 3	and B1-4
C_r[a] and C_{tr}[b] for crushed aggregates or recycled aggregates							
C_r[a]	NR[c]		NR[c]		C_{90} (or C_{50})[d]	NR[c]	C_{90} (or C_{50})
C_{tr}[b]					C_3 (or C_{30})[d]		C_3 (or C_{30})
Resistance to fragmentation of coarse aggregate (Los Angeles value)							
LA	NR[c]	≤50% or ≤60%	≤50%	≤50% or ≤60%	≤50%	NR[c]	≤50%
Acid-soluble surface content							
AS			Air-cooled blast furnace slag: ≤$AS_{1.0}$. Other aggregates: ≤$AS_{0.2}$				
Total sulfur content							
S			For air-cooled blast furnace slag: S_2. Other aggregates: S_1				
Fines quality (Plasticity index)							
PI	NR[c]			Non-plastic (NP)		NR[c]	NP
Maximum glass content (Class R_g)							
R_g				≤40%			
Maximum impurities (Class X)							
X	≤5%			≤3%		≤5%	≤3%

Source: Highways Agency, *The Manual of Contract Documents for Highway Works (MCDHW), Volume 1: Specification for Highway Works, Series 800: Road pavements – Unbound, cement and other hydraulically bound mixtures*, London: Department for Transport, Highways Agency, 2009b (© Highways Agency).

[a] C_r = crushed or broken particles.
[b] C_{tr} = totally rounded particles.
[c] NR = no requirement.
[d] No requirement if FABM 1 contains at least 3% cement by dry mass and trafficking is prevented for 7 days.

three are further sub-divided into category G1 (narrow range) and category G2 (slightly wider range).

The same applies with the aggregate gradings for the other HBMs such as slag bound mixtures (SBM), fly ash bound mixtures (FABM) and hydraulic road binder bound mixtures (HRBBM). They should fall within the grading envelopes specified by CEN EN 14227-2 (2013) for SBM, CEN EN 14227-3 (2013) for FABM and CEN EN 14227-5 (2013) for HRBBM.

Table 10.12 gives the aggregate grading requirements for the most commonly used type of HBM, the cement-bound granular mix, as specified by CEN EN 14771-1.

10.5.8.3 Requirements of treated soil by hydraulic binders

The soil requirements when treated by hydraulic binder are specified in the appropriate specification depending on the binder used: CEN EN 14227-10 (2006) for soil treated by cement, CEN EN 14227-11 (2006) for soil treated by lime, CEN EN 14227-12 (2006) for soil treated by slag, CEN EN 14227-14 (2006) for soil treated by fly ash and CEN EN 14227-13 (2006) for soil treated by hydraulic road binder.

Table 10.12 Grading envelopes for cement-bound granular mixtures

Sieve (mm)	CBGM A	CBGM B	CBGM C 0/31.5	CBGM C 0/20 GI	CBGM C 0/14 GI	CBGM C 0/10 GI
63	100	100				
40	—	—	100	—	—	—
31.5	85–100	85–100	85–100	100	—	—
25	—	—	75–100	—	100	—
20	—	—	65–94	85–100	—	—
16	—	—	—	—	—	100
14	—	—	—	—	85–100	—
10	—	—	44–78	55–80	68–90	85–100
6.3	—	—	—	42–66	50–72	62–83
4	—	—	26–61	32–56	38–60	48–71
2	15–100	15–50	18–50	23–43	26–46	33–54
0.5	—	—	8–30	11–26	13–27	17–31
0.25	—	—	6–22	8–19	10–20	12–23
0.063	0–15	5–15	3–11	3.5–9	4.5–10	6.5–12

Source: Adapted from CEN EN 14227-1, *Hydraulically bound mixtures – Specifications – Part 1: Cement bound granular mixtures*, Brussels: CEN, 2013. With permission (© CEN).

In the same specifications, details of the requirements for their composition and laboratory performance classification can also be found.

These HBMs are normally used as a capping or as a sub-base layer.

10.5.8.4 Laboratory mixture design procedure

Prior to the commencement of the work, the target proportions of the constituents, including the water, and the properties of the HBM should be determined.

The properties required to be determined depend on the requirements of the national specification. In general, it is required to determine the compressive strength after 28 days, tested in compliance with CEN EN 13286-41 (2003), or when required, the immediate bearing index, tested in compliance with CEN EN 13286-47 (2012). The latter is almost the same as the CBR test.

When the modulus of elasticity (E) and the tensile modulus (R_t) are required to be determined, the tests shall be carried out in compliance with CEN EN 13286-43 (2003) (modulus of elasticity), with CEN EN 13286-40 (2003) (direct tensile test) or with CEN EN 13286-42 (2003) (indirect tensile test).

Some countries may also require the determination of the resistance to water strength after immersion, the compacity (compaction/density) for SBMs only, the workability period and the resistance to frost damage.

The resistance to water strength after immersion is assessed by comparing the average compression strength of specimens cured and immersed for 14 days with that of specimens cured for 28 days (Highways Agency 2009b).

The compacity is determined as defined in CEN EN 14227-2 (2013) and the workability period is measured in conformity with CEN EN 13286-45 (2003).

The resistance to frost damage (heave) is judged by the magnitude of the compressive strength or indirect tensile strength and is achieved after 28 days of curing at 20°C. An

HBM shall be deemed resistant to frost heave where the compressive strength class is C3/4 or greater or R_{it} is >0.25 MPa (Highways Agency 2009b).

It must be noted that the typical compressive strength required for HBM is C3/4 or C5/6. HBMs with higher strengths, such as C8/10 and up to C16/20, are selected for pavement with high traffic volume, >80 million standard axles over a long design life period (≥30 years).

It is clarified that the first value of C refers to the 28-day strength determined from cylinders with a height-to-diameter ratio equal to 2, while the second value refers to the 28-day strength determined from cubes or cylinders with a height-to-width (or diameter) ratio between 0.8 and 1.21 (normally 1.0). The compressive strength is expressed in megapascals (MPa).

10.5.8.5 General requirements for mixing and layer construction of HBM

The HBMs may be produced/mixed in batch plants or in-place. The first method eliminates mix properties variations.

Construction of layers, including multiple lift layers, should be completed at a certain period depending on the type of HBM used. The duration should be specified and it depends on the hydraulic binder used and air temperature during construction.

According to British practice (Highways Agency 2009b), the minimum compacted lift should be 150 mm thick. The compaction should be completed without drying out and before setting of any part of the layer, meeting the requirements for density specified. Compaction of HBM, other than FABM 5, should be carried out by a vibrating roller or a pneumatic roller. When a vibrating roller is used, the last passes should always be with a pneumatic roller (at least eight passes when wheel load is ≥30 kN). Laying should be carried out at air temperatures above 5°C and should be ceased when it drops below 3°C and in the case of heavy or persistent rain. HBMs should never be laid on a frozen surface.

Curing before trafficking and protection of the surface during curing are important and instructions given should be strictly followed. Protection of the surface during curing can be carried out not only by the application of mist/fog/light spray of water but also with the application of concrete curing compound or by spraying bitumen emulsion. When bitumen emulsion is used, a quantity of 0.2 kg/m² of residual bitumen is sufficient but the membrane should be protected from any damage until the construction of overlaying layer (Highways Agency 2006).

Properly induced cracking, when required, is vital in order to avoid the development of thermal cracking of the HBM layer, which eventually will be reflected to the surface of the pavement. Methods used are as follows: (a) formation of transverse cracking at a certain spacing (normally every 3 m apart to a depth between one-half to two-thirds of the layer thickness) induced in fresh material after initial compaction and (b) induce random cracks in the layer after setting. All HBM layers that are expected to reach a compressive strength of 10 MPa at 7 days must have transverse cracks induced (Highways Agency 2006).

10.5.8.6 Testing and quality control of HBM

Testing, quality control and checking of the HBM should be carried out in accordance to the requirements set by the national specifications.

Testing of the constituent materials and of the mixtures should be carried out periodically to ensure compliance with requirements and imply control of quality. The tests may include determination of the following: (a) grading, water content and plasticity of the aggregate or soil sources; (b) mixture grading and hydraulic binder grading; (c) moisture content value (CEN EN 13286-46 2003); (d) degree of pulverisation (only if mixture is cohesive) (CEN EN 13286-48 2005); (e) in situ wet density; and (f) laboratory mechanical properties such as compressive strength, tensile strength, modulus of elasticity or strength after immersion or as requested.

A comprehensive table with the tests and quality control checks and their frequencies required in the United Kingdom is given in Highways Agency (2009b).

10.5.9 American practice for HBMs

The HBMs in the United States are called chemically stabilised material (CSM). They can also be referred to as cementitious stabilised materials. The latter is a general terminology that also covers soil stabilisation for subgrade improvement.

The distinct types of CSM used for the construction of base/sub-base layer are as follows: (a) cement-treated aggregate (CTA) or cement-treated base (CTB), (b) lean concrete (LC), (c) soil cement (base course), (d) lime–cement–fly ash base or pozzolanic-stabilised mixture, (e) lime-stabilised soil (LSS) and (f) open-graded cement stabilised.

However, apart from the cement binder (ASTM C 150 2012 or AASHTO M 85 2012), the hydraulic binders may be blended hydraulic cement (ASTM C 595 2013 or AASHTO M 240 2013), coal fly ash and raw or calcined natural pozzolan (ASTM C 618 2012 or AASHTO M 295 2011), ground granulated blast-furnace slag (ASTM C 989 2013 or AASHTO M 302 2013), silica fume (AASHTO M 307 2013) and lime (ASTM C 977 2010 or AASHTO M 216 2013).

The Federal Aviation Administration (FAA 2011) distinguishes the CSMs for base/sub-base into cement-treated base course (CTBC), econocrete base course (EBC) and soil cement base course (SCBC). The binder in the first and third material (CTBC and SCBC) is cement, while in the second material, the binder can be cement conforming to ASTM C 150 (2012) or blended hydraulic binder conforming to ASTM C 595 (2013), or with cementitious additives such as pozzolanic, meeting the requirements of ASTM C 618 (2012), or ground granulated slags, which conform to ASTM C 989 (2013).

Some states may use slightly different terminologies for CSMs. The state of California, for example, distinguishes CSMs for base/sub-base into CTB, lean concrete base (LCB) and LSS (Caltrans 2010).

The CTA mixtures and the LC are considered to be higher-quality materials and have higher strength than materials such as soil cement or LSS.

10.5.9.1 Aggregate and CSM requirements

The aggregate material for the CSMs may be select granular material, crushed or uncrushed gravel or stone, recycled aggregate, recycled crushed and graded Portland cement concrete, crushed and graded iron blast furnace slag or a combination thereof, as well as approved select soil.

The aggregate requirements for CSMs, such as their mechanical and physical properties and the grading of the aggregate mix, are basically determined from federal or state relevant specifications.

Generally speaking, the CSMs are required to have (a) some minimum compressive strength depending on the type of overlaying layers and hence type of pavement (flexible or rigid) and the importance of the layer in the pavement structure (base course or sub-base) and (b) freeze–thaw durability.

Table 10.13 Properties of aggregates and cementitious stabilised material requirements by FAA specifications

Property	CTBC[a]		EBC[b]	
Properties of aggregates				
Sieve size	Gradation A	Gradation B	37.5 mm	25 mm
50.8 mm (2")	100[c]	100[c]	—	—
37.5 mm (1 1/2")	—	—	100	—
25.4 mm (1")	—	—	70–95	100
19 mm (3/4")	—	—	55–85	70–100
4.75 mm (No. 4)	45–100	55–100	30–60	35–65
1.80 mm (No. 10)	37–80	45–100	—	—
0.425 mm (No. 40)	15–50	25–80	10–30	15–30
0.210 mm (No. 80)	0–25	10–35	—	—
0.075 mm (No. 200)	—	—	0–15	0–15
LA value[d]	≥40		Compliance with ASTM C 33 (2013)	
S-MS[e]	≤13			
LL and PI[f]	≤25 and ≤6			
Properties of cementitious stabilised material				
Compressive strength (7 days) (ASTM D 1633 2007 for CTBC, ASTM C 39 2012 for EBC)	3447–6895 kPa[g] (500–1000 psi)[g] or 5170–6895 kPa[h] (750–1000 psi)[h]		3445–5516 kPa (500–750 psi)	
W–D or F–T[i]	≤14[j]		≤14, if required	

Source: Adapted from FAA, *Advisory Circular, AC 150/5370-10F*, U.S. Department of Transportation, Federal Aviation Administration, 2011.

[a] CTBC = cement-treated base course.
[b] EBC = econocrete base course.
[c] Maximum size of aggregate is 25.4 mm (1") when used as base course under Portland cement concrete pavement.
[d] LA = resistance to degradation in Los Angeles (LA) machine, ASTM C 131 (2006).
[e] S-MS = soundness by use of magnesium sulfate, ASTM C 88 (2013).
[f] LL and PI = liquid limit and plasticity index.
[g] When placed under Portland cement concrete pavement.
[h] When placed under HMA pavement.
[i] W–D or F–T = wet–dry or freeze–thaw tests, ASTM D 559 (2003) and ASTM D 560 (2003), respectively.
[j] Not necessary if the 7-day compressive strength is ≥5170 kPa (750 psi).

Table 10.13 gives details on aggregate requirements according to the FAA specifications (FAA 2011) and Table 10.14 provides details on such requirements according to California state specifications (Caltrans 2010).

10.5.9.2 Laboratory mixture design procedure

For more details regarding the laboratory mixture design procedure, see the appropriate specification used.

10.5.9.3 General requirements for mixing and layer construction of CSM

For more details regarding the general requirements for mixing and layer construction of CSM, see the appropriate specification used.

Table 10.14 Properties of aggregates and cementitious stabilised material requirements by caltrans specifications

Property	CTB[a]		LCB[b]	
Properties of aggregates				
Sieve size	Class A[c]	Class B[c]	38.1 mm[d]	25.4 mm[d]
75 mm (3")	—	100	—	—
63.5 mm (2 1/2")	—	90–100	—	—
50.8 mm (2")	—	—	100	—
38.1 mm (1 1/2")	—	—	90–100	100
25.4 mm (1")	100	—	—	90–100
19 mm (3/4")	90–100	—	50–85	50–100
9.5 mm (3/8")	—	—	40–75	40–75
475 mm (No. 4)	40–70	35–70	26–60	35–60
0.600 (No. 30)	12–40	—	10–30	10–30
0.075 mm (No. 200)	3–15	3–20	0–12	0–12
Sand equivalent	≥21		≥21	
R value, California test 301[d]	≥60[e] ≥80[f]		—	
Properties of cementitious stabilised material				
Compressive strength	≥5170 kPa (≥750 psi)	—	≥4825 kPa[g] (≥700 psi)[g]	

Source: Caltrans, 2010. *Standard Specifications 2010 Division IV.* Sacramento, CA: Department of Transportation.

[a] CTB = cement-treated base.
[b] LCB = lean concrete base.
[c] Operating range. For contract compliance range, see Caltrans (2010), Article 27.
[d] Operating range. For contract compliance range, see Caltrans (2010), Article 28.
[e] Before aggregates mixed with cement.
[f] After aggregates mixed with cement ≤2.5% by weight of dry aggregate.
[g] The Portland cement content must be at least 160 kg/m³ (270 lb/cu yd).

10.5.9.4 Testing and quality control of CSM

For more details regarding the testing and quality control of CSM, see the appropriate specification used.

10.5.9.5 Elastic modulus, resilient modulus and flexural strength for design purpose

In a pavement design methodology such as the AASHTO methodology (AASHTO 1993), the elastic modulus, E (ASTM C 469 2010), or alternatively, the unconfined compressive strength (7 days) (ASTM D 1633 2007) of the CTA base, needs to be determined. With either value, the structural coefficient (a_2) is derived and the thickness of the corresponding layer as well as of all layers of a flexible pavement is determined (see Section 13.4.4.3). For a rigid pavement design, the elastic modulus of the sub-base is used (see Section 14.11.1).

In the Mechanistic–Empirical Pavement Design Guide (MEPDG) (AASHTO 2008), for a new pavement design, when CTA or LC material is used, the elastic modulus is required to be determined (ASTM C 469 2010). In the case of a flexible pavement only, its flexural strength is also required to be determined, according to AASHTO T 97 (2010).

When MEPDG is employed and lime–cement–fly ash material or soil cement is going to be used again, the flexural strength needs to be determined, according to AASHTO T 97

Table 10.15 Equations for estimating properties of chemically stabilised materials

Property	Material	Relationship[a]	Test method
Elastic modulus (E) or resilient modulus (M_r)	Lean concrete or cement-treated aggregate	$E = 57{,}000 \times f_c'^{0.5}$	AASHTO T 22 (2011)
	Lime–cement–fly ash	$E = 500 + q_u$	ASTM C 593 (2011)
	Soil cement	$E = 1200 \times q_u$	ASTM D 1633 (2007)
	Lime-stabilised soil	$M_r = 0.124 \times q_u + 9.98$	ASTM D 5102 (2009)
Flexural strength (required only for flexible pavements)	Use 20% of the compressive strength as an estimate of the flexural strength for all chemically stabilised materials		

Source: From AASHTO, *Mechanistic–Empirical Pavement Design Manual (PEPGD-1)*, Washington, DC: American Association of State Highway and Transportation Officials, 2008.

[a] f_c' or q_u unconfined compressive strength in psi of laboratory samples or extracted cores.

Table 10.16 Typical values of properties for chemically stabilised materials

Material	Property	Typical value
Lean concrete	E (psi)	2,000,000
Cement-treated aggregate		1,000,000
Open-graded cement-stabilised aggregate		750,000
Soil cement		500,000
Lime–cement–fly ash, E		1,500,000
Lime-stabilised soils, M_r	M_r (psi)	45,000
Chemically stabilised material placed under flexible pavement	Modulus of rupture (M_R) (psi)	750
Chemically stabilised material used as sub-base, select material or subgrade under flexible pavement		250

Source: AASHTO, *Mechanistic–Empirical Pavement Design Manual (PEPGD-1)*, Washington, DC: American Association of State Highway and Transportation Officials, 2008. With permission.

(2010) and ASTM D 1635 (2012), respectively. Finally, when LSS is used, the resilient modulus needs to be determined according to AASHTO T 307 (2007).

When there is limited or no testing capability, the abovementioned properties for CSMs may be estimated by using the equations shown in Table 10.15. Alternatively, typical values, as shown in Table 10.16, may also be used.

10.6 ASPHALT LAYERS

The asphalt layers in a flexible pavement consist of the asphalt base, the binder course and the surface layer or wearing course.

10.6.1 Asphalt base

The asphalt base is the first and most important layer of a flexible pavement, constructed over the unbound or hydraulically bound base. Together with the other overlying asphalt layers, it distributes the traffic load to the underlying layers and dignifies the strength of the pavement and its resistance to fatigue. The asphalts used should be of the best quality and possess the following:

a. Sufficient stiffness
b. Sufficient fatigue resistance
c. Acceptable resistance to permanent deformation
d. Very low water permeability, particularly when the surfacing material is of semi-open or open texture and the binder course is not used or is not distinguished as layer
e. Good resistance to the detrimental effect of the water

10.6.2 Binder course

The binder course is an intermediate layer between the asphalt base and the surface layer. Its purpose is to provide an even platform for the construction of the surface layer. The asphalts used should have the same properties as the asphalt base.

The binder course in some countries is not distinguished as a separate layer, since the type of asphalt used is usually the same as the asphalt base.

10.6.3 Surface layer

The surface layer is the layer that comes in direct contact with the tyres of the vehicles and provides the surface over which the traffic should roll with comfort and safety. The main functions and requirements of the surface layer are as follows:

a. To provide a surface with very good and long-lasting skid resistance ability
b. To sustain the destructive action of the traffic and the weather conditions
c. To be resistant to permanent deformation
d. To be resistant to surface thermal cracking
e. To be even for the provision of high-quality and safe driving
f. To provide a surface with acceptable level of traffic noise

10.6.3.1 Types of asphalts for asphalt base, binder course and surface layers

The asphalts (hot, warm, semi-warm mixed or cold mixed) used for the construction of asphalt base, binder course and surface layers together with their mechanical and other properties are described in detail in Chapters 5, 6 and 7. In addition, details on their production, laying, compaction and control of quality can be found in Chapters 8 and 9.

For the surface layer, in particular, the asphalts most commonly used nowadays are (a) asphalt concrete for very thin layers (AC-VTL), (b) porous asphalt (PA), (c) open-graded friction course mixture (OGFC), (d) stone mastic asphalt (SMA), (e) single or double surface dressing (S-SD or D-SD) and (f) micro-surfacing (MS). Table 10.17 gives the comparative properties/characteristics for all the abovementioned surfacing materials.

With regard to durability, Nicholls et al. (2011), after 9 years of monitoring, found that the life expectancy or durability of the surfacing materials examined, such as AC-VTL, SMA, D-SD and MS, can be higher than the values shown in Table 10.17.

Table 10.17 Comparative table of asphalts used in surface courses

Property	AC-VTL	PA (OGFC)	SMA	S-SD	D-SD	MS
Thickness	25–30 mm	40 mm	30 mm	15–20 mm	20 mm	10 mm
Traffic limitations	None	None	None	a	a	a
Structural equivalency[b]	0.8	0.5	1.0	0.0	0.0	0.0
Skid resistance:						
– Initial (after 3 months)	Excellent	Excellent	Very good	Excellent	Excellent	Excellent
– After 5 years	Good	Good	Good	Good	Good	Average
Sensitivity to weather conditions during laying (wind temperature)	Yes	Yes	Yes	No	No	No
Water permeability (proofing ability)	Very good	Excellent	Poor	NA	NA	NA
Noise reduction	Very good	Excellent	None	Negative	Negative	Negative
Spray reduction	Very good	Excellent	None	Negative	Negative	Negative
Ability to improve evenness	Minor	Yes	Minor/Yes	No	No	No/Minor[d]
Required quantity of aggregates (approximate index related to dense AC)	0.65	0.9	0.75	0.15	0.2	0.12
Cost index[c]	0.65 to 0.70	1.1 to 1.2	1.2 to 1.3	0.35 to 0.4	0.5 to 0.6	0.5 to 0.55
Typical life expectancy (years)	7 to 9	5 to 6	7 to 12	4 to 5	4 to 6	4 to 6

Note: NA = not applicable.

[a] Some countries impose a restriction to be used only in medium- to low-traffic roads.
[b] Structural equivalency with respect to dense asphalt concrete (AC) 40 mm thick.
[c] Range based on typical European unit price per unit area (m²) related to dense AC.
[d] If multiple layers are laid.

10.7 POISSON'S RATIO OF PAVING AND SUBGRADE MATERIALS

Poisson's ratio (v or μ) is the ratio of transverse strain to longitudinal strain in an axial loaded specimen.

Poisson's ratio is a basic parameter used in analytical pavement design methodologies (based on stress/strain calculations).

On paving and subgrade (soil) materials, Poisson's ratio is affected, among other parameters, by the elastic modulus of the material, the temperature, the cohesion, coarseness and roughness of the particles and the degree of saturation.

Table 10.18 gives the typical range values of Poisson's ratio for most materials used in pavements.

The typical values usually assumed for calculating stresses/strains in the pavement's layers are as follows: 0.35 or 0.4 for asphalts, 0.30 or 0.35 for the unbound base/sub-base material and 0.30 or 0.35 for the subgrade.

Table 10.18 Typical ranges of Poisson's ratio for paving and soil materials

Material	Poisson's ratio (v or μ)
Asphalts	0.35 to 0.40
Concrete	0.10 to 0.20
Unbound base/sub-base material	0.30 to 0.35
Lean concrete and HBM (CSM)	0.10 to 0.20[a]
Soil cement	0.15 to 0.35[a]
Lime–fly ash material	0.1 to 0.15[a]
Lime-stabilised soil	0.15 to 0.2[a]
Soil material	0.30 to 0.50

[a]   AASHTO, *Mechanistic–Empirical Pavement Design Manual (PEPGD-1)*, Washington, DC: American Association of State Highway and Transportation Officials, 2008.

REFERENCES

AASHTO. 1993. *Guide for design of pavement structures.* Washington, DC: American Association of State Highway and Transportation Officials.

AASHTO. 2008. *Mechanistic–empirical pavement design manual (PEPGD-1).* Washington, DC: American Association of State Highway and Transportation Officials.

AASHTO M 85. 2012. *Portland cement.* Washington, DC: American Association of State Highway and Transportation Officials.

AASHTO M 145. 2012. *Classification of soils and soil-aggregate mixtures for highway construction purposes.* Washington, DC: American Association of State Highway and Transportation Officials.

AASHTO M 147. 2012. *Materials for aggregate and soil-aggregate sub-base, base and surface courses.* Washington, DC: American Association of State Highway and Transportation Officials.

AASHTO M 216. 2013. *Lime for soil stabilization.* Washington, DC: American Association of State Highway and Transportation Officials.

AASHTO M 240M/M 240. 2013. *Blended hydraulic cement.* Washington, DC: American Association of State Highway and Transportation Officials.

AASHTO M 295. 2011. *Coal fly ash and raw or calcined natural pozzolan for use in concrete.* Washington, DC: American Association of State Highway and Transportation Officials.

AASHTO M 302. 2013. *Slag cement for use in concrete and mortars.* Washington, DC: American Association of State Highway and Transportation Officials.

AASHTO M 307. 2013. *Silica fume used in cementitious mixtures.* Washington, DC: American Association of State Highway and Transportation Officials.

AASHTO T 22. 2011. *Compressive strength of cylindrical concrete specimens.* Washington, DC: American Association of State Highway and Transportation Officials.

AASHTO T 97. 2010. *Flexural strength of concrete (using simple beam with third-point loading).* Washington, DC: American Association of State Highway and Transportation Officials.

AASHTO T 180. 2010. *Moisture-density relations of soils using a 4.54-kg (10-lb) rammer and a 457-mm (18-in.) drop.* Washington, DC: American Association of State Highway and Transportation Officials.

AASHTO T 190. 2013. *Resistance R-value and expansion pressure of compacted soils.* Washington, DC: American Association of State Highway and Transportation Officials.

AASHTO T 193. 2013. *The California bearing ratio.* Washington, DC: American Association of State Highway and Transportation Officials.

AASHTO T 307. 2007. *Determining the resilient modulus of soils and aggregate materials.* Washington, DC: American Association of State Highway and Transportation Officials.

Asphalt Institute MS-1. 1999. *The Thickness Design, Asphalt Pavements for Highways & Streets.* Manual Series No. 1. Lexington, KY: Asphalt Institute.

ASTM C 33/C 33M. 2013. *Standard specification for concrete aggregates.* West Conshohocken, PA: ASTM International.

ASTM C 39/C 39M-12a. 2012. *Standard test method for compressive strength of cylindrical concrete specimens.* West Conshohocken, PA: ASTM International.

ASTM C 88. 2013. *Standard test method for soundness of aggregates by use of sodium sulfate or magnesium sulfate.* West Conshohocken, PA: ASTM International.

ASTM C 131. 2006. *Standard test method for resistance to degradation of small-size coarse aggregate by abrasion and impact in the Los Angeles machine.* West Conshohocken, PA: ASTM International.

ASTM C 150/C 150M. 2012. *Standard specification for Portland cement.* West Conshohocken, PA: ASTM International.

ASTM C 469/C 469M. 2010. *Standard test method for static modulus of elasticity and Poisson's ratio of concrete in compression.* West Conshohocken, PA: ASTM International.

ASTM C 593. 2011. *Specification for fly ash and other pozzolans for use with lime for soil stabilization.* West Conshohocken, PA: ASTM International.

ASTM C 595/C 595M. 2013. *Standard specification for blended hydraulic cements.* West Conshohocken, PA: ASTM International.

ASTM C 618-12a. 2012. *Standard specification for coal fly ash and raw or calcined natural pozzolan for use in concrete.* West Conshohocken, PA: ASTM International.

ASTM C 977. 2010. *Standard specification for quicklime and hydrated lime for soil stabilization.* West Conshohocken, PA: ASTM International.

ASTM C 989/C 989M. 2013. *Standard specification for slag cement for use in concrete and mortars.* West Conshohocken, PA: ASTM International.

ASTM D 559. 2003. *Standard test methods for wetting and drying compacted soil-cement mixtures (Withdrawn 2012).* West Conshohocken, PA: ASTM International.

ASTM D 560. 2003. *Standard test methods for freezing and thawing compacted soil-cement mixtures (Withdrawn 2012).* West Conshohocken, PA: ASTM International.

ASTM D 806. 2011. *Standard test method for cement content of hardened soil-cement mixtures.* West Conshohocken, PA: ASTM International.

ASTM D 1557. 2012. *Standard test methods for laboratory compaction characteristics of soil using modified effort.* West Conshohocken, PA: ASTM International.

ASTM D 1633. 2007. *Standard test methods for compressive strength of molded soil-cement cylinders.* West Conshohocken, PA: ASTM International.

ASTM D 1635/1635M. 2012. *Standard test method for flexural strength of soil-cement using simple beam with third-point loading.* West Conshohocken, PA: ASTM International.

ASTM D 1883-7e2. 2007. *Standard test method for CBR (California bearing ratio) of laboratory-compacted soils.* West Conshohocken, PA: ASTM International.

ASTM D 2487. 2011. *Classification of soils for engineering purposes (Unified Soil Classification System).* West Conshohocken, PA: ASTM International.

ASTM D 2844/D 2844M. 2013. *Standard test method for resistance R-value and expansion pressure of compacted soils.* West Conshohocken, PA: ASTM International.

ASTM D 2940/D 2940M. 2009. *Standard specification for graded aggregate material for bases or sub-bases for highways or airports.* West Conshohocken, PA: ASTM International.

ASTM D 4318-10e1. 2010. *Standard test methods for liquid limit, plastic limit, and plasticity index of soils.* West Conshohocken, PA: ASTM International.

ASTM D 4759. 2011. *Standard practice for determining the specification conformance of geosynthetics.* West Conshohocken, PA: ASTM International.

ASTM D 4792/4792M. 2013. *Standard test method for potential expansion of aggregates from hydration reactions.* West Conshohocken, PA: ASTM International.

ASTM D 5102. 2009. *Standard test method for unconfined compressive strength of compacted soil lime mixtures.* West Conshohocken, PA: ASTM International.

ASTM D 5821. 2013. *Standard test method for determining the percentage of fractured particles in coarse aggregate.* West Conshohocken, PA: ASTM International.

Black W.P.M. and N.W. Lister. 1979. *The Strength of Clay Fill Sub-Grades, Its Prediction in Relation to Road Performance.* TRL Laboratory Report LR 889. Crowthorne, UK: Transport Research Laboratory.

Brown S.F. and A.R. Dawson. 1992. Two-stage mechanistic approach to asphalt pavement design. *Proceedings of 7th International Conference on Asphalt Pavements*, Vol. 2, p. 16. Nottingham.

Brown S.F. and E.T. Selig. 1991. Cyclic loading of soils, from theory to design, Chap. 6. In *The Design of Pavement and Rail Track Foundations.* (M.P. O'Reilly and S.F. Brown eds.) p. 249. Glasgow, Scotland: Blackie and Son Ltd.

BS 1377-4. 1990. *Methods of tests for soils for civil engineering purposes – Part 4: Compaction related tests.* British Standards Institution.

Caltrans. 2010. *Standard Specifications 2010, Division IV.* Sacramento, CA: Department of Transportation.

CEN EN 197-1. 2011. *Cement – Part 1: Composition, specifications and conformity criteria for common cements.* Brussels: CEN.

CEN EN 933-3. 2012. *Tests for geometrical properties of aggregates – Part 3: Determination of particle shape-Flakiness index.* Brussels: CEN.

CEN EN 933-5:1998/A1. 2004. *Tests for geometrical properties of aggregates – Part 5: Determination of percentage of crushed and broken surfaces in coarse aggregate particles.* Brussels: CEN.

CEN EN 933-8. 2012. *Tests for geometrical properties of aggregates – Part 8: Assessment of fines – Sand equivalent test.* Brussels: CEN.

CEN EN 933-9:2009+A1. 2013. *Tests for geometrical properties of aggregates – Part 9: Assessment of fines – Methylene blue test.* Brussels: CEN.

CEN EN 933-11:2009/AC. 2009. *Tests for geometrical properties of aggregates. Classification test for the constituents of coarse recycled aggregate.* Brussels: CEN.

CEN EN 1097-2. 2010. *Tests for mechanical and physical properties of aggregates – Part 2: Methods for the determination of the resistance to fragmentation.* Brussels: CEN.

CEN EN 1367-2. 2009. *Tests for thermal and weathering properties of aggregates – Part 2: Magnesium sulfate test.* Brussels: CEN.

CEN EN 1744-1:2009+A1. 2012. *Tests for chemical properties of aggregates. Chemical analysis.* Brussels: CEN.

CEN EN 13242: 2002+A1. 2007. *Aggregates for unbound and hydraulically bound materials for use in civil engineering work and road construction.* Brussels: CEN.

CEN EN 13249. 2014. *Geotextiles and geotextile-related products – Characteristics required for use in the construction of roads and other trafficked areas.* Brussels: CEN.

CEN EN 13251. 2014. *Geotextiles and geotextile-related products – Characteristics required for use in earthworks, foundations and retaining structures.* Brussels: CEN.

CEN EN 13285. 2010. *Unbound mixtures – Specification.* Brussels: CEN.

CEN EN 13286-2:2010/AC. 2012. *Unbound and hydraulically bound mixtures – Part 2: Test methods for the determination of the laboratory reference density and water content – Proctor compaction.* Brussels: CEN.

CEN EN 13286-3. 2003. *Unbound and hydraulically bound mixtures – Part 3: Test methods for laboratory reference density and water content – Vibrocompression with controlled parameters.* Brussels: CEN.

CEN EN 13286-4. 2003. *Unbound and hydraulically bound mixtures – Part 4: Test methods for laboratory reference density and water content – Vibrating hammer.* Brussels: CEN.

CEN EN 13286-5. 2003. *Unbound and hydraulically bound mixtures – Part 5: Test methods for laboratory reference density and water content – Vibrating table.* Brussels: CEN.

CEN EN 13286-40. 2003. *Unbound and hydraulically bound mixtures – Part 40: Test method for the determination of the direct tensile strength of hydraulically bound mixtures.* Brussels: CEN.

CEN EN 13286-41. 2003. *Unbound and hydraulically bound mixtures – Part 41: Test method for the determination of the compressive strength of hydraulically bound mixtures.* Brussels: CEN.

CEN EN 13286-42. 2003. *Unbound and hydraulically bound mixtures – Part 42: Test method for the determination of the indirect tensile strength of hydraulically bound mixtures.* Brussels: CEN.

CEN EN 13286-43. 2003. *Unbound and hydraulically bound mixtures – Part 43: Test method for the determination of the modulus of elasticity of hydraulically bound mixtures.* Brussels: CEN.

CEN EN 13286-45. 2003. *Unbound and hydraulically bound mixtures – Part 45: Test method for the determination of the workability period of hydraulically bound mixtures.* Brussels: CEN.

CEN EN 13286-46. 2003. *Unbound and hydraulically bound mixtures – Part 46: Test method for the determination of the moisture condition value.* Brussels: CEN.

CEN EN 13286-47. 2012. *Unbound and hydraulically bound mixtures – Part 47: Test method for the determination of California bearing ratio, immediate bearing index and linear swelling.* Brussels: CEN.

CEN EN 13286-48. 2005. *Unbound and hydraulically bound mixtures – Part 48: Test method for the determination of degree of pulverisation.* Brussels: CEN.

CEN EN 14227-1. 2013. *Hydraulically bound mixtures – Specifications – Part 1: Cement bound granular mixtures.* Brussels: CEN.

CEN EN 14227-2. 2013. *Hydraulically bound mixtures – Specifications – Part 2: Slag bound granular mixtures.* Brussels: CEN.

CEN EN 14227-3. 2013. *Hydraulically bound mixtures – Specifications – Part 3: Fly ash bound granular mixtures.* Brussels: CEN.

CEN EN 14227-4. 2013. *Hydraulically bound mixtures – Specifications – Part 4: Fly ash for hydraulically bound mixtures.* Brussels: CEN.

CEN EN 14227-5. 2013. *Hydraulically bound mixtures – Specification – Part 5: Hydraulic road binder bound mixtures.* Brussels: CEN.

CEN EN 14227-10. 2006. *Hydraulically bound mixtures – Specifications – Part 10: Soil treated by cement.* Brussels: CEN.

CEN EN 14227-11. 2006. *Hydraulically bound mixtures – Specifications – Part 11: Soil treated by lime.* Brussels: CEN.

CEN EN 14227-12. 2006. *Hydraulically bound mixtures – Specifications – Part 12: Soil treated by slag.* Brussels: CEN.

CEN EN 14227-13. 2006. *Hydraulically bound mixtures – Specifications – Part 13: Soil treated by hydraulic road binder.* Brussels: CEN.

CEN EN 14227-14. 2006. *Hydraulically bound mixtures – Specifications – Part 14: Soil treated by fly ash.* Brussels: CEN.

FAA. 2011. *Advisory Circular, AC 150/5370-10F.* Washington, DC: U.S. Department of Transportation, Federal Aviation Administration.

Highways Agency. 2006. *Design Manual for Roads and Bridges (DMRB), Volume 7: Pavement Design and Maintenance.* Section 2, Part 3, HD 26/06, Pavement Design. London: Department for Transport, Highways Agency.

Highways Agency. 2009a. *The Manual of Contract Documents for Highway Works (MCDHW), Volume 1: Specification for Highway Works, Series 600: Earthworks.* London: Department for Transport, Highways Agency.

Highways Agency. 2009b. *The Manual of Contract Documents for Highway Works (MCDHW), Volume 1: Specification for Highway Works, Series 800: Road pavements – Unbound, cement and other hydraulically bound mixtures.* London: Department for Transport, Highways Agency.

Highways Agency. 2009c. *Design Manual for Roads and Bridges (DMRB), Volume 7: Pavement Design and Maintenance.* Section 2, Part 2, IAN 73/06 Rev. 1 (2009), Design Guidance for Road Pavement Foundations. London: Department for Transport, Highways Agency.

Nicholls J.C., I. Carswell, and D.J. James. 2011. Durability of thin surfacing systems after nine years monitoring. *Proceedings of 5th International Conference 'Bituminous Mixtures and Pavements'.* Thessaloniki, Greece.

Nikolaides A.F. 2009. *Detailed Determination of Maintenance/Rehabilitation Works for the Year 2009 as Part of the Revised Pavement Management Program of Maliakos-Kleidi Motorway – Sections GU 6, 7 & 8.* Technical Report. Athens, Greece: MKC JV.

Potter J.F. and E.W.H. Currer. 1981. *The Effect of a Fabric Membrane on the Structural Behaviour of a Granular Road Pavement.* TRRL Report LR 996. Crowthorne, UK: Transport Research Laboratory.

Powell W.D., J.F. Potter, H.C. Mayhew, and M.E. Nunn. 1984. *The Structural Design of Bituminous Roads*. TRRL Laboratory Report LR1132. Crowthorne, UK: Transport Research Laboratory.

Robnett Q.L. and J.S. Lai. 1982. *Fabric-Reinforced Aggregate Roads – Overview*. Transportation Research Record 875, Transport Research Board, p. 42. Washington, DC: TRB.

Shell International. 1985. *Shell Pavement Design Manual*. London: Shell International Petroleum Company Ltd.

Chapter 11

Methods determining stresses and deflections

11.1 GENERAL

The determination of stresses developed in a pavement constitutes a fundamental prerequisite and is achieved by implementation of various methods depending on the number of distinct pavement layers.

By definition, the subgrade is considered as one layer. Thus, in the typical flexible pavement that consists of unbound layer (base/sub-base) and bitumen-bound layers all consisting of asphalts with the same mechanical properties, the number of distinct layers is three.

Generally, each of the constructed layers consisting of materials having different mechanical properties is considered as a distinct structural layer.

This chapter presents methods for determination of stresses, strains and deflections developed in a one-layer, two-layer, three-layer and multi-layer system.

11.2 ONE-LAYER SYSTEM – BOUSSINESQ THEORY

Boussinesq's theory (Boussinesq 1885) was developed at the end of the 19th century. Even though Boussinesq's theory is not currently used in multi-layer pavement engineering, it was the basis for the development of all subsequent stress/strain distribution theories. Boussinesq's theory refers to only one layer of uniform and homogeneous material such as, by assumption, soil material-subgrade.

According to Boussinesq, the stresses applied to an elastic, homogeneous and isotropic material extended to infinity at both directions (horizontal and vertical), and the stresses developed at any depth, z, below the surface under the influence of a *point load* (Figure 11.1a) can be calculated by the following fundamental equations:

$$\sigma_z = -\frac{3p}{2\pi}\left[\frac{z^3}{(r^2+z^2)^{5/2}}\right]$$

$$\sigma_r = -\frac{p}{2\pi}\left[\frac{1}{(r^2+z^2)^{3/2}} - (1-2\mu)\frac{(r^2+z^2)^{1/2}}{z+(r^2+z^2)^{1/2}}\right]$$

$$\sigma_\theta = \frac{p}{2\pi}(1-2\mu)\frac{1}{(r^2+z^2)}\left[\frac{z}{(r^2+z^2)^{1/2}} - \frac{(r^2+z^2)^{1/2}}{z+(r^2+z^2)^{1/2}}\right]$$

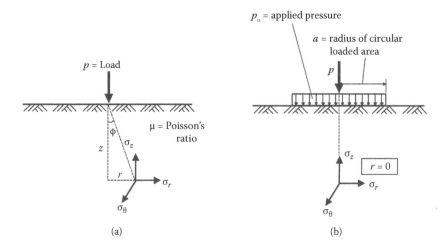

Figure 11.1 Developing stresses in accordance with Boussinesq's theory. (a) Point load. (b) Uniformly distributed load.

$$\tau_{r\theta} = \frac{3p}{2\pi}\left[\frac{rz^2}{(r^2+z^2)^{5/2}}\right],$$

where σ_z is the vertical compressive stress; σ_r, σ_θ are the horizontal tensile stresses (radial and tangential); $\tau_{r\theta}$ is the shear stress; and μ is Poisson's ratio.

If a *uniformly distributed load* is applied, the respective equations for stresses are the integral of the above equations. When the loaded area of the uniform distributed load is circular (the most usual consideration taken in pavement loading/calculations), the compressive stress, σ_z, and the horizontal tensile stresses, σ_r and σ_θ, at any depth, z, on the axis of symmetry of the loading surface (Figure 11.1b) are determined from the following equations:

$$\sigma_z = p_0\left[1 - \frac{z^3}{(a^2+z^2)^{3/2}}\right]$$

$$\sigma_r = \sigma_\theta = \frac{p_0}{2}\left[1 + 2\mu - \frac{2z(1+\mu)}{(a^2+z^2)^{1/2}} + \frac{z^3}{(a^2+z^2)^{3/2}}\right],$$

where p_0 is the applied pressure, a is the radius of applied circle of loading (or radius of loading), z is the distance from the surface and μ is Poisson's ratio.

It is noted that all the above equations are independent of the modulus of elasticity, since the layer was considered elastic and uniform, and the vertical compression stress is independent of Poisson's ratio.

11.2.1 Foster–Ahlvin diagrams

Foster and Ahlvin (1954), using Boussinesq's theory, developed diagrams from which not only the stresses but also the elastic deflections caused by a circular, uniformly applied load can be calculated much easier, at any depth from the surface and distance from the axis of the loading area.

The vertical compressive stress, σ_z, at any depth, z, and distance, r, from the axis of loading area can be calculated from the diagram shown in Figure 11.2. Similarly, the horizontal radial stress, σ_r, and the horizontal tangential stress, σ_θ, can be calculated from the diagrams shown in Figures 11.3 and 11.4, respectively.

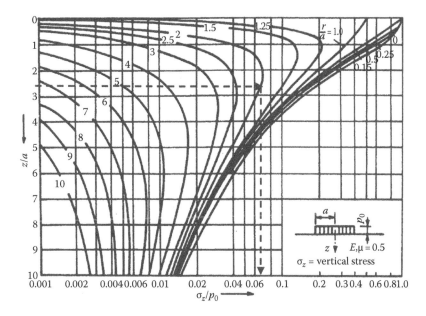

Figure 11.2 Determination of the vertical compressive stress σ. (From Foster C.R. and R.G. Ahlvin, Stresses and deflections induced by a uniform circular load. *Proceedings of the Highway Research Board*, Vol. 33, pp. 467–470, 1954.)

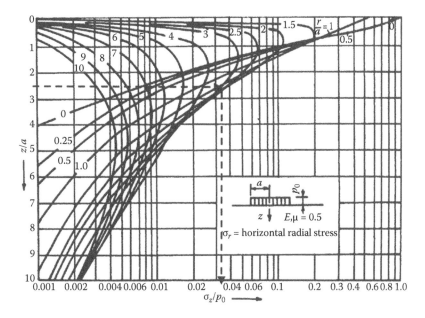

Figure 11.3 Determination of the horizontal radial stress σ_r. (From Foster C.R. and R.G. Ahlvin, Stresses and deflections induced by a uniform circular load. *Proceedings of the Highway Research Board*, Vol. 33, pp. 467–470, 1954.)

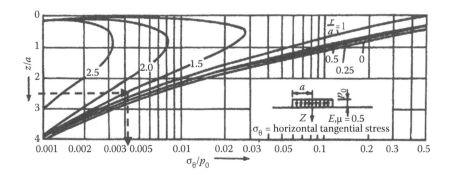

Figure 11.4 Determination of the horizontal radial stress σ_θ. (From Foster C.R. and R.G. Ahlvin, Stresses and deflections induced by a uniform circular load. *Proceedings of the Highway Research Board*, Vol. 33, pp. 467–470, 1954.)

The relevant stress, σ_z, σ_r or σ_u, is calculated from the ratio stress/applied pressure value obtained from the x-axis, since the applied pressure p_0 is known.

11.2.2 Elastic deflection

Boussinesq's theory/analysis can also be used to derive the elastic deflection of a homogeneous material subjected to uniform circular loading. The elastic surface deflection, Δ, at the centre of the loading area under the influence of a uniformly distributed load can be calculated from the following equation:

$$\Delta = \frac{2p_0 \times \alpha}{E}(1 - \mu^2),$$

where p_0 is the applied stress, a is the radius of loading, μ is Poisson's ratio and E is the modulus of elasticity of the subgrade.

For a uniformly distributed load, there are two types of deformation of the subgrade under the loading area: the uniform and non-uniform deformation.

Non-uniform deformation results when the load is applied though a non-rigid medium such as the tyre of a vehicle. In this case, the elastic deflection, w, developed is greater at the centre of the loading area than that developed at the circumference of the loading area (Figure 11.5a).

Uniform deflection results when the load is applied through a rigid medium (metal plate, concrete block, etc.). In this case, the elastic deflection is the same under the whole loading area (Figure 11.5b).

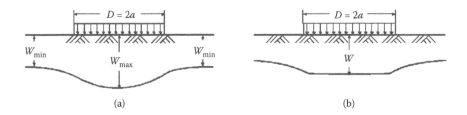

Figure 11.5 Deflection distribution forms. (a) Non-uniform deformation. (b) Uniform deformation.

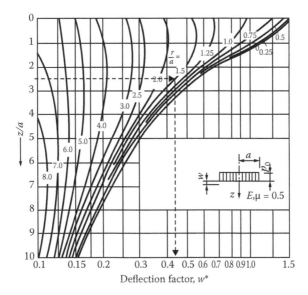

Figure 11.6   Determination of deflection factor, w^*. (From Foster C.R. and R.G. Ahlvin, Stresses and deflections induced by a uniform circular load. *Proceedings of the Highway Research Board*, Vol. 33, pp. 467–470, 1954.)

In case of non-uniform deformation, the maximum elastic deflection (i.e. at the centre of the loading area), w, at any depth, z, for Poisson's ratio $\mu = 0.5$ can be calculated from the following equation:

$$w = \left(\frac{p_0 \times a}{E} \right) \times w^*,$$

where p_0 is the applied pressure, a is the radius of loading, E is the modulus of elasticity of the subgrade and w^* is the deflection factor or constant.

The deflection factor, w^*, is determined from the nomograph of Figure 11.6.

It should be mentioned that values of deflection obtained by the above equation and Foster and Ahlvin's nomograph are valid only when Poisson's ratio μ is equal to 0.5. A supplementary work by Ahlvin and Ulery (1962) allowed stresses and deflections to be calculated for any value of Poisson's ratio. The results of this work were presented in tabular form instead of diagrams. The relevant tables as well as the equations developed can be found in Ahlvin and Ulery (1962) and Yoder and Witczak (1975).

11.3 TWO-LAYER SYSTEM – BURMISTER THEORY

Forty years after the development of Boussinesq's analysis, Westergaard (1926) developed his own analysis for the determination of stresses in a two-layer system, where the first layer consisted of a concrete slab. Even though this theory did not cover the case of flexible pavements, it triggered the development of analysis for flexible pavements with two layers.

Burmister (1943, 1945) was amongst the first researchers who developed theories on the determination of stresses and strains in flexible pavements. Other researchers, such as Hank and Scrivner (1948), Huang (1969) and Jelinek and Ranke (1970), also followed.

The theories/analyses for determining the stresses in a two-layer system developed were based on certain assumptions/prerequisites. These were the following: (a) the material of each layer is considered to be elastic, homogeneous and isotropic; (b) the first layer has infinite dimension in the horizontal level and finite dimension in the vertical level; (c) the second layer (subgrade) has infinite dimensions in both horizontal and vertical levels; (d) the upper surface of the first layer is not subject to any other horizontal or vertical loading, except the traffic loading; and (e) the first layer has a modulus of elasticity higher than or equal to that of the second layer.

A diagrammatic representation of a two-layer system is given in Figure 11.7.

It should be mentioned that the above assumptions were also used for theories/analyses for determining the stresses in three or more layer systems.

The numbering of the layers always begins from top to bottom. Thus, all properties bear the number '1', for the first layer and so forth. In order to determine the precise interface level, the letters 'u' (for under) and 'o' (for over) were used. The letter 'u' indicates the bottom surface of the layer, whereas the letter 'o' indicates its upper surface. Thus, for instance, the designation, σ_{r1u}, refers to the horizontal radial tensile stress developed at the bottom surface of the first layer. The same principle is also followed for systems with three or more layers.

According to the Burmister theory, the compressive stress at any depth, σ_z (usually less than $3a$, where a is the radius of loading), along the axis of symmetry of the circular loading area, assuming friction or bonding exists at the interface of the layers, can be determined from the nomograph of Figure 11.8.

The limitation of the depth, $z = 3a$, was set since it was found that developed stresses are too low (less than 10% of the applied stress) at greater depths. The ratio E_1/E_2 is the ratio of moduli of elasticity of the two layers.

The nomographs developed by Hank and Scrivner (1948) were similar to those developed by Burmister but included also the case where no friction exist at the interface of the layers. A sample of nomographs developed for determining the vertical stress at the bottom of the first layer (almost equal to the stress at the top of the subgrade) and the horizontal radial stress at the bottom of the first layer is given in Figures 11.9 and 11.10, respectively.

Other stresses, such as the tangential and the shear stress can be determined using similar nomographs.

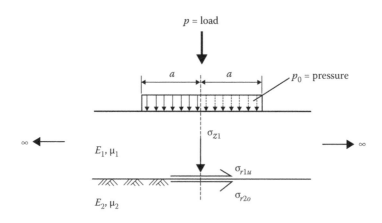

Figure 11.7 Two-layer system.

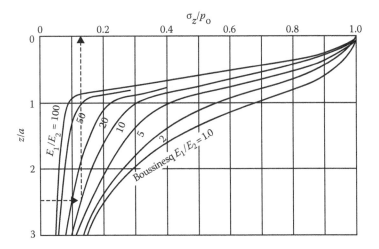

Figure 11.8 Determination of vertical stress according to Burmister's theory. (From Burmister D.M., Theory of stresses and displacements in layered systems and application to the design of airport runways. *Proceedings of the Highway Research Board*, Vol. 23, pp. 126–144, 1943; Burmister D.M., The general theory of stresses and displacements in layered systems. *Journal of Applied Physics*, Vol. 16, pp. 89–96, 1945.)

In general, the addition of one layer with better mechanical properties, higher modulus of elasticity, than the subgrade results in reduction of stresses exerted at the subgrade. This can clearly be seen in Figure 11.11.

Stress distribution curves in a one-layer system (Boussinesq analysis) are given in the left part, while the stress distribution curves in a two-layer system (Burmister analysis) are given in the right part of the Figure 11.11. In both systems, the same pressure is applied and friction was assumed to exist at the interface.

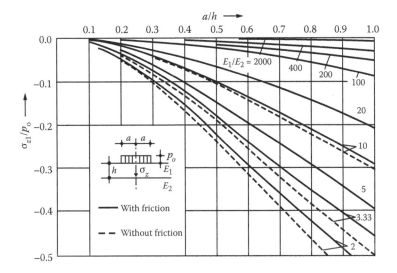

Figure 11.9 Determination of vertical stress by Hank and Scrivner analysis. (From Hank R.J. and F.H. Scrivner, Some numerical solutions of stresses in two- and three-layered systems. *Proceedings of Highway Research Board*, Vol. 28, pp. 457–468, 1948.)

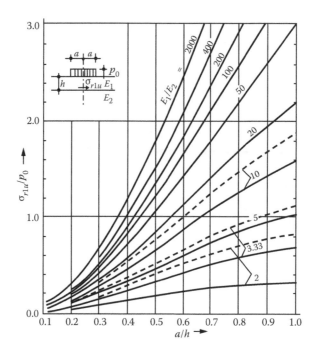

Figure 11.10 Determination of radial horizontal stress by Hank and Scrivner analysis. (From Hank R.J. and F.H. Scrivner, Some numerical solutions of stresses in two- and three-layered systems. *Proceedings of Highway Research Board*, Vol. 28, pp. 457–468, 1948.)

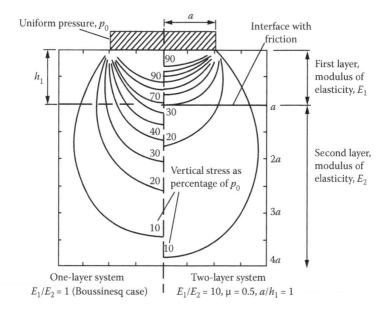

Figure 11.11 Comparison of vertical stress distribution in a one-layer and in a two-layer system. (From Davis E.H., *Pavement Design for Roads and Airfield*. Road Research Technical Paper No. 20. London: HMSO, 1951.)

In the particular example, the elasticity modulus of the first layer, E_1, is 10 times higher than the elasticity modulus of the subgrade, E_2; the layer thickness is equal to the radius of the loading area and Poisson's ratio is the same for both layers, $\mu_1 = \mu_2 = 0.5$.

As shown in Figure 11.11, the vertical stress, as a percentage of applied pressure, p_0, at a depth, h_1, equal to the radius of loading is only 30% of the applied pressure in the case of a two-layer system and approximately 70% in the case of a one-layer system.

11.3.1 Determination of surface deflection

The maximum surface deflection in the case of a two-layer system, Δ_{2-1}, can also be determined using Burmister's analysis. The surface deflection, Δ_{2-1}, is calculated using the following equation:

$$\Delta_{2-1} = \left(\frac{1.5 \times p_0 \times a}{E_2} \right) \times F,$$

where p_0 is the applied pressure, a is the radius of loading, E_2 is the modulus of elasticity of the second layer and F is the deflection factor, from Figure 11.12.

When the loading medium is a rigid plate, a coefficient of 1.18 is used instead of the coefficient 1.5 in the above equation.

The maximum surface deflection determined using the nomograph of Figure 11.12 is for Poisson's ratios $\mu_1 = \mu_2 = 0.5$ or $\mu_1 = 0.2$, $\mu_2 = 0.4$. When other Poisson's ratios are to be used, the deflection factor F should be determined from another appropriate nomograph.

The above inconvenience can be overcome by using Odemark's equivalent thickness theory, which converts the two-layer system into a one-layer system, and then applying Boussinesq's analysis.

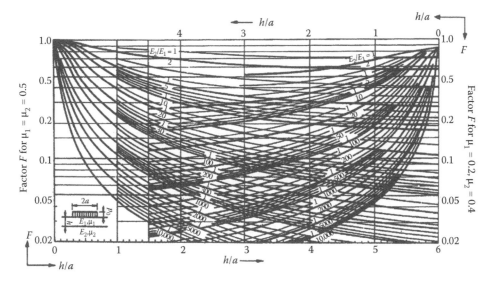

Figure 11.12 Determination of deflection factor, F. (From Burmister D.M., The general theory of stresses and displacements in layered systems. *Journal of Applied Physics*, Vol. 16, pp. 89–96, 1945.)

11.3.2 Odemark's equivalent thickness concept

Odemark's equivalent thickness concept consists of the transformation of a two or more layer system with different characteristic properties, E and μ, into an equivalent one-layer system with equivalent thickness but one elastic modulus, that of the bottom (last) layer. Thus, one elastic, isotropic and homogeneous layer results and calculations of stresses and strains are easier. The transformation a two-layer system into an equivalent one-layer system is explained in Figure 11.13.

Odemark's method is based on the assumption that the stresses and strains below a layer depend only on the stiffness of that layer. If the thickness, modulus and Poisson's ratio of a layer are changed, but the stiffness remains unchanged, the stresses and strains below the layer should also remain (relatively) unchanged.

According to Odemark (1949), the stiffness of the layer is proportional to the term $\dfrac{I \times E}{1 - \mu^2}$; hence, two layers are structurally equivalent when

$$\frac{E_n \times I_n}{1 - \mu_n^2} = \frac{E_{n+1} \times I_{n+1}}{1 - \mu_{n+1}^2},$$

where E_n, E_{n+1} are moduli of elasticity of layer n and $n + 1$; I_n, I_{n+1} are moments of inertia of layer n and $n + 1$; and μ_n, μ_{n+1} are Poisson's ratios of layer n and $n + 1$.

For a unit width of layer, the above equation converts to

$$\frac{E_n \times h_n^3}{1 - \mu_n^2} = \frac{E_{n+1} \times h_{n+1}^3}{1 - \mu_{n+1}^2},$$

where h_n, h_{n+1} are thicknesses of layer n and $n + 1$.

If a two-layer system is to be transformed into an equivalent one-layer system, the above equation becomes

$$\frac{E_1 \times h_1^3}{1 - \mu_1^2} = \frac{E_2 \times h_e^3}{1 - \mu_2^2} \text{ or } h_e = h_1 \sqrt[3]{\frac{E_1 \times \left(1 - \mu_2^2\right)}{E_2 \times \left(1 - \mu_1^2\right)}},$$

where h_e denotes the equivalent thicknesses, h_1 is the thickness of the first layer and μ_1, μ_2 are Poisson's ratios of the first and second layer.

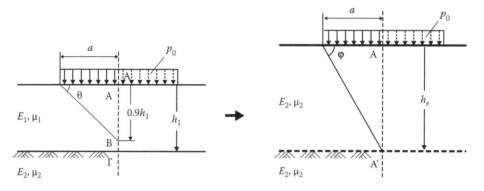

Figure 11.13 Odemark's equivalent thickness concept.

In order to achieve better agreement between the stresses and strains calculated using Odemark's concept and those from the theory of elasticity, a correction factor f is applied to the above equation. Thus,

$$h_e = h_1 \sqrt[3]{\frac{E_1 \times (1 - \mu_2^2)}{E_2 \times (1 - \mu_1^2)}} \times f,$$

where f is the correction factor; for non-rigid loading media, such as a tyre, f is usually taken equal to 0.9.

A recent study has confirmed that there is good agreement between the vertical stresses at the interface between the two layers, in a two-layer system, calculated using the theory of elasticity and Odemark's concept when using a correction factor (f) in the range of 0.8 to 0.9 (El-Badawy and Kamel 2011).

Odemark's transformation method has been widely used for pavement response analyses (Ullidtz 1987). Comparisons of measured and calculated stresses, strains and deflections have shown that the simple combination of Odemark's transformation with Boussinesq's equations (modified for nonlinearity) results in an agreement between measured and calculated values that is as good as that obtained with the finite element method. Linear elastic methods normally result in rather poor agreement (Ullidtz 1999).

Furthermore, Odemark's method has also been used for FWD backcalculation (Ullidtz et al. 2006) and the M-E pavement design guide implemented this method to transform a multi-layer pavement system into an equivalent one-layer system (NCHRP 1-37-A 2004).

Odemark's method of equivalent layer thickness (MET) is also applicable to a multi-layered base or sub-base, as well as to a multi-layered hot mix asphalt system (Zhou et al. 2010).

11.3.2.1 Surface deformation using Odemark's equivalent thickness concept

Odemark also developed an equation for calculating the elastic surface deflection, w, in a two-layer system. The equation developed is as follows (the elastic deflection, w, at the centre of the loading area):

$$w = \left(\frac{p_0 \times a}{E_1}\right) \times \left[\begin{array}{l} 2(1 - \mu_1^2) - (1 + \mu_1) \times \left[\cos\theta + (1 - 2\mu_1) \times \tan\left(45° - \frac{\theta}{2}\right) \right] \\ + \frac{E_1}{E_2} \times (1 - \mu_2) \times \left[\cos\varphi + (1 - 2\mu_2) \times \tan\left(45° - \frac{\varphi}{2}\right) \right] \end{array} \right],$$

where w is the surface deflection at the centre of loading area (max. deflection), p_0 is the applied pressure, a is the radius of loading, E_1 is the modulus of elasticity of layer 1 (top layer), E_2 is the modulus of elasticity of layer 2 (bottom layer, subgrade) and θ, φ are angles, as explained in Figure 11.13.

The surface deflection in a two-layer system can also be calculated by a nomograph given in Odemark (1949).

11.4 THREE OR MORE LAYER SYSTEM

A three or more layer system (multi-layer system) simulates better the real situation of flexible pavements, since there are more than two distinct layers in a typical pavement.

In a three-layer system, all asphalt layers are considered as one layer. The same applies for the unbound layers of base and sub-base. The third layer is the subgrade.

The increase of the number of layers to more than three represents the real situation of the pavement but requires more complicated calculations for the determination of stresses and strains developed at critical interfaces. When also considering variations of the properties of the materials that constitute the layer owing to the duration of loading or temperature, variation of loading area and applied load and non-truly elastic behaviour of the materials, the use of nomographs or tables for the determination of stresses and strains is almost impossible. Today, such calculations are carried out by software programs.

11.4.1 Determination of stresses/strains from tables or nomographs

Despite the complexity of three or more layer systems, several researchers tried to resolve the problem without the aid of computers. Burmister (1943) was the first to extend his theory to a three-layer system. A few years later, Acum and Fox (1951) presented the first tables for determining the vertical and horizontal stresses developed on the axis of symmetry of the circular loading area and at the interface of the layers.

Almost 15 years later, other researchers (Jones 1962, 1964; Peattie 1962) developed extensive tables that included more parametric values. These tables were used by many pavement engineers until the advent of computer software programs.

In 1973, another researcher (Iwanow 1973) suggested the use of nomographs for the determination of stresses in a three-layer system. Two of these nomographs for the calculation of vertical and radial stress are given in Figures 11.14 and 11.15 (Section 11.4.1.1).

At the beginning of the 1960s, Schiffman (1962) introduced a general solution to analyse stresses/deformations in a multi-layer system. The suggested equations required extensive calculations and effectively were never used extensively. However, the development of computing Schiffman's analysis for an n-layer system constituted the base of many software programs for the determination of pavement stresses and strains.

11.4.1.1 Determination of stresses and deflection using Iwanow's nomographs

The determination of vertical and horizontal stresses at the interfaces of the layers in a three-layer system may be carried out by the use of Iwanow's (1973) nomographs. Figures 11.14 through 11.16 show samples of the nomographs developed.

With the use of the nomograph in Figure 11.14, the vertical compressive stress at the interface of base/sub-base and subgrade along the axis of symmetry of the loaded area can be determined. Similarly, the radial horizontal stress in the bottom of the asphalt layers (asphalt base) may be determined from the nomograph in Figure 11.15. As for the determination of the deflection at the surface of the subgrade, the nomograph in Figure 11.16 is used.

It must be stated that Iwanow's nomographs are valid for the Poisson ratio of the first and second layer equal to 0.25 ($\mu_1 = \mu_2 = 0.25$) and for the Poisson ratio of the third layer equal to 0.35 ($\mu_3 = 0.35$).

The *vertical stress* developed at the bottom surface of the second layer (almost equal to the stress developed at the top of the subgrade), σ_{z2}, is calculated using the following equation:

$$\sigma_{z2} = \sigma_z^* \times k \times p_0,$$

where σ_z^* is the vertical stress factor from Figure 11.14, k is a constant value that ranges from 0.9 to 1.0 and p_0 is the applied pressure.

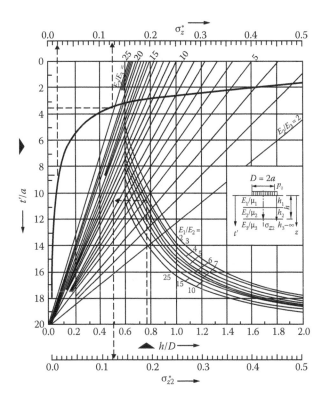

Figure 11.14 Iwanow's nomograph for the determination of vertical stress σ_{z2}. (From Iwanow N.N., *Konstruirowanije i restschot neschostkich doroschnich odeschd*. Moskau: Verlag Transport, 1973.)

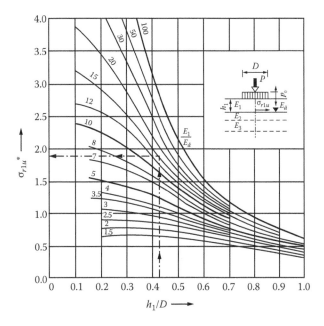

Figure 11.15 Iwanow's nomograph for the determination of the radial horizontal stress σ_{r1u}. (From Iwanow N.N., *Konstruirowanije i restschot neschostkich doroschnich odeschd*. Moskau: Verlag Transport, 1973.)

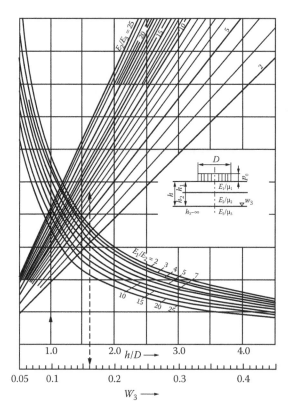

Figure 11.16 Iwanow's nomograph for the determination of deflection at the surface of the subgrade. (From Iwanow N.N., *Konstruirowanije i restschot neschostkich doroschnich odeschd.* Moskau: Verlag Transport, 1973.)

From the nomograph in Figure 11.14, the vertical stress, σ_z, at any depth, t', can also be calculated provided the three-layer system is transformed into a one-layer system using Odemark's equivalent thickness concept. In this case, the left y-axis and the top x-axis in Figure 11.14 are used.

The *radial horizontal stress* at the bottom of the first layer (asphalt layers), σ_{r1u}, is calculated using the following equation:

$$\sigma_{r1u} = p_0 \times \sigma^*_{r1u},$$

where p_0 is the applied pressure and σ^*_{r1u} is the horizontal stress factor from Figure 11.15.

In Figure 11.15, E_α is the modulus at the surface of the unbound layers.

The *deflection* at the surface of the subgrade, w_3, and at the centre of the loading surface can be calculated by the following equation:

$$w_3 = \frac{p_0 \times D}{E_3} \times F \times w'_3,$$

where E_3 is the modulus of elasticity of the third layer (subgrade), F is the correction coefficient from Table 11.1 and w'_3 is the deflection factor from Figure 11.16.

Table 11.1 Correction coefficient F

E_1/E_2	$h_1/h = 0.3$	$h_1/h = 0.5$	$h_1/h = 0.7$
	F values		
2	1.00	1.00	1.00
5	1.07	1.00	0.93
10	1.15	1.00	0.85
15	1.15	1.00	0.85
25	1.25	1.00	0.78

11.5 DETERMINATION OF STRESSES, STRAINS AND DISPLACEMENTS IN A MULTI-LAYER SYSTEM BY COMPUTER PROGRAMS

The determination of stresses, strains and displacements in a multi-layer system is conducted by computer programs using, typically, three methods: (a) the multi-layered elastic analysis, (b) two-dimensional (2D) finite element analysis and (c) three-dimensional (3D) finite element analysis.

The most popular method is the first one. In the multi-layered elastic analysis method, the system (pavement) is divided into a number of distinct layers where each layer has its own thickness and mechanical properties. Each of these distinct layers is considered as homogeneous with linear-elastic behaviour.

Some of the most widely used computer software developed to calculate stresses, strains and displacements in flexible pavements with the use of the multi-layered elastic theory are CHEVRON, BISAR, ELSYM5, KENLAYER and WESLEA.

The 2D analysis simulates better the real conditions since it takes into consideration material anisotropy, material non-linearity and a variety of additional restrictions. Unfortunately, the 2D analysis cannot include the non-uniform wheel loading and the multiplicity of the axial loads.

The 3D analysis overcomes all weaknesses of the 2D analysis and is becoming more popular. Of course, it requires more computational time for obtaining the results.

11.5.1 CHEVRON software

Chevron software was developed at the beginning of the 1960s by the Chevron company (Michelow 1963). The initial software (Chev5L) provided the ability to calculate stresses up to a five-layer system with one circular load area. The revised software edition accepts 10 layers with 10 different loading patterns (NHI 2002; Tu 2007).

11.5.2 BISAR software

The original BISAR mainframe computer program was developed by Shell (De Jong et al. 1973) in the early 1970s, which was used in developing the design charts of the Shell Pavement Design Manual. An abbreviated version for use on a personal computer was issued in 1987 (Koole et al. 1989) and the final release of BISAR 3.0 for use in the Windows environment replaced the DOS version BISAR-PC 2.0 (Shell International 1998).

BISAR 3.0 is a computer program for the calculation of stresses, strains and deflections for a variety of loading patterns. It is also able to deal with horizontal forces and slip between the pavement layers.

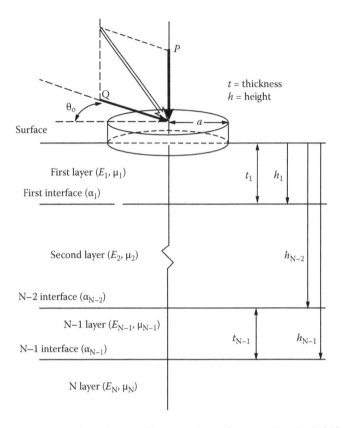

Figure 11.17   Typical notations and numbering of layers and interfaces used by the BISAR software.

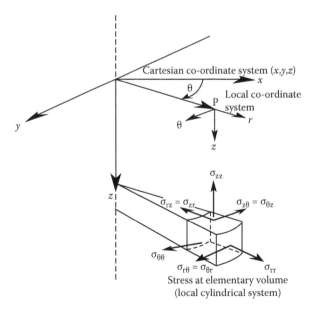

Figure 11.18   Principles of cylindrical coordinate system.

The typical notations and numbering of layers and interfaces used are as shown in Figure 11.17 (De Jong et al. 1973).

Figure 11.18 gives the principles of cylindrical coordinates and the assumption used for the positive or negative sign. The arrows in x, y, z direction, or r, θ, z, signify the positive sign.

11.5.3 ELSYM5, KENLAYER and WESLEA software

The ELSYM5 computer program was developed by the Federal Highway Washington Administration, and it has the ability to analyse up to five different layers under 20 different wheel loading configurations (Kopperman et al. 1986).

The KENLAYER computer program can analyse a multi-layer system being linear elastic, non-linear elastic or viscoelastic. It is capable of analysing up to 19 layers but with only one load pattern, circular loaded area (Huang 1993). An improved version of the KENLAYER software can be found in Huang (2004).

WESLEA is a multi-layer linear elastic computer program developed by the US Army Corps of Engineers Waterways Experiment Station. The original WESLEA version could handle up to five layers with varying interface conditions and a maximum of 20 loads (Van Cauwelaert et al. 1989).

A modified version of WESLEA, referred as JULEA, has been integrated into the Mechanistic Empirical Pavement Design Guide – MEPDG (NCHRP 2004).

A new layered elastic analysis program, called MNLAYER, is claimed to be more accurate than JULEA, especially at evaluation points located close to the surface or at large depths. Since, for the same structural system, it requires less computational time than JULEA and BISAR, MNLAYER is an attractive candidate for incorporation into the MEPDG software (Wang 2008).

11.5.4 DAMA software

The DAMA software was developed by Hwang and Witczak (1979) for the Asphalt Institute and has integrated into the pavement design program of the Asphalt Institute (Asphalt Institute SW-1 2005).

DAMA software not only calculates stresses, strains and deflections in a multi-layer system with elastic behaviour but also allows the user to analyse the influence of monthly variations in the temperature of the pavement. In the calculation procedures over a period, the mean monthly air temperatures are required as an input.

Stresses, strains and deflections are calculated at each layer interface and at the pavement surface in three distinct points: the centre of one tyre (point 1), the edge of one tyre (point 2) and at the midpoint of the dual wheels (point 3). All of these points lie on the centre-to-centre bisector of the dual wheels. The total number of layers is limited to five.

DAMA also computes the cumulative pavement damage due to excessive subgrade deformation leading to surface rutting and fatigue cracking of the asphalt-stabilised layer. More details can be found in the SW-1 user's guide (Asphalt Institute SW-1 2005).

11.5.5 Other software

During the 1980s and later on, various other computer programs of multi-layer system for flexible pavements were developed using 2D or 3D analysis of finite elements.

Two of these programs are ILLIPAVE (Raad and Figueroa 1980) and 3DMOVE (Siddharthan et al. 2000).

REFERENCES

Acum W.E. and L. Fox. 1951. Computation of load stresses in a three-layer elastic system. *Geotechnique*, Vol. 2, No. 4, pp. 293–300.

Ahlvin R.G. and H.H. Ulery. 1962. Tabulated values for determining the complete pattern of stresses, strains and deflections beneath a uniform circular load on a homogeneous half space. *Highway Research Bulletin*, Vol. 342, pp. 1–13.

Asphalt Institute SW-1. 2005. *Asphalt Pavement Thickness Design Software for Highways, Airports, Heavy Wheel Loads and Other Applications, User's Guide*. Lexington, KY: Asphalt Institute.

Boussinesq J.V. 1885. *Application des potentials a l'edude de l'equilibre et du mouvement des solides elastiques*. Paris: Gauthier-Villars.

Burmister D.M. 1943. Theory of stresses and displacements in layered systems and application to the design of airport runways. *Proceedings of the Highway Research Board*, Vol. 23, pp. 126–144.

Burmister D.M. 1945. The general theory of stresses and displacements in layered systems. *Journal of Applied Physics*, Vol. 16, pp. 89–96.

Davis E.H. 1951. *Pavement Design for Roads and Airfield*. Road Research Technical Paper No. 20. London: HMSO.

De Jong D.L., M.G.F. Peatz, and A.R. Korswagen. 1973. *Computer Program BISAR Layered Systems under Normal and Tangential Loads*. External Report AMSR.0006.73. Amsterdam: Konin Klijke Shell-Laboratorium.

El-Badawy S.M. and M.A. Kamel. 2011. Assessment and improvement of the accuracy of the Odemark transformation method. *International Journal of Advanced Engineering Sciences and Technologies*, Vol. 5, No. 2, pp. 105–110. Available at http://www.ijaest.iserp.org.

Foster C.R. and R.G. Ahlvin. 1954. Stresses and deflections induced by a uniform circular load. *Proceedings of the Highway Research Board*, Vol. 33, pp. 467–470.

Hank R.J. and F.H. Scrivner. 1948. Some numerical solutions of stresses in two- and three-layered systems. *Proceedings of Highway Research Board*, Vol. 28, pp. 457–468.

Huang Y.H. 1969. Computation of equivalent single wheel loads using layered theory. *Highway Research Record*, Vol. 291, pp. 144–155. Washington, DC: HRB, National Research Council.

Huang Y.H. 1993. *Pavement Analysis and Design*, 1st Edition. Englewood Cliffs, NJ: Prentice Hall.

Huang Y.H. 2004. *Pavement Analysis and Design*, 2nd Edition. Upper Saddle River, NJ: Pearson Prentice Hall.

Hwang D. and M.W. Witczak. 1979. *Program DAMA (Chevron), User's Manual*. Maryland, MD: Department of Civil Engineering, University of Maryland.

Iwanow N.N. 1973. *Konstruirowanije i restschot neschostkich doroschnich odeschd*. Moskau: Verlag Transport.

Jelinek R. and A. Ranke. 1970. Berechnung der Spannungsverteilung in einem Zweischichtsystem. *Bautechnik*, Vol. 47.

Jones A. 1962. Tables of stresses in three-layer elastic system. *Highway Research Board, Bulletin 342: Stress Distribution in Earth Masses*. Washington, DC: National Academy of Sciences.

Jones A. 1964. *The Calculation of Surface Deflection for Three-Layer Elastic Systems*. Chester, England: Shell Research Ltd, Thornton Research Centre.

Koole R.C., C.P. Valkering, and F.D.R. Stapel. 1989. Development of pavement design program for use on personal computer. *5th Conference of Asphalt Pavements for Southern Africa*, Swaziland.

Kopperman S., G. Tiller, and M. Tseng. 1986. *ELSYM5, Interactive Microcomputer Version User's Manual*. Report No. FHWA-TS-87-206. Washington, DC: Federal Highway Administration.

Michelow J. 1963. *Analysis of Stress and Displacements in an n-layered Elastic System under a Load Uniformly Distributed on a Circular Area*. Richmond, CA: California Research Corporation.

NCHRP. 2004. *Guide for Mechanistic-Empirical Design of New and Rehabilitated Pavement Structures*. Final Report for NCHRP Project 1-37A, Appendix CC-1. Transportation Research Board. Washington, DC: National Research Council.

NHI. 2002. *Analysis of New and Rehabilitated Pavement Performance with Mechanistic-Empirical Design Guide Software*. Arlington, VA: National Highway Institute, Course No. 131109 Participant Workbook, June.

Odemark N. 1949. *Investigations as to the Elastic Properties of Soils and Design of Pavements According to the Theory of Elasticity.* Stockholm: Staten Vagininstitut. Mitteilung Nr. 77.

Peattie K.R. 1962. Stress and strain factors for three-layer elastic system. *Highway Research Board, Bulletin*, Vol. 342.

Raad L. and J.L. Figueroa. 1980. Load response of transportation support systems. *Transportation Engineering Journal*, ASCE, Vol. 106, No. TE1, pp. 111–128.

Schiffman R.L. 1962. General solution of stresses and displacements in layered elastic systems. *Proceedings of the International Conference on the Structural Design of Asphalt Pavement*. Ann Arbor, MI: University of Michigan.

Shell International. 1998. *BISAR 3.0, User Manual*. Bitumen Business Group. The Hague: Shell International Oil Products B.V.

Siddharthan R.V., N. Krishnamenon, and P.E. Sebaaly. 2000. Finite-layer approach to pavement response evaluation. *Conference on Geotechnical Aspects of Pavements 2000*. Transportation Research Record No. 1709. Washington, DC: National Academy Press.

Tu W. 2007. Response modelling of pavement subjected to dynamic surface loading based on stress-based multi-layered plate. Columbus, OH: Ph.D. Thesis, The Ohio State University.

Ullidtz P. 1987. *Pavement Analysis*. Amsterdam, NY: Elsevier.

Ullidtz P. 1999. Deterioration models for managing flexible pavements. *Transportation Research Record, Journal of the Transportation Research Board*, Vol. 1655, pp. 31–34. Washington, DC: National Academies of Science.

Ullidtz P., J.T. Harvey, B.W. Tsai, and C.L. Monismith. 2006. *Calibration of Incremental-Recursive Flexible Damage Models in CalME Using HVS Experiments*. Sacramento, CA: Report prepared for the California Department of Transportation (Caltrans) Division of Research and Innovation by the University of California Pavement Research Center, Davis and Berkeley. UCPRC-RR-2005-06.

Van Cauwelaert F.J., D.R. Alexander, T.D. White, and W.R. Baker. 1989. *Multilayer Elastic Program for Backcalculating Layer Moduli in Pavement Evaluation*. STP 1026: Nondestructive testing of pavements and backcalculation of moduli. Philadelphia, PA: American Society for Testing and Materials (also ASTM STP19806S by ASTM International).

Wang Q. 2008. *Improvement of structural modeling of flexible pavements for mechanistic-empirical design*. Ph.D. Thesis, University of Minnesota. Ann Arbor, MI: ProQuest Information & Learning Company.

Westergaard H.M. 1926. Stresses in concrete pavements computed by theoretical analysis. *Public Roads*, Vol. 7, pp. 25–35.

Yoder E.J. and M.W. Witczak. 1975. *Principals of Pavement Design*. Hoboken, NJ: John Wiley & Sons, Inc.

Zhou F., E. Fernando, and T. Scullion. 2010. *Development, Calibration, and Validation of Performance Prediction Models for the Texas M-E Flexible Pavement Design System*. FHWA/TX-10/0-5798-2. Springfield, VA: National Technical Information Service.

Chapter 12

Traffic and traffic assessment

12.1 GENERAL

One of the major parameters involved in pavement design is the assessment of traffic over a period known as design traffic. The pavement should be designed to sustain the damage caused by the design traffic during its service life.

Various types of vehicles move on roads carrying different loads; hence, different types of axles and loads stress the pavement. To simplify the variability of axial loads and calculations, the equivalent single axial load (ESAL) has been introduced and used internationally as a unit of measure of traffic volume.

In the following paragraphs, a detailed description will be given on how the traffic volume is converted to ESALs, giving first a detailed account to vehicular loads, vehicle classification and ways of measuring axle loads. Reference to the tyre parameters, such as contact area and effect of tyre pressure, will also be made.

12.2 AXLE LOADS

The vehicles moving in a road is a mixture of vehicles such as private cars, buses, vans, trucks or lorries and articulated trucks or articulated vehicles. It is clear that the vehicular loads vary, as well as the number of axles and wheels each vehicle carries. The vehicular load is distributed to the axles and, in turn, to their wheels.

The vehicle load and its distribution are directly related to the damaging ability of each vehicle to the pavement. Thus, the damage ability of a private car, whose gross weight does not usually exceed 1.5 tonnes, is far less (almost negligible) than trucks or articulated trucks, whose gross weight may reach 40 tonnes or more.

All vehicles except private cars are characterised as commercial vehicles (CVs). Some organisations exclude from the category of CVs the vans or light goods vehicles with gross weight up to 3.5 tonnes. The structural wear caused by private cars or light goods vehicles is considered by many methodologies to be negligible.

The type of axles in CVs can be single, tandem or tri-axles, with a single or dual pneumatic tyre (wheels) on either side. The dimensions of the wheel and its other characteristics, pressure and so on are related to the vehicle type.

In order to protect pavements from excessive damage, authorised (permitted) maximum axle weights per type of axle and permitted maximum gross weights per type of vehicle have been enforced. The axle weight limit set is directly related to pavement engineering while the limit of vehicular gross weight is related to bridge engineering, road safety and competition.

The European Council has set the maximum authorised (permitted) axle weight per type of axle, distinguished between single and tandem driving or non-driving axle and tri-axles

Table 12.1 Maximum authorised axle weights in various countries and the EU

| Country | Maximum permitted axle weight (tonnes) | | | | |
| | Single axle | | Tandem[a] | | |
	Non-driving	Driving	Non-driving	Driving	Tri-axles or tridem[a]
Australia[b]	6/10	6	11–16.5	10–11	15–20
Austria	10	11.5	11–20	11.5–19	21–24
Belgium	10	12	10–20	12–22	10–30
Canada[c]	5.5	7.25	9.1–17	9.1–17	21–23
Denmark	10	11.5	11–20	10–16	21–24
Finland	10	11.5	11–20	11.5–16	11–24
France	13	13	13–21	13–21	24
Germany	10	11.5	16	16	—
Greece	10	11.5	20	—	—
Holland	10	11.5	10–20	11.5–20	24–30
Hungary	10	11	10–16	10–16	10–24
Ireland	10	10.5/11	11–20	19	21–24
Italy	12	12	19	19	—
Japan	—	10	20	—	—
Luxembourg	10	12	20	—	—
Norway	10	11.5	16	—	—
Portugal	11	12	16	—	—
Serbia	10	11.5	16	16	—
Slovenia	10	11.5	11–20	20–20	21–24
Spain	10	11.5	11–20	11.5–20	21–30
Sweden	10	11.5	11–20	11–20	21–24
Switzerland	10	11.5	18	18	—
Turkey	10	11.5	19	—	—
United Kingdom	10	11.5	16–20	16–19	21–24
United States[d]	9	9	15.4	15.4	23.2
European Union (EC)[e]	10	11.5	11–20[f]	11.5–19[g]	21–24

Sources: Adapted from Directive 96/53/EC, *Laying down for certain road vehicles circulating within the community the maximum authorized dimensions in national and international traffic and the maximum authorized weights in international traffic*, Official Journal of the European Communities, Brussels: The Council of the European Union, 1996; Directive 2002/7/EC, *Amending Council Directive 96/53/EC laying down for certain road vehicles circulating within the community the maximum authorised dimensions in national and international traffic and the maximum authorised weights in international traffic*, Official Journal of the European Communities, Brussels: The Council of the European Union, 2002; OECD, *Heavy trucks, climate and pavement damage, Road Transport and Intermodal Linkages Research Programme*, Paris: OECD Publishing, 1988; COST 323, *Collection and analysis of requirements as regards weighing vehicles in motion*, EURO-COST/323/2E/1997, European Co-operation in the Field of Scientific and Technical Research. Brussels, 1997; and De Ceuster et al., *Effects of adapting the rules on weights and dimensions of heavy commercial vehicles as established within Directive 96/53/EC*, TREN/G3/318/2007, Belgium: European Commission, 2008.

[a] For values expressed in range, the permitted axle weight depends on distance between axles, single/dual tyres, tyre width, load-sharing suspension and so on.
[b] VSC (2013) and RTA (2001).
[c] MOU (2011).
[d] TRB (2002).
[e] Directive 96/53/CE (1996) and Directive 2002/7/EC (2002).
[f] Tandem axles of trailers and semi-trailers.
[g] Tandem axles of motor vehicles.

(Directive 2002/7/EC 2002; Directive 96/53/EC 1996). Although limits have been set, axle weights have not yet been fully agreed upon across European Union (EU) countries.

The maximum authorised axial weights set by a number of countries and EU, per type of axle, are as shown in Table 12.1.

12.3 VEHICLE CLASSIFICATION

The CVs are classified into motor vehicles or trucks lorries with two, three or four axles; articulated trucks or tractors with three, four, five or six axles; road trains with four, five or six axles; and articulated trains with seven, eight or nine axles.

The maximum gross weight of each vehicle and the load distribution per axle type depend on the number, type of axles and distances between axles.

The maximum authorised vehicle gross weight legislated in the EU (Directive 96/53/EC 1996, Directive 2002/7/EC 2002) and recommended in United States, together with an indicative load distribution per axle type, is given in Table 12.2.

Needless to say, other countries outside the EU or the United States have imposed their own maximum permitted CV gross weights.

Table 12.2 Types of commercial vehicles, permitted vehicle gross weight and indicative axle-weight distribution in the EU and the United States

Type of commercial vehicle	Indicative weight distribution per axle (tonnes)	Permitted maximum vehicle gross weight (tonnes)
Two-axle motor vehicle or truck (MV2 or T2)[a]	6.5 11.5	18 EU
Three-axle motor vehicle or truck (MV3 or T3)[a]	6 20	25/26 EU (19–36.4 USA[f])
Four-axle motor vehicle or truck (MV4 or T4)[a]	14 18	32 EU (19–36.4 USA[f])
Three-axle articulated (Tr2-ST1)[b]	6 8 10	Not specified, EU (19–36.4 USA[f])
Four-axle articulated vehicle or tractor (Tr2-ST2)[b]	6 10 20	36 or 38[c] EU 19–36.4 USA[f]

(continued)

Table 12.2 Types of commercial vehicles, permitted vehicle gross weight and indicative axle-weight distribution in the EU and the United States (Continued)

Type of commercial vehicle	Indicative weight distribution per axle (tonnes)	Permitted maximum vehicle gross weight (tonnes)
Five-axle articulated vehicle or tractor (Tr2-ST3)[b]	8 10 22	40 EU (36.6–52.0 USA[f])
Five-axle articulated vehicle or tractor (Tr3-ST2)[b]	6 16 18	40 or 44[d] EU (36.6–52.0 USA[f])
Six-axle articulated vehicle or tractor (Tr3-ST3)[b]	10 13 21	40 or 44[d] EU (36.6–52.0 USA[f])
Four-axle road train (truck with trailer) (T2-Trl2)[e]	10 10 10 10	36 EU (19–36.4 USA[f])
Five-axle road train (truck with trailer) (T2-Trl3)[e]	6 10 10 10	40 EU (36.6–52.0 USA[f])
Five-axles road train (truck with trailer) (T3-Trl2)[e]	6 14 10 10	40 EU (36.6–52.0 USA[f])
Six-axle road train (truck with trailer) (T3-Trl3)[e]	6 12 11 11	40 EU (36.6–52.0 USA[f])
Five-axle articulated train (2T-ST1-Trl2)[b,e] (or six-axle articulated train)	6 9 8.5 7.3 7.3	Not specified, EU (36.6–52.0 USA[f])
Seven-axle articulated train (T3-ST2-Trl2)[b,e] (rocky mountain double)	4 14 13.5 7.5 7.0	Not specified, EU (37.0–52.2 USA[f])

(continued)

Table 12.2 Types of commercial vehicles, permitted vehicle gross weight and indicative axle-weight distribution in the EU and the United States (Continued)

Type of commercial vehicle	Indicative weight distribution per axle (tonnes)	Permitted maximum vehicle gross weight (tonnes)
Eight-axle articulated train (T3-ST2-Trl3)[b,e]	5.5 15.5 15 9.5 10	Not specified-EU (39.5–57.3 USA[f])
Nine-axle articulated train (T3-ST2-Trl4)[b,e] (turnpike double)	5.5 15 12.7 12.7 12.7	Not specified, EU (42.3–59.5 USA[f])

Sources: Directive 96/53/EC, *Laying down for certain road vehicles circulating within the community the maximum authorized dimensions in national and international traffic and the maximum authorized weights in international traffic*, Official Journal of the European Communities, Brussels: The Council of the European Union, 1996; Directive 2002/7/EC, *Amending Council Directive 96/53/EC laying down for certain road vehicles circulating within the community the maximum authorised dimensions in national and international traffic and the maximum authorised weights in international traffic*, Official Journal of the European Communities, Brussels: The Council of the European Union, 2002; TRB, *Regulation of Weights, Lengths and Widths of Commercial Motor Vehicles*. Special Report 267. Washington, DC: Transportation Research Board, 2002.

[a] MV, motor vehicle; T, truck; the number denotes the number of axles.
[b] Tr, tractor; ST, semi-trailer; the number denotes the number of axles.
[c] If the distance between the axles of the semi-trailer is greater than 1.8 m.
[d] With the semi-trailer carrying a 40-foot ISO container as a combined transport operation.
[e] T, truck; Trl, trailer; the number denotes the number of axles.
[f] TRB (2002); recommended bridge formula limits.

12.4 OVERLOADED VEHICLES

Even though maximum permitted gross vehicle weights and permitted weights per type of axle of CVs have been implemented, it is a quite common fact that a certain percentage of CVs moving along the highways are overloaded. This accelerates pavement wear and reduces pavement service life.

Early studies conducted by TRRL (Shane and Newton 1988) between 1980 and 1986 on motorways and trunk roads found that 8% of CVs had a gross weight greater than the permitted maximum and a further 5.4% that did not exceed the gross weight limit had overloaded axles. In the same study, it was concluded that overloading contributed directly to 5.7% of the road wear attributable to all goods vehicles.

Similar studies conducted in the United States found that approximately 25% of all CVs exceed the maximum permitted gross weight (FHWA 1989).

A study carried out in Greece (Mintsis et al. 1992) confirmed that the percentages of overloaded axles varied among different types of axles. On average, the percentage of overloaded axles, per type of axle, found was 16% for single non-driving axle, 6% for single driving axle and 30% for tandem non-driving axle. Regarding the percentage of overloaded CVs, it was found to range from 26% to 49% depending on the type of CV. During subsequent studies (Mintsis et al. 2002a,b, 2007; Proios et al. 2005), similar percentages of overloaded CVs were reported for some motorway sections.

A lower percentage of overloaded CVs, on average 10.7%, was reported in some motorway sections with an average daily flow of approximately 1000 CV/day. The highest percentage (16.8%) observed was in the vehicle category four or more axles (Egnatia Odos 2006).

Comparing the effect of overloaded vehicles to the number of ESALs, it was found that an increase of permissible axle load by 20% resulted in an increase of ESALs by 27.5% (Tsohos and Nikolaides 1989).

A study carried out between 1998 and 2000 in Australia to establish the distribution of heavy vehicles that have overloaded axle groups provided interesting results. The percentage of non-drive axle, in all vehicle types, was found to be higher than that of the drive axle (NTC 2005).

Overloaded axles or vehicles apart from anything else certainly cause excessive wear and damage to road pavements at the expense of the taxpayer.

In an extensive study to estimate the cost of overloaded (overweight) vehicles travelling on Arizona state highways (Straus and Semmens 2006), it was stated that overweight vehicles impose somewhere between $12 million and $53 million per year in uncompensated damage to Arizona roadways. The percentage of vehicles that are overweight was found to range from less than 0.5% to 30%. The type of vehicles found to have the highest rate of interstate overweight violations were those with five or more axles. Other similar studies have also been carried out in other states.

Pais et al. (2013) investigated the impact of overloaded vehicles on road pavements by studying the truck factors for different vehicle cases, applied to a set of pavements composed of five different asphalt layer thicknesses and five different subgrade stiffness moduli. The study revealed that the presence of overloaded vehicles can increase pavement costs by more than 100% compared to the cost of the same vehicles with legal loading.

The problem of overloaded vehicles or axles is serious and could be eliminated by introducing identification technologies and enforcing CV weight legislation and fines when the offence is confirmed.

The weigh-in-motion (WIM) technology is very effective and has been implemented by various countries in Europe. Weighing vehicles in motion on chosen routes makes it possible to monitor weights and to assess the 'aggressivity' of traffic. Relevant information can be found in the International Technology Scanning Program, sponsored by the FHWA, AASHTO and NCHRP (Honefanger et al. 2007).

12.5 MEASUREMENT OF VEHICLE AXLE LOAD

The measurement of every vehicle axle load (weight) in its simplest form is carried out by placing the vehicle axle on an appropriate device and weighing it. This procedure is slow and time-consuming. It also requires the participation of police or other authorised personnel to stop the vehicles for inspection. Additionally, with this method, it is impossible to record a sufficient number of axial loads over a period for the sample to be representative with the aim of drawing results and enforcing regulations. The enforcement of fines to every overloaded vehicle is also an impossible task.

For the inspection of the road network regarding vehicle axial loads and effectively vehicle gross weight devices, systems known as WIM are used.

12.5.1 WIM systems

WIM, according to the European WIM specification COST 323 (1997) (Jacob et al. 2002), is the process of estimating the weight of a moving vehicle, and the portion of the weight that is carried by each of its wheels or axles, by measurement and analysis of dynamic vehicle tyre forces. The terminology used for WIM by ASTM E 867 (2012) is similar to this.

The WIM system is a set of mounted sensor(s) and electronics with software that measures dynamic vehicle tyre forces and vehicle presence of a moving vehicle with respect to time and provides data for calculating wheel or axle loads and gross weights, as well as other parameters such as speed, axle spacing, silhouettes, and so on (Jacob et al. 2002).

Unlike static scales (weigh station scales), WIM systems are capable of measuring vehicles travelling at a normal or reduced traffic speed and do not require the vehicle to come to a stop. This makes the weighing process for legal purpose more efficient, since it allows the CVs under the weight limit to bypass static scales measurement.

WIM systems are distinguished into high-speed and low-speed systems. High-speed WIM systems (HS-WIM) are those installed on one or more traffic lane(s) and operate automatically under normal traffic conditions. Low-speed WIM systems (LS-WIM) are those installed in a specific weighing area, outside of the traffic lane(s), on which the vehicles to be weighed are diverted usually by police. They are also operated automatically, but the speed is limited (5 to 20 km/h maximum) and the travelling conditions of the vehicles are controlled. The WIM systems can be fixed or portable systems.

WIM systems are classified with respect to their accuracy, each of which corresponds to a range of applications or requirements, such as (a) legal purposes (for enforcement of legal weight limits), (b) infrastructure and pre-selection purposes (for detailed traffic analysis, design/maintenance of roads and bridges, vehicle classification, pre-selection for enforcement, etc.) and (c) statistics (for economic and technical studies of freight transport, general traffic evaluation, collecting statistical data, etc.).

According to the European WIM specification COST 323 (1997) (Jacob et al. 2002), there are six accuracy classes: Class A (5) and Class B+ (7) for legal purposes; Class B (10), Class C (15) and Class D+ (20) for infrastructure and pre-selection purposes; and Class D (25) for statistical purposes. The number enclosed in parentheses corresponds to the tolerance limit ($\pm$) of results for a confidence level (usually 95%). In addition to the above classes, other classes such as Class E (>25) characterise WIM systems that do not meet Class D (25) (systems installed in poor WIM sites). They are useful in that they give indications about the traffic composition and the load distribution and frequency.

The classification of WIM systems according to ASTM E 1318 (2009) is based on the needs of the user for intended applications. There are four classes or types distinguished: Type I (high accuracy) and Type II (lower cost) for WIM systems designed for installation at a traffic data collection site and capable of accommodating highway vehicles moving at speeds from 16 to 130 km/h, Type III for WIM systems designed for installation at weight-enforcement stations and capable of accommodating highway vehicles moving at speeds from 16 to 130 km/h and Type IV (low-speed WIM) designed for use at weight-enforcement stations to detect weight-limit or load-limit violations and capable of accommodating vehicles moving at speeds from 3 to 16 km/h. This type of system, as stated in the 2013 edition of standards, has not yet been approved for use in the United States. The accuracy of each type of WIM system and other performance requirements is provided in detail in ASTM E 1318 (2009).

The accuracy of WIM systems generally depends on a number of factors. Some of them depend on the system selected, particularly on the weight sensor used. Others are consequent to WIM technology and are related to the road geometry, pavement characteristics and environmental characteristics or requirements.

Apart from calculating the axial, tyre (wheel) and vehicle weight, WIM systems provide data such as vehicle speed, vehicle classification, axle spacing, wheelbase distance (front-most to rear-most axle), sequential vehicle record number, lane and direction of travel, date and time of passage, ambient temperature record or conversion to equivalent single-axle loads (ESALs).

12.5.2 Components of the WIM system

In general, WIM systems (LS-WIM or HS-WIM) consist of the following components:

a. Weight (mass) sensor
b. Vehicle classification sensor or identification sensor
c. Data storage and processing unit
d. User communication unit

A typical schematic arrangement of a WIM system is given in Figure 12.1.

12.5.2.1 Weight (mass) sensor

The weight sensor is the most fundamental and important component of a WIM system. The sensor or sensors are placed on or within the pavement, perpendicular to the direction of traffic. Depending on how they are placed, the sensors can be temporary, semi-permanent/temporary or permanent.

Temporary weight sensors are surface mounted and fixed so that they can be removed and re-installed at another location. Semi-temporary weight sensors are similar to the temporary ones, but only one component of the sensor is removed and re-installed (the gauge, transducer or signal processor). Permanent weight sensors are installed to the surface (or at a certain depth) once and are not removed until they are damaged.

Weight (mass) sensors are of various types, such as bending plate, capacitance pad, capacitance strip, piezo-electric cable, load cell or strain gauge mounted to an existing road structure such a culvert (Culway system).

12.5.2.2 Vehicle classification sensor or identification sensor

The classification of vehicles is carried out through various types of sensors, which are placed close to the weight sensor. The typical vehicle classification sensors include loops, piezo-electric cables, mechanical switches and tubes.

Some systems also bear a vehicle identification sensor in the form of a picture or a video image. In this case, the camera or video is placed off road in a specially built housing and placed in a position to capture the vehicle's registration number and other features.

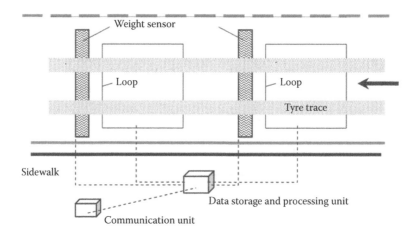

Figure 12.1 Typical arrangement of the WIM system with loops.

Nowadays, as the technology develops, laser scanners as well as 'smart cards' may also be used for the abovementioned purposes. Laser scanners are placed on a metal pole with a cantilever or a bridge covering the whole width of the road, while the 'smart cards' are carried onboard the vehicle.

12.5.2.3 Data storage and processing unit

The data storage and processing unit is placed at the roadside and in a protective cabinet. This unit is connected to the weight sensors, the vehicle classification sensor and the vehicle identification sensor. With the aid of software, it has the ability to present brief and analytical results. The data storage and processing unit also powers the WIM system; the power is supplied by mains, rechargeable electric or solar batteries.

12.5.2.4 User communication unit

Even though the data storage and processing unit provides the user with all the information needed, there is usually a communication link with the user for data recovery/downloading from the storage/processing unit. The communication link may be a particularly designed data recovery device, a personal computer or a modem with a telemetry link. In the latter case, the data are sent and stored into an offsite user unit.

12.5.3 Factors affecting performance of the WIM system

There are various factors affecting the performance of a WIM system. Some of them are related to the selected device, particularly the weight sensors used and related technology. However, the performance of a WIM system is greatly affected by the choice of the WIM site and the environmental characteristics at the selected site. In relation to the choice of site, the location and road geometry, the pavement characteristics are of great importance as well.

12.5.3.1 Location and road geometry at weigh station

The WIM system should be installed away from acceleration or deceleration or gear change areas (traffic lights, toll stations or slope sections). This is to ensure weighing the vehicles travelling at uniform speed.

It is also desirable to avoid areas where lane change is encouraged.

With regard to the road geometry, the longitudinal and transverse slope and radius of curvature before and after the weigh station are vital.

According to European specifications for WIM systems in COST 323 (1997) (Jacob et al. 2002), the longitudinal slope should be less than 1% for the class I site or less than 2% for other class sites in the road section between 50 m ahead of and 25 m beyond the system. Similarly, at all class sites, the transverse slope should be less than 3% and the radius of curvature should be greater than 1000 m. However, in all cases, it is preferable to place the system in a straight road segment.

The requirements set by ASTM E 1318 (2009) are similar to these. The Australian WIM specifications (Austroads 2010) specify road geometry requirements per type of weighing system used.

In the case of low-speed WIM systems, the characteristics of the location are determined by the manufacturer and should be fulfilled.

12.5.3.2 Pavement characteristics

Pavement characteristics such as rutting, evenness and deformation directly influence the signal recorded by any WIM sensor and limit the accuracy of the measurements. Cracking may also reduce the WIM sensor durability and affect its response.

With respect to rutting, measured by a 3 m beam, COST 323 (1997) classifies the sites to excellent, good and acceptable when the rut depth values are ≤4, ≤7 or ≤10 mm, respectively.

The surface evenness, determined using the International Roughness Index, for the abovementioned site classes is 0–1.3, 1.3–2.6 and 2.6–4.0 m/km, respectively. Evenness may be expressed in terms of APL rating (French profilometer) and respective range values are given in COST 232.

Surface evenness or smoothness may also be measured in accordance to AASHTO M 331 (2013) so that the WIM system complies with the performance tolerance required by ASTM E 1318 (2009).

As for deformation in the case of flexible or semi-flexible pavement, it can be measured with a deflectograph with a 13 t axial load at 2.0 to 3.5 km/h (quasi-static deflection) or a falling weight deflectometer with a test load of 5 t, at a reference temperature of 20°C (dynamic deflection). The range of deflection values per site class is given in COST 323 (1997) (Jacob et al. 2002).

12.5.3.3 Environmental characteristics

The air and pavement temperature, ice, salt and water directly affect the performance of a WIM system.

The air temperature directly affects the function of sensors and electronics and the pavement temperature affects the pavement modulus (asphalt layers).

The sensors and all electronic parts should be able to operate at temperatures ranging from –20°C to 60°C. The pavement modulus change owing to pavement temperature should be taken into consideration following the recommendations of COST 323 (1997). Guidelines regarding ambient temperature working range and the provision of a temperature-compensating mechanism in the electronic hardware are also given in ASTM E 1318 (2009) and Australian WIM specifications (Austroads 2010).

Furthermore, the sensors should not be insensitive to water and salt exposure. If the pavement is not well drained, deflection measurements may increase after rainfall.

Finally, the weight of vehicles and the use of studded tyres, in some countries, may also affect the performance of the sensors. According to the European WIM specification COST 323 (1997) (Jacob et al. 2002), the sensors should withstand 60 tonne vehicles and resist the destructive action of studded tyres.

12.5.4 Calibration of the WIM system

Calibration of the WIM system is necessary before measurements are taken. A calibration procedure for type approval is described in ASTM E 1318 (2009) and COST 323 (1997) (Jacob et al. 2002).

12.5.5 Available WIM systems

There seem to be plenty of WIM systems commercially available (more than 20). A list of suppliers and vendors, in addition to Cross Zlin (2014), Mettler Toledo (2014), Tollman (2014) and Traffic Tech (2014), can be found in Austroads (2000) and International Society for Weigh in Motion (2014).

12.6 CONTACT AREA AND PRESSURE

The shape of the contact area between the tyre and the pavement surface varies depending on the applied load and the internal tyre pressure. Lister and Nunn (1968) found that, by keeping the internal tyre pressure (P_o) constant and equal to the pressure recommended by the manufacturer, the shape of the contact area is almost cyclic when the applied load (P) is approximately 1/3 of the maximum allowable (P_n) by the manufacturer.

As the applied load increases, the contact area takes the shape of a 'barrel' (see Figure 12.2).

Similarly, by keeping the load constant to 1/3 of the maximum allowable and decreasing the tyre pressure below the typical value recommended by the manufacturer, the shape changes to approximately elliptical.

Because of the great variability in load and tyre pressure, even for a given type of tyre, and for simplicity reasons, the representative contact area is assumed to be circular.

Hence, the radius of the contact area is determined from the following equation:

$$a = \left(\frac{P}{\pi \times P_o} \right)^{1/2},$$

where a is the radius of contact area (m), P is the applied load (kN) and P_o is the tyre pressure (kN/m^2).

The load distribution in this cyclic area is affected by the actual tyre load (P)/maximum permissible load (P_n) ratio. When this ratio is lower than 1, the load distribution is parabolic, and when it is equal to 1, the distribution is uniform.

In case the above ratio is greater than 1, that is, the tyre load is greater than the maximum permissible load, the load distribution is non-uniform and obtains its maximum value at the perimeter of the contact area (Lister and Jones 1967). For simplicity, the load distribution is always assumed to be uniform across the surface.

Other factors that affect the tyre behaviour and have a direct effect on both distribution and magnitude of applied pressure are the following: type of tyre (radial or x-ply, high- or low-pressure tyre, etc.), surface and tyre condition (smooth or treaded, worn or new), temperature developed in the tyre, tyre resilience, type and condition of vehicle suspension and pavement surface irregularities. A detailed description of the effect of the above factors is beyond the scope of this book.

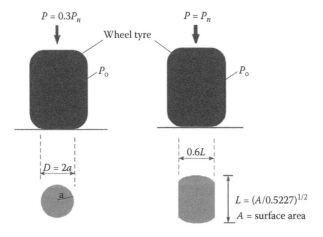

Figure 12.2 Schematic representation of contact area with respect to applied load.

In conclusion, it is indicated that the type and condition of tyre affect the internal tyre pressure and in turn the contact pressure. An increase in tyre temperature results in an increase in internal pressure, while surface irregularities cause an instantaneous load increase.

12.7 THE CONCEPT OF EQUIVALENT STANDARD AXLE LOADING – EQUIVALENCY FACTORS FOR FLEXIBLE PAVEMENTS

The great variety of CVs, axle configurations and loads causes different distress to the pavement structure. This alone creates a huge problem to the pavement designer. Many years ago, converting every axle configuration and load to an axial load was considered, causing an equivalent structural distress.

The first real-life pavement test undertaken in this sector was the AASHTO road test (HRB 1961, 1962) conducted in the United States from 1957 to 1961. Many different pavement structures, both flexible and rigid, were subjected to traffic loads, and equivalence between loads was established with respect to the load of a standard (or equivalent) axial chosen.

The standard axle and load chosen was the single axle with dual wheels carrying a total mass of 18,000 lb (80 kN).

The outcome of the road test was the establishment of complex equations (AASHTO 1986) from which load equivalency factor could be determined for any load of single, tandem or tridem axle. The structural distress caused by the traffic (all kinds of distresses cracking, rutting, etc.) was expressed in terms of pavement serviceability performance and is called terminal serviceability (p_t). The variation of flexible pavement structures was expressed in terms of pavement structural number (SN), while the variation of rigid pavement structures was expressed in terms of thickness of slab, D (AASHTO 1993; HRB 1962).

The load equivalency factors were derived from terminal serviceability factor (p_t) values equal to 2.0, 2.5 or 3.0 and pavement structural number (SN) values equal to 2, 3, 4, 5 or 6. The lower p_t value represents a pavement with serious structural distress where maintenance is unavoidable. The upper value of 3.0 represents a pavement with structural distresses needing maintenance, if high level of service is to be provided. The SN values represent different pavement structures. A sample of the equivalency factors derived from flexible pavements, for $p_t = 2.5$ and SN = 5, for the three types of axles (single, tandem and triaxial), is given in Table 12.3. All equivalency factors for flexible pavements can be found in tabular form in the AASHTO pavement design guide (AASHTO 1993).

When multiplied by the number of axles of a given axle weight, these equivalency factors provide the number of standard axle load applications that will have an equivalent effect on the performance of the pavement structure. This number of standard axle load applications is expressed as equivalent standard axle loads or loading (ESAL).

Similar studies were also conducted in various European countries during the 1980s. To study the structural distress (fatigue cracking in bound courses and permanent deformation of bituminous courses, unbound layers and subgrade), the notion of equivalence between loads in terms of the damage they cause has been developed. This equivalence was expressed in the form of a law:

$$\frac{N_i}{N_j} = \left(\frac{P_j}{P_i}\right)^{\gamma},$$

Table 12.3 Equivalency factors for converting axle loads to ESALs, for pt = 2.5 and SN = 5

Axle load		Equivalency factor per type of Axle (α_j)		
kN[a]	kips[a]	Single	Tandem	Triaxial
8.9	2	0.002	0.0000	0.0000
17.8	4	0.002	0.0003	0.0001
26.7	6	0.010	0.001	0.0003
35.6	8	0.034	0.003	0.001
44.5	10	0.088	0.007	0.002
53.4	12	0.189	0.014	0.003
62.3	14	0.360	0.027	0.006
71.2	16	0.623	0.047	0.011
80.0	18	1.00	0.077	0.017
89.0	20	1.51	0.121	0.027
97.9	22	2.18	0.180	0.040
106.8	24	3.03	0.260	0.057
115.7	26	4.09	0.364	0.080
124.6	28	5.39	0.495	0.109
133.5	30	7.0	0.658	0.145
142.4	32	8.9	0.857	0.191
151.3	34	11.2	1.09	0.246
160.2	36	13.9	1.38	0.313
169.1	38	17.2	1.70	0.393
178.0	40	21.1	2.08	0.487
186.8	42	25.6	2.51	0.597
195.8	44	31.0	3.00	0.723
204.7	46	37.2	3.55	0.868
213.6	48	44.5	4.17	1.033
222.5	50	53.0	4.86	1.22
231.4	52		5.63	1.43
240.3	54		6.47	1.66
249.2	56		7.4	1.91
258.1	58		8.4	2.20
267.0	60		9.6	2.51
275.9	62		10.8	2.85
284.8	64		12.2	3.22
293.7	66		13.7	3.62
302.6	68		15.4	4.05
311.5	70		17.2	4.52
320.4	72		19.2	5.03
329.3	74		21.3	5.57
338.2	76		23.7	6.15
347.1	78		26.2	6.78
356.0	80		29.0	7.45
364.9	82		32.0	8.2
373.8	84		35.3	8.9
382.7	86		38.8	9.8
391.6	88		42.6	10.6
400.5	90		46.8	11.6

Source: Adapted from AASHTO, *AASHTO Guide for design of pavement structures*, Washington, DC: American Association of State Highway and Transportation Officials, 1993.

[a]   1 kps is approximately equal to 4.44976 kN.

which is converted to

$$N_i = \left(\frac{P_j}{P_i}\right)^\gamma \times N_j,$$

where N_i is the number of loads of magnitude P_i to cause failure, N_j is the number of loads of magnitude P_j to cause failure and γ is the coefficient depending on the pavement structure, structural distress and axle configurations.

When P_i is the load of the standard axle, the above equation converts to

$$N_{SA} = \left(\frac{P_j}{P_{SA}}\right)^\gamma \times N_j$$

or

$$N_{SA} = \alpha \times N_j,$$

where N_{SA} is the number of standard axle load of magnitude P_{SA}, P_{SA} is the standard axle load, P_j is the axle load to be converted and α is the equivalency factor, equal to $\left(\frac{P_j}{P_{SA}}\right)^\gamma$.

From various experimental studies considering different structural distresses, axle configurations and standard axle loads, different values of coefficient γ were derived (see Table 12.4).

The OCED study (OECD 1988) concluded that an appropriate value for coefficient γ to be considered for all types of pavement distress is 4. Hence, the equivalency factor, α, can be determined from the equation

$$\alpha = \left(\frac{P_j}{P_{SA}}\right)^4,$$

where P_j is the load of axle to be converted and P_{SA} is the load of standard axle.

With respect to standard single axle load, the values used vary among countries. Most of them seem to adopt the value of 80 kN, while others adopt the value of 100 or 130 kN.

Table 12.4 Variability of coefficient γ from various experimental investigations

Country	Axle type	Criterion	Coefficient γ
Italy (Battiato et al. 1983)	Different axle types[a]	Cracking of asphalt layers	1.2–3.0
Finland (Huhtala 1986)	Single twin wheel[b] and Tandem twin wheel[b]	Cracking of asphalt layers	3.3S, 4.0T
		Subgrade deformation	4.1S, 4.0T
France (Autret et al. 1987)	Two arms test device, 100 kN and 130 kN, twin wheel[c]	Rutting (10, 15, 40 mm)	9.6, 8.1, 8.2
		Cracking of asphalt layers (20%, 50%, first crack)	2.1, 1.7, 1.3

[a] Pavement structure: 10 cm asphalt layers, 20 cm unbound layers (base/base).
[b] Pavement structure: 8 cm asphalt layers, 15 cm base and 40 subbase.
[c] Pavement structure: 5 cm asphalt layer and 45 cm unbound (base/subbase).

When a tandem or a tridem axle has to be converted to a standard single axle using the above equation, it has been suggested (OECD 1988) that they be considered as two or three independent single axles.

Another approach is to specify the standard load for a tandem or tridem axle. The Australian pavement design guide (Austroads 2012), for example, uses a standard load of 135 and 181 kN for the dual and tridem axle, respectively, to determine the ESALs using the above equation.

However, parameters with regard to vehicle characteristics (the type of tyres, introduction of more than three axle systems, vehicle suspension, vehicle dynamics, etc.) have changed ever since the AASHTO experiment or the European experiments have been executed. Research studies carried out by Bonaquist (1992) and Huhtala and Pihlajamäki (1992) have shown that the use of new wide-based single tyres (445 or 385/65R 22.5 and 350/75R 22.5 types) and the use of different types of trucks have a greater damaging effect than conventional tandem axles.

Considering the above plus the fact that pavements are constructed with a greater variety of materials, perhaps there is a need for a new extensive study to re-establish axle load equivalency factors.

12.7.1 Determination of the cumulative number of ESAL

Most current pavement design methodologies use the cumulative number of ESAL over the design period as one of the major input parameters in the calculation process to determine the thickness of pavement layers.

To calculate the cumulative number of ESAL over a period of n years (design period), the daily number of ESAL and the average annual increase of commercial traffic must be known.

The daily number of ESAL is calculated from the following equation:

$$\text{ESAL}_{\text{daily}} = \sum_{j=1}^{j=n} (\alpha_j \times N_j),$$

where $\text{ESAL}_{\text{daily}}$ is the total number of standard axle loads per day, α_j is the equivalency factor of axle load j to the standard axle load and N_j is the number of axles with load j per day.

The cumulative number of ESAL over the design period is calculated using the following equation:

$$\text{ESAL}_{\text{cum}} = \text{ESAL}_{\text{daily}} \times 365 \times \left[\frac{(1+r)^n - 1}{r} \right],$$

where ESAL_{cum} is the cumulative number of ESAL over a period of n years, r is the average annual increase of commercial traffic (in 0.01%) and n is the number of years of design life.

After extensive WIM studies, some countries, apart from using equivalency factors per axle load, have also developed standard axle equivalency factors per CV category or group of CVs. These equivalency factors are known as equivalent track factors.

The above eases collection of traffic data since only traffic composition and frequency need to be recorded.

The calculation of the daily ESAL is carried out by multiplying the number of vehicles per category type or group by the truck equivalency factor.

Table 12.5 gives the equivalent truck factors developed in Greece (Nikolaides 1996), which have been implemented in the Greek design methodology (Nikolaides 2005).

Table 12.5 Recommended truck equivalency factors

Vehicle category	Number of axles	Truck equivalency factor
Buses	2 or 3	1.3
Motor vehicles (truck)	2w	0.34
Motor vehicles or articulated trucks	3	1.5
Motor vehicles, articulated trucks or road trains	4 or more	3.025
All other small commercial vehicles (mini-buses, ambulances, agriculture vehicles, etc.)	2	0.026
Private cars	2	0.00

Source: Nikolaides A., Conversion of traffic on national road network into standard axle loads. *Proceedings of 2nd National Conference 'Bituminous Mixtures and Pavements'*. Thessaloniki, Greece, 1996.

12.8 EQUIVALENCY FACTORS FOR RIGID PAVEMENTS

For rigid pavements, the conversion of vehicle axial loads to ESAL is carried out with the similar equivalency coefficients derived from the AASHTO road test. The values of serviceability index, p_t, used were the same as in flexible pavements, but instead of the pavement structural number variable, the thickness, D, of the slab was used. The values of slab thickness considered were 6, 8, 10, 12 and 14 inches.

The magnitude of the equivalency factors for rigid pavements derived, in all cases, was higher than that for flexible pavements, indicating the different damaging effect of axle load and axle configuration on concrete slabs. In fact, the axle load is more significant in terms of damaging power than load configuration.

The equivalency factors derived from complex equations (AASHTO 1986) have been tabulated and are given in the AASHTO pavement design manual (AASHTO 1993).

However, as a rule of thumb, the 1993 AASHTO design guide (Part III, Chapter 5, paragraph 5.2.3) recommends the use of a multiplier of 0.67 to convert rigid pavement ESAL to flexible pavement ESAL (or 1.5 for the opposite). It is noted that using load spectra as proposed in the guide for the design of new and rehabilitated pavement structures (NCHRP 2004) eliminates the need for flexible to rigid ESAL conversions.

Studies carried out in Europe during the 1980s on rigid pavements for deriving coefficients using the equivalency law equation found that the value of coefficient γ ranges from 4.2 up to 33. The variability was due to the usage (or non-usage) of dowelled reinforcement, the thickness of the concrete slabs and the materials used (OECD 1988).

The higher value of coefficient γ confirms the predominant role of axle load magnitude and, hence, heavy loads or overloads in rigid pavements. This in turn signifies that rigid pavements are relatively insensitive to load repetitions.

No unique value of coefficient γ has yet been proposed for the calculation of the daily, or cumulative, ESAL using the equivalency law.

However, in the Australian pavement design methodology for rigid pavements (Austroads 2012), the axle loads determined as in flexible pavement (using the equivalency law) are multiplied by a load safety factor (LSF). The LSF is related to project reliability, which varied from 80% to 97.5%. For a reliability of 95%, the LSF for an unreinforced slab is 1.3, while that for a doweled or continuously reinforced slab is 1.25.

For rigid pavement design calculations, the designer for the calculation of design traffic flow, in general, is advised to follow the related guidelines given in the methodology being used.

REFERENCES

AASHTO. 1986. *AASHTO guide for design of pavement structures*, Vol. 2. Washington, DC: American Association of State Highway and Transportation Officials.

AASHTO. 1993. *AASHTO guide for design of pavement structures*. Washington, DC: American Association of State Highway and Transportation Officials.

AASHTO M 331. 2013. *Smoothness of pavement in weigh-in-motion (WIM) systems*. Washington, DC: American Association of State Highway and Transportation Officials.

ASTM E 867. 2012. *Standard terminology relating to vehicle-pavement systems*. West Conshohocken, PA: ASTM International.

ASTM E 1318. 2009. *Standard specification for highway weigh-in-motion (WIM) systems with user requirements and test methods*. West Conshohocken, PA: ASTM International.

Austroads. 2000. *Weigh-in-Motion Technology*. Publication No. AP-R168/00. Sydney: Austroads Inc.

Austroads. 2010. *Weigh-in-Motion Management and Operation Manual*. Austroads Publication No. AP–T171/10. Sydney: Austroads Ltd.

Austroads. 2012. *Guide to Pavement Technology Part 2: Pavement Structural Design*. Publication No. AGPT02-12. Sydney: Austroads Inc.

Autret P., A. Baucheron de Boissoudy, and J.C. Gramsammer. 1987. The circular test track of the LCPC Nantes: First results. *Proceedings of the 6th International Conference on the Structural Design of Asphalt Pavements*. Ann Arbor, MI: The University of Michigan.

Battiato G., G. Camomillia, M. Malgarini, and G. Scapaticci. 1983. *Measurement of the 'Aggressiveness' of Goods Traffic on Road Pavements*. Essai de Nardo, Report No. 3, Tandem effect evaluation. *Autostrade*, Vol. 1. Rome, Italy.

Bonaquist R. 1992. An assessment of the increased damage potential of wide based single tires. *Proceedings of the 7th International Conference on the Structural Design of Asphalt Pavements*, Vol. 3, p. 1. Nottingham, UK.

COST 323. 1997. *Collection and Analysis of Requirements as Regards Weighing Vehicles in Motion*. EURO-COST/323/2E/1997. Brussels: European Co-operation in the Field of Scientific and Technical Research.

Cross Zlin. 2014. Available at http://www.cross.cz.

De Ceuster G., T. Breemersch, B. Van Herbruggen, K. Verweij, I. Davydenko, M. Klingender, B. Jacob, H. Arki, and M. Bereni. 2008. *Effects of Adapting the Rules on Weights and Dimensions of Heavy Commercial Vehicles as Established within Directive 96/53/EC*. TREN/G3/318/2007. Belgium: European Commission.

Directive 2002/7/EC. 2002. Amending Council Directive 96/53/EC laying down for certain road vehicles circulating within the Community the maximum authorised dimensions in national and international traffic and the maximum authorised weights in international traffic. *Official Journal of the European Communities*, Vol. 67, pp. 47–49. Brussels: The Council of the European Union.

Directive 96/53/EC. 1996. Laying down for certain road vehicles circulating within the Community the maximum authorized dimensions in national and international traffic and the maximum authorized weights in international traffic. *Official Journal of the European Communities*, Vol. 235, pp. 59–75. Brussels: The Council of the European Union.

Egnatia Odos. 2006. Traffic measurements along Egnatia motorway. *Newsletter*, Vol. 2, p. 2. Thessaloniki, Greece: Egnatia Odos S.A.

FHWA. 1989. *Overweight Vehicles-Penalties and Permits: An Inventory of State Practices for Fiscal Year 1987*. Washington, DC: U.S. Department of Transportation, Federal Highway Administration.

Honefanger J., J. Strawhorn, R. Athey, J. Carson, G. Conner, D. Jones, T. Kearney, J. Nicholas, P. Thurber, and R. Woolley. 2007. *Commercial Motor Vehicle Size and Weight Enforcement in Europe*. FHWA-PL-07-002. Washington, DC: FHWA-HPIP.

HRB (Highway Research Board). 1961. *The AASHO Road Test: History and Description of the Project*. Special Report 61A. National Academy of Science. Washington, DC: National Research Council.

HRB (Highway Research Board). 1962. *The AASHO Road Test*. Report 5, Special Report 61E, Publication 954. National Academy of Science. Washington, DC: National Research Council.

Huhtala M. 1986. The effect of different trucks on road pavements. *Proceedings of International Symposium on Heavy Vehicles Weights and Dimensions.* Kelowna, Canada.

Huhtala M. and K. Pihlajamäki. 1992. New concepts on load equivalency measurements. *Proceedings of the 7th International Conference on the Structural Design of Asphalt Pavements*, Vol. 3, p. 194. Nottingham, UK.

International Society for Weigh in Motion. 2014. Available at http://www.iswim.free.fr.

Jacob B., E. O'Brien, and S. Jehaes. 2002. *COST 323, Weigh-in-Motion of Road Vehicles.* Final Report, Appendix 1: European WIM Specification. Paris: COST/LCPC.

Lister N.W. and D.F. Nunn. 1968. *Contact Areas of Commercial Vehicle Tires.* TRRL Report LR 172. Crowthorne, UK: Transport Research Laboratory.

Lister N.W. and R. Jones. 1967. The behavior or flexible pavements under moving wheel loads. *Proceedings of 2nd International Conference on Structural Design of Asphalt Pavements.* pp. 1021–1035. Ann Arbor, MI: University of Michigan.

Mettler Toledo. 2014. Available at http://www.mtwim.com.

Mintsis G., A. Nikolaides, C. Taxiltaris, S. Basbas, P. Patonis, A. Filaktakis, and S. Lambropoulos. 2002a. Determination and assessment of the dynamic characteristics of HGVs and their impact to National highway network. *Proceedings of the 3rd International Conference on Bituminous Mixtures and Pavements*, pp. 761–770. Thessaloniki, Greece: Aristotle University of Thessaloniki.

Mintsis G., A. Nikolaides, and G. Tsohos. 1992. Count and analysis of heavy vehicle characteristics in sections of National road. *Proceedings of 1st National Conference of Asphalt Concrete and Flexible Pavements*, p. 134. Thessaloniki, Greece: Aristotle University of Thessaloniki.

Mintsis G., C. Taxiltaris, S. Basbas, and A. Filaktakis. 2002b. *Analyzing HGV Data Collected on Main Road Network in Greece.* Transportation Research Record (TRR), No. 1809, pp. 110–115. Transportation Research Board. Washington, DC: National Research Council.

Mintsis G., C. Taxiltaris, S. Basbas, A. Filaktakis, K. Koutsoukos, S. Guy, and E. Viskos. 2007. Temporal evolution of HGV traffic data along Egnatia Odos motorway. *Proceedings of the 4th International Conference on Bituminous Mixtures and Pavements*, pp. 591–601. Thessaloniki, Greece: Aristotle University of Greece.

MOU (Memorandum of Understanding). 2011. *Summary of the Memorandum of Understanding on vehicle weights and dimensions: Heavy truck weight and dimension limits for interprovincial operations in Canada: Summary information.* Council of Ministers Responsible for Transportation and Highway Safety. Ottawa: Council of Ministers.

NCHRP. 2004. Mechanistic. *Empirical design of new and rehabilitated pavement structures.* National Cooperative Highway Research Program, NCHRP Project I-37A Report, Part 2, Chapter 2. Washington, DC: National Research Council.

Nikolaides A. 1996. Conversion of traffic on national road network into standard axle loads. *Proceedings of 2nd National Conference 'Bituminous Mixtures and Pavements'.* Thessaloniki, Greece.

Nikolaides A. 2005. *Flexible Pavements: Pavement Design Method, Asphalts and Surfacing Mixtures.* Thessaloniki: A. Nikolaides. ISBN 960-91849-1-X.

NTC. 2005. *Review of Heavy Vehicle Axle Load Data: Information Data.* National Transport Commission. Melbourne: NTC.

OECD. 1988. *Heavy Trucks, Climate and Pavement Damage. Road Transport and Intermodal Linkages Research Programme.* Paris: OECD Publishing.

Pais J., S. Amorin, and M. Minhoto. 2013. Impact of traffic overload on road pavement performance. *Journal of Transportation Engineering*, Vol. 139, No. 9, pp. 873–879. ASCE.

Proios A., G. Mintsis, C. Taxiltaris, S. Basbas, and F. Filaktakis. 2005. The issue of the overloaded HGVs in the Greek National road network. *Proceedings of the 2nd Pan-Hellenic Conference on Highways.* Volos, Greece.

RTA. 2001. *National Heavy Vehicle Reform: Heavy Vehicle Mass, Loading and Access.* RTA/Pub. 01.029. Cat no. RTA45070666E. Sydney: Road and Traffic Authority.

Shane B.A., and W.H. Newton. 1988. *Goods Vehicle Overloading and Road Wear: Results from Ten Roadside Surveys (1980–1986).* RR 133. Wokingham, UK: TRL Limited.

Straus S.H. and J. Semmens. 2006. *Estimating the Cost of Overweight Vehicle Travel on Arizona Highways.* FHWA-AZ-06-52B. Springfield, VA: National Technical Information Service.

Tollman. 2014. Available at http://www.tollman.co.in.
Traffic Tech. 2014. Available at http://www.traffic-tech.com.
TRB. 2002. *Regulation of Weights, Lengths and Widths of Commercial Motor Vehicles*. Special Report 267. Washington, DC: Transportation Research Board.
Tsohos G. and A. Nikolaides. 1989. Overloaded vehicles: Investigation on their effect on pavements' design and performance. *Bulletin of Central Research Laboratory of Greece*, Vol. 103–104, p. 149. Athens: KEDE.
VSC (Vehicle Standards and Compliance). 2013. *Permit guidelines for oversize and overmass vehicles, Version 4*. Transport Regulation and Compliance. Northern Territory Government. Department of Transport. Australia: Northern Territory Government of Australia.

Chapter 13

Flexible pavement design methodologies

13.1 GENERAL

Pavement design consists of the determination of thickness of every layer composing the pavement, to ensure that the stresses/strains caused by the traffic loads and transmitted through the pavement to the subgrade do not exceed the supportive capacity of each layer and the subgrade.

The determination of thickness of each layer depends directly on the volume and composition of the traffic, the mechanical characteristics of each layer constructed and the subgrade and the temperature and moisture environment prevailing in each layer.

Additionally, a set of other factors should be always taken into consideration to determine the final pavement design. These factors are construction cost, ease and cost of future maintenance and service life. As a consequence, pavement design is not a merely computational procedure following a particular methodology, but a rather complicated techno-economic procedure.

Pavement design was almost unknown in the beginning of the century. The determination of usually one layer constructed over the subgrade was based on engineer's experience and assessment, since the requirements were not as demanding. After the Second World War and with the rapid growth of the automotive industry, there was an intense need for economic design and construction of pavements in road networks and airports. Thus, various design methodologies developed were based, originally, on fundamental theories of soil mechanics. These methodologies gave satisfactory results until the 1960s.

The economic crisis in the 1970s, the rapid development of highways and the appearance of early failures forced authorities and institutions of various countries to review, modify and develop new pavement design methodologies, capable of designing adequately economical pavements suited to the new requirements. A better understanding of the mechanical performance of constituent material and the development of computing contributed to this effort.

Today, flexible pavement design is carried out by modern analytical or semi-analytical methodologies. They are all based on a multi-layer structure having isotropic linear elastic behaviour and numerous research studies. The basic design criteria in all methodologies are the same: fatigue cracking of asphalt layers and excessive deformation of the subgrade.

The *analytical methodologies* enable to design flexible pavements consisting of, theoretically, any number of different asphalt and base/sub-base layers. Layer thickness determination may be carried out using any configuration of axle loading and at any environmental temperature; some methodologies have the ability to examine partial bonded or un-bonded interfaces apart from the fully bonded interfaces. The determination of the thickness of the layers is carried out by the use of appropriate software developed.

The *semi-analytical methodologies* provide limited number of solutions in terms of type of asphalt and base/sub-base layers and use only standard axle loading configuration and a fixed environmental temperature; they all have been developed assuming a fully bonded interface condition. The determination of the layer's thickness is carried out by the use of appropriate nomographs or graphs developed. The related software available in some cases is simply to ease design procedure and increase accuracy of interpretation of the result. The nomographs or graphs developed have resulted from employing theories and models similar to those in analytical methodologies and incorporating, in some cases, results obtained from full-scale experiments. Few semi-analytical methodologies provide recipe-type solutions.

In most pavement design methodologies for roads and highways, the traffic load is expressed in terms of equivalent standard axle load (ESAL). The bearing capacity or strength of the subgrade material is expressed in terms of CBR or resilient modulus, M_R, whereas that of the unbound or hydraulically bound layers is expressed in terms of modulus of elasticity and that of the asphalt layers is expressed in terms of stiffness modulus.

This chapter describes in detail some of the semi-analytical methodologies developed, such as Asphalt Institute (1999), AASHTO (1993), the British methodology (Highways Agency 2006a) and the Aristotelian University of Thessaloniki (AUTh) methodology (Nikolaides 2005). Reference is also made to various other semi-analytical and analytical methodologies developed such as the Australian methodology (Austroads 2012) and others.

With regard to analytical flexible pavement design methodologies, a brief description is given to the Shell method (Shell International 1985) and the ALIZE-LCPC (2011) method.

A brief description is also given to the recently developed *AASHTO Mechanistic–Empirical Pavement Design Guide* (MEPDG) (AASHTO 2008).

13.2 DESIGN CRITERIA FOR FLEXIBLE PAVEMENTS

The design criteria or distress modes adopted by most analytical and semi-analytical methodologies for flexible pavement design are as follows: (a) fatigue of treated layers, that is, the asphalt layer or the hydraulically bound layer not to crack under the influence of traffic; and (b) deformation of the subgrade, that is, the subgrade to be able to sustain the traffic without excessive deformation.

The first is controlled by the horizontal tensile strain, ε_r, at the underside of the asphalt base course, or hydraulically bound, layer. The second is controlled by the vertical compressive strain, ε_z, at the top of the subgrade (formation layer) (see Figure 13.1).

Repetitive tensile strain applied during the pavement's service life will initiate the first crack at the underside surface of the bottom asphalt layer (or hydraulically bound layer), which will propagate to the surface causing surface cracking. Similarly, repetitive compressive strain at the surface of the subgrade will cause subgrade deformation, which will eventually spread to the surface of the pavement, resulting in surface or structural deformation.

The expression 'repetition of loading' during the pavement's service life implies that the structural condition of the pavement deteriorates with time. In fact, the deterioration mechanism of the pavement is a fatigue phenomenon, related to the number of load repetitions and the magnitude of the exerted strain.

The number of load repetitions for failure may be determined from fatigue equations by knowing the magnitude of the exerted strain, ε. Similarly, from a fatigue equation, the allowable strain may be determined by knowing the number of load repetitions. A fatigue equation has the following mathematical expression:

$$N = c \times \varepsilon^{-m},$$

where c and m are material coefficients determined empirically.

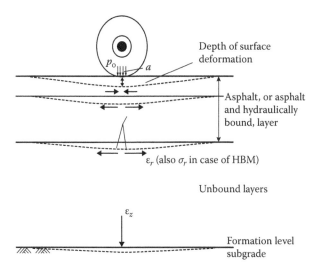

Figure 13.1 Representation of typical design criteria for flexible pavements.

Fatigue equations for *asphalts* have been developed by various organisations and research institutes and more information can be found in Section 7.7.

Fatigue equations for *hydraulically bound materials* have also been developed; however, limited information is available. As an example, the fatigue equation used by the Australia pavement design methodology (Austroads 2012) is as follows:

$$N = \text{RF}\left[\frac{11,300/E^{0.804} + 191}{\varepsilon}\right]^{12}$$

where N is the cumulative number of load repetitions of 80 kN standard axle loading (allowable number), RF is the reliability factor (e.g., for project reliability 97.5%, RF = 0.5), E is the modulus of cemented material (MPa) and ε is the strain developed, expressed in microstrain (10^{-6}).

The above fatigue equation (criterion) is valid for cemented materials with moduli within the range 2000 to 10,000 MPa.

The current British pavement design methodology (Highways Agency 2006a) for hydraulically bound mixture (HBM) uses as design criterion the predicted tensile stress, σ_r, at the underside of the hydraulic bound base, which, for heavy traffic design (>80 × 10⁶ ESAL), has to satisfy the following relationship:

$$\sigma_r \leq f_f \times K_{\text{Hyd}} \times K_{\text{safety}},$$

where f_f is the 360-day flexural strength of hydraulically bound material (MPa); K_{Hyd} is the calibration factor depending on the compressive strength of the HMB, including temperature effects, curing behaviour and transverse cracking characteristics; and K_{safety} is the factor controlling the inherent risk in pavement design.

It must also be stated that the current British pavement design, after the introduction of the foundation layer concept, no longer uses the subgrade compressive strain criterion. Details of all the above can be found in Nunn (2004) and Highways Agency (2006a).

Cracks could also initiate from the top asphalt layer and propagate down to the lower asphalt layers. These cracks are non-load related and caused by repeated tensile strains developed at the surface owing to thermal stresses (surface contraction) or, in some cases, by pavement uplift (upheave) owing to frost or swelling, or even by subgrade shrinkage.

Thermal cracking is not considered as a design criterion by any of the published design methodologies for flexible pavements. Nevertheless, the AASHTO mechanistic–empirical pavement design methodology, MEPDG (AASHTO 2008), considers thermal cracking as a prediction performance indicator of the pavement.

Similarly, the Asphalt Institute's pavement thickness design software (Asphalt Institute 2005) also takes into consideration thermal cracking but only as behaviour forecast of the bituminous mixture used.

With respect to surface cracking causing pavement uplift attributed to frost or swelling, the AASHTO methodology (AASHTO 1993) considers it indirectly in the determination and usage of loss of serviceability index in the design procedure.

As for the surface cracking attributed to the shrinkage of the subgrade material, no methodology considers it since such a material (clayey material) is not permitted to be utilised.

Regarding fatigue for the *subgrade*, causing surface structural deformation (normally less than 20 mm), various institutions and research centres have developed relevant fatigue equations. Some of the most widely used fatigue equations for the prediction of structural surface deformation, subgrade design criterion, are as follows:

Asphalt Institute (Asphalt Institute 1990):

$$\varepsilon_z = 1.047 \times 10^{-2} \times N^{-0.2234} \text{ or } N = 1.365 \times 10^{-9} \times \varepsilon_z^{-4.477}$$
(for surface deformation from 13 to 19 mm)

Shell International Petroleum (Shell 1985):

$$\varepsilon_z = 2.1 \times 10^{-2} \times N^{-0.25} \text{ or } N = 1.945 \times 10^{-7} \times \varepsilon_z^{-4.0} \text{ (for 85\% reliability)}$$

$$\varepsilon_z = 1.8 \times 10^{-2} \times N^{-0.25} \text{ or } N = 1.0498 \times 10^{-7} \times \varepsilon_z^{-4.0} \text{ (for 95\% reliability)}$$

Nottingham (Brown 1980):

$$\varepsilon_z = 2.16 \times 10^{-2} \times N^{-0.28} \text{ or } N = 1.1324 \times 10^{-6} \times \varepsilon_z^{-3.57}$$

AASHTO (AASHTO 1993):

$$\varepsilon_z = 2.80 \times 10^{-2} \times N^{-0.25} \text{ or } N = 0.6147 \times 10^{-6} \times \varepsilon_z^{-4.0} \text{ (for PSI = 2.5)}$$

TRRL (Powell et al. 1984):

$$\varepsilon_z = 1.495 \times 10^{-2} \times N^{-0.2535} \text{ or } N = 6.1659 \times 10^{-8} \times \varepsilon_z^{-3.95}$$
(for 85% possibility of survival to design life)

Austroads (Austroads 2012):

$$\varepsilon_z = 0.93 \times 10^{-2} \times N^{-0.1428} \text{ or } N = 6.017 \times 10^{-15} \times \varepsilon_z^{-7}$$
(for 90% confidence limit)

where, in all cases, ε_z is the vertical strain on the subgrade, induced by 80 kN standard axle loading, and N is the cumulative number of load repetitions of 80 kN standard axle loading (allowable number).

Some analytical methodologies (AASHTO 2008; Asphalt Institute 2005; Shell 1985) consider surface pavement deformation owing to deformation of asphalt layers (rutting). However, the estimated rut depth is taken as a performance indicator of the bituminous mixtures used. In case of excessive estimated rut depth, it is recommended to alter the composition of the bituminous mixture to avoid excessive rutting during the pavement's service life.

The deformation of unbound layers in almost all methodologies, apart from the MEPDG (AASHTO 2008), is not taken into consideration either as a design criterion or as a layer performance indicator.

In conclusion, the aim of the abovementioned design criteria is to design a pavement that reduces the stresses exerted on the subgrade to such a magnitude ensuring that there will be only limited surface deformation at the end of the design life of the pavement. Additionally, the aim is to determine the thickness of bound layers so that the developed tensile strain at the underside of the bottom bitumen bound or hydraulically bound layer is less than or equal to the permissible value for fatigue cracking.

13.3 ASPHALT INSTITUTE PAVEMENT DESIGN METHODOLOGY

The current semi-analytical flexible pavement design methodology of the Asphalt Institute (1999) was originally published in 1981 and is based on the application of elastic theory in a multi-layer system. The design criteria considered are cracking of the asphalt layer owing to horizontal tensile strain and deformation of the subgrade owing to vertical compressive strain.

The methodology can be used to design flexible pavements composed of the following: (a) asphalt layers and unbound layers; (b) asphalt layers, exclusively (full-depth pavement); (c) asphalt layer and cold asphalt layers; and (d) asphalt layer, cold asphalt layer and unbound layer.

The thickness of the individual layer is determined from appropriate design charts, developed with the aid of the multi-layered elastic analysis software called DAMA.

The steps in the design procedure are illustrated in the flow diagram shown in Figure 13.2.

This mechanistic/empirical flexible pavement methodology of the Asphalt Institute for highways and streets, together with pavement methodology for airport pavements and for industrial facilities supporting heavy wheel loads, is provided in a computer program known as SW-1 Asphalt Institute software (Asphalt Institute 2005).

13.3.1 Determination of cumulative ESAL

The cumulative number of ESALs (18 kps) over the design period is determined once the following are estimated: (a) average annual traffic composition during the first year of opening, in terms of axle weight composition or vehicle-type composition; (b) percentage of truck traffic in the design lane; and (c) percentage of annual growth rate for trucks.

The traffic composition may be estimated from past surveys or representative area traffic data or may be determined from new counts. The conversion to ESAL may be carried out by using equivalency factors or truck factors provided. The equivalency factors provided by the methodology are those given in Table 12.3.

The percentage of trucks in the design lane depends on the total number of traffic lanes, considering the percentage of traffic in both directions. For the determination of traffic in the design lane, if no other information is available, the methodology proposes to use values ranging between 50%, for sections with one lane in each direction, and 25%, for sections with three or more lanes in each direction.

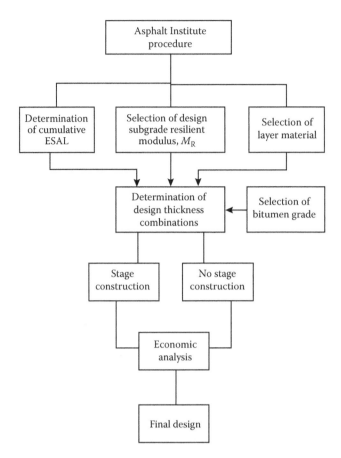

Figure 13.2 Flow diagram for Asphalt Institute flexible pavement design methodology procedure. (From Asphalt Institute, MS-1, *The Thickness Design, Asphalt Pavements for Highways & Streets*, Manual Series No. 1 [MS-1], 9th Edition, Lexington, USA: Asphalt Institute, Reprinted 1999. With permission.)

The percentage of annual growth is estimated and should be representative for the whole design period. Former traffic studies in the project area, in conjunction with the national annual growth in urban or rural regions, help predict more accurately this growth.

The cumulative number of ESALs over the design period is determined with the use of the equation given in Section 13.5.1 and the value determined multiplied by the estimated percentage of trucks in design lane.

13.3.2 Selection of subgrade resilient modulus

The selection of the resilient modulus, M_R, of the subgrade, the other necessary data for the determination of layer thickness, is carried out after representative sampling and laboratory testing.

Sampling and testing of soil materials are carried out according to established procedures. It is recommended to test all subgrade material within 600 mm of the planned elevation and in embankment areas to test the expected source of the fill material. Apart from the resilient modulus test (AASHTO T 307 2007; Asphalt Institute 1997), all other tests required for subgrade material determination also need to be carried out (liquid and plastic limit, plasticity index, sieve analysis, etc.).

If a sample from a location gives low M_R, indicating an extremely weak soil, it is recommended to take and test additional samples to determine the boundaries of the area with weak soil. Such areas require replacement with improved subgrade material.

The design M_R value is not the average of the values determined but the value that is lower than a certain percentile of all test values. This percentile depends on the level of traffic volume and is taken as 60% for <10^4 ESAL, 75% for 10^4–10^6 ESAL or 87.5% for >10^6 ESAL.

The procedure for selecting the design M_R value that corresponds to a certain percentile is explained by an example in Asphalt Institute (1999).

With the procedure adopted, effectively a lower design M_R value results for sites with high traffic volume in relation to sites with low traffic volume. This results in a more conservative design for the high traffic volume situation, ensuring in a way the expected performance of the pavement during the design period.

To facilitate the use of the design charts, the resilient modulus may be approximated from CBR or R test values, using the correlations given in Section 1.9.2.

The procedure also sets compaction requirements. For *cohesive material*, the compaction for the top 300 mm layer should be at least 95% of the laboratory maximum density determined (ASTM D 1557 2012 or AASHTO T 180 2010), and for the remaining layer, it should be below at least 90%. For *non-cohesive material*, compaction for the top 300 mm layer should be at least 100% of the laboratory maximum density and at least 95% of the laboratory maximum for the layer below.

13.3.3 Selection of layer materials

13.3.3.1 Unbound aggregate for base and sub-base

The aggregate material for unbound or untreated base or sub-base layers should be in accordance to ASTM D 2940 (2009) and should meet the requirements of Table 13.1.

The compaction base and sub-base should be conducted at optimum moisture content, with a permissible deviation of ±1.5%, to achieve a minimum density of 100% of the maximum density determined in the laboratory, in accordance to ASTM D 1557 (2012) or AASHTO T 180 (2010). At least three field density tests should be conducted on each 2700 t of material.

Table 13.1 Unbound aggregate base and sub-base quality requirements

Test	Test requirements	
	Sub-base	Base
CBR, minimum[a]	20	80
R value, minimum[a]	55	78
Liquid limit, maximum	25	25
Plasticity index, maximum	6	NP
Sand equivalent, minimum	25	35
Passing through No. 200 sieve, maximum	12	7

Source: Asphalt Institute, MS-1, *The Thickness Design, Asphalt Pavements for Highways & Streets*, Manual Series No. 1 (MS-1), 9th Edition, Lexington, USA: Asphalt Institute, Reprinted 1999. With permission.

[a] The formulae relating CBR and R value to subgrade resilient modulus do not apply to unbound aggregate base and sub-base.

13.3.3.2 Asphalts for surface and base mixtures

The asphalts used for surface, binder and asphalt base are asphalt concrete and cold asphalt.

13.3.3.2.1 Asphalt concrete

The asphalt concrete is designed in accordance to the Marshall (Asphalt Institute 6th Edition) or Superpave design (Asphalt Institute 2001).

Emphasis is given to the proper compaction of the asphalt layers. In core sampling or nuclear determination, the average of five density determinations should be equal to or greater than 96% of the average density of laboratory-prepared specimens, and no individual specimen should be lower than 94%. When compacted density is compared to the theoretical maximum specific gravity, the above percentage values become 92% and 90%, respectively.

13.3.3.2.2 Cold asphalt mixtures – emulsified asphalt base mixtures

The cold asphalts (emulsified asphalt mixtures) recommended to be used are divided into three characteristic types:

1. Type I, cold mixtures made with processed and dense-graded aggregates
2. Type II, cold mixtures made with semi-processed crusher-run, pit-run or bank-run aggregates
3. Type III, cold mixtures made of sands and silty sands

The cold asphalts and the binder, which, in all cases, is bitumen emulsion, should be designed based on and meet the requirements of the asphalt emulsion manual (Asphalt Institute 2008).

Regarding compaction requirements, the average of five dry density field measurements is compared to the target dry density determined in the laboratory. The average value obtained should be ≥95% and no individual determination should be <92%. More details regarding field and laboratory dry densities are given in the asphalt emulsion manual (Asphalt Institute 2008).

13.3.4 Selection of bitumen grade

The bitumen grade is selected according to mean annual air temperatures (MAATs). Three different temperature conditions are distinguished: cold (≤7°C), warm (7°C to 24°C) and hot (≥24°C). The bitumen grades recommended are 120/150 pen or 85/100 pen for cold conditions, 85/100 pen or 60/70 pen for warm conditions and 60/70 pen or 40/50 pen for hot conditions.

13.3.5 Thickness determination (procedure)

Once the traffic volume (in cumulative ESAL), the representative value of the resilient modulus M_R of the subgrade and the materials to be used have been determined, the thickness of each layer is determined from appropriate nomographs. The nomographs developed are for cold, warm and hot temperature conditions, noted as for 7°C, 15.5°C and 24°C, respectively, and for two alternative thickness of unbound base course, 150 or 300 mm.

13.3.5.1 Thickness determination for unbound flexible pavement

The thickness determination of a flexible pavement with asphalt concrete layers over untreated aggregate base/sub-base is obtained using the appropriate nomograph. A sample of those nomographs, for 15.5°C MAAT and for 300 mm thickness of base/sub-base, is given in Figure 13.3.

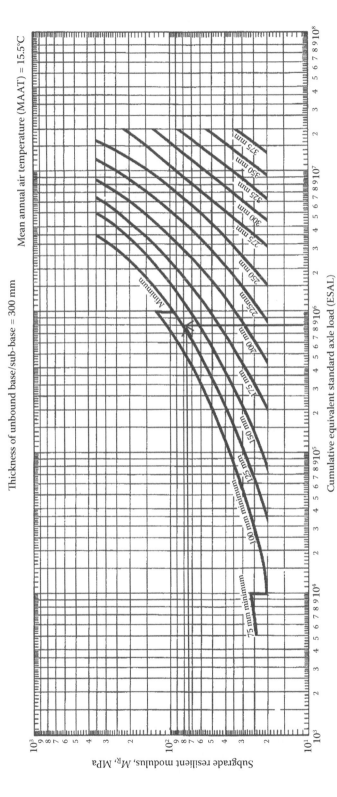

Figure 13.3 Nomograph for thickness determination of pavements with asphalt concrete over 300 mm untreated aggregate base. (From Asphalt Institute, MS-1, *The Thickness Design, Asphalt Pavements for Highways & Streets*, Manual Series No. 1 [MS-1], 9th Edition, Lexington, USA: Asphalt Institute, Reprinted 1999. With permission.)

Table 13.2 Minimum thickness of asphalt concrete over untreated aggregate base/sub-base

Traffic level, ESALs	Traffic condition	Minimum thickness of asphalt concrete
<10^4	Light traffic, parking lots, driveways and light rural roads	75 mm[a]
10^4–10^6	Medium truck traffic	100 mm
>10^6	Heavy truck traffic	>125 mm

Source: Asphalt Institute, MS-1, *The Thickness Design, Asphalt Pavements for Highways & Streets,* Manual Series No. 1 (MS-1), 9th Edition, Lexington, USA: Asphalt Institute, Reprinted 1999. With permission.

[a] For full-depth asphalt concrete or emulsified asphalt pavements, a minimum of 100 mm applies in this traffic region.

The methodology requires minimum thickness of asphalt layers depending on the level of traffic. These minimum requirements are as shown in Table 13.2.

13.3.5.2 Thickness determination for full-depth pavement

The determination of the thickness of a full-depth asphalt concrete pavement is carried out from a similar to a flexible pavement with unbound layer nomographs. A sample of the nomographs used for 15.5°C MAAT is shown in Figure 13.4.

The resulting total pavement thickness is, in all cases, less than that of any alternative flexible-type pavement.

The first asphalt layer is laid directly over the subgrade. The only requirement by methodology is the provision of a sub-base layer in case construction equipment experience difficulties in moving over the subgrade material. The author proposes to consider full-depth pavement only when the subgrade resilient modulus is greater then 100 MPa (approximate CRB value, 10).

13.3.5.3 Thickness determination for pavement with cold asphalt

For the determination of the thickness of the pavement consisting of cold asphalt, there are two alternatives. One is for all the layers to be constructed with cold asphalts, and the other is for the cold asphalt to be laid over an unbound aggregate layer.

In the first case, the total thickness of the cold asphalt layers is determined from relevant nomographs, a sample of which, for Type I cold asphalt and MAAT = 15.5°C, is given in Figure 13.5. Similar nomographs are provided for Type II and Type III cold asphalts and for temperatures 7°C, 15.5°C and 24°C.

When cold asphalt Type I is used, a surface treatment is required. When cold asphalt Type II or III is used, the top layer is required to be from hot asphalt concrete. The minimum required thickness of asphalt concrete is determined by the traffic volume and is as shown in Table 13.3. However, it is common to replace the top 50 mm with hot asphalt concrete even if Type I mix is used.

In the second case where the cold asphalt will be laid over an unbound aggregate layer, the thickness of the cold asphalt layer is determined as follows:

1. Determine the thickness of the asphalt layers, T_A, such as in a full-depth pavement (use Figure 13.4 or a similar nomograph). Assuming that the top asphalt concrete layers will have a certain thickness, say 50 mm, the thickness of the remaining asphalt concrete layers is (T_A – 50).
2. Determine the thickness of cold asphalt layers, T_{CA}, as if laid over the subgrade (use Figure 13.5 or a similar nomograph). Considering that the top 50 mm is going to be from asphalt concrete, the thickness of cold asphalt layers will be T_{CA} – 50.

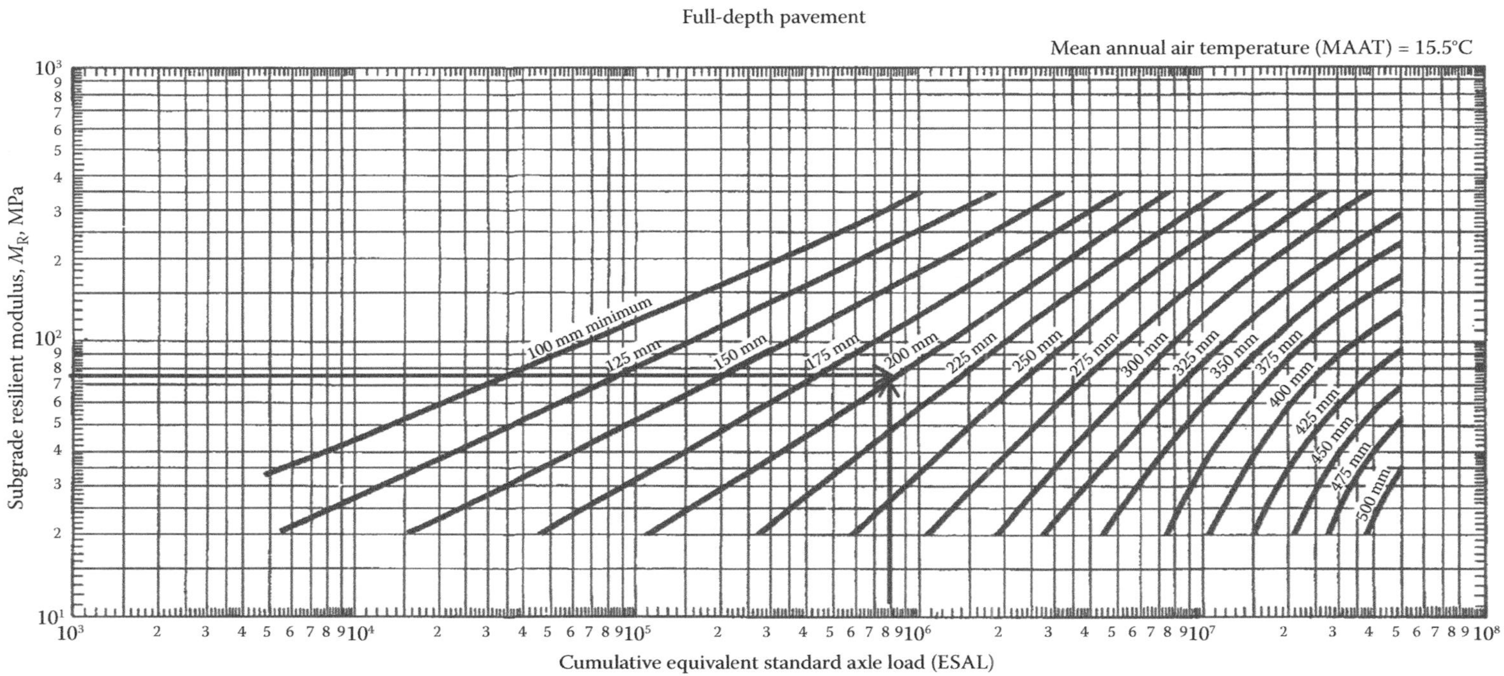

Figure 13.4 Nomograph for thickness determination of full-depth asphalt concrete pavements. (From Asphalt Institute, MS-1, *The Thickness Design, Asphalt Pavements for Highways & Streets*, Manual Series No. 1 [MS-1], 9th Edition, Lexington, USA: Asphalt Institute, Reprinted 1999. With permission.)

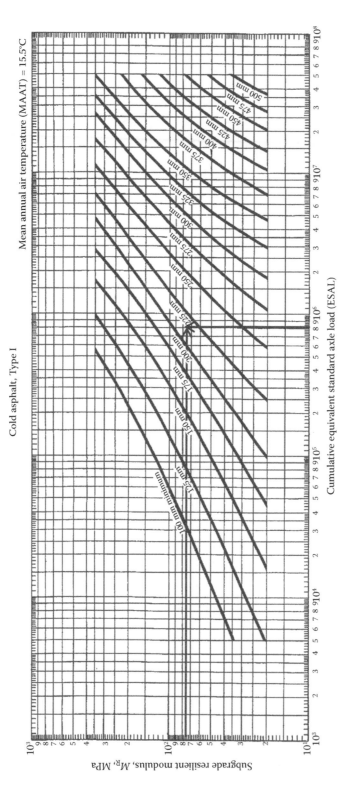

Figure 13.5 Nomograph for thickness determination of pavements with Type I cold asphalt in all layers. (From Asphalt Institute, MS-1, *The Thickness Design, Asphalt Pavements for Highways & Streets*, Manual Series No. 1 [MS-1], 9th Edition, Lexington, USA: Asphalt Institute, Reprinted 1999. With permission.)

Table 13.3 Minimum thickness of asphalt concrete over cold
asphalt, Type II or Type III

Traffic level, ESALs	Minimum thickness of asphalt concrete (mm)
10^4	50
10^5	50
10^6	75
10^7	100
$>10^7$	130

Source: Asphalt Institute, MS-1, *The Thickness Design, Asphalt Pavements for Highways & Streets*, Manual Series No. 1 (MS-1), 9th Edition, Lexington, USA: Asphalt Institute, Reprinted 1999. With permission.

3. Determine the thickness of the asphalt concrete layers over unbound aggregate base/sub-base material, T_U, of 150 or 300 mm in thickness (use Figure 13.3 or a similar nomograph).
4. Determine the thickness of hot asphalt concrete to be replaced by cold asphalt. This is equal to $T_U - T_{U\,min}$. The thickness of $T_{U\,min}$ is 50 mm or a value taken from Table 13.6, in case cold asphalt Type II or Type III is used.
5. Determine the thickness of the layer that will be replaced by emulsified asphalt ($T_{CA\text{-base}}$) using the following equation:

$$T_{CA\text{-base}} = (T_U - T_{U\,min}) \times \left(\frac{T_{CA} - 50}{T_A - 50} \right).$$

Hence, the flexible pavement with cold base and unbound aggregate base/sub-base will consist of top 50 mm asphalt concrete, a layer of cold asphalt of thickness $T_{CA\text{-base}}$ and a base/sub-base of thickness 150 or 300 mm, as selected.

13.3.6 Planned stage construction

Planned stage construction is the construction where the asphalt layers are constructed at stages according to a predetermined time schedule.

A key prerequisite for a planned stage construction is that all subsequent stages (usually one plus one) should be applied according to schedule and before pavement fatigue signs appear.

The planned stage construction may be preferred in cases where the entire fund is not available for the construction of the full design thickness or there are difficulties in estimating traffic for a long design period (say 20 or 30 years).

It should be highlighted that when planned stage construction is decided, regular traffic counts during the first period after construction to re-adjust, if needed, the time scheduled for the second stage of construction.

13.3.6.1 Design method

According to Asphalt Institute (1999), the design method involves three stages: (1) first-stage design, (2) preliminary design of the second-stage overlay and (3) final design of the second-stage overlay.

The *first-stage* design method is based on the remaining life concept (Asphalt Institute 1999). In this concept, the first stage is for a design period less than that which would produce fatigue failure. To use this concept, the estimated first stage design ESAL, for the first stage design period, is adjusted by a factor of 1.67 (100/60). Then, first stage design traffic, $ESAL_1$, is 1.67 × $ESAL_1$. The pavement is constructed at this thickness, where the thickness of the asphalt layers is h_1.

In the *preliminary second-stage* overlay design, the remaining design $ESAL_2$ is multiplied by a factor of 2.5 (100/40) and the thickness of the asphalt layers, h_2, is determined. The thickness of the overlay to be constructed at the second stage of construction, h_s, is equal to $(h_2 - h_1)$.

The *final design of the second-stage overlay* is carried out normally 1 year before the end of the first-stage period utilising the traffic counts data and conducting a condition survey. Depending on the pavement condition, alteration of the preliminary second design or of the period of construction of the second stage may be needed. Otherwise, preliminary second-stage design is implemented. A working example can be found in Asphalt Institute (1999).

13.3.7 Economic analysis

An economic analysis may be carried out before the final design decision is taken, by making economic comparisons between alternative pavement designs.

It incorporates the concept of present worth for evaluating future expenditure and the procedure of life cycle cost analysis (LCCA). The basic factors required for LCCA are as follows: (a) initial cost of pavement structure; (b) cost of future overlays, major maintenance or reconstruction, or other interventions; (c) time, in years, from initial construction up to each intervention; (d) salvage value of the structure at the end of the analysis period; (e) interest rate; and (f) determination of the analysis period.

Details for LCCA using the concept of present worth value can be found in reference (Asphalt Institute 1999) or elsewhere.

13.4 AASHTO PAVEMENT DESIGN METHODOLOGY

13.4.1 General

The current AASHTO pavement design methodology (AASHTO 1993) is based on the methodology developed in 1972 (AASHTO 1972), as an output of the AASHTO road test (HRB 1962), which, in turn, was revised in 1981 and 1986.

In contrast to all other design methodologies for flexible pavements, the AASHTO methodology uses a pavement performance criterion, which, apart from cracking and subgrade deformation, includes other parameters affecting the performance of the pavement. This criterion is called serviceability, expressed as present serviceability index (PSI).

Thus, the pavement is designed in such a way that its deterioration, manifested as cracking, deformation, surface irregularities, potholes and so on, until the end of its service life, results in the pavement offering a minimum tolerable level of serviceability.

The above constitutes a totally different design concept, somewhat more practical, provided the user is not interested in whether the pavement has been cracked or deformed, or whether it has one or two patches. He is interested in whether or not the pavement provides some tolerable serviceability level.

The subjectivity included in the definition of 'tolerable serviceability level' was overridden by defining various values for the PSI.

The design requirements in this methodology for flexible pavement structural design can be generally classified into four categories.

The first category includes *design variables – data*, such as the following: traffic volume, design life, reliability of data/results and environmental effects (temperature and moisture).

Pavement performance criteria related to the user belong to the second category. These criteria are expressed in terms of initial PSI, p_o and terminal PSI, p_t.

Properties of the materials and layers, expressed in resilient modulus or subgrade, M_R or moduli for sub-base, E_{SB}, base, E_{BS} and asphalt layers constructed from asphalt concrete, E_{AC}, belong to the third category. The structural performance of each layer is expressed in layer coefficients (a_1, a_2, etc.).

Finally, *pavement structural characteristics* and, particularly for flexible pavements, drainage belong to the fourth category.

The AASHTO design procedure for a new flexible pavement is illustrated in the flow chart shown in Figure 13.6.

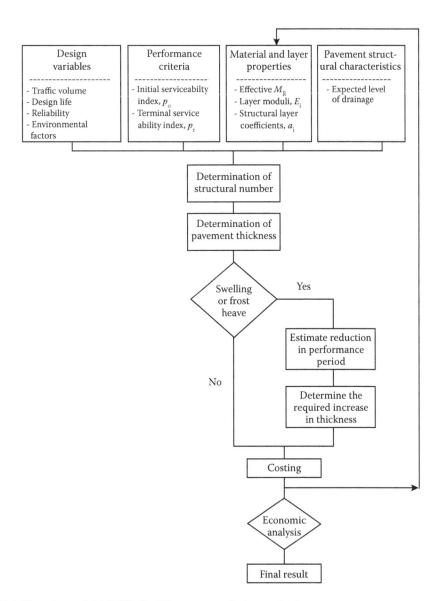

Figure 13.6 Flow chart of AASHTO flexible pavement design method.

13.4.2 Design variables

13.4.2.1 Traffic volume

Traffic volume is expressed in ESALs. The conversion of mixed axial loads to ESAL is conducted with the use of equivalency coefficients related to the terminal serviceability value, p_t, and the structural number, SN, of the pavement. For the design of new flexible pavements, it is suggested that equivalency coefficients corresponding to $p_t = 2.5$ and SN = 5 be used (see Table 12.3).

To calculate the cumulative number of ESALs over a design period of n years, the equations given in Section 13.5.1 are used.

For the determination of traffic in the design lane (design traffic), $W_{18\text{-des}}$, if the cumulative ESAL in both directions (e.g. North–South, East–West, etc.) is known, the following equation is used:

$$W_{18\text{-des}} = D_D \times D_L \times w_{18},$$

where $W_{18\text{-des}}$ is the cumulative ESAL (18,000 lb) in the design lane; D_D is the directional distribution factor, expressed as a ratio (percentage) that accounts for distribution of ESAL by direction; D_L is the lane distribution factor, expressed as a ratio (percentage) that accounts for distribution of traffic when two or more lanes are available per direction; and w_{18} is the cumulative two-directional ESAL during the analysis period.

The directional distribution factor, D_D, is usually 0.5 (50%). However, there are instances where more traffic is moving in one direction than the other. If this is the case, the pavement is designed with a greater number of ESAL.

As for the lane distribution factor, D_L, if no other information is available, the methodology proposes to use values ranging between 100%, for sections with one lane in each direction, and 50%, for sections with four lanes in each direction.

13.4.2.2 Design life

The design life is distinguished by the AASHTO methodology into 'analysis period' and 'performance period' (in years).

The *analysis period* refers to the period for which the analysis is to be conducted and is analogous to the term *design life*. During the analysis period, surface maintenance, or some kind of rehabilitation or surfacing, is unavoidable. The methodology recommends considering the analysis period for high-volume roads longer than low-volume roads. The guidelines are given in AASHTO (1993).

The *performance period* refers to how long a new pavement structure will last before it needs rehabilitation. The performance period is distinguished into minimum and maximum periods.

The minimum performance period is the shortest amount of time an initial pavement structure, or stage construction, lasts before some intervention is performed.

The maximum performance period is the maximum practical amount of time expected from a given pavement structure or stage construction. Theoretically, the maximum performance period should be equal to the analysis period, but in practice, this rarely happens (taking into consideration the effect of environmental factors, surface deterioration, etc.).

13.4.2.3 Reliability

Reliability is a means of incorporating some degree of certainty into the design process to ensure that the design or various alternatives will last throughout the analysis period.

The reliability design factor accounts for change in traffic prediction, $W_{18\text{-des}}$, and is associated with the functional classification of the project (road category). High reliability levels, R, are used for highways and low reliability levels are used for local roads. The greater the value of reliability, the more pavement structures required. The recommended range of values is given in AASHTO (1993).

For a given reliability level, R, the reliability factor is a function of the overall standard deviation, S_o, which accounts for both change variation in the traffic prediction and normal variation in pavement performance prediction for a given design traffic. The standard deviation selected should be representative of the local conditions.

In the AASHTO road test, it was found that the resulting standard deviation, S_o, was 0.35 and 0.45 for rigid and flexible pavements, respectively.

13.4.2.4 Environmental factors

The environmental factors that are considered by the methodology to affect the performance of new pavement are the *subgrade swelling* and *frost heave*. Either of these factors, or both, can lead to a significant loss of serviceability or ride quality during the analysis period.

This loss is expressed in terms of the PSI index. The swelling loss, ΔPSI_{SW}, or the frost heave loss, ΔPSI_{FH}, can be quantified using relevant serviceability models and calculation procedures given in the methodology (AASHTO 1993).

The objective of this step is to produce a graph of serviceability loss versus time. The serviceability loss owing to environmental factors is added to that resulting from the cumulative axle loads (see Figure 13.7). In general, if this loss is significant, measures are taken

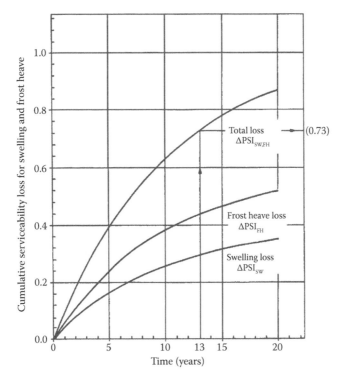

Figure 13.7 Conceptual example of pavement serviceability loss attributed to swelling and frost heave. (From AASHTO, *AASHTO Guide for Design of Pavement Structures*, Washington, DC: American Association of State Highway and Transportation Officials, 1993. With permission.)

to reduce it by replacement of the subgrade material with non-expansive material or replacement of the frost-susceptible material with non-susceptible materials to a depth more than the estimated frost depth. More details can be found in Parts I and II and Appendix G of the AASHTO pavement design guide (AASHTO 1993).

13.4.3 Pavement performance criteria

The pavement performance criterion, which is taken into account in this methodology, is the pavement serviceability level.

Pavement serviceability is defined as the ability of the pavement to serve the user/traffic. This is expressed with the PSI, which obtains values ranging from 0 (impossible road) to 5 (perfect road). Since an impossible road cannot be travelled on by a user and a perfect road is almost impossible to be constructed, the terms terminal serviceability index (p_t) and initial serviceability index (p_o) have been introduced.

The *terminal serviceability index* (p_t) expresses the lowest tolerable serviceability level before pavement resurfacing, rehabilitation or even reconstruction becomes necessary. The terminal index value is related to everything that is characterised as tolerable level by the user, depending on the road significance. AASHTO suggests a p_t value equal to 2.5 or higher for highways or main road arteries and 2.0 for all other cases.

The *initial serviceability index* (p_o) expresses the initial serviceability index when pavement opens to traffic. The maximum value of this index is 5.0, but given that is impossible to have an absolutely perfect construction, a lower value should always be used. The p_o values observed at the AASHTO road test were 4.2 for flexible pavements and 4.5 for rigid pavements.

Having established the p_o and p_t values, the change in PSI (ΔPSI) is determined by the following equation:

$$\Delta\mathrm{PSI} = p_o - p_t,$$

where p_o is the initial serviceability index and p_t is the terminal serviceability index.

13.4.4 Material and layer properties

13.4.4.1 Subgrade material – effective resilient modulus

The subgrade material is characterised by the elastic or resilient modulus, M_R, determined by AASHTO T 307 (2007) (also see Section 1.9.1). This property changes with seasonal variation of moisture content. The seasonal resilient moduli must be determined and from that an effective subgrade resilient modulus must be established.

The AASHTO (1993) pavement design methodology describes a procedure on how to determine the effective resilient modulus. In brief, this consists of determining the relative damage factor, u_f, for each M_R value determined in the laboratory by simulating seasonal moisture conditions. The relative damage factor, u_f, is determined using the following equation:

$$u_f = 1.18 \times 10^8 \times M_R^{-2.32}.$$

Having determined the relative damaging factor for each M_R value, the average value of u_f is calculated and from that the effective resilient modulus, using the above equation.

13.4.4.2 Pavement layer materials – layer moduli

The materials with which the sub-base and base course layers constructed are characterised also by elastic or resilient modulus (AASHTO T 307 2007), using different designations: E_{SB} and E_{BS}, for sub-base and base course, respectively.

The bound or higher stiffness materials such as stabilised bases and asphalt concrete are characterised by the elastic modulus, E. The elastic modulus, E, of the cement-treated aggregates may be determined according to ASTM C 469 (2010) or, alternatively, from the compressive strength, f'_c (AASHTO T 22 2011 or ASTM C 39 2012), and using the correlation equation

$$E = 57,000 \times \left(f'_c\right)^{0.5}.$$

As for the asphalt concrete, the elastic modulus, E, also known nowadays as M_R (instantaneous or total resilient modulus) is determined according to ASTM D 7369 (2011) (see also Section 7.4.8).

13.4.4.3 Structural layer coefficients

The structural layer coefficients (a_i) are necessary to convert actual layer i thickness (D_i) into layer i structural number (SN_i) using the following empirical relationship:

$$SN_i = a_i \times D_i.$$

The *structural layer coefficients* (a_i) depend on the type and function of layer material. These are asphalt concrete, granular base, granular sub-base, cement-treated and bituminous base. In order to estimate the structural layer coefficients, different charts have been developed. A sample of charts for asphalt concrete and granular base is shown in Figures 13.8 and 13.9. For other materials such as lime, lime fly ash and cement fly ash, the methodology suggests each agency to develop relevant charts.

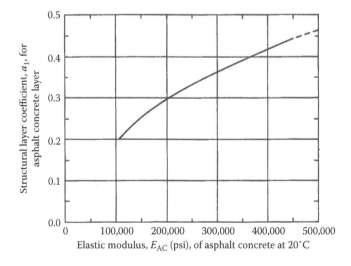

Figure 13.8 Chart for estimating structural layer coefficient of dense-graded asphalt concrete (a_1) based on the elastic (resilient) modulus. (From AASHTO, *AASHTO Guide for Design of Pavement Structures*, Washington, DC: American Association of State Highway and Transportation Officials, 1993. With permission.)

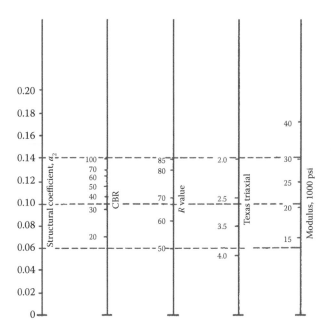

Figure 13.9 Determination and variation of granular base layer coefficient (a_2) with various base strength parameters. (From AASHTO, *AASHTO Guide for Design of Pavement Structures*, Washington, DC: American Association of State Highway and Transportation Officials, 1993. With permission.)

For the determination of the asphalt concrete structural coefficient, the methodology recommends the elastic (resilient) modulus (E_{AC}) at 20°C to be less than 450,000 psi (3100 MPa), since asphalt concretes with higher elastic modulus values are more susceptible to thermal and fatigue cracking.

As it can be seen, the structural layer coefficients for the granular base or sub-base layers may be derived from different laboratory tests including resilient modulus (E_{BS} or E_{SB}). Similarly, the structural layer coefficients for the cement-treated bases or bituminous-treated bases may be derived from unconfined compressive strength or Marshall stability, respectively, including elastic modulus.

13.4.5 Pavement structural characteristics

The methodology considers the effects of certain levels of drainage on predicted pavement performance. The drainage level (quality) of construction affects the behaviour of all layers apart from asphalt layers. There are five drainage levels (quality) considered: excellent (water removed within 5 h), good (water removed within 1 day), fair (water removed within 1 week), poor (water removed within 1 month) and very poor (water will not drain). For each drainage quality, drainage coefficients, m_i, have been proposed that are intergraded into the structural number (SN) equation for a three-layered pavement:

$$\text{SN} = a_1 D_1 + a_2 D_2 m_2 + a_3 D_3 m_3,$$

where a_1, a_2 and a_3 are structural coefficients of asphalt concrete, sub-base and base layer, respectively; D_1, D_2 and D_3 are the thickness (in inches) of asphalt concrete, sub-base and base layer, respectively; and m_2, m_3 are drainage coefficients.

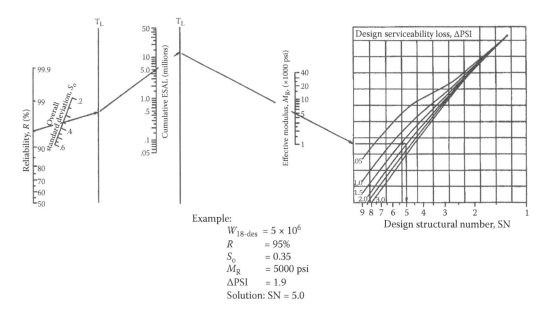

Example:
$W_{18\text{-des}} = 5 \times 10^6$
$R = 95\%$
$S_o = 0.35$
$M_R = 5000$ psi
$\Delta PSI = 1.9$
Solution: SN = 5.0

Figure 13.10 Design chart for flexible pavement based on using mean values for each input parameter. (From AASHTO, *AASHTO Guide for Design of Pavement Structures*, Washington, DC: American Association of State Highway and Transportation Officials, 1993. With permission.)

Details of the drainage coefficients for the five drainage levels considered are given in AASHTO (1993).

The estimation of the proper drainage coefficient (m_i) is subjective and depends on the judgment and experience of the designer. The value of the coefficient $m_i = 1.00$ is taken when drainage quality is fair and pavement is exposed to moisture levels approaching saturation, that is, 5% of its service life.

13.4.6 Determination of structural number

The determination of the design structural number (SN) for specific conditions including cumulative traffic over design or performance period $(W_{18\text{-des}})$, reliability (R), overall standard deviation (S_o), effective modulus (M_R) and design serviceability loss $(\Delta PSI = p_o - p_t)$, when the pavement carries more than 50,000 ESALs of 18,000 lb, is carried out by using the nomograph given in Figure 13.10.

The equation from which the nomograph has been derived is given in AASHTO (1993).

13.4.7 Determination of pavement thickness

The determination of the thickness of a new pavement and effectively the thickness of each layer may be carried out by two alternative ways: (a) by selection of layer thickness or (b) by layered design analysis.

13.4.7.1 Selection of layer thickness

Once the structural number of the pavement (SN) is determined, using the equation given in Section 13.4.5, the thickness of each layer (D_i) is determined by selecting layer thicknesses so that the right-hand side of the equation is equal to SN. This implies that this method does not have a single and unique solution. For a cost-effective solution and to avoid impractical design, minimum thicknesses for asphalt concrete and aggregate base layer are suggested by AASHTO (1993).

13.4.7.2 Layered design analysis

In this case, the pavement structure is considered as a layered system and is designed accordingly. The structure is designed according to the principle shown in Figure 13.11. First, the structural number required over the subgrade is computed from the nomograph given in Figure 13.10 using the strength (equivalent modulus, M_R) of the subgrade. In the same way, the structural number required over the sub-base layer and base layer is also computed, using the applicable strength values (M_R values).

By working the differences between the computed structural numbers required over each layer, the maximum allowable thickness of any given layer can be computed. For example, the maximum allowable structural number of the sub-base material would be equal to the structural number required over the sub-base subtracted from the structural number required over the subgrade (roadbed soil). In a similar manner, the structural numbers of other layers may be computed.

The procedure normally adopted is as follows:

The minimum thickness of the first layer (bituminous layers) is determined by the following equation:

$$D_1 \geq \frac{SN_1}{a_1}.$$

In case the thickness of the first layer is rounded up to the nearest 1/2 inch, say D_1^*, then the structural number is modified as

$$SN_1^* = D_1^* \times a_1 \text{ and } SN_1^* > SN_1.$$

The thickness of the second layer (base course) is determined by the following equation:

$$D_2 \geq \frac{SN_2 - SN_1^*}{a_2 \times m_2}.$$

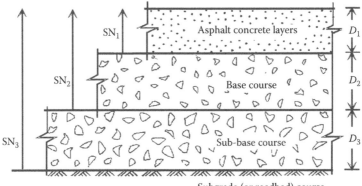

Figure 13.11 Determination of structural number in a multiple-layer system. (From AASHTO, *AASHTO Guide for Design of Pavement Structures*, Washington, DC: American Association of State Highway and Transportation Officials, 1993. With permission.)

In case the thickness of the second layer is rounded up to the nearest 1/2 inch, say D_2^*, then the structural number is modified as

$$SN_2^* = D_2^* \times a_2 \times m_2 \text{ and } (SN_1^* + SN_2^*) > SN_2.$$

Thus, the third layer thickness (sub-base) is determined by the following equation:

$$D_3 \geq \frac{SN_3 - (SN_1^* + SN_2^*)}{a_3 \times m_3}.$$

It is stated that this procedure is not applicable to determine SN values of sub-base or base materials having a modulus greater than 40,000 psi. For such cases, layer thickness of materials above a 'high' modulus layer should be established on the basis of cost-effectiveness and minimum practical thickness considerations (AASHTO 1993).

13.4.8 Swelling or frost heave

The above design procedures define the layer thicknesses of the pavement without taking into account the negative effect of swelling or frost heave. If one or both conditions exist, the calculated layer thicknesses provide a pavement that will not last as many years as anticipated (design life). In this case, the pavement is redesigned considering reduction in performance owing to swelling or frost heave. The procedure is described in detail in AASHTO (1993).

13.4.9 Costing and economic analysis

The methodology recommends to cost the pavement structure determined and also to carry out an economic analysis among alternative pavements using different materials. Stage construction may also be selected. For more information, see AASHTO (1993).

13.5 UK FLEXIBLE PAVEMENT DESIGN METHODOLOGY

The UK pavement design methodology (Highways Agency 2006a) is a semi-analytical methodology based on the methodology suggested in 1970 (RRL 1970). The method took its present form following two reviews, one in 1987 based on the 20-year research work by TRRL (Powell et al. 1984) and the second one in 2006 based on a TRL research study (Nunn 2004). With the methodology, the thickness of a flexible pavement, flexible composite as well as of a rigid pavement can be determined. The methodology for a rigid pavement is described in Chapter 13.

The methodology has introduced the concept of the foundation and of the pavement structure being either flexible or rigid. The design criteria used to develop the design charts are as follows: for the foundation design, the deflection of the foundation surface and the minimum thickness of the upper foundation layer (Chaddock and Roberts 2006), and for the pavement design, the strain of the asphalt layer and the stress of the hydraulically bound layer (Nunn 2004).

The foundation, constructed over the subgrade, consists of the sub-base layer and the capping layer, if used. The foundation is distinguished into four different classes of materials, known as foundation classes. The four foundation classes are defined by their surface modulus, known as foundation surface modulus. The materials for the foundation of the pavement may be unbound or hydraulically bound material. The latter are distinguished into fast-setting and low-setting material.

The pavement structure consists of upper and lower layers. The upper layers are asphalt layers and the lower (base) layers are either asphalt layers or hydraulic bound layers.

The thickness of the foundation depends on the subgrade CBR or subgrade stiffness modulus, while the thickness of the pavement structure depends on the foundation classes and the traffic volume.

This revised/modified methodology gives a wider choice of material and design configurations. It also enables a stronger foundation, incorporating hydraulically bound materials to be used, which results in reduction of thickness of the more expensive asphalt layers.

The UK design procedure for flexible pavement is illustrated in the flow chart shown in Figure 13.12.

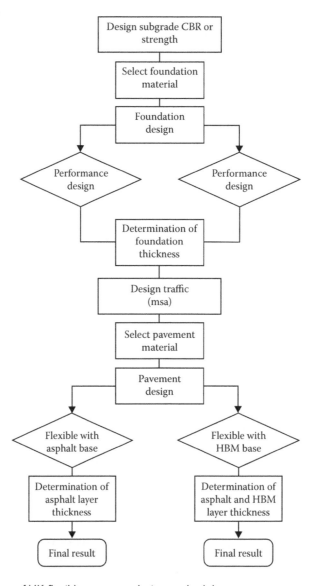

Figure 13.12 Flow chart of UK flexible pavement design methodology.

13.5.1 Determination of design traffic

The design traffic is the commercial vehicle loading over the design period expressed as the number of equivalent standard (80 kN) axles ($ESA_{80\text{-}kN}$). The design traffic is calculated using the commercial vehicle flow, vehicle wear factors and traffic growth.

A commercial vehicle (cv) is defined as that having a gross weight of more than 3.5 tonnes. Lighter vehicles are not taken into consideration since the structural wear caused is considered negligible.

Commercial vehicles are classified into eight classes and three categories, as shown in Table 13.4.

The pavement structural wear caused by a class of commercial vehicle is expressed by its *wear factor* (W). For the determination of the structural wear factor, the fourth power law was used to equate the wear caused by each vehicle type to the number of equivalent standard axles. A standard axle was defined as an axle applying a force of 80 kN.

The methodology proposes different wear factors to be used for the pavement maintenance (W_M) and new pavement design (W_N) cases. These factors are given in Table 13.4.

The wear factors for the new design case are higher than for the maintenance case. This is, as stated, in order to allow for the additional risk that arises from the additional uncertainty with traffic predictions for new designs (Highways Agency 2006b).

In case there are no analytical traffic data per type of vehicle, a combined wear factor for each of the two distinguished categories (OGV1 + PSV) and OGV2 can be used. These factors are as shown at the bottom of Table 13.4.

Additionally, for new road designs, the percentage of OGV2 vehicles obtained by calculation or modelling should not be less than the percentage determined from a graph provided in the source (Highways Agency 2006b).

The *traffic growth* is expressed by the growth factor (G), which is a function of the design period and the annual traffic percentage increase. The growth factor represents the proportional difference between the average vehicle flow over the entire design period and the present flow (or flow at opening). The growth factor for future traffic in the United Kingdom is calculated using a graph provided in the source (Highways Agency 2006b).

The *design traffic* (T) is the sum of the future cumulative flow, in terms of million standard axles (msa), of each commercial vehicle class, T_i, that is,

$$T = \sum T_i,$$

and the T_i is determined according to the following equation:

$$T_i = 365 \times F \times Y \times G \times W \times P \times 10^{-6}, \text{ in million standard axles,}$$

where T_i is the cumulative flow, in terms of million standard axles for commercial vehicle class i; F is the average annual daily flow of traffic for each traffic class at opening; Y is the design period (in years); G is the growth factor; W is the wear factor for each traffic class (W_N for new pavement design and W_M for maintenance design case), from Table 13.4; and P is the percentage of vehicles in the heaviest loaded lane, from Figure 13.13.

The design period (life) is recommended to be taken as 40 years for all sections where traffic is heavy; for less heavily trafficked sections or major maintenance schemes a 20 years design period may be appropriate.

Table 13.4 Classes of commercial vehicles and wear factors per commercial vehicle category

Class no.	Commercial vehicle (CV)	Schematic representation of CVs	Category	Wear factors Maintenance W_M	New W_N
1	Buses and coaches		PSV[a]	2.6	3.9
2	2-Axle rigid		OGV1[b]	0.4	0.6
3	3-Axle rigid			2.3	3.4
4	4-Axle rigid		OGV2[b]	3.0	4.6
5	3-Axle articulated			1.7	2.5
6	4-Axle articulated				
7	5-Axle articulated			2.9	4.4
8	6 (or more)-Axle articulated			3.7	5.6
	OGV1 + PSV			0.6	1.0
	OGV2			3.0	4.4

Source: Adapted from Highways Agency, *Design Manual for Roads and Bridges (DMRB), Vol. 7: Pavement design and maintenance*, Section 2, Part 1, HD 24/06: Traffic assessment, Department for Transport. London: Highways Agency, 2006b (© Highways Agency).

[a] PSV, public service vehicle.
[b] OGV, other goods vehicles.

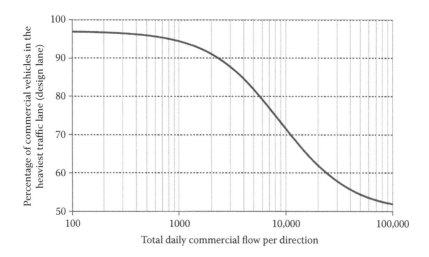

Figure 13.13 Percentage of commercial vehicles in heaviest loaded lane (design lane). (From Highways Agency, *Design Manual for Roads and Bridges [DMRB], Vol. 7: Pavement design and maintenance*, Section 2, Part 1, HD 24/06: Traffic assessment, Department for Transport. London: Highways Agency, 2006b [© Highways Agency].)

13.5.2 Subgrade design CBR and surface stiffness modulus

The design CBR of the subgrade is determined by executing CBR tests in the laboratory, over a range of conditions to reproduce, as far as possible, the conditions of moisture content and density, which are likely to be experienced during construction and in the completed pavement.

Where it is not possible to collect material samples for laboratory assessment of CBR, the design CBR may be estimated from Table 10.1.

The design CBR, for performance design, must be converted to subgrade surface modulus, E. The subgrade surface modulus is an estimated value of stiffness modulus based on subgrade CBR and is estimated by the following equation, derived from work on certain soils (Powell et al. 1984):

$$E = 17.6 \times \text{CBR}^{0.64},$$

where E is the subgrade surface stiffness modulus (MPa) and CBR is the California bearing ratio (% value).

The validity of the above equation is restricted to fine soil material with laboratory CBR values ranging from 2% to 12%. For coarser materials, the plate bearing test may also be appropriate.

The subgrade surface modulus used in the design must be the lower value of the long-term and short-term CBR obtained.

13.5.3 Foundation classes

The foundation consists of the sub-base layer and the capping layer (if used) and is constructed with unbound or hydraulically bound materials. Depending on the type of materials used, the methodology distinguished four foundation classes, on the basis of foundation surface modulus.

Table 13.5 Foundation classes according to Highways Agency (2009a) and materials to be used

Foundation class	Materials	Foundation surface modulus (MPa)
Class 1	Unbound selected granular material or stabilised granular material as per Table 10.2	≥50
Class 2	Unbound crushed, selective granular or with asphalt arisings material as per Tables 10.4[a] and 10.5, as well as bound materials such as CBGM A or B as per Table 10.10, or soil cement achieving compressive strength of at least C3/4	≥100
Class 3	Cement-bound granular material CBGM A or B as per Table 10.10 achieving compressive strength of at least C8/10	≥200
Class 4	Cement-bound granular material CBGM A or B as per Table 10.10 achieving the required minimum foundation surface modulus	≥400

Source: Highways Agency, *Design Manual for Roads and Bridges (DMRB)*, Vol. 7: *Pavement Design and Maintenance*, Section 5, Part 2, IAN 73/06 Revision 1: Design guidance for road pavement foundations, Department for Transport. London: Highways Agency, 2009a (© Highways Agency).

[a] Type II material is used only when the cumulative ESAL is $\leq 5 \times 10^6$.

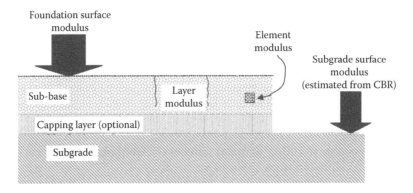

Figure 13.14 Modulus definitions on the surface of foundation and of subgrade. (From Highways Agency, *Design Manual for Roads and Bridges [DMRB], Vol. 7: Pavement Design and Maintenance*, Section 5, Part 2, IAN 73/06 Revision 1: Design guidance for road pavement foundations, Department for Transport. London: Highways Agency, 2009a [© Highways Agency].)

The classes are characterised by the materials used and types of unbound or hydraulically bound materials (or bound mixtures). The bound mixtures are distinguished into fast-setting and slow-setting types. The fast-setting bound mixtures achieve more than 50% of their compressive strength after 28 days at 20°C, while the slow-setting bound mixtures achieve 50% or less of their compressive strength within the same period and under the same curing condition.

Table 13.5 gives details of the foundation classes and the expected foundation surface modulus, while Figure 13.14 explains the foundation and surface modulus concepts.

13.5.3.1 Definitions to expressions used in Figure 13.14 and in the methodology

According to IAN 73/06 (Highways Agency 2009a), *foundation surface modulus* is defined as a measure of 'stiffness modulus' on the basis of the application of a known load at the top of the foundation; it is a composite value with contributions from all underlying layers.

Subgrade surface modulus is defined as an estimated value of 'stiffness modulus' based on subgrade CBR and used for foundation design.

Layer modulus is defined as a measure of 'stiffness modulus' assigned to a given foundation layer; usually, this is a long-term estimate that takes account of degradation caused by factors such as cracking.

Element modulus is defined as a measure of 'stiffness modulus' assigned to a discrete sample of material and usually characterised by a laboratory test; it does not normally take account of degradation owing to factors such as cracking.

13.5.4 Foundation design

The thickness determination of the foundation, per foundation class, is carried out by two alternative design approaches: the restricted foundation design and the performance foundation design (Highways Agency 2009a). Both design methods use as design variable the subgrade CBR or the subgrade stiffness modulus.

The *restricted foundation design* approach is intended for use in cases where it is inappropriate to carry out the range of compliance testing required by the performance-related specification for foundations. The designs are conservative, making allowances for uncertainty in material performance and in layer thickness. Hence, restricted foundation design is intended for use on schemes of limited extent.

The foundation classes are limited to classes 1, 2 and 3. Foundation class 4 is excluded since it is considered essential to measure the properties of such a foundation during construction to give adequate assurance that the appropriate long-term foundation surface modulus is likely to be achieved.

The test required to be carried out are limited to the following: (a) CBR value at the top of the exposed subgrade, immediately prior to placement of the overlying foundation layers, (b) material density and the actual thickness for each stage of foundation construction and (c) compliance with the relevant material specifications.

The *performance foundation design* approach offers greater flexibility to the designer, since a wide range of resources, incorporating natural, secondary and recycled materials, may be utilised. Additionally, the mechanical properties of the materials used are utilised more efficiently and the performance foundation design provides some assurance that the material performance assumptions made at the design stage are being, or are likely to be, achieved.

Apart from the laboratory tests on materials acceptance, it is required to carry out an in situ test for subgrade and foundation acceptance.

The in situ test for subgrade acceptance is the CBR test and the value obtained must be equal to, or greater than, the design CBR. The in situ CBR may also be determined using a dynamic cone penetrometer (see Section 1.6.3) or the plate bearing test (see Section 1.7 and Highways Agency 2009a).

The in situ test for foundation acceptance is the determination of foundation surface modulus immediately prior to the construction of the overlaying pavement layers. Table 13.6 gives the unadjusted mean foundation surface modulus and the minimum foundation surface modulus values, for each foundation class and for different categories of materials, to be achieved or exceeded.

The in situ foundation surface modulus is measured by dynamic plate loading devices, such as the falling weight deflectometer (FWD) or the light weight deflectometer (see Sections 1.9.3 and 16.4.1.3 and Highways Agency 2009a).

The methodology also imposes maximum permissible layer stiffness values for each foundation class, regardless of the foundation design method employed. This is to minimise the

Table 13.6 Top of foundation surface modulus requirements

		Surface modulus (MPa)			
		Foundation class			
		1	*2*	*3*	*4*
Long Term in Service Surface Modulus		≥50	≥100	≥200	≥400
Mean foundation surface modulus	Unbound mixture types	40[a]	80[b]	[c]	[c]
	Fast-setting mixture types	50[a]	100	300	600
	Slow-setting mixture types	40[a]	80	150	300
Minimum foundation surface modulus	Unbound mixture types	25[a]	50[b]	[c]	[c]
	Fast-setting mixture types	25[a]	50	150	300
	Slow-setting mixture types	25[a]	50	75	150

Source: Highways Agency, *Design Manual for Roads and Bridges (DMRB), Vol. 7: Pavement Design and Maintenance*, Section 5, Part 2, IAN 73/06 Revision 1: Design guidance for road pavement foundations, Department for Transport. London: Highways Agency, 2009a (© Highways Agency).

[a] Only permitted on trunk roads including motorways that are designed for not more than 20 msa.
[b] Not permitted for pavements designed for 80 msa or above.
[c] Unbound materials are unlikely to achieve the requirements for Classes 3 and 4.

risk of selecting very thin or very stiff foundation layers at lower subgrade CBR values. The maximum permissible layer stiffnesses to be used are as follows: 100 MPa for class 1, 350 MPa for class 2, 1000 MPa for class 3 and 3500 MPa for class 4 (Highways Agency 2009a).

The layer stiffness (modulus) can be estimated from the element modulus of the foundation materials. The element modulus is assessed by (a) the triaxial testing for unbound materials (CEN EN 13286-7 2004), (b) the modulus of elasticity testing in compression for fast-setting and slow-setting mixtures (CEN EN 13286-43 2003), (c) the dynamic plate testing of compacted trial layers for both unbound and fast- and slow-setting mixtures (see Section 1.9.3) or (d) the Springbox testing of laboratory specimens for unbound mixtures (see Section 1.9.4). More details can be found in Highways Agency (2009a).

To conclude, the choice as to which approach and which foundation class is selected is usually made on economic grounds on the basis of the materials that are available, the size of the scheme and relevant costing information. The designers should give full consideration to the use of local and secondary materials.

13.5.4.1 Determination of foundation thickness for restricted design

The determination of the foundation thickness in the case of restricted design is carried out with the use of design charts shown on Figures 13.17 and 13.18. In all cases, the designer chooses the foundation class, and for a given subgrade CBR or subgrade stiffness modulus, the total foundation layer thickness is determined.

The chart in Figure 13.15 is used when a single layer of foundation is to be used, that is, capping layer or sub-base.

The chart in Figure 13.16 is used when two foundation layers are to be used, that is, capping layer and sub-base.

13.5.4.2 Determination of foundation thickness for performance design

There are a large number of possible designs for the various combinations of subgrade surface modulus and foundation material, in order to achieve the desired foundation class.

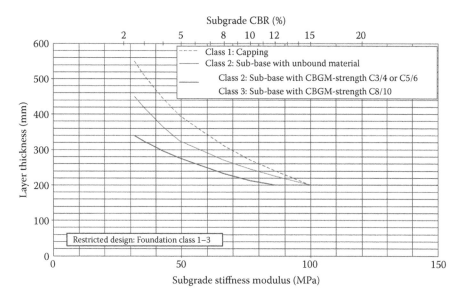

Figure 13.15 Thickness determination of single foundation (sub-base or capping). (From Highways Agency, *Design Manual for Roads and Bridges [DMRB], Vol. 7: Pavement Design and Maintenance*, Section 5, Part 2, IAN 73/06 Revision 1: Design guidance for road pavement foundations, Department for Transport. London: Highways Agency, 2009a [© Highways Agency].)

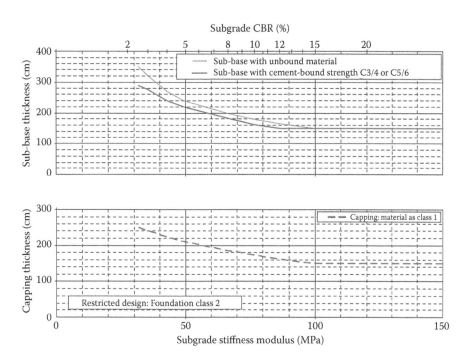

Figure 13.16 Thickness determination of foundation with sub-base and capping layer. (From Highways Agency, *Design Manual for Roads and Bridges [DMRB], Vol. 7: Pavement Design and Maintenance*, Section 5, Part 2, IAN 73/06 Revision 1: Design guidance for road pavement foundations, Department for Transport. London: Highways Agency, 2009a [© Highways Agency].)

Design charts to determine the foundation thickness in case a performance foundation design is employed have been developed. Each chart refers to a different foundation class. The equations used to derive these charts are provided in Highways Agency (2009a). An example for class 1 designs, for a single foundation layer, is given in Figure 13.17.

When two-layer foundation design is chosen and the foundation class is 2, the chart in Figure 13.18 can be used for the determination of the layer's thicknesses.

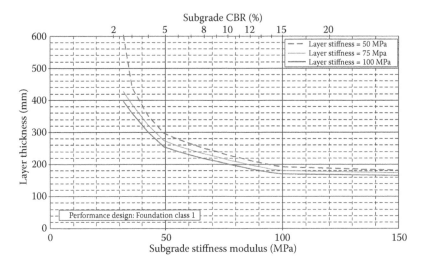

Figure 13.17 Class 1 designs – single foundation layer. (From Highways Agency, *Design Manual for Roads and Bridges [DMRB], Vol. 7: Pavement Design and Maintenance,* Section 5, Part 2, IAN 73/06 Revision 1: Design guidance for road pavement foundations, Department for Transport. London: Highways Agency, 2009a [© Highways Agency].)

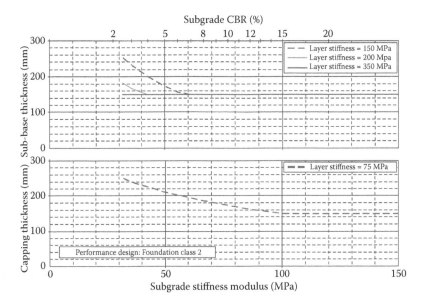

Figure 13.18 Class 2 designs – sub-base on capping. (From Highways Agency, *Design Manual for Roads and Bridges [DMRB], Vol. 7: Pavement Design and Maintenance,* Section 5, Part 2, IAN 73/06 Revision 1: Design guidance for road pavement foundations, Department for Transport. London: Highways Agency, 2009a [© Highways Agency].)

Table 13.7 Categories of hydraulic bound mixtures (HBM)

HBM category	A	B	C	D
Crushed rock coarse aggregate[a]	—	CBGM B-C8/10	CBGM B-C12/15	CBGM B-C16/20
		SBM BI-C9/12	SBM BI-C12/16	SBM BI-C15/20
		FABMI-C9/12	FABMI-C12/16	FABMI-C15/20
Gravel coarse aggregate[b]	CBGM B-C8/10	CBGM B-C12/15	CBGM B-C16/20	—
	SBM BI-C9/12	SBM BI-C12/16	SBM BI-C15/20	
	FABMI-C9/12	FABMI-C12/16	FABMI-C15/20	

Source: Highways Agency, *Design Manual for Roads and Bridges (DMRB), Vol. 7: Pavement design and maintenance*, Section 2, Part 3, HD 26/06: Pavement design, Department for Transport. London: Highways Agency, 2006a (© Highways Agency).

[a] With coefficient of thermal expansion <10 × 10^{-6} per degree Celsius (typically limestone).
[b] With coefficient of thermal expansion ≥10 × 10^{-6} per degree Celsius.

The determination of the foundation layer thickness is carried out having determined the layer stiffness to be used and the subgrade CBR or subgrade stiffness modulus.

In all the design charts, the foundation layer is constructed of the same material (single foundation layer); materials are as those described in Table 13.7. However, it must be noted that the use of granular sub-base, Type II (for details, see Section 10.5.2), is not recommended for any pavement design for more than 5 msa.

The methodology recommends using the performance design only for foundation classes 1 and 2 and not for classes 3 and 4. As it is stated (Highways Agency 2009a), the structural contribution of capping materials with low layer stiffness values is limited when compared with the stiffness of sub-base materials required to achieve foundation classes 3 and 4. Their inclusion in the design model does not, therefore, demonstrate a significant reduction in the thickness of sub-base required but designers should consider the practical advantages of including capping materials in the foundation design. More information is available in TRL Report PPR 127 (Chaddock and Roberts 2006).

13.5.5 Cases where subgrade CBR is low (CBR < 2.5%)

When the subgrade has a CBR value less than 2.5%, its bearing capacity is considered unsuitable to support a pavement foundation. In these cases, the subgrade must be permanently improved. This is obtained by removing the subgrade material and replacing it with a more suitable material or, if the soil is cohesive, by treating (stabilising) it with lime (or similar).

The thickness removed is typically between 0.5 and 1.0 m. Suitable materials for subgrade improvement are those used for capping layer (see Section 10.3.1; for soil stabilisation, see Section 10.3.2).

In all cases where the subgrade has been improved, the new design subgrade CBR is assumed to be equal to 2.5%.

The incorporation of geosynthetic material into the foundation, in certain conditions, may be advantageous. In this case, an alternative design CBR value is necessary to be determined.

13.5.6 Flexible pavement design

Flexible pavement, in this methodology, is considered the structure consisting of all layers above foundation. The upper layers of the flexible pavement are bound in bitumen and the lower (base) layers are bound in either bitumen or hydraulic binder.

In the first case, the pavement is called flexible pavement with asphalt base, while in the second, it is called flexible pavement with HBM base. The latter was previously known as flexible composite pavement.

13.5.7 Selection of pavement material

The designer may choose from a range of asphalt mixtures for asphalt base and binder course, all with graded bitumen, and a range of HBMs, having a 28-day compressive cube strength ranging from 10 to 20 MPa.

As for the top upper layer (surfacing), the recommended material is asphalt concrete for very thin layers. Other asphalts, such as porous asphalt (PA) and hot rolled asphalt (HRA) with coated chippings, may also be used in new pavements.

13.5.7.1 Asphalt base and binder course material

The methodology offers the use of four different types of dense asphalts. These asphalts all contain penetration grade bitumen and are as follows: (a) dense bitumen macadam (DBM125), (b) hot rolled asphalt (HRA50), (c) heavy duty macadam (HDM50) or dense bitumen macadam (DBM50) and (d) enrobe a module eleve (EME2) (high stiffness modulus) material based on French practice. The stone mastic asphalt (SMA) may be used but only as binder course.

It is noted that the DBM material is similar to the dense asphalt concrete. DBM125 is the least stiff material and EME2 is the stiffest of all.

DBM125 for base and binder course must contain 100/150 penetration grade binder. HRA50, DBM50 and HDM50 for base and binder course must contain 40/60 penetration grade binder. EME2 must target a penetration grade of 15/20, which can be achieved using 10/20 or 15/25 penetration grade binder.

It is noted that EME2 is recommended to be laid over a foundation class 3 or 4; or foundation class 2 that has a surface stiffness modulus of at least 120 MPa at the time of construction.

13.5.7.2 HBMs for base layer

The types of HBM permitted to be used are distinguished into four categories, from A to D, on the basis of their strength, and are as shown in Table 13.7. Distinction is made with regard to the type of coarse aggregates used. Better performance is expected for those mixtures made with a crushed rock coarse aggregate that has a coefficient of thermal expansion less than 10×10^{-6} per degree Celsius (typically limestone) (Highways Agency 2006a).

The symbolism of their compressive class (C) is in accordance to CEN EN 14227-1 (2013). The first value corresponds to the characteristic compressive strength (R_{ck}) obtained from a cylindrical specimen with H/D (ratio between height and diameter) equal to 2, while the second value corresponds to the characteristic compressive strength (R_{ck}) from cubes or cylinders with H/D equal to 1.0 (in fact, 0.8 to 1.21). The characteristic strength (R_{ck}) is the 28-day compressive strength expressed in megapascals (MPa).

HBM can practically be cement-bound granular mixtures (CBGMs) or slug bound mixtures (SBMs) or fly ash bound mixtures (FABMs). The aggregate mixture may be crushed rock materials or gravels. Further information for these mixtures is provided in Section 10.5.3.

13.5.7.3 Surface course material

With regard to the surface course material in all new construction or major maintenance works, the Highways Agency proposes the use of thin surface course system or thin surfacings. The HRA with coated chippings as a surfacing material is permitted only for works in Scotland (Highways Agency 2006c). Other surfacing materials such as PA, for new construction works, or HRA, PA, surface dressing and slurry/microsurfacing for maintenance works may be used under approval (Highways Agency 1999, 2006c, 2011).

The typical thickness of the thin surfacing is >25 to <40 mm (or <50 mm) and is made of asphalt concrete for very thin layer according to CEN EN 13108-2 (2008).

In the case where PA (CEN EN 13108-7 2008) is to be used, the surface layer thickness should be 50 mm and the asphalt must be modified with polymer or fibre additive. Its contribution to the material design thickness is only 20 mm; hence, an increase by 30 mm of the total asphalt thickness is necessary. Additionally, a 60 mm dense binder course is required to be laid beneath PA surfacing.

More information for the above asphalts for surfacing, binder course and asphalt base can be found in Chapter 5.

13.5.8 Determination of flexible pavement with asphalt base

The total thickness of the flexible pavement structure, comprising the surface course, binder course and base, is obtained from the right-hand portion of the nomograph shown in Figure 13.19, and it depends on the type of base material.

Thus, by knowing the cumulative number of ESAL, the foundation class and the type of asphalt to be used in the base and binder course, the total asphalt thickness is determined (see Figure 13.19, right-hand portion).

The thickness of the individual layers (surface, binder and base course) is determined by the designer. The method assumes and recommends the use of the same type of asphalt for the base and binder course.

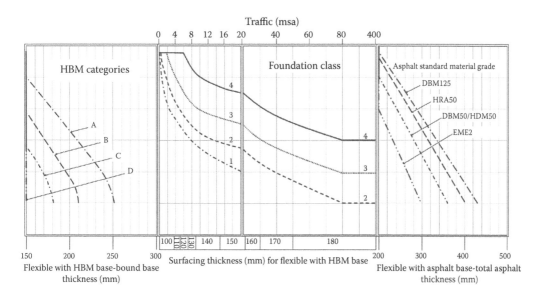

Figure 13.19 Asphalt layers' thickness determination of flexible and flexible with HBM base pavements. (From Highways Agency, *Design Manual for Roads and Bridges [DMRB], Vol. 7: Pavement design and maintenance*, Section 2, Part 3, HD 26/06: Pavement design, Department for Transport. London: Highways Agency, 2006a [© Highways Agency].)

13.5.9 Determination of flexible pavement with HBM base

In a composite pavement structure, the upper layer part is from asphalt and the lower part is from HBMs.

The thickness of the HBM layer, also called hydraulically bound base layer, depends on the type and strength of the mixture and is determined using the left portion of the nomograph shown in Figure 13.19.

The thickness of the overlying asphalt layer is determined by the bottom axis in the central portion of the same nomograph, knowing the cumulative traffic over the design period expressed in million standard axles. An example is shown in Figure 13.19.

For the choice of the surfacing and binder course on base with HBMs, all those mentioned in Sections 13.5.7.3 and 13.5.7.1 can be applied.

When constructing a flexible pavement on HBM base, all HBM layers that are expected to reach a compressive strength of 10 MPa at 7 days must have cracks induced (Highways Agency 2006a). The cracks must be formed in accordance with the specification MCHW 1 Clause 818 (Highways Agency 2009b).

The induced transverse cracks should be 3 m apart and must be aligned (maximum 100 mm tolerance) with any induced cracks in the underlying construction. The laying width of HBM is recommended to be not more than 4.75 m. The above minimise the possibility of having contraction or reflective cracks on the surface of the pavement in the transverse and longitudinal direction.

13.5.10 Frost protection

All material within 450 mm of the road surface, where the mean annual frost index (MAFI) of the site is ≥50, must be non-frost susceptible in the long term. Where the MAFI is <50, the thickness of non-frost-susceptible material may be 350 mm. The frost index is defined as the product of the number of days of continuous freezing and the average amount of frost (in degrees Celsius) on those days. For slower-curing HBM, appropriate measures must be taken to prevent frost damage in the short term. Further guidance is provided in Highways Agency (2006a, 2009a).

13.6 AUTH FLEXIBLE PAVEMENT DESIGN METHODOLOGY

13.6.1 General

The AUTh flexible pavement design methodology is a semi-analytical procedure originally developed in 1997 by a group of scientists (Nikolaides et al. 1997) for the needs of Egnatia Odos S.A. Soon, it was endorsed and incorporated into Egnatia Odos S.A Design Guidelines for Highway Works (DGHW) (Egnatia Odos 2001).

The methodology was published in 2005 after incorporating additional nomographs, to cover a wider range of temperatures and amendments regarding material requirements, and so on (Nikolaides 2005). Since 2005, it has been used by many designers and is considered as the Greek methodology for designing flexible pavement.

The AUTh pavement design methodology consists of nomographs and tables that allow determining the thickness of all layers of flexible pavement consisting of dense asphalt concrete with penetration grade bitumen (40/50 or 60/70) and unbound base/sub-base layers. Provision is also made for the use of modified bitumen in the surfacing and binder layer or the use of 80/100 pen grade bitumen.

The surfacing layer is chosen from the following bituminous mixtures: asphalt concrete for very thin layers, PA, SMA, dense asphalt concrete (for low traffic areas) or micro-surfacing.

The pavement thickness can be determined for three MAATs, 13°C, 16°C or 19°C, representing warm environments.

The nomographs were developed using the principles of the elastic layered system where materials are characterised by their stiffness and Poisson's ratio. The stresses and strains developed were calculated using BISAR software (de Jong et al. 1973) incorporated in Shell's pavement design software SPDM-PC (Valkering and Stapel 1992).

The design criteria used were the horizontal strain in the asphalt layer and the vertical strain on the subgrade, as outlined in Section 13.7. Rutting of the asphalt layers was also considered so that the thickness of the asphalt layers does not reach more than 25 mm (maximum) during the design life of the pavement.

At the end of the design life, the pavement is expected to fail owing to fatigue cracking or surface deformation. During the lifetime of the pavement, the surfacing material is expected to be rehabilitated to meet surface characteristic requirements. The frequency of the rehabilitation depends entirely on the life expectancy of the surfacing material used in relation to traffic volume.

Finally, the method also provides a means for stage construction and recommendations for economic analysis.

The design process of this method is illustrated in the flow chart shown in Figure 13.20.

13.6.2 Determination of design traffic ESAL

The determination of the cumulative traffic in equivalent standard (80 kN) axle loading (ESAL) is carried out as outlined in Section 12.7.1.

The conversion of the commercial vehicle (all except private cars) to standard axles is carried out using the AASHTO equivalency factors (see Table 12.3) or, alternatively, the equivalency truck factors given in Table 12.5.

The traffic distribution to the design lane is estimated graphically using the curve proposed by the old UK pavement design methodology (Highways Agency 1996), which is similar to the one shown in Figure 13.13.

13.6.3 Determination of design subgrade CBR

The CBR test is carried out on soaked specimens (for 4 days) and in accordance to CEN EN 13286-47 (2012). Samples are taken within the first 600 mm below formation level. The CBR value is determined at 90% of the maximum dry density determined by proctor (modified) compaction (see Section 1.4.7). When undisturbed samples are taken from naturally cemented soil material, where water table is low (>1 m), the value of 90% changes to 95%.

The CBR tests are required to be carried out on samples taken at the design stage and also during construction. The number of samples tested at the design stage should be six to eight, and during construction, three per soil category are encountered.

The design CBR is the value that is greater by 90% of the CBR values determined, for a project with design ESAL $\geq 1 \times 10^6$, and by 75% of the CBR values, for a project with design ESAL $< 1 \times 10^6$.

Design CBR values greater than 20% are considered as being equal to 20%.

13.6.4 Determination of MAAT

The MAAT is the mean of all monthly average air temperatures. Its determination requires air (ambient) temperature data recorded over long periods, available at any central metrological office.

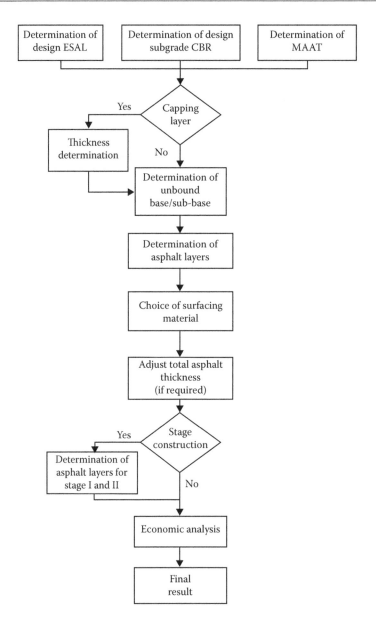

Figure 13.20 Flow chart of AUTh flexible pavement design methodology. (From Nikolaides, A.F., *Flexible Pavements: Pavement Design Methodology, Bituminous Mixtures, Antiskidding Layers*, 1st Edition. Thessaloniki, Greece: A.F. Nikolaides. ISBN 960-91849-1-X, 2005.)

For the needs of the methodology, a table is provided in the source containing the MAAT of most major cities in the Greek territory (Nikolaides 2005). The values were determined by utilising metrological data from 1965 to 1995. It is noted that the MAAT in the Greek territory was found to range from 13°C to 19°C.

However, although the MAAT is required as data, the design charts have been developed on the basis of weighted mean annual pavement temperature, using the method published in Shell methodology (Shell 1978).

13.6.5 Thickness determination of capping layer

The methodology requires the provision of a capping layer when the design CBR value is less than 5%.

The thickness of the capping layer is determined in terms of the design CBR of the subgrade. For simplicity, the methodology distinguishes two limiting CBR values and therefore only two capping layer thicknesses. In particular:

- When design CBR is ≤2.5%, the thickness of the capping layer should be ≥600 mm.
- When design CBR is >2.5% and ≤5%, the thickness of the capping layer should be 300 mm.

The abovementioned capping layer thicknesses are also shown in Table 13.10, aside from the recommended base/sub-base thickness.

If the design subgrade CBR is lower than 3%, the value of CBR used to determine the asphalt layer thickness is taken as being equal to 3%. In all other cases, the design CBR is that determined from laboratory tests.

The materials for capping layer may be selected granular material or cement- and lime-stabilised material, as those described in Section 10.3.1. Soil stabilisation is also recommended to be considered, when highly cohesive soil material is encountered. For details, see Section 10.3.2.

The use of geotextiles or geotextile-related materials is encouraged before laying the capping layer or when construction is carried out at high subgrade moisture content, to ensure proper compaction of the overlaying layers. Information about geotextile materials is given in Section 10.3.4.

13.6.6 Thickness determination of base/sub-base

The base and sub-base are dealt as one unified layer. The thickness of this layer can be 200, 300 or 400 mm, which simplifies construction works (base/sub-base is constructed in 100 mm thick layers).

The determination of the required thickness of the base/sub-base layer is based on design subgrade CBR and simply follows the recommendation of Table 13.8.

The base/sub-base materials are unbound crushed aggregates, Type I, or, if the thickness is 400 or 300 mm, selective granular material, Type II (see Tables 10.4 and 10.6 for

Table 13.8 Thicknesses determination of base/sub-base layer

Design subgrade CBR (%)	Capping layer (mm)	Base and sub-base total thickness (mm)
Required thickness		
≤2.5	600	400
2.6–5.0	300	400
5.1–10.0	0	400
Recommended thickness[a]		
10.1–20.0	0	300
>20.0	0	200

Source: Nikolaides, A.F., *Flexible Pavements: Pavement Design Methodology, Bituminous Mixtures, Antiskidding Layers*, 1st Edition. Thessaloniki, Greece: A.F. Nikolaides. ISBN 960-91849-1-X, 2005.

[a] The thickness of 400 mm, if desired, is not excluded.

gradation and requirements). When the thickness of the base/sub-base is 400 or 300 mm, the top 200 mm is always constructed with unbound crushed aggregates Type I.

HBMs may be used as a replacement part of the total base/sub-base thickness and are always placed over an unbound aggregate layer. However, in case of using HBM, the methodology does not cater to any expected reduction of the overlaying asphalt layers. Details on HBM are given in Sections 10.5.7, 10.5.8 and 10.5.9.

13.6.7 Thickness determination of bituminous layers

The determination of bituminous layer thickness is carried out with respect to design ESAL and design CBR value. A number of nomographs (design charts) have been developed, for each distinct base/sub-base thickness (400, 300 and 200 mm) and for each MAAT (13°C, 16°C and 19°C), for two penetration grade bitumen (50/70 and 40/50 pen).

A sample of the nomographs developed are given in Figures 13.21, 13.22 and 13.23 (for a base/sub-base thickness of 200, 300 and 400 mm, MAAT = 16°C and 50/70 grade bitumen). All the other nomographs are provided in Nikolaides (2005).

By choosing the appropriate nomograph, the total thickness of asphalt layers is easily determined.

It is noted that when MAAT is between the values taken for the development of the nomographs, the thickness of the asphalt layers is determined by linear extrapolation.

In all cases, thickness of the asphalt layers determined is rounded up to the next 5 mm.

The type of bituminous mixture for which the thickness is determined is dense asphalt concrete mixture with 40/50 or 60/70 pen graded bitumen.

The designer is assisted on his decision of where and when to choose the type of bitumen in the asphalt layers (surfacing, binder or asphalt base) by the protocol provided in Table 13.9. This effectively ensures that rutting caused by the permanent deformation of

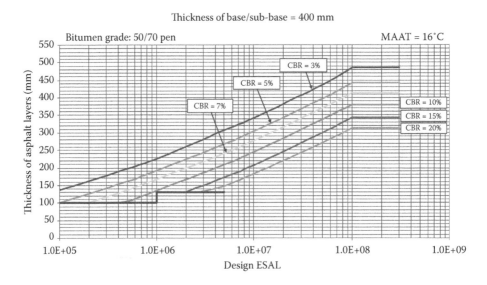

Figure 13.21 Nomograph determining the thickness of asphalt layers with 50/70 grade bitumen, base/sub-base thickness of 400 mm and MAAT = 16°C. (From Nikolaides, A.F., *Flexible Pavements: Pavement Design Methodology, Bituminous Mixtures, Antiskidding Layers*, 1st Edition. Thessaloniki, Greece: A.F. Nikolaides. ISBN 960-91849-1-X, 2005.)

Thickness of base/sub-base = 300 mm

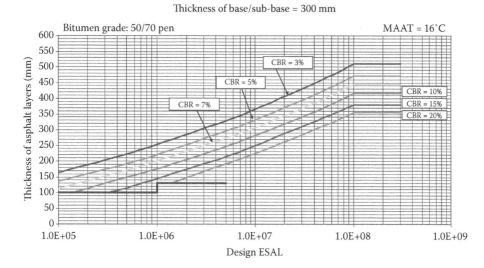

Figure 13.22 Nomograph determining the thickness of asphalt layers with 50/70 grade bitumen, base/sub-base thickness of 300 mm and MAAT = 16°C. (From Nikolaides, A.F., *Flexible Pavements: Pavement Design Methodology, Bituminous Mixtures, Antiskidding Layers*, 1st Edition. Thessaloniki, Greece: A.F. Nikolaides. ISBN 960-91849-1-X, 2005.)

Thickness of base/sub-base = 200 mm

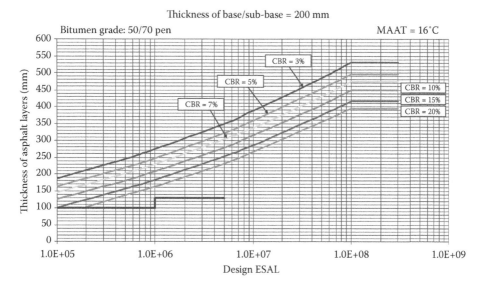

Figure 13.23 Nomograph determining the thickness of asphalt layers with 50/70 grade bitumen, base/sub-base thickness of 200 mm and MAAT = 16°C. (From Nikolaides, A.F., *Flexible Pavements: Pavement Design Methodology, Bituminous Mixtures, Antiskidding Layers*, 1st Edition. Thessaloniki, Greece: A.F. Nikolaides. ISBN 960-91849-1-X, 2005.)

Table 13.9 Protocol for choosing the appropriate grade binder

Cumulative traffic (ESAL)	Mean annual air temperature (°C)	Penetration grade binder (pen)	Layer with a certain penetration grade binder
≤1 × 10⁷	13–19	50/70	All
1 × 10⁷ to 5 × 10⁷	13–14	50/70	All
	15–19	40/50	Surface layer only
5 × 10⁷	13–14	40/50	Surface layer only
	15–19	40/50	All

Source: Nikolaides, A.F., *Flexible Pavements: Pavement Design Methodology, Bituminous Mixtures, Antiskidding Layers*, 1st Edition. Thessaloniki, Greece: A.F. Nikolaides. ISBN 960-91849-1-X, 2005.

the asphalt layers will not be excessive during the pavement's design life. When 40/50 grade bitumen is used in the surface layer instead of 50/70 grade bitumen, the total thickness of the asphalt layers determined remains unchanged. The same applies when modified bitumen is used in the surface layer mixture.

The use of 80/100 pen bitumen is recommended only for the construction of the asphalt base layer and only at locations with MAAT < 15°C or at locations with design ESAL < 1 × 10⁷ irrespective of MAAT. In such cases, the asphalt base thickness resulting from the nomograph used is increased by 7% to cater to the use of 80/100 grade bitumen.

13.6.8 Selection of surfacing material

The surfacing material should provide a pavement surface with long-term good surface characteristics, particularly anti-skidding properties. The asphalts recommended to be used by the methodology are the following:

a. Asphalt concrete for very thin layers (AC-VTL) (see Section 5.5).
b. PA (see Section 5.6)
c. SMA (see Section 5.8).
d. Micro-surfacing (MS) (see Section 6.8).

Some of the surfacing materials above have structural strengths and load distribution abilities different from those of dense asphalt concrete, which was considered in the calculations for the development of the design charts.

Therefore, when any from these mixtures is chosen, the total thickness of the bituminous layers should be adjusted, if needed. The necessary adjustments are summarised in Table 13.10.

Table 13.10 Required modifications for total thickness of asphalt layers

Surfacing material	Increase total asphalt thickness by:
AC-VTL	+5 mm, for 2.5 to 3.0 mm layer
PA	+20 mm, for 40 mm layer (or +25 mm for 50 mm, or +30 mm for 60 mm)
SMA	None (for any thickness)
MS	None; the thickness of MS is simply added to the determined total thickness of the asphalt layers
AC[a]	None

Source: Nikolaides, A.F., *Flexible Pavements: Pavement Design Methodology, Bituminous Mixtures, Antiskidding Layers*, 1st Edition. Thessaloniki, Greece: A.F. Nikolaides. ISBN 960-91849-1-X, 2005.

[a] Used in low traffic areas.

Table 13.11 Use of AC-VTS as surface material – thickness adjustment and recommendations for thickness and type of dense asphalt concrete (AC)

Total thickness of asphalt layers (mm)	Surfacing Thickness (mm)	Binder course Thickness (mm)	Binder course Type of AC	Asphalt base Thickness (mm)	Asphalt base Type of AC
100	25	40 + 40	AC12.5	0	AC12.5
150	25	50	AC12.5 or AC20	80	AC 20
200	25	40	AC12.5	60 + 80	AC 20
250	25	60	AC20 or AC12.5	80 + 90	AC 20
300	25	60		90 + 130	AC20+ AC31.5
350	25	60		(80 + 80) + 110	AC20+ AC31.5
400	25	60		(70 + 70 + 80) + 100	AC20+ AC31.5
450	25	60		(80 + 80 + 80) + 130	AC20+ AC31.5

Source: Adapted from Nikolaides, A.F., *Flexible Pavements: Pavement Design Methodology, Bituminous Mixtures, Antiskidding Layers*, 1st Edition. Thessaloniki, Greece: A.F. Nikolaides. ISBN 960-91849-1-X, 2005.

Having determined the total thickness of the asphalt layers and after choosing the surfacing material, the methodology provides analytical tables with the required thickness adjustment, recommended thicknesses per layer to be constructed and type of dense asphalt concrete to be used. Table 13.11 provides an example of the above for an AC-VTL surfacing material and various depths of asphalt layers determined, at increments of 50 mm; in Nikolaides (2005), the increment used is 10 mm.

13.6.9 Planned stage construction and economic analysis

The methodology offers the capacity of pavement design in case of stage construction of the bituminous layers.

The process for the determination of the layer thickness in case of stage construction is similar to the one proposed by Asphalt Institute (1999). A working example is given in Nikolaides (2005).

13.6.10 Economic analysis

The methodology recommends performing an economic analysis on the alternative designs. The analysis is based on LCCA and explained with a working example (Nikolaides 2005).

13.7 SHELL FLEXIBLE PAVEMENT DESIGN METHODOLOGY

The Shell flexible pavement design method initially presented in 1963 and published as a design manual in 1978 is known as the Shell pavement design manual (Shell 1978). In 1985, there was an addendum to the manual giving guidelines on the use of safety factors or confidence levels into the design. It was one the first integrated semi-analytical detailed methodology developed.

The method uses the elastic theory in multi-layered systems and distinguishes up to five distinct layers, namely, (a) bituminous layers, which may be subdivided into two layers; (b) the unbound layer or two unbound layers with different moduli; and (c) subgrade.

The design criteria are fatigue cracking of the asphalt or hydraulically bound layer and subgrade deformation, as mentioned in Section 13.2. However, the method offers the ability to estimate surface permanent deformation (rutting) caused exclusively by the asphalt layers. It is thus possible to forecast rutting depth of the given asphalt mixture considered in the design.

The traffic volume is expressed in equivalent standard (80 kN) axles. The process of converting axial loads into typical axles uses equivalency factors, determined from the equivalence law to the power of 4. The strength of all layers is expressed by their elastic moduli and Poisson ratios. The method also considered environmental temperatures, that is, monthly average air temperatures converted to weighted layer temperature, since they directly affect the stiffness of the asphalt considered in the design.

For the determination of the critical strains in the asphalt layer and subgrade, the equations given in Sections 13.2 and 7.7.3 have been used. For the strains developed in the pavement structure, the in-house BISAR software was utilised to develop the design charts.

The pavement design for a given traffic, subgrade strength and average monthly environmental temperature is carried out by using a series of appropriate diagrams, nomographs and design charts (almost 250 in total are contained in the manual).

In 1992, the first release of software SPDM-PC (Valkering and Stapel 1992) was issued, followed the philosophy of the manual. The computer program allowed the use of a wide variety of temperatures, bitumen grades and material properties.

The last Windows version release of SPDM 3.0 (Shell 1998) contains modules for thickness design, rutting calculations and asphalt overlay design. Additionally, more emphasis was given to material properties incorporating the use of polymer-modified bitumen. The rutting calculations are now based on creep characteristics obtained from dynamic and not static tests, since it appeared not to be applicable to modified bitumen (Lijzenga 1997).

The Shell pavement design method has been and is still used by various organisations and agencies around the world.

13.8 FRENCH FLEXIBLE PAVEMENT DESIGN METHODOLOGY

The French-LCPC method for flexible road pavements (Corte 1997; Destombes 2003) provides a catalogue of designs for different classes of roads (from low traffic volume [minor roads] to high traffic volume [motorways]). The road classes are classified with respect to average daily commercial vehicle of gross weight ≥3.5 tonnes (35 kN); heavy commercial vehicles weighing 5.0 tonnes may also be used for road classification. For the determination of the catalogue of designs, the multi-layer elastic theory, fatigue of bound layers and subgrade deformation as design criteria and the environmental effect (temperature) were used. The catalogue of designs provides the thickness of each layer with respect to 20, 50, 120 and 200 MPa subgrade modulus and cumulative standard axle (design traffic) classes, TC_i. The pavement may consist of asphalt concrete with unbound aggregate layers, asphalt-treated bases or cement-stabilised aggregates.

In addition to French flexible pavement design methodology, ITECH Co. on behalf of LCPC developed 'Alize-LCPC Routes', also designated as 'Alize-LCPC' or 'Alize' software. Alize-LCPC software provides a rational mechanical design method for pavement structures, as developed by the LCPC and SETRA French organisations. It now constitutes the regulatory approach for designing pavements throughout the French national road network and moreover has been adopted by many other road project development agencies.

The integral version of Alize-LCPC includes three main modules: (a) the mechanical computation module based on the determination of the stresses and strains created in the road materials by the traffic loads, named 'Alize-mechanical module'; (b) the module dedicated to the verification of the design as regards the pavement frost–thaw behaviour, named 'Alize-frost thaw module'; and (c) the back-calculation module used for the computation of the elastic modules of the pavement materials from the measured deflection bowls, named the 'Alize-back calculation module' (LCPC 2011).

The procedure for pavement design (use of the Alize-mechanical module) consists of the following:

i. Specify the loading; any axle load and load configuration can be used including aircraft type of loading.
ii. Estimate the thickness of components.
iii. Consider the material's fatigue properties and traffic level for allowable values; a library of a wide variety of material is provided; however, the user may input a new material giving its characteristic properties.
iv. Carry out a structure analysis; bonded, half-bonded or un-bonded layer interfaces may be chosen.
v. Compare critical stresses, strains or deflections with allowable values.
vi. Make adjustments to material or geometry until a satisfactory design is achieved.

The Alize-frost thaw module calculates the diffusion of frost into the pavement structure and the module may be used for the analysis of positive temperature conditions.

The back-calculation module used to compute layers' elastic modulus requires data obtained from a Benkelman beam, a Lacroix deflectograph, or a FWD.

13.9 AUSTRALIAN FLEXIBLE PAVEMENT DESIGN METHODOLOGY

The Australian flexible pavement design methodology published by Austroads (2012) has been developed as a general guide to Australian road agencies. The agencies may publish design manuals or supplements that translate the design guidance provided by Austroads into design practice reflecting local materials, environment, loading and pavement performance.

With the use of the design guide, flexible and rigid pavements can be designed. The flexible pavements may consist of asphalt, unbound, hydraulically bound or hydraulically bound/unbound layers.

The design procedure for a flexible pavement design is as shown in Figure 13.24.

In the analysis, pavement materials are considered to be homogeneous, elastic and isotropic, except for unbound granular materials and subgrades, which are considered to be anisotropic.

Stress analysis is carried out by using a linear elastic model, such as the computer program CIRCLY (Mincad Systems 2009).

The design criteria used are as follows: the horizontal tensile strain at the bottom of the asphalt layer or of the HBMs and the vertical compressive strain at the top of the layer; no consideration was given for the unbound aggregate layer.

For the selection of the materials, it is recommended to also consider subsequent publications related to the manual (Austroads 2013a,b). The first is related to modified granular materials (unconfined compressive strength <2.0 MPa) and bound materials (hydraulically or bitumen bound) for base/sub-base course. The second is related to selected subgrade and lime-stabilised subgrade material.

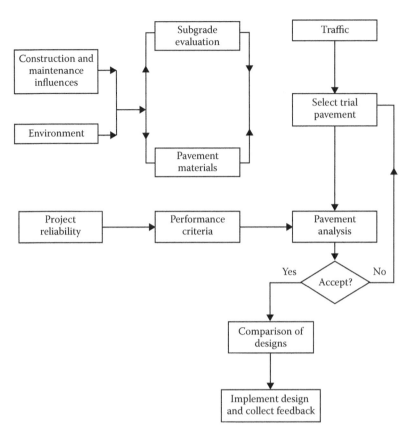

Figure 13.24 Austroads design procedure for flexible pavement design. (From Austroads, *Guide to Pavement Technology Part 2: Pavement Structural Design.* Publication No. AGPT02-12. Sydney: Austroads Inc., 2012. With permission.)

Regarding the environmental factors, subgrade moisture and temperature have been considered to significantly affect pavement performance. Freeze/thaw conditions were not considered since, as stated, they rarely occur in Australia.

With respect to traffic, the initial average daily number of heavy vehicles required is estimated, in descending order of accuracy, (a) from WIM surveys, (b) from vehicle classification counters or (c) from data obtained from single tube axle counters or manual traffic count surveys, together with an estimate of the proportion of heavy vehicles (two or more axle trucks, three or more axle rigid vehicles or articulated vehicles or six or more road trains; for details, see Table 7.1 in Austroads 2012).

The standard axle loading consists of a dual-wheeled single axle, applying a load of 80 kN. However, loads on axle groups with dual tyres causing the same damage as standard axle are also provided.

The design traffic is expressed as cumulative equivalent standard loads (ESAL) over the design period. To convert daily traffic to ESAL, two different approaches related to the equivalency factors are considered. The first approach is applied to the empirical design method, for flexible pavements with only a thin asphalt layer (25 to 40 mm), and the second approach is applied to the mechanistic design method, for flexible pavements with asphalt, unbound or HBM layers.

In the first approach, the exponent 'γ' (noted as *m* in the reference manual) for calculating the equivalency factor using the power law is 4 (see Section 12.7). In this case, one type of damage is considered, that of the overall deterioration of the pavement. With this approach, the equivalent standard axle per heavy vehicle per axle group is determined.

In the second approach, the values of the exponent 'γ' takes three different values depending on the type of fatigue damage. For fatigue damage to asphalt, γ = 5; for fatigue damage to cemented material, γ = 12; and for surface deformation (rutting) attributed to subgrade deformation, γ = 7. This implies that three different ESAL values are determined, called standard axle repetitions per equivalent standard load for each type of fatigue damage. This gives the ability to determine the pavement failure type for given layer thicknesses.

The typical distribution factors of the traffic per lane (design lane or other) for rural and urban locations are provided in the design manual. The recommended design period for flexible pavement is 20 to 40 years (for rigid pavement, 30 to 40 years).

The design manual provides design charts as examples to determine the thickness of asphalt, unbound or HBM layer for given subgrade values (30, 50 and 120 MPa). These design charts, as stated, cannot be used for thickness design. They serve as an illustration of the application of mechanistic design procedures for flexible pavements and may be useful in selecting a trial pavement configuration (Austroads 2012).

The design methodology also provides procedures for the evaluation of pavement damage attributed to specialised vehicles as well as the procedure for pavement design for very light traffic locations (design ESAL less than 1×10^5).

13.10 THE NOTTINGHAM UNIVERSITY/MOBIL DESIGN METHOD

The University of Nottingham developed a flexible pavement design methodology (Brown 1980), which was the basis for the Mobil method published for UK roads and conditions (Mobil 1985). The University of Nottingham method used multi-layer elastic theory, fatigue of bound layers and subgrade deformation as design criteria (distress modes); the environmental effect of temperature; and HRA and unbound materials as pavement materials. Design chats were developed with the use of the computer program ANPAD for stress analysis. The University of Nottingham method was later revised by Brown and Dawson (1992). However, this design method is not in use in the United Kingdom ever since the introduction of the British methodology (Highways Agency 2006a,b, 2009).

13.11 OTHER PAVEMENT DESIGN METHODS

Apart from the above methods, many other semi-analytical methodologies have been developed by various institutions in various countries.

Some of them are the German methodology (RStO 01 2001), the Illinois State methodology (IDOT 2010), the New York State methodology (CPDM 2002) and the Washington State methodology (WSDOT 2011).

In Canada, there is no single pavement design methodology. The provinces and the federal government use various methodologies, primarily the AASHTO (1993) methodology (C-SHRP 2002).

13.12 AASHTO MECHANISTIC–EMPIRICAL PAVEMENT DESIGN GUIDE

13.12.1 General

The MEPDG is the outcome of years of work, carried out by the AASHTO Joint Task Force on Pavements in cooperation with NCHRP (2004) and FHWA, to eliminate the limitations of the AASHTO (1993) methodology and provide a pavement design methodology on the basis of engineering mechanistic–empirical properties. The design and analysis procedure developed calculates the pavement stresses, strains and deflections and uses them to compute incremental damage over time. The procedure then empirically relates the cumulative damage to observe pavement distresses.

The MEPDG (AASHTO 2008) provides a uniform set of procedures for the analysis and design of new and rehabilitated flexible as well as rigid pavements. It employs common design parameters for all pavement types to develop alternative designs by using a variety of materials. Recommendations are also provided for the new and rehabilitated structure (layer materials and thickness), including procedures to select pavement thickness, rehabilitation treatments, subsurface drainage, foundation improvement strategies and other design features.

The outputs from the procedure of pavement design using the MEPDG are predicted pavement distresses and smoothness (International Roughness Index [IRI]), at a selected reliability level. Thus, it is not a direct thickness design procedure, but rather an analysis tool to evaluate a trial design for a given set of site conditions and failure criteria at a specified level of reliability. The trial design is a combination of layer thickness, layer material types and other design features and may be determined by AASHTO (1993) or other pavement design methods.

Data input and computational analysis are carried out by specially built powerful software known as AASHTOWare Pavement ME design (AASHTOWare 2013). Users are advised to build their own material library and traffic library, set up field sections for local calibration and so on, with the aim of continually upgrading the software.

The MEPDG design approach consists of three major steps and multiple steps, as shown in Figure 13.25 for new pavement and rehabilitation design and analysis.

The *first stage* consists of the evaluation/determination of the input values for the trial design. During this stage, inputs are provided and strategies are identified for consideration in the design stage. In the case of new pavement design, the inputs concerned are as follows: site investigation/foundation analysis, paving materials characterisation, traffic analysis, climate/environment analysis and design criteria or threshold values.

The site investigation mainly concerns resilient modulus determination, evaluation of the shrink–swell potential of high-plasticity soils, frost heave–thaw weakening potential of frost-susceptible soils and soils' drainage potential.

The paving materials characterisation concerns resilient modulus determination of all unbound layers, dynamic modulus for all asphalt layers and elastic modulus for Portland cement or chemically stabilised layers.

Traffic analysis consists of estimating the axle load distribution spectrum (single, tandem, tridem and squad axle loadings) applied to the pavement structure. These data are obtained from WIM studies; default values for axle load distribution are also available in case limited WIM data are available. The MEPDG does not use ESAL and hence does not require load equivalency factors. The MEPDG procedure, apart from average annual daily truck traffic, lane distribution, truck growth factor and axle load distribution, also utilises some new input data. These are the operational speed, the monthly and hourly distribution of traffic and the lateral wander of axle loads.

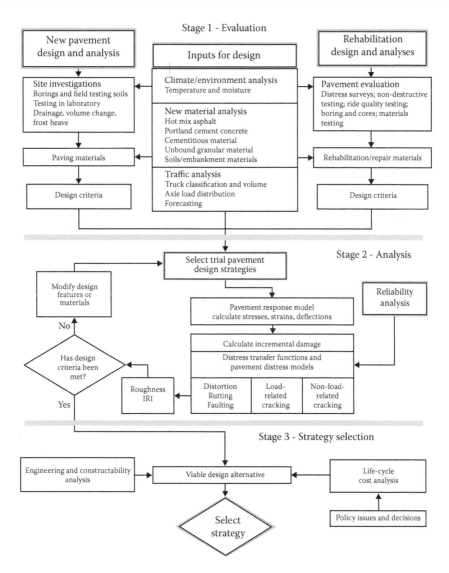

Figure 13.25 Conceptual flow chart of the MEPDG three-stage design/analysis process. (From AASHTO, *Mechanistic–Empirical Pavement Design Guide [MEPDG]*, Washington, DC: American Association of State Highway and Transportation Officials, 2008. With permission.)

In the climate/environment analysis, detailed climatic data (hourly temperature, precipitation, wind, relative humidity, etc.) are required for predicting temperature and moisture content in each of the pavement layers and then pavement distress. Climatic data are available from weather stations but MEPDG has an extensive number of weather stations embedded in its software for ease of use.

The design performance criteria or threshold values, together with the design reliability level, greatly affect construction costs and performance. Design performance criteria or threshold values for flexible pavements concern alligator cracking, rut depth, transverse cracking length (thermal cracks) and IRI (smoothness). With regard to reliability level, it is selected with respect to functional classification of the roadway and its location (urban or

Table 13.12 Design criteria for acceptability of trial designs (flexible and rigid pavements and overlays)

Pavement type	Performance criteria	Maximum value at end of design life
Hot mix asphalt pavements (flexible pavements) and overlays	Alligator cracking (HMA bottom up cracking)	Interstate: 10% lane area Primary: 20% lane area Secondary: 35% lane area
	Rut depth (permanent deformation in the wheel paths)	Interstate: 0.40 in Primary: 0.50 in Others (<45 mph): 0.65 in
	Transverse cracking length (thermal cracks)	Interstate: 500 ft/mi Primary: 700 ft/mi Secondary: 700 ft/mi
	IRI (smoothness)	Interstate: 160 in/mi Primary: 200 in/mi Secondary: 200 in/mi
JPCP new, CRP and overlays	Mean joint faulting	Interstate: 0.15 in Primary: 0.20 in Secondary: 0.25 in
	Percent transverse slab cracking	Interstate: 10% Primary: 15% Secondary: 20%
	IRI (smoothness)	Interstate: 160 in/mi Primary: 200 in/mi Secondary: 200 in/mi

Source: AASHTO, *Mechanistic–Empirical Pavement Design Guide (MEPDG)*, Washington, DC: American Association of State Highway and Transportation Officials, 2008. With permission.

rural), for example, 95% for urban interstate/freeways, 85% for rural principal arterials and so on.

The *second stage* of the design process is the structural analysis and predictions of performance indicators. It is an iterative process that begins with an initial trial design obtained from a design procedure or from general catalogue. The trial section is analysed over time using the pavement response and distress models. If the trial section does not meet the design criteria, modifications are made and the analysis is repeated until a satisfactory result is obtained.

The design criteria or threshold values recommended to be used in judging the acceptability of trial design for new flexible pavements or overlays, as well as for rigid pavements and overlays, are shown in Table 13.12.

The *third stage* of the design process concerns activities required to strategically select the structurally viable alternatives. These include engineering constructability analysis and life cost analysis of the alternatives.

13.12.2 MEPDG for flexible pavements

The mixture characterisation of hot mix asphalt (HMA) also differs in MEPDG. The typical differences between the *Guide for Design of Pavement Structures* (AASHTO 1993) and the *Mechanistic–Empirical Pavement Design Guide* (AASHTO 2008) in terms of HMA mixture characterisation are shown in Figure 13.26.

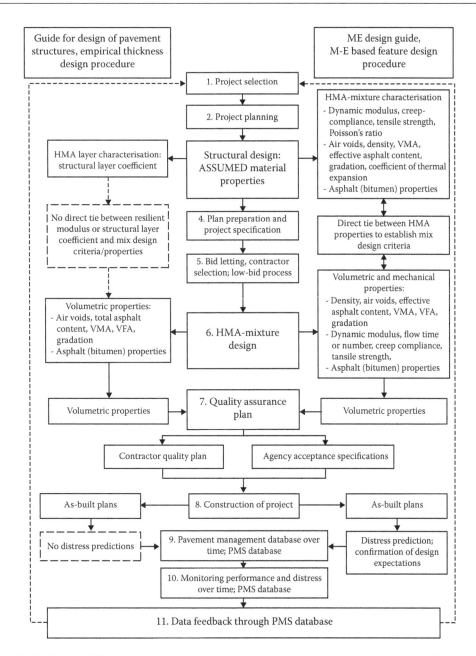

Figure 13.26 Typical differences between empirical design procedures and an integrated M-E design system in terms of mixture characterisation of HMA. (From AASHTO, *Mechanistic–Empirical Pavement Design Guide [MEPDG]*, Washington, DC: American Association of State Highway and Transportation Officials, 2008. With permission.)

The MEPDG analyses expected performance of new and reconstructed flexible pavements and asphalt overlays. The design strategies that can be simulated with MEPDG for new flexible pavement are shown in Figure 13.27.

A similar figure to Figure 13.27 but for HMA overlay design strategies of flexible, semi-rigid (flexible with HMB base) and rigid pavements that can be simulated with the MEPDG is given in AASHTO (2008).

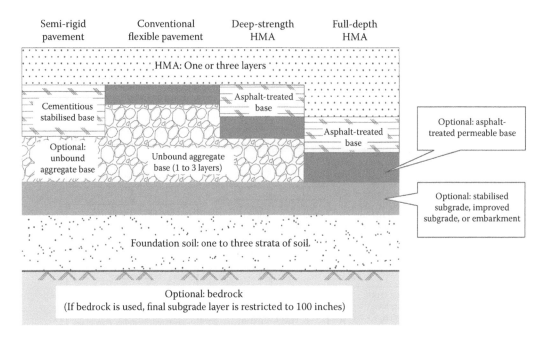

Figure 13.27 Design strategies that can be simulated with the MEPDG for new flexible pavement design. (From AASHTO, *Mechanistic–Empirical Pavement Design Guide [MEPDG]*, Washington, DC: American Association of State Highway and Transportation Officials, 2008. With permission.)

More details about MEPDG and the MEPDG software can be found in AASHTO (2008) and AASHTOWare (2013).

REFERENCES

AASHTO. 1972. *AASHTO interim guide for design of pavement structures.* American Association of State Highway and Transportation Officials. Washington, DC: AASHTO.

AASHTO. 1993. *AASHTO guide for design of pavement structures.* Washington, DC: American Association of State Highway and Transportation Officials.

AASHTO. 2008. *Mechanistic-empirical pavement design guide (MEPDG).* Washington, DC: American Association of State Highway and Transportation Officials.

AASHTO T 22. 2011. *Compressive strength of cylindrical concrete specimens.* Washington, DC: American Association of State Highway and Transportation Officials.

AASHTO T 180. 2010. *Moisture-density relations of soils using a 4.54 kg rammer and a 457 mm drop.* Washington, DC: American Association of State Highway and Transportation Officials.

AASHTO T 307. 2007. *Determining the resilient modulus of soils and aggregate materials.* Washington, DC: American Association of State Highway and Transportation Officials.

AASHTOWare. 2013. *AASHTOWare pavement ME design.* Washington, DC: American Association of State Highway and Transportation Officials.

ALIZE-LCPC. 2011. *ALIZE-LCPC Software Version 3.1-User Manual.* Montreuil, France: LCPC.

Asphalt Institute. 1990. *Computer Program CAMA, CP-6, Version 2.0.* Lexington, KY: Asphalt Institute.

Asphalt Institute MS-1. 1999. *The Thickness Design, Asphalt Pavements for Highways & Streets*, 9th Edition. Manual Series No. 1, Reprinted 1999. Lexington, KY: Asphalt Institute.

Asphalt Institute MS-2. *Mix Design Methods-For Asphalt Concrete and Other Hot-Mix Types*, 6th Edition. Manual Series 2 No. 6. Lexington, KY: Asphalt Institute.

Asphalt Institute MS-10. 1997. *Soils Manual*. Manual Series No. 10. Lexington, KY: Asphalt Institute.

Asphalt Institute MS-19. 2008. *Basic Asphalt Emulsion Manual*, 4th Edition. Manual Series No. 19. Lexington, KY: Asphalt Institute.

Asphalt Institute SP 2. 2001. *Superpave Mix Design*, 3rd Edition. Superpave Series No. 2. Lexington, KY: Asphalt Institute.

Asphalt Institute SW-1. 2005. *Asphalt Pavement Thickness Design Software for Highways, Airports, Heavy Wheel Loads and Other Applications*. Lexington, KY: Asphalt Institute.

ASTM C 39/C 39M-12a. 2012. *Standard test method for compressive strength of cylindrical concrete specimens*. West Conshohocken, PA: ASTM International.

ASTM C 469/C 469M. 2010. *Standard test method for static modulus of elasticity and Poisson's ratio of concrete in compression*. West Conshohocken, PA: ASTM International.

ASTM D 1557. 2012. *Standard test methods for laboratory compaction characteristics of soil using modified effort (56,000 ft-lbf/ft³ (2,700 kN-m/m³))*. West Conshohocken, PA: ASTM International.

ASTM D 2940/D 2940M. 2009. *Standard specification for graded aggregate material for bases or sub-bases for highways or airports*. West Conshohocken, PA: ASTM International.

ASTM D 7369. 2011. *Standard test method for determining the resilient modulus of bituminous mixtures by indirect tension test*. West Conshohocken, PA: ASTM International.

Austroads. 2012. *Guide to Pavement Technology Part 2: Pavement Structural Design*. Publication No. AGPT02-12. Sydney: Austroads Inc.

Austroads. 2013a. *Review of Definition of Modified Granular Materials and Bound Materials*. Publication No. AP-R434-13. Sydney: Austroads Ltd.

Austroads. 2013b. *Proposed Procedures for the Design of Pavements on Selected Subgrade and Lime-Stabilised Subgrade Materials*. Publication No. AP-R435-13. Sydney: Austroads Ltd.

Brown S.F. 1980. *The Analytical Design of Bituminous Pavements*. Nottingham, UK: Nottingham University.

Brown S.F. and A.R. Dawson. 1992. Two-stage mechanistic approach to asphalt pavement design. *7th International Conference on Asphalt Pavements*, Vol. 1, p. 16. Nottingham, UK.

CEN EN 13108-2:2006/AC. 2008. *Bituminous mixtures – Material specifications – Part 2: Asphalt concrete for very thin layers*. Brussels: CEN.

CEN EN 13108-7:2006/AC. 2008. *Bituminous mixtures – Material specifications – Part 7: Porous asphalt*. Brussels: CEN.

CEN EN 13286-7. 2004. *Unbound and hydraulically bound mixtures – Part 7: Cyclic load triaxial test for unbound mixtures*. Brussels: CEN.

CEN EN 13286-43. 2003. *Unbound and hydraulically bound mixtures – Part 43: Test method for the determination of the modulus of elasticity of hydraulically bound mixtures*. Brussels: CEN.

CEN EN 13286-47. 2012. *Unbound and hydraulically bound mixtures – Part 47: Test methods for the determination of the California bearing ratio, immediate bearing index and linear swelling*. Brussels: CEN.

CEN EN 14227-1. 2013. *Hydraulically bound mixtures. Specifications – Part 1: Cement bound granular mixtures*. Brussels: CEN.

Chaddock B. and C. Roberts. 2006. *Road foundation design for major UK highways*. Published Project Report PPR 127. Crowthorne, UK: TRL Limited.

Corte J.-L. 1997. *French Design Manual for Pavement Structures* (Translation of the December 1994 French version of the technical guide). Paris: Laboratoire Central des Ponts et Chausees.

CPDM. 2002. *Comprehensive Pavement Design Manual, Chapter 4 – New Construction/Reconstruction*. Design Division & Technical Services Division. Albany: New York State Department of Transportation.

C-SHRP. 2002. *Pavement Structural Design Practices across Canada*. C-SHRP Technical Brief No. 23. Ottawa, Ontario: Canadian Strategic Highway Research Program.

de Jong D.L., M.G.F. Peutz, and A.R. Korswagen. 1973. *Computer Program BISAR-Layered Systems under Normal and Tangential Surface Loads*. External Report AMSR.0006.73. Amsterdam: Koninklijke/Shell Laboratorium.

Destombes M.A. 2003. *Catalogue des Structures de Chaussees: Guide Technique Pour l'utilisation des Materiaux Regionaux d'ile-de-France*. Paris: Laboratoire régional de l'ouest Parisien.

Egnatia Odos. 2001. *Design Guidelines for Highway Works (DGHW), Chapter 3: Pavement Design, 1988/Revision 2001*. Thessaloniki: Egnatia Odos S.A.

Highways Agency. 1999. *Design Manual for Roads and Bridges (DMRB), Vol. 7: Pavement Design and Maintenance*. Section 5, Part 2, HD 37/99: Bituminous Surfacing Material and Techniques. Department for Transport. London: Highways Agency.

Highways Agency. 2006a. *Design Manual for Roads and Bridges (DMRB), Vol. 7: Pavement Design and Maintenance*. Section 2, Part 3, HD 26/06: Pavement Design. Department for Transport. London: Highways Agency.

Highways Agency. 2006b. *Design Manual for Roads and Bridges (DMRB), Vol. 7: Pavement Design and Maintenance*. Section 2, Part 1, HD 24/06: Traffic Assessment. Department for Transport. London: Highways Agency.

Highways Agency. 2006c. *Design Manual for Roads and Bridges (DMRB), Vol. 7: Pavement Design and Maintenance*. Section 5, Part 1, HD 36/06: Surfacing Materials for New and Maintenance Construction. Department for Transport. London: Highways Agency.

Highways Agency. 2009a. *Design Manual for Roads and Bridges (DMRB), Vol. 7: Pavement Design and Maintenance*. Section 5, Part 2, IAN 73/06 Revision 1: Design Guidance for Road Pavement Foundations. Department for Transport. London: Highways Agency.

Highways Agency. 2009b. *Manual of Contract Documents for Highway Works (MCHW), Vol. 1: Specification for Highway Works, Series 800: Road Pavements – Unbound, Cement and Other Hydraulically Bound Mixtures, Clause 818*. Department for Transport. London: Highways Agency.

Highways Agency. 2011. *Design Manual for Roads and Bridges (DMRB), Vol. 7: Pavement Design and Maintenance*. Section 5, IAN 157/11: Thin Surface Course Systems – Installation and Maintenance. Department for Transport. London: Highways Agency.

HRB (Highway Research Board). 1962. *The AASHO Road Test*. Report 5, Special Report 61E, Publication 954. National Academy of Science. Washington, DC: National Research Council.

IDOT. 2010. *Bureau of Design and Environment Manual, Part 6, Chapter 54, Pavement Design*. Revised August 2013. Springfield, IL: Illinois Department of Transportation.

LCPC. 2011. *ALIZE-LCPC Software, v 1.3.0*. Montreuil, France: LCPC.

Lijzenga J. 1997. On the prediction of pavement rutting in the Shell pavement design method. *2nd European Symposium on Performance and Durability of Bituminous Materials*. pp. 175–193. Leeds.

Mincad Systems. 2009. *CIRCLY 5 Users' Manual*. Richmond, Victoria, AU: MINCAD Systems Pty Ltd. Available at http://www.mincad.com.au.

Mobil. 1985. *Asphalt Pavement Design Manual for U.K. Ltd*. London: Mobil Oil Company.

NCHRP. 2004. *Guide for Mechanistic-Empirical Design of New and Rehabilitated Pavement Structures, 1-37A*. National Cooperative Highway Research Program, Transportation Research Board, Washington DC: National Research Council.

Nikolaides A., G. Tsohos, and A. Papavasileiou. 1997. *Pavement Design Guide*. Submitted Report. Thessaloniki: Egnatia Odos S.A.

Nikolaides A.F. 2005. *Flexible Pavements: Pavement Design Methodology, Bituminous Mixtures, Antiskidding Layers*, 1st Edition. Thessaloniki, Greece: A.F. Nikolaides. ISBN 960-91849-1-X.

Nunn M. 2004. *Development of a More Versatile Approach to Flexible and Flexible Composite Pavement Design*. Report TRL615. Crowthorne, UK: TRL Limited.

Powell W.D., J.F. Potter, H.C. Mayhew, and M.E. Nunn. 1984. *The Structural Design of Bituminous Roads*. TRRL Laboratory Report LR 1132. Crowthorne, UK: Transport Research Laboratory.

RRL. 1970. *A Guide to the Structural Design of Pavements for New Roads, Road Note 29*, 3rd Edition. Road Research Laboratory. Department of Environment. London: HMSO.

RStO 01. 2001. *Recommendations for Standardizing the Construction of Road Pavements* (translated). Köln, Germany: FGSV Verlag.

Shell. 1978. *Shell Pavement Design Manual: Asphalt Pavements and Overlays for Road Traffic*. London: Shell International Petroleum Company Ltd.

Shell. 1985. *Addendum to the Shell Pavement Design Manual: Asphalt Pavements and Overlays for Road Traffic*. London: International Petroleum Company Ltd.

Shell. 1998. *SPDM version 3.0*. Bitumen Business Group. The Hague: Shell International Oil Products B.V.

Valkering C.P. and F.D.R. Stapel. 1992. The Shell pavement design method on a personal computer. *7th International Conference on Asphalt Pavements*, pp. 351–374. Nottingham, UK.

WSDOT. 2011. *WSDOT Pavement Policy*. Environmental and Engineering Programs Division. Olympia: Washington State Department of Transportation.

Chapter 14

Rigid pavements and design methodologies

14.1 GENERAL

Rigid pavements are high-stiffness pavements with the main structural layer being Portland cement concrete (PCC). Because of their high stiffness in contrast to flexible pavements, the deformation of the underlying layer to the concrete layer that may appear is not reflected to the pavement surface. However, after a certain period, surface cracks appear on the surface of the concrete slab.

A typical rigid pavement consists of a PCC layer and a sub-base layer. Recent construction techniques and methodologies also require the use of an asphalt surfacing layer, certainly in the case of continuously reinforced concrete layer. The use of continuously reinforced concrete pavements (CRCPs) with an asphalt layer on top can provide a 'long life' with all the advantages offered by the noise-reducing properties of the surfacing (Highways Agency 2006b).

Depending on the existence (or non-existence) of reinforcement, rigid pavements are divided into four types of pavements: (a) unreinforced concrete pavements (URCPs), (b) jointed reinforced concrete pavements (JRCPs), (c) CRCPs and (d) CRCPs with asphalt surfacing layer, formerly known as composite rigid pavements.

The CRCPs with asphalt surfacing layer may be further distinguished into CRCPs with thin asphalt overlay (minimum thickness of 30 mm) or CRCPs with an asphalt overlay of 100 mm. The latter is known as rigid pavement with continuously reinforced concrete base (CRCB) and asphalt overlay.

The characteristic types of the abovementioned rigid pavements are presented in Figure 14.1.

The basic difference among URCP, JRCP and CRCP is that in the first two types, joints are constructed in both directions (transverse and longitudinal), whereas in the last type, joints are constructed only along the longitudinal direction (construction joints). It should be noted that in the jointed reinforced pavements, the number of transverse joints is less than that in the unreinforced pavements, whereas in the case of continuously reinforced pavement, there are no transverse joints.

Nowadays, there is a tendency to construct continuously reinforced rigid pavements with an asphalt overlay. In some countries, like the United Kingdom, when rigid pavement is to be constructed in motorways or heavily trafficked areas, it is advised to use exclusively continuously reinforced pavements with an asphalt overlay.

The construction of a URCP is simpler and cheaper in absolute cost values when compared to jointed reinforced pavement. However, the disadvantage of the URCP is the greater number of transverse joints that reduces comfort during driving and increases the possibilities of construction failure and differential slab settlement in the future.

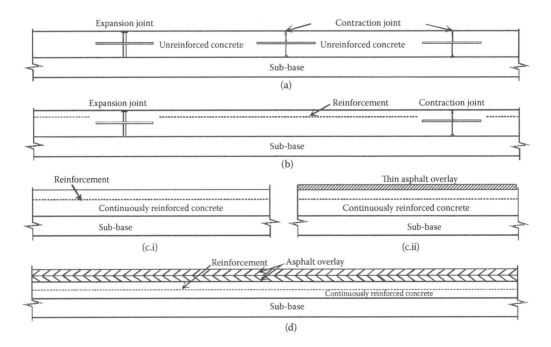

Figure 14.1 Characteristic types of rigid pavements. (a) Unreinforced concrete pavement (URCP); (b) jointed reinforced concrete pavement (JRCP); (c.i) continuously reinforced concrete pavement (CRCP), without asphalt overlay; (c.ii) CRCP, with thin asphalt overlay and (d) rigid pavement with continuously reinforced concrete base (CRCB) and asphalt overlay. (Adapted from Highways Agency, *The Manual of Contract Documents for Highway Works [MCDHW], Volume 3: Highway Construction Details*, Department for Transport. London: Highways Agency, 2006c.)

The disadvantage of transverse joints is eliminated by the use of continuous reinforcement, without increasing the cost when compared to JRCP.

During the recent years in certain countries, the continuously reinforced pavements prevail, particularly those with an asphalt overlay. These pavements are more expensive than all the other abovementioned types but are more cost-effective, offer better serviceability level, are easily maintained and are ideal to the application of further asphalt overlays at stages during the future pavement life.

It must be mentioned that in a small number of projects, pre-stressed reinforcement has been used. The use of pre-stressed concrete has some advantages, but the construction cost tends to be higher and it requires specialised staff and construction equipment. The advantages of using pre-stressed concrete slabs are as follows: (a) reduction of slab thickness (usually 10 to 15 cm slab is required), (b) ability to apply greater loads and (c) substantial reduction of total number of transverse joints, when compared to URCP.

The present chapter discusses the basic principles of rigid pavements, provides some information on concrete material, gives constructional details for the joints and describes in detail two well-known rigid pavement design methodologies: the UK methodology (Highways Agency 2006a) and the American methodology (AASHTO 1993). It also outlines Australian (Austroads 2012) and AASHTO's MEPDG (AASHTO 2008) rigid pavement design methodologies used.

14.2 SUBGRADE AND LAYERS OF RIGID PAVEMENTS

14.2.1 Subgrade

The main concern in the design of a rigid pavement is whether the subgrade is able to provide a uniform and stable platform for the concrete slab throughout the design life and the entire length of the pavement. The main parameters that directly affect the bearing capacity (strength) of the subgrade are moisture and low-temperature (freeze–thaw) conditions.

To face the problem of moisture, especially when the soil material is sensitive to moisture with a plasticity index (PI) <25, the lowering of the water table level is necessary to maintain it at a pre-determined depth below formation level (typically greater than 300 mm). This is achieved by the provision of an effective sub-drainage system or by raising the pavement structure to construct an embankment.

With regard to freeze–thaw conditions, the overlying pavement structure should have a thickness greater than the pre-determined depth of frost penetration.

The bearing capacity, or strength, of the subgrade may be expressed by the California bearing ratio (CBR) value or the resilient modulus (M_R). Details for both parameters and their correlation are provided in Sections 1.6.1 and 1.9.1. Some methodologies may use, instead, the modulus of subgrade reaction (k value) (see Section 1.7).

In general, it could be said that the quality of the subgrade affects to a lesser extent the design of rigid pavements compared to flexible pavements. However, this does not mean to overlook the fact that a pavement is as good as the subgrade material it sits on.

14.2.1.1 Swelling/shrinkage of subgrade

Swelling and shrinkage of the subgrade material certainly negatively affect the performance of the rigid pavement, in particular the unreinforced or jointed reinforced type.

Swelling and shrinkage of high-plasticity subgrade soil material are avoided by taking the same measures as in flexible pavements. These measures are (a) to provide effective sub-drainage and run-off drainage, to keep subgrade moisture at equilibrium; (b) to stabilise the soil material and (c) to construct an embankment of an appropriate height.

14.2.1.2 Frost protection of subgrade

If the subgrade is frost susceptible, the same measures are taken as in flexible pavements, that is, to make sure that the overlying layers (sub-base and concrete slab), consisting of non-frost-susceptible material, have a thickness greater than or equal to the frost heave (frost penetration depth). A thickness of 350 mm of non-frost-susceptible material over the subgrade for coastal areas, or of 450 mm anywhere else, is considered sufficient for UK environmental conditions (Highways Agency 2006b).

14.2.1.3 Drainage of the subgrade

It is of vital importance to keep water out of the subgrade both during compaction of sub-base/construction of sub-base and during the service life of the rigid pavement. The measures taken against high moisture content attributed to natural water table are similar to the ones applied for flexible pavements. As a rule of thumb, keep the water table at least 300 mm below formation level.

14.2.2 Sub-base

The sub-base layer provides a uniform and even surface to support the concrete slab and assist construction works. Its thickness certainly contributes to the strength of the pavement but not as much as the sub-base/base layer in flexible pavements.

The materials used for the construction of the sub-base are almost exclusively hydraulic (cement) bound material (HBM), as those used for sub-base/base in flexible pavements (see Section 10.5.7). The minimum compressive strength of the cement-bound material required for a rigid pavement design varies from country to country and may range from 3 to 10 MPa (7-day compressive strength).

The use of HBM eliminates pumping of fines most certainly to appear in jointed concrete pavements after a period of time, if unbound aggregate material had been used and joint maintenance is delayed. Pumping causes edge or corner cracking at the joints.

In case sub-base is constructed with unbound aggregates, the aggregate material should not contain high percentage of fines (less than 10% passing through a 0.063 or 0.075 mm sieve) and its PI should be less than 6. Greater insurance is provided when the material used is cement bound (Lister and Maggs 1982).

The compacted surface of the sub-base should be even with the required transverse slope to facilitate the construction of the uniform thickness concrete slab and economise on construction cost.

Most rigid pavement construction methods enforce the use of separation and waterproofing between the upper surface of the sub-base and the reinforced or unreinforced slabs. This membrane prevents water loss from the concrete during curing. It also acts as a frictionless surface, particularly useful for unreinforced-type pavements. It also eliminates the friction developed at the interface between the sub-base and concrete slab owing to the temperature changes during curing.

Generally, the role of the sub-base in rigid pavements is not to increase the structural strength of the pavement but to provide a layer of uniform bearing capacity to avoid local failures. Additionally, the sub-base provides a good surface for construction vehicle traffic.

14.2.3 Concrete layer and constituents of concrete

The concrete layer in a rigid pavement is the main structural element that receives and distributes the traffic loads to the underlying layer and provides a comfortable and safe surface to the traffic.

An equally important function of the concrete layer is to withstand the thermal stresses attributed to expansion, contraction or warping; the friction stresses that developed with the underlying layer; and the stresses attributed to moisture changes of the concrete.

Although the concrete layer is more durable than an asphalt layer, it requires frequent joint maintenance and restoration of the skid resistance property of the surface. The frequency of intervention depends on the quality/suitability of the material used and workmanship achieved.

Joint maintenance, although time consuming, is generally effective; restoration of the concrete surface's skid resistance though is a difficult task and needs special consideration and not always as effective.

14.2.4 Constituents of concrete

The materials used for the production of concrete are cement, aggregates, water and chemical additives. The aim of this chapter is not to give a detailed description of the concrete technology but to provide the reader the basic information and characteristics of concrete that will help in the design of a rigid pavement.

14.2.4.1 Aggregates

The aggregates used for concrete production are crushed or natural, clean and durable aggregates, the properties of which conform to national specifications, such as CEN EN 12620 (2008) for countries in the European Union (EU) and ASTM C 33 (2013) or AASHTO M 6 (2013) and AASHTO M 80 (2013) for the United States.

It is noted that blast furnace slags may also be used as aggregates for the production of concrete, after they have been crushed, and provided their properties conform to the requirements of the abovementioned standards.

The gradation of the aggregate mixture is not so strict as in the case of bituminous mixtures. Usually, the ratio of sand (particles passing through the 4, 5 or 4.75 mm sieve) to coarse aggregates (particles retained on the 4, 5 or 4.75 mm sieve) is determined. Institutions and countries use their own gradation curves that are more or less similar.

The maximum particle size used is defined by the size of the slab. As a general rule for optimal strength and workability of the concrete mixture, the maximum particle size should not exceed 1/4 of the slab thickness. The most common mixtures used are those with 40 or 20 mm maximum nominal particle size.

The aggregates, especially the sand in the concrete exposed to traffic, should consist of hard and durable rock (not from limestone) for the achieved surface micro-texture to last.

14.2.4.2 Cement

The cement used for the production of concrete is the common cement (Portland) or cement with fly ash. The cement should be in compliance with national specifications. For Europe and the United States, the relevant specifications for cement are the following: CEN EN 197-1 (2011) and CEN EN 450-1 (2012) if with pulverised fuel ash, and CEN EN 197-2 (2014), ASTM C 150 (2012), ASTM C 595 (2013), ASTM C 1157 (2011), AASHTO M 85 (2012) and AASHTO M 240 (2013).

In certain cases, when the concrete layer is to be seated on a sub-base or subgrade that contains water-soluble sulfides, the cement should have high resistance to sulfides.

14.2.4.3 Water and admixtures

Water should be free of substances that may affect concrete quality. It can be water from any water supply company or other source provided it conforms to CEN EN 1008 (2002) for the EU.

Chemical additives (admixtures) may be used to improve workability, to reduce permeability, to vary the hydration temperature so as to be mixed and placed at low temperatures, to accelerate the development of strength during the first stages of curing, to increase frost–thaw resistance and to improve durability. In all cases, the addition of any additive is allowed, provided the required strength and other properties of the concrete are satisfied.

The chemical additives of concrete should be in compliance with CEN EN 934-1 (2008) and CEN 934-2 (2012), for the EU, and with ASTM C 260 (2010), ASTM C 226 (2012), ASTM C 688 (2008), ASTM C 465 (2010), AASHTO M 194 (2013) and AASHTO T 157 (2012), for the United States.

With respect to air-entraining admixture, the British manual of contract documents for highway works (Highways Agency 2006d) impose that the air-entraining admixture in concretes be placed at the top 50 mm of the concrete layer, except for the following cases: (a) for pavements with an exposed aggregate concrete surface where at least the top 40 mm of the surface slab is air entrained, (b) for surface slabs of pavements with at least a class C40/50 concrete, (c) for surface slabs of pavements with a class C32/40 concrete that are to be overlaid by a 30 mm minimum thickness thin surface course system and (d) for surface slabs of pavements with a class C35/45 concrete that are to be overlaid by a 20 mm minimum thickness thin surface course system.

The incorporation of entrained air into a mix enables the concrete to withstand better the action of frost and de-icing salts. The use of air-entraining admixture also improves workability.

It is noted that the addition of certain additives slightly decreases the concrete strength, despite its positive effects. Early studies found that the addition of air-entraining admixture reduces the concrete's compressive and tensile strength (HRB 1971; Teychenne et al. 1975). The average reduction on the concrete's compressive and tensile strength was found to be at the 5% and 4% level, respectively, for every 1% by volume of additive in the mixture (Teychenne et al. 1975).

14.2.5 Concrete and concrete strength

The structural requirements and performance of concrete should comply with national requirements, such as CEN EN 13877-2 (2013) for the EU and ASTM C 94 (2013) or AASHTO M 157 (2013) for the United States. The constituents of the concrete should also comply with national specifications such as CEN EN 206 (2013) and CEN EN 13877-1 (2013) for the EU.

The strength of concrete is a defining property for rigid pavement design. However, the strength of concrete may be expressed as compressive (f_c), flexural (f_f) or tensile strength (f_t). The determination of the compressive strength may be carried out according to CEN EN 12390-3 (2011), ASTM C 39 (2012), AASHTO T 22 (2011) or AASHTO T 107. The flexural strength may be determined according to CEN EN 12390-5 (2009), ASTM C 78 (2010) or AASHTO T 97 (2010), and the tensile strength may be determined according to ASTM C 496 (2011) or AASHTO T 198 (2009) or CEN EN 12390-6 (2009).

The majority of specifications nowadays for the design of concrete pavements consider flexural strength rather than compressive strength as the criterion to determine the slab thickness.

The compressive strength characterises the concrete. The concrete classes, according to CEN EN 206 (2013), are denoted by the letter C and two numbers corresponding to compressive strength obtained from testing a cylinder or a cube, for instance C32/40.

The first value refers to the characteristic compressive strength of a cylinder with height $H = 300$ mm diameter and diameter $D = 150$ mm ($f_{c\text{-cylinder}}$) whereas the second refers to the characteristic compressive strength of a cube with an edge length of 150 mm ($f_{c\text{-cube}}$).

The characteristic compressive and flexural strength, as well as the characteristic tensile strength, are typically determined after 28 days of curing and are expressed in megapascals (MPa).

In case the characteristic flexural strength of concrete is to be estimated by its characteristic compressive strength, the following equations were derived by Hassan et al. (2005):

for siliceous gravel: $f_f = 0.45 \times (f_{c,\text{cube}})^{0.62}$

for limestone aggregate: $f_f = 0.87 \times (f_{c,\text{cube}})^{0.49}$,

where f_f is the 28-day flexural strength (MPa) and $f_{c,\text{cube}}$ is the 28-day compressive cube strength (MPa).

From the above equations, it can be seen that for the same compressive strength, the flexural strength is higher when limestone aggregates are used.

In general, the predominant factor affecting the strength of the concrete is the amount of cement and water in the mixture, expressed as water/cement ratio (w/c). Other factors

Table 14.1 Values of material constants

Concrete aggregate	Values of constants		
	a	b	c
Gravel	0.773	0.0301	0.11
Crushed rock (crushed granite or limestone aggregate)	0.636	0.0295	0.16

Source: Nunn, *Development of a More Versatile Approach to Flexible and Flexible Composite Pavements Design*, TRL Report TRL615, Crowthorne, UK: TRL Limited, 2004.

such as mineralogy, surface texture and particle shape of the aggregates also influence the strength of concrete. The strength of concrete aggregates does not affect the strength of concrete (Hassan et al. 1998; Viorin et al. 2001).

14.2.6 Elastic modulus and strength

In some cases, the elastic modulus of concrete needs to be known. The elastic modulus is determined by the dynamic compressive test in accordance to CEN EN 13412 (2006) or ASTM C 469 (2010).

The elastic modulus is related to the strength of concrete. The relationships suggested (Nunn 2004) may be used, if the elastic modulus is to be estimated from the strength of cement-bound mixtures (CBMs):

$$E = \frac{\log f_f + a}{b}$$

$$f_f = c \times f_c,$$

where E is the elastic modulus (dynamic) (GPa), f_f is the flexural strength (MPa), f_c is the compressive strength (MPa) and a, b are material constants, obtained from Table 14.1.

According to AASHTO (2008), the modulus of elasticity of lean concrete and cement-treated aggregate may be estimated by using the following equation:

$$E = 57,000 \times f_c^{0.5},$$

where E is the elastic modulus (psi) and f_c is the compressive strength (psi).

The elastic modulus, in contrast to compressive strength, does not noticeably increase with curing time, as it has been found by Croney and Croney (1991) testing specimens produced with crushed aggregates or gravels and cured for up to 520 and 104 weeks, respectively. Moreover, Croney and Croney (1991) state that the computed stresses in the concrete slab are not very sensitive to small changes in the elastic modulus.

14.3 CONSIDERATION OF STRESSES THAT DEVELOPED ON A CONCRETE SLAB

The stresses that developed on the concrete slab of a rigid pavement result from various sources, the major ones being the traffic load (axial loads), the temperature changes, the slab's moisture changes and the frictions developed on the interface with the underlying layer.

The abovementioned factors cause tensile, compressive or flexural stress. A detailed analysis of all the stresses that developed on a concrete slab is beyond the scope of this book. However, in the present section, some general information will be provided regarding the way these stresses are developed and how they are handled.

14.3.1 Stresses attributed to traffic loads

Undoubtedly, as the axial loads pass over the rigid slab, stresses are developed. The severity of stresses that developed depends on the position of the wheels (loading point) with respect to the slab. In addition, axle configuration, frequency of loading (intensity of traffic), quality/strength of sub-base and subgrade and Poisson's ratio should also be considered.

Westergaard (1926, 1933) carried out an initial and fundamental study for rigid unreinforced slabs that rested on subgrade. In his analysis, Westergaard examined three critical conditions of loading (circular contact area): when the slab was loaded at the interior (centre), at the edge and at the corner (see Figure 14.2). The equations developed for calculating the stresses can be found in Westergaard's original work (Westergaard 1926, 1933) as well as in other references (O'Flaherty 2002).

Of these conditions, the stresses produced at the corner (tension σ_c at the top) are by far the most severe and hence the most important. Edge loading produces stresses (tension σ_e at the bottom) that are slightly less than those caused by corner loading, whilst a load at the interior of the slab generates the least stress (tension σ_i at the bottom).

Thus, the slab is expected to crack first at the corners, then at the edges and at the end at the interior. In practice, this does not always happen, as the loading conditions differ. It should be highlighted that the loading conditions at the corners and the edges of the concrete slab are further enhanced by the stresses that developed owing to temperature changes of the slab and lack of support attributed to possible pumping of fines at the joints.

After Westergaard, who in 1948 (Westergaard 1948) also examined the case of elliptical loading surface, other researchers have also investigated the stresses that developed on concrete slabs, such as Pickett (1951), Pickett and Ray (1951), Wang et al. (1972), Shi et al. (1994) and Fwa et al. (1996).

The detrimental effect of stresses induced by an applied load substantially increases with the loading configuration (other than single wheel load), at rest or during movement. The impact may also be considerable when the adjacent slabs that make up the pavement are 'non-flushed' with each other. To handle this impact, the transverse joints provided have effective load transfer devices such as dowels.

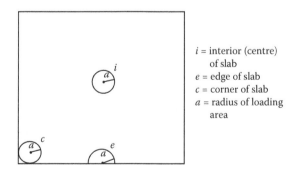

Figure 14.2 Critical conditions of loading according to Westergaard analysis.

14.3.2 Thermal stresses

The thermal stresses that developed owing to temperature changes in the concrete slab have a greater impact during the early stage of concrete curing rather than at later stages, after the concrete has hardened.

14.3.2.1 Contraction stresses

Contraction stresses develop when slab/environmental temperature decreases. Contraction tensile stresses should be defused; otherwise, contraction cracks will develop. To avoid contraction stresses, adequately spaced contraction joints are provided.

14.3.2.2 Expansion stresses

The expansion stresses develop when slab/environmental temperatures increase.

The expansion stresses that developed may be estimated by using the following equation:

$$\sigma_c = E \times \varepsilon \times \delta T,$$

where E is the elastic modulus of concrete (MPa), ε is the coefficient of thermal expansion of concrete per degree Celsius and δT is the increase in temperature (°C).

The expansion (compressive) stresses that developed, when the slab is constrained at its end, should also be defused; otherwise, there will be crack formation. This is achieved by the provision of adequately spaced expansion joints.

14.3.2.3 Warping stresses

Warping stresses develop as a result of the differential temperature (temperature gradient) between the upper and lower surface of the slab. The non-uniform distribution of temperatures within the depth of the concrete slab may cause upward or downward curling. Concrete slab curling has an effect on the pavement's behaviour, which can be attributed to the consequent loss of support (Shi et al. 1993; Tang et al. 1993).

The problem of warping was addressed by Westergaard (1926) who developed equations for three different cases (with the assumption that the temperature gradient from the top to the bottom of the slab was linear), the simplest of which is given below:

$$\sigma_0 = \frac{E \times \varepsilon \times t}{2 \times (1-\mu)},$$

where E is the concrete modulus of elasticity (lb/in^2), ε is the coefficient of linear thermal expansion of concrete per degree Fahrenheit, t is the temperature difference between the two surfaces (°F) and μ is Poisson's ratio.

Since then, it has been shown that the temperature gradient is closer to a curve line. This results in calculated stress values that are much lower than those derived by Westergaard (O'Flaherty 2002). Work carried out by Zhang et al. (2003) concluded that warping stresses caused by the assumption of linear temperature distribution could be as high as 30% or more.

For a temperature gradient of 5°C between the top and bottom of a 150 mm slab, the stresses induced by temperature warping are not as detrimental as might be expected (O'Flaherty 2002).

14.3.3 Temperature–friction stresses

As the concrete slab expands or contracts owing to temperature changes, the friction developed on the interface with the subgrade causes tensile or compressive stress, respectively. These stresses when combined with thermal stresses may lead to serious slab damage.

In case of slab expansion without the provision of sufficient joint spacing, slabs may blow up. For this reason, it is recommended that plastic membranes be placed on unreinforced pavements so as to eliminate the thermal–friction stresses that developed.

14.3.4 Stresses attributed to moisture change

The concrete slab tends to expand as moisture content in its mass increases; otherwise, it contracts. Thus, tensile or compressive stresses are developed. The development of such stresses, attributed to non-uniform moisture distribution, causes slab warping, as in the case of warping owing to temperature gradient.

It could be said that the stresses that developed as a result of moisture change actually oppose the stresses that developed owing to thermal contraction/expansion. This happens because during the summer, the moisture of the concrete slab is reduced, resulting in the development of compressive stresses, whereas at the same time, because of high temperature, tensile stresses are developed.

Moisture-induced stresses are more important at the early stage of curing and in hot regions with wet and dry seasons.

14.4 CRACKING OF FRESH AND HARDENED CONCRETE

Cracking, other than fatigue cracking, is a common phenomenon in any concrete structure, including concrete pavements. Cracking of fresh or hardened concrete is caused, to a great extent, by the moisture-induced volume changes of the concrete during the early stages of curing, as well as after concrete has hardened. External factors also cause cracking of hardened concrete.

In general, concrete cracking may be distinguished into two: cracking that developed when the concrete is fresh (plastic stage) and cracking that developed when the concrete is hardened.

Cracking that developed in fresh concrete can be further distinguished into the following:

1. Plastic settlement cracking
2. Plastic shrinkage cracking

Cracking that developed in hardened concrete can be further distinguished into the following:

1. Crazing cracking
2. Early thermal contraction cracking or thermal stress cracking
3. Drying shrinkage cracking or long-term drying shrinkage cracks
4. Chemical reaction cracking (aggregates)
5. Cracking attributed to poor construction practice
6. Cracking attributed to errors in design and detailing
7. Cracking attributed to externally applied loads
8. Weathering cracking or defrost cracking
9. Cracking attributed to corrosion of reinforcement
10. Cracking attributed to overloading

Finally, cracking in hardened concrete may be caused by propagation of exiting cracks (reflective cracks) owing to lack of subgrade support or swelling/shrinkage of soil material.

14.4.1 Cracks in fresh concrete

The cracks in fresh concrete develop during its initial stage of curing, approximately 10 min to 6 h after concreting.

14.4.1.1 Plastic settlement cracking

Plastic settlement cracking normally appears on reinforced elements and is caused by excess bleeding and rapid early drying conditions in conjunction with some obstruction such as reinforcing bars or large particles of aggregates. They are formed in the longitudinal or perpendicular direction and they can be of variable width and of variable length. Their typical time of appearance is 10 min to 3 h. Plastic settlement cracking can be avoided by reducing bleeding or using vibration during concreting.

14.4.1.2 Plastic shrinkage cracking

Plastic shrinkage cracking is caused by rapid early drying and low rate of bleeding. Plastic shrinkage cracks are formed in random and roughly parallel to each other, spaced from a few centimetres up to 3 m. They can be very deep (typical depth, 25 to 100 mm), ranging in width from 0.1 to 3.0 mm and of varying lengths (short to as long as 1.5 m or longer). The typical time of appearance is 30 min to 6 h.

Plastic shrinkage cracking is generally believed to likely occur when the rate of evaporation exceeds the rate at which the bleeding water rises to the surface, but it has been observed that cracks also form under a layer of water and merely become apparent on drying (Lerch 1957). Evaporation rate that approaches the rate or 1 kg/m²/h is considered to be critical and 'precautions against plastic shrinkage are necessary' (ACI 2000). Furthermore, drying rates greater than 0.3 kg/m²/h may present a problem for paving concrete, depending on bleeding rates during the same time (Pool 2005). In effect, as concrete hardens, the cement and water begin to shrink, and the stresses created by this shrinking cannot be overcome by the small amount of strength developed by the young concrete.

Environmental conditions such as ambient temperature, relative humidity and wind velocity during curing are very critical. An increase in ambient temperature, a drop in relative humidity and an increase in wind velocity encourage the development of shrinkage cracking.

The rate of water evaporation can be determined from the nomograph given in Figure 14.3, for certain weather conditions (ambient temperature, relative humidity and wind velocity) and concrete temperature at delivery. The rate of evaporation obtained should always be less than the recommended limit of 1 kg/m²/h.

To avoid or to eliminate the effect of all the above factors affecting shrinkage, it is necessary during curing to give particular emphasis and attention to details with respect to moisture retention and concrete temperature. Both are affected by the timing of concreting, as well as by the environmental conditions during early curing.

Figure 14.4 illustrates the influence of timing of concreting with respect to ambient temperature. When ambient temperatures are high, concreting time should be delayed to stagger the daily temperature peak from the time of heat generation during hydration (Jeuffroy and Sauterey 1996). High ambient temperature is considered to be >25°C, or lower, if humidity is lower than 50%.

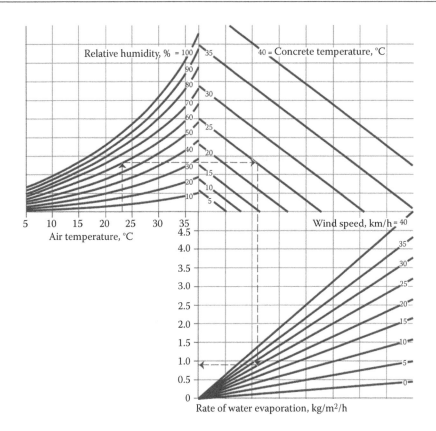

Figure 14.3 Determination of rate of water evaporation during curing. (From ACI, *Hot Weather Concreting*. ACI 305R-99. Farmington Hills, MI: American Concrete Institute, 2000; and CCAA, *Plastic Shrinkage Cracking*. Australia: Cement Concrete and Aggregates, 2005.)

Plastic shrinkage cracks do not seem to cause problems in some situations, but in other cases, they provide an entry for de-icing salts and may contribute to freezing and thawing damage (Pool 2005).

Plastic shrinkage cracking can be avoided by taking extra care during early curing.

14.4.2 Cracks in hardened concrete

14.4.2.1 Crazing cracking

Crazing cracking is caused by over-towelling, poor curing or usage of mixes rich in cement and occurs when the surface zone of the concrete has higher water content than deeper in the interior.

The pattern of crazing looks like irregular network (the term *map cracking* is often used) with spacing of up to approximately 15 cm. The cracks are very shallow, very fine (less than 0.15 mm in width) and barely visible except when the concrete is drying after the surface has been wet. The typical time of appearance is after 1 to 7 days from concreting and sometimes later.

Crazing cracks, apart from being unsightly and can collect dirt, are of very little importance. Crazing is not structurally serious and does not ordinarily indicate the start of a future deterioration.

Crazing can be avoided by improving curing and finishing procedures of concrete.

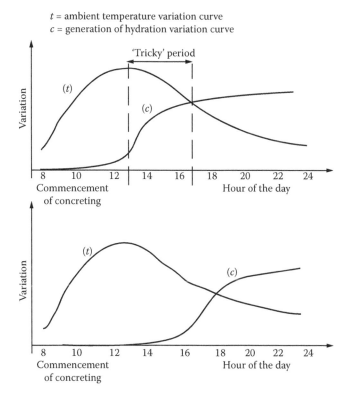

t = ambient temperature variation curve
c = generation of hydration variation curve

Figure 14.4 Influence of timing of concreting. (From Jeuffroy, G. and Sauterey, R., *Cement Concrete Pavements [Chaussées en betón de ciment]*, published by arrangement with Presses de l'Ecole Nationale des Ponts et Chaussées. Brookfield, WI: A.A. Balkema Publishers, 1996.)

14.4.2.2 Early thermal contraction cracking

Early thermal contraction cracking, also known as thermal stress cracking or curling, is caused by excess temperature gradients (deferential) or moisture content between the top and bottom of a slab, attributed to rapid cooling when ambient temperature drops and, in the case of thick (mass) concrete elements, to excess heat generation during hydration of the cement.

The early thermal contraction cracks are linear and formed in the slab's corners or edges. The cracks may be deep and of variable length depending on the size of the concrete element. The typical time of appearance is after 1 day to 2–4 weeks.

To prevent early thermal contraction cracking, reduce heat during hydration or insulate the concrete element. Early thermal contraction cracking can be reduced by using properly spaced joints, creating uniform moisture content and temperature of the slab from top to bottom, using a low-shrinkage concrete mix, using thickened slab edges and, in the case of reinforced slab, using large amount of reinforcing steel 50 mm down from the surface.

14.4.2.3 Drying shrinkage cracking

Drying shrinkage cracking, also known as long-term drying shrinkage cracking, is a common type of cracking in hardened concrete. Drying shrinkage cracking is caused by inefficient joints, excess shrinkage and inefficient curing.

The drying shrinkage cracks are almost linear cracks and are formed, when single, either transversely or longitudinally to the longer length of the slab, or when branched into a three-crack pattern, at an angle approximately 120°. They can be very deep, quite long and as wide as the plastic shrinkage cracks (0.1 to 3 mm).

The typical time of their appearance is several weeks or months after concreting. Some movements have been observed even after 29 years but part of this long-term shrinkage is likely to be attributed to carbonation (Neville 2011).

Drying shrinkage is caused by the withdrawal of water from concrete exposed in unsaturated air. The change in volume of drying concrete is not equal to the volume of water removed, owing to internal restrain provided by the aggregate particle. It is the combination of shrinkage and restraint that causes tensile stresses to develop, and when the tensile strength of the concrete is exceeded, concrete will crack. More about the complex mechanism of drying shrinkage can be found in ACI (2007) and Neville (2011).

The main factors affecting drying shrinkage are the amount of total water, the amount of coarse aggregate, the elastic properties of the aggregate (compressibility or stiffness) and the type (nature) of the aggregates. The higher water content, the greater the amount of shrinkage is. As the quantity of coarse aggregate increases, the shrinkage decreases. The higher the stiffness of the aggregate, the more effective it is in reducing the shrinkage of the concrete (Brooks and Neville 1992; Neville 2011). As for the type of aggregates, the use of shrinking aggregates increases the shrinkage. The use of calcium chloride admixtures with shrinking aggregates increases even further the shrinkage (PCA 2001).

The effect of aggregate type: Within the range of aggregates used, there is a considerable variation in shrinkages of the resulting concrete. The usual natural aggregates are not subject to shrinkage, but there exist rocks that shrink on drying about the same magnitude as shrinkage of concrete made with non-shrinking aggregates. These aggregates are called shrinking aggregates. Shrinking aggregates are those produced from some dolerites/diabases, basalts and also some sedimentary rocks such as greywacke and mudstones. On the other hand, granite, limestone and quartzite have been found to be non-shrinking (Neville 2011).

When shrinking aggregates are used, even plastic cracking takes place very early and the time of sawing the joints should be adapted to the specific conditions of the work (percentage of shrinking aggregate used in the mix and daily temperature variations) (Jeuffroy and Sauterey 1996). In the case of using shrinking aggregates (basically non-calcareous aggregates), the spacing distance of the contraction joints should be decreased, if the original spacing distance had been determined for non-shrinking (calcareous) aggregates, and vice versa.

Additionally, when non-calcareous aggregates are used, special precautions should be taken in order to retard irregular cracking. These precautions are as follows (Jeuffroy and Sauterey 1996):

- Concreting time should be delayed to stagger the ambient temperature peak from the time of heat generation during hydration (see Figure 14.4).
- Joints should be made deeper.
- Transverse joints should be made in fresh concrete (only in the case of pavements not requiring a very high level of riding quality).
- One out of two or three joints should be sawn as early as possible while the others should be sawn with minimal delay.

Similar precautions are also recommended by Highways Agency (2006d).

Other factors: Other factors influencing drying shrinkage cracking are as follows: (a) use of cement deficient in gypsum exhibits a significant increase in shrinkage, (b) inclusion of either fly ash or granulated blast furnace slag in the mix increases shrinkage, (c) use of water-reducing admixtures may cause a small increase in shrinkage and (d) entrainment of air has been found to have no effect on shrinkage (Keene 1960; Brooks and Neville 1992; Neville 2011). The increase in cement content, within the range of usual concrete mixtures, 280 to 445 kg/m³ cement, has little to no effect on shrinkage as long as the water content is not increased significantly and non-shrinking aggregates are used (PCA 2001).

Drying shrinkage cracking can be prevented by (a) reducing the amount of the water content, (b) using non-shrinking aggregates, (c) maximising the size and amount of coarse aggregates, (d) improving curing, (e) providing contraction joints that are effectively spaced, (f) using shrinkage compensating additives and (g) avoiding the use of excessive amount of cementitious materials.

The expected drying shrinkage cracking is compensated by planned joint cutting. However, if significant drying of the hardened concrete occurs within the first few days/weeks after placement, the drying shrinkage that developed will result in cracking at closer intervals than that expected and planned for by the contraction joints. Stresses attributed to drying and temperature loss may be additive, leading to increased potential for uncontrolled cracking (Pool 2005).

14.4.2.4 Chemical reaction cracking

Chemical reaction cracking is caused by deleterious chemical reactions. These reactions may be attributed to materials used to make the concrete or materials that come into contact with the concrete after it has hardened (sulfates, de-icing salts, etc.) (ACI 2007).

Concrete may crack with time as a result of slowly developing expansive reactions between aggregates containing active silica and alkalis derived from cement hydration, admixtures or external sources (such as curing water, groundwater, de-icing chemicals and alkaline solutions stored or used in the finished structure) (Brooks and Neville 1992).

Certain carbonated rocks participate in reactions with alkalis that, in some instances, produce detrimental expansion and cracking. These detrimental alkali–carbonate reactions are usually associated with argillaceous dolomitic limestones that have a very fine-grained structure (ACI 2007).

Chemical reaction cracking due to alkali–aggregate reaction typically appears after more than 5 years.

14.4.2.5 Poor construction practice cracking

A wide variety of poor construction practice results in cracking in concrete structures. Among these, the most common practice is to add water to concrete to increase its workability. Others are lack of curing, inadequate consolidation, inadequate formwork supports and placement of construction joints at points of high stress. Additionally, by adding cementitious material, even if the water/cement ratio remains constant, more shrinkage will occur because the paste volume is increased (ACI 2007).

Inadequate curing before opening the pavement to traffic may result in concrete with poor skid resistance owing to loss of surface mortar. Areas with heavy traffic or lanes for commercial vehicles are going to suffer first.

To avoid poor construction cracking, simply follow precisely and strictly the instructions and specifications provided.

14.4.2.6 Cracking attributed to errors in design and detailing

Errors in design and detailing a pavement slab that will result in cracking include inefficiently spaced joints, improper foundation design, lack of reinforcement (in case of a reinforced slab) and lack of detailing of reinforcement.

Improper foundation design refers to the sub-base, over which the concrete slab is going to be constructed, with respect to its thickness, related also to the bearing capacity of the subgrade. Special consideration should be given when constructing a slab over an existing sub-base. Any cracks that may develop in the sub-base must be investigated before suitable measures are taken. If not detected, they will certainly cause additional cracking (reflective cracking).

The time of appearance of the reflective cracking, or cracking owing to settlement of the subgrade, as well as cracking owing to improper joint spacing or lack of reinforcement, can be several weeks to months.

14.4.2.7 Cracking attributed to externally applied loads

This type of cracking is caused by load-induced tensile stresses and is not a common phenomenon in concrete pavement slabs. However, an early use of the pavement slab without the concrete having achieved its required strength may cause cracking, primarily, near the edges of the joints.

14.4.2.8 Weathering cracking or defrost cracking

This type of cracking requires the existence of freezing and thawing periods over a certain period. It is exclusively caused by the use of frost-damaged aggregates.

Typical time of appearance of defrost cracking is more than 10 years. To avoid defrost cracking, use non-frost-damaged aggregates (normally aggregates with low percentage water absorption) or reduce aggregate size.

14.4.2.9 Cracking attributed to corrosion of reinforcement

This type of cracking is caused by the corrosion of the steel and it appears more than 2 years after construction.

Corrosion of the reinforcing steel occurs if the alkalinity of the concrete is reduced through carbonation or if the passivity of the steel is destroyed by aggressive ions (usually chlorides) (ACI 2007).

Corrosion cracks are formed in the longitudinal direction parallel to the reinforcing bars. To avoid corrosion cracking, use concrete of low permeability and adequate cover (ACI 2007; Neville 2011).

14.4.2.10 Cracking attributed to overloads

Cracking attributed to overloads is caused either by inadequate design (failure to predict the loads going to be applied) or by construction not conforming to the specification with respect to the early stages of handling and the concrete gaining stiffness.

For inadequate design (under-designed structure), the typical time for cracking to appear is after some years, depending on the original design life assumed.

For not conforming to the specifications with regard to handling, cracking will appear very early. A typical case for this type of cracking can be seen in pre-cast members when they are not properly supported during transportation, lifting or erection.

However, cracking attributed to overloads can appear on cast-in-place concrete, such as pavement slabs. When the concrete has not gained its sufficient strength and the pavement is opened to traffic, the stresses imposed are most likely to cause cracking.

14.4.3 The effect of surface cracking and its repair

The development of cracking on pavement slabs, for reasons described above, may impair the durability of concrete. It makes the concrete more permeable, which may contribute to more rapid deterioration from scaling after the use of de-icing salts. Additionally, it mars the appearance of the structures (Pool 2005; Neville 2011).

Cracking does not really affect the load-bearing capacity because the weakened zone is usually confined to the top 50 mm of concrete. Load-bearing capacity could be a problem if full-depth cracking has occurred or if concrete is cured for an insufficient amount of time (Pool 2005).

Successful repair of cracking is often very difficult to attain. Cracking in fresh or hardened concrete is a type of failure that is far easier to prevent and almost impossible to repair efficiently and successfully.

It is better to leave most of the concrete cracking unrepaired than to attempt an inadequate or improper repair (Smoak 2012).

However, prior to any attempt to repair cracking, it is necessary to determine the width and cause of the cracks, as well as whether they are still active or stabilised. The alternatives to repairing cracks range from doing nothing (if cracks are ≤0.25 mm) up to reconstructing the whole cracked area. In between, the cracks can be routed and sealed or they can be gravity filled. Surface dressing or micro-surfacing can also be used.

Routing and sealing are used when cracks are linear or almost linear, so that the saw cutter can operate easily. Sealing, after thorough cleaning, is carried out using gravity fill polymers.

Gravity fill is also used when cracking is random or patterned. The material normally used is gravity polymer covered with hard and durable sand. Epoxy resin–based materials instead of elastomer/plastomer-based materials are preferred when a longer-lasting surface skid resistance is required. However, this method of repair can be very expensive.

Surface dressing is cheaper than micro-surfacing but both do no not have anywhere near the same life expectancy as epoxy resin treatment.

14.4.4 The role of joints

Since the contraction of concrete engenders cracking, to avoid self-propagation of cracks in an anarchic manner, the strength of the pavement is deliberately reduced at certain selective positions, so that cracking is induced and localised. This is achieved by reducing the slab thickness, cutting it with a saw at selective locations, that is, introducing a transverse contraction joint in the longitudinal direction.

The spacing distance between the contraction joints is very important and critical. When it is greater than required, cracks will develop, and hence, the joint does not serve its purpose.

The spacing between transverse contraction joints depends on the slab thickness, the aggregate type used and, in the case of reinforced slab, the quantity of reinforcement. The use of 'shrinking aggregates' requires lesser joint spacing in comparison to 'non-shrinking' aggregates. The UK pavement design methodology (Highways Agency 2006b) considers that the distance determined for concrete with aggregate having a coefficient of thermal expansion greater than 10×10^{-6} per degree Celsius (use of shrinking aggregate) may be increased by 20% if concrete is used with aggregate having a coefficient of thermal expansion less than 10×10^{6} per degree Celsius (use of non-shrinking aggregates).

Table 14.2 gives coefficients of thermal expansion of some parent rock aggregate, cement paste under saturated conditions, concrete and steel.

The sawing time is of importance, especially when concreting in hot weather and even more when aggregates with high coefficient of thermal expansion are used. Sawing should start after the concrete has hardened to that degree that track marks cannot be left by the

Table 14.2 Coefficients of thermal expansion

	Coefficient of thermal expansion	
	10^{-6} per °C	10^{-6} per °F
Aggregate		
Granite	7–9	4–5
Basalt	6–8	3.3–4.4
Limestone	6	3.3
Dolomite	7–10	4–5.5
Sandstone	11–12	6.1–6.7
Quartzite	11–13	6.1–7.2
Marble	4–7	2.2–4
Cement paste (saturated)		
w/c = 0.4	18–20	10–11
w/c = 0.5	18–20	10–11
w/c = 0.6	18–20	10–11
Concrete	7.4–13	4.1–7.3
Steel	11–12	6.1–6.7

Source: FHWA, *Portland Cement Concrete Pavements Research: Thermal Coefficient of Portland Cement Concrete.* Available at http://www.fhwa.dot.gov/publications/research/infrastructure/pavements/pccp/thermal.cfm, 2011.

sawing machine. Spalling or breaking at the edges of the slabs under the rotation of the saw must also be safeguarded.

To judge when exactly sawing should start, one should monitor the variations in properties of the cement supplied and the changes in the conditions, while simultaneously observing the rate of opening of the joints already constructed. This is the reason for having a temperature and humidity recording system at the job site. It is also recommended that a meteorological station be consulted for forecasts for the following day (minimum and maximum temperature, humidity and wind). This information will assist the engineer not only to estimate the interval after which sawing may be carried out but also to decide whether it is necessary to delay the commencement of concreting.

14.4.5 Hot weather concreting

When concreting in hot weather, special attention should be paid to the risks of surface drying causing the development of cracks. The higher the initial concrete temperature owing to high ambient temperatures, the more open the cracks at the joints properly spaced and the greater the risk of developing other random surface cracks owing to shrinkage. The first, in particular, may entail a low load transfer, especially with non-limestone aggregates as they have a high coefficient of expansion, higher than limestone. Hence, any measure that will lower the temperature of the concrete is favourable.

The risks of early crack development, when concreting in hot weather, are also high because of the time of concreting. If concreting is carried out in the morning, the phase of maximum liberation of heat during the chemical reactions coincides with the peak afternoon temperatures and the final phase coincides with peak temperatures towards the end of the night. This greatly fosters the development of cracking the following morning. It is thus advisable to commence concreting in hot weather in the afternoon, so that the maximum liberation of the heat of setting compensates for the fall in temperature at the end of the night (Jeuffroy and Sauterey 1996).

According to many specifications, the temperature of fresh concrete before placement should normally be less than 30°C. However, it is recommended (Jeuffroy and Sauterey 1996) that, when the ambient temperature is higher than 20°C and the relative humidity of the air is less than 50%, the content of curing compound should be increased and, if needed, two successive applications, each with the dosage initially specified, should be carried out to ensure proper waterproofing. In certain cases, cooling of the constituents of the concrete may even be considered (cement, water and aggregates). An effective solution is simply to cool the mixing water of the concrete.

14.5 JOINTS IN CONCRETE PAVEMENTS

In order to address the stresses described above, joints are provided, which are distinguished into three main categories: (a) contraction joints, (b) expansion joints and (c) warping joints (see Figure 14.5).

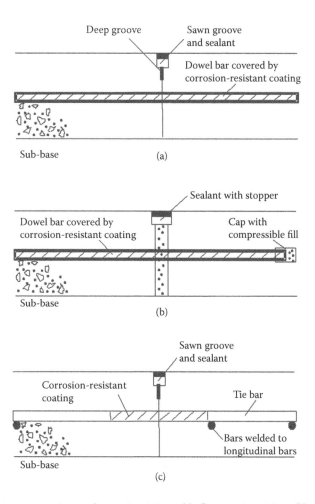

Figure 14.5 Contraction, expansion and warping joints. (a) Contraction joint; (b) expansion joint and (c) warping joint.

For unreinforced and jointed reinforced type of concrete pavements, all types of joints are provided, while in the continuously reinforced pavements, only longitudinal (warping) joints are provided (see Figure 14.6).

There are also joints that, because of the interruption of works or emergency joints (as will be explained later), do not constitute a separate joint type. The longitudinal construction joints are warping joints.

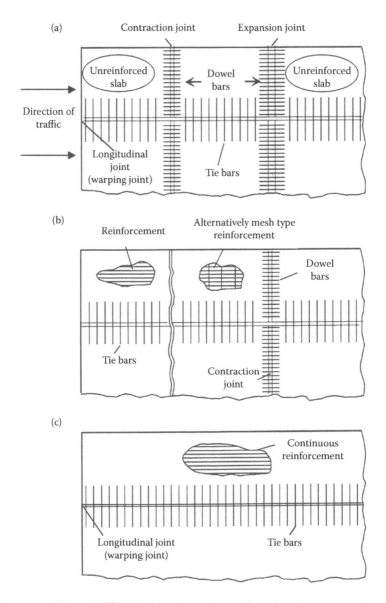

Figure 14.6 Schematic plant view of unreinforced, jointed reinforced and continuously reinforced type of rigid pavements. (a) Unreinforced rigid pavement with contraction and expansion joints; (b) jointed reinforced rigid pavement with contraction and expansion joints (expansion joint is not shown in the plan view) and (c) continuously reinforced rigid pavement with longitudinal joint.

14.5.1 Contraction joints

The contraction joints are constructed in the transverse direction to relieve tensile stresses that developed because of thermal contraction and force the slab to crack at a pre-determined position.

The latter is obtained by reducing the effective depth of the slab, inducing a sawn groove. The depth of the groove is related to the thickness of the slab; the width of the groove may be determined, but in any case its minimum value is 3–4 mm. The groove induced is filled with a suitable sealant material. Details of a typical contraction joint are shown in Figure 14.7.

In case the width of the groove (ΔL) for a contraction joint is to be calculated, the equation below may be used (AASHTO 1993):

$$\Delta L = \left[\frac{C \times l \times (\varepsilon \times \Delta T + Z)}{S} \right] \times 100,$$

where ΔL is the width of the groove (inches); C is the adjustment factor owing to sub-base/slab friction restraint, 0.65 for stabilised sub-base and 0.80 for non-stabilised sub-base; l is the contraction joint spacing (inches); ε is the coefficient of contraction (°F); ΔT is the temperature range expected (°F); Z is the drying shrinkage coefficient of the concrete slab (can be neglected for a re-sealing project) (inches/inches); and S is the allowable strain of joint sealant material; most current sealants are designed to withstand strains of 25% to 35%.

In an unreinforced concrete slab, the contraction joints are spaced approximately every 4 to 5 m, depending on the slab thickness and type of aggregate. In a jointed reinforced concrete slab, contraction joints are typically spaced every 20 to 25 m, depending on the amount of reinforcement. The designer is advised to strictly follow the instructions of the pavement methodology used with regard to the spacing of the contraction joints.

It is common to place dowel bars at the position of the contraction joint. The *dowel bars* assist the transfer of loads over the contraction joint and are placed approximately in the middle of the slab. Dowel bars, since no bonding with concrete is required, are smooth and covered by corrosion-resistant coating. Their diameter is typically 20 or 25 mm and they are usually spaced every 300 to 400 mm.

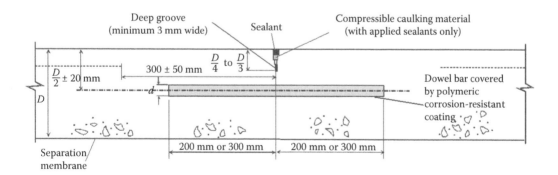

Figure 14.7 Construction details of a typical contraction joint with sawn groove. (Adapted from Highways Agency, *The Manual of Contract Documents for Highway Works [MCDHW], Volume 3: Highway Construction Details*, Department for Transport. London: Highways Agency, 2006c [© Highways Agency].)

Dowel bars may not be used when the joint spacing is smaller than 4 to 4.5 m or when traffic volume is low.

Although contraction joints are constructed perpendicularly to the direction of traffic in most cases, some designers choose to construct them at an angle, approximately 10° to 15° off the transverse direction, skew joints. By using skew joints, contraction joints are not stressed simultaneously along the full length and the repetitive noise produced by the joints is minimised.

14.5.2 Expansion joints

The expansion joints are constructed in the transverse direction to provide space for the thermal expansion of the slab, thus preventing cracking owing to compressive stresses that developed when restricted.

At the contraction joint, there is a complete discontinuity of concrete slab. The gap width introduced depends on the slab thickness, the coefficient of thermal expansion of concrete, the developed friction with the sub-base and the spacing between two expansion joints. The gap width is usually 20 to 25 mm. The top part of this gap is always filled with appropriate sealant. The spacing between expansion joints is determined by the pavement design methodology followed.

The use of dowel bars is necessary in the expansion joints, because of the gap created between slabs, ensuring smooth load transfer from one slab to the other and increasing concrete's resistance to bending and shear.

Because of the expected expansion, one end of the dowel bar is free to move owing to the provision of a waterproofing cap filled with compressible material. A typical cross section of an expansion joint is shown in Figure 14.8.

The dowel bars for expansion joints are slightly longer than those used in the contraction joints, typically 500–600 mm, and are usually placed every 300 mm. The bar diameter is always greater than that of the dowel bars placed in the contraction joints, usually 25 or 32 mm. As in contraction joints, the dowel bars are made of smooth steel and coated by a corrosion-resistant material.

For the exact determination of the diameter, length, spacing length, steel quality of the dowel bars or any other related construction detail, the designer should follow precisely the instructions of the methodology used.

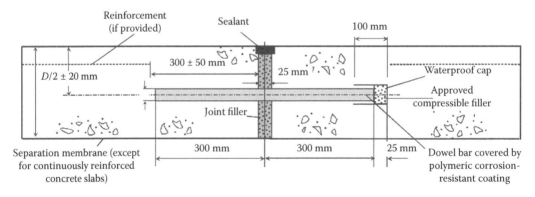

Figure 14.8 Construction details of a typical expansion joint. (Adapted from Highways Agency, *The Manual of Contract Documents for Highway Works [MCDHW], Volume 3: Highway Construction Details,* Department for Transport. London: Highways Agency, 2006c [© Highways Agency].)

Expansion joints, in particular, together with contraction joints, are the weakest points of rigid pavements in terms of early maintenance. Therefore, particular attention must be given during construction and sealing of the joints. The joint filling material and the sealant should be appropriate and long lasting.

14.5.2.1 Sealants and joint fillers

The sealants and joint fillers should be impervious to water, should be capable of being compressed and recovered, should have good adhesion properties to concrete, should not soften or harden as temperature varies and should be age resistant.

The sealants are usually modified bituminous materials applied hot or cold. The hot applied sealant materials should comply with the requirements of specifications such as CEN EN 14188-1 (2004) or ASTM D 6690 (2012). The cold applied sealants should comply with the requirements of specifications such as CEN EN 14188-2 (2004) or ASTM D5893 (2010).

The quantity of joint sealant to be poured into the joint depends on the depth and width of the gap or groove to be filled. As a rule of thumb, for expansion joints, a depth-to-width ratio of 1:1 to 1:1.5 is used, and generally, the depth should always be lower than or equal to the width. To limit the amount and depth of sealant applied to the joint, a sealant backer material (compressible caulking material) conforming to ASTM D 5249 (2010) or other standards is used.

Apart from the poured joint sealants, particularly for the expansion type of joints, there are also preformed joint seals that are placed after compressing them into the expansion joint. These materials are such that they are always compressed, even when there is a complete expansion of the joint. The preformed joint seals should comply with specifications such as CEN EN 14188-3 (2006), ASTM D 2628 (2011), ASTM 1752 (2013), AASHTO T 42 (2010), AASHTO M 33 (2012), AASHTO M 153 (2011) or AASHTO M 213 (2010).

14.5.3 Warping joints

The warping joints are constructed to control cracks that may occur owing to the development of warping stresses in the longitudinal or transverse direction. The warping joints are constructed along the longitudinal direction between slabs of any type of concrete pavement; hence, they are known as longitudinal or hinge joints. The only case that warping joints are constructed in the transverse direction is in unreinforced pavements. In the reinforced pavements, the warping that developed along the transverse direction is compensated by the longitudinal reinforcement.

The function of warping joints is to stop the warping of slabs attributed to the differential temperature of the slab with respect to its depth. By no means are the warping joints created to withstand any slab expansion or contraction. Anchorage of the steel bars placed between the slabs is of vital importance. Because of the different functions of steel bars placed, they are called tie bars.

The proper anchorage of the tie bars is achieved either by welding the tie bars on longitudinal supportive steel bars to be placed at a certain depth (or height) or by bending both ends. Suffice it to say that the tie bars are not covered with a corrosion-resistant coating at their full length, in contrast to the dowel bars in the expansion or contraction joints.

As the tie bars are not designed to act as a load transfer mechanism, they have a smaller diameter and usually are longer than the dowel bars.

A typical cross section of a warping joint for unreinforced pavements along the transverse direction is shown in Figure 14.9.

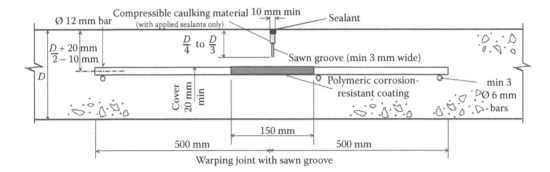

Figure 14.9 Construction details of a typical warping joint in URCP, in the transverse direction. (Adapted from Highways Agency, *The Manual of Contract Documents for Highway Works [MCDHW], Volume 3: Highway Construction Details*, Department for Transport. London: Highways Agency, 2006c [© Highways Agency].)

In URCPs, usually every third contraction joint is replaced by a warping joint.

The longitudinal joints are needed not only to control cracks along the longitudinal direction but also to provide the ability to construct concrete slabs with an appropriate width. The maximum allowable slab width is usually 4.2 m for unreinforced slab and 6 m for reinforced slab. A typical cross section of a longitudinal joint between two separately constructed unreinforced or jointed reinforced slabs is shown in Figure 14.10a.

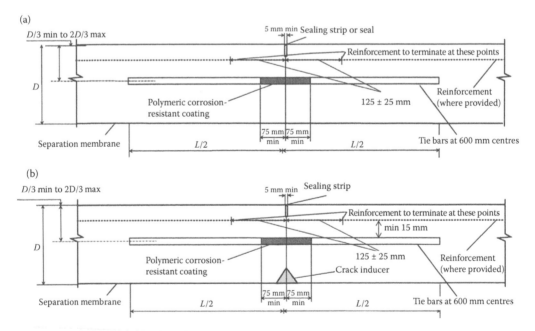

Figure 14.10 Construction details of a longitudinal joint in URCPs and CRCPs. (a) Longitudinal construction joint between two separately constructed unreinforced or jointed reinforced slabs and (b) wet formed longitudinal joint for slabs more than one lane width constructed in one operation. (Adapted from Highways Agency, *The Manual of Contract Documents for Highway Works [MCDHW], Volume 3: Highway Construction Details*, Department for Transport. London: Highways Agency, 2006c [© Highways Agency].)

The machinery used nowadays for the construction of rigid pavements is capable of laying concrete at widths two or three times greater than the abovementioned maximum widths. In these cases, the longitudinal joint is again constructed at a width equal to the maximum allowable width. The formation of the longitudinal joint is obtained by placing a crack inducer on the subgrade surface before concreting and inducing a groove at the same position while the concrete is in its plastic stage. A typical cross section of a longitudinal joint for slabs with more than one lane constructed in one operation is shown in Figure 14.10b.

In case of continuously reinforced pavements, the longitudinal construction joints are formed in a manner similar to the abovementioned ways.

14.5.4 Joints attributed to interruption of works

The joints attributed to the interruption of works are not in fact a different kind of joints. They are necessarily created as a result of work interruption. Under normal working conditions, work should stop at a pre-determined joint (contraction or expansion) so that an additional unnecessary joint is not created. However, in case of sudden and unscheduled work interruptions (mechanical failure, etc.), the vertical concrete surface is formed to a trapezoidal shape and tie bars 1 m long and 12 mm in diameter are placed at a spacing of 600 mm.

14.6 THE FUNCTION OF STEEL REINFORCEMENT

The steel reinforcement in rigid pavements does not have the same function as any other concrete structure. In all structures apart from pavements, the tensile stresses that developed are much greater than those the concrete can sustain. Therefore, the reinforcement is placed to counteract the tensile stresses caused by applied external loads.

In rigid pavements, the concrete slab is sited over the cement-bound layer and the tensile stresses that developed owing to traffic loading or ground reaction are very low. Even if they are high, they cannot cause sudden catastrophic failure of the pavement, as in other structures. The surface cracks may appear to affect the aesthetics and certainly accelerate pavement deterioration, but the structure will still be in use for a certain period (years).

The functions of steel reinforcement in rigid pavements are primarily to counteract the temperature- and moisture-induced stresses, to reduce the number of transverse joints (jointed reinforced slabs) or eliminate the transverse joints (continuously reinforced slabs), to minimise future maintenance cost and, to a certain extent, to reduce the thickness of the slab. The reduction or elimination of transverse joints has a direct impact on the improvement of the riding quality offered. Some countries like the United Kingdom recommend to use, almost exclusively, continuously reinforced slabs on new constructions, hence the absence of transverse joints, with an thin (30 mm) or thick (100 mm) asphalt layer on top.

14.6.1 Amount and position of reinforcement

14.6.1.1 Jointed reinforced concrete pavements

Extensive studies have been carried out in the past to determine the amount of steel required in reinforced slabs. The principle adopted was that an increase of the amount of steel reinforcement limits the severity of cracking that may appear and reduces the number of transverse joints.

The amount of reinforcement required in reinforced concrete slabs is best determined by the formula or procedure given in the pavement design methodology adopted.

For information purposes, the minimum steel quantity required can be calculated by the following equation (Yoder and Witczak 1975):

$$A_s = \frac{W \times L \times f}{f_s},$$

where A_s is the required area of steel per unit width (mm²/m), W is the weight of slab per unit area (kg/m²), L is the length of slab (m), f is the coefficient of sliding resistance between slab and underlying layer (equal to 1.5) and f_s is the allowable tensile stress in steel (MPa).

From the above equation, it is evident that the quantity of steel is dependent on the geometric dimensions of the slab, as well as the tensile strength of the steel. As the slab length (as well as the width) increases, greater amount of steel is required.

The quantity of steel reinforcement in the case of jointed reinforced pavements is placed either in the longitudinal direction (the most common case nowadays) or in both directions in the form of a mesh. In the first case, some steel bars are also placed in the transverse direction simply to withhold the reinforcement in the longitudinal direction.

If the reinforcement is in the form of a mesh, the bars placed transversely are usually 20% to 30% of the total steel area required, A_s.

The position of the reinforcement has been established to be above the mid-depth of the slab and near the top surface of the slab, leaving a cover typically ranging from 50 to 70 mm.

14.6.1.2 Continuously reinforced concrete pavements

In the case of continuously reinforced pavements, the quantity of steel used normally ranges from 0.5% to 0.7% of the slab cross-sectional area, depending on the design method used.

The reinforcement almost always is placed in the longitudinal direction and at the mid-depth of the specified thickness of the slab. When siliceous aggregates are used, the reinforcement is recommended to be located at a depth equal to one-third of the slab thickness, since it has been found to improve the crack pattern of the siliceous aggregate CRCPs (Hassan et al. 2005).

The reinforcement is usually high-yield deformed steel bars (500 MPa) with diameter and spacing as defined by the design method employed.

14.7 PRE-STRESSED CONCRETE

The use of pre-stressed concrete in rigid pavements is not very common. However, the main advantage of pre-stressed concrete is the reduction of the thickness of the concrete slab. The typical thickness of a pre-stressed concrete slab ranges from 100 to 150 mm.

Thickness reduction results in saving concrete material and the ability to construct a pavement where height clearance restriction is a problem (tunnels, under bridges, etc.) in conjunction with the inability to dig deeply to accommodate the thickness of a typical concrete slab.

However, these advantages are not sufficient to establish the pre-stressed concrete as an alternative in pavement construction. Problems are also encountered during construction (anchoring, use of special equipment, construction of special type of joints, etc.); the rate of daily output is lower and the construction cost is higher than that of a typical rigid pavement.

Because of all the above, pre-stressed concrete pavements are rarely used and are only used in very specific locations.

14.8 FIBRE-REINFORCED CONCRETE

Fibre-reinforced concrete is a mixture of concrete with a dispersion of discontinuous fibres. The fibres may be steel fibres, synthetic fibres (micro-synthetic or macro-synthetic), glass fibres (alkali-resistant only, AR glass fibres) or natural fibres.

The addition of fibres affects the plastic and hardened properties of concrete. The properties that improved are plastic shrinkage cracking and tensile and flexural strength. The addition of fibres also enhances the ability to resist cracking and spalling and to reduce crack propagation.

The most useful parameters describing the fibres are length, length/diameter ratio (aspect ratio), fibre tensile strength and shape. In general, fibre length varies from 6 to 65 mm and the diameter varies from 0.5 to 1.0 mm.

Steel fibres have a diameter and length in the 0.5 to 1.0 mm and 12.5 to 65 mm range, respectively, and a specific gravity of 7.85. They should preferably have an aspect ratio of 30 to 100 (ACI 2008). The typical tensile strengths of steel fibres are in the 500 to 2500 MPa range. High tensile strength fibres should be used when the compressive strength of concrete is high (more than 50 MPa).

Steel fibres are produced in a variety of types (cold-drawn wire, cut sheet, melt extracted into shaved cold-drawn wire and milled from blocks) and shapes (straight, shaped, round, oval, polygonal or crescent).

Micro-synthetic fibres are defined as those having diameters less than 0.3 mm and macro-synthetic fibres are those with diameters greater than 0.3 mm. Their specific gravity is 0.91 (polypropylene fibres) or 1.14 (nylon fibres).

AR glass fibres, also distinguished into micro- or macro-fibres, are either 13 or 18 mm in diameter with a specific gravity 2.7. Their tensile strength should range from 1 to 1.7 GPa (ACI 2008).

Fibre-reinforced concrete may be used in unreinforced as well as reinforced concrete pavements. The use of steel fibre-reinforced usually does not cause reduction of reinforcement required, it simply allows to increase the spacing between joints, which is advantageous.

Fibres may be added last to the transit mixer, or added to the aggregates before mixing, and in the case of AR glass fibres, they are chopped and added directly to the mixer or conveyor.

Although addition and mixing of fibres are an easy process, special attention must be given to the mix design since the addition of fibres affects workability, slump and water content. The use of entrained air compound is almost unavoidable, certainly if steel fibres are used. More information about mix proportioning, batching, mixing, delivery and sampling can be found in ASTM C1116 (2010) and an ACI relevant guide (ACI 2008).

Steel fibres should comply with CEN EN 14889-1 (2006), ASTM A 820 (2011) or other national specifications. Polymer fibres should comply with CEN 14889-2 (2006) or ASTM C 1116 (2010), and glass fibres should comply with CEN EN 15422 (2008), ASTM C 1666 (2008) or other national specifications.

Despite the benefits of using fibres and the possibility of reducing the amount of reinforcement, their cost should always be considered and economic analysis should be carried out before a decision is made.

14.9 RIGID PAVEMENT DESIGN METHODOLOGIES

Various methodologies have been developed for the design of rigid pavements, the majority of which are semi-empirical and mainly based on results from experimental pavements.

The design criteria of rigid pavements are far more complicated than those of flexible pavements. The horizontal tensile stresses causing cracking of the slab are generated by the combined effect of wheel loading and thermally induced internal and warping stresses.

Deformation of the subgrade is not considered as a design criteria in rigid pavements since the stresses transmitted to the subgrade are low. However, poor subgrade associated with improper drainage will certainly cause further cracking to the slab. In addition, the absence of transverse joints in the sub-base will also contribute to the cracking of the slab.

All pavement design methodologies assume that rigid pavement failure due to load related distresses occurs at the end of the design period.

The type of load-related distresses considered in the design of jointed reinforced or CRCPs are longitudinal, transverse and corner cracks, crack width and punchouts.

The distress level considered as failure, per type of distress, varies among methodologies.

14.10 UK RIGID PAVEMENT DESIGN METHODOLOGY

The revised 2006 UK methodology for rigid pavements considers rigid concrete construction as a permitted option for trunk roads including motorways if it has an asphalt surfacing. This requirement generally makes jointed construction unsuitable for consideration, since reflection cracking of the surfacing at the joints will occur (Highways Agency 2006b).

Thus, the preferred rigid pavement construction is either (a) CRCP with a thin asphalt overlay of minimum thickness 30 mm (see Figure 14.1c.ii) or (b) CRCB pavement with an asphalt overlay of 100 mm (see Figure 14.1d). The latter is used to be called rigid composite pavement.

As for the CRCP, it is stated that 'the use of continuously reinforced concrete pavements with a thin surface course system (TSCS) can provide a 'long life' with all the advantages offered by the noise reducing properties of the surfacing. Such pavements are ideally suited to the application of further asphalt overlays at stages during the future pavement life' (Highways Agency 2006b).

The other types of rigid pavements, unreinforced or jointed reinforced, are not excluded by the methodology. They can be used upon approval and certainly when an existing old JRCP is to be widened or rehabilitated.

The methodology uses the foundation concept as in flexible pavements. Out of the four foundation classes established, class 1 is excluded for rigid pavements.

The determination of the slab thickness is related to the traffic expressed in ESAL, the foundation class and the flexural strength of concrete.

The design life of any new construction and of locations where design traffic is heavy in relation to the capacity of the layout is recommended to be taken as 40 years. In less heavily trafficked sites or for major maintenance, 20 years is considered sufficient.

For continuously reinforced pavements (CRCP or CRCB), it has been assumed that no concrete slab maintenance or reinforcement will be required over the design period. As for the maintenance of thick or thin asphalt layers, it depends on the nature of the traffic.

14.10.1 Design parameters

14.10.1.1 Design traffic

The design traffic is determined in cumulative equivalent standard axle loads (ESAL), expressed in million standard axles (msa). The process for the determination of the ESAL is identical to that followed for flexible pavements and described in Section 13.5.1.

14.10.1.2 Subgrade and foundation layers

The strength of the subgrade is expressed in CBR or stiffness modulus, and from this, the foundation thickness is determined for the foundation class chosen to be used. Out of the four classes established, foundation class 1 cannot be used for rigid pavements and bound foundation class 2 can be used only upon approval of the overseeing organisation.

The foundation classes and the foundation design procedure are the same as those in flexible pavements and are outlined in Section 13.5.3.

The separation membrane used in jointed slabs is omitted from CRCP/CRCB construction in order to give a higher level of friction between the concrete slab and the sub-base.

14.10.1.3 Moisture and frost

To prevent problems associated with high moisture of the subgrade, it is recommended to maintain the water table at least 300 mm below the formation level. This can be achieved with the provision of an effective drainage system.

Additionally, in case the subgrade material is frost susceptible, the total pavement thickness (slab and sub-base) is recommended to be greater than 450 mm, unless the estimated mean annual frost index is less than 50 (see Section 13.5.10).

14.10.2 Thickness determination of slab with continuous reinforcement

Knowing the design traffic and after selecting the foundation class to be used, the determination of the total thickness of the continuously reinforced slab (excluding any asphalt surfacing) is carried out using the design nomograph shown in Figure 14.11.

The thickness of the CRCP for a selected 28-day flexural strength of concrete is determined from the right-hand portion of the nomograph, and the thickness of the CRCB is determined from the left-hand portion of the nomograph in Figure 14.11. In all cases, the result is rounded up to the next 10 mm.

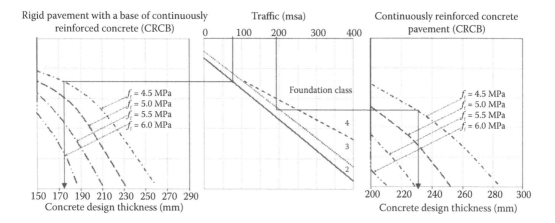

Figure 14.11 Determination of a rigid pavement's thickness with continuous reinforcement. (Adapted from Highways Agency, *The Manual of Contract Documents for Highway Works [MCDHW], Volume 3: Highway Construction Details*, Department for Transport. London: Highways Agency, 2006c [© Highways Agency].)

The concrete thickness design value determined assumes the presence of a minimum 1 m edge strip or tied shoulder; otherwise, the concrete design thickness should be increased by 30 mm.

It is also pointed out that, if concrete with a flexural strength ≥5.5 MPa is to be used, the concrete aggregates should have a coefficient of thermal expansion less than 10×10^{-6} per degree Celsius.

14.10.2.1 Thickness and type of asphalt overlay

The thickness of the asphalt overlay in the case of CRCP is recommended to be 30 mm (minimum) to 40 mm (maximum).

The type of asphalt is usually asphalt concrete for very thin layers in accordance to CEN EN 13108-2 (2008).

The thickness of the asphalt layer in the case of CRCB is recommended to be at the minimum, 100 mm (the minimum value is normally used). This overlay usually consists of a binder course of dense asphalt concrete (CEN EN 13108-1 2008) with a surfacing layer of 30 mm.

If a porous asphalt (CEN EN 13108-7 2008) surface course is used over CRCB, it must be modified with a polymer or fibre additive and laid over a dense binder course. The porous asphalt should be 50 mm thick over a 90 mm binder course or 50 mm thick over a 60 mm binder course with the CRCB thickness increased by 10 mm.

14.10.2.2 Steel reinforcement

The quantity of steel reinforcement in CRCP required is 0.6% of the concrete slab cross-sectional area, comprising 16 mm diameter deformed steel bars. The transverse steel to be used must be 12 mm diameter deformed bars at 600 mm spacings.

In CRCB, the longitudinal steel reinforcement is required to be 0.4% of the concrete slab cross-sectional area, comprising 12 mm diameter deformed steel bars. The transverse steel must be 12 mm diameter deformed bars at 600 mm spacings.

14.10.3　Thickness determination of unreinforced and jointed reinforced concrete slab

In special cases where a URCP slab is to be designed, the concrete slab thickness is determined by the following equation (Highways Agency 2006b):

$$\mathrm{Ln}(H_1) = \left[\frac{\mathrm{Ln}(T) - 3.466 \times \mathrm{Ln}(R_c) - 0.484 \times \mathrm{Ln}(E) + 40.483}{5.094}\right],$$

where H_1 is the thickness of the concrete slab without a tied shoulder or 1 m edge strip (minimum slab thickness, 150 mm) (mm), T is the design traffic (maximum 400 msa) ($\times 10^6$ msa), R_c is the mean compressive cube strength at 28 days (N/mm^2 or MPa) and E is the foundation class stiffness (MPa) (200 MPa for foundation class 3 or 400 MPa for foundation class 4).

As for jointed reinforced pavement, the thickness of the reinforced concrete slab is determined by the following equation:

$$\mathrm{Ln}(H_1) = \left[\frac{\mathrm{Ln}(T) - R - 3.171 \times \mathrm{Ln}(R_c) - 0.326 \times \mathrm{Ln}(E) + 45.150}{4.786}\right],$$

where H_1, T, R_c and E are as in the equation for unreinforced slab; $R = 8.812$ for 500 mm²/m reinforcement (minimum quantity); $R = 9.071$ for 600 mm²/m reinforcement; $R = 9.289$ for 700 mm²/m reinforcement; and $R = 9.479$ for 800 mm²/m reinforcement.

In case there is a tied shoulder or 1 m edge strip of thickness ≥150 mm, the concrete slab thickness, in both cases, is reduced and determined by the following equation:

$$H_2 = 0.934 \times H_1 - 12.5,$$

where H_2 is the thickness of the concrete slab with a tied shoulder or 1 m edge strip (mm) and H_1 is the thickness of the concrete slab without a tied shoulder or 1 m edge strip (minimum thickness 150 mm) (mm).

More information regarding the tied shoulder can be found in TRRL RR 87 (Mayhew and Harding 1987).

14.10.3.1 Maximum transverse joint spacings for URC pavements

The maximum transverse contraction joint spacing is 4 m, when slab thickness is <230 mm, while the maximum transverse expansion joint spacing is usually 40 m.

When the slab thickness is ≥230 mm, the maximum transverse contraction joint spacing is 5 m, and the maximum transverse expansion joint spacing is usually 60 m.

14.10.3.2 Maximum transverse joint spacings for JRC pavements

When the quantity of slab reinforcement is ≥600 mm²/m, the maximum transverse joint spacing is recommended to be 25 m.

When the slab reinforcement is <600 mm²/m, the maximum joint spacing depends on the slab thickness. The recommended spacing is as follows:

25 m, if the slab thickness is <280 mm
24 m, if the slab thickness is <290 mm
23 m, if the slab thickness is <300 mm
22 m, if the slab thickness is <310 mm
21 m, if the slab thickness is <320 mm
20 m, if the slab thickness is <330 mm

Every third transverse joint is recommended to be an expansion joint.

The above are valid for concretes with aggregates having a coefficient of thermal expansion ≥10 × 10⁻⁶ per degree Celsius (sand gravels).

If concrete is used with aggregates having a coefficient of thermal expansion <10 × 10⁻⁶ per degree Celsius (limestone aggregates), the maximum transverse joint spacings may be increased by 20%.

Further information on the design of a rigid pavement with jointed slabs or continuously reinforced slabs is given in Highways Agency (2006c).

14.10.4 Concrete surfacing and materials

When the concrete slabs are not covered with asphalt overlay, the concrete surface must have sufficient texture to provide good resistance to skidding.

The usual techniques used are to brush, burlap drag or tine the concrete at its plastic stage. They all texture the surface in the transverse direction.

Details for the above techniques can be found in Highways Agency (1999).

Apart from the above techniques, exposed aggregate concrete mix may also be used in the top concrete layer. The technique of using exposed aggregate concrete mix was first developed in Belgium in the late 1960s and has been adopted and improved by many countries ever since.

The exposure of the surface aggregate is a two-stage process including retarding of the surface mortar and brushing afterwards the surface to expose the aggregate.

A suitable retarder is applied (sprayed) to the surface of the concrete to retard the action of the cement at the surface of the top layer so that the targeted texture depth can be achieved by brushing operation. The initiation of the second stage, brushing, is very critical. Brushing too early will result in chipping loss, and brushing too late will entail difficulty in achieving the desired surface texture.

The size and shape of the hard and durable aggregate, as well as proper mix design and the curing process, play an important role in this technique. The cost of exposed aggregate concrete surfacing is slightly higher than that of a similar pavement with conventional brushed concrete finishes.

More information on the exposed aggregate concrete technique with regard to UK design methodology for rigid pavements can be found in Highways Agency (1999, 2006d).

14.11 AASHTO RIGID PAVEMENT DESIGN METHODOLOGY

The AASHTO rigid pavement design methodology (AASHTO 1993) was developed at the same time as the AASHTO flexible pavements design methodology and is described analytically in the same design guide (AASHTO 1993).

With the use of the methodology, jointed (unreinforced or reinforced) and continuously reinforced pavements can be designed. Additionally, it gives general guidelines on the design of pre-stressed concrete pavement. Unlike the UK methodology, in the AASHTO methodology, there is no requirement or obligation to use asphalt overlay in CRCP.

The design data required for the determination of the slab thickness for an assumed sub-base thickness (>150 mm) are as follows: resilient modulus of subgrade (M_R), elastic modulus of sub-base (E_{SB}), modulus of rupture of concrete (S_c) (or flexural strength of concrete), elastic modulus of concrete (E_c), cumulative ESAL over the design period (W_{18}), overall standard deviation (S_o) and design serviceability (ΔPSI) as in flexible pavement design.

In addition to the above, the effective modulus of subgrade reaction (k) and the load transfer coefficient (J) must be determined and the expected level of drainage must be considered, expressed as drainage coefficient C_d.

The methodology allows the use of other hydraulically bound material for the construction of the sub-base.

A detailed procedure for designing rigid concrete pavements with jointed or continuously reinforced concrete slabs can be found in the AASHTO manual (AASHTO 1993). However, in the following paragraphs, a brief description for determining the thickness of the concrete slab in all cases (unreinforced or reinforced slabs) and the required quantity of reinforcement are given for the benefit of the reader.

14.11.1 Thickness determination of slab (all cases)

The thickness determination of the concrete slab in a rigid pavement regardless of being unreinforced or reinforced is carried out as follows:

a. Determine the seasonal roadbed soil (subgrade) resilient modulus, M_R, usually for every month of the year.

b. Determine the seasonal sub-base elastic modulus, E_{SB}, usually for every month of the year.

c. Select a sub-base thickness, D_{SB}, and with the use of Figure 3.3 of AASHTO (1993), determine the composite modulus of subgrade reaction, k value, for each month, assuming semi-infinite subgrade depth, k_∞.

   If the slab is going to be placed directly on the subgrade (i.e. no sub-base), the composite modulus of subgrade reaction, k value, is defined by using the equation

$$k = \frac{M_R}{19.4}.$$

d. Correct the composite modulus of subgrade reaction, k value, for each month, to consider the effect of rigid foundation near the surface (within 10 ft), by using Figure 3.4 of AASHTO (1993). This step is disregarded if the depth to a rigid foundation is greater than 10 ft.

e. Determine the relative damage of the slab, u_r, for each month, by using Figure 3.5 of AASHTO (1993), assuming a slab thickness (projected slab thickness).

f. Determine the average relative damage, $\bar{u}_r$, from the monthly relative damage factors determined in (e). Then, for the average relative damage factor, determine the effective modulus of subgrade reaction, k value, using again Figure 3.5 of AASHTO (1993).

g. Adjust the effective modulus of subgrade reaction to account for the potential loss of support, LS, arising from sub-base erosion, using Figure 3.6 of AASHTO (1993). The loss of support depends on the sub-base material and its value is estimated from Table 2.7 of AASHTO (1993).

Calculation of slab thickness

h. The determination of the slab depth (unreinforced or reinforced) for the adjusted effective k value determined in (g) is carried out using the nomographs in Figure 14.12a and b.

Other input data for the determination of the concrete slab are as follows: elastic modulus of concrete, E_c, mean concrete modulus of rupture, S_c (three-point bending test), load transfer coefficient, J (from Table 2.6 of AASHTO 1993), drainage coefficient, C_d (coefficient similar to the one used in flexible pavements), design serviceability loss (ΔPSI), estimated future traffic, W_{18} (as determined in flexible pavements but using equivalency coefficients for rigid pavements), the overall standard deviation, S_o (usually 0.35), and the reliability, R (as determined in flexible pavements).

The thickness of the concrete slab may also be determined from the equation given at the top of Figure 14.12a representing the nomographs in Figure 14.12a and b.

Consideration of road swelling and frost heave

The approach to consider the effects of swelling and frost heave in rigid pavement design is almost identical to that for flexible pavements. In fact, the total environmental serviceability loss caused by swelling and frost heave (ΔPSI$_{SW,FH}$) is calculated and is then used to predict its effect on performance period. For more details, see AASHTO (1993).

Finally, it is noted that, in this methodology, unlike the UK methodology, there is no adjustment (increase) of the slab thickness owing to the absence of concrete shoulder.

(a)

Nomograph solves:

$$\log_{10} W_{18} = z_R * S_o + 7.35 * \log_{10}(D+1) - 0.06 + \frac{\log_{10}\left[\dfrac{\Delta PSI}{4.5 - 1.5}\right]}{1 + \dfrac{1.624 * 10^7}{(D+1)^{8.46}}} + (4.22 - 0.32 p_t) * \log_{10}\left[\dfrac{S'_c * C_d \left[D^{0.75} - 1.132\right]}{215.63 * 3 \left[D^{0.75} - \dfrac{18.42}{(E_c/k)^{0.25}}\right]}\right]$$

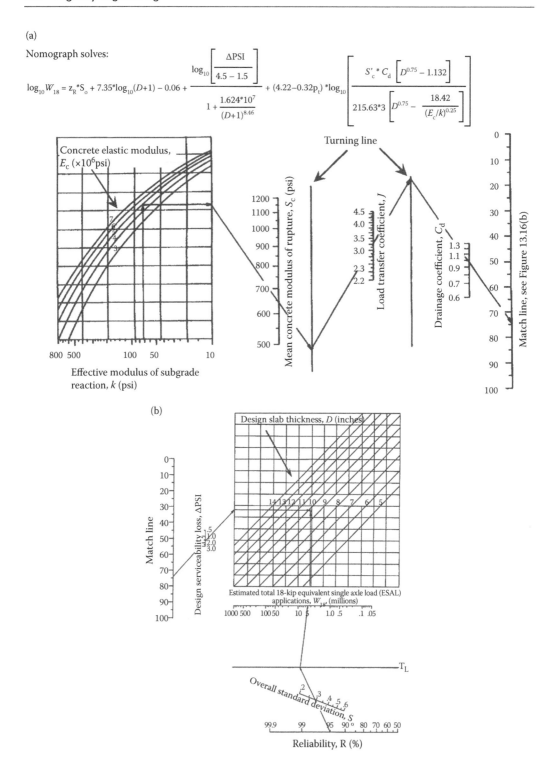

Figure 14.12 Design charts for slab thickness determination in rigid pavements. (From AASHTO, *Guide for Design of Pavement Structures [GDPS]*, Washington, DC: American Association of State Highway and Transportation Officials, 1993. With permission.)

Table 14.3 Recommended friction factors

Type of material beneath slab	Friction factor, F
Surface treatment	2.2
Lime stabilisation	1.8
Asphalt stabilisation	1.8
Cement stabilisation	1.8
River gravel	1.5
Crushed stone	1.5
Sandstone	1.2
Natural subgrade	0.9

Source: AASHTO, *Guide for Design of Pavement Structures (GDPS)*, Washington, DC: American Association of State Highway and Transportation Officials, 1993. With permission.

14.11.2 Determination of reinforcement in jointed reinforced slabs

In jointed reinforced slabs, the amount of reinforcement required in either transverse or longitudinal direction is determined using the following equation:

$$P_s = \frac{L \times F}{2 \times f_s} \times 100,$$

where P_s is the amount of steel reinforcement quantity (% of the slab cross-sectional area), L is the slab length (ft), F is the friction factor from Table 14.3 and f_s is the steel working stress (typically 75% of the steel yield strength) (psi).

It is noted that the above equation can be used to determine the amount of steel reinforcement (if required) along the transverse direction of the continuously reinforced slab.

14.11.3 Determination of tie bar spacing

The determination of the tie bar spacing depends on the distance of the joint from the free edge of the rigid pavement (distance between transverse joints) and the slab thickness. The tie bar spacing when a 1/2-inch-diameter bar is used is determined from Figure 3.13 of AASHTO (1993); when a 5/8-inch-diameter bar is used, the tie bar spacing is determined from Figure 3.14 of AASHTO (1993).

The minimum length of tie bars is recommended to be 25 inches (635 mm) for 1/2-inch-diameter bars and 30 inches (750 mm) for 5/8-inch-diameter bars.

The above apply to all jointed reinforced rigid pavements.

14.11.4 Determination of dowel bars

The determination of the spacing and diameter of dowel bars is left to the designer's experience. However, the following general guide is given: the bar diameter should be 1/8 of the slab thickness and the spacings should be 12 to 18 inches.

14.11.5 Joint spacing

The spacing of both transverse and longitudinal joints is again left to the designer's experience in relation to the materials used and environmental conditions. As a general guide, it is stated that 'the joint spacing (in feet) for unreinforced pavement should not greatly exceed twice the maximum slab thickness (in inches)'. Additionally, the ratio of slab width to length should not exceed 1.25 (AASHTO 1993). According to the author's experience, the above guides are too loose to prevent the development of thermal cracking.

As for the expansion joints, it is stated that 'the use of expansion joints is generally minimized on a project due to cost, complexity and performance problems'.

Finally, with regard to the longitudinal construction joints, it is simply stated that 'they should be placed at lane edges to maximize pavement smoothness and minimize load transfer problems' (AASHTO 1993).

14.11.6 Determination of reinforcement in CRCPs

The determination of the longitudinal reinforcement in continuously reinforced pavements is carried out by a procedure on the basis of three limiting criteria. These limiting criteria are as follows: (a) crack spacing considering spalling and punchouts, (b) crack width considering spalling and water penetration and (c) steel stress to guard against steel fracture and excessive permanent deformation. To satisfy the limiting criteria mentioned above and determine the amount of longitudinal reinforcement, three different nomographs are used. Details regarding the design procedure can be found in the AASHTO pavement design guide (AASHTO 1993).

14.11.7 Pre-stressed concrete pavement

The AASHTO provides only some general guidelines on the design of pre-stressed concrete pavements. Details can be found in the manual (AASHTO 1993).

14.12 AUSTRALIAN RIGID PAVEMENT DESIGN METHODOLOGY

The Australian rigid pavement design methodology published by Austroads (2012) has been developed together with the pavement design for flexible pavements as a general guide to Australian road agencies.

The methodology covers the design of jointed (unreinforced and reinforced), continuously reinforced and fibre-reinforced concrete pavements. The case of pre-stressed rigid pavement is not covered.

Unlike UK rigid pavement design methodology, the CRCP with asphalt overlay is not enforced to be used in all new constructions. However, the use of lean concrete, cement-stabilised crushed aggregate or dense asphalt concrete as a sub-base material in moderate to heavily trafficked roads is necessary; for lightly trafficked roads, unbound granular sub-base may also be used.

The procedure for the determination of the thickness of rigid pavements, and of the concrete slabs in particular, is based on the USA 1984 Portland Cement Association method (Packard 1984). Additionally, the method assumes that the base and sub-base layers are not bonded.

For the determination of the concrete slab thickness, two distress modes have been considered: (a) the flexural fatigue cracking of the slab and (b) the subgrade/sub-base erosion arising from repeated deflections at joints and planned cracks.

The design method is based on assessments of the following (design data): (a) predicted traffic volume and composition over the design period (design traffic), (b) strength of the subgrade in terms of its CBR and (c) flexural strength of the slab concrete.

The design traffic for rigid pavement design is the cumulative number of axle groups over the design period, referred to as heavy vehicle axle group (HVAG), classified according

Table 14.4 Minimum sub-base thickness requirements for rigid pavements

Design traffic (HVAG)	Sub-base type
Up to 10^6	125 mm bound
Up to 5×10^6	150 mm bound or 125 lean concrete[a]
Up to 1×10^7	170 mm bound or 125 mm lean concrete[a]
$>1 \times 10^7$	150 mm lean concrete[a]

Source: Austroads, *Guide to Pavement Technology Part 2: Pavement Structural Design*, Publication No. AGPT02-12, Sydney: Austroads Inc., 2012. With permission.

[a] Lean concrete characteristic 28-day compressive strength ≥5 MPa.

to the type of axle group (single/single tyre, tandem/single tyre, single/dual tyre, tandem/dual tyre, triaxial/dual tyre and quad-axle/dual tyre) and the load on the specific axle group type. The process of determining the design traffic is best explained in Austroads (2012).

The assessment of design CBR is similar to the one for flexible pavement design and is basically based on laboratory testing. The subgrade material is required to have a minimum CBR value of 5%. Additionally, the depth of the subgrade material below sub-base should be increased to 600 mm over highly expansive subgrade materials.

The methodology requires minimum sub-base thickness as shown in Table 14.4.

Because of the provision of the sub-base, there is an increase in effective subgrade strength that can be determined using Figure 14.13.

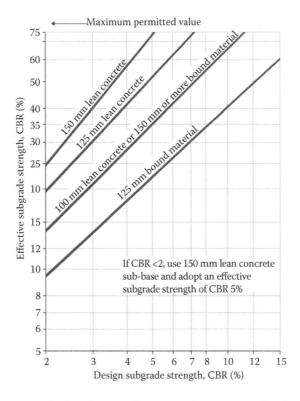

Figure 14.13 Effective increase of subgrade strength owing to the provision of a sub-base. (From Austroads, *Guide to Pavement Technology Part 2: Pavement Structural Design*. Publication No. AGPT02-12, Sydney: Austroads Inc., 2012. With permission.)

Table 14.5 Minimum concrete slab thickness

Type of concrete slab	Design traffic, heavy vehicle axle group (HVAG)		
	1×10^6 to 1×10^7	1×10^7 to $<5 \times 10^7$	$\geq 5 \times 10^7$
Unreinforced	150	200	250
Jointed reinforced and dowelled	150		
Steel fibre-reinforced concrete	125	180	230
Continuously reinforced concrete	150		

Source: Austroads, *Guide to Pavement Technology Part 2: Pavement Structural Design*, Publication No. AGPT02-12, Sydney: Austroads Inc., 2012. With permission.

14.12.1 Determination of concrete slab thickness

The procedure for the determination of the thickness of the concrete slab is based on the two distress modes, as mentioned earlier, the flexural fatigue cracking of the slab and the subgrade/sub-base erosion. The two distress modes are expressed by two fatigue equations, provided in the reference manual, from which the allowable load repetitions can be calculated.

For an assumed slab thickness, the minimum required values are given in Table 14.5; the allowable number of load repetitions is determined in each case and for each axle load group type. Then, the ratio of the expected fatigue repetitions to the allowable load repetitions is calculated and multiplied by 100 to determine the fatigue percentage, for each axle load group type, per type of distress.

The sum of all percentages of fatigue is then determined per type of distress. If any of the sums of percentage of fatigue is greater than 100%, the assumed thickness is not sufficient and the calculations should be repeated using a greater number corresponding to slab thickness.

The required minimum characteristic design concrete flexural strength for concrete pavements with a design traffic of 10^6 HVAG or more is 4.5 MPa at 28 days, except for steel fibre–reinforced concrete, where the requirement is 5.5 MPa.

The methodology also requires the provision of concrete shoulders. These can be either integral or structural.

Integral concrete shoulders are made up of the same concrete and are the same thickness as the concrete slab, and they are cast integrally with the slab with a minimum width of 600 mm. The minimum width for integral cast shoulders in the median lane may be reduced to 500 mm.

A structural shoulder is connected with a tied corrugated joint and has a minimum width of 1.5 m or is a 600 mm integral widening outside of the traffic lane.

14.12.2 Determination of reinforcement

The amount of reinforcement for JRCP and CRCP is calculated by two equations provided in the reference manual.

The use of steel fibre-reinforced concrete does not affect the amount of reinforcement required. It is simply used to increase flexural strength when cracking in odd-shaped slabs needs to be controlled and where increased abrasion resistance is required for durability.

This type of pavement is often used for toll plazas, roundabouts and bus stops. Steel fibres are typically between 15 and 50 mm in length with enlarged ends that act as anchorages or crimps to improve bond. However, limits apply to the ratio of fibre length to the minimum dimension of test specimens (such as test cylinders and beams). Typically, fibres are added to the concrete at a rate of approximately 45 to 75 kg/m^3 (Austroads 2012).

14.12.3 Dowel and tie bars

The dowel bars are recommended to be plain round steel bars of grade 250 N and 450 mm long, placed at 300 mm centres. More than half of the length of the dowel bar should be coated with a debonding agent to ensure effective debonding from the concrete on that side of the joint. The dowel diameters vary from 24 to 36 mm, depending on the slab thickness.

Tie bars are recommended to be 12 mm in diameter, grade 500 N deformed steel bars, 1 m long and placed centrally in the joint. The spacing is determined accordingly depending on the thickness of the concrete slab, the interlayer friction and the distance to the nearest free edge of pavement.

More details about reinforcement design, joint spacing and joint construction can be found in Austroads (2012).

14.12.4 Concrete surface finishes

The surface of the concrete slab should have good texture to provide sufficient and lasting skid resistance. Other factors such as noise generation may be of concern. The provision of good texture is obtained by usual techniques such as tining, hessian drag, broomed, exposed aggregate concrete, stamped concrete and so on, the details of which can be found in Austroads (2009).

14.13 MEPDG FOR RIGID PAVEMENTS

The set of procedures for the analysis and design of new and rehabilitated rigid pavements is the same as that for flexible pavements, outlined in Section 13.12.1.

However, the mixture characterisation of the PCC is different in MEPDG (AASHTO 2008) when compared to the empirical thickness design procedure given in the guide for design of pavement structures (AASHTO 1993). The typical differences between them are shown in Figure 14.14.

The MEPDG analyses the expected performance of new and reconstructed rigid pavements, as well as PCC overlays and concrete pavement restoration. The design strategies that can be simulated with MEPDG for new rigid pavement are shown in Figure 14.15.

A figure similar to Figure 14.15, but for PCC overlay design strategies for flexible, semi-rigid (flexible with HMB base) and rigid pavements that can be simulated with MEPDG, is given in AASHTO (2008).

More details about the MEPDG and MEPDG software can be found in AASHTO (2008) and AASHTOWare (2014).

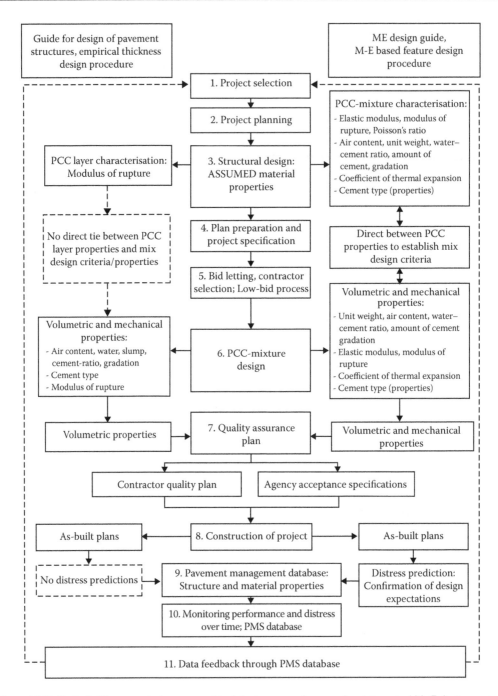

Figure 14.14 Typical differences between empirical design procedures and an integrated M–E design system in terms of PCC-mixture characterisation. (From AASHTO, *Mechanistic–Empirical Pavement Design Guide [MEPDG]*, American Association of State Highway and Transportation Officials, Washington, DC: AASHTO, 2008. With permission.)

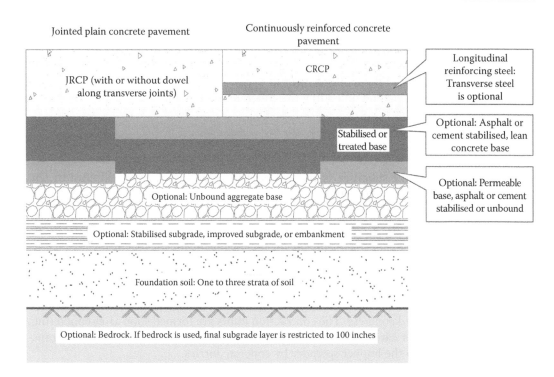

Figure 14.15 Design strategies that can be simulated with the MEPDG for new rigid pavement design. (From AASHTO, *Mechanistic–Empirical Pavement Design Guide [MEPDG]*, Washington, DC: American Association of State Highway and Transportation Officials, 2008. With permission.)

14.14 CONSTRUCTION OF RIGID PAVEMENTS

The construction of rigid pavements is characterised by the use of specially designed machinery, each of which performs a specific task. Thus, it becomes obvious that there should be good coordination among them to produce quality results with minimum delays.

There are two basic types of concrete paving equipment: the fixed-form paving and the slip-form plant.

The advantages and disadvantages of the slip-form plants compared to the fixed-form paving are as follows:

a. Higher daily output (usually 2–3 km/day)
b. Engagement of fewer machinery and staff
c. Requirement of greater investment and more specialised staff
d. Need for greater quantities of concrete to be supplied
e. There is always a risk of concrete slumping as the forms slide off
f. Machine breakdown results in complete interruption of works

During the construction of a rigid pavement, the basic stages involved are as follows:

1. Preparation of the subgrade or the sub-base
2. Form laying (if fixed-form paving is used)
3. Placing the joints and the reinforcement (if used)
4. Mixing, transporting and unloading the concrete
5. Placing the concrete
6. Compacting and finishing
7. Curing of concrete

The preparation for the subgrade or the sub-base mainly consists of providing sufficient compaction. This is achieved by using the same technique and machinery used as in flexible pavements.

Form laying, in the case of fixed-form paving, is a time-consuming but a necessary process. The forms normally used are steel forms to avoid warping of the formwork. Greater ease and higher speed of construction are achieved when slip-form paving is used. On large projects, slip-form paving is almost exclusively used.

Placing of the expansion, warping and longitudinal joints and slab reinforcement (if used) are carried out manually before concreting and after laying the separation membrane, in the case of jointed reinforced concrete. The contraction joints are formed soon after concreting.

Mixing of concrete is carried out in a concrete mixing plant and the concrete is transported and unloaded to the hopper spreader by concrete lorries.

Placing the concrete is carried out by self-propelled machines consisting of (apart from the hopper) a spreader, a vibratory compactor, an oscillating-beam finisher and spraying equipment for the application of the curing compound to be used. Some concrete pavers may be equipped with a guillotine-type component for forming the transverse contraction joints or equipment for the mechanical placement of joint tie bars or dowel bars. The concrete pavers move on steel rail tracks to ensure evenness and uniform thickness of the concrete slab.

Compaction and initial finishing are carried out by using the concrete paver, while the final finishing of the concrete is performed by a separate unit. Surface texturing is also carried out by a separate appropriate unit. Surface texturing may be carried out using one of the following techniques: tining, hessian drag, hessian drag and tine, wire brushing, grooving and stamped (impressed) or exposed aggregate technique.

Curing of concrete is arguably the most important stage of construction since formation of early thermal cracking may occur during this stage depending on the environmental conditions (temperature in particular), concrete temperature, type of concrete and type of aggregate used.

In particular, rapid loss of moisture will result in drying shrinkage, and plastic shrinkage cracking will develop. Conditions that mainly favour plastic shrinkage are high air temperature, low humidity, wind, high concrete temperature, exposure of concrete to the sun and high evaporation rate. A more analytical description of cracking of fresh and hardened concrete and precautions that need to be taken are given in Section 14.4.

REFERENCES

AASHTO. 1993. *Guide for design of pavement structures (GDPS)*. Washington, DC: American Association of State Highway and Transportation Officials.

AASHTO. 2008. *Mechanistic–empirical pavement design guide (MEPDG)*. Washington, DC: American Association of State Highway and Transportation Officials.

AASHTO M 6. 2013. *Fine aggregate for hydraulic cement concrete*. Washington, DC: American Association of State Highway and Transportation Officials.

AASHTO M 33. 2012. *Preformed expansion joint filler for concrete (bituminous type)*. Washington, DC: American Association of State Highway and Transportation Officials.

AASHTO M 80. 2013. *Coarse aggregate for hydraulic cement concrete*. Washington, DC: American Association of State Highway and Transportation Officials.

AASHTO M 85. 2012. *Portland cement*. Washington, DC: American Association of State Highway and Transportation Officials.

AASHTO M 153. 2011. *Preformed sponge rubber and cork expansion joint fillers for concrete paving and structural construction*. Washington, DC: American Association of State Highway and Transportation Officials.

AASHTO M 157. 2013. *Ready-mixed concrete*. Washington, DC: American Association of State Highway and Transportation Officials.

AASHTO M 194M/M 194. 2013. *Chemical admixtures for concrete*. Washington, DC: American Association of State Highway and Transportation Officials.

AASHTO M 213. 2010. *Preformed expansion joint fillers for concrete paving and structural construction (nonextruding and resilient bituminous types)*. Washington, DC: American Association of State Highway and Transportation Officials.

AASHTO M 240/M 240. 2013. *Standard specification for blended hydraulic cement*. Washington, DC: American Association of State Highway and Transportation Officials.

AASHTO T 22. 2011. *Compressive strength of cylindrical concrete specimens*. Washington, DC: American Association of State Highway and Transportation Officials.

AASHTO T 42. 2010. *Preformed expansion joint filler for concrete construction*. Washington, DC: American Association of State Highway and Transportation Officials.

AASHTO T 97. 2010. *Flexural strength of concrete (using simple beam with third-point loading)*. Washington, DC: American Association of State Highway and Transportation Officials.

AASHTO T 157. 2012. *Air-entraining admixtures for concrete*. Washington, DC: American Association of State Highway and Transportation Officials.

AASHTO T 198. 2009. *Splitting tensile strength of cylindrical concrete specimens*. Washington, DC: American Association of State Highway and Transportation Officials.

AASHTOWare. 2014. *Pavement*. Washington, DC: AASHTO, http://www.aashtoware.org.

ACI. 2000. *Hot weather concreting*. ACI 305R-99. Farmington Hills, MI: American Concrete Institute.

ACI. 2007. *Causes, evaluation and repairs of cracks in concrete structure*. ACI 224.1R-07. Farmington Hills, MI: American Concrete Institute.

ACI. 2008. *Guide for specifying, proportioning, mixing, placing, and finishing fiber-reinforced concrete*. ACI 544.3R-93 (Reapproved 1998). Farmington Hills, MI: American Concrete Institute.

ASTM A 820/A 820M. 2011. *Standard specification for steel fibers for fiber-reinforced concrete*. West Conshohocken, PA: ASTM International.

ASTM C 33/C 33M. 2013. *Standard specification for concrete aggregates*. West Conshohocken, PA: ASTM International.

ASTM C 39/C 39M-12a. 2012. *Standard test method for compressive strength of cylindrical concrete specimens*. West Conshohocken, PA: ASTM International.

ASTM C 78/C 78M-10e1. 2010. *Standard test method for flexural strength of concrete (using simple beam with third-point loading)*. West Conshohocken, PA: ASTM International.

ASTM C 94/C 94M-13a. 2013. *Standard specification for ready-mixed concrete*. West Conshohocken, PA: ASTM International.

ASTM C 150/C 150M. 2012. *Standard specification for Portland cement*. West Conshohocken, PA: ASTM International.

ASTM C 226. 2012. *Standard specification for air-entraining additions for use in the manufacture of air-entraining hydraulic cement*. West Conshohocken, PA: ASTM International.

ASTM C 260/C 260M-10a. 2010. *Standard specification for air-entraining admixtures for concrete*. West Conshohocken, PA: ASTM International.

ASTM C 465. 2010. *Standard specification for processing additions for use in the manufacture of hydraulic cements*. West Conshohocken, PA: ASTM International.

ASTM C 469/C 469M. 2010. *Standard test method for Static modulus of elasticity and Poisson's ratio of concrete in compression*. West Conshohocken, PA: ASTM International.

ASTM C 496/C 496M. 2011. *Standard test method for splitting tensile strength of cylindrical concrete specimens*. West Conshohocken, PA: ASTM International.

ASTM C 595/M. 2013. *Standard specification for blended hydraulic cements*. West Conshohocken, PA: ASTM International.

ASTM C 688-8e1. 2008. *Standard specification for functional additions for use in hydraulic cements*. West Conshohocken, PA: ASTM International.

ASTM C 1116/C 1116M-10a. 2010. *Standard specification for fiber-reinforced concrete*. West Conshohocken, PA: ASTM International.

ASTM C 1157/C 1157M. 2011. *Standard performance specification for hydraulic Cement*. West Conshohocken, PA: ASTM International.

ASTM C 1666/C 1666M. 2008. *Standard specification for alkali resistant (AR) glass fiber for GFRC and fiber-reinforced concrete and cement*. West Conshohocken, PA: ASTM International.

ASTM D 1752-04aR13. 2013. *Standard specification for preformed sponge rubber cork and recycled PVC expansion joint fillers for concrete paving and structural construction*. West Conshohocken, PA: ASTM International.

ASTM D 2628-91R11. 2011. *Standard Specification for preformed polychloroprene elastomeric joint seals for concrete pavements*. West Conshohocken, PA: ASTM International.

ASTM D 5249. 2010. *Standard specification for Backer material for use with cold- and hot-applied joint sealants in Portland-cement concrete and asphalt joints*. West Conshohocken, PA: ASTM International.

ASTM D 5893/5893M. 2010. *Standard Specification for cold applied, single component, chemically curing silicone joint sealant for Portland cement concrete pavements*. West Conshohocken, PA: ASTM International.

ASTM D 6690. 2012. *Standard specification for joint and crack sealants, hot applied, for concrete and asphalt pavements*. West Conshohocken, PA: ASTM International.

Austroads. 2009. *Guide to Pavement Technology Part 3: Pavement Surfacings*. Publication No. AGPT03-09. Sydney: Austroads Inc.

Austroads. 2012. *Guide to Pavement Technology Part 2: Pavement Structural Design*. Publication No. AGPT02-12. Sydney: Austroads Inc.

Brooks J.J. and A. Neville. 1992. Creep and shrinkage of concrete affected by admixtures and cement replacement materials, ACI SP-135. *Creep and Shrinkage of Concrete: Effect of Materials and Environment*. Detroit, MI: American Concrete Institute.

CCAA. 2005. *Plastic Shrinkage Cracking*. Australia: Cement Concrete and Aggregates.

CEN EN 197-1. 2011. *Cement – Part 1: Composition, specifications and conformity criteria for common cement*. Brussels: CEN.

CEN EN 197-2. 2014. *Cement – Part 2: Conformity evaluation*. Brussels: CEN.

CEN EN 206. 2013. *Concrete – Part 1: Specification performance, production and conformity*. Brussels: CEN.

CEN EN 450-1. 2012. *Fly ash for concrete – Part 1: Definition, specifications and conformity criteria*. Brussels: CEN.

CEN EN 934-1. 2008. *Admixtures for concrete, mortar and grout – Part 1: Common requirements*. Brussels: CEN.

CEN EN 934-2:2009+A1. 2012. *Admixtures for concrete, mortar and grout – Part 2: Concrete admixtures. Definitions, requirements, conformity, marking and labeling*. Brussels: CEN.

CEN EN 1008. 2002. *Mixing water for concrete. Specification for sampling, testing and assessing the suitability of water, including water recovered from processes in the concrete industry, as mixing water for concrete*. Brussels: CEN.

CEN EN 12390-3:2003/AC. 2011. *Testing hardened concrete – Part 3: Compressive strength of test specimens*. Brussels: CEN.

CEN EN 12390-5. 2009. *Testing hardened concrete – Part 5: Flexural strength of test specimens*. Brussels: CEN.

CEN EN 12390-6. 2009. *Testing hardened concrete – Part 6: Tensile splitting strength of test specimens.* Brussels: CEN.

CEN EN 12620:2002+A1. 2008. *Aggregates for concrete.* Brussels: CEN.

CEN EN 13108-1:2006/AC. 2008. *Bituminous mixtures – Material specifications – Part 1: Asphalt concrete.* Brussels: CEN.

CEN EN 13108-2:2006/AC. 2008. *Bituminous mixtures – Material specifications – Part 2: Asphalt concrete for very thin layers.* Brussels: CEN.

CEN EN 13108-7:2006/AC. 2008. *Bituminous mixtures – Material specifications – Part 7: Porous asphalt.* Brussels: CEN.

CEN EN 13412. 2006. *Products and systems for the protection and repair of concrete structures – Test methods – Determination of modulus of elasticity in compression.* Brussels: CEN.

CEN EN 13877-1. 2013. *Concrete pavements – Part 1: Materials.* Brussels: CEN.

CEN EN 13877-2. 2013. *Concrete pavements – Part 2. Functional requirements for concrete pavements.* Brussels: CEN.

CEN EN 14188-1. 2004. *Joint fillers and sealants – Part 1: Specifications for hot applied sealants.* Brussels: CEN.

CEN EN 14188-2. 2004. *Joint fillers and sealants – Part 2: Specifications for cold applied sealants.* Brussels: CEN.

CEN EN 14188-3. 2006. *Joint fillers and sealants – Part 3: Specifications for preformed joint seals.* Brussels: CEN.

CEN EN 14889-1. 2006. *Fibres for concrete steel fibres: Definitions, specifications and conformity.* Brussels: CEN.

CEN EN 14889-2. 2006. *Fibres for concrete – Polymer fibres: Definitions, specifications and conformity.* Brussels: CEN.

CEN EN 15422. 2008. *Precast concrete products – Specification of glassfibres for reinforcement of mortars and concretes.* Brussels: CEN.

Croney D. and P. Croney. 1991. *The Design and Performance of Road Pavements*, Chap. 15, 2nd Edition. London: McGraw-Hill International.

FHWA. 2011. *Portland Cement Concrete Pavements Research: Thermal Coefficient of Portland Cement Concrete.* Washington, DC: Federal Highway Administration. Available at http://www.fhwa.dot .gov/publications/research/infrastructure/pavements/pccp/thermal.cfm.

Fwa T.F., X.P. Shi, and S.A. Tan. 1996. Analysis of concrete pavements by rectangular thick-plate model. *Journal of Transportation Engineering*, Vol. 122, No. 2, pp. 146–154. ASCE.

Hassan K.E., J.G. Cabrera, and M.K. Head. 1998. The effect of aggregate type on the properties of high performance, high strength concrete: Material properties, structural behaviour and field applications. *Proceedings of the International Conference 'High Performance High Strength Concrete'.* Perth, WA: School of Civil Engineering, Curtin University of Technology.

Hassan K.E., J.W.E. Chandler, H.M. Harding, and R.P. Dudgeon. 2005. *New Continuously Reinforced Concrete Pavement Designs.* TRL Report TRL 630. Crowthorne, UK: TRL Limited.

Highways Agency. 1999. *Design Manual for Roads and Bridges (DMRB), Vol. 7, Pavement Design and Maintenance.* Section 5, Part 3: Surfacing and surfacing materials, HD 38/97: Concrete surfacing and materials, Amendment No. 1. Department for Transport. London: Highways Agency.

Highways Agency. 2006a. *Design Manual for Roads and Bridges (DMRB), Vol. 7: Pavement Design and Maintenance.* Section 2. Part 2. Ian 73/06. Design guidance for road pavement foundations (Draft HD 25). Department for Transport. London: Highways Agency.

Highways Agency. 2006b. *Design Manual for Roads and Bridges (DMRB), Vol. 7: Pavement Design and Maintenance.* Section 2. Part 3. HD 26/06: Pavement design. Department for Transport. London: Highways Agency.

Highways Agency. 2006c. *The Manual of Contract Documents for Highway Works (MCDHW), Vol. 3: Highway Construction Details.* Department for Transport. London: Highways Agency.

Highways Agency. 2006d. *The Manual of Contract Documents for Highway Works (MCDHW), Vol. 1: Specification for Highway Works.* Series 1000: Road pavements – Concrete materials. Department for Transport. London: Highways Agency.

HRB. 1971. *Admixtures in Concrete: Accelerators, Air Entrainers, Water Reducers, Retarders, Pozzolans*. Special Report 119. Washington, DC: Highway Research Board.

Jeuffroy G. and R. Sauterey. 1996. *Cement Concrete Pavements (Chaussées en béton de ciment)*, published by arrangement with Presses de l'Ecole Nationale des Ponts et Chaussées. Brookfield, WI: A.A. Balkema Publishers.

Keene P.W. 1960. *The Effect of Air-Entrainment on the Shrinkage of Concrete Stored in Laboratory Air*. Technical Report TRA/331. London: Cement Concrete Association.

Lerch W. 1957. Plastic shrinkage. *Journal of American Concrete Institute (ACI), Proceedings*, Vol. 53, No. 8, pp. 797–802. Farmington Hills, MI: American Concrete Institute.

Lister N.W. and M.F. Maggs. 1982. Research and development in the design of concrete pavements. *International Symposium on Concrete Roads*. London: ICE.

Mayhew H.C. and H.M. Harding. 1987. *Thickness Design of Concrete Roads*. TRRL Research Report RR87. Crowthorne, UK: Transport Research Laboratory.

Neville A.M. 2011. *Properties of Concrete*, 5th Edition. Essex, England: Pearson Education Limited.

Nunn M. 2004. *Development of a More Versatile Approach to Flexible and Flexible Composite Pavements Design*. TRL Report TRL615. Crowthorne, UK: TRL Limited.

O'Flaherty C.A. 2002. *Highways: The Location, Design, Construction and Maintenance of Pavements*, 4th Edition. Oxford, UK: Butterworth-Heinenmann.

Packard R.G. 1984. *Thickness Design for Concrete Highway and Street Pavements*. Ottawa, Ontario: Canadian Portland Cement Association.

PCA. 2001. *Concrete Information – Concrete Slabs Surface Defects: Causes, Prevention, Repair*. PCA R&D Serial No. 2155, IS 177.07. Skokie, IL: Portland Cement Association.

Pickett G. 1951. *A Study of Stresses in the Corner of Concrete Pavement Slabs under Large Corner Loads*. Chicago: Portland Cement Association.

Pickett G. and G.K. Ray. 1951. Influence charts for concrete pavements. *Transaction*, Vol. 116, pp. 49–73. ASCE.

Pool T.S. 2005. *Guide for Curing of Portland Cement Concrete Pavements*. Report No. FHWA-RD-02-099, Vol. 1. McLean, VA, USA, Washington, DC: Federal Highway Administration.

Shi X.P., T.F. Fwa, and S.A. Tan. 1993. Warping stresses on concrete pavements on Pasternak foundation. *Journal of Transportation Engineering*, Vol. 119, No. 6. ASCE.

Shi X.P., S.A. Tan, and T.F. Fwa. 1994. Rectangular thick plate with free edges on Pasternak foundation. *Journal of Engineering Mechanics*, Vol. 120, No. 5, pp. 971–988.

Smoak G. 2012. *Guide to Concrete Repair*. Denver, CO: US Department of Interior, Bureau of Reclamation, Technical Service Centre.

Tang T.X., D.G. Zollinger, and S. Sandheera. 1993. Analysis of concave curling in concrete slabs. *Journal of Transportation Engineering*, Vol. 119, No. 4, pp. 18–32. ASCE.

Teychenne D.C., R.E. Franklin, and H.C. Erntroy. 1975. *Design of Normal Concrete Mixes*. Building Research Establishment & Transport and Road Research Laboratory. London: HMSO.

Viorin J., D. Desmoulin, and A. Lecomte. 2001. Predicting the long term strength of road mixtures treated with hydraulic binders. *Bulletin des Labatoire des Ponts et Chaussées*, Vol. 231, pp. 3–16.

Wang S.K., M.A. Sargious, and Y.K. Cheung. 1972. Advance analysis of rigid pavements. *Transportation Engineering Journal*, Vol. 98, pp. 37–44. ASCE.

Westergaard H.M. 1926. Stress in concrete pavements computed by theoretical analysis. *Public Roads*, Vol. 7. No. 2, pp. 25–35.

Westergaard H.M. 1933. Analytical tools for judging results of structural tests of concrete pavements. *Public Roads*, Vol. 14. No. 10, pp. 185–188.

Westergaard H.M. 1948. New formulae for stresses in concrete pavements of airfields. *Transaction*, Vol. 113, pp. 425–444. ASCE.

Yoder E.J. and M.W. Witczak. 1975. *Principles of Pavement Design*, 2nd Edition. Somerset, NJ: John Wiley & Sons, Inc.

Zhang J., T.F. Fwa, and X. Shi. 2003. Model for nonlinear thermal effect on pavement warping stresses. *Journal of Transportation Engineering*, Vol. 129, No. 6, pp. 695–702. ASCE.

Chapter 15

Pavement maintenance rehabilitation and strengthening

15.1 GENERAL

Every newly constructed pavement, by the time it is opened to traffic, is subjected to the disastrous effect of various factors, such as traffic, weather conditions, solar radiation and so on. At the same time, a gradual deterioration of the pavement's functional and structural quality starts. This is attributed to the ageing and wear of the surfacing material and fatigue of materials composing the pavement.

The above factors, in combination with the reliability of the design, the compliance of materials used and the quality of the construction achieved, are the only reasons for the emergence of pavement surface distresses, fatigue failure and, finally, pavement disintegration.

The construction of a new pavement should always be considered as a social investment. The administrator of the public fund is responsible and obliged not only to preserve the capital invested but also to confer a benefit.

The benefit may be direct or indirect. The direct benefits include reduction of accidents, reduction of travelling time/costs and reduction (or no increase) of vehicle maintenance cost. The indirect benefit is the social benefit arising from comfortable and safe transportation of the users for social and commercial activities.

To preserve the capital and obtain the above benefits, the pavement should be regularly maintained in order to sustain a tolerable level of service throughout its service life. The profit may be maximised by setting the limit of tolerable level of service high.

As a consequence, it becomes clear that pavement maintenance is imperative. The term *maintenance* in this particular case is a broad term. Other more precise terms are best to be used, such as those specified below.

15.2 TERMINOLOGY

The terminology used for keeping the pavement at a tolerable level of service differs significantly from country to country. In most of the countries, the following terms are used: routine maintenance, preventive maintenance, corrective maintenance, major maintenance or pavement rehabilitation, strengthening and rejuvenation.

Routine maintenance is defined as the number of activities (works) carried out repeatedly, on a daily, weekly, monthly or annual basis on all elements of the road/highway in order to ensure serviceability at all times and under all weather conditions.

The main activities in routine maintenance are as follows: (a) the cleansing of carriageway, verges, ditches, drains, signs and signals and safety barriers, to name as few, as well as grass cutting and tree pruning; (b) repair of damaged areas around manhole covers; (c) replacement of damaged safety barriers, road signs and, generally, road furniture; and

(d) winter maintenance, such as clearance of snow and prevention of ice formation on the pavement surface. It is obvious that works that are directly related to the pavement structure are not included in routine maintenance.

Preventive maintenance is defined as the number of activities aiming to prevent the premature emergence of distresses and consequently premature pavement destruction.

Corrective maintenance is defined as the number of activities aiming to correct pavement surface imperfections, which affect the safety of the user.

The activities included in the preventive and corrective maintenance are not essentially independent of each other, except perhaps crack filling, and thus they will not be individually mentioned per case. Works for preventive and corrective maintenance include crack filling, pothole filling, patching, surface skid resistance restoration and surface evenness restoration.

Major maintenance or pavement rehabilitation strengthening may be defined as the number of activities aiming to fully restore the qualitative state of the pavement. The works consist of constructing an asphalt layer of a certain thickness (asphalt overlay) consisting of new or recycled materials, with or without levelling course or milling of the old pavement surface. This asphalt layer may be catered to extending a pavement's service life.

For a better clarification of the terms *maintenance* and *rehabilitation* and for the purpose of this book, the definitions proposed by the Asphalt Institute will be adopted.

The Asphalt Institute (Asphalt Institute MS-17 3rd Edition) defines *maintenance* as routine work to keep a pavement as close to its desired level of serviceability as possible. This includes the preservation of existing pavement surfaces, resurfacing of less than the nominal overlay thickness, resurfacing of a short length of pavement, patching and repair of minor failures and the undersealing of concrete slabs.

Rehabilitation is the extension of the pavement structure's life when maintenance techniques are no longer viable to maintain adequate serviceability. It requires structural evaluation, corrective action and at least a nominal hot mix asphalt (HMA) overlay. A nominal overlay has a thickness of three times the nominal maximum aggregate size. Since many agencies specify 12.5 mm nominal size aggregate for their surface mixtures, their minimum HMA overlay thickness (over a HMA pavement) should be at least 38 mm (Asphalt Institute MS-17 3rd Edition).

The thickness of the HMA overlay is determined according to the existing structural condition of the pavement and the required number of years the pavement will be of service in the future.

Generally, maintenance works are considered as those for maintaining the capital invested while rehabilitation works are considered as those for increasing capital efficiency.

15.3 MAINTENANCE, REHABILITATION AND PAVEMENT LIFE

Each pavement is designed to last a certain number of years (design life). On the date of opening to traffic, the pavement offers the highest level of service. Soon, the detrimental effect of traffic and environment results in a continuous decrease of the pavement's level of service. If the pavement is to reach the design life providing all the time an acceptable level of service, periodic maintenance needs to be carried out.

The time and frequency of pavement maintenance vary depending primarily on the surfacing material used, quality of construction achieved and the traffic volume.

Under normal circumstances, when proper materials were used, quality construction was very good, and for a medium/high level of traffic, the first pavement surface maintenance usually appears within the first 8 to 12 years after construction.

Theoretically, the time for rehabilitation is towards the end of the pavement's design life. How close it is towards the end is difficult to determine since a number of factors (such as

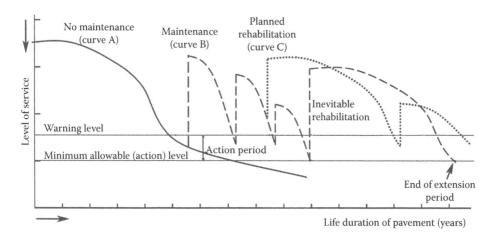

Figure 15.1 Effect of pavement maintenance and rehabilitation to pavement level of service level and pavement life duration.

compatibility of estimated loading with actual traffic loading, structural behaviour of layer material inclusive of the subgrade used, efficiency/insufficiency of drainage system provided, reliability of pavement design methodology used and unforeseen traffic and environmental conditions) affect the pavement's performance and hence the time when rehabilitation is needed. Rare is the case where rehabilitation is needed after the expiration of the design life.

One way to determine the rehabilitation time is by monitoring the provided level of service related to some surface distresses, mainly transversal/longitudinal evenness and cracking. Another way is by structural evaluation survey of the pavement.

The first approach, provided a minimum allowable level of service is set, can only give you the time when rehabilitation is inevitable. The second approach is more preferable since it is possible to determine the pavement's remaining life and decide for planned rehabilitation before the stage of inevitable rehabilitation. Planned rehabilitation is particularly useful when extension of pavement design life is required.

Planned rehabilitation always requires less capital investment in comparison to 'inevitable' rehabilitation, since it utilises the remaining life of the pavement found at the time of structural evaluation.

Figure 15.1 shows the effect of maintenance and rehabilitation actions to the pavement's level of service and pavement's life. It has been assumed that for a certain period, no maintenance is required for, say, a 20-year design period pavement. Figure 15.1 also shows the warning level and the minimum allowable level, or action, of service set.

When maintenance (curve A) is applied, it will be impossible for the pavement to reach its design life. With periodic maintenance (curve B), the pavement will require an 'inevitable' rehabilitation a few years before the end of its design life. When planned rehabilitation is elected (curve C), the rehabilitation cost is less than the cost of rehabilitation applied later on ('inevitable' rehabilitation), even if pavement is maintained once to reach the end of the decided extension period.

15.4 TYPICAL SURFACE DISTRESSES IN FLEXIBLE PAVEMENTS

All types of surface distresses occurring in flexible pavements can be classified into four categories: *cracking, distortion, disintegration* and *loss of skid resistance.*

A detailed description of all abovementioned distresses along with the possible causes and treatments is given below. It should be stressed that for applying the most appropriate treatment-maintenance procedure, the determination of the exact cause(s) of distress, among other things, is of vital importance.

15.5 CRACKING IN FLEXIBLE PAVEMENTS

The forms of surface cracking vary and are attributed to various reasons. They can be hairline cracks, line cracks, line cracks with branches and block cracks.

The extent and severity of cracking as well as their detrimental effect on the pavement structure determine the appropriate maintenance treatment.

In general, hairline cracks are best treated by surface dressing or thin surfacing. Line cracks of limited width are best treated by crack sealing without or with widening. Line cracks with branches may be best treated by removing the affected area and patching it. Block cracks may be best treated by removing the asphalt layers of the affected area to a desired depth and replacing them with new asphalt.

The above treatments are indicative. The best approach relies on determination of the cause(s), extent of cracking and average crack width.

15.5.1 Crack sealing/filling – general

The main purpose of crack sealing/filling is to prevent the intrusion of water through the crack into the underlying layers, causing further deterioration of the pavement. Additionally, crack sealing/filling prevents incompressible materials from entering the crack, causing pavement deterioration as it expands or contracts with temperatures.

Crack sealing/filling is only applied to crack widths greater than 3 mm. Cracks with width less than 3 mm (hairline cracks) are treated more effectively by micro-surfacing, surface dressing (chip seal), sand seal or fog seal.

The crack sealing procedure consists of either just cleaning and sealing the crack or sawing or routing, cleaning and sealing the crack with an appropriate sealant.

In particular, when the crack width is between 3 and 10 mm, the crack is usually sawn or routed to a width of 6 to 12 mm. The sawn crack is then cleaned and sealed with hot-applied sealant. If the cracks are more than 50 mm deep, a backer rod is recommended to be installed to conserve sealant.

When the crack width is between 10 and 20 mm, it is usually required to remove the possible loose parts, clean the crack and then seal it with hot-applied sealant. If the cracks are more than 50 mm deep, a backer rod should be installed to conserve sealant.

When the crack width is more than 20 mm, the crack is cleaned and filled with appropriate sealant.

Cleaning the crack is best carried out by hot air blasting since it not only cleans but also heats the walls of the crack. Cold air blasting as well as sandblasting may also be used for cleaning the cracks; sandblasting must always be followed by air blasting and is more expensive and labour intensive. Water blasting is not recommended for crack cleaning since it is time consuming (water should be dried out completely).

Sealing the crack with hot-applied sealant is carried out immediately after cleaning, creating an overbanding of approximately 20 to 40 mm in width. The hot-applied sealant should comply with the requirements of CEN EN 14188-1 (2004), ASTM D 6690 (2012) or other national specifications. A backer rod should be installed to conserve sealant. The backer rod should comply with ASTM D 5249 (2010) or other national specifications.

Cold applied single component, chemically curing, silicone sealant may also be used for crack filling, complying to CEN EN 14188-2 (2004), ASTM D 5893 (2010) or other specifications.

Cold bituminous emulsions or hot non-modified bitumen may be used for filling wide cracks (more than 20 mm crack width). Crack filling with these two types of sealants provides only temporary crack treatment.

An alternative to sealing/overbanding, cracks may be widened by milling out to a depth from 20 mm to the full depth of the surfacing layer (wearing course) and reinstated with a repair material (sealant or hot asphalt with small particle size aggregate).

The crack sealing procedure is recommended to comply with the guidelines provided by the producer of the sealant or contractual requirements.

When the overbanding width is more than 20 mm, particularly on the longitudinal joint or other cracking, some countries, such as the United Kingdom, require surfacing the overbanding with durable aggregates to provide an initial skid resistance value of at least 60 (Highways Agency 1998).

15.5.2 Alligator (fatigue) cracking

Alligator cracking is characterised by branched and interconnected cracks forming polygonal blocks, the pattern of which resembles an alligator's skin. In some cases, the blocks give the impression that they are ready to be detached. A typical form of alligator cracking is shown in Figure 15.2.

15.5.2.1 Causes

Alligator cracking, also known as fatigue cracking, is caused by excessive deflection of the asphalt layer. This can be caused by overloading the asphalt layer (fatigue cracking of asphalt layer) or by weakening of the subgrade or unbound layers because of moisture increase. The above are also related to insufficient pavement thickness and to thin asphalt layer.

(a)   (b)

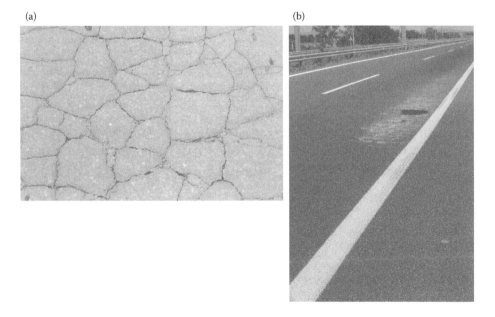

Figure 15.2 Alligator cracking. (a) Typical pattern of alligator cracking. (b) Localised alligator cracking.

When alligator cracking is extensive and has initiated from the wheel paths, and there is no surface depression, the cause is related to fatigue of the asphalt layer. The particular pavement section has failed and reached its service life. If, together with the alligator cracking, surface depression is observed, the pavement has also failed as a result of settlement failure of the subgrade. Which failure came first is a matter of investigation to be conducted.

When alligator cracking is not extensive but localised, the cause is most probably the local weakening of the subgrade or, to a lesser extent, of the unbound layers, owing to moisture increase. This implies that, locally, the subgrade material had very low bearing capacity (existence of medium- to high-plasticity material) or the unbound materials used do not comply with the requirements. Localised alligator cracking more often is accompanied by surface depression. Localised alligator cracking usually appear within the first few years after pavement construction (2 to 5 years).

15.5.2.2 Treatment

When the *alligator cracks are extensive*, the treatment may consist of (a) replacing all asphalt layers, (b) replacing part of asphalt layers or (c) placing an asphalt overlay, provided pavement elevation permits.

When it is time to replace part of the asphalt layers, or to place an overlay, the use of modified bitumen in the new asphalt hot mix is recommended in most cases. The use of asphalt-reinforcing grids, or other reflection cracking-reducing technique, is also advised to be considered. For details, see Section 15.6.

In cases where the formation of alligator cracking is at the beginning, and the width of the cracks is less than 3 mm, a temporary maintenance such as slurry surfacing, or surface dressing, may be considered. This will only extend the time before permanent repair is needed.

When *alligator cracking is localised*, the best repair is to remove and replace all asphalt layers and most probably the unbound layers or part of the subgrade (at least 300 to 600 mm, depending on the strength of the subgrade). The above procedure is known as full-depth patching. At the same time, the cause of moisture increase must be determined and appropriate measures (drainage maintenance or drainage installation) must be taken.

Full-depth patching in brief: The area that is to be repaired (patch area) should be extended approximately 500 mm inside the non-stressed area. The outline of the patch is cut by a pavement saw cutter. All materials are carefully removed, or excavated, down to the required depth; in the case of localised alligator cracking area, most probably down to part of the subgrade material. All cut and excavated faces should be straight and vertical.

The replaced unbound material should be laid in layers of not more than 100 to 150 mm, for the unbound material, or 50 to 100 mm, for asphalt. Compaction should also be sufficient using a vibratory plate compactor, for small patches, or a vibratory/static roller, for larger areas.

The vertical surfaces of the old asphalt layers should be tack coated, as well as the unbound base surface, with suitable bitumen emulsion.

Extra care should always be taken for the surface of the patch to be at the same level (grade) as the surrounding pavement.

After completion of compaction of the last asphalt layer, it is recommended that the edges of the patch be sealed with an appropriate joint sealant. This provides an excellent sealing protection against water infiltration. The sealant material is poured along the edges, forming a bitumen band of 30 to 40 mm wide. The sealant is the same as the one used for crack/joint sealing (modified bitumen).

Figure 15.3 Edge cracking.

15.5.3 Edge cracks

These cracks are longitudinal and appear at a distance of approximately 30–50 cm from pavement edges with or without branching cracks towards the shoulder (see Figure 15.3).

15.5.3.1 Causes

Edge cracking is caused by insufficient pavement support owing to one or more of the following reasons: poor compaction, shrinkage, poor drainage or frost action of surrounding/underlying material.

The crack width may vary from a few millimetres up to 50 mm or more.

15.5.3.2 Treatment

The treatment for the edge cracking depends on the severity of the distress. Single and short cracks may be filled with bituminous binder, sand–asphalt mix or slurry surfacing mix, depending on the width of the crack.

More than one long edge crack may require removal from the required depth of the unbound material not complying with the requirements and replacing it with well-compacted material complying with the requirements. New hot asphalt is placed over the compacted new base material.

In case water is also a contributing factor, installation or improvement of drainage is needed to eliminate the problem.

Any other treatment such as crack filling with sealant, slurry surfacing material or fine asphalt is only a temporary treatment.

15.5.4 Paving joint and widening cracking

This kind of cracking appears at construction joints between paving lanes or at pavement widening. At the beginning, cracking is a single linear crack developed in the longitudinal direction along the joint of the paving lanes; if left unattended, small branched cracks may also develop. Figure 15.4a shows a typical lane (joint) cracking at the initial stage of formation.

Lane cracking may develop at the transverse direction at the initiation or termination of asphalt works, at the joint between the old and new paved lanes.

(a)   (b)

Figure 15.4 Paving joint cracking and maintenance. (a) Initiation of joint cracking. (b) Treated joint cracking.

15.5.4.1 Causes

The cases are attributed only to poor construction, such as lack of material at the joint, poor or insufficient adhesion of the vertical surfaces of the joint (weak bond in the joint) or insufficient compaction at the joint. The temperature difference between lanes during paving may affect the development of joint cracking, and thermal stresses during the service life of the pavement should not be excluded, particularly on wide paved lanes.

In rare cases, paving joint cracking may be caused by insufficient compaction of underlying asphalt lanes where the construction joint is on top of each other. However, in these cases, minor longitudinal settlement will most certainly appear.

15.5.4.2 Treatment

The maintenance of a typical paving joint crack is carried out by using the crack sealing procedure described in Section 15.5.1. Figure 15.4b shows a joint crack after treatment.

15.5.5 Reflective cracks

Reflective cracks occur in overlays laid to rehabilitate old cracked flexible pavement (Figure 15.5). Typical reflective cracks also appear in asphalt overlays on rigid pavements, flexible pavements with cement-bound bases or asphalt overlays constructed over tied shoulders or widenings.

Figure 15.6 shows a transverse reflective crack of asphalt overlay over a cement-bound base flexible pavement. The crack was attributed to the absence of joint formation of the cement-bound base.

The pattern and direction of the reflective cracks vary from linear in the longitudinal, transverse or diagonal direction to block cracking depending on the old pattern of the cracks that developed before applying the overlay.

Figure 15.5 Reflective cracking initiated over an old paving joint in a flexible pavement.

(a)   (b)

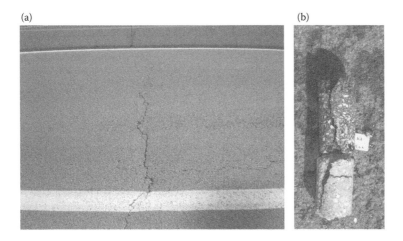

Figure 15.6 (a) Reflective crack on the transverse direction. (b) Core taken at the reflective crack.

15.5.5.1 Causes

Reflection cracks are caused by vertical and horizontal movements of the underlying parts of the pavement showing discontinuity (broken parts or joints). These movements are induced by traffic, expansion and contraction owing to temperature or moisture changes or swelling/shrinkage of the subgrade.

15.5.5.2 Treatment

The maintenance of reflective cracks depends on their size (crack width) and severity. The following are the usual techniques used:

a. When reflective cracks are linear and localised, crack sealing is applied (see Section 15.5.1).
b. When reflective cracks are extensive and non-linear, the most effective treatment would be to remove the asphalt layer in the distressed area and to replace it with a new asphalt layer. The use of asphalt-reinforcing grids, or other reflection cracking-reducing technique, may also be considered.

Figure 15.7 Slippage cracking. (From Asphalt Institute, *Asphalt in Pavement Preservation and Maintenance*, Manual Series No. 16 [MS-16], 4th Edition, Lexington, US: Asphalt Institute, 2009. With permission.)

15.5.6 Slippage cracking

Slippage cracks are crescent-shaped (half-moon) cracks formed owing to the slippage of the asphalt layer. They usually develop in areas where horizontal forces induced by traffic develop (heavy braking areas, downhills, uphills and junctions). Figure 15.7 shows typical slippage cracking.

15.5.6.1 Causes

Slippage cracking is exclusively caused by lack of bond (or poor bond) between two asphalt layers. Lack of bond is attributed to the absence of tack coat and to the presence of dust, traces of clayey soil material, oil, dirt or even water on the asphalt surface to be paved.

15.5.6.2 Treatment

The maintenance of these cracks is carried out by removing the affected area, cleaning thoroughly the exposed surface and applying tack coat and a surface patch.

15.5.7 Shrinkage cracking

Shrinkage cracks are block-patterned cracks of almost square shape formed over the whole area of the pavement. They can also be linear, running parallel to the longitudinal axis at any distance from the edge of the pavement.

15.5.7.1 Causes

Block-patterned cracks are usually caused by volume change of the asphalt mix owing to low environmental temperatures, while shrinkage longitudinal cracks are caused by volume change (shrinkage) of highly plastic subgrade material.

The block pattern shrinkage cracks are usually formed on old flexible pavements with low penetration grade bitumen and high stiffness modulus asphalt. The lack of traffic, the use of absorptive aggregates and possibly the use of crushed gravel (high silica content compared to limestone) encourage the development of shrinkage cracking. Thick asphalt layers

Figure 15.8 Shrinkage cracking, block patterned.

with high stiffness modulus asphalt regardless of traffic also encourage the development of shrinkage cracks caused by volume changes of the asphalt.

Figure 15.8 shows block shrinkage cracks (location and conditions: altitude 980 m, sub-zero and high air temperatures during winter/summer period, low traffic and dense asphalt concrete with crushed gravel).

The longitudinal shrinkage cracks are more likely to develop in pavements built on highly plastic and expansive subgrade material (Figure 15.9). The absence of pavement drainage or ineffective pavement drainage hastens this type of shrinkage cracking.

Figure 15.9 shows shrinkage cracks initiated owing to shrinkage of the subgrade material and reflected through the layers to the surface of the pavement, within less than a year (December to June). Details of the construction are as follows: plasticity index of subgrade material, 37; thickness of asphalt layers, 100 mm; thickness of unbound base/sub-base layers, 400 mm; a capping layer of 300 mm; between soil material and the capping layer, a filtration/separation woven geotextile had been placed, as shown in the last photograph of Figure 15.9.

Figure 15.9 Shrinkage cracking owing to subgrade shrinkage; cracks shown on each layer and on the surface of the subgrade.

15.5.7.2 Treatment

The maintenance of the block-patterned cracks is best carried out by replacing the asphalt layer in the stressed area with new asphalt. Crack sealing may also be chosen as a temporary treatment.

The maintenance of longitudinal shrinkage cracks is carried out by excavating the pavement down to the subgrade level, stabilising the soil material or increasing the thickness of the capping layer and then reconstructing the pavement. Alternatively, if pavement elevation permits, remove the asphalt layer, increase the thickness of the unbound layers to the required depth and lay new asphalt layers. In both alternative treatments, the provision of effective drainage is essential. The placement of a reinforcing geotextile at the formation level will be beneficial.

15.5.8 Linear wheel path cracks

Linear wheel path cracks develop along the wheel paths of the near side lane, not always simultaneously, and consist of a single longitudinal crack with small braches. Linear wheel path cracks are exclusively fatigue cracks and, when left untreated, soon change to an alligator cracking pattern.

15.5.8.1 Causes

The causes of linear wheel path cracking are the same as those of alligator (fatigue) cracking (Section 15.5.2).

15.5.8.2 Treatment

The maintenance of the linear wheel path crack is carried out by crack sealing procedure. The treatment will certainly delay the imminent formation of alligator cracking, provided the cracks are sealed as soon as they appear.

15.5.9 Helical or diagonal cracks

Helical or diagonal cracks usually start from the centre of the pavement and run away towards the pavement edge (Figure 15.10a) or diverge from the longitudinal axis (Figure 15.10b); in other cases, they may run diagonally across the pavement (Figure 15.10c). They are exclusively formed in pavements constructed on medium to high embankments or cut/fill. Helical or diagonal cracks are not load-related cracks.

15.5.9.1 Causes

Helical or diagonal cracks are caused by instability or settlement of the embankment.

15.5.9.2 Treatment

The maintenance of these cracks is carried out by crack sealing or by milling out a narrow zone to a certain depth and then filling it with hot asphalt.

After a crack emerges, an investigation aimed at determining the actual causes of embankment failure is recommended to establish proper and permanent treatment.

Figure 15.10 Helical or diagonal cracks. (a) Helical crack, after treatment; (b) Helical crack formation and (c) diagonal crack.

15.6 RETARDATION OF REFLECTIVE CRACKING TREATMENTS AND ASPHALT REINFORCEMENT TECHNIQUES

Retardation of reflective cracking, when an overlay is to be constructed over an area with cracks, can be obtained by applying one of the following treatments: (a) use of geosynthetics for asphalts, (b) use of stress-absorbing membrane interlayer (SAMI) and (c) use of fibre-reinforced membrane.

The effectiveness of each treatment depends on the type/extent of cracking, workmanship at placement and daily traffic. All treatments may be applied on milled surfaces (the most usual case) or on existing pavement surface.

15.6.1 Geosynthetics for asphalts

From the wide range of geosynthetic products, the ones used in asphalt rehabilitation works are the geotextiles, geogrids and geocomposites, all able to sustain the high temperature of the asphalt during application.

Geotextiles, non-woven textiles in particular, also known as paving fabrics, were the first to be used in asphalt paving. The original idea was for geotextiles to act as an asphalt reinforcement to improve the tensile properties of the asphalt and hence enhance asphalt tensile fatigue and thermal cracking. Soon, it was realised that a geotextile functions as a stress relief layer rather than as a reinforcement. When properly installed between the road surface and the asphalt overlay, it allows for slight movements between the two layers, which delays or arrests crack propagation in the asphalt overlay.

The use of geotextiles in conjunction with the bitumen layer sprayed also functions as an interlayer water barrier to the ingress of surface water and thus prevents or delays the deterioration of the pavement.

Geogrids for asphalts were developed after the introduction of paving fabrics and exclusively for reinforcing the asphalt; hence, they are called *asphalt-reinforcing geogrids*. Geogrids are polymeric structures consisting of a regular open network of integrally connected tensile elements bonded by needle-punching, agglutinating or interlacing (CEN EN ISO 10318 2005).

Geogrids are typically made of polyester, polypropylene, glass or carbon fibres. The aperture size of the grid is 40 mm by 40 mm or less.

Soon, it was realised that the bond between the geogrid and the asphalt layers was the most important factor for the geogrid to successfully function as an asphalt reinforcement. As a result, most of the geogrids nowadays are coated with bitumen to improve the bond with the asphalt mixture.

Geocomposites were the latest products developed in the mid-1990s that utilise the benefit of nonwoven geotextiles and geogrids or reinforcing woven textiles that – by needle-punching or heat setting – establish a single unit (CEN EN ISO 10318 2005).

Both parts of the geocomposite, reinforcement grid and textile, are usually coated with bitumen, or modified bitumen, for better bonding with the asphalt. The geocomposite, together with the tack coat applied, fulfils three functions: it provides asphalt reinforcement, a stress relief layer and a water barrier layer. Today, asphalt-reinforcing geocomposites are considered to be the ultimate products that simultaneously face retardation of reflective cracking and pavement strengthening.

It is noted that apart from the non-metallic geogrids, there are also metallic geogrids. Despite reinforcing the asphalt, the impossibility of bonding with the old asphalt layer (hence, they are pegged on the surface), the great difficulty as regards handling and placement and their higher cost make metallic geogrids less attractive than their non-metallic counterparts.

The following are the properties required for a geosynthetic to be used in pavement rehabilitation or new pavement asphalt works: tensile strength and elongation (in both directions), bitumen retention, durability, alkaline resistance (in case a geogrid or a geocomposite is used over concrete surface) and melting point, to name a few. Property requirements and other characteristics are specified in CEN EN 15381 (2008), ASTM D 7239 (2013) or other relevant standards.

The use of geogrids or geocomposites effectively results in the reduction of the thickness of the asphalt overlay. For the determination of the resulting reduction in thickness, special calculations are needed to be carried out. However, the thickness of the asphalt layer above the reinforcement should never be less than 50 mm or the minimum value required by the supplier. This minimises the horizontal tensile forces that may cause slippage of the geocomposite or the overlay.

In general, to avoid slippage failure in geosynthetics application, the following should be ensured: (a) the surface needs to be clean and free of dirt, sand, mud, oil and so on; (b) application of the proper amount of tack coat; and (c) laying the geosynthetic free of folds or waves. Typical quantities of bituminous emulsion required for tack coat range from 0.4 to 0.8 kg/m² of residual bitumen, depending on the conditions of the pavement surface. In all cases, for proper handling, application and avoidance of application failures of the

Figure 15.11 Slippage of asphalt overlay over a geotextile.

geosynthetic, it is recommended that the supplier's guidelines and instructions are strictly followed.

Figure 15.11 shows a common early slippage failure of the overlay over a geotextile owing to the inadequate quantity of tack coat.

From the author's experience, slippage failure is more likely to occur when a geotextile is used rather than a geocomposite product. This is the outcome of the author's involvement in seven projects since 1997. Early slippage failure had occurred in three out of seven projects where geotextiles were used. There was no slippage failure in the four projects where geocomposites were used. However, after corrective works, the performance of the geotextile-treated pavements in all cases was excellent, as well as the performance of the pavements treated with geocomposites until now.

With respect to choosing between geotextile and geocomposite material, it is advised to choose the latter as it turns out to be more cost-effective.

15.6.2 Stress-absorbing membrane

The stress-absorbing membrane (SAM) technique is similar to the single surface dressing. The only differences are that the quantity of the bituminous material sprayed is higher and the bituminous binder is exclusively modified bitumen.

The SAM technique was initially developed in the United States and was soon applied in Europe (Buchta and Nievelt 1989).

The technique consists of the creation of a membrane approximately 3.5 mm thick, capable of absorbing the tensile stresses that developed. Thus, it is known as the 'stress-absorbing membrane' (SAM) technique. On the membrane created, and immediately after spraying the modified bitumen, single-size aggregates of 8/10 or 10/16 mm are spread and slightly rolled with a pneumatic-tyred roller. The aggregates are applied hot for better adherence to the modified bitumen.

The spraying temperature of the modified bitumen is specified by the supplier, which usually ranges from 170°C to 200°C. The quantities of materials used are usually (a) 2.5 to 2.8 kg/m² for modified bitumen and approximately 18 kg/m² for aggregate size 10/16 mm or (b) 2.0 to 2.3 kg/m² for modified bitumen and approximately 15 kg/m² for aggregate size 8/10 mm.

The technique, as described above, is considered as temporary maintenance, protecting the pavement from disintegration and, to a certain extent, from further cracking.

When an asphalt layer with a thickness of 40 mm (or more) is laid on SAM, SAM acts as a stress-absorbing interlayer and the technique is called SAMI. The advantage of this technique is the excellent adhesion achieved between the old pavement surface, the absorbing interlayer and the asphalt overlay. SAMI is a much longer-lasting method of treating reflective cracking than SAM.

15.6.3 Fibre-reinforced membrane

This technique is similar to the SAM or SAMI technique. The only difference is that immediately after spraying the bitumen, suitable synthetic or natural fibres are laid, followed by new bitumen spraying. The procedure is best conducted with a modified bituminous sprayer capable of dispersing fibres as well.

On the reinforced bitumen membrane formed, single-size aggregates are spread and lightly compacted. The fibre-reinforced chipped surface may be exposed to traffic or covered soon by an asphalt overlay. The quantities required per type of material and the type of bitumen used are specified by the supplier/producer.

The results of this technique have been found to be as good as those achieved by the use geotextile and asphalt overlay, at a significantly lower cost (Yeates 1994). Useful information of a recent application of SAMI with fibres can be found in a paper by Sproule et al. (2012).

15.7 SURFACE DISTORTIONS IN FLEXIBLE PAVEMENTS

Pavement surface distortions are generally distresses related to pavement unevenness in the longitudinal or transverse direction. Surface evenness affects traffic safety and comfort. Distortions may be accompanied by cracking, which then affects the structural performance of the pavement.

Distortions may take one of the following forms: (a) rutting, (b) corrugations and shoving, (c) local depressions or (d) upheaval. Figure 15.12 shows schematically the different forms of surface distortion.

Surface distortions may be caused by one or more of the following reasons: (a) viscoelastic behaviour of the asphalt, (b) low resistance to deformability of the asphalt, (c) poor compaction of one or some of the pavement layers and (d) subgrade settlement.

For effective pavement distortion repair/maintenance, as in all pavement defects, it is absolutely necessary to determine the cause.

Generally, the repair/maintenance of surface distortions may range from hot asphalt filling, complete removal of the affected area and replacement with new material, thin or full-depth patching or application of micro-surfacing.

15.7.1 Rutting

Rutting is canalised surface depression along the wheel paths (see Figure 15.12a). Rutting may also occur with side uplift of the asphalt (see Figure 15.12b).

15.7.1.1 Causes

Rutting in the form of Figure 15.12a usually occurs after a long period (years) and may be attributed to one or more of the following reasons: permanent deformation of the asphalt layers, consolidation of pavement layers or settlement of the subgrade. The permanent

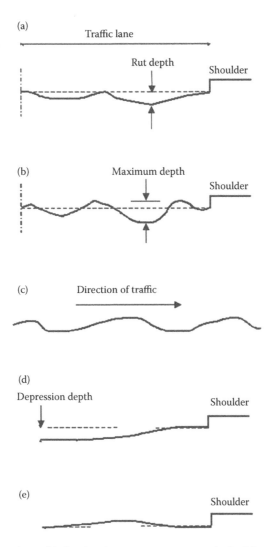

(a)

Traffic lane

Rut depth

Shoulder

(b)

Maximum depth

Shoulder

(c)

Direction of traffic

(d)

Depression depth

Shoulder

(e)

Shoulder

Figure 15.12 Surface distortions. (a) Rutting (transverse cross section), (b) rutting with side upheaval (transverse cross section), (c) corrugations (longitudinal cross section), (d) local depression and (e) local upheaval.

deformation of asphalt layers is exclusively attributed to the intrinsic property of the asphalt inherited from the bitumen. The consolidation of the layers is caused by insufficient compaction, and the settlement of the subgrade is caused by excessive loading of the subgrade.

Rutting, together with side uplift of the asphalt (Figure 15.12b), usually forms relatively early, even within the first summer months, and is exclusively attributed to low resistance in deformation (high deformability) of the asphalt laid (usually of the top asphalt layer[s]). Low void content in relation to aggregate gradation and high percentage of bitumen content, assisted by the use of inappropriate (soft) bitumen and possible use of uncrushed aggregate, are the main contributing factors.

Figures 15.13 and 15.14 show typical forms of rutting after long and short periods, respectively.

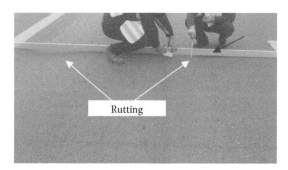

Figure 15.13 Typical rutting usually formed after a long period, as shown schematically in Figure 15.12a.

Figure 15.14 Typical rutting quite often formed very early, as shown schematically in Figure 15.12b.

15.7.1.2 *Treatment*

The treatment for rutting after a long period may consist of one of the following actions, depending on the cause:

a. If the rut depth is less than 20 mm, application of micro-surfacing, in one or two layers (temporary measure).
b. If rut depth is more than 20 mm and rutting is caused by permanent deformation of the asphalt layer(s), milling out to a certain depth to remove the asphalt layer with low resistance to permanent deformation and replacement with new asphalt having acceptable resistance to permanent deformation. If pavement elevation permits, a levelling course and an overlay, preferably with modified bitumen, may be a solution. Always consider the remaining life of the pavement.
c. If the rut depth is more than 20 mm and the cause is in the base/sub-base or subgrade, complete reconstruction of the pavement to the required depth is necessary. Examine effectiveness of the drainage system or provision of drainage, if water is a contributing factor. Figure 15.15 shows a typical example of rutting that developed because of weak subgrade (CBR = 3.2) settlement.

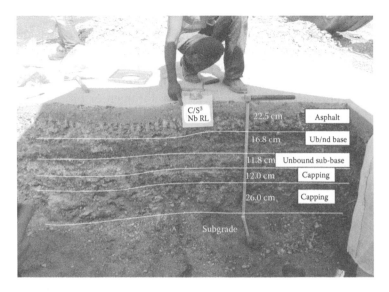

Figure 15.15 Rutting attributed to subgrade settlement.

In all the above treatments, thorough cleaning of the surface and application of the proper amount of tack coat before laying the new asphalt layer are essential; cationic emulsion is recommended to obtain 0.15 to 0.35 g/m^2 of bitumen residue on the surface (quantity depends on the surface's bitumen absorption).

15.7.2 Corrugations and shoving

Corrugations and shoving are ripples formed in the longitudinal direction of the asphalt pavement surface (Figure 15.12c) owing to plastic displacement of the asphalt. When the plastic displacement is local, the distress is called just shoving.

Corrugations and shoving usually appear in areas where high horizontal stresses develop such as in downhill or uphill areas or in vehicle braking areas (traffic lights or crossings). Typical corrugations and shoving are shown in Figure 15.16. Corrugation and shoving are not accompanied by cracking.

Figure 15.16 Corrugations and shoving.

15.7.2.1 Causes

Corrugation and shoving are exclusively attributed to lack of stability and shear resistance of the asphalt. Factors affecting the above are high bitumen content, use of soft bitumen and high percentage of sand size aggregate particles or round and smooth aggregate. Contamination of bitumen with oil substances (during transportation), oil spillage and lack of aeration in case of using cut-back bitumen or bitumen emulsion in the asphalt or tack coat may also cause corrugation and shoving. In some, rare, cases, the excessive moisture in the subgrade in conjunction with type of soil material may cause corrugations.

15.7.2.2 Treatment

The only effective treatment for corrugations and shoving is to remove the deficient asphalt from the whole affected area and replace it with well-designed and properly produced asphalt, avoiding the use of cut-back for tack coating.

In case of local shoving, remove the affected area and patch it.

In areas with very low traffic where the thickness of the asphalt layer may only be 50 mm, corrugations may be repaired by breaking and scarifying the asphalt layer and part of the base, adding a small amount of binder, mixing on site and, after compaction, applying a single surface dressing.

15.7.3 Local depressions

Local depressions are localised areas with slightly lower elevations than the rest of the surrounding pavement (Figure 15.12d). They are more easily noticeable after rain.

15.7.3.1 Causes

Local depressions are caused by local inadequate compaction of the unbound layers or failure of the subgrade. When accompanied by cracks, fatigue of asphalt layers has also occurred.

15.7.3.2 Treatment

Local depressions with less than 20 mm maximum depression depth may be treated by micro-surfacing (temporary maintenance).

When the depression depth is higher than approximately 20 mm, determine the cause and repair by removing and replacing the affected pavement to the required depth.

15.7.4 Upheaval

Upheaval is the upward pavement displacement, usually occurring locally. Soon, surface cracking is developed in the same area.

15.7.4.1 Causes

Upheaval is caused by swelling of the subgrade owing to the moisture effect on expansive soils. Sub-zero temperatures and frost-susceptible subgrade material will also cause pavement upheaval owing to expansion of ice formation.

15.7.4.2 Treatment

Upheaval is treated by excavating the pavement down to the subgrade level, stabilising or replacing the subgrade material, correcting moisture conditions at the subgrade level and reconstructing the pavement.

Temporarily, milling out the area may be applied, covered by surface dressing or micro-surfacing.

15.7.5 Utility cut depressions

Utility cut depression is a failure of utility installation (water, gas, electricity supply, etc.).

15.7.5.1 Causes

The cause of utility cut depressions is almost exclusively the inadequate compaction of the back fill material or the asphalt patch.

15.7.5.2 Treatment

Permanent repair of utility cut depressions requires removal of asphalt patch, proper re-compaction of the backfill material and placement of new asphalt patch compacted adequately.

15.8 DISINTEGRATION OF FLEXIBLE PAVEMENTS

Disintegration is the fragmentation of the pavement surface into small loose pieces. This also includes the detachment (loss) of aggregates from the surface of the pavement called ravelling. If disintegration is not treated at the early stage, the pavement will certainly require rehabilitation.

The main forms of pavement disintegration are ravelling and potholes.

15.8.1 Ravelling

Ravelling is the progressive dislodgment and loss of fine, first, and then coarse aggregate from the pavement surface. As the fine and coarse aggregates come off, the surface becomes rough and soon 'pock marks' or 'pocket holes' form, which, if left untreated, will develop into potholes. Figure 15.17a shows a ravelled surface and Figure 15.17b shows a ravelled surface with potholes after temporary patching.

15.8.1.1 Causes

Ravelling is caused by one or more or the following reasons: (a) low bitumen content in the mixture, (b) construction at low temperatures or during rain, (c) use of aggregates with poor affinity to bitumen (hydrophilic aggregates), (d) use of disintegrating or dirty aggregate, (e) bitumen ageing, (f) bitumen or asphalt overheating or (g) inadequate compaction. Ravelling is encouraged by the presence of water.

15.8.1.2 Treatment

Ravelling in its early stage is best treated by micro-surfacing or chip seal. When excessive ravelling has occurred, resulting in the formation of potholes, the best treatment is to mill off the surfacing and replace it.

(a)   (b)

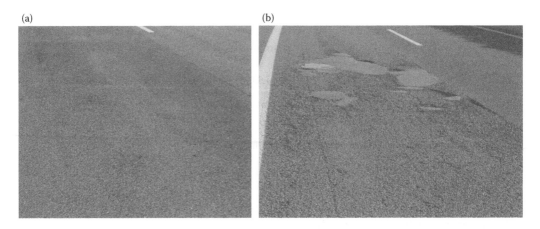

Figure 15.17 Ravelling. (a) Ravelling at early stage and (b) ravelling at late stage.

15.8.2 Potholes

Potholes are bowl-shaped holes resulting from local surface distress.

15.8.2.1 Causes

Potholes are caused by severe ravelling, alligator cracking, ill-treated patching, ill-treated utility cut repair or local deficiency of the asphalt laid.

15.8.2.2 Treatment

Potholes are treated by filling or patching. Filling consists of removing all loose parts; cleaning the pothole; applying tack coat (bitumen emulsion); applying usually premixed cold asphalt packed in bags or re-sealed pails, cold emulsion–based material blown in the pothole or hot asphalt; and compacting with a static or vibratory compactor.

Cold emulsion–based material containing chopped glass fibres may significantly increase the life of a pothole or reinstatement/repair (Woodside et al. 2011).

Patching consists of marking and saw cutting outside the distress area, removing the old asphalt and thorough cleaning, applying tack coat, backfilling with hot asphalt and compacting with a vibratory plate compactor or small vibratory roller compactor.

15.9 LOSS OF SURFACE SKID RESISTANCE

Loss of surface skid resistance is a skid hazard associated with smooth and slippery surface, which directly affects traffic safety. Unlike all other surface distresses, a smooth and slippery surface does not affect the structural deterioration of the pavement.

Slippery surface may result from polishing and abrasion of the surface aggregate or asphalt bleeding. Both factors reduce the macro- and micro-texture of the pavement surface and the surface skid resistance provided.

In the presence of water, the surface becomes more slippery; hence, the pavement's surface becomes more dangerous. An excess amount of water may result in aquaplaning, a situation where skid resistance is eliminated.

15.9.1 Polished aggregates

Polished aggregates are aggregates that have lost their inherited surface micro-texture and angularity.

15.9.1.1 Causes

The cause for aggregates to become polished is the polishing action by the traffic. The degree of aggregate polishing is related to the polish and abrasion resistance of the aggregates.

15.9.1.2 Treatment

The only effective treatment is to remove, with the use of a milling machine, the surfacing and to replace it with a surfacing material containing suitable crushed surface aggregate.

Crushed aggregates derived from sedimentary rocks such as limestone and dolomite are unsuitable for surfacing. The aggregates to be used in surfacing are those mainly produced from a majority of igneous rocks, sandstones from sedimentary rocks and hornfels from metamorphic rocks.

As surfacing material, the following may be considered: asphalt concrete for very thin layers, porous asphalt, SMA and micro-surfacing. Details can be found in Chapter 5. Surface dressing technique may also be considered as an alternative (see Section 15.15).

Surface texture restoration by mechanical means, such as sandblasting, glass bead blasting or micro-milling, provides a good but temporary treatment.

15.9.2 Bleeding or flushing

Bleeding or flushing of bitumen is the upward movement of bitumen and its appearance in the surface of the pavement. This results in the formation of a film of bitumen on the surface of variable thickness. Figure 15.18 shows a typical bleeding.

15.9.2.1 Causes

The only cause for bleeding is the excess amount of bitumen present in the asphalt layers or tack coat. Hot summer months accelerate bleeding as well as traffic if asphalt layers are overcompacted. Excessive sealant in crack filling may also cause bleeding.

Figure 15.18 Bleeding.

15.9.2.2 Treatment

Bleeding is usually treated by spreading hot sand or grit (heated to at least 140°C) over the affected area. Sometimes, the affected area may also be heated by hot air or infrared heaters before the application of the sand or grit. The rate of spread typically ranges from 5 to 7 kg/m². Immediately after spreading the sand, the surface is rolled with a pneumatic roller. After the surface cools down, the excess amount of aggregate is removed by a broom or a suction sweeping machine.

In severe and extensive bleeding, the only effective treatment is through the removal and replacement of surfacing.

15.10 TYPICAL DISTRESSES IN RIGID PAVEMENTS

The surface distresses in rigid pavements may be classified into *cracking*, *deformation* (*distortion*), *disintegration* and *loss of skid resistance*.

Before describing in detail the distresses and their causes, it should be mentioned that effective maintenance/treatment of the distresses in rigid pavements with cement mortar or concrete (unreinforced or reinforced) is too difficult and sometimes impossible. For this reason, bitumen and asphalt are the materials greatly used in maintenance of rigid pavements.

15.11 CRACKING OF RIGID PAVEMENTS

Cracking is the most common distresses of rigid pavements. It is caused, to a great extent, not only by the concrete's volumetric behaviour under temperature variations but also by the repetitive loading caused by traffic and some other external factors.

Cracks in concrete may develop at any direction, from the first hours of curing. Many cracks could be avoided or limited by provision of proper joint spacing, careful curing of concrete, good concrete mix and pavement design and good workmanship.

Timely maintenance sealing of the joints and of the cracks is of vital importance to prevent development of further distresses, which eventually accelerates the complete deterioration of the pavement.

The use of continuous reinforced concrete and, to a lesser extent, jointed reinforced concrete eliminates the development of some of the distresses described below.

15.11.1 Linear cracking

Linear cracking are formed in the transverse, longitudinal or diagonal direction that divides the slab into two to four parts. They extend down to a certain depth of the slab or to the full depth of the slab (Figure 15.19).

15.11.1.1 Causes

Linear cracking is caused by one or more of the following reasons: (a) traffic overloading (usually the stage where pavement construction reaches its excepted life), (b) improper transverse joint spacing (slab size, temperature related), (c) late sawing of the joint groove, (d) dowel restrain gross misalignment, (e) compression failure, (f) ingress of incompressible material into the joint, (g) edge restrain, (h) lack or loss of sub-base or subgrade support, (i) excessive swelling/shrinkage of the subgrade, (j) settlement of the subgrade owing to improper compaction and (k) propagation cracking (in unreinforced slabs) as a result of crack formed in the sub-base.

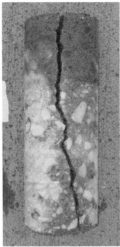

Figure 15.19 Transverse cracking (crack through the full depth of the slab).

15.11.1.2 Treatment

The treatment depends on the time of appearance of the linear cracking and its cause. If linear cracking appears early and there is no structural problem, cleaning and hot sealing are sufficient. The hot sealant should comply with the requirements of CEN EN 14188-1 (2004), ASTM D 6690 (2012) or other specifications.

Otherwise, if there are structural problems, full-depth repair (part of the slab or whole slab replacement) or pavement reconstruction at the affected area is necessary.

15.11.2 Corner cracking

Corner cracking appears at the corners and is a crack that, together with the two sides of the joints, forms a triangle. Corner cracking usually extends the full depth of the slab.

15.11.2.1 Causes

Corner cracking is caused by load repetitions combined with a loss of corner support (sub-base or subgrade), poor load transfer across the joint, dowel bar restrain near the edge of the slab and ingress of solids into the joint. The loss of support may be created by pumping or warping stresses.

15.11.2.2 Treatment

The temporary treatment, until such time that corner cracking is permanently repaired, consists of cleaning and sealing the crack. Permanent repair consists of full-depth concrete or asphalt patching.

15.11.3 Joint cracking and extrusion of joint seal

Joint cracking appears at the joint in the form of a crack (or cracks) either on the surface of the joint sealant or between the sealant and the wall of the joint.

In some cases, the joint may be safer from extrusion of the sealant.

15.11.3.1 Causes

A crack on the surface of the sealant is a cohesion failure resulting primarily from ageing of the sealant. Other factors such as inappropriate sealing material or incorrect sealing groove may also be the cause.

A crack between the sealant and the wall of the joint is an adhesion failure resulting from inadequate preparation of the sealing groove, inappropriate sealing material, the presence of moisture in the sealing groove, incorrect dimensions of the sealing groove or the chilling effect.

Extrusion of the sealant may be caused by overfill sealing groove and lack of compressibility of the backer material (caulking strip), if used.

15.11.3.2 Treatment

The only appropriate treatment is to remove the old joint sealant material, clean out the joint thoroughly, prepare the groove (if needed) and replace the backer material (if used) and reseal with appropriate sealant. The procedure is known as joint repair.

Sealant materials could be hot applied (complying with CEN EN 14188-1 2004 or ASTM D 6690 2012), cold applied (complying with CEN EN 14188-2 2004 or ASTM D 5893 2010) or preformed joint seal material applied by compression (complying with CEN EN 14188-3 2006, ASTM D 2628 2011 or other specifications). When backer material is used, it should comply with ASTM D 5249 (2010) or any other specification.

15.11.4 D-cracking

D-cracking is the formation of a series of cracks parallel and close to each other curving around corners of joints or cracks intersecting edges. D-cracking always starts from the bottom of the slab and progresses upwards until it reaches the surface, usually near the pavement joints.

15.11.4.1 Causes

D-cracking is caused by freezing and thawing cycles of the concrete containing expansive or poor-quality aggregates. Accumulation of water in the pavement structure (sub-base or subgrade) is the only contributing factor for the development of this type of cracking.

15.11.4.2 Treatment

The only appropriate treatment is full-depth patch or replacement of the slab.

15.11.5 Multiple cracking

Multiple cracking is the crack pattern that divides the slab into more than four parts. It can be likened to a situation where the slab is shattered.

15.11.5.1 Causes

Multiple cracking is caused by fatigue and other factors. When multiple cracking forms earlier than expected, the probable cause is poor support of the slab.

15.11.5.2 Treatment

The treatment, in case there are only few shattered slabs, is full replacement of the slabs. Otherwise, if the problem is widespread, pavement reconstruction is the only alternative.

Reconstruction of the pavement with an asphalt overlay, after cracking and seating of the concrete slabs, should always be considered as a proper and perhaps more cost-effective alternative solution, particularly in unreinforced concrete pavements.

15.11.6 Cracking of fresh and hardened concrete

Cracking of fresh and hardened concrete is non-fatigue-related cracking caused, to a great extent, by the moisture-induced volume changes of the concrete during the early stages of curing (plastic stage) as well as after when the concrete is hardened.

The most typical forms of cracking that developed on fresh and hardened concrete are plastic shrinkage cracking, drying (long term) shrinkage cracking, crazing cracking and early thermal contraction cracking.

The plastic shrinkage cracks are linear, formed randomly and usually parallel to each other; the drying shrinkage cracks are almost linear cracks and are formed either transversely or longitudinally to the longer length of the slab (single type) or into a three-crack pattern (branched type); the crazing cracks are of an irregular network pattern (map cracking); and the early thermal contraction cracks are linear and formed in the slab's corners or edges. The opening width of the cracks varies from 0.1 to 3 mm and their depth can be very shallow or very deep, depending on the form of cracking.

Analytical details about cracks that developed on fresh and hardened concrete can be found in Section 14.4.

Figure 15.20 shows typical crazing cracking; cracks are very fine (less than 0.5 mm in width), very shallow and barely visible.

Figure 15.21 shows a linear-branched drying shrinkage crack, extended to a certain depth.

Figure 15.22 shows a linear (longitudinal) drying shrinkage crack, extended to the full depth of the slab.

Figure 15.20 Crazing cracking.

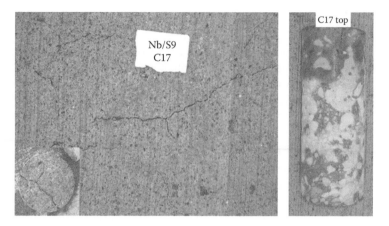

Figure 15.21 Drying shrinkage cracking, linear-branched into three cracks.

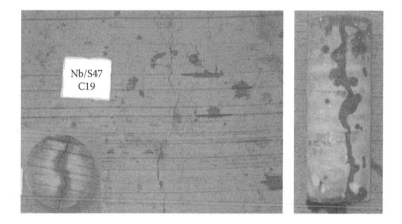

Figure 15.22 Drying shrinkage cracking.

15.11.6.1 Causes

Cracking of fresh and hardened concrete, in general, is caused by a combination of the following factors: environmental conditions during curing, rate of bleeding, rate of cooling, rate of water evaporation, temperature gradient or moisture content between the top and bottom of a slab, inefficient joints, excess shrinkage and inefficient curing.

The type of aggregate used is also an important factor, usually overlooked, for the development of shrinkage cracking. When shrinking aggregate is used, extra care during curing should be taken to avoid development of shrinkage cracking. For shrinking and non-shrinking aggregates, see Section 14.4.2.3.

Figure 15.21 shows the effect of using two different aggregates: diabase (shrinking aggregate) at the top and limestone (non-shrinking aggregate) at the bottom.

Figure 15.22 also shows the effect of using a shrinking aggregate; a drying shrinkage crack developed through the depth of the slab, under the same curing conditions as the slab in Figure 15.21.

15.11.6.2 Treatment

Successful repair of cracking in fresh or hardened concrete owing to reasons mentioned above is often very difficult to attain. It is a type of failure that is far easier to prevent and almost impossible to repair efficiently and successfully.

The alternatives to treating and repairing these cracks range from doing nothing (if cracks are ≤0.25 mm) up to reconstructing the whole cracked area. In between, the cracks can be routed and sealed or they can be gravity filled. Surface dressing or micro-surfacing can also be used. For more details, see Section 15.15 or Section 14.4.3, respectively.

15.11.7 Cracking of hardened concrete caused by other factors

External factors may also cause cracking of hardened concrete. These are cracking attributed to poor construction practice, cracking attributed to errors in design and detailing, cracking attributed to externally applied loads, weathering cracking or defrost cracking, cracking attributed to corrosion of reinforcement and cracking attributed to overloading.

15.11.7.1 Causes

The causes for the development of the above cracking are explained in Sections 14.4.2.4 through 14.4.2.10.

15.11.7.2 Treatment

The treatment depends on the cause, time of appearance and how hazardous the cracking may be.

15.12 SURFACE DEFORMATION IN RIGID PAVEMENTS

Surface deformation is another very common type of distress of jointed rigid pavements. Surface deformation is the difference in elevation across a joint or a crack resulting from vertical movements.

Vertical movement may develop either by permanent movement in the form of settlement (also known as faulting) or by dynamic movement caused by passing traffic usually associated with mud pumping. Evaluation of faulting may be carried out by using AASHTO R 36 (2013).

15.12.1 Causes

Settlement or faulting is caused by compaction/consolidation or shrinkage of the layers underlying the slab, movements in the underlying ground or lack of effective load transfer between slabs at joints (effectiveness of dowels and tie bars).

Dynamic movement is caused by lack of support from the sub-base, lack of effective load transfer at joints or poor pavement (surface) or subgrade drainage.

15.12.2 Treatment

Severe settlement is treated either by reconstruction of the slab(s) or by the construction of an asphalt overlay after applying a levelling course.

Localised settlement may be treated by means of slab lifting in conjunction with either pressure or vacuum grouting.

Dynamic movement, when it is not associated with mud pumping, is treated by pressure or vacuum grouting. When mud pumping is present, dynamic movement is treated by renewing or improving existing surface and subgrade drainage or by sealing the joints or cracks.

Pumping suitable hot bitumen for undersealing concrete pavements may also be a solution (usually temporary). The bitumen should comply with ASTM D 3141 (2009) or other specifications.

Blow-up is the localised upward movement (buckling) that can be considered as surface deformation of rigid pavements. It usually occurs at transverse joints owing to excessive expansion of the slabs during high-temperature periods. The main contributor to blow-ups is the infiltration of incompressible materials into the joints, provided the joint (groove) is sufficient.

To treat a blow-up, full-depth repair is necessary.

15.13 DISINTEGRATION OF RIGID PAVEMENTS

Disintegration, as in flexible pavements, is the fragmentation of the pavement surface into small loose pieces. This may also include the detachment (loss or dislodging) of aggregate particles from the surface of the rigid pavement. If disintegration is not treated at an early stage, the slab will certainly require complete reconstruction.

The main forms of disintegration in rigid pavements are scaling and joint (or edge of slabs) spalling.

15.13.1 Scaling

Scaling is the flaking or peeling of the surface of hardened concrete. It starts as localised patches that later on may merge. Light scaling does not expose coarse aggregates. Moderate scaling exposes the coarse aggregate, while in severe scaling, coarse aggregates are clearly exposed and stand out.

15.13.1.1 Causes

Scaling is caused by one of the following: over-finish (surface bleed water), improper mixing, improper/insufficient curing, use of unsuitable aggregate, chemical action of the de-icing material or exposure to freezing and thawing particularly when non-air-entrained concrete is used.

15.13.1.2 Treatment

When scaling is less than approximately 10 mm in depth, temporary repair is carried out by micro-surfacing; otherwise, the surface should be overlaid by asphalt. Application of the proper amount of tack coat and a clean surface are of vital importance.

15.13.2 Spalling

Joint (or edge) and corner of slabs spalling is the breaking up or disintegration of the concrete at these joint locations or cracks. Spalling may be distinguished into shallow or deep and impairs the effectiveness of joint seal.

15.13.2.1 Causes

Spalling is caused by one of the following: weak concrete, poor joint sealing (or crack sealing), infiltration of fine material or sand into the joint, improper forming of joint or damage caused by removal of formwork, penetration of stones into the joint groove, misplacement of dowels or dowel restrain. The last three factors cause deep spalling.

15.13.2.2 Treatment

Minor and shallow spalling is usually treated by removing the affected area by widening the joint groove and filling it with sealant or by removing the broken area (to the required width and depth of at least 10 mm) and filling it with concrete mortar, if the depth is approximately 20 mm, or with fine concrete, if it is deeper; joint sealant may also need to be replaced.

Deep spalling requires full-depth joint repair.

Generally, when the rigid pavement shows extensive deep spalling together with surface deformation or other cracking, strengthening the asphalt or concrete overlay is the only effective treatment.

15.14 LOSS OF SURFACE SKID RESISTANCE

Loss of surface skid resistance to levels below acceptable is associated with smooth and slippery concrete surface, which directly affects traffic safety. Unlike all other surface distresses, a smooth and slippery surface does not affect the structural deterioration of the rigid pavement.

15.14.1 Cause

The cause for loss of surface skid resistance is exclusively the wear of the surface texture provided, the polishing of the fine aggregate and the polishing of the exposed coarse aggregate.

15.14.2 Treatment

Restoration of skid resistance in rigid pavements may be carried out by one of the following techniques: (a) mechanical cutting–grooving, (b) mechanical roughening, (c) surface dressing, (d) micro-surfacing and (e) hot asphalt surfacing overlay. Some methods will restore only macro-texture or only micro-texture while others do both.

The choice of the most suitable method to be used depends on the type of road (traffic volume and speed of traffic) and characteristic properties and coarseness of the aggregate exposed.

Mechanical cutting–grooving, also known as diamond grinding technique, is carried out by a specialised machine carrying diamond saw blades. It restores macro-texture and is more suitable in high-speed roads.

Mechanical roughening techniques include grit or shot blasting, bush hammering and flailing. Grid or shot blasting is for the short-term improvement of micro-texture; bush hammer improves macro-texture but there is a risk to damaging the surface by creating micro-cracking. Flailing improves macro-texture.

Surface dressing restores both micro- and macro-texture; the type of surface dressing to be used depends on the road category, and extra care should be given to the selection of the binder. More details for surface dressing are given in Section 15.15.

Micro-surfacing also restores both micro- and macro-texture. For longer-lasting treatment, a double micro-surfacing layer is recommended to be used: first, a layer with minimal size aggregate 0/4 mm and then a layer with 0/8 or 0/10 mm.

Hot asphalt surfacing overlay restores micro- and macro-texture and is the most suitable treatment for restoration of skid resistance of continuously reinforced pavement surface. All types of hot asphalts used in restoration of skid resistance of flexible pavements can be used, namely: asphalt concrete for very thin layers, SMA and porous asphalt.

On jointed concrete pavements when hot asphalt surfacing overlay is used for texture restoration, reflective cracks at the joints are expected to appear after a certain period. The same applies when micro-surfacing or surface dressing technique is applied, but reflective cracks will appear sooner. To avoid the above, it is recommended that all joints be replaced after laying the hot asphalt, micro-surfacing or surface dressing.

15.15 SURFACE DRESSING

Surface dressing, also known as chip seal, is one of the oldest surface treatment techniques. It is applied in many countries to restore surface skid resistance and to seal the surface inhibiting pavement disintegration.

Surface dressing provides a simple and cost-effective pavement maintenance alternative solution contributing to road safety and protection of pavement. Today, considering proper design, careful application and use of appropriate material surface dressing can be used in any types of roads. It should be underlined, however, that surface dressing cannot restore the pavement's surface evenness nor can it contribute to the pavement's structural strengthening.

The typical surface dressing consists of spraying binder and immediately applying a single layer of chippings to 'dress' the binder.

The binder may be bituminous emulsion (modified or unmodified), penetration grade bitumen or polymer-modified bitumen. The use of modified bitumen emulsion or modified bitumen improves quality of construction and extends life expectancy of the treatment.

Crushed rock or slag chippings should be clean, hard and durable (appropriate PSV and AAV values). Crushed gravel chippings may be used provided they satisfy the requirements.

Even though single surface dressing is an effective and economical method for restoration of skid resistance, it increases the noise generated by the tyres and surface texture. However, the use of the double surface dressing, or sandwich surface dressing, or generally the use of smaller size aggregate reduces the noise and improves the relevant problem. Alternative single surface dressing may be covered by slurry seal. This technique is known as cape seal, which, apart from reducing the noise generated from single surface dressing, also minimises or eliminates the detachment of chippings particularly during early periods of trafficking and thereafter.

Surface dressing is sensitive to weather conditions. Applying surface dressing during hot months with pavement surface temperature higher than 40°C will certainly cause ravelling, if there is no strict speed traffic control. The same phenomenon may be observed when it starts raining shortly after completion.

Finally, after completion of surface dressing, the surface should always be cleaned from loose aggregates.

The following comprises the successful application of surface dressing: selection of appropriate type of surface dressing, selection of appropriate material, use of appropriate quantities of material, choice of favourable weather conditions, proper surface preparation, good workmanship, surface cleaning on completion and obedience of traffic to speed control applied for a certain period.

The visual assessment of the defects that may occur with a surface dressing is determined by CEN EN 12272-2 (2003). The assessment of the defects evaluates the condition or performance of the surface dressing and, thus, gives an indication on the durability of the surface dressing.

15.15.1 Types of surface dressing

There are several types of surface dressings varying according to the number of layer chippings and binder. According to CEN EN 12271 (2006), the different types of surface dressing are distinguished into the following: single surface dressing, racked-in surface dressing, double dressing, inverted double dressing, and sandwich surface, as shown in Figure 15.23.

The definition for each surfacing dressing technique, according to CEN EN 12271 (2006), and a brief description, based on UK practice (Highways Agency 1999; Roberts and Nicholls 2008), are given below.

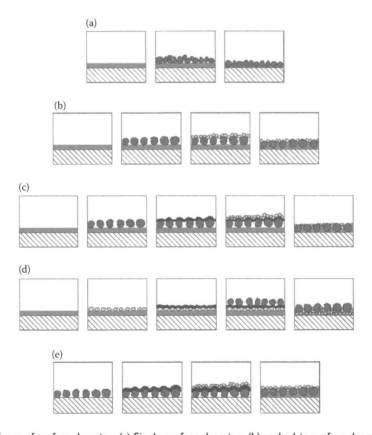

Figure 15.23 Types of surface dressing. (a) Single surface dressing; (b) racked-in surface dressing; (c) double dressing; (d) inverted double dressing and (e) sandwich dressing (example of pre-chipping dressing). (Reproduced from CEN EN 12271, *Surface dressing – Requirements*, Brussels: CEN, 2006. With permission [© CEN].)

15.15.1.1 Single surface dressing

Single surface dressing (Figure 15.23a) refers to the fundamental type of surface dressing technique, consisting of successive laying of one layer of binder and one layer of aggregates (chippings).

The upper size of chippings is usually 6.3 or 10 mm. The rate of spread of chippings should be such to achieve 100% to 105% shoulder-to-shoulder coverage (Heslop et al. 1982), as measured by CEN EN 12272-1 (2002).

The single surface dressing has the least number of operations and requires the least quantities of material (chippings and binder) than the rest of the types of surface dressing. The quantities of material required mainly depend, as in all cases, on the condition of the pavement surface (mainly hardness), traffic category and size of chippings.

15.15.1.2 Racked-in surface dressing

Racked-in surface dressing (RISD) (Figure 15.23b) refers to successive laying of one layer of binder and two layers of chippings, the second layer being of a smaller size.

The rate of spread of the bituminous binder applied is higher than that in single surface dressing to create a thicker layer of binder. The size of the primary chippings is typically 14 mm on fast heavily trafficked roads and is laid at a lower rate, 90% of that for a single dressing. The second layer of smaller chippings (usually 6 mm) fills gaps and achieves mechanical interlocking.

The advantages of this method are high initial texture depth, early stability of dressings and major reduction in the initial loss of large chippings. The RISD is mainly used where traffic is heavy or fast.

15.15.1.3 Double dressing

Double dressing (Figure 15.23c) refers to successive laying of a first layer of binder and a first layer of chippings followed by a second layer of binder and a second layer of chippings, the second layer of chippings being of a smaller size.

The rate of spread of first-layer chippings is such to achieve 95% shoulder-to-shoulder coverage, and that of the second is such to achieve 100%–105% shoulder-to-shoulder coverage as measured by CEN EN 12272-1 (2002).

Double dressing generally produces a slightly lower texture depth than RISD and is particularly suitable for road surfaces that are binder lean. It requires more rolling and curing (if emulsion is used) before opening to traffic. However, double dressing enhances durability, may be used in all road categories and is considered as a standard practice in some countries.

15.15.1.4 Inverted double dressing

Inverted double dressing (Figure 15.23d) refers to successive laying of a first layer of binder and a first layer of chippings followed by a second layer of binder and a second layer of chippings, the second layer of chippings being of a larger size.

Inverted double surface dressing is used on existing very hard (concrete) or hard surface with high and variable macro-texture resulting from surface disintegration. The first dressing with the small size chippings, also known as pad coat, provides a uniform and softer surface to which the coarser and main chipping applied is imbedded better, reducing chipping loss.

Both layers of chippings are applied at a rate of spread to achieve 100%–105% shoulder-to-shoulder coverage as measured by CEN EN 12272-1 (2002). The surplus of chippings from the first layer is removed before the second layer of chippings is applied. It is not uncommon to apply the second layer of chippings after a certain period. The size of the chippings for the first layer is usually 6 mm and that for the second layer is 10 or 14 mm.

15.15.1.5 Sandwich surface dressing

Sandwich surface dressing, also known as pre-chipping dressing (Figure 15.23e), refers to successive laying of one layer of chippings (pre-chipping layer) followed by surface dressing as part of the process.

Sandwich surface dressing is used in situations where the pavement surface is binder rich and works are carried out during hot weather.

The pre-chipping layer consists of large-size chippings (usually 14 mm) and the chippings of the single surface dressing consist of small-size chippings (usually 6 mm).

The rate of spread of chippings is as for double surface dressing.

15.15.2 UK surface dressing design methodology (RN 39)

The RN 39 surface dressing design methodology is incorporated in the design guide developed (Roberts and Nicholls 2008) for the design of surface dressing for roads throughout the United Kingdom.

It is the most complete surface dressing design methodology that may also be used outside the United Kingdom with minor modifications.

The input parameters required for determining the binder spread rate related to chipping size, road hardness and traffic category are as follows:

a. Surface temperature category
b. Road hardness
c. Traffic category and traffic speed
d. Highway layout
e. Aggregate (chipping) properties

The rate of spread of chippings is determined with respect to the percentage of shoulder-to-shoulder coverage required.

15.15.2.1 Surface temperature category

Since the climate is not uniform in the United Kingdom, the design guide divides the country into four pavement surface temperature categories depending on the location and altitude of the project position (geographical position of the project). The categories have been designated as A, B, C and D.

For instance, all areas of south England, south of Nottingham and Stoke-on-Trent, with an altitude of 200 m or less, are considered in surface temperature category A. More information on the rest of the surface temperature categories is given in the design guide (Roberts and Nicholls 2008).

15.15.2.2 Road hardness

The hardness of pavement surface constitutes a property representing the surface resistance to chipping embedment into the surface. As a consequence, this property is affected by climatic conditions and by the type and age of the surfacing material.

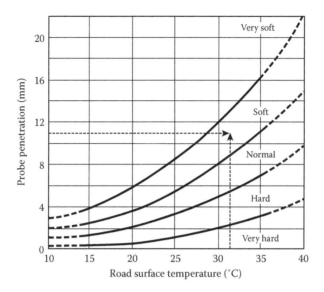

Figure 15.24 Determination of road hardness for surface temperature category A. (From Roberts, C. and Nicholls, J.C., *Design Guide for Road Surface Dressing.* TRL Road Note RN 39, 6th Edition. Crowthorne, UK: Transport Research Laboratory, 2008.)

The surface hardness is measured with a test method described in BS 598-112 (2004) using a hardness probe. During the test, a weight of 35 kg is applied through a metal probe on the road surface for 10 s. After 10 s, the probe penetration depth (in millimetres) is measured, and at the same time, the surface temperature is recorded. Measurements are recommended to be conducted at surface temperatures between 15°C and 35°C. The average of 10 measurements is considered to be the representative value for the determination of road hardness category.

The road hardness category distinguished as very soft, soft, normal hard and very hard is determined from four figures developed for each surface temperature category. A sample for surface temperature category A is shown in Figure 15.24.

All rigid pavement surfaces are designated as very hard.

15.15.2.3 Traffic category and traffic speed

The traffic volume is expressed in daily volume of medium and heavy vehicles per design lane. Medium and heavy vehicles are defined as vehicles of unladen weight greater than 1.5 tonnes.

For the purposes of surface dressing design, eight traffic categories are distinguished, which are as shown in Table 15.1.

Where surface dressing is likely to be subjected to regular high speeds (i.e. permitted speeds above 80 km/h), consideration should be given to a stronger surface dressing, such as double surface dressing and RISD.

Table 15.1 Traffic categories

Medium and heavy vehicles/lane/day	≤50	51–125	126–250	251–500	501–1250	1251–2000	2001–3250	>3250
Traffic category	H	G	F	E	D	C	B	A

Source: Roberts, C. and Nicholls, J.C., *Design Guide for Road Surface Dressing.* TRL Road Note RN 39, 6th Edition. Crowthorne, UK: Transport Research Laboratory, 2008.

15.15.2.4 Highway layout

The gradient, the radius of curvature and whether the project is at junction or crossing (approach and non-approach) are the main considerations for surface dressing design.

The gradient is distinguished into three categories: less than 5%, 5% to 10% and greater than 10% gradient.

Regarding the radius of curvature, three categories are distinguished: less than 100 m, 100 to 250 m and greater than 250 m radius.

15.15.2.5 Aggregate properties

Aggregates should be single sized and crushed from appropriate hard rocks, free of dust or soil, complying to CEN EN 13043 (2004).

The hardness of aggregate chippings is mainly determined by their resistance to polishing (PSV). However, the resistance to fragmentation by Los Angeles (LA) as well as the resistance to aggregate abrasion value (AAV) should also be taken into account.

According to the RN 39 design guide, the chippings for single surface dressing and for primary layer in multiple surface dressings should have PSV values as determined by Table 15.2. Additionally, the LA value is required to be less than 30 and the AAV should be less than 12.

For the secondary chippings, in case of double layers, the PSV value should be ≥50, in all cases.

On very hard substrates, particularly when the traffic category is A and B, a lower LA value than the abovementioned should be considered and the AAV value should be less than 10.

Size of chippings: For single surface dressing, the size of chippings is either 6.3/10 mm or 2.8/6.3 mm ($G_c85/35$), depending on the traffic category. Chippings of 8/14 mm are not recommended to be used because of the high risk of windscreen damage. If used, they can only be used in softer substrates and with utmost care.

For RISD, the size of chippings is 6.3/10 mm for the coarse ones and 2.8/6.3 or 2/4 mm for the fine ones.

For double surface dressing, the size of coarse chippings is 10/14 or 6.3/10 mm and the size for the fine chippings is 2.8/6.3 or 2/4 mm.

Finally, in inverted double dressing, the size of the first-layer chippings is usually 2.8/6.3 mm.

15.15.2.6 Selection of type of surface dressing

The selection of the type of surface dressing, suitable for consideration, is carried out by utilising two flow chart diagrams developed (Roberts and Nicholls 2008): one for lightly trafficked sites (categories G and H) and one for heavily trafficked sites (categories A to F). The flow chart diagram for heavily trafficked sites is given in Figure 15.25.

15.15.2.7 Type and quantity of bituminous binder

The binder to be used in surface dressings is cationic bituminous emulsion, unmodified or polymer modified. However, in most of the cases, polymer-modified bitumen emulsion is used.

The bitumen emulsion should have a nominal binder content greater than 69% and meet the requirements specified according to CEN EN 13808 (2013). The RN 39 design guide suggests the use of C69B2, C69B3, C69B4 or C69BP3 bituminous emulsions. Bituminous emulsions with flux oil or flux, such as C69BF2, C69BF3 and C69BF4, could be also used. More information about bituminous emulsions can be found in Section 3.6.

Great emphasis is given to the cohesion of bituminous emulsion residue by using the Vialit pendulum test according to CEN EN 13588 (2008). According to the RN 39 design guide,

Table 15.2 Minimum PSV for chippings or coarse aggregate in bituminous surfacing (excluding hot-applied thin surface course)

Site category	Site description	IL[a]	Min PSV for given IL[a], traffic level[b] and type of site									
			<250	251–500	501–750	751–1000	1001–2000	2001–3000	3001–4000	4001–5000	5001–6000	>6000
A1	Motorways where traffic is generally free-flowing in a relatively straight line	0.30	50	50	50	50	50	55	55	60	65	65
		0.35	50	50	50	50	50	60	60	60	65	65
A2	Motorways where some braking regularly occurs	0.35	50	50	50	55	55	60	60	65	65	65
B1	Dual carriageways where traffic is generally free-flowing in a relatively straight line	0.30	50	50	50	50	50	55	55	60	65	65
		0.35	50	50	50	50	50	60	60	60	65	65
		0.40	50	50	50	55	60	65	65	65	65	68+
B2	Dual carriageways where some braking regularly occurs	0.35	50	50	50	55	55	60	60	65	65	65
		0.40	55	60	60	65	65	68+	68+	68+	68+	68+
C	Single carriageways where traffic is generally free-flowing in a relatively straight line	0.35	50	50	50	55	55	60	60	65	65	65
		0.40	55	60	60	65	65	68+	68+	68+	68+	68+
		0.45	60	60	65	65	68+	68+	68+	68+	68+	68+
G1/G2	Gradients >5% longer than 50 m	0.45	55	60	60	65	65	68+	68+	68+	68+	68+
		0.50	60	68+	68+	HFS[c]	HFS	HFS	HFS	HFS	HFS	HFS
		0.55	68+	HFS[c]	HFS	HFS	HFS	HFS	HFS	HFS	HFS	HFS

Site category[d]	Description	IL[a]	\	\	\	\	Traffic level[b] (cv/lane/day)	\	\	\	\
K[d]	Approaches to pedestrian crossings and other high-risk situations	0.50	65	65	68+	68+	68+	HFS	HFS	HFS	HFS
		0.55	68+	HFS	HFS	HFS	HFS	HFS	HFS	HFS	HFS
Q[d]	Approaches to major and minor junctions on dual carriageways and single carriageways where frequent or sudden braking occurs but in a generally straight line	0.45	60	65	65	68+	68+	68+	68+	68+	HFS
		0.50	65	65	65	68+	68+	68+	HFS	HFS	HFS
		0.55	68+	68+	HFS	HFS	HFS	HFS	HFS	HFS	HFS
R	Roundabout circulation areas	0.45	50	55	60	60	65	65	68+	68+	68+
		0.50	68+	68+	68+	68+	68+	68+	68+	68+	68+
S1/S2	Bends (radius <500 m) on all types of road, including motorway link roads; other hazards that require combined braking and cornering	0.45	50	55	60	60	65	65	68+	68+	HFS
		0.50	68+	68+	68+	HFS[c]	HFS	HFS	HFS	HFS	HFS
		0.55	HFS	HFS	HFS	HFS	HFS	HFS	HFS	HFS	HFS

Source: Highways Agency, Revision of Aggregate Specification for Pavement Surfacing, Interim Advice Note (IAN) 156/12, London: Department for Transport, Highways Agency, 2012 (© Highways Agency).

Note: Where '68+' material is listed in this table, none of the three most recent results from consecutive PSV tests relating to the aggregate to be supplied must fall below 68. For site categories G1/G2, S1/S2 and R, any PSV in the range given for each traffic level may be used for any IL and should be chosen on the basis of local experience of material performance. In the absence of this information, the values given for the appropriate IL and traffic level must be used. Where designers are knowledgeable or have other experience of particular site conditions, an alternative PSV value can be specified.

[a] IL, investigatory level, defined in HD 28 (Highways Agency 2004).
[b] The traffic level is expressed in terms of the expected commercial vehicles (cv) per lane, per day (cv/lane/day).
[c] HFS, specialised high-friction surfacing, incorporating calcined bauxite aggregate.
[d] Site categories K and Q should not be applied to the circulatory parts of a roundabout.

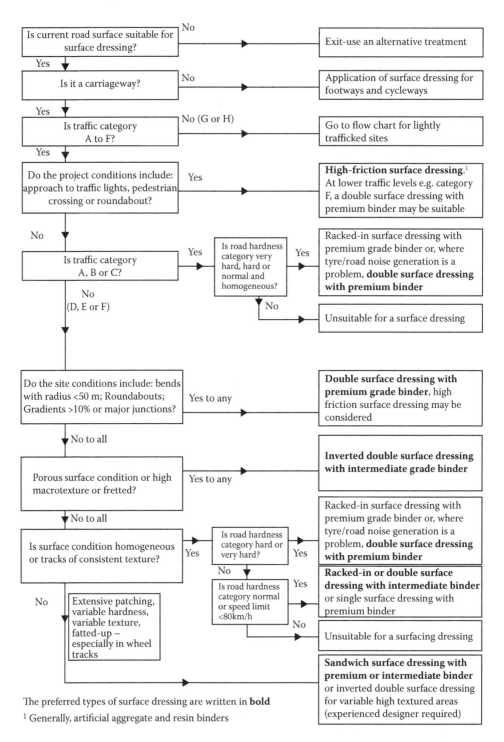

Figure 15.25 Selection of type of surface dressing for traffic categories A to F. (From Roberts, C. and Nicholls, J.C., *Design Guide for Road Surface Dressing*. TRL Road Note RN 39, 6th Edition. Crowthorne, UK: Transport Research Laboratory, 2008.)

the minimum peak cohesion value required for different grades of bitumen emulsion (UK classification) are as follows: for unmodified, 0.7 J/cm^2; for intermediate grade, 1.0 J/cm^2; for premium grade, 1.2 J/cm^2; and for super-premium grade, 1.4 J/cm^2.

The binder quantity to be sprayed per traffic category depends on the hardness category of the surface and the size of chippings. The recommended target rates of spread of binder for single surface dressing and double surface dressing are given in Tables 15.3 and 15.4.

Information for the target rates of spread of binder regarding the other types of surfacings can be found in the design guide (Roberts and Nicholls 2008).

The binder spread rates given in Tables 15.3 and 15.4 should be slightly adjusted to allow for different local conditions (season, aggregate type, surface condition in terms of richness in binder, gradient, etc.). The corrections recommended to be applied are as shown in Table 15.5. When the accumulated effect of adjustments is greater than +0.2 L/m^2, it is recommended that the original design be revised.

15.15.2.8 Quantity of chippings for surface dressing

The quantity of chippings to achieve 100% or other shoulder-to-shoulder coverage depends on the size, shape and density of the chippings; hence, it varies accordingly. However, to facilitate the engineer estimating the total amount of chippings required, Table 15.6 is provided for single surface dressing achieving 100% shoulder-to-shoulder coverage.

15.15.3 Austroads surface dressing design methodology

The Austroads surface dressing, or sprayed seal, design methodology (Austroads 2006, 2013) provides a guide to the determination of rates of spread of binder and aggregates (chippings). The design methodology covers the most common type of surface dressing, the single surface dressing, called single seal (Austroads 2006), and the double surface dressing type, called double seal (Austroads 2013).

Table 15.3 Recommended nominal size of chippings and target rates of spread binder at spraying temperature for single surface dressings

Traffic category	Hardness category of road surface									
	Very hard		Hard		Normal		Soft		Very soft	
	Chip. mm	Bin. L/m^2	Chip. mm	Bin. L/m^2	Chip. mm	Bin. L/m^2	Chip. mm	Bin. L/m^2	Chip. mm	Bin. L/m^2
A	a		a		a		b		b	
B	6.3	10 1.8c	a		a		a		b	
C	6.3	10 1.8c	6.3	10 1.6c	a		a		b	
D	2.8	6.3 1.5c	6.3	10 1.6c	a		a		a	
E	2.8	6.3 1.5c	6.3	10 1.6c	6.3	10 1.6c	6.3	10 1.6c	a	
F	2.8	6.3 1.5c	2.8	6.3 1.5c	6.3	10 1.6c	6.3	10 1.6c	a	
G	2.8	6.3 1.5	2.8	6.3 1.5	2.8	6.3 1.5	6.3	10 1.6	a	
The	2.8	6.3 1.5	2.8	6.3 1.5	2.8	6.3 1.5	2.8	6.3 1.4	2.8	6.3 1.4

Source: Roberts, C. and Nicholls, J.C., *Design Guide for Road Surface Dressing*. TRL Road Note RN 39, 6th Edition. Crowthorne, UK: Transport Research Laboratory, 2008.

Note: Chip., size of chipping, referring to D/d. Bin., binder rate of spread (bitumen emulsion).

[a] Multiple layer surface dressing is preferred.
[b] Conditions not suitable for single surface dressings.
[c] Polymer-modified bituminous emulsion is preferred in those conditions.

Table 15.4 Recommended nominal size of chippings and target rates of spread binder at spraying temperature for double surface dressings

Traffic category	Very hard Chip. mm	Bin. L/m² 1st	2nd	Hard Chip. mm	Bin. L/m² 1st	2nd	Normal Chip. mm	Bin. L/m² 1st	2nd	Soft Chip. mm	Bin. L/m² 1st	2nd	Very Soft Chip. mm	Bin. L/m² 1st	2nd
									Hardness category of road surface						
A	10 and 6	1.1	1.2[b]	14 and 6	1.2	1.3[b]	14 and 6	1.2	1.1[b]	a			a		
				10 and 6	1.0	1.2[b]	10 and 4	1.0	1.0[b]						
B	10 and 6	1.1	1.2[b]	14 and 6	1.2	1.3[b]	14 and 6	1.2	1.1[b]	a			a		
				10 and 6	1.0	1.2[b]	10 and 4	1.0	1.0[b]						
C	10 and 6	1.1	1.2[b]	10 and 6	1.0	1.2[b]	14 and 6	1.2	1.1[b]	a			a		
							10 and 4	1.0	1.0[b]						
D	10 and 6	1.1	1.2[b]	10 and 6	1.0	1.2[b]	14 and 6	1.2	1.2[b]	14 and 6	1.0	1.1	14 and 6	0.8	1.0
							10 and 4	1.0	1.1[b]						
E	10 and 6	1.1	1.2[b]	10 and 6	1.1	1.2[b]	10 and 6	1.0	1.1[b]	14 and 6	1.0	1.1	10 and 6	0.8	1.0
										10 and 6	0.8	1.0			
F	10 and 6	1.2	1.2[b]	10 and 6	1.1	1.2[b]	10 and 6	1.0	1.1[b]	14 and 6	1.0	1.2	10 and 6	0.8	1.1
										10 and 6	0.8	1.1			
G	10 and 6	1.2	1.3[b]	10 and 6	1.1	1.3[b]	10 and 6	1.0	1.2[b]	10 and 6	1.0	1.1	10 and 6	0.8	1.0
The	10 and 6	1.2	1.3[b]	10 and 6	1.1	1.3[b]	10 and 6	1.0	1.2[b]	10 and 6	1.0	1.1	10 and 6	0.8	1.0

Source: Roberts, C. and Nicholls, J.C., *Design Guide for Road Surface Dressing*. TRL Road Note RN 39, 6th Edition. Crowthorne, UK: Transport Research Laboratory, 2008.

Note: Chip., chipping size. Chipping size is given as D_1 and D_2, where D_1 is the upper sieve size of the first set of chippings and D_2 is the upper sieve size of the second set of chippings to be used. Bin., binder rate of spread (bitumen emulsion).

[a] Conditions not suitable for double surface dressings.
[b] Polymer-modified bituminous emulsion is preferred under these conditions.

The single surface dressing design methodology distinguishes two procedures for determining the rate of binder and the aggregate: one for large-size chippings (10 mm or larger) and one for small-size chippings (7 mm or smaller).

The small aggregate seals are often used as correction courses to provide an interim, even surfacing, with uniform texture prior to the placement of a more durable seal treatment. Small aggregate sizes are appropriate on low- to medium-traffic roads, particularly over existing seals with large aggregates and high surface texture. Additionally, small aggregate seals are used in situations that can tolerate, or only require, a reduced surfacing life, such as where a temporary surfacing is required (Austroads 2006).

The determination of the rate of spread of binder and chippings is based on the use of conventional bitumen as a binder. When polymer-modified bitumen or bitumen emulsion (unmodified or modified) is used, necessary adjustments to the values for conventional bitumen are made to determine the rate of spread of polymer-modified bitumen or bitumen emulsion and the rate of spread of chippings.

15.15.3.1 Single surface dressing (single seal)

15.15.3.1.1 Determination of binder application rate of spread

The design process for the determination of binder (conventional bitumen) application rate of spread for 10 mm or larger chippings or 7 mm or smaller chippings (when the average

Table 15.5 Secondary factors influencing the rate of spread of binder for traffic categories a to f and all types of surface dressing

Factor influence	Property	Change in binder (L/m^2)	Comments
Season	Early and mid season	0	Late season work is very risky
	Late season	+0.2	
Aggregate type	Crushed rock or slag	0	Gravel is only appropriate for categories F, G and H
	Gravel	+0.1	
Flakiness	FI <10%	+0.1	
	FI 10% to 25%	0	
	FI >25%	−0.1	
Chipping size	Size smaller	−0.1	Size appropriate to traffic category can be changed to the adjacent size if required
	'Design' size	0	
	Size larger	+0.2	
Shade	Un-shaded, open to sun	0	Shaded areas are cooler
	Partially shaded	+0.1	
	Fully shaded	+0.2	
Surface condition	Very binder rich	−0.1	The road condition will affect how much binder is required to provide similar conditions at the interface
	Binder rich and normal	0	
	Normal	0	
	Binder lean	+0.1	
	Very binder lean/porous	+0.2	
Gradient	>5% uphill	−0.1	The gradient affects the stresses applied to the surfacing
	<5%	0	
	>5% downhill	+0.1	
	>10% downhill	+0.2	
Speed of traffic	High speed (≥80 km/h)	+0.1	Roads subject to high traffic speeds induce greater surface stress
	Low speed (<80 km/h)	0	
Local traffic	Design range	0	Hard shoulders and sizeable areas with hatched lines to exclude traffic are effectively untrafficked
	Effectively untrafficked	+0.2	

Source: Roberts, C. and Nicholls, J.C., *Design Guide for Road Surface Dressing*. TRL Road Note RN 39, 6th Edition. Crowthorne, UK: Transport Research Laboratory, 2008.

Table 15.6 Typical range of rate of spread of chippings for single surface dressings

Nominal size of chippings (mm)	Range of spread rates	
	kg/m^2	$m^3/tonne$
2.8/6.3	7–10	143–100
6.3/10	9–13	111–77
8/14	12–16	83–62

Source: Roberts, C. and Nicholls, J.C., *Design Guide for Road Surface Dressing*. TRL Road Note RN 39, 6th Edition. Crowthorne, UK: Transport Research Laboratory, 2008.

least dimension is known) is summarised as follows: (a) determine the basic voids factor, (b) determine the design void factor and (c) determine the basic binder application rate.

The *basic voids factor*, Vf ($L/m^2/mm$), is related to traffic level and is determined for a traffic volume of 0–500 vehicles/lane/day from Figure 15.26 and for a traffic volume of 500 to 10,000 vehicles/lane/day from Figure 15.27. It is noted that the central target line is used to determine the basic voids factor in all cases.

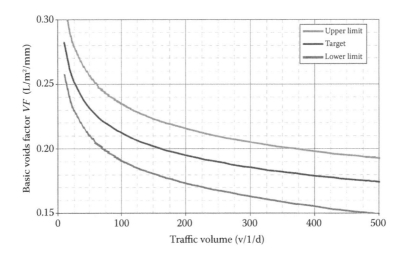

Figure 15.26 Basic voids factor for a traffic volume of 0 to 500 vehicles/lane/day. (From Austroads, *Update of the Austroads Sprayed Seal Design Method*, Austroads Publication No. AP-T68/06, Sydney: Austroads Incorporated, 2006. With permission.)

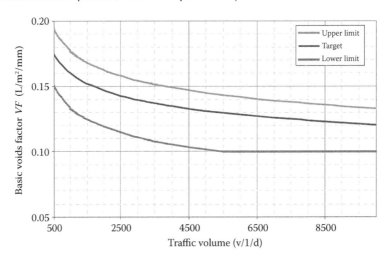

Figure 15.27 Basic voids factor for a traffic volume of 500 to 10,000 vehicles/lane/day. (From Austroads, *Update of the Austroads Sprayed Seal Design Method*, Austroads Publication No. AP-T68/06, Sydney: Austroads Incorporated, 2006. With permission.)

The *design void factor*, *VF*, is the basic voids factor adjusted for aggregate shape, using the shape adjustment factor, *Va*, and for traffic effects, using the traffic effect adjustment factor, *Vt*. The design void factor is then calculated using the following equation:

$$VF = Vf + Va + Vt.$$

The shape adjustment factor, *Va*, is obtained from Table 15.7, and the traffic effect adjustment factor, *Vt*, is obtained from Table 15.8.

The *basic binder application rate*, B_b, is then determined by the following formula:

$$B_b = VF + ALD,$$

Table 15.7 Aggregate shape adjustment factor

Aggregate type	Aggregate shape	Flakiness index (%)	Shape adjustment Factor, Va (L/m²/mm)
Crushed or partly crushed	Very flaky	>35	Not recommended
	Flaky	26 to 35	0 to −0.01
	Angular	15 to 25	0
	Cubic	<15	+0.01
	Rounded	—	
Not crushed	Rounded	—	+0.01

Source: Austroads, *Update of the Austroads Sprayed Seal Design Method*, Austroads Publication No. AP-T68/06, Sydney: Austroads Incorporated, 2006. With permission.

Table 15.8 Traffic effect adjustment factor

	Traffic adjustment factor, Vt (L/m²/mm)			
	Flat or downhill		Slow moving/climbing lanes	
Traffic	Normal	Channelised	Normal	Channelised
On overtaking lanes of multi-lane rural roads where traffic is mainly cars with ≤10% of HV	+0.01	0.00	n.a.	n.a
Non-traffic areas such as shoulders, medians, parking areas	+0.02	n.a.	n.a.	n.a.
EHV[a] (%) 0–15	0	−0.01	−0.01	−0.02
16–25	−0.01	−0.02	−0.02	−0.03
26–45	−0.02	−0.03	−0.03	−0.04[b]
>45	−0.03	−0.04[b]	−0.04[b]	−0.05[b]

Source: Austroads, *Update of the Austroads Sprayed Seal Design Method*, Austroads Publication No. AP-T68/06, Sydney: Austroads Incorporated, 2006. With permission.

Note: n.a., not applicable.

[a] EHV, equivalent heavy vehicle; EHV = heavy vehicles (>3.5 tonnes) +3 × large heavy vehicles (with seven or more axles).
[b] If adjustments for aggregate shape and traffic effects result in a reduction in basic voids factor of 0.4 L/m²/mm or more, special consideration should be given to the suitability of the treatment and possible selection of alternative treatments. Note that the recommended minimum VF is 0.10 L/m²/mm in all cases.

where B_b is the basic binder application rate for conventional bitumen (L/m²), VF is the design voids factor (L/m²/mm) and ALD is the aggregate least dimension (mm).

The aggregate least dimension, ALD, is defined as the smallest dimension of a particle when placed on a horizontal surface (effectively the height). ALD can be determined by either direct measurement (AS 1141.20.1 2000; AS 1141.20.2 2000) or by using a nomograph from the grading, median size and flakiness index (AS 1141.20.3). ALD is an important parameter in surface dressing design by Austroads methodology and can also be used to justify variations in the quality of the chippings produced related to the shape of chippings.

Having determined the basic binder application rate, B_b, for conventional bitumen, the basic binder application rate for modified bitumen, bitumen emulsion or polymer-modified bitumen emulsion can calculated using one of the following equations:

$$B_{pmb} = B_b \times PF$$

or

$$B_{be} = B_b \times EF$$

or

$$B_{pmbe} = B_b \times EF \times PF,$$

where B_{pmb} is the basic binder application rate for polymer-modified bitumen (L/m²), B_{be} is the basic binder application rate for bitumen emulsion (L/m²), B_{pmbe} is the basic binder application rate for polymer-modified bitumen emulsion (L/m²), PF is the polymer factor from Table 4.1 of Austroads (2006) (varies from 1.1 to 1.4) and EF is the emulsion factor, which is 1.0 for typical emulsions with 60% residual bitumen and 1.1 to 1.2 for bitumen emulsions with ≥67% residual bitumen.

The basic binder application rate is rounded to the nearest 0.1, and in the bitumen emulsion or polymer-modified bitumen emulsion, it refers to residual bitumen.

In all the above cases, to complete the design, allowances applied to the basic binder application rate should be considered. Allowances (in L/m²) are made for the following: (a) surface texture of existing surface, (b) potential aggregate embedment into the existing surface, (c) potential binder absorption into the existing pavement and (d) potential binder absorption into the sealing aggregate.

The allowances are determined to the nearest 0.1 L/m² and are cumulative. They must be added to or subtracted from the basic binder application rate. Hence, the *design binder application rate*, B_d, is determined by the following equation:

$$B_d = B_b + \text{Allowances,}$$

where B_d is the design binder application rate (L/m²) (rounded to nearest 0.1 L/m²), B_b is the basic binder application rate (L/m²) (rounded to nearest 0.1 L/m²) and Allowances is as determined (L/m²).

More information regarding the allowances can be found in Austroads (2006).

When the average least dimension (ALD) for size 7 mm and smaller aggregates is not available, the determination of the basic binder application rate is carried out using Table 15.9.

The lower of the basic application rate values is recommended to be selected for use with flaky aggregates or where traffic includes heavy vehicles of 15% or higher. If not certain of the conditions and traffic composition, but it appears to be 'normal', it is recommended that the mid-point basic binder application be selected (Austroads 2006).

The design binder application rate is then calculated by applying all the allowances mentioned above.

Table 15.9 Basic binder application rates for size 7 mm and smaller aggregates

Traffic (v/l/d)	Basic binder application rate, B_b (L/m²)
<100	1.0 to 0.8
100–600	0.9 to 0.7
601–1200	0.8 to 0.6
1201–2500	0.7 to 0.5
>2500	0.5

Source: Austroads, *Update of the Austroads Sprayed Seal Design Method*, Austroads Publication No. AP-T68/06, Sydney: Austroads Incorporated, 2006. With permission.

15.15.3.1.2 Determination of aggregate spread rate

The amount of 10 mm or larger aggregates required in single surface dressings (sprayed seals) is based on the ALD value of the aggregate.

The design aggregate spread rates for single seals with size aggregates 10 mm and larger and for any type of binder to be used are given in Table 15.10.

The design aggregate spread rates for single seals with size aggregates 7 mm and smaller and for any type of binder to be used are given in Table 15.11.

To convert the aggregate spread rate expressed in square metres per cubic metre to units of kilograms per square metre, the loose bulk density of the aggregate is divided by the value of the spread rate determined from the above tables; the loose bulk density of aggregates is determined by CEN EN 1097-3 (1998), ASTM C 29 (2009), AASHTO T 19 (2009), AS 1141.4 (2000) or any other specification.

15.15.3.2 Double surface dressing (double seal)

The Austroads double surface dressing (double seal, as it is called) design (Austroads 2013) distinguishes two different design procedures: design for little or no trafficking between applications, which is the most common case, and design for second application delayed.

Double surface dressing is designed to provide a robust seal able to cope with high traffic stress. To achieve the design outcome, the second application should have an aggregate size not larger than half the nominal size used in the first application. Recommended combinations are as follows (Austroads 2013):

Table 15.10 Design aggregate spread rate for single seals with size 10 mm and larger aggregates

	Type of binder		
	Bitumen	Polymer-modified bitumen	Bitumen emulsion
Traffic conditions	Aggregate spread rate (m²/m³)		
Traffic,[a] >200 v/l/d	900/ALD	750/ALD	750/ALD
Very low traffic,[b] <200 v/l/d	850/ALD	800/ALD	700/ALD

Source: Adapted from Austroads, *Update of the Austroads Sprayed Seal Design Method*, Austroads Publication No. AP-T68/06, Sydney: Austroads Incorporated, 2006.

[a] Category >200 v/l/d when modified bitumen is used changes to >300 v/l/d.
[b] Category <200 v/l/d when modified bitumen is used changes to <300 v/l/d.

Table 15.11 Design aggregate spread rate for single seals with size 7 mm aggregates

	Type of binder		
	Bitumen	Polymer-modified bitumen	Bitumen emulsion
Seal aim	Aggregate spread rate (m²/m³)		
Normal, small aggregate mosaic, ALD design	900/ALD	[a]	260–250
Normal, small aggregate mosaic, no ALD design	200–250	160–200	

Source: Adapted from Austroads, *Update of the Austroads Sprayed Seal Design Method*, Austroads Publication No. AP-T68/06, Sydney: Austroads Incorporated, 2006.

[a] Increase the rate when bitumen is used by 10%–20%, depending on the design binder application rate and design traffic.

- 20 and 10 mm, 20 and 7 mm or 20 and 5 mm
- 16 and 7 mm, 16 and 5 mm
- 14 and 7 mm, 14 and 5 mm
- 10 and 5 mm, mainly for urban-type locations where noise may be an issue and traffic speeds are relatively low

15.15.3.2.1 Design for little or no trafficking between applications

15.15.3.2.1.1 DETERMINATION OF BINDER APPLICATION RATE OF SPREAD

The determination of the design binder application rate for the first application is carried out in a similar way as for a single surface dressing (single seal) described above.

In particular, the basic binder application rate for first application for conventional bitumen, B_{b1}, is determined first and then the design binder application rate for first application for conventional bitumen, B_{d1}, considering allowances for surface texture, aggregate embedment and so on, as for single surface dressing.

The *basic binder application rate for first application*, B_{b1}, is determined by the following formula:

$$B_{b1} = VF_1 + ALD,$$

where B_{b1} is the basic binder application rate for first application for conventional bitumen (L/m^2), VF_1 is the design voids factor for first application ($L/m^2/mm$) and ALD is the aggregate least dimension (mm).

The design voids factor for first application, VF_1, is determined by the following equation:

$$VF_1 = Vf_1 + Va + Vt,$$

where Vf_1 is the basic voids factor for first application ($L/m^2/mm$) from Figure 15.28 or Figure 15.29, Va is the shape adjustment factor from Table 15.7 and Vt is the traffic effect adjustment factor from Table 15.8.

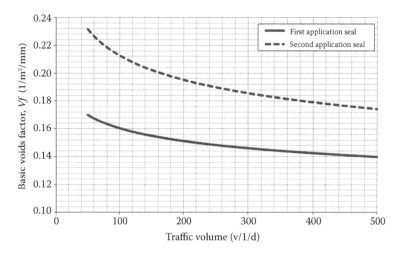

Figure 15.28 Double dressing basic voids factor for a traffic volume of 0 to 500 vehicles/lane/day. (From Austroads, *Update of Double/Double Design for Austroads Sprayed Seal Design Method*, Austroads Publication No. AP-T236-13, Sydney: Austroads Ltd., 2013. With permission.)

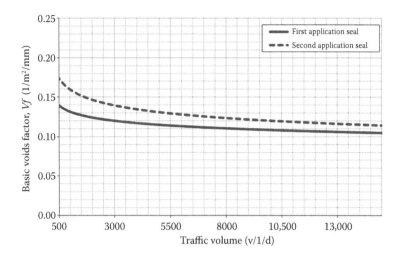

Figure 15.29 Double dressing basic voids factor for a traffic volume of 500 to 15,000 vehicles/lane/day. (From Austroads, *Update of Double/Double Design for Austroads Sprayed Seal Design Method*, Austroads Publication No. AP-T236-13, Sydney: Austroads Ltd., 2013. With permission.)

The *design binder application rate for first application* for conventional bitumen, B_{d1}, is calculated by the following equation:

$$B_{d1} = B_{b1} + \text{Allowances},$$

where B_{d1} is the design binder application rate for first application (L/m²) (rounded to the nearest 0.1 L/m²), B_{b1} is the basic binder application rate for first application (L/m²) (rounded to the nearest 0.1 L/m²) and Allowances is as in single surface dressing (L/m²).

The *basic binder application for second application*, B_{b2}, is determined in exactly the same way as the first application, except for the determination of aggregate shape factor, *Va*. In this case, the aggregate shape factor takes values from Table 15.9 considering that the lower of the basic binder application rates should be selected for use with flaky aggregates (FI > 25%); the higher of the basic binder application rates should be selected for use with more cubically shaped aggregates.

The *design binder application rate for second application*, B_{d2}, is equal to the basic binder application for second application, B_{b2}, since allowances for surface texture, embedment or pavement absorption are not applied for the second application seal binder. Hence, $B_{d2} = B_{b2}$ (in L/m²).

When instead of conventional bitumen, another type of binder such as polymer-modified bitumen, bitumen emulsion, or polymer-modified emulsion is used, the design binder application rate for the first and second application is determined in a similar way as above but considering first the polymer factor and emulsion factor, as explained in single surface dressing.

Details for the polymer and emulsion factors can be found in Austroads (2013).

15.15.3.2.1.2 DETERMINATION OF AGGREGATE SPREAD RATE

As for a single seal, the design aggregate spread rate is also based on ALD, but reduced by approximately 10% to provide a slightly more open mosaic to allow the second application of aggregate to firmly interlock.

Table 15.12 Double seal design aggregate spread rates for first application seal

	Type of binder		
	Bitumen	Polymer-modified bitumen	Bitumen emulsion
Design traffic	Aggregate spread rate (m²/m³)		
>200 v/l/d	950/ALD	[a]	[b]
<200 v/l/d	900/ALD	[a]	[b]

Source: Adapted from Austroads, *Update of Double/Double Design for Austroads Sprayed Seal Design Method*, Austroads Publication No. AP-T236-13, Sydney: Austroads Ltd., 2013.

[a] When polymer-modified binder is used, the aggregate spread rates are approximately 10% less than that for a normal single seal.

[b] When bitumen emulsion is used, the aggregate spread rate for the first application is the same as that for conventional bitumen. When modified bitumen emulsion is used, the aggregate spread rate for the first application is the same as that for polymer-modified bitumen.

Table 15.13 Double seal design aggregate spread rates for second application seal

	Type of binder	
	Bitumen or polymer-modified bitumen	Bitumen emulsion
Aggregate size (mm)	Aggregate spread rate (m²/m³)	
10	1050/ALD to 1100/ALD	[a]
7 (ALD known)	1100/ALD to 1150/ALD	[a]
7 or 5 (no ALD)	250 to 300	[a]

Source: Adapted from Austroads, *Update of Double/Double Design for Austroads Sprayed Seal Design Method*, Austroads Publication No. AP-T236-13, Sydney: Austroads Ltd., 2013.

[a] When bitumen emulsion is used, the aggregate spread rate for the second application is the same as that for conventional bitumen. When modified bitumen emulsion is used, the aggregate spread rate for the second application is the same as that for polymer-modified bitumen.

Aggregate spread rates for the first application of a double seal are as shown in Table 15.12.

Aggregate in the second application is normally no more than half the size of the first application, and the spread rate is just sufficient to fill the voids in the first application. Aggregate spread rates for the second application of a double seal are as shown in Table 15.13.

To convert the aggregate spread rate expressed in square metres per cubic metre to units of kilograms per square metre, the loose bulk density of the aggregate is divided by the value of spread rate determined from the above tables; the loose bulk density of aggregates is determined by CEN EN 1097-3 (1998), ASTM C 29 (2009), AASHTO T 19 (2009), AS 1141.4 (2000) or any other specification.

15.15.3.2.2 Design for second application delayed

When the second application of dressing (seal) is purposely delayed, it is recommended to be more than 12 months (Austroads 2013); the first application design is carried out as single dressing. The second application design is also carried out as single dressing.

However, if the time delay of the second application is less than 12 months, the allowance for texture depth of existing surface is only part of the surface texture allowances determined for single dressing. For details, see the relevant technical report (Austroads 2013).

15.15.4 AASHTO chip seal design

The AASHTO surface dressing (chip seal) design, particularly the emulsion-based material (AASHTO 2012), is the same as the Austroads design described above.

15.15.5 Asphalt institute surface dressing (chip seal) design

The Asphalt Institute cheap seal (surface dressing) design method (Asphalt Institute 2009) is much more simplified than the above-described design methods. It basically consists of tables providing the quantities of emulsion and aggregate (chippings) for single, double and triple surface treatment (dressing).

The quantities (rate of spread or application) of bitumen emulsion and aggregate required for single and double surface treatment are given in Tables 15.14 and 15.15, respectively.

The quantities may be varied according to local conditions and experience. It is also mentioned that the mass of aggregate shown in the tables is based on aggregate with a specific gravity of 2.65; for any variation, adjustment should be made to the values given by multiplying by the ratio specific gravity of the aggregate over 2.65.

As for the quantity of bitumen emulsion, the values refer specifically to the emulsion grades as given in Table 15.14. The quantities of emulsion cover the average range of conditions that include primed granular bases and old pavement surfaces.

The bitumen emulsion sprayed rate should be adjusted for the conditions of the road surface, such as smooth, absorbent or flushed surface. The corrections applied are given in Table 15.16.

Rates of spread (quantities) of the aggregate and bituminous emulsion similar to those recommended by the Asphalt Institute are provided by ASTM D 1369 (2012) for single,

Table 15.14 Quantities of emulsion and aggregate for single surface dressing

Nominal aggregate size	Aggregate quantity (kg/m²)	Emulsion quantity (L/m²)	Emulsion grade
19.0 to 9.5 mm	22–27	1.8–2.3	RS-2, CRS-2, CRS-2P
12.5 to 4.75 mm	14–16	1.4–2.0	CRS-2L, HFRS-2
9.5 to 2.38 mm	11–14	0.9–1.6	
4.75 to 1.18 mm	8–11	0.7–0.9	RS-1, MS-1, CRS-1,
Sand	5–8	0.5–0.7	HFRS

Source: Asphalt Institute, *Asphalt in Pavement Preservation and Maintenance*, Manual Series No. 16 (MS-16), 4th Edition, Lexington, US: Asphalt Institute, 2009. With permission.

Table 15.15 Quantities of emulsion and aggregate for double surface dressing

	Nominal aggregate size	Aggregate quantity (kg/m²)	Emulsion quantity (L/m²)
12.5 mm thick			
1st application	9.5 to 2.36 mm	14–19	0.9–1.4
2nd application	4.75 to 1.18 mm	5–8	1.4–1.8
15.9 mm thick			
1st application	12.5 to 4.75 mm	16–22	1.4–1.8
2nd application	4.75 to 1.18 mm	8–11	1.8–2.3
19.0 mm thick			
1st application	19.0 to 9.5 mm	22–27	1.6–2.3
2nd application	9.5 to 2.36 mm	11–14	2.3–2.7

Source: Asphalt Institute, *Asphalt in Pavement Preservation and Maintenance*, Manual Series No. 16 (MS-16), 4th Edition, Lexington, US: Asphalt Institute, 2009. With permission.

Table 15.16 Correction of bitumen emulsion sprayed rate owing to surface conditions

Pavement texture	Correction (L/m²)
Black, flushed asphalt	−0.04 to −0.27
Smooth, non-porous	0.00
Absorbent – slightly porous, oxidised	0.14
– slightly ravelled, porous, oxidised	0.27
– badly ravelled, porous, oxidised	0.40

Source: Asphalt Institute, *Asphalt in Pavement Preservation and Maintenance,* Manual Series No. 16 (MS-16), 4th Edition, Lexington, US: Asphalt Institute, 2009. With permission.

double and triple surface dressing (treatment). The rate of spread is given in cubic metres per square metre unit, which needs to be converted to kilograms per cubic metre by multiplying it by the loose unit weight of the aggregate used.

The properties of the aggregates for surface dressings (treatments) should comply with ASTM D 1139 (2009).

15.15.6 Equipment and construction sequence of surface dressing

The main equipment needed for surface dressing (chip seal) is as follows: the distributor, the aggregate spreader, the roller(s) and cleaning equipment.

The *distributor* is a truck- or trailer-mounted insulating tank that applies via a spray bar the binder material uniformly and in specified quantity (spread rate). Almost all distributors are equipped with a heating system.

Calibration of the distributor before initiation of works must always be carried out to ensure quality work. The key factors to consider to achieve a uniform application of the required spread rate of the binder are the correct size nozzles for the type of binder, the angle of nozzle opening and the height of the spray bar. For a selected size nozzle, the angle of the nozzle openings and the height of the spray bar must be adjusted to spray the binder at a uniform thickness and at the required quantity across the full width of spraying application.

The *aggregate spreader* applies uniformly and at a specified rate the aggregate (chippings) over the freshly spread surface with binder. The aggregate spreaders may simply attach to the tailgate of the truck (vane spreaders), mounted to the truck (hoppers on wheels) and propelled by reversing aggregate truck or self-propelled spreaders.

The spreaders mounted to the truck usually contain augers to distribute aggregate and provide better control of the rate of spread compared to vane spreaders. The self-propelled spreaders have a receiving hopper and belt conveyors to carry the aggregate to the hopper spreader and offer high daily production rates.

The *rollers* ensure good embedment of the aggregate particles and should always be pneumatic-type, rubber tyre rollers. A pneumatic roller with tyre pressure in the region of 500 kPa is sufficient for rolling dressings. The number of rollers required depends on the lane width, distributor speed and the appropriate speed of the roller to achieve the desired embedment. Rolling starts shortly after aggregate spreading.

The cleaning equipment, such as vacuum brooms, preferably, or rotary power brush brooms or sweepers, is necessary not only to clean the pavement surface before spraying the binder but also to remove the loose particles after the surface treatment is completed.

Cleaning the surface before binder application is essential to the durability of the surface and cleaning the surface after completion of works is essential to road safety.

The sequence of single surface dressing construction may be summarised as follows:

a. Prepare the existing surface (cleaning and, if needed, patching potholes or repair damaged areas).
b. Uniformly spray the bituminous binder at the specified rate and specified temperature.
c. Uniformly spread the chippings at a specified rate and immediately after spraying the binder.
d. Roll the chippings to orient and embed the particles.
e. Allow emulsion to cure or bitumen temperature to drop, and remove all loose chippings before opening the treated surface to traffic.

In the case of double dressing, steps (b) to (d) are repeated after single dressing is applied.

In the case of an RISD, the second layer of small-size chippings is spread before step (d).

In the case of inverted double dressing, the procedure is the same as for double surface dressing but the first layer of chippings consists of small-size aggregate.

Finally, in the case of sandwich dressing, the large-size chippings are spread on the clean surface first and then steps (b) to (e) follow.

The application of any type of surface dressing and its performance characteristics should comply with the requirements of CEN EN 12271 (2006).

Guidelines for the design and construction of surface treatments (dressings) are also provided in ASTM D 5360 (2009).

15.15.7 Adhesivity tests and quality control tests for surface dressing

The adhesion between binder and chippings is of vital importance for successful surface dressing.

The adhesivity test is carried out by the Vialit plate shock adhesion test according to CEN EN 12272-3 (2003) before construction to ensure that good bond is achieved for the given binder and chippings. The test method specifies the following methods of measurements: (a) the mechanical adhesion of the binder to the surface of aggregates, (b) the active adhesivity of the binder to chippings, (c) the improvement of the mechanical adhesion and active adhesivity either by adding an adhesion agent into the mass of the binder or by spraying the interface between binder and chippings, (d) the wetting temperature and (e) the fragility temperature. The test is briefly described in Section 3.7.3.7.

In addition to the above test, the pendulum (Vialit) test (CEN EN 13588 2008) may also be carried out for binder selection. The test gives a measure of the ability of a binder to resist traffic stress. Some details for the test are given in Section 3.7.3.8.

It has been suggested that the area under the graph above some arbitrary value, say 0.5 J/cm^2, would be an alternative criterion by which to compare binders. It should be noted that very high levels of cohesion (over 2 J/cm^2) are sometimes associated with poor adhesion (Highways Agency 1999).

The sweep test (ASTM D 7000 2011) can also be used to measure the performance characteristics of bituminous emulsion and aggregates by simulating the brooming of a surface dressing in the laboratory.

The test is particularly useful for classifying rapid-setting bituminous emulsion for surface dressings. It is intended to evaluate the potential curing characteristics of a binder–aggregate combination to ensure that the surface dressing is sufficiently cured before allowing traffic onto the dressing.

A brush (designed to closely replicate the sweeping action of a broom) exerts a force on the aggregate used on surface dressing. The bituminous emulsion is applied to an asphalt felt disk and then aggregate is applied and embedded into the bituminous emulsion. The sample is then conditioned at a prescribed temperature and period before testing. A mixer abrades the surface of the sample using a nylon brush. After 1 min of abrasion, the test is stopped, any loose aggregate is removed and the per cent mass loss is calculated. Further details can be found in ASTM D 7000 (2011).

During construction, two quality control tests are carried out for the determination of the rate of spread and accuracy of spread of the binder and the chippings.

The test methods are used on site to check the ability of binder sprayers and chipping spreaders to meet the intended rates of spread and tolerances and coefficients of variation.

For the determination of rate of spread of binder, according to CEN EN 12272-1 (2002), samples of the binder sprayed by the binder sprayer are collected to determine the average rate of spread. At least five trays, boards or tiles are used, each with a minimum area of 0.1 m² and having a minimum combined area of 0.5 m², spaced evenly across the full width of the road to be sprayed. Alternatively, a continuous strip of tiles or boards abutting each other is arranged across this entire width.

The *rate of spread of binder* is determined by the difference of the mass of the tray with the binder and mass of tray empty. The mean rate of spread of the binder is calculated, as well as the proportional range of spread.

The proportional range of spread is calculated using the following equation:

$$P_R = \frac{(d_{max} - d_{min})}{D},$$

where P_R is the proportional range, d_{max} is the highest rate of spread of binder found on an individual sampling device, d_{min} is the lowest rate of spread of binder found on an individual sampling device and D is the mean rate of spread of binder (expressed in kg/m² reported to the nearest 0.05 kg/m²).

The test is repeated if the proportional range is greater than 0.20.

The rate of spread of chippings is determined by collecting chippings in three calibrated boxes (known mass) laid on the road in front of the chipping spreader.

The *rate of spread of chippings* is expressed by three samples and is expressed by mass in kilograms per square metre or by volume in litres per square metre. More details can be found in CEN EN 12272-1 (2002).

When each box is laid flat, with its lid removed, it serves as a tray to collect the chippings from a known area. With its lid replaced, each box is stood on its end and used to directly measure the bulk volume of the collected chippings. Alternatively, each box is used to collect chippings for subsequent weighing.

The determination of the rate of spread of binder and rate of spread of chippings can also be carried out, by a test similar to the above, according to ASTM D 2995 (2009) and ASTM D 5624 (2013), respectively, or other national specifications.

15.15.8 Type of failures of surface dressing

The most common types of failures in surface dressings are loss of cover aggregates (or fretting), streaking, bleeding and fatting. Other failures may also occur such as tearing, tracking and scabbing.

15.15.8.1 Loss of cover aggregates

Loss of cover aggregates (chippings), or fretting, can be dangerous since loose aggregates thrown by the tyres can damage vehicle windshields, and if an excessive amount is lost, the surface becomes slippery.

There are several causes for this type of failure, namely, wrong combination of chippings and binder, use of dirty chippings, use of wrong size chippings, applying too little binder or too much chippings, improper rolling, improperly embedding the chippings owing to the hardness of the surface, application in adverse weather conditions, surface too dirty or wet, application of chippings after emulsion has broken, emulsion washing away if it rains soon after construction, sweeping too soon or improper control of traffic.

The remedy is to redress the surface having established the cause of failure.

15.15.8.2 Streaking

Longitudinal streaking occurs as a result of the non-uniform application of binder. Streaking leaves an unsightly appearance and reduces the service life.

Streaking is caused exclusively by malfunctioning of the sprayer and, in some cases, by improper spraying operation (too little or too much binder at joints between two applications. With regard to malfunctioning of the sprayer, the height of the spay bar; the size, angle and functioning of the nozzles; inconsistent operation of the pump delivering the binder; varying the sprayer's speed of travel; and improper viscosity of binder are all potential candidates to cause streaking.

The remedy is to redress the surface paying particular attention to spraying operations.

15.15.8.3 Bleeding

Bleeding or flushing of the binder results in a surface that is too rich in binder. Bleeding can cause hazardous conditions owing to loss of surface skid resistance.

The causes of bleeding may be improper application of binder (too much), improper amount of chippings, binder migration from underlying surface, low viscosity of binder used, high environmental temperatures or trapped water resulting in stripping.

The only treatment/remedy is to plane off all binder and redress.

15.15.8.4 Fatting

Fatting is similar to bleeding and is difficult to be distinguished from bleeding.

Fatting is due to the appearance of binder at the surface caused by the penetration of chippings into the underlying surface. The main reason for this phenomenon is the effect of traffic related to the size of chippings and the hardness of the underlying surface (too soft), which stemmed from poor design.

The remedy is not always easy. A temporary measure may be to redress the surface but a long-term solution is to remove the fatting area and redress it using correctly designed system.

15.15.8.5 Other failures

Other failures such as tearing, occurring in roundabouts or areas where heavy vehicles brake hard, and tracking, usually long-term failure along the wheel paths, are caused by unsuitability of the surface dressing technique in such sites. The only remedy is to remove the surface dressing and apply an alternative suitable material.

Scabbing occurs locally owing to inadequate surface preparation and the presence of mud or other contaminants on the existing surface.

The only remedy is to clean the surface and redress the area.

15.16 PAVEMENT STRENGTHENING

During pavement maintenance or rehabilitation, it is also recommended that the case of simultaneous pavement strengthening be examined. However, pavement strengthening may be decided at any time regardless of the need for maintenance or rehabilitation of the pavement. In any case, the thickness of the strengthening overlay is determined after examining the structural condition of the existing pavement.

A rational approach for the decision-making process for the selection of an appropriate treatment of a flexible pavement with asphalt base is shown in the flow chart in Figure 15.30.

Various procedures have been developed for determining the required thickness of the strengthening asphalt layer (overlay or inlay). All can be classified into either deflection procedure or effective thickness procedure.

The deflection procedures for asphalt overlay design are based on pavement deflection data collected by dynamic, static or vibratory non-destructive testing (NDT) devices. Examples of such devices are as follows: Benkelman beam, deflectograph, falling weight deflectometer (FWD), Dynaflect or Road Rater; for details of the testing devices, see Section 16.4.

Two of the most known deflection procedures requiring the use of the Benkelman beam or the deflectograph will be described below.

The design procedures employing the use of FWDs or other devices require extensive analytical calculations and fall outside the scope of this book.

The effective thickness procedure is a thickness deficiency approach on the basis of the physical condition of the pavement structure, determined by conditional survey, and the thickness and composition of the pavement layers.

The most well-known effective thickness procedure developed and still in use, in the absence of availability of NDT devices, is the Asphalt Institute procedure, which will also be described in this chapter.

15.17 ASPHALT OVERLAY DESIGN METHODS OVER FLEXIBLE PAVEMENTS

15.17.1 Asphalt Institute deflection method

The Asphalt Institute deflection method for asphalt overlay design (Asphalt Institute MS-17 3rd Edition) requires surface deflection measurements for the determination of pavement structural condition. The method can be used for asphalt overlay design for either flexible or rigid pavements.

The procedure for determination of the asphalt overlay thickness consists of the following steps:

a. Determination of the representative rebound deflection (RRD) using preferably Benkelman beam data, as described in Section 16.4.1.1. When other NDT devices are used, such as Dynaflect, FDW or Road Rater, it is necessary to convert the deflection results obtained to equivalent Benkelman beam rebound deflections, using correlation equations developed (see Section 16.4.1.2).

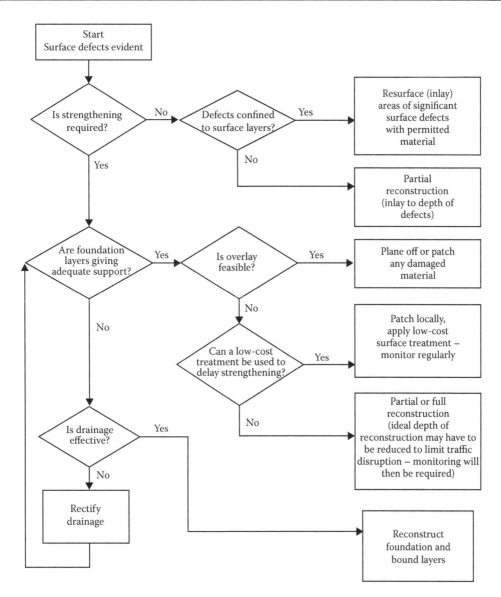

Figure 15.30 Treatment options for flexible pavements with asphalt base. (From Highways Agency, *Design Manual for Roads and Bridges [DMRB], Volume 7: Pavement Design and Maintenance, Section 3, Part 3, HD 30/08, Maintenance Assessment Procedure*, London: Department for Transport, Highways Agency, 2008b [© Highways Agency].)

b. Determination of the design traffic, expressed in equivalent standard axle loading (ESAL), over the number of years that the pavement is designed to last after overlaying. The procedure for calculating design ESAL is as explained in Section 13.3.1.

c. Determination of overlay thickness, using the above data, RRD and design ESAL and the chart shown in Figure 15.31.

The asphalt overlay consists of dense asphalt concrete, and before laying, any surface defects (potholes, minor cracking, surface depressions, etc.) should be repaired.

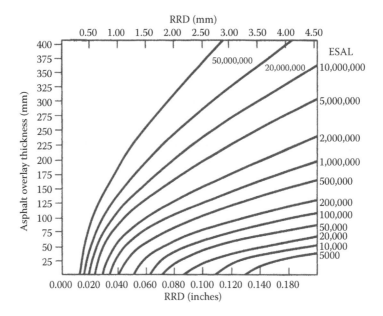

Figure 15.31 Asphalt concrete overlay thickness required to reduce pavement deflections from a measured to a design deflection value. (From Asphalt Institute MS-17 3rd Edition, *Asphalt Overlays for Highway and Street Rehabilitation*, Manual Series No.17 [MS-17], Lexington, US: Asphalt Institute. With permission.)

More information about this overlay design method can be found in Asphalt Institute MS-17 3rd Edition.

15.17.2 UK highways agency deflection method

The UK Highways Agency method (Highways Agency 2008a) for asphalt overlay design uses deflection data collected by the deflectograph automated deflection measuring system (Highways Agency 2008b). The old overlay design method based on Benkelman beam deflection data (Kennedy and Lister 1978) is not used anymore.

The procedure for determination of the asphalt overlay consists of the following steps:

a. Collection and electronic recording of the deflection measurements
b. Data processing with the use of software, known as PANDEF
c. Determination of asphalt overlay thickness using PANDEF or HAPMS software

Relevant information about PANDEF or HAPMS can be found in Highways Agency (2008b).

Highways Agency (2008a) recommends strengthening by overlay or partial reconstruction to be designed to extend the pavement life for a further 20 years. This procedure is known as pavement strengthening/upgrading.

The thickness of the overlay may increase to convert the pavement from 'determinate life pavement' to 'life pavement'. Provided deflection measurements allow it, this can be achieved by using the graph shown in Figure 15.32.

The total thickness of bituminous material (TTBM) shown in Figure 15.32 is the combined thickness of all the contiguous intact asphalt layers present in the pavement, subject to the following criteria:

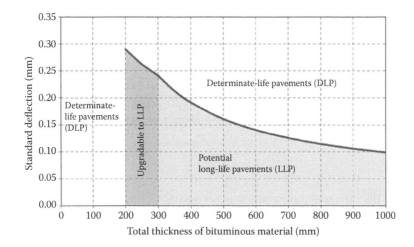

Figure 15.32 Deflectograph-based pavement life categories. (From Highways Agency, *Design Manual for Roads and Bridges [DMRB], Volume 7: Pavement Design and Maintenance, Section 3, Part 3, HD 30/08, Maintenance Assessment Procedure*, London: Department for Transport, Highways Agency, 2008b [© Highways Agency].)

a. Asphalt surfacing layers (i.e. those within the top 100 mm of the existing pavement) are included in TTBM regardless of their condition.
b. Asphalt layers that are known to be severely deteriorated and whose upper surface is at a depth greater than 100 mm are not included in TTBM.
c. Any intact asphalt (or deteriorated surfacing material) that is separated from other intact asphalt materials by either a severely deteriorated asphalt layer or any granular layer (either of which must be greater than 25 mm thick and have their upper surface at a depth greater than 100 mm) is not to be included in TTBM (Highways Agency 2008a).

Highways Agency (2008a) also recommends that asphalt overlay design should not be based on deflection data alone, since deflectograph analysis does not take into account all factors related to pavement performance.

It has also been stated (Highways Agency 2008a) that another approach to an overlay design could be to design the pavement as new, to carry both past and future traffic, and then to compare the result with the existing pavement structure. The difference in asphalt layer thickness is an estimate of the necessary overlay, provided allowances are made for any deterioration or initial deficiencies, as well as the varying material in the existing pavement.

15.17.3 Asphalt institute effective depth method

The effective depth method developed by the Asphalt Institute (Asphalt Institute MS-17 3rd Edition) is an alternative asphalt overlay design method to the deflection method. It is based on the concept that the pavement deteriorates and at any time there is a 'remaining life' that could be utilised in designing the pavement for future conditions. This 'remaining life' is provided by a pavement with a depth less than the original, called 'effective depth'.

To convert the existing pavement into an effective depth pavement (full-depth pavement), conversion factors are used related to the condition of each layer. The determination of the condition of each layer is carried out by visual inspection/survey.

The effective method can be used for asphalt overlay design for either flexible or rigid pavements.

The procedure for determining the thickness of the asphalt overlay for flexible pavements consists of the following steps:

a. Determination of the effective thickness of the total existing pavement structure, T_e
b. Determination of thickness for a theoretical full-depth asphalt concrete pavement, T_n, for the expected future traffic and existing subgrade strength
c. Determination of asphalt overlay thickness (T_o) using the following equation:

$$T_o = T_n - T_e$$

To calculate the effective pavement, it is necessary to know the condition, composition and thickness of each pavement layer.

Once the condition of each distinct layer is determined, its equivalency performance is estimated, expressed by conversion factors (see Table 15.17). The range of conversion factor values entails a degree of subjectivity. However, experience has shown that they are reasonable and useful for overlay design (Asphalt Institute MS-17 3rd Edition).

The equivalent thickness of each layer is determined by multiplying the layer thickness with its conversion factor. Hence, the equivalent thickness of the total existing pavement is the sum of all equivalent layers; that is,

$$T_e = \sum_{n=1}^{n=i} T_{e_i} = \sum_{n=1}^{n=i} C_i \times T_i,$$

where T_e is the effective thickness of total existing pavement (mm), T_{e_i} is the effective thickness of layer i (mm), C_i is the conversion coefficient for layer i, T_i is the thickness of layer i (mm) and n is the number of layers.

The determination of full-depth asphalt concrete pavement thickness, T_n, of a theoretical new pavement, is carried out from the nomograph given in Figure 13.4, by knowing the subgrade strength (resilient modulus, M_R) and the cumulative future traffic for which the overlay is designed for (in equivalent standard load, ESAL).

Available subgrade strength data from initial design records may be used, but it is advisable to execute a limited number of laboratory or non-destructive tests (mainly FWD).

When subgrade strength is determined other than by the resilient modulus value, M_R, such as CBR or R value, the correlation relationships given in Section 1.7.1 may be used.

When the determination of subgrade strength is not possible by laboratory or non-destructive tests, the subgrade strength can be estimated from soil type by grouping the subgrade into three general classes (Asphalt Institute MS-17 3rd Edition): poor, fair and good.

Poor soil material is defined as soil that softens and becomes plastic when it moistens. In other words, it includes soil materials with appreciable amount of clay and silt. The typical strength values considered are as follows: $M_R = 30$ MPa, CBR = 3 or R value = 6.

Fair soil material is described as soil that retains a moderate degree of hardness under adverse moisture conditions. Such soils may be loams, silty sands and sand-gravels containing moderate amount of clay and fine silt. The typical strength values considered are as follows: $M_R = 80$ MPa, CBR = 8 or R value = 20.

Table 15.17 Conversion factors for converting thickness of existing pavement components to effective thickness

Classification of materials	Description of material	Conversion factors
I	a. Native subgrade in all cases b. Improved subgrade – predominately granular materials – may contain some silt and clay but have a plasticity index (PI) ≤ 10 c. Lime-modified subgrade constructed from high-plasticity soils, materials with PI > 10	0.0
II	Granular sub-base or base – reasonably well-graded, hard aggregates with some plastic fines and CBR > 20. Use upper part of range if PI ≤ 6 and lower part of range if PI > 6	0.1–0.2
III	Cement or lime–fly ash-stabilised sub-bases and bases constructed from low-plasticity soils with PI ≤ 10	0.2–0.3
IV	a. Emulsified or cut-back asphalt surfaces and bases that show extensive cracking, considerable ravelling or aggregate degradation, appreciable deformation in the wheel paths and lack of stability. b. Portland cement concrete (PCC) pavements (including those under asphalt surfaces) that have been broken into small pieces (which have maximum dimension ≤ 0.6 m) prior to overlay construction. Use upper part of range when sub-base is present; lower part of range when slab is on subgrade. c. Cement or lime–fly ash-stabilised bases that have developed pattern cracking, as shown by reflected cracks. Use upper part of range when cracks are narrow and tight; lower part of range with wide cracks, pumping or evidence of instability.	0.3–0.5
V	Fractured slab techniques a. Rubblisation b. Crack/break and seat (When unstabilised base is present, use lower end of range. When stabilised base is present, use upper end of range.)	0.4–0.7 0.5–0.7
VI	a. Asphalt concrete surface and base that exhibit appreciable cracking and crack patterns b. Emulsified or cut-back asphalt surface and base courses that, although remain stable, exhibit some fine cracking, ravelling or aggregate degradation, and slight deformation in the wheel paths c. Appreciably cracked and faulted PCC pavement (including those previously overlaid with HMA) that cannot be effectively undersealed. Slab fragments that range in size from 1 to 4 m^2 and have been well seated on the subgrade by heavy pneumatic-tyred rolling	0.5–0.7
VII	a. HMA surfaces and bases that exhibit some fine cracking, have small intermittent cracking patterns and have slight deformation in the wheel paths but remain stable b. Emulsified or cut-back asphalt surface and bases that are stable and generally uncracked, show no bleeding and exhibit little deformation in the wheel paths c. PCC pavements (including those under HMA) that are stable and generally undersealed, have some cracking but contain no pieces smaller than approximately 1 m^2	0.7–0.9
VIII	a. Hot mix asphalt, including HMA base, generally uncracked and with little deformation in the wheel paths b. Portland cement concrete base, under HMA surface, that is stable and non-pumping and exhibits little reflected surface cracking	0.9–1.0

Source: Asphalt Institute MS-17 3rd Edition, *Asphalt Overlays for Highway and Street Rehabilitation*, Manual Series No.17 (MS-17), Lexington, US: Asphalt Institute. With permission.

Good soil material is defined as soil that retains a substantial amount of its load-support capacity when is wet. Such soils may be clean sands, sand-gravels and soils relatively free of plastic fines. The typical strength values considered are as follows: M_R = 170 MPa, CBR = 17 or R value = 43.

The traffic volume in ESAL is calculated as described in Section 13.3.1.

15.18 ASPHALT OVERLAY OVER RIGID PAVEMENT

The asphalt overlay laid on worn rigid pavements involves (entails) the concept of correcting functionally related surface defects rather than exclusively strengthening the pavement to carry future traffic load. In all cases, asphalt overlay on rigid pavements is very much affected by reflective cracking.

Asphalt overlay is rarely laid directly on rigid pavement without any pre-treatment or maintenance of the existing surface. Even if the surface is almost free of distresses, reflective cracking at the joints are expected to appear after a certain period.

More often, the existing concrete surface shows (displaces) some kind of distress (settlement, joint failures, cracking, etc.). In this very common case, it is vital to first determine the causes and then repair the distress areas before applying the asphalt overlay.

In order to control or minimise reflective cracking, one of the following techniques are employed, before laying the asphalt overlay:

a. Rubblisation
b. Break (or crack) and seat
c. Saw-cut and seal
d. Saw-cut, crack and seat
e. Provision of crack-relief layer
f. Use of dense asphalt concrete with SBS-modified bitume
g. Use of interlayers (geosynthetics/geogrids or geotextiles or SAMI)

Rubblisation is a technique that involves breaking the pavement into fragments varying in size from 25 mm at the surface to more than 300 mm at the bottom of the layer or below the reinforcement.

Rubblisation requires the use of special equipment. The rubblising equipment can be distinguished into multiple head breakers (Figure 15.33a) and resonant pavement breakers (Figure 15.34d). The rubble is further pulverised by a vibratory grid roller (Figure 15.33b) and then seated using a steel or pneumatic roller. The rubblisation technique is applied to either unreinforced or reinforced rigid pavements.

Break (or crack) and seat is a technique to reduce the effective length of the concrete slab so that the horizontal strains resulting from the thermal expansion are reduced and evenly distributed. The above reduces the potential for reflecting cracking to occur. The break and seat technique is normally applied to unreinforced pavements but it may be applied to jointed reinforced pavements as well.

The crack patterns vary and typically range from 0.5 by 2.0 m (Coley and Carswell 2006) or 0.3 m by 1.2 m to 1.8 m by 1.8 m (Asphalt Institute MS-17 3rd Edition).

The equipment used are drop hammers, using guillotine action (Figures 15.33a and 15.35a), or whiphammers (Figure 15.35b), using chisel-type impact head powered by a spring-loaded action.

Useful information regarding the performance of the crack and seat technique that was applied in airfield pavements can be found in the TRL Report 590 (Langdale et al. 2003).

(a)   (b)

Figure 15.33 Equipment for rubblisation. (a) Multiple head breaker and (b) grid roller. (From Coley, C. and Carswell, I., *Improved Design of Overlay Treatments to Concrete Pavements. Final Report on the Monitoring of Trials and Schemes.* TRL Report TRL 657. Crowthorne, UK: Transport Research Laboratory, 2006.)

(a)   (b)   (c)   (d)

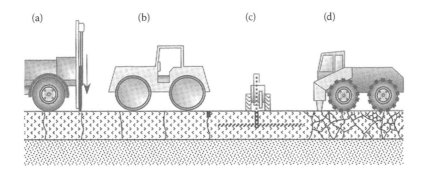

Figure 15.34 Some equipment for concrete pre-treated options prior to asphalt overlay. (a) Cracking (guillotine action); (b) seating (roller); (c) saw-cutting and (d) rubblisation (resonant breaker). (From Thom, N., *Principles of Paving Engineering*, Thomas Telford, 2008.)

The *saw-cut and seal* technique consists of saw-cutting the asphalt overlay by introducing joints into the asphalt overlay directly above the joints in the concrete slab. The joints created are then filled with an approved sealant. Thus, contraction and expansion of the underlying jointed concrete slab will be accommodated in the joints created and will not cause cracking of the overlay.

The equipment used for the saw-cut and seal technique is a saw-cutter (Figure 15.34c), which cuts the asphalt overlay to a depth of 25 to 35 mm and creates a joint approximately 20 mm.

This technique is applied to unreinforced or jointed reinforced concrete slabs that show no major surface distresses.

The *saw-cut, crack and seat* technique is a combination of the saw-cut and the crack and seat techniques described above. It is applied to jointed reinforced slabs with length greater than 6 m. The equipment used are those used in the saw-cut and crack and seat techniques.

The provision of *crack-relief layer* was pioneered in the United States for use in jointed and continuously reinforced concrete pavements to minimize reflection cracking.

(a)

(b)

Figure 15.35 Equipment for crack and seat. (a) Guillotine action (drop hammer) and (b) Whiphammer action. (From Coley, C. and Carswell, I., *Improved Design of Overlay Treatments to Concrete Pavements. Final Report on the Monitoring of Trials and Schemes.* TRL Report TRL 657. Crowthorne, UK: Transport Research Laboratory, 2006.)

The crack-relief layer is typically a 90 mm thick layer of coarse and open-graded hot asphalt, containing 25% to 35% voids and made up of 100% crushed aggregate (Asphalt Institute MS-17 3rd Edition). The recommended aggregate grading limits are as shown in Table 15.18.

The crack-relief concept requires three layers. The first is the crack-relief layer, followed by a well-graded intermediate levelling or binder course and then a dense-graded surface course. The binder course is a typical 18 mm mixture placed to a thickness of 50 mm and the

Table 15.18 Aggregate grading limits for the crack-relief layer

Sieve size (mm)	Blend A[a]	Blend B[a]	Mixture C[b]
	Percentage passing (%)		
75	100	—	—
63	95–100	100	—
50	—	—	100
37.5	30–70	35–70	75–90
19.0	3–20	5–20	50–70
9.5	0–5	—	—
4.75	—	—	8–20
2.36	—	0–5	—
0.150	—	—	0–5
0.075	—	0–3	—
Asphalt content 40/50 pen	1.5% to 3.0% per weight of mix		

Source: Asphalt Institute MS-17 3rd Edition, *Asphalt Overlays for Highway and Street Rehabilitation*, Manual Series No.17 (MS-17), Lexington, US: Asphalt Institute. With permission.

[a] This aggregate blend is recommended for highly expansive PCC pavements, such as those built with crushed silica gravels or with joint spacing exceeding 6 m.
[b] This aggregate blend may be used for less expansive PCC pavements.

third is surface course asphalt, at least 38 mm thick, composed of a nominal size aggregate of 19 mm. When a 90 mm crack-relief system is used, the total overlay thickness, inclusive of the levelling and surface course, will be from 175 to 225 mm. The crack-relief layer should be compacted with a 3.5 to 9 tonne static roller; over-compaction should be avoided and tack coating is generally not required before placing the crack-relief layer (Asphalt Institute MS-17 3rd Edition).

In the United Kingdom, TRL trials using porous friction course mixture gradation as a crack-relief layer on military airfields have shown good performance in resisting reflection cracking, even with a crack-relief layer as thin as 20 mm (Coley and Carswell 2006).

The use of *SBS-modified bitumen* in asphalt overlays is expected to delay the propagation of reflection cracking and should always be considered as an option.

Similarly, the use of *SAMI* or geogrids/geotextiles between the concrete slab and the overlay is also expected to delay propagation of reflection cracking (Ogundipe et al. 2011).

15.18.1 Asphalt Institute effective thickness method for asphalt overlay over rigid pavement

In the effective thickness asphalt overlay design method, the components of the rigid pavement are evaluated so that representative effective thickness is assigned and used in assessing structural adequacy (Asphalt Institute MS-17 3rd Edition).

The determination of asphalt overlay thickness on rigid pavements is conducted in exactly the same way as for asphalt overlay on a flexible pavement: use of conversion factors, determination of equivalent thickness pavement and determination of full-depth pavement (see Section 15.17.3).

The overall effective thickness of the asphalt overlay computed should be checked for conformation to the minimum thickness requirements based on slab length and local temperature conditions. Figure 15.36 provides the minimum thicknesses assuming the slab has been fractured.

The temperature differential is the typical maximum and minimum daily temperature of the hottest and the coldest month, respectively.

Figure 15.36 is divided into three sections (sections A, B and C). In section A, the minimum thickness of 100 mm should reduce the deflection by 20% and minimise the reflection. Overlay values in section B may be used as given, but when rubblisation, or crack and seat of the pavement, is selected, the minimum overlay value may be reduced to 125 mm, in all

Slab length (m)						
3	100	100	100	100	100	100
4.5	100	100	100	100	100	100
6	100	100	100	100	130	140
7.5	100	100	100	130	150	180
9	100	100	130	150	180	200
10.5	100	115	150	180	215	T1 or T2
12	100	140	180	200	T1 or T2	T1 or T2
13.5	115	150	190	230	T1 or T2	T1 or T2
15	130	180	215	T1 or T2	T1 or T2	T1 or T2
18	150	200	T1 or T2	T1 or T2	T1 or T2	T1 or T2
	17	22	28	33	39	44

Temperature differential (°C)

Figure 15.36 Minimum thickness of asphalt overlay over Portland cement concrete pavement. (From Asphalt Institute MS-17 3rd Edition, *Asphalt Overlays for Highway and Street Rehabilitation*, Manual Series No.17 [MS-17], Lexington, US: Asphalt Institute. With permission.)

cases. In section C, the crack (brake) and seat (T1) or the rubblisation (T2) technique is recommended to be selected. This is because an overlay thickness of more than 200 mm creates serious problems by increasing pavement elevation substantially (grade changes, shoulder reconstruction, grade-crossing clearance, etc.).

15.18.2 Asphalt Institute deflection procedure for asphalt overlay over rigid pavements

The Asphalt Institute deflection procedure for asphalt overlay design over rigid pavements is considered as a supplement to the effective thickness design method. It is used to determine if corrective actions to minimise pavement deflection, particularly at the joints, should be initiated prior to overlay construction; this is only in the case when rubblisation, or crack (brake) and seat, has not been selected.

The deflection measurements may be carried out by any of the devices shown in Table 15.19. Deflection measurements are conducted along the edge of the edge of the concrete slab.

For jointed concrete pavements, the differential deflection $(D_1 - D_2)$ and the mean deflection $[(D_1 + D_2)/2]$ determined from the collected data should have a maximum value as that given in Table 15.19, depending on the device used.

Similarly, for continuously reinforced concrete pavements, the critical deflection (maximum deflection), D_3, should have a maximum value as that given in Table 15.19, depending on the device used.

The maximum values given in Table 15.19 act as criteria for the need of stabilisation and undersealing of the slabs (usually at the joints or at the corners). When the measured deflection is greater than the maximum permissible values, stabilisation and undersealing are required. Undersealing is carried out by pumping bitumen under the slab; the bitumen to be used should meet ASTM D 3141 (2009) requirements.

When the measured deflection is above the maximum permissible value and since the asphalt overlay reduces the measured deflection (the Asphalt Institute assumes a deflection reduction of 0.2% per millimetre of asphalt overlay), the theoretical deflection after the application of the overlay of certain thickness can be determined. When this theoretical deflection is still greater than the maximum permissible value, it is more economical to apply undersealing than to increase the thickness of the overlay. The above is best illustrated by an example given in Asphalt Institute MS-17 3rd Edition.

Table 15.19 Maximum deflection criteria

Criteria	Maximum deflection (mm)				
	Benkel.	Dynafl.	RR 4&5	RR 2000	FWD
JRC					
Differential deflection	0.05	0.005	0.005	0.13	0.08
Mean deflection	0.36	0.019	0.019	0.52	0.57
CRCP					
Critical deflection	0.27	0.015	0.015	0.41	0.44

Source: Asphalt Institute MS-17 3rd Edition, *Asphalt Overlays for Highway and Street Rehabilitation*, Manual Series No.17 (MS-17), Lexington, US: Asphalt Institute. With permission.

Note: JRC, jointed concrete pavements; CRCP, continuously reinforced concrete pavement; Benkel., Benkelman beam; Dynafl., Dynaflect; RR 4&5, Road Rater 400 and 510; measurements were made at 4.5 kN and 8–10 Hz; RR 2000, Road Rater 2000; measurements were made at 36 kN and 15 Hz; FWD, falling weight deflectometer; measurements were made at 40 kN.

Table 15.20 General guidelines for evaluating load transfer efficiency using deflection ratio

Deflection ratio	Load transfer efficiency
>0.75	Adequate
0.6–0.75	Fair
<0.60	Poor

Source: Asphalt Institute MS-17 3rd Edition, *Asphalt Overlays for Highway and Street Rehabilitation*, Manual Series No. 17 (MS-17), Lexington, US: Asphalt Institute. With permission.

An alternative to the above, the load transfer efficiency (LTE) between concrete slabs can be measured by computing the deflection ratio, DR, as

$$DR = d_2/d_1,$$

where d_1 is the deflection on the unloaded side of the joint, or crack, measured 300 or 450 mm away from the centre of a 300 or 450 mm load plate, respectively, and d_2 is the deflection under load plate.

For NDT devices such as a FWD with a 300 mm diameter load plate, the edge of the load plate should be 75 mm from the joint or crack; for a 450 mm diameter load plate, the edge of the plate should be placed 50 mm from the joint or crack (Asphalt Institute MS-17 3rd Edition).

Having determined the deflection ratio, the LTE between the slabs can be evaluated from Table 15.20.

Pavements with poor LTE are considered candidates for undersealing, cracking and seating or rubblising prior to overlay. The repair option depends on the condition of the slab.

More information and solved examples are given in Asphalt Institute MS-17 3rd Edition. Relevant information regarding the deflection devices can also be found in Section 16.4.

15.18.3 British practice regarding overlays on rigid pavements

The British practice regarding overlays on rigid pavements is covered by DMRB HD 29 (Highways Agency 2008a) and DMRB HD 30 (Highways Agency 2008b). DMRB HD 29 describes methods of assessing residual life of rigid pavements and DMRB HD 30 covers the selection (and basic thickness design) of maintenance measures.

Useful information regarding improved design of overlay treatments to rigid pavements based on monitoring performance on trials and schemes in the United Kingdom can be found in TRL 657 (Coley and Carswell 2006).

Information regarding maintenance and strengthening of rigid pavements, in general, can be found in DMRB HD 32 (Highways Agency 1994).

Treatments are generally suggested depending on the nature of distresses and condition of the concrete pavement. With respect to overlays, the instructions and recommendations are summarised as follows:

a. Asphalt overlay over rigid pavements is recommended when cut and seal technique is employed.
b. Asphalt overlay of minimum thickness of 150 mm is recommended when crack (break) and seat technique is employed.

c. When concrete overlay is selected, it is recommended to use continuously reinforced concrete covered by a thin asphalt layer.

d. Full-depth reconstruction should be considered if concrete pavement is severely damaged.

There is currently no standard British method for assessing the thickness of an asphalt overlay required to strengthen rigid pavements.

15.19 CONCRETE OVERLAY OVER RIGID OR FLEXIBLE PAVEMENT

Concrete overlay on rigid or flexible pavement is not a common practice as asphalt overlay. However, when concrete overlay is used, it is distinguished into thin overlay (50 to 100 mm) and thick overlay (>150 mm) construction.

Thin concrete overlay over either flexible or rigid pavement presupposes intact pavement surface and ensures good bond between the old surface and the concrete overlay.

Thick concrete overlay is also applied over damaged surfaces and is most commonly applied over existing rigid pavements.

Good bond in the case of thin concrete overlay is achieved by thoroughly cleaning the asphalt surface and washing the asphalt surface prior to placement of the concrete overlay, or in the case of concrete surface, by milling, grit or sand blasting and thorough cleaning, wetting of the existing clean surface, or by applying cement grout immediately before the overlay is placed. The use of epoxy resins applied over the intact existing concrete surface is an effective but expensive solution when applied to large-scale strengthening/rehabilitation schemes.

In the case of thick concrete overlay, the new slab needs to be free to move horizontally to deter the appearance of reflection cracks. The above can be achieved with the provision of an interlayer, which can be a polythene membrane or an asphalt interlayer. The use of an asphalt interlayer is preferred if the underlying concrete pavement has a significant remaining integrity (Thom 2008).

In thin-bonded concrete overlay over asphalt surface, transverse joints should be provided at very close distance (no more than approximately 1 m; Thom 2008) and the slab should be either fibre reinforced or steel reinforced, if slab thickness allows it.

In thin concrete overlay over jointed concrete pavement, the joints in the overlay should be in exactly the same position as the existing joints. In thick concrete overlay over an existing concrete pavement, such a requirement does not exist.

When thick concrete overlay is placed over an existing concrete surface, the concrete material and in particular the aggregates must be compatible with the aggregates used in the existing concrete. This is to avoid shrinkage cracking (see Section 15.11.6 and Figure 15.21).

As for the design of the thickness of concrete overlay, the thickness of the thin concrete overlay over intact surface seems to rely on proven past experience and in any case ranges from 50 to 100 mm.

Thickness determination of the thick concrete overlay may be carried out by considering the failed concrete pavement as a sub-base layer. Usually, the minimum thickness considered for an unreinforced concrete slab is 150 mm and that for a reinforced slab is 200 mm.

Additional information regarding bonded and unbonded concrete overlay over rigid pavements according to UK standards can be found in DMRB HD 32/94 (Highways Agency 1994).

15.20 AASHTO OVERLAY DESIGN METHOD

15.20.1 According to the AASHTO guide of pavement structures

The AASHTO guide of pavement structures (AASHTO 1993) provides the most comprehensive overlay design methods for rehabilitating/strengthening flexible or rigid pavements.

It covers asphalt or concrete overlays of flexible or rigid pavements as well as asphalt overlays of rigid pavements with an existing asphalt overlay and concrete overlay over a flexible pavement. In particular, it covers the following:

a. Asphalt overlay of flexible pavement
b. Asphalt overlay of fractured concrete slab (break/seat or rubblised/compact) rigid pavement
c. Asphalt overlay of jointed unreinforced (JPCP)/reinforced (JRCP) or continuously reinforced pavement (CRCP)
d. Asphalt overlay of rigid pavements with existing asphalt overlay
e. Bonded concrete overlay of jointed unreinforced/reinforced or continuously reinforced pavements
f. Unbonded jointed unreinforced/reinforced or continuously reinforced overlay of jointed unreinforced/reinforced or continuously reinforced pavements
g. Jointed unreinforced/reinforced or continuously reinforced overlay of flexible pavement

In all cases, to determine the overlay thickness, the following is required: functional evaluation of the existing pavement, structural evaluation of the existing pavement, determination of design subgrade resilient modulus, M_R, and traffic analysis (past and future traffic).

All existing pavement design and construction details (thickness and material type of each layer, subgrade soil information, etc.) are also required.

Despite the collected information on the existing pavement, coring and materials testing are strongly recommended to determine the resilient modulus of the subgrade, M_R, (or CBR), the elastic modulus of the asphalt, E_{AC}, and the concrete modulus of rupture, S_c', in most cases.

The functional evaluation of the existing pavement is based on condition surveys for certain distress types and severities, depending on the overlay selected.

The structural evaluation of the existing pavement, with the aim to determine the structural capacity of the existing pavement, is carried out based on visual survey and materials testing, non-destructive deflection testing or fatigue damage from traffic (remaining life procedure). The structural capacity of the existing pavement for flexible pavements is expressed as effective structural number, SN_{eff}, while that for rigid and composite pavements is expressed as effective slab thickness, D_{eff}.

Having obtained all the necessary information as described above, the thickness of an asphalt overlay of flexible pavement or of fractured concrete slab rigid pavement, cases (a) and (b) above, is determined by the following equation:

$$D_{ol} = \frac{SN_f - SN_{eff}}{a_{ol}},$$

where D_{ol} is the required asphalt overlay thickness (inches); SN_f is the structural number for future traffic, computed after determining the appropriate design inputs from the nomograph in Figure 13.10; SN_{eff} is the effective structural number of the existing pavement, determined from the calculus procedure (NDT method), condition surveys or remaining life concept procedure; and A_{ol} is the structural coefficient of asphalt overlay (suggested values are given in Table 5.2 of Part III of AASHTO 1993).

Respectively, the thickness of an asphalt overlay of jointed unreinforced/reinforced or continuously reinforced pavement, or of rigid pavements with existing asphalt overlay, cases (c) and (d) above, is determined by the following equation:

$$D_{ol} = A \times (D_f - D_{eff}),$$

where D_{ol} is the required asphalt overlay thickness (inches); A is the factor to convert concrete thickness deficiency to asphalt overlay thickness (determined by an equation based on D_f and D_{eff} values); D_f is the slab thickness to carry future traffic (inches), computed after determining the appropriate design inputs from the nomograph in Figure 14.12; and D_{eff} is the effective thickness of slab (inches), determined from condition surveys or remaining life concept procedure.

The thickness of bonded concrete overlay of any type of rigid pavement, case (e), is determined by the following equation:

$$D_{ol} = D_f - D_{eff},$$

where D_{ol} is the required concrete overlay thickness (inches) and D_f and D_{eff} are as determined above.

The thickness of unbonded jointed unreinforced/reinforced or continuously reinforced overlay of jointed unreinforced/reinforced or continuously reinforced pavements, case (f), is determined by the following equation:

$$D_{ol} = \sqrt{D_f^2 - D_{eff}^2},$$

where D_{ol} is the required concrete overlay thickness (inches) and D_f and D_{eff} are as determined above.

Finally, the thickness of jointed unreinforced/reinforced or continuously reinforced overlay of flexible pavement, case (g), is determined by the following equation:

$$D_{ol} = D_f,$$

where D_{ol} is the required concrete overlay thickness (inches) and D_f is the slab thickness to carry future traffic (inches), computed after determining the appropriate design inputs from the nomograph in Figure 14.12.

All the detailed information regarding the determination of the thickness of the asphalt or concrete overlays of flexible or rigid pavements can be found in the AASHTO *Guide for Design of Pavement Structures* (AASHTO 1993).

15.20.2 According to the AASHTO mechanistic–empirical pavement design guide

The AASHTO *Mechanistic–Empirical Pavement Design Guide* (MEPDG) (AASHTO 2008) rehabilitation (overlay) design process is completely different from the overlay design procedure of the AASHTO *Guide for Design of Pavement Structures* (AASHTO 1993), which has been developed to determine the thickness of the overlay.

The MEPDG design process is an iterative hands-on approach by the designer, starting with a trial rehabilitation strategy. The trial rehabilitation design (overlay) may be initially determined using the guide for the design of pavement structure outlined in the previous paragraph, a rehabilitation design catalogue or an agency-specific procedure. The MEPDG software may then be used to analyse the trial design to ensure that it will meet the user's performance expectations (AASHTO 2008).

The MEPDG provides detailed guidance on the use and design of rehabilitation strategies, depending on the type and conditions of the existing pavement. It also provides specific details on the use of material-specific overlays for existing flexible and rigid pavements. The steps that are suggested for use in determining a preferred rehabilitation strategy are shown in Figure 15.37.

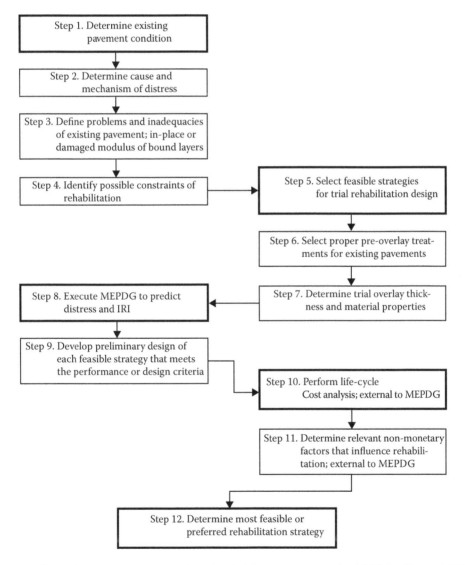

Figure 15.37 Steps for determining a preferred rehabilitation strategy by MEPDG. (From AASHTO, *Mechanistic–Empirical Pavement Design Guide [MEPDG]*, American Association of State Highway and Transportation Officials, Washington, DC: AASHTO, 2008. With permission.)

For the rehabilitation design with asphalt (HMA) overlays according to MEPDG, seven options are covered, which are as follows:

a. HMA overlay of existing flexible pavements
b. HMA overlay of semi-rigid pavements
c. HMA overlay of intact concrete (PCC) slabs
d. HMA overlay of fractured concrete (PCC) slabs
e. HMA overlay of existing intact concrete (PCC) pavements including composite pavements (one or more HMA overlays of existing JPCP and CRCP)
f. HMA overlay of existing intact concrete (PCC) pavements including composite pavements
g. HMA overlay of fractured concrete (PCC) pavements

A generalised flow chart for pavement rehabilitation with asphalt (HMA) overlays together with the pre-overlay treatments suggested is shown in Figure 15.38.

A detailed rehabilitation (overlay) design process for asphalt of flexible and rigid pavements for each of the above cases is covered in the MEPDG (AASHTO 2008).

As for the rehabilitation design with concrete (PCC) overlays that can be analysed by MEPDG, again seven options are covered, which are shown in Table 15.21.

The general design processes for major concrete (PCC) rehabilitation strategies considered by MEPDG are shown in Figure 15.39.

A detailed rehabilitation (overlay) design process for concrete (PCC) of flexible and rigid pavements for each of the above cases is covered in the MEPDG (AASHTO 2008).

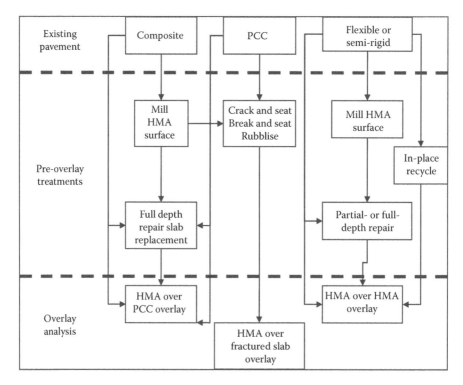

Figure 15.38 Flow chart of rehabilitation design options using HMA overlays according to MEPDG. (From AASHTO, *Mechanistic–Empirical Pavement Design Guide [MEPDG]*, American Association of State Highway and Transportation Officials, Washington, DC: AASHTO, 2008. With permission.)

Table 15.21 Rehabilitation options/strategies to correct surface and structural deficiencies of all types of existing pavement

Type of PCC overlay	Existing pavement	Rehabilitation of existing pavement	Separation layer and surface preparation
Unbonded JPCP overlay	JPRP, JRCP and CRCP	Repair by slab replacement or full-depth repair (FDR).	Place HMA layer for level up and separation. Do not diminish bonding between PCC overlay and HMA.
	Fractured JPCP, JRCP and CRCP	Fracture and seat existing pavement if concerns over rocking slabs exist.	Place HMA layer for level up and separation. Do not diminish bonding between PCC overlay and HMA.
	Composite pavement (HMA/PCC)	Mill off portion or all of existing HMA for level up (all if stripping exists), FDR existing PCC pavement or fracture and seat existing pavement.	Place HMA layer for level up and separation. Do not diminish bonding between PCC overlay and HMA.
Unbonded CRCP overlay	JPCP, JRCP and CRCP	Repair by FDR or fracture and seat existing pavement if concerns over poor transverse joint load transfer or rocking slabs exist.	Place HMA layer for level up and separation. Increase thickness if poor joint and crack load transfer efficiency. Maximise bonding between CRCP overlay and HMA layers.
	Fractured JPCP, JRCP and CRCP	Fracture existing pavement if concerns over rocking slabs or reflection cracking exist (poor existing joint LTE).	Place HMA layer for level up and separation. Maximise bonding between CRCP overlay and HMA layers.
Bonded PCC overlay	JPCP and CRCP in fair or better condition only	FDR deteriorated joints and cracks.	Preparation of existing surface to maximise bond with PCC overlay.
JPCP and CRCP overlays	Existing flexible pavement	Mill portion of existing HMA material for level up and removal of deterioration. Patch as needed.	Place HMA layer for level up and separation. Maximise bonding between PCC overlay and HMA layers.

Source: AASHTO, *Mechanistic–Empirical Pavement Design Guide (MEPDG)*, American Association of State Highway and Transportation Officials, Washington, DC: AASHTO, 2008. With permission.

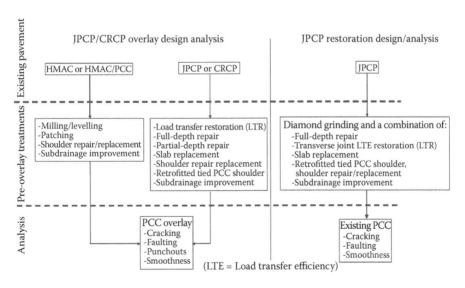

Figure 15.39 Overall design processes for major PCC rehabilitation strategies of all pavement types. (From AASHTO, *Mechanistic–Empirical Pavement Design Guide [MEPDG]*, American Association of State Highway and Transportation Officials, Washington, DC: AASHTO, 2008.)

REFERENCES

AASHTO R 36. 2013. *Standard practice for evaluating faulting of concrete pavements*. Washington, DC: American Association of State Highway and Transportation Officials.

AASHTO T 19M/T 19. 2009. *Bulk density ("unit weight") and voids in aggregate*. Washington, DC: American Association of State Highway and Transportation Officials.

AASHTO. 1993. *Guide for design of pavement structures (GDPS)*. Washington, DC: American Association of State Highway and Transportation Officials.

AASHTO. 2008. *Mechanistic–empirical pavement design guide (MEPDG)*. Washington, DC: American Association of State Highway and Transportation Officials.

AASHTO. 2012. *Manual for emulsion-based chip seals for pavement preservation*. Washington, DC: American Association of State Highway and Transportation Officials.

AS 1141.4. 2000. *Methods for sampling and testing aggregates – Bulk density of aggregate*, Standards Australia. Sydney: Standards Australia Ltd.

AS 1141.20.1. 2000. *Methods for sampling and testing aggregates – Average least dimension – Direct measurement (nominal size 10 mm and greater)*, Standards Australia. Sydney: Standards Australia Ltd.

AS 1141.20.2. 2000. *Methods for sampling and testing aggregates – Average least dimension – Direct measurement (nominal sizes 5 mm and 7 mm)*, Standards Australia. Sydney: Standards Australia Ltd.

AS 1141.20.3-2000 (R2013). 2013. *Methods for sampling and testing aggregates - Average least dimension - Calculation (nomograph)*, Standards Australia. Sydney Australia Ltd.

Asphalt Institute MS-17. *Asphalt overlays for highway and street rehabilitation*. Manual Series No. 17 (MS-17), 3rd Edition. Lexington, US: Asphalt Institute.

Asphalt Institute. 2009. *Asphalt in pavement preservation and maintenance*, Manual Series No. 16 (MS-16), 4th Edition. Lexington, US: Asphalt Institute.

ASTM C 29/C29M. 2009. *Standard test method for bulk density ("unit weight") and voids in aggregate*. West Conshohocken, PA: ASTM International.

ASTM D 1139/1139M. 2009. *Standard specification for aggregate for single or multiple bituminous surface treatments*. West Conshohocken, PA: ASTM International.

ASTM D 1369. 2012. *Standard practice for quantities of materials for bituminous surface treatments*. West Conshohocken, PA: ASTM International.

ASTM D 2628. 2011. *Standard specification for preformed polychloroprene elastomeric joint seals for concrete pavements*. West Conshohocken, PA: ASTM International.

ASTM D 2995. 2009. *Standard practice for estimating application rate of bituminous distributors*. West Conshohocken, PA: ASTM International.

ASTM D 3141/D3141M. 2009. *Standard specification for asphalt for undersealing Portland-cement concrete pavements*. West Conshohocken, PA: ASTM International.

ASTM D 5249. 2010. *Standard specification for backer material for use with cold- and hot-applied joint sealants in Portland-cement concrete and asphalt joints*. West Conshohocken, PA: ASTM International.

ASTM D 5360. 2009. *Standard practice for design and construction of bituminous surface treatments*. West Conshohocken, PA: ASTM International.

ASTM D 5624. 2013. *Standard test method for determining the transverse-aggregate spread rate for surface treatment applications*. West Conshohocken, PA: ASTM International.

ASTM D 5893/D5893M. 2010. *Standard specification for cold applied, single component, chemically curing silicone joint sealant for Portland cement concrete pavements*. West Conshohocken, PA: ASTM International.

ASTM D 6690. 2012. *Standard specification for joint and crack sealants, hot applied, for concrete and asphalt pavements*. West Conshohocken, PA: ASTM International.

ASTM D 7000. 2011. *Standard test method for sweep test of bituminous emulsion surface treatment*. West Conshohocken, PA: ASTM International.

ASTM D 7239. 2013. *Standard specification for hybrid geosynthetic paving mat for highway applications*. West Conshohocken, PA: ASTM International.

Austroads. 2006. *Update of the Austroads sprayed seal design method*, Austroads Publication No. AP-T68/06. Sydney: Austroads Incorporated.

Austroads. 2013. *Update of double/double design for Austroads sprayed seal design method*, Austroads Publication No. AP-T236-13. Sydney: Austroads Ltd.

BS 598-112. 2004. *Sampling and examination of bituminous mixtures for roads and other paved area: Method for the use of road surface hardness probe.* London: British Standards Institution.

Buchta H. and G. Nievelt. 1989. *Stress Absorbing Membranes (SAM) and chip seals for medium maintenance of flexible roads.* 4th Eurobitume, p. 596, Madrid, Spain.

CEN EN 1097-3. 1998. *Tests for mechanical and physical properties of aggregates, Part 3: Determination of loose bulk density and voids.* Brussels: CEN

CEN EN 12271. 2006. *Surface dressing – Requirements.* Brussels: CEN.

CEN EN 12272-1. 2002. *Surface dressing – Test methods – Part 1: Rate of spread and accuracy of spread of binder and chippings.* Brussels: CEN.

CEN EN 12272-2. 2003. *Surface dressing – Test methods – Part 2: Visual assessment of defects.* Brussels: CEN.

CEN EN 12272-3. 2003. *Surface dressing – Test method – Part 3: Determination of binder aggregate adhesivity by the Vialit plate shock test method.* Brussels: CEN.

CEN EN 13043:2004/AC. 2004. *Aggregates for bituminous mixtures and surface treatments for roads, airfields and other trafficked areas.* Brussels: CEN.

CEN EN 13588. 2008. *Bitumen and bituminous binders – Determination of cohesion of bituminous binders with pendulum test.* Brussels: CEN.

CEN EN 13808. 2013. *Bitumen and bituminous binders – Framework for specifying cationic bituminous emulsions.* Brussels: CEN.

CEN EN 14188-1. 2004. *Joint fillers and sealants – Part 1: Specifications for hot applied sealants.* Brussels: CEN.

CEN EN 14188-2. 2004. *Joint fillers and sealants – Part 2: Specifications for cold applied sealants.* Brussels: CEN.

CEN EN 14188-3. 2006. *Joint fillers and sealants – Part 3: Specifications for preformed joint seals.* Brussels: CEN.

CEN EN 15381. 2008. *Geotextiles and geotextile-related products – Characteristics required for use in pavements and asphalt overlays.* Brussels: CEN.

CEN EN ISO 10318. 2005. *Geosynthetics – Terms and definitions.* Brussels: CEN.

Coley C. and I. Carswell. 2006. *Improved design of overlay treatments to concrete pavements. Final Report on the monitoring of trials and schemes.* TRL Report TRL 657. Crowthorne, UK: Transport Research Laboratory.

Heslop M.W., M.J. Elborn and G.R. Pooley. 1982. Recent developments in surface dressing. *The Highway Engineer*, pp. 6–19. London: Institution of Highways and Transportation.

Highways Agency. 1994. *Design Manual for Roads and Bridges (DMRB), Volume 7: Pavement design and maintenance, Section 4, Part 2, HD 32/94, Maintenance of concrete roads.* London: Department for Transport, Highways Agency.

Highways Agency. 1998. *Design Manual for Roads and Bridges (DMRB), Volume 7: Pavement design and maintenance. Section 4. Pavement maintenance methods, Part 1, HD 31/94, Amendment No. 2: Maintenance of bituminous roads.* Department for Transport. London: Highways Agency.

Highways Agency. 1999. *Design Manual for Roads and Bridges (DMRB), Volume 7: Pavement Design and Maintenance, Section 5, Surfacing and surfacing materials, Part 2, HD 37/99, Amendment No. 1, Bituminous surfacing materials and techniques.* London: Department for Transport, Highways Agency.

Highways Agency. 2004. *Design Manual for Road and Bridges (DMRB), Volume 7: Pavement design and maintenance, Section 3, Part 1, HD 28/04.* London: Department for Transport, Highways Agency.

Highways Agency. 2008a. *Design Manual for Roads and Bridges (DMRB), Volume 7: Pavement design and maintenance, Section 3, Part 2, HD 29/08, Data for pavement assessment.* London: Department for Transport, Highways Agency.

Highways Agency. 2008b. *Design Manual for Roads and Bridges (DMRB), Volume 7: Pavement design and maintenance, Section 3, Part 3, HD 30/08, Maintenance assessment procedure.* London: Department for Transport, Highways Agency.

Highways Agency. 2012. *Revision of Aggregate Specification for Pavement Surfacing, Interim Advice Note (IAN)156/12*. London: Department for Transport, Highways Agency.

Kennedy C. K. and N.W. Lister. 1978. *Prediction of pavement performance and the design of overlays*. TRRL Report LR 833. Crowthorne, UK: Transport Research Laboratory.

Langdale P.C, J.F. Potter and S.J. Ellis. 2003. *The use of the crack and seat treatment in the refurbishment of airfield pavements*. TRL Report TRL 590. Crowthorne, UK: Transport Research Laboratory.

Ogundipe O.M., A.C. Collop and N.H. Thom. 2011. *Mitigating reflective cracking using stress absorbing membrane interlayers (SAMIs)*. Proceedings, 5th International Conference Bituminous Mixtures and Pavements. Thessaloniki, Greece: AUTh.

Roberts C. and J.C. Nicholls. 2008. *Design guide for road surface dressing*. TRL Road Note RN 39, 6th Edition. Crowthorne, UK: Transport Research Laboratory.

Sproule D., T. Filson and D. Nunn. 2012. *Stress absorbing membrane interlayer (SAMI) using the FiberMat™ process*. 2012 Conference of the Transportation Association of Canada, 'Innovations in pavement preservation' Session, Fredericton, New Brunswick, Canada.

Thom N. 2008. *Principles of paving engineering*. Thomas Telford.

Woodside A.R., W.D.H. Woodward and A. Graham, 2011. *The use of glass fibre addition to reinforce a road patch*. Proceedings, 5th International Conference Bituminous Mixtures and Pavements. Thessaloniki, Greece: AUTh.

Yeates C. 1994. *An evaluation of the use of a fibre-reinforced membrane to inhibit reflective cracking*. 2nd International Symposium on Highway Surfacing. Ulster: Ulster University.

Chapter 16

Pavement evaluation and measurement of functional and structural characteristics

16.1 PAVEMENT EVALUATION

The evaluation of the pavement condition is a fundamental parameter for the determination of the timing and type of intervention (maintenance, rehabilitation or reconstruction) and the overall pavement management practice. Usually, two types of pavement condition evaluation are conducted: the functional and the structural evaluation.

Functional evaluation considers the surface characteristics of a pavement and is user related. Surface characteristics include longitudinal evenness (smoothness), skid resistance, rutting, cracking or any other surface distress that affects riding quality and safety. Functional evaluation is used to decide whether the pavement needs to be maintained, rehabilitated or reconstructed; essentially, the necessity for intervention and its type is decided.

Structural evaluation considers layer thickness, materials and strength and is load related. Structural evaluation is used to determine the ability of the pavement structure to carry traffic loading; essentially, the remaining life of the pavement is determined, and from that, it is decided whether rehabilitation or strengthening of the pavement structure is required for the anticipated future traffic loading.

Functional evaluation and structural evaluation are complimentary to each other and quite often are best to be executed together, particularly when more precise determination of pavement rehabilitation strategy is required.

The functional evaluation of pavements is carried out by visual condition surveys or purposely built mobile devices.

The structural evaluation of the pavement is carried out by non-destructive testing using deflection measuring devices, supported by detailed information data of the pavement structure. Limited number of coring is necessary to verify primarily pavement layer thickness, even if a ground penetration radar (GPR) device in most cases is used.

16.2 FUNCTIONAL EVALUATION BY VISUAL CONDITION SURVEY

The visual condition survey, also known as distress survey, is the first step to functional evaluation of the pavement and perhaps the most useful of all types of surveys. In some cases, it is a stand-alone procedure for decision making.

The visual condition survey of pavements consists of (a) recording of pavement distresses, (b) pavement rating and (c) detailed presentation of pavement condition.

The visual *recording of pavement distresses* is usually conducted by crews of at least two trained people. The collection of pavement distress information may be carried out by viewing the pavement surface from a slow-moving vehicle or by walking on the pavement surface. The first may require stop and start for taking representative photographs or for better identification of the distress. When a survey is carried out, walking closer along the pavement lane is necessary and measurements are taken related to the extent of the distress (area or linear length) for assessing more accurately the distress level.

Visual condition surveys require the use of GPS facility and the use of survey forms and distress identification manual. Table 16.1 provides some details on the type of distresses to be recorded during a visual condition survey of flexible pavements carried out by the Highway Engineering Laboratory of the Aristotle University of Thessaloniki (AUTh).

Portable data recording devices (data capture devices [DCDs]) may also be used during surveying, enabling faster data processing by the use of specially developed software. The information on DCDs may also be downloaded to pavement management software. Such an example is the software developed by UK Highways Agency known as Highways Agency Visual Surveys package, suitable for flexible pavements and rigid pavements with asphalt surfacing. The data are subsequently downloaded to the Highways Agency Pavement Management System (HAPMS) (Highways Agency 2008b).

Pavement rating may be descriptive or quantitative.

In case of descriptive rating of the pavement condition, the terms *good*, *fair* and *bad* are usually used. The meaning of the above terms could be as follows:

- Good: pavement section requires no intervention
- Fair: pavement section requires some kind of maintenance or rehabilitation of the surface layer
- Bad: pavement section has structurally failed and requires rehabilitation of all bituminous layers or even reconstruction of the pavement structure

Intermediate ratings may also be used, such as 'fair to good' or 'fair to bad'. These terms provide the immediacy aspect of the intervention, namely, if the intervention is to be performed after a certain period or immediately.

In the case of quantitative rating of the pavement condition, the relevant condition index or distress level is determined.

The most well-known pavement condition index (PCI) after visual survey is the one described in ASTM D 6433 (2011). PCI values range from 0 to 100. In particular, when PCI is >85, the pavement is characterised as good; when PCI is 70 to 85, the pavement is characterised as satisfactory; when PCI is 55 to 69, the pavement is characterised as fair; when

Table 16.1 Proposed types of surface distresses to be recorded during visual inspection of flexible pavements

Code	Distress type	Code	Distress type
1	Ravelling	12	Reflection cracking
2	'Pocket' holes	13	Shrinkage cracking
3	Potholes	14	Edge cracking
4	Severe surface disintegration	15	Slippage cracking
5	Linear joint cracking	16	Shoving
6	Linear joint cracking with branching off cracks	17	Corrugation
7	Linear joint cracking with disintegration	18	Rutting
8	Wheel path linear cracking	19	Bleeding
9	Wheel path cracking with branching off cracks	20	Depression
10	Alligator cracking	21	Depression at utility cuts
11	Transverse cracking	22	Other surface distresses

PCI is 40 to 54, the pavement is characterised as poor; when PCI is 25 to 39, the pavement is characterised as very poor; when PCI is 10 to 24, the pavement is characterised as serious; and when PCI is 0 to 9, the pavement failed. More details regarding the procedure can be found in ASTM D 6433 (2011). An example for calculating the PCI can be found in Shahin (2005). It is also noted that the MicroPAVER software developed for pavement management studies (Shanin 2005) is based on PCI determination.

Other rating methods for determining pavement condition based on visual condition surveys have also been developed and reported (Chong et al. 1977; Nikolaides et al. 1992; Swiss Road Standards 1989–1992). The methods basically use coefficients of severity, frequency of distress occurrence and weighted coefficients in terms of risk hazard per type of distress.

After completion of pavement condition rating, the results are usually presented graphically on a linear scale representing the length of the road section surveyed, using different colours for each descriptive or quantitative rate.

All pavement rating methods mainly help determine the order of priority of the pavement sections for maintenance or rehabilitation.

16.2.1 Present serviceability ratio and present serviceability index concept

The first expression of the functional evaluation of the pavement was related to the 'rideability' or acceptability of the pavement by road users. The determination of the ride quality was carried out by a group of people who drove over various pavement sections of excellent to unacceptable quality level and then rated the pavement sections on a scale of 0 to 5. Such pavements were those constructed during the AASHTO experiment (AASHTO 1962; Carey and Irick 1960).

During the rating, the drivers were asked to take into account the surface irregularities along the transverse and longitudinal direction, cracking and in general any pavement distress affecting the ride quality. From the average rating obtained, the term *present serviceability ratio* (PSR) was defined, where 0 to 1 indicates a pavement in very poor condition (unacceptable), 1 to 2 denotes a pavement in poor condition, 3–4 indicates a pavement in good condition and 4 to 5 denotes a pavement in very good condition.

The subjectivity of rating the pavement condition by users in terms of PSR has led to the development of the present serviceability index (PSI). The same sections of the AASHTO study were also surveyed at the same time and physical measurements (slope variance in wheel paths, cracking, area of patching and rut depth) were carried out (AASHTO 1962). The PSR estimated from objective physical measurements was termed PSI. The equations derived to determine PSR in an objective way in terms of PSI were as follows:

For flexible pavements:

$$\text{PSI} = 5.03 - 1.91 \times \log(1 + \overline{\text{SV}}) - 0.01(C + P)^{0.5} - 1.38 \times \overline{\text{RD}^2}.$$

For rigid pavements:

$$\text{PSI} = 5.41 - 1.8 \times \log(1 + \overline{\text{SV}}) - 0.09(C + P)^{0.5},$$

where PSI is the present serviceability index; $\overline{\text{SV}}$ is the average slope variance on both wheel paths, as obtained by AASHTO profilometer (this is an expression of surface irregularities); C is the major cracking in linear feet per 1000 ft^2 of pavement area; P is the asphalt patching in square feet per 1000 ft^2 of pavement area; and RD is the average rut depth of both wheel paths based on a 4 ft straightedge in inches (this is an expression of permanent deformation).

The determination of PSI using the above equations provided the ability to quantify more objectively the pavement condition from condition surveys. Many organisations in the United States and in other countries have adopted the PSI approach for periodical evaluation of the pavement condition (road network) aiming to prioritise and organise maintenance and rehabilitation works. Others have modified the PSI equations according to their own findings and needs, for example, inclusion of International Roughness Index (IRI) measurements (Hernán de Solminihac et al. 2003).

In the United States, the PSI values for new pavements are considered to range from 4.2 to 4.7, depending on the quality of the construction. The PSI value is decreased as pavement is in use, and usually when the PSI value is below 2.0, that is, the pavement is in poor condition and at an unsatisfactory level, immediate rehabilitation should be considered. A PSI value of 2.5 is usually considered as a warning level for future pavement rehabilitation works.

PSI assessment is a useful tool for decision making with regard to maintenance, rehabilitation or even reconstruction of the pavement.

It is mentioned that the PSI concept is also used in the thickness design of new flexible pavements or asphalt overlays using the AASHTO *Guide for Design of Pavement Structures* (AASHTO 1993).

16.2.2 Structural adequacy evaluation by visual distress survey by MEPDG

The AASHTO *Mechanistic–Empirical Pavement Design Guide* (MEPDG) (AASHTO 2008) recommends an assessment of structural adequacy of flexible or rigid pavements based on visual distress survey.

Table 16.2 Distress types and levels recommended for assessing flexible pavement structural adequacy

Distress type	Highway classification	Current distress level regarded as:		
		Inadequate	Marginal	Adequate
Fatigue cracking, percent of total lane area	Interstate f/w	>20	5 to 20	<5
	Primary	>45	10 to 45	<10
	Secondary	>45	10 to 45	<10
Longitudinal cracking in wheel path (ft/mi)	Interstate f/w	>1060	265 to 1060	<265
	Primary	>2650	530 to 2650	<530
	Secondary	>2650	530 to 2650	<5
Reflection cracking, percent of total lane area	Interstate f/w	>20	5 to 20	<5
	Primary	>45	10 to 45	<10
	Secondary	>45	10 to 45	<10
Transverse cracking length (ft/mi)	Interstate f/w	>800	500 to 800	<500
	Primary	>1000	800 to 1000	<800
	Secondary	>1000	800 to 1000	<800
Rutting, mean depth, maximum between both wheel paths (in)	Interstate f/w	>0.45	0.25 to 0.45	<0.25
	Primary	>0.6	0.35 to 0.60	<0.35
	Secondary	>0.8	0.40 to 0.80	<0.40
Shoving, percent of wheel path area	Interstate f/w	>10	1 to 10	None
	Primary	>20	10 to 20	<10
	Secondary	>50	20 to 45	<20

Source: AASHTO, *Mechanistic–Empirical Pavement Design Guide (MEPDG)*, American Association of State Highway and Transportation Officials, Washington, DC: AASHTO, 2008. With permission.

Note: f/w, freeway.

Table 16.3 Distress types and levels recommended for assessing rigid pavement structural adequacy

Distress type	Highway classification	Current distress level regarded as:		
		Inadequate	Marginal	Adequate
JPCP deteriorated cracked slabs (medium- and high-severity transverse and longitudinal cracks and corner breaks) (% slabs)	Interstate f/w	>10	5 to 10	<5
	Primary	>15	8 to 15	<8
	Secondary	>20	10 to 20	<10
JRCP deteriorated cracked slabs (medium- and high-severity transverse cracks and corner breaks) (#/lane-mi)	Interstate f/w	>40	15 to 40	<15
	Primary	>50	20 to 50	<20
	Secondary	>60	25 to 60	<25
JPCP mean transverse joint/crack faulting (in)	Interstate f/w	>0.15	0.1 to 0.15	<0.10
	Primary	>0.20	0.12 to 0.20	<0.12
	Secondary	>0.30	0.15 to 0.30	<0.15
CRCP punch outs (medium and high severity) (#/lane-mi)	Interstate f/w	>10	5 to 10	<5
	Primary	>15	8 to 15	<8
	Secondary	>20	10 to 20	<10

Source: AASHTO, *Mechanistic–Empirical Pavement Design Guide (MEPDG)*, American Association of State Highway and Transportation Officials, Washington, DC: AASHTO, 2008. With permission.

Note: f/w, freeway.

The assessment relates the condition of the pavement surface as to whether the pavement is structurally adequate, marginal or inadequate. The distresses considered together with the distress levels for the pavement to be regarded as adequate, marginal or inadequate are as shown in Table 16.2 for flexible pavements and in Table 16.3 for rigid pavements.

'Adequate' implies that the surface condition or individual distresses would not trigger any major rehabilitation activity and the existing pavement has some remaining life. 'Marginal' distress level implies that the existing pavement has exhibited distress levels that do require maintenance or some type of minor repairs, and 'inadequate' implies that the pavement has distresses that require immediate major rehabilitation and has no remaining life (AASHTO 2008).

16.3 FUNCTIONAL EVALUATION BY DEVICES MEASURING SURFACE CHARACTERISTICS

The most objective assessment of functional pavement condition evaluation is conducted using purposely built devices that record the pavement's surface characteristics, in particular evenness, skid resistance and texture depth. These devices are self-mobile, mobile or portable when a short length of road pavement has to be surveyed.

The additional advantages when using self-mobile or mobile devices are as follows: much faster data collection, less traffic interruption and quicker data processing. Needless to say, the cost of mobile devices is considerably higher than that of the portable devices.

Before referring to and describing the devices used for functional evaluation of the pavement, general issues regarding pavement surface characteristics will be discussed in the following paragraph.

16.3.1 Pavement surface characteristics

The pavement surface characteristics that are closely related to the pavement functionality are the ones related to its geometric profile on the vertical level.

Depending on the magnitude of the wavelength of the surface profile, the range of each surface property is determined. Thus, the micro-texture has a wavelength up to approximately 5×10^{-4} m; the macro-texture, from 5×10^{-4} m to 5×10^{-2} m; the mega-texture, from 5×10^{-2} m to 5×10^{-1} m; and roughness, from 5×10^{-1} m to 5×10^{1} m (see Figure 16.1).

The pavement surface characteristics affect not only driving safety and comfort but also the environmental pollution (mainly noise) and the vehicle operating cost (maintenance cost, tyre replacement cost, wasted fuel cost).

A pavement surface should primarily have a good friction skid resistance (good micro- and macro-texture) for the avoidance of accidents, as well as good evenness in the transverse and longitudinal direction for comfort during driving and additional driving safety.

Additionally, the pavement surface should not generate high noise levels and should have good light reflective properties. For the latter, the micro- and macro-texture in conjunction with the water withholding at the surface are the basic affecting factors.

Figure 16.2 provides the proportional influence of the abovementioned characteristics on the safety, comfort, environment and the reduction of the operational cost.

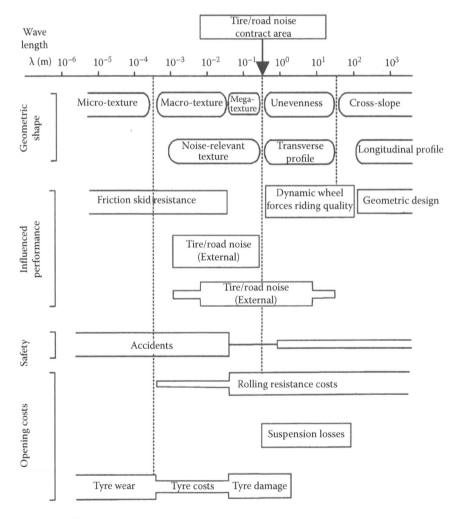

Figure 16.1 Road surface wavelength spectrum and its effects on driving safety, comfort and vehicle operational cost. (From PIARC, Surface characteristics. In: *Proceedings of the XIXth World Road Congress*, Technical Committee on the Surface Characteristics, Report 19.01B. Marrakech, 1991.)

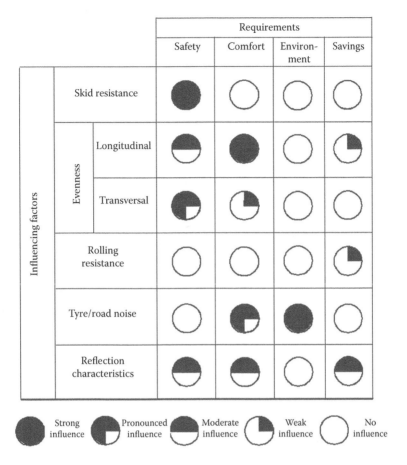

		Requirements			
		Safety	Comfort	Environ-ment	Savings
Influencing factors	Skid resistance	●	○	○	○
	Evenness — Longitudinal	◒	●	○	◔
	Evenness — Transversal	◔	◔	○	○
	Rolling resistance	○	○	○	◔
	Tyre/road noise	○	◔	●	○
	Reflection characteristics	◒	◒	○	◒

● Strong influence   ◔ Pronounced influence   ◒ Moderate influence   ◔ Weak influence   ○ No influence

Figure 16.2 Influence of surface characteristics in safety, comfort, environment and savings. (From OECD, *Road Transport Research, Road Surface Characteristics: The Interaction and Their Optimisation.* Road Transport and Intermodal Linkages Research Programme. Organisation for Economic Co-operation and Development. Paris: OECD, 1984.)

16.3.2 Surface skid resistance

The pavement surface skid resistance is directly related to the safety of the vehicles driving primarily on wet surface.

It is known that the presence of water between the tyre and the riding surface reduces the skid resistance and that skid resistance increases as the dry contact area between the tyre and the pavement increases. The micro- and macro-texture of the pavement surface primarily contributes to the latter.

At high speeds, it is more difficult for the tyre to displace the water. The contact surface becomes smaller, and in the area a (Figure 16.3), there is no contact because of the water film. In area b, the water film is locally interrupted by some surface protuberances and friction forces can develop owing to the deformation of the tyre. It is only area c where contact is still adequate since water can evacuate. The size of area c is determined by the water film, the tyre pattern and the surface texture. At the moment where contact area c is nil, the well-known dangerous phenomenon of aquaplaning occurs (OECD 1984). Greater surface texture depth increases area b by providing rapid drainage routes between the tyre and the surface.

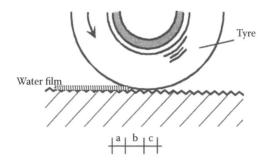

Figure 16.3 Contact areas between tyre and road surface in wet surface. (From OECD, *Road Transport Research, Road Surface Characteristics: The Interaction and Their Optimisation*. Road Transport and Intermodal Linkages Research Programme. Organisation for Economic Co-operation and Development. Paris: OECD, 1984.)

The micro-texture is related to the aggregate surface roughness, whereas the macro-texture is related to the pavement surface roughness owing to the protrusion of aggregates from the surface. The coexistence or lack of either of the above characteristics defines the surface as rough and harsh (coexistence of micro- and macro-texture), rough and polished (existence of macro-texture only), smooth and harsh (existence of micro-texture only) or smooth and polished (lack of micro- and macro-texture). The above characterisations, together with the terms *mega-texture*, *micro-texture length* and *texture depth* are explained in Figure 16.4.

The effect of the above surface texture characteristics on the skid resistance related to vehicle speed is shown in Figure 16.5. As it can be seen, as the speed increases, the surface skid resistance decreases; the decrease is more distinct when the surface lacks macro-texture (curves C and D). The best skid resistance value is provided when the surface exhibits both macro- and micro-texture (curve A) and the worst is provided when the surface lacks both macro- and micro-texture (curve D). The existence of macro-texture alone provides good skid resistance (curve C) at speeds higher than 50 km/h.

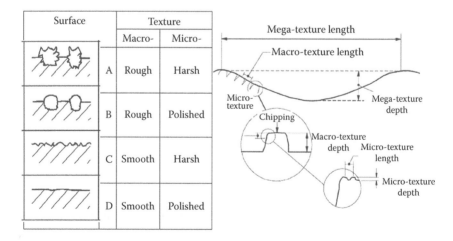

Figure 16.4 Terms and explanatory notes on surface texture. (Adapted from Sabey, B.E., *The Road Surface Texture and the Change in Skidding Resistance with Speed*. Ministry of Transport, Road Research Laboratory, Report LR 20. Harmondsworth [Crowthorne], UK: Transport Research Laboratory, 1966.)

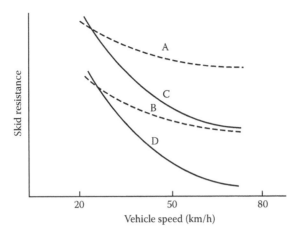

Figure 16.5 Skid resistance on wet road surfaces (measured with smooth tyre). (From Sabey, B.E., *The Road Surface Texture and the Change in Skidding Resistance with Speed*. Ministry of Transport, Road Research Laboratory, Report LR 20. Harmondsworth [Crowthorne], UK: Transport Research Laboratory, 1966.)

The surface texture (macro- and micro-texture) is expressed in terms of texture depth. Figure 16.6 shows the effect of the texture depth to the braking force coefficient (BFC) when measuring speed increased from 50 to 130 km/h (different surfacing materials were tested).

The reduction of BFC when speed increases is more noticeable in asphalts that cannot provide an initial texture depth higher than 0.8 mm, such as asphalt concrete.

Evidence similar to the effect of texture depth in skid resistance related to speed is given in Table 16.4. As it can be seen, skid resistance (SCRIM [Sideway-force Coefficient Routine

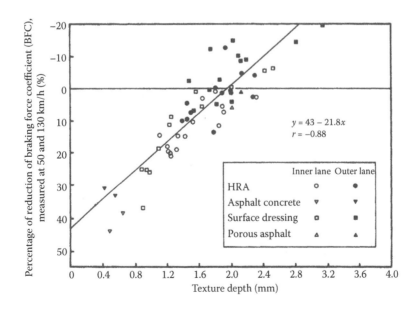

Figure 16.6 Effect of texture depth in BFC when measuring speed increased from 50 to 130 km/h. (From Jacobs, F.A., *M40 High Wycombe By-Pass: Results of a Bituminous Surface Texture Experiment*. Department of the Environment, Department of Transport. TRRL Report, LR 1065. Crowthorne, UK: Transport Research Laboratory, 1983.)

Table 16.4 Typical reduction in skid resistance (SC) compared with 20 km/h value

Speed	Texture depth (mm)		
50 km/h	40%	30%	25%
120 km/h	70%	60%	50%

Source: Highways Agency, *Design Manual for Roads and Bridges* (DMRB), *Volume 7: Pavement Design and Maintenance, Section 3, Part 1, HD 28/04*, London: Department for Transport, Highways Agency, 2004. © Highways Agency.

Investigation Machine] measurements) decreases with reduction of texture depth; the effect becomes apparent even at speeds as low as 50 km/h and is increasingly significant at higher speeds (120 km/h).

Other factors that directly affect the skid resistance are tyre pattern and tyre resilience (high or low percentage of plastic/elastic material). The smoother the tyre surface, the greater the reduction of skid resistance as vehicle speed increases. The higher the percentage of plastic incorporated in the tyre, the lesser the skid resistance on rough surfaces (Sabey 1966).

Finally, another factor affecting skid resistance variation is the seasonal factor. The pavement surface provides higher skid resistance during winter months, compared to summer months (Giles et al. 1964). This is attributed to the surface cleanness during the winter months. More details on skid resistance are provided in Hosking (1992), Hosking and Woodford (1976b) and Nikolaides (1978).

16.3.3 Measurement of surface skid resistance

Surface skid resistance is measured by self-powered mobile devices or towed mobile devices. Spot measurements or measurements over short road sections are usually carried out by the British pendulum or ether portable devices.

The mobile devices are distinguished into two categories. The first category includes devices where the skid resistance is measured by an independent wheel moving at zero angles to the direction of travel. Measurements are taken while the wheel is fully locked (blocked),

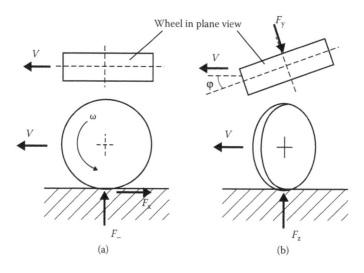

Figure 16.7 Illustration of skid resistance coefficients. (a) Braking force coefficient (BFC). (b) Side force coefficient (SFC).

hence the locked wheel-type device, or partially locked, hence the fixed slip ratio-type device (Figure 16.7a).

The second category includes devices where the independent measuring (test) wheel is at an angle to the direction of travel and partially blocked (Figure 16.7b).

In the first case, the horizontal friction force, F_x, and vertical reaction force, F_z, developed during braking are recorded (electronically). The ratio of the two forces (F_x/F_z) determines the BFC, or the friction coefficient, along the longitudinal direction (Figure 16.7a).

In the second case, the side horizontal friction force, F_y, and the vertical reaction force developed are measured (Figure 16.7b). The ratio of the two forces (F_y/F_z) determines the sideway force coefficient (SFC).

16.3.4 Mobile devices measuring skid resistance

A substantial number of mobile devices for measuring surface skid resistance have been developed globally. The majority of them have been manufactured in European Union (EU) countries. A relevant EU survey identified and reported 21 mobile devices (TYROSAFE 2008). Table 16.5 presents 19 of those devices together with their basic characteristic features.

The most popular friction devices according to the TYROSAFE report are the SCRIM (used by nine EU countries) and the GripTester, towed device (used by eight EU countries). Two EU countries use the Skiddometer BV11 and two EU countries use the SRM (Stuttgarter Reibungsmesser) or RoadSTAR (Road Surface Tester of Arsenal Research). All other devices are used by only one EU country (TYROSAFE 2008). An overview of highway and runway friction devices is also provided by NCHRP web-only document 108 (Hall et al. 2009).

Another popular device is the ASTM skid trailer (towed device), known as ASTM skiddometer.

Apart from the abovementioned devices, there are also the following skid resistance measuring devices: the Japanese Skid Tester (locked wheel type; operating speed during measurements, 30–90 km/h), DBV of NASA (locked wheel type; operating speed, usually 65 km/h), Komatsu Skid Tester (Japan) (fixed slip ratio type; between 10% and 30% slip ratio, operating speed between 30 and 60 km/h) and SRT-3 (Japan) (locked wheel type; operating speed between 30 and 90 km/h). More information is provided in NCHRP (2000).

Other devices to be mentioned are the Mu-meter (United Kingdom) (fixed slip ratio type 13% or 7.5%; operating speed between 20 and 80 km/h) and the Runway Friction Tester (United States) (fixed slip ratio type 15%; operating speed between 30 and 91 km/h). These devices have initially been developed for runway skid resistance measurements.

All devices have the ability to wet the pavement surface, creating a water film, usually approximately 0.5 mm thick, just before surface skid resistance is measured. All measurements are taken on a wet surface and readings are recorded continuously at pre-determined intervals, usually between 10 to 50 m length.

Finally, apart from the mobile devices, portable devices such as the British pendulum (SRT), the Japanese DFT (Dynamic Friction Tester) and the Swedish T2GO and VTI are in use. From the above, the British pendulum has been widely used in many countries around the world and the portable friction testers T2GO and VTI are slowly gaining popularity.

Apart from the braking system (locked wheel or partially locked [fixed slip ratio]), other factors influence the skid resistance measurements. These factors are (a) the speed of the device during measurements, (b) the tyre type of test wheel (the smooth type is the most common, or ribbed), (c) the load applied on the test wheel, (d) the amount of water applied on the surface (water film created) before the measurement and (e) the temperature of the air, tyre and pavement.

Table 16.5 Skid resistance devices used in European union countries

Name of device/vehicle	Tech. spec. CEN/TS	Type	Operating typical speed (km/h)	Method (% of slip ratio)	Load wheel (N)	Length for the mean value (m)	Country
ADHERA	15901-3 (2009)	LFC[a]-Trailer	40–120	LW[b] (100%)	2500	20	France
BV-11 or SFT	15901-12 (2011)	LFC-trailer or vehicle	70	FSR[c] (17%)	1000	20	Finland and Sweden
GripTester	15901-7 (2009)	LFC-Trailer	5–100	FSR (15%)	250	10 or 20	United Kingdom
RoadSTAR	15901-1	LFC-Vehicle	60 (min 30)	FSR (18%)	3500	50	Austria
ROAR DK	15901-5 (2009)	LFC-Trailer	60 (60 or 80)[d]	FSR (20%)	1200	min 5	Denmark
ROAR NL	15901-2 (2009)	LFC-Vehicle	70 and 50	FSR (86%)	1200	100 (min 5)	Netherlands
RWS NL	15901-9 (2009)	LFC-Trailer	70 and 50	FSR (86%)	1962	100 (min 5)	
SCRIM	15901-6 (2009)	SFC[e]-Vehicle	50	FSR (34%)	1960	10	United Kingdom
Skiddometer BV-8	15901-10 (2009)	LFC-Trailer	40, 60, 80	LW or FSR (14%)	3500	30–50	Switzerland
SKM	15901-8 (2009)	SFC-Vehicle	40, 60, 80	FSR (20%)	1960	100	Germany
SRM	15901-11 (2011)	LFC-Vehicle	40, 60, 80	LW or FSR (15%)	3500	20	Germany
TRT	15901-4 (2009)	LFC-Vehicle	40–140	FSR (25%)	1000	20	Czech Republic
IMAG	—	LFC-Vehicle	65 (up to 140)	FSR (15%)	1500	—	France
ODOLIOGRAPH	15901-13 (2011)	SFC-Vehicle	80	FSR (20%)	2700	—	Belgium
OSCAR	—	LFC-Trailer	—	FSR (18%)	4826	—	Norway
PFT (TRL)	ASTM[f]	LFC-Trailer	20–130	Almost 100%	—	—	United Kingdom

Source: Adapted from TYROSAFE, D04 Report on state-of-the-art of test methods, FEHRL, 7th Framework Programme, 2008.

[a] LFC, longitudinal friction coefficient.
[b] LW, locked wheel type (100%).
[c] FSR, fixed slip ratio type (percentage of slip).
[d] For new pavements control, V = 60 or 80 km/h.
[e] SFC, side force coefficient.
[f] ASTM E 274, locked wheel tester.

As far as the temperature is concerned, in almost all cases, it is recommended that measurements should be conducted when minimum air temperature is greater than 4°C and maximum pavement temperature is lower than 50°C.

The measurements taken by each device are not completely comparable. For this reason, the limiting values of skid resistance values set by national specifications should be differentiated for every device recommended to be used. This problem can be resolved by expressing the skid resistance value obtained with respect to the Skid Resistance Index (SRI) (see Section 16.3.6).

16.3.4.1 SCRIM

SCRIM originated from research by TRL Ltd (United Kingdom) and was developed by W.D.M. Limited. W.D.M. Limited is the sole licensed manufacturer worldwide. SCRIM is a registered trademark of W.D.M. Limited.

SCRIM consists of a purposely built vehicle equipped with a side test (measuring) smooth wheel positioned between the front and back axis, a water tank, usually 3 m³ in capacity, and a data measuring/recording/processing system (see Figure 16.8).

The measurements are taken with a standardised – not entirely blocked (34% fixed slip ratio) – smooth wheel positioned at a 20° angle to the direction of travel, loaded with a 1962 N load. The speed at which the measurements are taken is either 80 or 50 km/h (test speed). The 80 km/h test speed is used in motorways or other roads where the speed limit is greater than 80 km/h, whereas the 50 km/h test speed is used in all other cases.

Skid resistance data are recorded continuously and stored as an average, usually for each 10 m section of road; this length is called 'length for the mean value'.

The wet-road skid resistance is determined by the SFC known as SCRIM coefficient, SC. The result is expressed in two decimal points and it ranges, theoretically, from 0 to 1.00 units (the greater the value, the higher the skid resistance provided by the pavement surface).

The measurements are directly affected by the test speed and to a lesser extent by the temperature of the air or road at which the measurements are taken.

When measurements are taken at speeds other than the standard test speed, 50 km/h, the data collected are corrected to the standard test speed before use. The correction according to HD 28/04 (Highways Agency 2004) for data collected within the speed range 25 to 85 km/h is carried out using the following equation:

$$SC_{(50)} = SC_{(s)} + (s \times 2.18 \times 10^{-3} - 0.109),$$

Figure 16.8   SCRIM and test wheel assembly (wheel in raised position). (Courtesy of WDM Limited.)

where $SC_{(50)}$ is the SCRIM coefficient corrected to 50 km/h, $SC_{(s)}$ is the SCRIM coefficient measured at speed s and s is the test speed (km/h).

Studies carried out by TRL on behalf of Highways Agency (Brittain 2011) proposed the use of the following revised correction equation, when data are collected within the speed range 30 to 80 km/h:

$$SC_{50} = SC_{(s)} \times (-0.015 \times s^2 + 4.765 \times s + 799.25)/1000.$$

With respect to the temperature of the air or pavement, it has been found that under normal UK conditions, the influence of temperature is not of practical significance in comparison with other factors affecting the measurements (Highways Agency 2004). Hence, temperature correction is not necessary for surveys carried out under the conditions set out in HD 28/04 (Highways Agency 2004). One of the conditions is to take measurements within the testing period defined as the summer period (1 May to 30 September); when yearly periodic measurements are carried out, it is recommended to be carried out during the same month.

The range of taking measurements with SCRIM is determined by the capacity of the water tank equipped; with a 3 m³ capacity tank, the range of action is approximately 60 km.

Further information on SCRIM is provided in Highways Agency (2004) and Hosking and Woodford (1976a,b).

The operational procedure for determining the skid resistance of a pavement surface by SCRIM or other similar SFC measuring devices is described and should comply with the CEN technical specification CEN/TS 15901-6 (2009).

16.3.4.2 GripTester

The GripTester device was developed in the United Kingdom (by Findlay Irvine, Scotland) at the end of the 1980s and consists of a trailer carrying a standard smooth (ASTM E 524 2008) test wheel (see Figure 16.9) and a data measuring/recording/processing system. The wheel is loaded with a 250 N load.

The trailer can be towed by any vehicle that carries the water tank (of sufficient capacity) and the portable data recording/processing unit.

Skid resistance measurements are taken by the independent test wheel moving parallel to the direction of travel and with a fixed slip ratio of 15%.

The GripTester can conduct measurements with a speed within the 5 to 130 km/h range, which gives it the advantage of being used not only on road and highway surveys but also

Figure 16.9 GripTester device and field measurements. (Courtesy of Findlay Irvine Ltd.)

on airport runways as well as pedestrian areas. However, the typical measurement speed for highway surveys is 50 km/h.

The skid resistance (longitudinal friction coefficient) measured by the GripTester is called GripNumber (GN), is reported to two decimal points and ranges from 0 to 1.0.

GripNumber (GN) can be converted to SCRIM coefficient (SC) by using the conversion equation established after TRL studies (Dunford 2010). The equation derived, and recommended to supersede all previous versions, is as follows:

$$SC = 0.89 \times GN.$$

The GripTester is considerably cheaper than SCRIM and the ASTM skid trailer, has lower operating and maintenance cost and is easy to carry (length, 1010 mm; width, 790 mm; height, 510 mm; and weight, 85 kg). The abovementioned advantages and the fact that it performs measurements throughout the speed range up to 130 km/h render the GripTester as an alternative to a skid measurement device. More information on the GripTester is provided by Findlay Irvine Ltd.

The operational procedure for determining the skid resistance of a pavement surface by the GripTest is described and should comply with the CEN technical specification CEN/TS 15901-7 (2009).

16.3.4.3 BV-11 skiddometer and Saab friction tester

The BV-11 skiddometer is a towed trailer device with the measuring wheel between the two reference wheels (Figure 16.10) and was developed in Finland. Also available is the model BV-11 VI where the trailer is vehicle integrated and sets out automatically by a telescopic mechanism (Moventor Oy Inc.).

The Sarsys friction tester (SFT), which has the measuring wheel built into a Saab 9-5 vehicle (Figure 16.11) and hence is known as the Saab friction tester, was developed in Sweden.

Other models are also available, on the basis of the SFT principle, such as SVFT (the measuring wheel built into a Volvo V70 vehicle), SFTT (the measuring wheel built into a Volkswagen transporter) and STFT (the measuring wheel on a trailer) (Sarsys AB).

The measuring wheel in BV-11 and SFT devices moves parallel to the direction of travel (longitudinal friction principal coefficient) and carries a load of 1000 N (TYROSAFE 2008).

The measurements are usually taken with a constant slip ratio of 13%–17% and at a typical speed of 45 to 96 km/h.

Figure 16.10 Moventor BV-11 device. (Courtesy of Moventor Oy Inc.)

Figure 16.11 Sarsys SFT device. (Courtesy of Sarsys AB.)

Both devices are popular in friction measurements at airport runways. More information about the Moventor BV-11 skiddometer or SFT can be obtained from Moventor Oy Inc. and Sarsys AB, respectively.

The operational procedure for determining the skid resistance of a pavement surface by BV-11 or SFT is described and should comply with the CEN technical specification CEN/TS 15901-12 (2011).

16.3.4.4 RoadSTAR

The RoadSTAR device was developed in Austria device was developed in Austria (AIT, formerly known as Arsenal Research) and consists of a modified truck that carries a measuring ribbed wheel mounted at the rear of the chassis, a water tank, a storage cabinet and control equipment for data recording/processing (Figure 16.12). The wheel moves parallel to the direction of travel and is loaded with 3500 N load, and measurements are taken, usually, at a slip ratio of 18% (slip ratio may vary, if desired).

1 = Measuring wheel including braking torque measurement
2 = Pneumatic cylinder
3 = Wetting unit
4 = Pre-wetting system
5 = Gearbox
6 = Water tank
7 = Sorage cabinet
8 = Drivers cabin – digital data acquisition

Figure 16.12 RoadSTAR vehicle. (From TYROSAFE, *D04 Report on State-of-the-Art of Test Methods*, FEHRL, 7th Framework Programme, 2008.)

The measurements are usually conducted at 60 km/h speed and the longitudinal friction coefficient is the mean values over 50 m. The vehicle may also be equipped with a laser device to record surface macro-texture.

More information on RoadSTAR is provided in Maurer (2004) and TYROSAFE (2008).

The operational procedure for determining the skid resistance of a pavement surface by RoadSTAR is described and should comply with the CEN technical specification CEN/TS 15901-1 (2009).

The latest model of RoadSTAR, owned by the Austrian Institute of Technology (AIT), is equipped with laser sensors and cameras to measure other surface parameters such as transverse evenness (rut depth), longitudinal evenness, macro-texture, road layout parameters, surface distress (potholes, fretting, etc.) and road environment (AIT 2014).

16.3.4.5 SRM

The SRM was developed in Germany and consists of a specially modified truck that carries two ribbed (165 R15) test wheels at the rear of the vehicle, a storage cabinet and control equipment for data recording/processing (Figure 16.13).

The longitudinal friction coefficient is measured by locked wheel with a slip ratio of 0% (locked wheel: standard) or a slip ratio of 15% $\pm$ 1%, or ABS and a controlled speed. The control speeds typically used are 40, 50 or 60 km/h and the friction coefficient is the mean values over 20 m.

More information on the SRM vehicle is given in FKFS and TYROSAFE (2008).

The operational procedure for determining the skid resistance of a pavement surface by SRM is described and should comply with the CEN technical specification CEN/TS 15901-11 (2011).

16.3.4.6 Mu-meter

The Mu-meter is one of the first side-force friction-measuring devices developed in the United Kingdom in the late 1960s, purposely built for airport runways and taxiways; today, it can also be used in highways and other roads.

Figure 16.13 SRM of IVT ETH Zürich. (From TYROSAFE, *D04 Report on State-of-the-Art of Test Methods*, FEHRL, 7th Framework Programme, 2008; and http://www.ivt.ethz.ch.)

Figure 16.14 Mu-meter device. (Courtesy of Douglas Equipment.)

The Mu-meter consists of a three-wheeled trailer, and unlike other side-force devices, two of its wheels are measuring wheels. The third wheel provides distance measurement and helps keep the trailer operating on a straight line.

The Mu-meter measures SFC on the basis of forces developed on the two linked test wheels, each angled at 7.5° to the direction of travel; the slip ratio is fixed at approximately 13%. Figure 16.14 shows a Mu-meter device.

The operational procedure for determining the skid resistance of a pavement surface by Mu-meter is described and should comply with ASTM E 670 (2009).

16.3.4.7 ASTM locked wheel skid tester

The ASTM locked wheel skid tester or the pavement friction tester (PFT) was developed in America and is one of the first pavement surface skid resistance measuring devices.

The device consists of a trailer with two wheels towed by a suitable vehicle; one or both wheels are test wheels. The apparatus contains a transducer (or transducers), instrumentation, a water supply and proper dispensing system and actuation controls for the brake of the test wheel (usually one at time in case both are test wheels). The test wheel is equipped with a standard smooth or ribbed tyre according to ASTM E 524 (2008) or ASTM E 501 (2008), respectively.

The apparatus shall be of such a design as to provide an equal static load of 4800 ± 65 N to each test wheel.

Figure 16.15 ASTM locked wheel skid tester. (Courtesy of Dynatest International A/S.)

The water tank (usually of 1 m³ capacity) and the data recording/processing system are located inside the cabinet of the towing vehicle (Figure 16.15).

The measurements are taken with the test wheel fully locked for a short period (usually 2 s). Readings are taken at regular intervals, usually every 200 m, and at a standard test speed of 65 km/h.

The measurements can be carried out at ambient temperatures between 4°C and 40°C with the tyre inflation pressure at 165 ± 3 kPa and after 10 km warm-up drive.

The skid resistance of the pavement surface is determined from the resulting force or torque record and reported as skid number (SN), which is determined from the force required to slide the locked test tyre at a stated speed, divided by the effective wheel load and multiplied by 100.

The skid number (SN) expressed as an integer may range from 1 to 100 (the greater the value, the better the skid resistance provided by the pavement surface).

More details on the operation and calibration of the device are given in ASTM E 274 (2011) and ASTM E 2793 (2010).

16.3.4.8 Continuous reading, fixed slip skid testers

The continuous reading, fixed slip technique is similar to the abovementioned locked wheel skid tester technique but it provides a record of skid resistance along the whole length of the test track surface. It also enables averages to be obtained for specified test segments. Hence, the instantaneous friction reading is recorded and the BFC for each friction length is calculated and recorded.

The devices/equipment used are known as continuous reading, fixed-slip measuring equipment. The testing procedure and method of measurement are carried out in accordance to ASTM E 2340 (2011).

The measuring apparatus consisting of a test wheel with a smooth tyre (ASTM E 1551 2008), force or torque transducers and a watering system is usually built into a vehicle or into a trailer towed by a vehicle.

The vehicle carries an appropriate capacity water tank, usually 1000 L, for watering the surface before taking measurements. A typical vehicle with the measuring system built into the back of a vehicle is shown in Figure 16.16.

Measurements are taken on a wet surface at a constant slip ratio of usually 15% and at a standard test speed within the 20 to 80 km/h range.

More information regarding the operation of the device and how measurements are implemented is provided in the ASTM E 2340 (2011) standard.

Figure 16.16 Continuous slip friction tester. (Courtesy of Dynatest International A/S.)

16.3.5 Portable devices measuring skid resistance

16.3.5.1 British pendulum

The British pendulum was designed with the aim of providing a low-cost and portable apparatus for stationary measurements of slip/skid resistance of a surface in the field or in the laboratory. This test method provides a measure of frictional property, micro-texture in particular, of surfaces and may be used to determine the relative effects of various polishing processes on materials or material combinations.

The device consists of a pendulum arm that has a spring-loaded standard rubber slider to its end (Figure 16.17). On release of the pendulum from a horizontal position, the loss of energy as the slider assembly passes over the test surface is measured by the reduction in length of the upswing using a calibrated scale.

The arm is released from its horizontal position to swing over a contact path length of 126 ± 1 mm (for horizontal site surfaces, not for laboratory curved specimens for polished stone determination). The contact path length is obtained by adjusting the height of the pendulum arm.

Prior to the start of taking measurements, the surface is wetted with a sufficient amount of water and then the arm is released without recording the reading. This operation is repeated five times, re-wetting the surface copiously just before releasing the pendulum and recording the result each time. The average of the five values recorded is the pendulum test value (PTV). If the five values recorded differ by more than three units, further measurements are taken until three consecutive values are the same; otherwise, the test is repeated in an adjacent location.

The PTV is defined as the loss of energy as the standard rubber-coated slider assembly slides across the test surface and provides a standardised value of slip/skid resistance (CEN EN 13036-4 2011).

The PTV determined at a pavement surface temperature other than 20°C is corrected using the correction values given in Table 16.6.

It has also been proposed (Oliver 1980) that the following equation for correcting PTV (or skid resistance value [SRV] measured by British pendulum) be used when measuring at a pavement surface temperature other than 20°C:

$$SRV(PTV)_{20} = \frac{SRV_t}{\left[1 - 0.00525 \times \left(T_{0_c} - 20\right)\right]},$$

where T_{0_c} is the test temperature (°C).

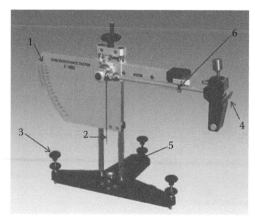

1 = Calibrated scale
2 = Pointer
3 = Levelling screw
4 = Rubber slider
5 = Spirit level
6 = Release button

Figure 16.17 British pendulum.

Table 16.6 Correction of PTV when the test is carried out at temperatures other than 20°C

Measured temperature of pavement surface (°C)	Correction to measured PTV
5	−5
10	−3
15	−2
20	0
30	+2
40	+3

Source: Reproduced from CEN EN 13036-4 (2011), *Road and airfield surface character-istics – Test methods – Part 4: Method for measurement of slip/skid resistance of a surface – The Pendulum Test*, Brussels: CEN, 2011. With permission. © CEN.

The test is performed according to CEN EN 13036-4 (2011) or ASTM E 303 (2008) standards. The PTV according to the ASTM E 303 (2008) test method is noted as the BPN (British pendulum number).

The slip/skid resistance values determined by the British pendulum do not necessarily agree or directly correlate with those obtained utilising other methods of determining friction properties or skid resistance.

To convert PTV to SCRIM coefficient (SC), the following equation may be used (NZTA 2012):

$$SC = 0.0071 \times PTV + 0.033.$$

16.3.5.2 Other portable devices

Other portable-push devices were developed for skid resistance measurements in areas with certain restrictions, such as road markings, or in general where the skid resistance measurements with mobile devices are not possible.

Two of these devices are the T2GO device and the VTI device (Figure 16.18a and b, respectively).

In both cases, the measurements are taken at walking speed as the user pushes the device on the pavement surface.

(a) (b)

Figure 16.18 Portable devices: (a) T2GO. (From TYROSAFE, *D04 Report on State-of-the-Art of Test Methods*, FEHRL, 7th Framework Programme, 2008.) (b) VTI 9. (Courtesy of VTI/Hejdlösa bilder AB.)

(a)　　　　　　　　　　　　　　　　　(b)

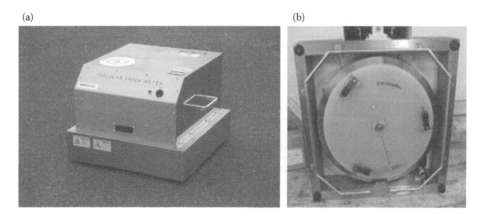

Figure 16.19 Dynamic friction tester portable device. (a) Overall view. (b) Bottom view. (Courtesy of Nippo Sangyo Co. Ltd.)

The T2GO device takes measurements with a 20% fixed slip ratio. Similarly, the VTI device, works on the fixed slip ratio principal.

More information is provided in ASFT Industries AB (2014), Bergström et al. (2003) and Wallman and Astrom (2001).

In Japan, a DFT has been developed. The DFT (Figure 16.19) is a static device operating on the rotating-head rubber-slider principle.

The DFT consists of a horizontal spinning disk fitted with three spring-loaded rubber sliders that contact the surface. A water supply unit delivers water to the surface being tested. The torque generated by the slider forces measured during the spin down is used to calculate the friction as a function of speed.

This device may be used for laboratory investigations and measurements on site. This test is performed according to ASTM E 1911 (2009).

16.3.6 Skid resistance index

The skid resistance of the pavement surface may be measured by a considerably great number of devices on the basis of different principles. This fact renders the direct comparison of the results almost impossible.

For harmonising the results obtained from all permitted skid resistance devices, the SRI has been invented. SRI is an objective estimate of skid resistance that is independent of the friction-measuring device used.

According to CEN EN 13026-2, the permitted devices for friction measurements are those that have their measuring principle and procedure described in CEN/TS 15901-1 (2009) to CEN/TS 15901-13 (2011) (Table 16.2).

For the determination of SRI, it is necessary to also measure the surface texture depth profile, according to CEN EN ISO 13473-1. This is achieved by the use of appropriate laser devices or other optical non-contact devices (high-resolution three-dimensional [3D] devices may also be used) placed at the front or at the back of a survey vehicle.

The SRI can be determined by using the following equation (CEN/TS EN 13036-2 2010):

$$SRI = B \times F \times e^{\left[(S-30)/S_0\right]},$$

where $S_0 = a \times MPD^b$; a, b, B are parameters specific to the friction-measuring device used, determined in the course of the calibration of the device; F is the measured friction coefficient at slip speed, S; S is the slip speed (km/h) and MPD is the mean profile depth of pavement surface, determined from surface profile depth measurements according to CEN EN ISO 13473-1.

In the case that the mean texture depth (MTD) is known by the volumetric technique (sand patch method) according to CEN EN 13036-1 (2010), the mean profile depth (MPD) may be estimated by the following equations:

$$MPD = (5 \times MTD - 1)/4, \text{ for } MTD > 0.2$$

or

$$MPD = 0, \text{ for } MTD < 0.2.$$

It is noted that since the precision of the equation for determining the SRI has not yet been determined, the SRI cannot yet be used for setting limit values in national specifications.

More information about the SRI is given in CEN EN 13036-2 (2010).

16.3.7 International friction index

The International Friction Index (IFI) is similar to the SRI discussed above and was developed in the PIARC international experiment to compare and harmonise texture and skid resistance measurements. The index allows for the harmonising of friction measurements with different devices that use smooth tread test tyres to a common calibrated index. The procedure for determining the IFI is covered in ASTM E 1960 (2011).

IFI comprises two parameters, the F60 and the S_p. These parameters refer to the calibrated friction index at a 60 km/h speed (F60) and the speed constant S_p, during measurements on a wet pavement surface.

The speed constant, S_p, is calculated from the MPD. The MPD is determined according to ASTM E 1845 (2009), using laser or other optical non-contact methods.

The speed constant of the pavement (S_p) with the measured friction (FRS) at some slip speed (S) is used to calculate the friction at 60 km/h (FR60), and a linear regression is used on FR60 to find the calibrated friction value at 60 km/h (F60).

The F60 and S_p are reported as IFI (F60, S_p).

The relevant equations determining the above parameters and other information are provided in ASTM E 1960 (2011).

16.3.8 Limit skid resistance values

In order to assure road safety, countries have implemented 'intervention' or 'threshold' skid resistance limit values (levels) that, when reached, would require immediate action for surface skid resistance restoration.

Certain countries have also implemented 'investigatory' or 'warning' limit values. When the skid resistance reaches the assigned level, the pavement segment is systematically monitored, so that the restoration of its skid resistance is programmed in a timely manner.

The intervention or investigatory levels depend on the road site category and its geometry and differ from country to country, even if skid resistance is monitored with the same device. Indicative limit values used in some countries are given in Table 16.7.

In the United Kingdom, skid resistance monitoring is treated with greater rigor than some other countries. UK standards (Highways Agency 2004) use only investigatory levels in

Table 16.7 Indicative warning and threshold values of skid resistance

Country	Instruments	Warning value (investigatory level)	Threshold value (immediate action limit)
Belgium	SCRIM	—	$SC_{60} = 0.4$
Canada	ASTM LWT	$SN_{40} = 40$	$SN_{40} = 30$
France	SCRIM	$SFC_{60} = 0.55$	$SC_{60} = 0.25$–0.45 (depending on the road category)
Germany	Stuttgarter Reibungsmesser-SRM	—	$\mu_{60} = 0.33$, $\mu_{50} = 0.42$ or $\mu_{40} = 0.26$, if >10% of values are lower than the threshold value
Greece	ASTM LWT	$SN_{40} = 40$[a]	$SN_{40} = 35$ dangerous sites[a] $SN_{40} = 30$ other sites[a]
Ireland	SCRIM	—	$SC_{50} = 0.5$
Italy	SCRIM	—	$SC_{60} = 0.31$–0.40 for bends, $SC_{60} = 0.28$ for straight sections
Spain	SCRIM	$SFC_{60} = 0.5$–0.4	$SC_{60} = 0.4$
Switzerland	Skiddometer BV-8	—	$\mu_{80} = 0.36$–0.32, $\mu_{60} = 0.43$–0.39 or $\mu_{40} = 0.52$–0.48, if 10%–6% of the values are less than the threshold value
United Kingdom		See Table 15.5	
United States	ASTM LWT	$SN_{40} = 35$–40	$SN_{40} = 30$

Source: OECD, *Road Transport Research, Road Surface Characteristics: The Interaction and Their Optimisation*, Road Transport and Intermodal Linkages Research Programme, Organisation for Economic Co-operation and Development. Paris: OECD, 1984; and Private Communication.

[a] Recommended by the author and used nationwide, but non-statutory.

Table 16.8 Investigatory levels of CSC values for the United Kingdom

Category	Site definition	Investigatory levels of CSC at 50 km/h							
		0.30	0.35	0.40	0.45	0.50	0.55	0.60	0.65
A	Motorway	*	†						
B	Dual carriageway non-event	*	†	†					
C	Single carriageway non-event		*	†	†				
Q	Approaches to and across minor and major junctions, approaches to roundabouts				†	†	†		
K	Approaches to pedestrian crossings and other high-risk situations					†	†		
R	Roundabout				†	†			
G1	Gradient 5%–10% longer than 50 m				†	†			
G2	Gradient >10% longer than 50 m				*	†	†		
S1	Bend radius <500 m – dual carriageway				†	†			
S2	Bend radius <500 m – single carriageway				*	†	†		

Source: Highways Agency, *Design Manual for Roads and Bridges (DMRB), Volume 7: Pavement Design and Maintenance, Section 3, Part 1, HD 28/04*, London: Department for Transport, Highways Agency, 2004. © Highways Agency.

Note: CSC, characteristic SCRIM coefficient.

Table 16.9 Minimum pendulum test values

Category	Type of site	Minimum PTV
A	Difficult sites such as: Roundabouts, bends with radius <150 m Gradients 1:20 or steeper of lengths >100 m, Approaches to traffic lights	65
B	Motorways and heavily traffic roads (>2000 commercial vehicles)	55
C	All other sites	45

Source: RRL, *Instruction for using the portable skid-resistance tester*, Road Note No. 27, Road Research Laboratory, Department of Science and Industrial Research. London: HMSO, 1969.

terms of SCRIM values. When the set investigatory level is achieved, a decision is made, depending on the possibility of accident occurrence, whether an immediate restoration of surface skid resistance is needed or a specific section such as frequent monitoring is required.

The investigatory limit values of the characteristic SCRIM coefficient (CSC) at 50 km/h per site category are given in Table 16.8. The CSC is an estimate of the underlying skid resistance once the effect of seasonal variation has been taken into account (Highways Agency 2004).

The dagger in Table 16.8 indicates the range of investigatory levels that will generally be used for road sections carrying significant traffic levels. The asterisk indicates a lower investigatory level that will be appropriate in low-risk situations, such as low traffic levels or where the risks present are well mitigated and a low incidence of accidents has been observed. Exceptionally, a higher or lower investigatory level may be assigned if justified by the observed accident record and local risk assessment (Highways Agency 2004).

It should be said that when restoration of skid resistance is decided and until works begin, it is vital that the road authorities put appropriate warning sings regarding the low skid resistance provided by the pavement surface.

Finally, when PTVs are used to justify skid resistance safety limits, the values given in Table 16.9 may be used.

16.3.9 Surface texture

The pavement surface texture, that is, surface macro-texture, can be determined in terms of MTD, using a volumetric technique (sand patch method), or in terms of MPD, using laser sensor devices.

The surface texture changes with time and may be detected nowadays by 3D high-resolution devices (laser imaging profilers) (see Section 16.3.13).

16.3.9.1 Measurement of texture depth by volumetric technique

This method was formerly known as the sand patch method. The name changed after the introduction of fine glass spheres instead of fine sand. The test procedure is carried out by CEN EN 13036-1 (2010) of ASTM E 965 (2006).

A known volume of at least 25 ml graded glass spheres (90% passing through the 0.25 mm sieve and retained on the 0.18 mm sieve) is poured on the dry and clean pavement surface. The quantity is then spread on the surface, filling the surface voids flushed with the glass spheres, to form a circular patch (Figure 16.20). A slight pressure on the hand is used, just enough to ensure that the disc will spread out the material so that the disc touches the surface aggregate particle tips.

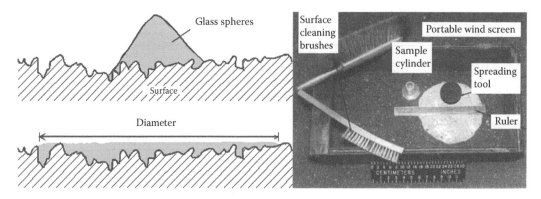

Figure 16.20 Explanation of texture depth determination and measuring device. (Adapted from CEN EN 13036-1, *Road and Airfield Surface Characteristics – Test Methods – Part 1: Measurement of Pavement Surface Macro-Texture Depth Using a Volumetric Patch Technique.* Brussels: CEN, 2010. With permission [© CEN]; and ASTM E 965, *Standard Test Method for Measuring Pavement Macro-Texture Depth Using a Volumetric Technique.* West Conshohocken, PA: ASTM International, 2006.)

The MTD is calculated by the following equation:

$$\text{MTD} = 4 \times V/(\pi \times D^2),$$

where MTD is the mean texture depth (mm), V is the volume of spheres (mm^3) and D is the average diameter of the area covered by the material (mm).

A detailed description of the method is provided in CEN EN 13036-1 (2010) or ASTM E 965 (2006).

16.3.9.1.1 Limit values of texture depth

Texture depth directly affects the skid resistance of the pavement surface. In many countries, the minimum allowed texture depth is 0.5 mm. Pavement segments with values lower than 0.5 mm should be investigated immediately, along with skid resistance measurements, and restoration of the surface skid resistance should be programmed.

Minimum surface texture depth is also required after completion of asphalt works. The minimum values required depend on the type of asphalt used in surfacing. In the case of new asphalt layer with asphalt concrete for very thin layers (AC-VTL), the minimum MTD usually required is 1.00 mm; in addition to that, no individual value should be less than 0.8 mm.

Some countries, such as the United Kingdom, require greater texture depth values after construction. According to the British specifications (Highways Agency 2008a), the minimum mean surface texture depth required after construction for a surface layer of AC-VTL, where the posted speed limit is >80 km/h, is 1.3 mm per 1000 metre section and the average of a set of 10 consecutive texture depth measurements should greater than 1.0 mm. For locations with posted speed limit <80 km/h, the above minimum values are 1.0 and 0.9 mm, respectively.

16.3.9.2 Measurement of texture profile using laser sensors

The use of laser sensor devices provides the ability of rapid and safe determination of the pavement macro-texture in terms of MPD without traffic disruption. The test procedure and determination of MPD are carried out according to CEN EN ISO 13471-1 (2004) or ASTM E 1845 (2009).

The laser devices are attached either at the front or at the rear of the vehicle and scan the pavement surface. These vehicles, apart from recording the pavement texture profile, also record other surface characteristics such as longitudinal profile, transverse profile (rut depth) (see Figure 16.30) or even surface cracking (Figure 16.36).

The laser sensor may also be placed on small portable surface texture measuring devices. Figure 16.21 shows a portable-push device measuring MPD.

Moreover, a portable stationary laser device has been developed, known as a circular track meter (CTM), which also measures the MPD (Figure 16.22). The test procedure and determination of MPD are carried out according to ASTM E 2157 (2009). The CTMeter can be used for both laboratory investigations and field measurements.

The MPD measured in accordance to CEN EN ISO 13471-1 (2004) or ASTM E 1845 (2009) may be transformed to an estimated texture depth (ETD) by using the following equation:

$$ETD = 0.2 + 0.8 \times MPD,$$

where ETD is the estimated texture depth (mm) and MPD is the mean profile depth (mm).

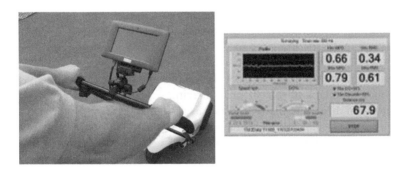

Figure 16.21 Portable-push laser device, TM2. (Courtesy of WDM Limited.)

(a)   (b)

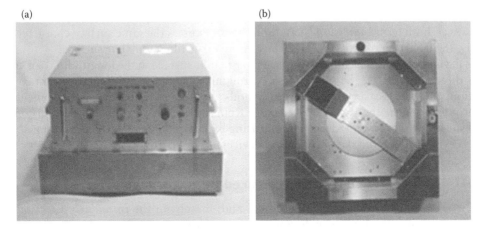

Figure 16.22 Portable CTM. (a) Front view. (b) Bottom view. (Courtesy of Nippo Sangyo Co. Ltd.)

As for the CTMeter, it has been found that the average of the MPD values for the eight segments using the CTMeter is extremely highly correlated with the MTD and can replace the volumetric measurement for determination of the MTD (ASTM E 2157 2009). The recommended relationship for the estimate of the MTD from the MPD by the CTMeter is as follows (ASTM E 2157 2009):

$$MTD = 0.947 \times MPD + 0.069,$$

where MTD is the texture depth (mm) determined by the volumetric technique and MPD is the mean profile depth (mm) determined by CTM.

It is reminded that the mean (texture) profile depth (MPD) is a necessary parameter for calculating the SRI (see Section 16.3.6).

16.3.10 Unevenness and mega-texture of pavement surface

Unevenness is defined as the deviation of a pavement surface from a true filtered planar surface in a wavelength range of 0.5 to 50 m (CEN EN 13036-6 2008; PIARC 1991).

Mega-texture is defined as the deviation of a pavement surface from a filtered true planar surface in a wavelength range of 50 to 500 mm.

The above wavelength ranges in relation to the spectrum of wavelengths of other surface characteristics are explained in Figure 16.2.

The unevenness and mega-texture affect the dynamics of the vehicle, the driving quality, the dynamic loading, the surface water drainage and the driving safety.

Unevenness and mega-texture refer to the transverse and longitudinal direction and are related to the pavement profile along these directions. Transverse and longitudinal profiles affect the vertical and lateral movement of the vehicle and create the feeling of poor driving quality, 'wobbling' and lack of safety during driving.

In the longitudinal direction, distresses such as settlement and swelling, or paving defects, present high values of wavelength, on the order of 10 to 50 m. On the other hand, corrugations, shoving and potholes present low values of wavelength, from 0.5 to 10 m. In the transverse direction, ruttings have a wavelength value of approximately 1 m.

Apart from the term *unevenness*, usually used by EN standards, the term *roughness* is also used, usually by US standards. According to ASTM E 867 (2012), travelled surface roughness is the deviations of a surface from a true planar surface with characteristic dimensions that affect vehicle dynamics, ride quality, dynamic loads and drainage, for example, longitudinal profile, transverse profile and cross slop.

However, both terms, *unevenness* and *roughness*, express the same surface characteristic, namely, the deviation of a pavement surface from a true filtered planar surface.

The deviation of the pavement from true filter planar surface in the longitudinal (or transverse) direction, that is, the unevenness or roughness in the longitudinal (or transverse) direction, could be measured by various methods and devices measuring the pavement surface profile. These methods and devices range from rod and level equipment used in conventional surveying works to modern methods using high-speed moving vehicles equipped with the necessary sensors and data acquisition devices.

16.3.11 Surface irregularities measured by static device

The surface irregularities in the longitudinal and, mainly, in the transverse direction (rut depth) may be measured by a simple static apparatus known as straightedge.

The test method is slow and labour intensive and requires the use of a straightedge and a gauge graduated to 1 mm or a calibrated wedge. The straightedge is a rectangular beam, usually 3.0 or 4.0 m long by 25 mm horizontal width, of rigid construction such that when suspended at the end points, its measurement edge shall not deviate from a true plane by more than ±0.5 mm at any point. The straightedge should also be straight along its length and should not deviate from straight by more than 1.5 mm (requirements by CEN EN 13036-7 2003).

The straightedge is placed on the road surface free from detritus and the distance between the measurement edge of the straightedge and the pavement surface is measured. The straightedge is placed perpendicular to the centreline of the road for rut depth measurements or parallel to the centreline of the road for longitudinal surface irregularities measurements. Figure 14.13 shows typical measurement of rutting by straightedge.

Measurements are made to ascertain the greatest distance between the straightedge and the surface to be measured and are recorded to the nearest 1 mm.

The test procedure is carried out by CEN EN 13036-7 (2003), ASTM E 1703 (2010) or other national specifications.

This test method is not applicable to providing information on profile or general unevenness and the rut depth value obtained may not correlate well with values obtained using other methods.

16.3.12 Profile-measuring devices

Pavement surface profile, particularly its unevenness (or roughness), was of concern to road engineers centuries ago. In the early 1900s, an Irish engineer, J. Brown, invented the first device measuring surface roughness, known as Viagraph. It was a straightedge, 12 feet long and 9 inches wide, applied continuously to the road surface along which it is drawn. Brown's device furnished virtually all the information obtained from the most modern profilometers units today, except that he measured roughness in feet instead of inches per mile (Hveem 1960).

Ever since and until the end of the 1960s, many other devices were developed, namely, the Benkelman profilometer (1922), the Hveem profilograph (1929), BPR (Bureau of Public Roads) road roughness–trailer unit device (1941), the AASHTO profilometer (1960s), CHLOE profilometers (the initials come from the names of the inventors) (1960s) and others. Details of all the devices mentioned can be found in Hveem (1960) and Less (1974).

A much greater variety of pavement profile-measuring devices were developed after the late 1960s and early 1970s from which the roughness (unevenness) of the surface can be determined. They may be distinguished into four categories according to the horizontal reference datum level used to determine roughness. These categories are (a) the true profile-measuring devices, (b) the moving reference datum level devices, (c) the dynamic relative displacement or response-type devices and (d) the dynamic inertial devices (profilometers). A dynamic device is one that executes measurements running in a normal traffic flow at a certain speed (low to high speed).

Table 16.10 summarises the above four categories.

Today, pavement surface profile or pavement unevenness (or roughness) is usually measured by rolling devices or by dynamic devices, mounted on a specially designed trailer and pulled by the vehicle or mounted on the test vehicle.

Rolling devices are used over short road sections for quality control and quality acceptance (QC/QA), while the dynamic devices are used, primarily, for pavement evenness surveys in the longitudinal and, depending on the device used, in the transverse direction.

Table 16.10 Classification of some surface roughness (evenness) measuring devices developed after the 1970s

Device category	Characteristics and typical devices
True profile devices	Measure the true surface profile of the pavement. Reference elevation is the instrument height or position. Roughness is determined in the fundamental index IRI. Typical devices: rod and level, Dipstick and walking profiler.
Moving reference datum level devices (profilographs)	Measure the surface profile in relation to the reference level of the road, which is constantly changing (test tyre constantly in contact with the road surface). Roughness is determined in number of irregularities, expressed in millimetres. Typical devices: rolling straightedge, travelling beam and profilographs such as California profilograph, Rainhart profilograph and CHLOE profilometers.
Dynamic relative displacement or RTRRMS (roughness meters)	Measure the relative vertical displacement of a wheel, or the wheel's axle, by a transducer summing up the vertical displacements at fixed distances. Surface profile cannot be determined as with other devices. Roughness is determined in relevant indices usually expressed in centimetres per kilometre, which may be converted to IRI. Typical devices: TRRL bump integrator on trailer, TRLL bump integrator on vehicle, NAASRA roughness meter, Mays ride meter and Road meter PCA.
Dynamic inertial profilometers	Measure surface profile to an inertial reference datum. The profile is determined by either a. Inertial accelerometer and at least one of three types of sensors: lasers, acoustic or infrared transducer. Non-contact determination of profile. Roughness is determined in IRI or RN or other ride quality index. Typical device: GM inertial profiler (the first inertial device built), Dynatest RSP-III, Furgo ARAN 7000, AMES 8200 high profiler, SSI-High speed profiling systems, VTI laser RST, Greenwood LaserProf, NCAT at Auburn University ARAN inertial profiler and ROMDAS profilometers. Multi-functional vehicles: Dynatest MFV, Highways Agency TRACS vehicle, Furgo ARAN 9000, etc. b. Inertial pendulum. Determination of profile by contact (a measuring wheel, where the angular displacement of the inertia pendulum axle to a horizontal level is measured). Roughness is determined in APL25 coefficient, NBO and IRI. Typical devices: Longitudinal profile analyser (APL in French) and KJ Law Profilometer (United States).

16.3.12.1 True profile-measuring devices

The rod and level equipment method is labour-intensive with respect to all other methods measuring the longitudinal (or transverse) profile and, as a result, is mainly used for either validating other profile-measuring methods or calibrating response-type roughness-measuring systems. More details on the static level method measuring roughness can be found in ASTM E 1364 (2012) and AASHTO R 40 (2010), and some details can be found in CEN EN 13036-6 (2008).

However, an auto rod and level profiling device, consisting of a rotary laser and a specially designed rod on a three-wheel push trailer, has been developed and is claimed to measure and produce an accurate and sufficiently fast pavement profile (Figure 16.23).

A purposely built patented device for obtaining pavement profile measurements with a reported accuracy of 0.01 mm (0.0004 inches), known as the Dipstick (FACE Construction Technologies Inc. 2014), has been developed (Figure 16.24a). The device consists of a high-resolution inclinometer mounted on a 0.3 m beam frame, a handle and a microcomputer mounted on the Dipstick. The Dipstick is 'walked' along the line being profiled. The operator leans the device to the leading leg and pivots it 180° along the line being profiled. The computer monitors the sensor, and when the device is stabilised, it records the change in elevation and beeps, signalling for the next step to be taken in exactly the same way. The data

Figure 16.23 Auto rod and level profiling device. (Courtesy of APR Consultants.)

(a)   (b)   (c)

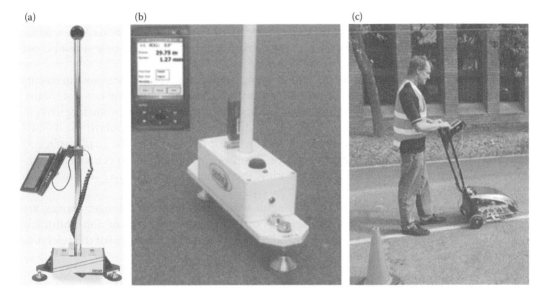

Figure 16.24 The (a) Dipstick, the (b) reference profiler (ROMDAS Z250) and the (c) walking profiler
(ARRB walking profiler G2) (FACE Construction [ROMDAS; ARRB Group Ltd.] Technologies).
(Photos courtesy of the abovementioned companies.)

are analysed by the microcomputer and the surface roughness expressed in IRI is printed.
The daily productivity can be several kilometres depending on the operator. Measurements
carried out by Dipstick comply with AASHTO R 41 (2010) and those carried out by incli-
nometers in general comply with ASTM E 2133 (2009).

A device similar to the Dipstick, known as the ROMDAS Z250 reference profiler, has
been developed (Figure 16.24b).

Another device based on the same principle as the Dipstick is also available, except that
the walking process is automated by the operator pushing the machine along a wheel path
at walking speed. The device is known as ARRB walking profiler (ARRB Group Ltd. 2014)
(Figure 16.24c). As the readings are taken at a slow but continuous walking speed, the time
taken to profile the wheel path is less than that for the Dipstick or ROMDAS Z250 devices.

Overall, the advantages of the Dipstick and walking profiles are that they are lower-cost devices, easy to operate and accurate (Class 1 resolution ASTM E 950), and can be used to calibrate other profile devices. However, they are not ideal for measuring long lengths.

16.3.12.2 Moving reference datum level devices – rolling devices

The moving reference datum level devices, known as rolling devices, are simpler to operate and the cheapest to acquire. They are pushed at a walking speed (approximately 2 km/h) and consequently have very low daily output.

The rolling straightedge was developed by TRRL UK and consists of a 3 m long and 25 cm wide aluminium frame, with rigid wheels attached at the bottom edge of both sides and a wheel at the middle that is able to move vertically (Figure 16.25a). The vertical movement of the test wheel in relation to the reference straight line, determined by the fixed wheels, can be continuously measured on a dial, visible to the operator, as the device is pushed along. More details can be found in Young (1977).

The travelling beam (Figure 16.25b) is a similar but a lighter version device to rolling straightedge. It consists of a 3 m beam with only three wheels, one rigid wheel at each end and a wheel in the middle that is able to move vertically. The vertical deviation of the surface from the reference straight line between the two fixed wheels is recorded in a special chart paper (one roll of paper can record up to 1 km).

The rolling straightedge or the travelling beam is used for compliance measurements of surface longitudinal regularity of asphalt surface and binder courses or concrete slabs. According to UK specifications (Highways Agency 2009), the number of surface irregularities should be within the relevant limits given in Table 16.11. An irregularity is defined as a variation of not less than 4 mm or not less 7 mm of the profile of the road surface as measured by the rolling straightedge set at 4 or 7 mm as appropriate or an equivalent apparatus capable of measuring irregularities within the same magnitudes over a 3 m length (Highways Agency 2009).

Devices similar to the rolling straightedge but with much larger and longer frames are known as profilographs (7 m long). They consist of 12 or more uniformly or non-uniformly spaced wheels, with an extra sensing wheel and recorder located at the centre of the reference

(a)   (b)

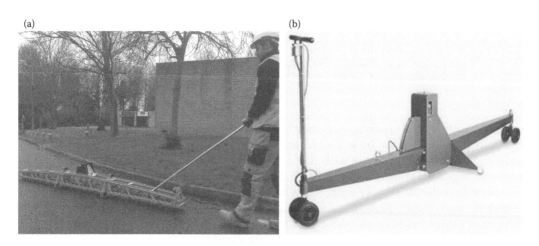

Figure 16.25 Rolling beam devices. (a) Rolling straightedge. (Courtesy of T&J Farnell Ltd.) (b) Travelling beam device. (Courtesy of Controls Srl.)

Table 16.11 Maximum permitted number of surface irregularities

	Surfaces of each lane, hard strip, hard shoulder for each irregularity limit				Surfaces of lay-bys, service areas and associated bituminous binder courses for each irregularity limit	
	Surface courses		Bituminous binder courses			
Irregularity limits	4 mm	7 mm	4 mm	7 mm	4 mm	7 mm
Length (m)	300 75	300 75	300 75	300 75	300 75	300 75
Category A[a] roads	20 9	2 1	40 18	4 2	40 18	4 2
Category B[b] roads	40 18	4 2	60 27	6 3	60 27	6 3

Source: Highways Agency, *The Manual of Contract Documents for Highway Works (MCDHW), Volume 1: Specification for Highway Works, Series 700: Road pavements – General*, London: Department for Transport, Highways Agency, 2009. © Highways Agency.

[a] Category A = motorways.
[b] Category B = dual carriageway (all purpose).

platform. Typical devices are the California profilograph (Caltrans 2012), the Rainhart profilograph (Rainhart Co. 2014) and the CHLOE profilometer (Carey et al. 1962), which is rarely used.

The surface record is analysed to determine the rate of roughness and to identify bumps that exceed a specified threshold.

When roughness is measured by a profilograph, it is defined as the height of each continuous excursion of the surface rounded to the nearest 1 mm, except those less than 0.8 mm vertically and 0.6 m longitudinally (ASTM E 867 2012).

More information about measuring pavement roughness by a profilograph can be found in ASTM E 1274 (2012).

Overall, the moving reference datum level devices are low-cost devices, are easy to operate and can easily locate bumps. However, the operating speed is, as in Dipstick and walking profilers, slow, does not provide a true profile and lacks precision.

16.3.12.3 Relative displacement or response-type devices

The relative displacement or response-type devices were the first to be introduced for measuring surface evenness at driving speed. These devices have a low purchase cost and are still in use in many countries. They measure the relative displacement of the measuring wheel, or of the vehicle's rear axle, and have the ability to rapidly record surface evenness since measurements are taken at 30 km/h or higher speeds.

A response-type device developed in the 1970s to measure the relative displacement of the test wheel was the TRRL bump integrator on a trailer (Keir 1974) (Figure 16.26).

The TRRL bump integrator was based on BPR road roughness device principles, has standard leaf springs and shock absorbers and is designed to be independent of the vertical movement of the towing vehicle. The device incorporates a unit known as the 'integrator unit' that is capable of detecting downward movement in relation to the trailer frame. The accumulated movement is measured as well as the distance travelled. The minimum wavelength of irregularities that can be measured is limited by the length of the contact area of the test wheel.

The roughness index determined by the TRRL bump integrator at a standard operating speed of 32 km/h, BI_{r32}, is the integrated vertical movements, in centimetres, divided by the distance travelled, in kilometres. The determination of BI_{r32} is usually carried every 1 km length interval; other length intervals may also be used (0.25 or 0.5 km).

Figure 16.26 TRRL bump integrator on a trailer. (From Jordan, P.G. and Young, J.C., *Developments in the Calibration and Use of the Bump-Integrator for Ride Assessment*. TRRL Supplementary Report, SR 604. Crowthorne, Berkshire, UK: Transport Research Laboratory, 1980.)

When the test speed is other than 32 km/h and within the range of 20 to 65 km/h, the following conversion equations can be used (Jordan and Young 1980):

- For uneven surfaces and operating speeds between 20 and 65 km/h and for even surfaces and operating speeds between 20 and 32 km/h, the equation is as follows:

$$BI_{r32} = \sqrt{\frac{V}{32}} \times (BI_{rv} - 47) + 47.$$

- For even surfaces and operating speeds between 32 and 65 km/h, the equation is as follows:

$$BI_{r32} = \left(\frac{V}{32}\right) \times (BI_{rv} - 47) + 47,$$

where BI_{r32} is the bump integrator index at a standard speed of 32 km/h (cm/km), V is the test speed (km/h) and BI_{rv} is the bump integrator index at test speed V (cm/km).

To improve the accuracy of measurement on short sub-sections of a road, a microprocessor-based integrator unit has been developed and has replaced the mechanical integrator fitted to the trailer of the standard bump integrator. The new equipment enabled the unevenness index of sub-sections as short as 30 m to be measured with an accuracy of better than 3%. This accuracy on short sub-sections made the bump integrator suitable for use with the CHART highway maintenance system (Still et al. 1983).

The bump integrator measuring device can also be fitted at the rear axle of a vehicle (Figure 16.27); however, measurements are dependent on the suspension system of the vehicle. Measurements are taken at a standard speed of 32 km/h, and the downward movement of the axle is detected with reference to the chassis of the vehicle. The roughness index, BI_{r32-v}, is determined in exactly the same way as described above.

The dynamic properties of each vehicle are unique and also change with time (springs and shock absorbers wear). It is therefore essential that the roughness values obtained from a response-type road roughness meter, particularly bump integrator fitted to a vehicle, are converted to units of IRI by regularly calibrating it with the MERLIN device (Figure 16.28) or

(a)

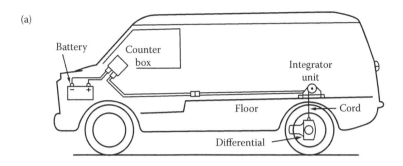

(b)   (c)

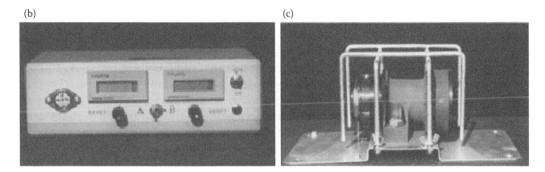

Figure 16.27 TRRL bump integrator mounted on a vehicle. (a) Test vehicle. (b) Counter box. (c) Integrator unit (Highway Engineering Laboratory of AUTh devices bought from CNS Farnell Ltd. 2014).

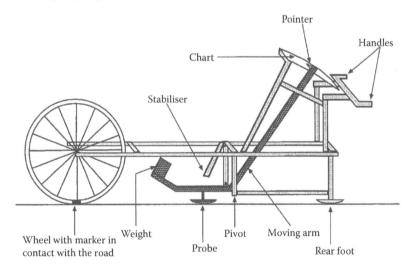

Figure 16.28 MERLIN Mark 2. (From Cundill, *The MERLIN Road Roughness Machine: User Guide, TRL Report 229*. Crowthorne, UK: Transport Research Laboratory, 1996.)

other profilometric method. The calibration procedure is described, together with other operational information, in TRRL RR 301 (Cundill 1991), Mrawira and Haas (1996) and TRL Report 229 (Cundill 1996). Relevant information can also be found in TRL DFID (1999).

MERLIN is considered as a low-cost instrument that evaluates pavement roughness on the basis of the measurement of longitudinal deformities of pavement surface and its results are correlated to the IRI.

The raw roughness data collected by the bump integrator on a vehicle when fed to a data acquisition system, such as in the ROMDAS bump integrator, manufactured by Data Collection Ltd., can be converted into a calibrated roughness index such as IRI using user-supplied roughness equations, which are determined during roughness calibration (ROMDAS).

Other response-type roughness measuring meters developed are as follows: the Mays ride meter (Walker and Hudson 1973) and the Road meter PCA (Brokaw 1973), both developed in the United States, as well as the NAASRA roughness meter developed in Australia (Priem 1989; Scala and Potter 1977). In all devices, the relative displacement between the rear axle and the body of the vehicle is measured by a mechanical device mounted on the vehicle. The roughness meters basically differ on the way the measured displacement is weighted in the accumulation process.

Measurements of surface roughness from all the abovementioned devices are calibrated so that they are expressed in IRI values.

Relevant US and EN standards for response-type devices measuring surface roughness are ASTM E 1082 (2012) (for measurement of vehicular response), ASTM E 1215 (2012) (trailers used), ASTM E 1448 (2009) (for calibration of systems) and CEN EN 13036-6 (2008).

Overall, the response-type road roughness measuring system (RTRRMS) consists of relatively low-cost devices, is relatively easy to operate, is reasonably accurate and operates at normal traffic speed. However, roughness measurements are affected by where the system is mounted (vehicle or trailer), do not provide surface profile and need calibration certainly over a range of speeds.

16.3.12.4 Dynamic inertial profilometers

Dynamic inertial profilometers are the most common devices used nowadays in road network surveys to determine pavement surface roughness. The inertial profiling system is mounted on a vehicle; hence, they are called high-speed profilometers.

All inertial profilometers available today, except one, the longitudinal pavement analyser, APL, use accelerometers, which provide the horizontal plane of reference in a non-contact manner.

16.3.12.4.1 Inertial accelerometer profilometers

The accelerometer is a transducer designed to measure vertical acceleration and used to establish the inertial reference and effectively the response of the vehicle to the road surface (instant height of accelerometer in the host vehicle). The vertical distance between the accelerometer and the travelled surface is measured by displacement transducers or sensors (laser, acoustic or infrared sensors) and the distance travelled by distance transducers. Figure 16.29 shows an accelerometer and a single sensor mid-mount on a vehicle.

The accelerometer's and sensor's signals are all combined by computer software and produce the longitudinal profile of the travelled surface.

The great majority of inertial (accelerometer) profilometers use laser sensors rather than acoustic or infrared sensors. A study has reported that 44 out of 51 agencies in the United States used laser profilometers; 3 agencies used profilometers with acoustic sensors and 4 used infrared profilometers (McGhee 2004). Ultrasonic sensors have also been used, but they are not as popular owing to the disadvantage of not being sensitive to high-speed measurements.

One laser sensor determines the longitudinal pavement profile in one location (path), whereas any number may be used to capture several longitudinal locations and hence the transverse profile may be determined as well. The laser sensors are usually mounted at the front of the vehicle.

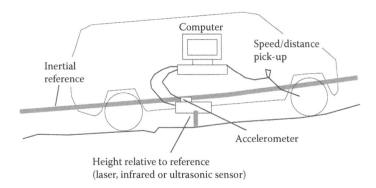

Figure 16.29 The principal of inertial reference profiler. (From Sayers, M.W. and Karamihas, S.M., *The Little Book of Profiling*. Michigan: The Regent of the University of Michigan, 1998. Courtesy of the University of Michigan Transportation Research Institute.)

Today, most laser profilometer devices determine not only the longitudinal profile and roughness indices but also other surface characteristics such as transverse profile, macro-texture, crossfall, gradient or radius of curvature. The availability depends on the manu-facturer and the model purchased. There are a number of high-speed profilometers on the market. A sample of the devices, also known as road surface profilometers (RSPs) or road surface testers (RSTs), is shown in Figure 16.30.

Apart from laser road surface profilometers, there are vehicles equipped with additional sensors and devices to determine other surface characteristics, such as laser rut measure-ments and forward-facing and backward-facing video record of the road being surveyed. These vehicles are known as integrated road analysers; Figure 16.31a shows the TRAffic speed Condition Survey (TRACS) vehicle used by surveys carried out in the United Kingdom under the Highways Agency contract, and the ARAN 9000 integrated vehicle is shown in Figure 16.31b.

16.3.12.4.2 Inertial pendulum profilometer

The high-speed profilometer that uses *inertial pendulum* for the determination, by contact, of the horizontal plane of reference has been developed by Laboratoire Central des Ponts et Chaussees (LCPC, France). It is known as the APL profilometer (lengthwise profile analyser in French). The device is a specially designed trailer with a combination of instruments and built-in mechanical properties that allow profile measurement.

In particular, the APL profilometer device usually includes two single-wheel trailers towed at constant speed by a vehicle and employs a data acquisition system. A ballasted chassis supports an oscillating beam, holding a feeler wheel that is kept in permanent contact with the pavement by a suspension and damping system. Vertical movements of the wheel(s) resulting in angular travel of the beam are measured with respect to the horizontal arm of an inertial pendulum, independent of the movements of the towing vehicle.

Figure 16.32 shows in detail the APL system and Figure 16.33 shows an APL profilometer on the road.

The test speed is normally 72 km/h and pavement evenness may be expressed in CAPL25 index, wavelength band notes (NBO) index (typical indices used in France), IRI or other indices. More information can also be found in Lucas and Viano (1979).

The APL trailer is also used by the Centre for Road Research in Belgium; however, even-ness for Belgian use is expressed in CP index.

(a)

(b)

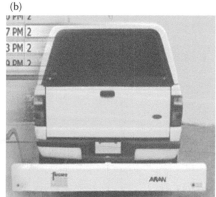

(c)

(d)

Figure 16.30 Some RSPs. (a) Dynatest RST-II (Dynatest International A/S). (b) Furgo ARAN 7000 (Furgo Roadware). (c) Ames 8200 high profiler (Ames Engineering Inc.). (d) VTI laser RST (VTI/ Hejdlösa bilder AB). (Photos courtesy of the abovementioned companies.)

However, a new high-speed vehicle system for measuring longitudinal evenness in France has been reported. The vehicle consists of laser sensors, an accelerometer and a gyroscope. The vehicle is called MuLtiProfilometer, and information can be found in Martin et al. (2012).

Relevant standards to the abovementioned vehicles, depending on their measuring capabilities, are as follows: CEN EN 13036-6 (2008) for longitudinal and transverse surface profile, ASTM E 950 for longitudinal profile, AASHTO M 328 (2010) for inertial profilers,

(a)   (b)

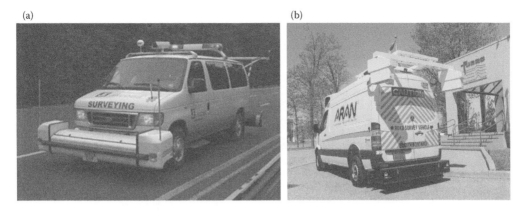

Figure 16.31 (a) TRACS vehicle (Highways Agency) and (b) ARAN 9000 integrated vehicle (Furgo Roadware). (Photos courtesy of the abovementioned organisation and company.)

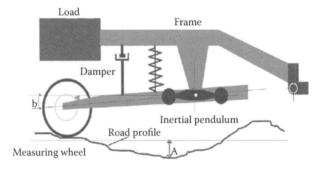

Figure 16.32 Longitudinal profile analyser – APL. (From Imine et al., road profile input estimation in vehicle dynamics simulation. *Vehicle System Dynamics*, Vol. 44, No. 4, Taylor & Francis, 2006.)

Figure 16.33 APL profilometer. (Courtesy of Vectra France.)

(a)

(b)

Figure 16.34 Lightweight profilometers. (a) Ames 6200 lightweight profiler (Ames Engineering Inc.). (b) SSI CS8700 lightweight profiler (Surface Systems & Instruments Inc.). (Photos courtesy of the abovementioned companies.)

AASHTO R 57 (2010) for operating inertial profiling systems, AASHTO R 56 (2010) for certification of inertial profiling systems, CEN EN 13036-8 (2008) for transverse unevenness (rut depth, etc.), CEN EN ISO 13473-1 for mean macro-texture profile depth, ASTM E 1926 (2008) for the determination of IRI and ASTM E 1489 (2008) for the determination of the ride number (RN).

An inventory of high-speed road evenness (longitudinal and transverse) measuring equipment in Europe has been carried out under the FILTER (FEHRL Investigation on Longitudinal and Transverse Evenness of Roads) project and reported by Descornet (1988). The models listed have been changed ever since, as well as their capability and accuracy of measurements; however, there are still a great number of devices available on the world market.

Overall, the inertial high-speed profilometers measure true road surface profile, operate at high traffic speeds, have very good repeatability and are the most suitable to network roughness surveys. However, they are also the most costly to purchase.

16.3.12.4.3 Lightweight inertial profilometers

The need for having a contactless device for quality control and quality acceptance of newly constructed pavements or surfaces led to the development of lightweight profilometers. These devices use the same principle as the inertial accelerometer-type vehicle profilometers, but the system is mounted on small vehicles (golf carts) (see Figure 16.34).

The lightweight profilometers cost less than high-speed profilometers, measure true profile, are flexible to use and are ideal for quality control measurements during paving operations. However, they require certain traffic control during operation, lack reproducibility and are not really suitable for network roughness surveys. Some relevant information can be found in Mondal et al. (2000).

16.3.13 3D laser imaging profilers

More than 10 years ago, the two-dimensional (2D) laser imaging technology was introduced to pavement surface analysis, particularly for cracking surveying (Wang 2000; Wang

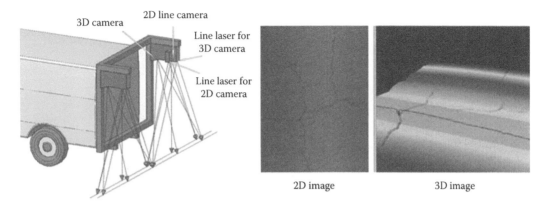

2D image   3D image

Figure 16.35 Schematic representation of 3D laser image system and comparison of 2D and 3D images. (From Wang, K., Prototyping Automated Distress Survey Based on 2D to 3D Laser Images. In: *Proceedings of the 5th International Conference 'Bituminous Mixtures & Pavements'*. Thessaloniki, Greece, 2011.)

and Gong 2005). The evolution of 2D/3D laser imaging technique during the last 5 years offered the development of 3D laser profiles that can accurately survey pavement surface for cracking and ravelling at high traffic speeds.

The 3D laser profilers consist of high-speed cameras and line lasers to obtain both 2D images and high-resolution 3D road profile. The data collected are analysed with specially designed software to detect and analyse cracks, rutting, ravelling, macro-texture and other surface defects such as potholes, patches and depressions. The process is known as image processing and pattern recognition.

Figure 16.35 shows a schematic representation of 3D laser image system and the difference between 2D and 3D images of cracking with a resolution of 1 mm.

The 3D laser image system is mounted at the back of a vehicle that usually carries an inertial profilometer at the front. Figure 16.36 shows a sample of three different 3D laser profilers.

It has been demonstrated that the 3D data may also be used for macro-texture measurement (determination of MPD) (Wang et al. 2011).

The Dynatest Multi-Functional Vehicle (MFV) combines the functionality of the RSP with the Laser Rut Measurement System, the Laser Road Imaging System or the Laser Crack Measurement System from Pavemetrics. The MFV measures the IRI/RN, longitudinal and transverse profile, macro-texture, geometrics (crossfall, gradient and radius of curvature) plus 2D or 3D pavement imagery and provides photo logging by up to five right-of-way cameras. The MFV brings safety to the forefront, allowing surveys of roads and airports to be performed from a vehicle at normal traffic speeds, day or night.

16.3.14 Highways agency road research information system (HARRIS-2)

HARRIS-2 is a compact vehicle, developed by TRL for the Highways Agency, capable of carrying out pavement condition assessment surveys at traffic speed over the full range of conditions encountered on the network.

HARRIS-2 employs high-resolution systems for the measurement of road shape and visual condition. Pavement shape measurements are provided by a scanning laser measurement system that offers unparalleled capability for measurement of transverse unevenness, such as rutting and longitudinal deformation. HARRIS-2 also provides a state-of-the-art image collection system that provides high-speed/high-resolution linescan cameras in conjunction

(a)

(b)

(c)

Figure 16.36 Pavement surface distress survey vehicles based on 3D laser imaging technology. (a) 3D laser image system and digital highway data vehicle. (Courtesy of WayLink Systems Co.) (b) Multifunctional vehicle (MFV). (Courtesy of Dynatest.) (c) ICC's 3D profiler. (Courtesy of International Cybernetics Corp.)

Figure 16.37 HARRIS 2. (Courtesy of TRL UK.)

with high-power laser line projectors to image 4 m transverse road sections with 1 mm resolution. Highly accurate location measurements are assured by the differential GPS system combined with a survey grade inertial measurement system (TRL website). Figure 16.37 shows a HARRIS 2 vehicle.

More information can be obtained from Transport Research Laboratory UK (http://www.trl.co.uk/facilities).

16.3.15 Roughness indices

There are several roughness indices, all based on specific profilers, measuring different ranges of wavelengths. Hence, each of them has a different correlation with highway user response.

The most common roughness indices developed and still in use by some agencies are as follows: the RN, computed from inertial measuring devices; the Profile Index (PI), derived from profilograph output (California profilograph or Rainhart profilograph); the bump integrator index, derived from response-type meter trailer-mounted data (TRRL bump integrator trailer [BI_{r32}]); the bump integrator index, derived from response-type meter vehicle-mounted data (TRRL bump integrator vehicle mounted [BI_{r32-v}], NAASRA index [Priem 1989] or ROMDAS index); and the French CAPL25 index computed from inertial pendulum trailer-type meter.

Other roughness indices have also been introduced, such as moving average, power spectral density, half roughness index, Mays response meter, root-mean-square acceleration, root-mean-square vertical acceleration, Texas MO and Brazil QI, to name a few (Sayers and Karamihas 1996).

However, the unique, internationally recognised roughness indicator is the IRI.

The RN is a profile index intended to indicate rideability on scale similar to PSI and was developed when IRI was not well known (Sayers and Karamihas 1996).

In particular, RN is the rideability index of a pavement using a scale of 0 to 5, with 5 being perfect and 0 being impassable, and a rideability index is an index derived from controlled measurements of longitudinal profile in the wheel tracks and correlated with panel ratings of rideability (ASTM E 867 2012).

The mathematical procedure to calculate RN was developed by NCHRP Report 275 (Janoff et al. 1985) and the software for computing the RN from mean panel rating was developed by Sayers and Karamihas (1996). Details on the RN rating scale, the intermediate points and the equation to compute RN from inertial profile-measuring devices can be found in ASTM E 1489 (2008).

The PI is derived from the profilograph output (California profilograph or Rainhart profilograph). PI is expressed in units of millimetres per kilometre (or inches per mile) and represents the total accumulated excursion of the strip chart trace (which is the profilograph output) beyond a tolerance zone (blanking band) (Mondal et al. 2000).

Some details on the roughness *indices based on bump integrators* can be found in Section 16.3.12.3, and for the French CAPL25 index, see Section 16.3.12.4.

16.3.15.1 International roughness index

The IRI is defined as an index computed from a longitudinal profile measurement using a reference mathematical RTRRMS (a quarter-car simulation) for a standard simulation speed of 80 km/h (ASTM E 867 2012; Sayers 1986b).

The IRI and its algorithm to be computed from profile data were the outcome of the International Road Roughness Experiment conducted in Brazil and are described in detail in the technical reports by Sayers et al. (1986a,b).

The World Bank experiments have validated that IRI can be measured by an extensive range of equipment including rod and level, Mays meter cars, NAASRA car, French APL, BPR roughometer, TRRL bump integrator, TRRL beam, Swedish Road Surface Tester, ARAN, Face Dipstick, GMR-type profilometers, South Dakota-type profiling system and others. For each type of hardware, required procedures have been defined along with the level of accuracy to be expected, and these principles have been tested throughout the world. The rod and level method for measuring profiles from which to calculate IRI is now reduced to the standard method. Thus, the IRI allows data from different devices/instruments and different countries to be directly compared and enables historical trends to be determined with confidence.

IRI can be interpreted as the output of a response-type measuring system where the physical vehicle and instrumentation are replaced with a mathematical model. The units of slope correspond to accumulated suspension motions (e.g. metres), divided by the distance travelled (e.g. kilometres).

The IRI is portable in that it can be obtained from longitudinal profiles obtained with a variety of instruments. The IRI is stable with time because true IRI is based on the concept of a true longitudinal profile, rather than the physical properties of a particular type of instrument (ASTM E 1926 2008). The instruments used for the collection of roughness data are characterised into four classes as defined by Sayers et al. (1986b).

When profiles are measured simultaneously for both travelled wheel tracks, then the mean roughness index (MRI) is considered to be a better measure of road surface roughness than the IRI for either wheel track. The MRI scale is identical to the IRI scale. Relevant guidance on the calculation of IRI and MRI can be found in ASTM E 1926 (2008) or AASHTO R 43 (2013).

The IRI scale starts at zero for a road with no roughness and covers positive numbers that increase in proportion to roughness. Sayers et al. (1986b) provided a series of descriptors for selected levels on the roughness (IRI) scale, applied to paved roads with asphalt concrete or surface treatment surfacing (chipseal or surface dressings) and for unpaved roads with gravel or earth surfaces. For example, for paved roads with asphalt concrete or surface treatment surfacing:

When roughness ranges from 1.5 to 2.5 (m/km IRI): Ride is comfortable over 120 km/h. Undulation is barely perceptible at 80 km/h in the range 1.3 to 1.8. No depressions, potholes or corrugations are noticeable; depressions <2 mm/3 m. Typical high-quality asphalt surface, 1.4 to 2.3, and high-quality surface treatment, 2.0 to 3.0.

When roughness ranges from 4.0 to 5.3 (m/km IRI): Ride comfortable up to 100–120 km/h. At 80 km/h, moderately perceptible movements or large undulations may be felt. Defective surface: occasional depressions, patches or potholes (e.g. 5–15 mm/3 m or 10–20 mm/5 m with frequency 1–2 per 50 m) or many shallow potholes (e.g. on surface treatment showing extensive ravelling). Surface without defects: moderate corrugations or large undulations.

When roughness ranges from 7.0 to 8.0 (m/km IRI): Ride comfortable up to 70–90 km/h, strongly perceptible movements and swaying. Usually associated with defects: frequent moderate and uneven depressions or patches (e.g. 15–20 mm/3 m or 20–40 mm/5 m with frequency 3–5 per 50 m) or occasional potholes (e.g. 1–3 per 50 m). Surface without defects: strong undulations or corrugations.

When roughness ranges from 9.0 to 10.0 (m/km IRI): Ride comfortable up to 50–60 km/h, frequent sharp movements or swaying. Associated with severe defects: frequent deep and uneven depressions and patches (e.g. 20–40 mm/3 m or 40–80 mm/5 m with frequency 3–5 per 50 m) or frequent potholes (e.g. 4–6 per 50 m).

When roughness ranges from 11.0 to 12.0 (m/km IRI): Necessary to reduce velocity below 50 km/h; many deep depressions, potholes and severe disintegration (e.g. 40–80 mm deep with frequency 8–16 per 50 m).

Road roughness estimation scale for unpaved roads with gravel or earth surfaces can be found in Sayers et al. (1986b) or ASTM E 1926 (2008).

In addition to the above, Figure 16.38 shows IRI ranges represented by different classes of roads.

Acceptable IRI values are influenced by the normal traffic speed on the facility, since roughness is less acceptable at higher traffic speeds. The US Federal Highway Administration (FHWA), for example, uses an IRI value of 2.68 m/km (170 in/mi) as the dividing line between good and bad interstate pavements. Higher IRI values would typically be allowed on non-interstate pavements with lower traffic speed (AASHTO 2012).

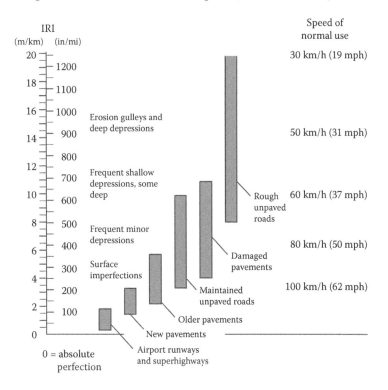

Figure 16.38 IRI ranges represented by different classes of roads. (From Sayers, M.W. and Karamihas, S.M., *The Little Book of Profiling*. Michigan: The Regent of the University of Michigan, 1998. Courtesy of the University of Michigan Transportation Research Institute.)

Countries or states around the world have set their own rideability QC/QA limits for newly constructed pavements in terms of IRI or other roughness indices. Some have, rightly, also introduced payment adjustment procedures so that projects receive bonuses or penalties. Relevant information for the latter can be found in references such as Baus and Hong (2004), Bowman et al. (2002) and others.

Rideability threshold limits in terms of roughness have also been set for maintenance rehabilitation decision-making works, which slightly vary from country to country. The reader is advised to follow the relevant national standard.

Finally, when IRI needs to be converted to PSR, the following equation may be used (Al-Omari and Darter 1994):

$$PSR = 5 \times e^{(-0.26 \times IRI)},$$

where IRI is in millimetres per metre (or metres per kilometre).

16.3.15.2 Prediction of evenness in terms of IRI

A lot of work has been carried out over the last 30 or more years regarding the prediction of evenness of flexible and rigid pavements. One of the first IRI prediction models developed with good reliability was the Paterson model (Paterson 1987).

From all prediction models developed, only the AASHTO MEPDG (AASHTO 2008) prediction model for new asphalt pavement (HMA) and asphalt (HMA) overlays of flexible pavement is included in this book as an example.

The AASHTO MEPDG prediction model for evenness (smoothness) degradation of new HMA pavements and HMA overlays of flexible pavements is as follows:

$$IRI = IRI_0 + 0.0150 \times (SF) + 0.400 \times (FC_{Total}) + 0.0080 \times (TC) + 40 \times (RD),$$

where IRI is the predicted IRI (in/mile), IRI_0 is the initial IRI after construction (in/mile), SF is the site factor (see equation below), FC_{Total} is the area of fatigue cracking (combined alligator, longitudinal and reflection cracking in the wheel path) (percent of total area), TC is the length of transverse cracking (including the reflection of transverse cracks in existing HMA pavement) (ft/mile) and RD is the average rut depth (in).

The site factor is calculated in accordance with the following equation:

$$SF = Age \times [0.02003 \times (PI + 1) + 0.00794 \times (P_{recip} + 1) + 0.000636 \times (FI + 1)],$$

where Age is the pavement age (years), PI is the plasticity index of the soil (%), FI is the average annual freezing index (°F days) and P_{rcip} is the average annual precipitation or rainfall (in).

Similar predictive equations for smoothness degradation have also been developed for HMA overlays of rigid pavements, jointed reinforced concrete pavements (JRCPs) and continuously reinforced concrete pavements (CRCPs), which can be found in the MEPDG manual (AASHTO 2008).

16.3.16 TRACS vehicle for pavement surface condition evaluation

The TRACS vehicle (Figure 16.31) is used for surveys carried out under UK Highways Agency to evaluate the pavement's surface condition.

The TRACS survey vehicle measures (Highways Agency 2008b) the following:

- Texture profile in the nearside wheel track at approximately 1 mm longitudinal intervals
- Transverse profile across a 3.2 m width at approximately 0.15 m longitudinal intervals
- Longitudinal profile in the nearside wheel track at approximately 0.1 m longitudinal intervals
- Cracking over a width of 3.2 m (continuous monitoring)
- Vehicle geographical position (Northing, Easting and altitude) as well as road geometry (gradient, crossfall and curvature) at discrete 5 m intervals
- A forward-facing video record of the road being surveyed
- A downward-facing video record of the road being surveyed

All data collected by the TRACS survey vehicle are referenced to the network sections to a longitudinal accuracy of ±1 m, processed by specially designed software and fed, if required, to HAPMS.

The pavement surface condition is assessed in terms of riding quality, rut depth, texture depth and surface cracking, and, if hot rolled asphalt surfacing, fretting.

The assessment of ride quality or profile unevenness is carried out using three Enhanced Longitudinal Profile Variance values that indicate the level of profile unevenness within wavelength ranges less than or equal to 3, 10 and 30 m.

The rut depth (in millimetres) is measured on each wheel track every 10 m, from which the average rut depth and then the maximum rut depth is determined.

The texture depth (in millimetres) is calculated using the Sensor Measured Texture Depth method.

Threshold values for ride quality per category of road are given in Table 16.12, and the threshold values for rut depth and texture depth for all road categories are given in Table 16.13.

All the threshold values and guidance levels are based on characteristic values associated with 100 m lengths and are for the assessment of TRACS data collected from in-service roads, as opposed to newly constructed roads.

The definition of condition categories 1 to 4 is as follows:

Category 1: Sound – no visible deterioration.

Category 2: Some deterioration – lower level of concern. The deterioration is not serious and more detailed (project level) investigations are not needed unless they extend over long periods; several parameters at this category are at isolated positions.

Category 3: Moderate deterioration – warning level of concern. The deterioration is becoming serious and needs to be investigated. Priorities for more detailed (scheme level) investigations depend on the extent and values of the condition parameters.

Category 4: Severe deterioration – intervention level of concern. This condition should not occur very frequently on the motorway and the all-purpose trunk road network as earlier maintenance should have prevented this state from being reached. At this level of deterioration, more detailed (scheme level) investigations should be carried out on the deteriorated lengths at the earliest opportunity and action should be taken if and as appropriate.

More details as well as details on surface cracking and fretting measurements/evaluation can be found in Highways Agency (2008b).

Table 16.12 Ride quality criteria of all types of construction for TRACS measurements (enhanced longitudinal profile variance [E-LPV], 100 m lengths)

Road classification	Threshold value						
	Condition category						
	1	↓	*2*	↓	*3*	↓	*4*
Motorways and rural dual carriageways							
E-LPV (mm²)							
3 m		0.7		2.2		4.4	
10 m		1.6		6.5		14.7	
30 m		22		66		110	
Urban dual carriageways							
E-LPV (mm²)							
3 m		0.8		2.2		5.5	
10 m		2.8		8.6		22.8	
30 m		30		75		121	
Rural single carriageway roads							
E-LPV (mm²)							
3 m		0.8		2.2		5.5	
10 m		2.8		8.6		22.8	
30 m		30		75		121	
Urban single carriageway roads							
E-LPV (mm²)							
3 m		1.4		3.8		9.3	
10 m		6.1		18.3		36.6	
30 m		48		97		193	

Source: Highways Agency, *Design Manual for Roads and Bridges (DMRB), Volume 7: Pavement Design and Maintenance, Section 3, Part 2, HD 29/08*, London: Department for Transport, Highways Agency, 2008b. © Highways Agency.

Table 16.13 Rutting and texture depth criteria for TRACS measurements (100 m lengths)

	Threshold value						
	Condition category						
	1	↓	*2*	↓	*3*	↓	*4*
Maximum rut (mm)		11		10		20	
Texture depth (mm)		1.1		0.8		0.4	

Source: Highways Agency, *Design Manual for Roads and Bridges (DMRB), Volume 7: Pavement Design and Maintenance, Section 3, Part 2, HD 29/08*, London: Department for Transport, Highways Agency, 2008b. © Highways Agency.

16.4 STRUCTURAL EVALUATION OF PAVEMENTS

The structural evaluation of pavements is conducted primarily by suitably designed devices and is a non-destructive procedure. However, destructive tests are often necessary to be carried out to confirm the results of the non-destructive structural evaluation of the pavement. On a project level, destructive testing alone may be sufficient for pavement structural evaluation.

The destructive tests comprise core extraction and trench opening. Core extraction is necessary for the determination of (a) the thickness of the asphalt layers, (b) the degree of compaction, (c) the composition of the asphalt, (d) the mechanical properties of the asphalt and (e) the extension/initiation of cracking within the asphalt layers.

Trench opening down to the formation level is necessary for the determination of (a) the strength of the subgrade, (b) the strength and stiffness of all layers beneath the asphalt layers, (c) the thickness and degree of compaction of the layers, (d) moisture content, (e) particle gradation, (f) plasticity and other properties of the constituent materials of the unbound layers, (g) crack propagation and (h) layer deformation.

The devices used in the non-destructive structural evaluation of the pavements are, in general, instruments imposing a load to the pavement surface and measuring the oncoming surface deflection; so-called deflection measuring devices. The magnitude of the deflection is an indicator of the pavement capacity to withstand further traffic loading. The greater the oncoming deflection, the lesser the structural capacity of the pavement to withstand more loads. The surface deflection measured is practically the sum of all layer deflections, including the subgrade, caused by test loading.

The structural evaluation of the pavement, after correlating the oncoming deflection at a given type and magnitude load with the equivalent standard axle loading (ESAL), is expressed in remaining life in terms of ESAL. Furthermore, by knowing the remaining ESAL, the average daily traffic in ESAL and the average annual increase of traffic, the remaining life can be estimated in years and the required overlay thickness is determined for the pavement to last a certain number of years.

The non-destructive structural evaluation of a pavement may be carried out by a wide variety of deflection measuring devices. The devices may be classified into three categories: (a) static devices, (b) semi-static devices and (c) moving or rolling devices.

Static devices are those that have to stop to take measurements and the loading is rolling, vibrational or impulse.

Semi-static devices are those moving slowly during measurements, the loading is rolling or vibrating mass and the measuring apparatus is stationary and in contact with the pavement surface.

Moving or rolling devices are those taking deflection measurements as they move, the loading is rolling and the measuring apparatus is non-stationary and has no contact with the pavement surface.

On the basis of the above classification, the most common devices used today to measure surface deflection of flexible or concrete pavements are listed in Table 16.14.

A general description of the various types of static and semi-static deflection testing devices and procedures for deflection measurement is provided by ASTM D 4695 (2008) or AASHTO T 256 (2011).

16.4.1 Static deflection measuring devices

16.4.1.1 Benkelman beam

The Benkelman beam is the simplest and the oldest deflection test device, developed in the United States in the mid-1950s, that has been extensively used worldwide for many years because of its simplicity in taking measurements and its lowest purchasing cost. Its use today is limited only to the project-level structural evaluation of the pavement.

The device is composed of an aluminium beam approximately 3.60 m in total length, having a 2.44 or 2.5 m long probe. The cross section of the probe is such to fit in the space between the dual tyres. The device is placed behind a truck between the dual tyres of the

Table 16.14 Deflection measuring devices

Type of deflection measuring device	Type of equipment	Type of loading	Typical speed at testing (km/h)
Static	Benkelman beam	Rolling wheel	0
	Dynaflect	Vibration load	0
	Road rater		
	FWDs	Impulse loading	
Semi-static (moving measurement vehicles with stationary measurement apparatus)	LaCroix deflectograph and other deflectograph versions (UK, PaSE Australian, etc.)	Rolling wheel load	2.5 or 3.5
	Flash deflectograph (Vectra France)		3.5 or 7.0
	Curviameter (Euro Consulting Group)		18
Moving (moving measurement vehicles with non-stationary measurement apparatus)	Rolling dynamic deflectometer (RDD) (Texas)	Vibrating mass	5
	Airfield rolling weight deflectometer (ARWD)	Rolling wheel	35
	Rolling wheel deflectometer (RWD) (ARA for FHWA)		Up to 80
	Traffic speed deflectometer (TSD) (Greenwood, DK)		60 to 80

rear single axle with the tip of the probe positioned approximately 1.38 m in front of the loading wheel axle (back axle). Details of the Benkelman beam and deflection taking readings are shown in Figures 16.39 and 16.40.

The truck is loaded so that the rear single axle imparts a standard axle load, typically 80 kN (18,000 lb) with the dual tyres inflated to 480 to 550 kPa (70 to 80 psi); loading conditions to the previous one may vary in some countries.

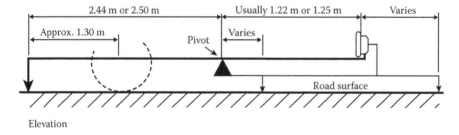

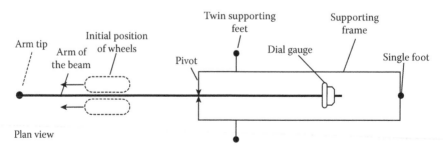

Figure 16.39 Diagrammatic representation of the principle of the Benkelman beam. (Adapted from Norman, P.J. et al., *Pavement Deflection Measurements and Their Application to Structural Maintenance and Overlay Design*, TRRL Report LR 571. Transport and Road Research Laboratory. Crowthorne, UK: Transport Research Laboratory, 1973.)

Figure 16.40 Benkelman beam device. (Courtesy of Controls Srl.)

After positioning the device, the truck is slowly driven forward (creep speed) and the maximum deflection is recorded as the tyres pass the tip of the arm. The vehicle continues to move forward for some metres (approximately 7 more metres) and the minimum deflection is recorded.

The surface deflection is calculated using the manufacturer's recommended formula, which is based on the configuration of the pivot on the beam. When a ratio of 2:1 is used (i.e. length of arm, 2.44 m; length of beam from pivot to the gauge, 1.22 m), the difference of the maximum and minimum reading is multiplied by 2 and this is the rebound deflection (RD) caused by the loaded wheel.

Deflection measurements should preferably be made when the road temperature is close to 20°C; measurements outside the range 10°C–30°C should be avoided because of the large temperature correction likely to be necessary. The tyre pressure and tyre dimensions should be in accordance with the specification followed.

The spacing between measurements is usually between 20 and 50 m.

More information can be found in Norman et al. (1973), ASTM D 4695 (2008), AASHTO T 256 (2011) or other related national guidelines/specifications.

16.4.1.1.1 Data processing and analysis

The aim of Benkelman beam testing, like any other deflection testing procedure, is to estimate the integrity, or remaining life, of the pavement and then to determine the thickness of the overlay, if required. This is obtained by correlating measured deflection with remaining life, expressed as number of standard loads.

One of the procedures to follow is the one described in the Asphalt Institute MS-17 (Asphalt Institute 3rd Edition). According to this procedure, the representative rebound deflection (RRD) value needs to be determined first. The RRD value is the mean of the deflection measurements (rebound reflections) adjusted for temperature and period of executing the measurements. The RRD can be computed from the following equation:

$$\text{RRD} = (\bar{x} + 2s) \times c,$$

where $\bar{x}$ is the arithmetic mean of the individual values that have been adjusted for temperature, $\bar{x} = \dfrac{\sum x_i f_i}{n}$; s is the standard deviation, $s = \sqrt{\dfrac{\sum x^2 - \bar{x} \sum x}{n-1}}$; c is the critical period adjustment factor; $c = 1$ when tests are carried out during spring; otherwise it is estimated based on experience (values range between 0.5 and 1.0); x_i is the individual deflection values; f_i is the temperature adjustment factor of individual deflections (see Figure 16.39); and n is the number of individual deflection test values.

The temperature adjustment factor for a three-layered asphalt concrete pavement (asphalt, unbound aggregate layer and subgrade) and for a full-depth pavement can be determined from Figure 16.41.

Having determined the RRD value, the structural adequacy of the pavement in terms of remaining ESAL (18 kps) can be estimated from Figure 16.42.

To calculate the overlay thickness, having determined the remaining ESAL, see Section 15.17.1.

Similar to the Asphalt Institute MS-17 procedure for temperature adjustment of deflection measured by Benkelman beam and estimation of remaining life is the UK procedure. Details can be found in TRRL Report LR 571 (Norman et al. 1973).

16.4.1.1.2 Estimation of SNP from Benkelman measurements

The Benkelman beam measurements can also be used to determine the adjusted structural number (SNP), a useful parameter for estimating pavement deterioration in pavement management studies, used in WDM-4. The SNP is basically the structural number (SN) used by AASHTO as a measure of pavement strength (AASHTO 1993) but adjusted for the reduced contribution of each layer with depth (Morosiuk et al. 2004).

The equations used to estimate the SNP distinguish between flexible pavements with granular (unbound) and cement bound base. Details can be found in the WDM-4 Manual (Morosiuk et al. 2004). However, the SNP may also be estimated from Benkelman rebound deflection (80 kN with 520 kPa tyre pressure) using the curves shown in Figure 16.43.

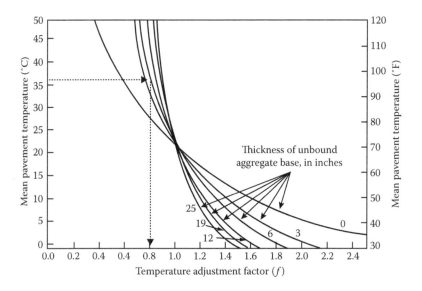

Figure 16.41 Benkelman beam deflection adjustment factors owing to temperature variation for full-depth and three-layered asphalt concrete pavement. (From Asphalt Institute 3rd Edition, *Asphalt Overlays for Highway and Street Rehabilitation*. Manual Series No.17 [MS-17]. Lexington, US: Asphalt Institute. With permission.)

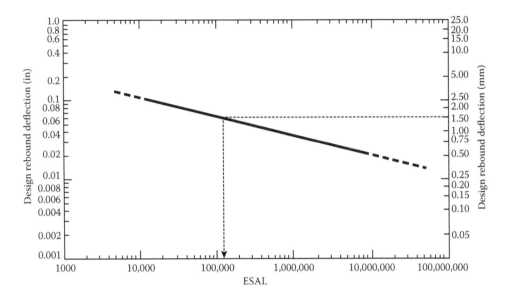

Figure 16.42 Calculation of remaining ESAL from design rebound deflection. (From Asphalt Institute 3rd Edition, *Asphalt Overlays for Highway and Street Rehabilitation*. Manual Series No.17 [MS-17]. Lexington, US: Asphalt Institute. With permission.)

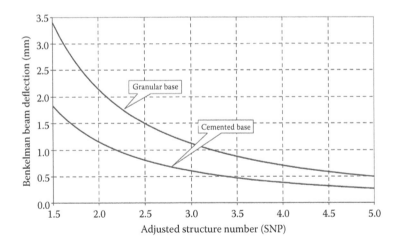

Figure 16.43 Relationship of Benkelman beam deflection and SNP. (From Morosiuk, G. et al., *Modelling Road Deterioration and Works Effects, Version 2, HDM-4*. The Highway Development and Management Series, Vol. 6. Paris: World Road Association, 2004.)

16.4.1.1.3 Correlation between Benkelman beam and other devices

Numerous comparisons have been made between Benkelman beam and other deflection devices in order to develop correlation equations. This offers the possibility to determine the overlay thickness required by following the Asphalt Institute procedure based on the RRD, outlined above.

The approximate correlation equations proposed by Asphalt Institute MS-17 (Asphalt Institute 3rd Edition) between Benkelman beam deflection and other deflection devices such as Dynaflect, Road Rater and falling weight deflectometer (FWD) are listed below. A short description of the above deflection devices is given in the following paragraphs.

Correlation between Benkelman beam and Dynaflect device:

BB = 22.3 × D − 2.73

Correlation between Benkelman beam and Road Rater device:

BB = 2.57 + (1.27 × RR)

Correlation between Benkelman beam and FWD device:

BB = 1.61 × FWD

where BB is the Benkelman beam deflection (inches, $\times 10^{-3}$), D is the Dynaflect centre deflection (inches, $\times 10^{-3}$), RR is the Road Rater model 2000 deflection (inches, $\times 10^{-3}$ at 36 kN and 15 Hz) and FWD is the FWD deflection (inches, $\times 10^{-3}$ corrected to a 40 kN applied load on a 300 mm diameter plate).

As mentioned in the bibliography, the validity of the above equations depends on whether the measurements are taken under conditions similar to those when correlation tests were performed. For this reason, when overlay thickness is determined using the correlation equations, it is recommended to be compared with other design procedures such as the effective thickness procedure described in Section 15.17.3.

16.4.1.2 Dynaflect and road rater

Dynaflect and Road Rater are vibrating load devices developed in the United States in the 1970s. They measure dynamic deflection while they are stationary, using a sinusoidal force. Unlike the Benkelman beam and semi-static devices, they are also capable of measuring the deflection basin since deflection is recorded by velocity transducers (usually five). Both devices comply with ASTM D 4695 (2008) and ASTM D 4602 (2008).

In the Dynaflect, the sinusoidal load, 455 kg peak to peak, is applied through eccentric masses that counter-rotate at a fixed frequency, through two rubber-coated steel wheels.

Dynaflect deflection measurements may be converted to Benkelman deflection using the correlation equation given in Section 16.4.1.1 or a similar equation. The main drawback of Dynaflect is that since the applied load is low, it is not really suitable for thick asphalt pavements.

More details about the Dynaflect can be found in Asphalt Institute MS-17 (3rd Edition), Eagleson et al. (1982) and Pavement Interactive.

In the Road Rater, the sinusoidal load may range from 455 to 3650 kg depending on the model and is generated by a hydraulic acceleration of steel mass; this is basically the main difference compared to Dynaflect. Figure 16.44 shows a Road Rater mounted on a trailer.

The Road Rater deflection measurements may also be converted to Benkelman deflection using the correlation equation given in Section 16.4.1.1 or a similar equation.

More details about the Road Rater can be found in Asphalt Institute MS-17 (3rd Edition), Wang (1985) and Rutherford (1993).

Figure 16.44 The Road Rater device. (From Asphalt Institute 3rd Edition, *Asphalt Overlays for Highway and Street Rehabilitation*. Manual Series No. 17 [MS-17]. Lexington, US: Asphalt Institute. With permission.)

16.4.1.3 FWDs – impulse devices

The FWD was initially developed in France in 1963 (Bretonniere 1963) and then evolved in Denmark (Ullidtz 1987) as an alternative non-destructive device for structural evaluation of pavements. Today, the following FWDs exist on the market: Dynatest (Dynatest International A/S 2014), KUAB (ERI 2014), PRIMAX (Grontmij A/S) and JILS (JILS) FWD.

All FWDs are based on the same principle; they apply an impulse load using a falling system and all comply with ASTM D 4695 (2008) and ASTM D 4694 (2009) standard guidelines. Figure 16.45 shows a typical FWD capable of measuring road and airport pavements, known also as heavy weight deflectometer.

An FWD has the ability to provide information for the structural condition of all pavement layers, including the subgrade. In particular, with FWD deflection measurements, the following may be obtained: (a) estimation of modulus of elasticity or stiffness modulus of each layer, (b) estimation of remaining life of pavement, (c) detection of weak points in the pavement, (d) estimation of the severity of potential cracks in stabilised layers, (e) determination of the effectiveness of the load transfer at the joints of rigid pavements and (f) detection of cracks on rigid pavements.

The determination of the overlay thickness is usually performed by analytical methods similar to those applied in the analytical pavement design methodologies. However, by converting deflection measurements obtained by FWD to Benkelman beam, or to deflectograph deflection, the established methodologies described in Section 15.17, may also be used.

FWD in most cases is mounted on a trailer, rather than on a vehicle. The load is applied through a 300 mm diameter circular metal plate for road pavements or through a 450 mm

Figure 16.45 FWD, heavy weight deflectometer. (Courtesy of Dynatest International A/S.)

diameter plate for airport pavements. The applied load for flexible pavements is usually 50 kN, whereas that for a rigid pavement is usually 75 kN.

The resulting deflections are measured at the centre of the applied load (through a hole in the centre of the loading plate) and at various distances away from the load with sensors. The sensors may be displacement transducers, velocity transducers, accelerometers, geophones or even seismometers.

The deflection is measured with the aid of seven sensors placed on a straight line passing through the centre of the loading area. The first sensor is placed exactly at the centre of the loading plate (having a hole), whereas the others are placed at certain distances, as given in Table 16.15. In rigid or flexible pavements with stabilised base, the above distances may be smaller for crack detection.

The central deflection, d1, gives an indication of the overall pavement performance whilst the deflection difference (d1–d4) relates to the condition of the bound pavement layers. Deflection d6 is an indication of subgrade condition (see also Figure 16.46, deflection bowl).

The measurements are usually taken every 20 m or more, and even every 200 m in the case of surveying a long road network. When measurements are taken at long distances apart, there may be a necessity, in some sections, to also take measurements every 20 m, to detect precisely the suspected pavement weakness.

Table 16.15 Recommended geophone positions for FWD measurements

| Type of pavement | Geophone distance from centre of loading plate (geophone number and distance in mm) | | | | | | |
	d1	d2	d3	d4	d5	d6	d7
Flexible and flexible composite	0	300	600	900	1200	1500	2100
Rigid and rigid composite (CRCB)	0	300	600	900	1350	1800	2250

Source: Highways Agency, Design Manual for Roads and Bridges (DMRB), Volume 7: Pavement Design and Maintenance, Section 3, Part 2, HD 29/08, London: Department for Transport, Highways Agency, 2008b.

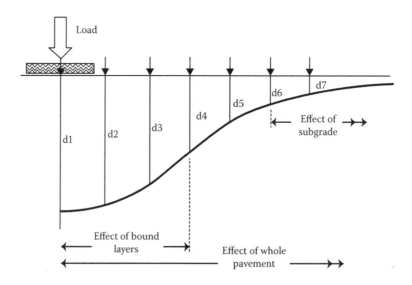

Figure 16.46 FWD deflection bowl. (From Highways Agency, Design Manual for Roads and Bridges [DMRB], Volume 7: Pavement Design and Maintenance, Section 3, Part 2, HD 29/08, London: Department for Transport, Highways Agency, 2008b [© Highways Agency].)

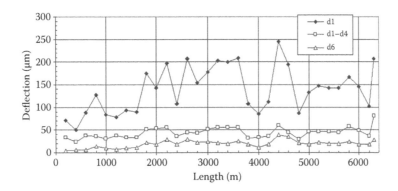

Figure 16.47 Typical FWD deflection results.

Given that the temperature affects the mechanical behaviour of the asphalt layers, pavement temperature should also be considered and measured during testing. This is achieved by taking a measurement at approximately 100 mm below the pavement surface using an appropriate electronic thermometer.

All data collected are processed by the software provided by the FWD producer. A sample of printout of deflection results, in graphical presentation, is given in Figure 16.47.

16.4.1.3.1 Analysis and evaluation of results

Considering the curve corresponding to central deflection, d1, in Figure 16.46, it can be seen that the pavement for the first length of 1600 m behaves better than the rest of the surveyed section, apart from some localised exceptions (between 3800 and 4200 m, at around 4800 m and at around 6200 m length). At these locations, additional FWD measurements, closely spaced (every 20 m), are recommended to better detect the exact locations where the pavement behaviour changes.

FWD provides the ability to estimate the stiffness moduli of each individual layer after back-calculation analysis of the deflection data. This is carried out using specially developed computer programs, almost all using the principles and assumptions of the multi-layer system linear elastic analysis. Some guidance for calculating in situ equivalent elastic moduli of pavement materials using layered elastic theory is provided by ASTM D 5858 (2008).

The curves in Figure 16.48 show typical results of layer stiffness after back-calculation analysis along a project that is 21 km long; the variation of the stiffness of the bituminous layers (E_1) is shown in Figure 16.48a, that of the base/sub-base layer (E_2) is shown in Figure 16.48b and that of the formation layer (E_3) is shown in Figure 16.48c.

Having estimated the stiffness moduli and knowing the thickness of each layer, the developed strain at critical interfaces can be calculated, from which, using fatigue equations, it is possible to estimate the remaining life of the pavement, in ESAL.

Figure 16.48d gives the estimated remaining life of the pavement along the same stretch of road length (21 km).

However, results from back-analysing FWD data should always be considered with a degree of scepticism and using engineering judgement. There are many factors, apart from the assumptions used in the back-calculation analysis that may affect stiffness estimation. In all cases, certainly when low stiffness values are concerned, it is always advisable that the results be compared with results from core testing or field material testing.

However, stiffness determination from back-calculation analysis of FWD data provides the possibility of also assessing the quality of pavement layers. To assess the quality of the

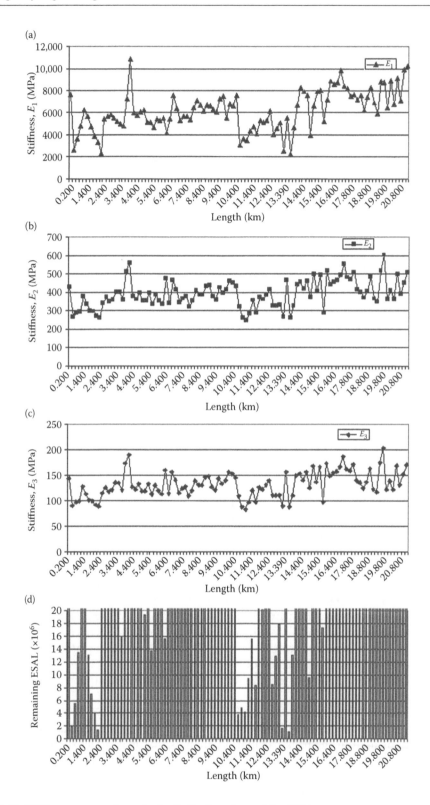

Figure 16.48 FWD measurement results for E_1, E_2, E_3 and remaining ESAL.

Table 16.16 Conditions related to bound layer stiffness

| | Bound layer stiffness[a] at 20°C derived from FWD | | |
Pavement type	Poor integrity throughout	Some deterioration	Good integrity
Asphalt	<3 GPa	3–7 GPa	>7 GPa
HBM[b]	<8	8–15	>15
PQ concrete	<20	20–30	>30

Source: Highways Agency, Design Manual for Roads and Bridges (DMRB), Volume 7: Pavement design and maintenance, Section 3, Part 3: HD 30/08, London: Department for Transport, Highways Agency, 2008c. © Highways Agency.

[a] The stiffness applies to layers consisting of only one material type.
[b] HBM, hydraulically bound mixture.

layers from FWD back-calculated stiffness, Highways Agency (2008c) proposes the use of the values given in Table 16.16.

Using the condition given in Table 16.16, the results of stiffness of the asphalt layers from Figure 16.48a and the results of remaining life (Figure 16.48d), it is obvious that the sections with asphalt stiffness less than 3 GPa, that is, sections with poor integrity, have indeed limited remaining life, less than 2×10^6 ESAL.

16.4.1.3.2 Estimation of SNP from FWD measurements

The FWD measurements can be used for determining the SNP, a useful parameter for estimating pavement deterioration for pavement management studies, as mentioned in Section 16.4.1.1. The equations developed by various researchers are outlined in the WDM-4 Manual (Morosiuk et al. 2004).

16.4.2 Semi-static deflection measuring devices

Many semi-static devices have been developed ever since the first semi-static deflection measuring device, the California Travelling deflectograph in the early 1950s, was presented in Hveem's paper (1955). A good historical review of the evolution of semi-static deflection measuring devices can be found in the Andrén report (Andrén 2006).

This paragraph covers most of the semi-static deflection measuring devices currently in use.

16.4.2.1 LaCroix deflectograph

The LaCroix deflectograph was developed in France (late 1950s), and soon similar devices, based on exactly the same principle, were developed in other countries. All later developed devices bear the name deflectograph. Such devices are namely, the UK deflectograph (Highways Agency 2008b), the PaSE deflectograph (Austroads 2007, 2011) and the Flash deflectograph (Vecrta France). All abovementioned deflectographs carry out deflection measurements at 3.5 km/h while the Flash deflectograph is capable of taking measurements also at 7.0 km/h.

The deflectograph is a two-axle truck carrying a beam assembly and an associated recording system between its axles. Deflection measurements are based on the same principle as in Benkelman beam measurements. The deflectograph provides a more rapid method of measuring pavement deflection than the Benkelman beam since setting arrangements are much faster.

The beam assembly is lowered, rests on the road and is suitably aligned between the front and rear axles of the vehicle, and deflections are measured as the rear wheel assemblies approach and pass the tips of the beam (see Figure 16.49).

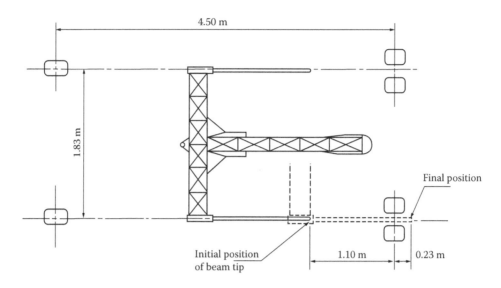

Figure 16.49 Diagrammatic representation of the deflectograph. (From Norman, P.J. et al., *Pavement Deflection Measurements and Their Application to Structural Maintenance and Overlay Design*, *TRRL Report LR 571*. Transport and Road Research Laboratory. Crowthorne, UK: Transport Research Laboratory, 1973.)

The rear axle is loaded to a standard weight similar to the weight used in the Benkelman beam test. The maximum deflection is recorded by electrical transducers located near the beam pivots. When measurements are taken, the beam assembly is pulled forward at approximately twice the speed of the vehicle, to a new position ready for the next measurement. The working speed of the deflectograph is approximately 2 km/h (walking speed) and measurements are taken every 3.8 to 4.0 m. A deflectograph vehicle is shown Figure 16.50.

The processing of the data measurements with the necessary (temperature) corrections is performed by the PC software and the result is expressed as average deflection per point of measurement.

Further information regarding the measurement collection process with the TRL UK Deflectograph is provided in Norman et al. (1973), Highways Agency (2008b) and WDM Limited.

Figure 16.50 WDM Limited deflectograph. (From Highways Agency, *Design Manual for Roads and Bridges [DMRB]*, *Volume 7: Pavement Design and Maintenance*, Section 3, Part 2, HD 29/08, London: Department for Transport, Highways Agency, 2008b [© Highways Agency].)

16.4.2.2 Curviameter

The curviameter device originally developed by France (Paquet 1978) operates at a speed of 18 km/h and is capable of not only measuring deflection but also determining pavement deflection bowl. The curviameter measuring system (see Figure 16.51a) is mounted on a two-axle truck with dual rear wheels (see Figure 16.51b).

The deflection measurements are carried out by three geophones mounted on a 15 m chain 5 m apart. The chain is positioned on the ground and passes between the dual wheels. As the truck moves forward, it constantly places the chain down in front of the rear wheels. The chain remains at a fixed location on the pavement as the truck wheels roll over it.

Deflection and radius of curvature data, usually corrected for pavement temperature, are processed by specially developed software. Curviameter test standards and documents of CEDEX (2006), AFNOR (1991) and MDRW (2002) include a detailed description of the measuring system and the test preparation, sensor calibration and the measurement principle and procedure. More details on the curviameter device can also be found in Ramos et al. (2013).

Correlations have been established between curviameter results and FWD, Benkelman beam or deflectograph results. Relevant details for the developed correlation equations over the last years can be found in Van Geem (2010).

The data obtained from the curviameter have also been correlated with remaining life in $ESAL_{80}$ (Gorski 1999).

(a)

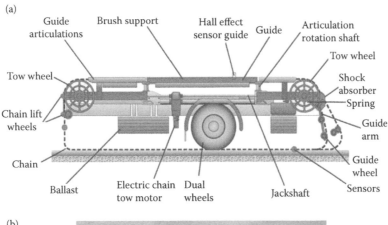

(b)

Figure 16.51 Curviameter. (a) Detailed diagram of the measuring system. (From Sánchez, D.F., new regulations for the measuring of deflections with a curviameter. *Symposium on the Measuring of Texture and Structural Condition of Pavements and Road Surface of Spain's Road Network, CEDEX,* May, Madrid, Spain, 2008.) (b) Curviameter vehicle. (Courtesy of Euroconsult.)

16.4.3 Moving deflection measuring devices

Moving deflection measuring devices measure deflection continuously as they travel and, hence, are called continuous, rolling or traffic speed deflectometers. They have started to emerge in the 1990s with the aim of performing deflection measurements while moving at a certain speed ranging from 5 to 80 km/h. In fact, they are non-destructive devices measuring pavement response at traffic speeds with almost no traffic control from which the bearing capacity of the pavement can be estimated.

Although the continuous deflectometers do not provide as much information as obtained from the FWD deflection bowl, they can effectively evaluate pavements for pavement management purposes. The devices that operate at high traffic speeds such as 70 or 80 km/h are more favourable to be used for network-level evaluation compared to those operating at lower speeds.

At the time of writing (2013), four continuous deflection measuring devices were found to be in operation: the rolling dynamic deflectometer (RDD), the airfield rolling weight deflectometer (ARWD), the rolling wheel deflectometer (RWD) and the traffic speed deflectometer (TSD).

16.4.3.1 Rolling dynamic deflectometer

The RDD developed by the Center for Transportation Research at the University of Texas at Austin and the current version are a modification of a Vibroseis truck originally used for oil exploitation (Bay and Stokoe 1998).

The dynamic load system (a combination of static and dynamic force, hence servo-vibrating load) is located at the middle of a two-axle truck. The applied dynamic load is applied to the pavement surface via two rollers placed parallel to the track wheel axles. The deflection is measured by vertical accelerometers.

The operating speed of the RDD is 5 km/h, which is a disadvantage when it is compared to other rolling profilometers operating at much higher speeds 60 to 80 km/h.

However, the RDD demonstrated good potential for providing continuous profiles of flexible (and rigid) pavement structures. A comparison between RDD and FWD data also showed very good correlation. The RDD may be used for other testing purposes, such as estimating pavement depth and measuring pavement resistance to fatigue cracking (Bay and Stokoe 1998).

Additional information regarding the RDD can be found in Bay et al. (1999).

A new-generation RDD has been announced to be under development as a joint effort of the Texas A&M Transportation Institute (TTI), the Texas Department of Transportation, the Center for Transportation Research at The University of Texas and Industrial Vehicles International. The device is called Total Pavement Acceptance Device (TPAD), also known as the RDD. The TPAD is a multi-function device that integrates a deflection-measuring system with GPR and high-definition video imaging of pavements (TTI 2013) (see Figure 16.52).

16.4.3.2 Airfield rolling weight deflectometer

The ARWD was originally developed by Quest Integrated, Inc., to test the deflection of airfield pavements.

The ARWD system is composed of a towed trailer that is capable of travelling with speeds of 35 km/h. A long loading platform is used to load the pavement through the trailer's rear double axle with a single wheel total load of 80 kN (18 kips). A horizontal 10 m (33 ft) long

Figure 16.52 TPAD rolling dynamic deflectometer. (From Texas A&M Transportation Institute [TTI], *Using New Technologies to Solve Old Problems*, Texas Transportation Researcher, Vol. 49, No. 4, http://www.tti.tamu.edu, 2003.)

beam is positioned at the central axis of the trailer, which carries four equally spaced optical sensors (2.74 m apart); one is positioned near the load and the other three are placed ahead of the loading axle.

The deflection is measured by using a laser triangulation and parallax method. The laser sensors are responsible for measuring the distance to the pavement surface, and the deflection is calculated from the difference between the slope of the beam and the slope of the pavement at the initial point and after the device moves to the next point. The device is provided with a data acquisition system used for data collection, storage and analysis (Elseifi et al. 2012).

Although the ARWD device is suitable for airfields, the device needs to be modified for highways before it can be implemented (Arora et al. 2006).

16.4.3.3 Rolling wheel deflectometer

The RWD was developed by ARA (Applied Research Associates Inc.) in collaboration with FHWA Office of Asset Management in the early 2000s.

The RWD consists of a 16.15 m (53 ft) long semitrailer applying a standard 80 kN (18,000 lb) axle load on the pavement structure by means of a regular dual-tyre assembly over the rear single axle. The trailer is specifically designed to be long enough to separate the deflection basin, owing to the 80 kN rear axle load from the effect of the front axle load. (Steele et al. 2009).

The deflection is measured by four laser triangulation sensors; one is placed between the dual tyres and the other three are placed in front of the back axle. The laser between the dual tyres measures the pavement profile (maximum deflection) under axle load and the other three, spaced every 2.44 mm (8 ft), measure the unloaded pavement profile. The three lasers are mounted at approximately 1.1 m (3.6 ft) above the pavement surface into the truck body in a temperature-controlled enclosure (see Figure 16.53a). Two additional sensors are placed in front of the wheels to measure a secondary pavement deflection (Elseifi et al. 2012).

The RWD measures wheel deflections at the pavement surface by means of a spatially coincident method, which compares the profiles of the surface in both undeflected and deflected states (Gedafa et al. 2008). Figure 16.53b illustrates the spatially coincident method.

The RWD data are used to provide a 'structural map' of an entire highway network and to target areas for detailed inspection and testing using an FWD, coring or other static types of testing (FHWA 2011). Since the mid-2000s, numerous pavement evaluation studies have been conducted in the United States, together with FWD measurements, which demonstrated that RWD results could be included in pavement management systems.

(a)   (b)

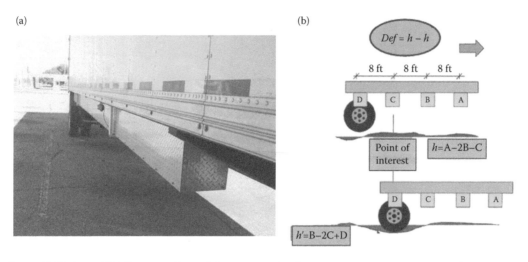

Figure 16.53 Part of RWD semitrailer and illustration of deflection measurement principle. (a) Laser sensors mounted into the trailer body, with enclosure. (b) Illustration of spatially coincident method. (From Elseifi, M. et al., *Implementation of Rolling Wheel Deflectometer [RWD] in PMS and Pavement Preservation.* FHWA/11.492. Baton Rogue, LA: Department of Civil and Environmental Engineering, Louisiana State University, 2012.)

In a recent work on the implementation of RWD in PMS and pavement preservation (Elseifi et al. 2012), it was concluded that RWD deflection measurements were in general agreement and had similar trends to FDW measurements; however, the mean centre deflections from the RWD and the FWD were statistically different for 15 of the 16 sites.

In the same study (Elseifi et al. 2012), a model was developed to estimate SN based on RWD deflection data. Although the developed SN expression based on RWD data was independent of the pavement thickness and layer properties, it was concluded that it provides promising results as an indicator of structural integrity of pavement structure at the network level; at the project level, further evaluation of the proposed model was recommended.

Similar observations were also made in another study assessing continuous pavement deflection measuring technologies (Flintsch et al. 2013).

16.4.3.4 Traffic speed deflectometer

The TSD is the latest developed moving deflectometer. The TSD was developed in Denmark in the early 2000s by Greenwood (Greenwood Engineering A/S) where it was used to be called high-speed deflectometer.

It is the only moving deflection measuring device that uses Doppler laser sensors to measure deflection. In fact, the main difference of the TSD from all other deflectometers is that it measures the velocity of deflection rather than displacement.

TSD is an articulated truck with a rear axle load of 100 kN (22 kips) that uses Doppler lasers mounted on a servo-hydraulic beam to record the deflection velocity of a loaded pavement. The number of Doppler lasers in the first TSD devices was four, while in the second-generation TSD devices, seven Doppler lasers are used (Figure 16.54). However, it has been reported that the device delivered to South Africa in 2013 had 10 lasers and included laser profiling and right-of-way imaging (Greenwood Engineering A/S).

TSD can take deflection measurements continuously at a survey speed up to 80 km/h; to prevent thermal distortion of the steel beam, a climate control system maintains the trailer temperature at a constant temperature of 20°C (see Figure 16.54).

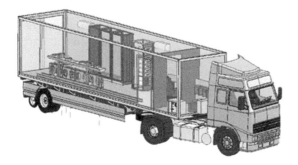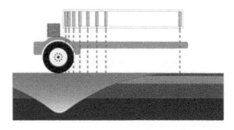

Figure 16.54 Second-generation TSD vehicle and deflection bowl. (From Krarup, J., *Traffic Speed Deflecto-meter [TSD]*. Measuring systems for infrastructure – For roads and for railroads, Sacramento, CA: FWD User's Group, 2012.)

The TSD, over the last 10 years, has undergone extensive evaluation, namely, in the United Kingdom by TRL, in France by LCPC, in Australia by the road authorities in New South Wales and Queensland and in the United States by TRB. In all assessment studies, good repeatability of TSD results was found, as well as good agreement/comparability of TSD results with FWD, deflectograph, RTA deflectograph (Australian model), or Flash deflecto-graph results (Baltzer et al. 2010; Flintsch et al. 2013; Simonin et al. 2005).

In an assessment study of a continuous pavement measuring device (Flintsch et al. 2013), it was also found that, at least for the section investigated, the strains at the bottom of the asphalt layer estimated with measurements using the FWD and TSD resulted in an approxi-mately one-to-one relationship. Similarly, it was concluded that the effective SN estimated with measurements obtained from TSD testing at two sites broadly matched the expected SN calculated from the layer composition and surface condition.

The above findings are encouraging since not only does TSD provide accurate bearing capacity results continuously at traffic speeds, the results could be utilised for the estimation of residual life and strengthening requirements of roads as well.

16.4.4 Image deflection measuring technique

The image deflection measuring technique for surface deflection measurements has recently been tried on pavements. This non-destructive method consists of the projection of one light pattern (line, grid, etc.) on the pavement surface. The surface is captured by a camera from a different viewpoint, and software analyses the pattern. This allows the measurement of surface deflection and its gradient.

The first experiments have been carried out by the LCPC putting the projector and the camera on a static fatigue device. The first applications showed interesting results, which validate the choice of measuring deflection basin with fringe projection. However, this tech-nique needs to be improved in order to obtain an operational device. An ongoing project aims to mount the system on a heavy truck (Muzet et al. 2009).

The imaging technique, although promising, is still in the early stage of development.

16.4.5 Other devices

Another device used in pavement evaluation studies is the ground-penetrating RADAR (RAdio Detection And Ranging), known as GPR.

The GPR device is used for rapid and non-destructive determination of pavement layer thickness. It may also be used to detect cracks and determine crack depths, discontinuities within layers in general, presence of moisture, voids within unbound materials, particularly below concrete slabs, and position of reinforcement.

GPR uses radar pulses to image a pavement's subsurface. It transmits short electromagnetic pulses into the ground, and by receiving and analyzing the reflected images of these pulses, layer thickness and other layer characteristics (cracks, etc.) are determined.

The GRP device is usually mounted on an FWD vehicle or any other moving deflection measuring vehicle.

Figure 16.55 shows a typical GPR device mounted on a vehicle and typical partial processed data.

(a)

(b)

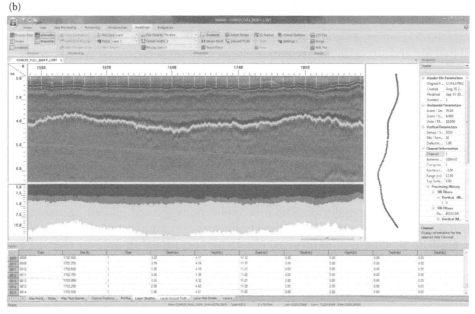

Figure 16.55 (a) GPR device and (b) sample of processed data showing thickness variation (1 = overlay, 2 = original asphalt layer, 3 = base/sub-base layer). (Courtesy of GSSI Inc.)

When short lengths or small areas are to be surveyed, the GPR device can also be operated by hand such as on a trolley or pulled on a sled across the ground. Figure 16.56a shows a GPR device with the GPS facility on a trolley.

For spot measurements of the total thickness of bound material, a new device has been developed called the e-Spott device (see Figure 16.56b). It has been developed not to replace traditional coring but to compliment core sampling and target limited coring resources.

The performance of the e-Spott device has been assessed by TRL (Cook 2011). It was found that is a very simple device, gives repeatable results with reasonable accuracy and it compares reasonably well with traditional GPR data that have been calibrated by coring, but its accuracy is insufficient to negate the needed coring.

In fact, filed coring is needed in any GPR measurements not only for calibrating the devices but also to confirm GRP thickness results. Apart from the usual coring technique in which a cylindrical core, usually 100 or 150 mm in diameter, is extracted, a new method for measuring pavement thickness has been developed. The novel method is based on creating a 20, 30 or 50 mm hole and using a borescope to obtain images from the exposed wall of the hole; layer boundaries can be observed in the images (Gopaldas et al. 2009). The theory behind the calculations of layer thickness using these images and recommendations for additional work to further develop the technique are provided by Gopaldas et al. (2009).

(a)

(b)

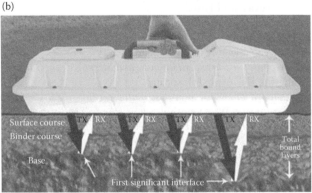

Figure 16.56 (a) GPR on trolley. (Courtesy of MALÅ.) (b) e-Spott device. (Courtesy of PipeHawk Plc.)

REFERENCES

AASHTO. 1962. *The AASHO road test: Report 5, Pavement research.* Special Report No. 61E. Washington, DC: Highway Research Board.

AASHTO. 1993. *Guide for design of pavement structures (GDPS).* Washington, DC: American Association of State Highway and Transportation Officials.

AASHTO. 2008. *Mechanistic–empirical pavement design guide (MEPDG).* Washington, DC: American Association of State Highway and Transportation Officials.

AASHTO. 2012. *Pavement management guide,* 2nd Edition. Washington, DC: American Association of State Highway and Transportation Officials.

AASHTO M 328. 2010. *Inertial profiler.* Washington, DC: American Association of State Highway and Transportation Officials.

AASHTO R 40. 2010. *Measuring pavement profile using a rod and level.* Washington, DC: American Association of State Highway and Transportation Officials.

AASHTO R 41. 2010. *Measuring pavement profile using a dipstick.* Washington, DC: American Association of State Highway and Transportation Officials.

AASHTO R 43. 2013. *Quantifying roughness of pavements.* Washington, DC: American Association of State Highway and Transportation Officials.

AASHTO R 56. 2010. *Certification of inertial profiling systems.* Washington, DC: American Association of State Highway and Transportation Officials.

AASHTO R 57. 2010. *Operating inertial profiling systems.* Washington, DC: American Association of State Highway and Transportation Officials.

AASHTO T 256. 2011. *Pavement deflection measurements.* Washington, DC: American Association of State Highway and Transportation Officials.

AFNOR. 1991. *Measure de la deflexión engendrée par une charge roulante. Détermination de la déflexion et du rayon de courbure avec le curviamètre.* NPF-98-200-7. Paris: Association Française de Normalisation.

AIT. 2014. *Austrian Institute of Technology.* Available at http://www.ait.ac.at/research-services/research-services-mobility.

Al-Omari B.H. and M. Darter. 1994. Relationships between International Roughness Index and present serviceability rating. *Transportation Research Record,* No. 1435, pp. 130–136. Washington, DC: TRB, National Research Council.

Ames Engineering Inc. 2014. Available at http://www.amesengineering.com.

Andrén P. 2006. Development and results of the Swedish road deflection tester. Licentiate thesis from Royal Institute of Technology Department of Mechanics, Stockholm, Sweden.

APR Consultants. 2014. Available at http://www.aprconsultans.com.

Arora J., V. Tandon, and S. Nazarian. 2006. *Continuous Deflection Testing of Highways at Traffic Speeds.* FHWA/TX Report No. 06/0-4380-1. El Paso, Texas: Center for Transportation Infrastructure Systems, University of Texas at El Paso.

ARRB Group Ltd. 2014. Available at http://www.arrb.com.au.

AIT (Austrian Institute of Technology). 2014. Available at http://www.ait.ac.at.

ASFT Industries AB. 2014. Available at http://www.asft.se.

Asphalt Institute MS-17. *Asphalt Overlays for Highway and Street Rehabilitation,* 3rd Edition. Manual Series No. 17. Lexington, KY: Asphalt Institute.

ASTM D 4602. 2008. *Standard guide for non-destructive testing of pavements using cyclic-loading dynamic deflection equipment.* West Conshohocken, PA: ASTM International.

ASTM D 4694. 2009. *Standard test method for deflections with a falling-weight-type impulse load device.* West Conshohocken, PA: ASTM International.

ASTM D 4695. 2008. *Standard guide for general pavement deflection measurements.* West Conshohocken, PA: ASTM International.

ASTM D 5858. 2008. *Standard guide for calculating in situ equivalent elastic moduli of pavement materials using layered elastic theory.* West Conshohocken, PA: ASTM International.

ASTM D 6433. 2011. *Standard practice for roads and parking lots pavement condition index surveys.* West Conshohocken, PA: ASTM International.

ASTM E 274/E 274M. 2011. *Standard test method for skid resistance of paved surfaces using a full-scale tire.* West Conshohocken, PA: ASTM International.

ASTM E 303. 2008. *Standard test method for measuring surface frictional properties using the British Pendulum tester.* West Conshohocken, PA: ASTM International.

ASTM E 501. 2008. *Standard specification for standard rib tire for pavement skid-resistance tests.* West Conshohocken, PA: ASTM International.

ASTM E 524. 2008. *Standard specification for standard smooth tire for pavement skid-resistance tests.* West Conshohocken, PA: ASTM International.

ASTM E 670. 2009. *Standard test method for testing side force friction on paved surfaces using the Mu-meter.* West Conshohocken, PA: ASTM International.

ASTM E 867. 2012. *Standard terminology relating to vehicle–pavement systems.* West Conshohocken, PA: ASTM International.

ASTM E 965. 2006. *Standard test method for measuring pavement macro-texture depth using a volumetric technique.* West Conshohocken, PA: ASTM International.

ASTM E 1082. 2012. *Standard test method for measurement of vehicular response to travelled surface roughness.* West Conshohocken, PA: ASTM International.

ASTM E 1215. 2012. *Standard specification for trailers used for measuring vehicular response to road roughness.* West Conshohocken, PA: ASTM International.

ASTM E 1274. 2012. *Standard test method for measuring pavement roughness using a profilograph.* West Conshohocken, PA: ASTM International.

ASTM E 1364. 2012. *Standard test method for measuring road roughness by static level method.* West Conshohocken, PA: ASTM International.

ASTM E 1448/E 1448M. 2009. *Standard practice for calibration of systems used for measuring vehicular response to pavement roughness.* West Conshohocken, PA: ASTM International.

ASTM E 1489. 2008. *Standard practice for computing ride number of roads from longitudinal profile measurements made by an inertial profile measuring device.* West Conshohocken, PA: ASTM International.

ASTM E 1551. 2008. *Standard specification for special purpose, smooth tread tire, operated on fixed braking slip continuous friction measuring equipment.* West Conshohocken, PA: ASTM International.

ASTM E 1703/E 1703M. 2010. *Standard test method for measuring rut-depth of pavement surfaces using a straightedge.* West Conshohocken, PA: ASTM International.

ASTM E 1845. 2009. *Standard practice for calculating pavement macro-texture mean profile depth.* West Conshohocken, PA: ASTM International.

ASTM E 1911-09ae01. 2009. *Standard test method for measuring paved surface frictional properties using the dynamic friction tester.* West Conshohocken, PA: ASTM International.

ASTM E 1926. 2008. *Standard practice for computing International Roughness Index of roads from longitudinal profile measurements.* West Conshohocken, PA: ASTM International.

ASTM E 1960. 2011. *Standard practice for calculating International Friction Index of a pavement surface.* West Conshohocken, PA: ASTM International.

ASTM E 2133. 2009. *Standard test method for using a rolling inclinometer to measure longitudinal and transverse profiles of a traveled surface.* West Conshohocken, PA: ASTM International.

ASTM E 2157. 2009. *Standard test method for measuring pavement macro-texture properties using the circular track meter.* West Conshohocken, PA: ASTM International.

ASTM E 2340/E 2340M. 2011. *Standard test method for measuring the skid resistance of pavements and other trafficked surfaces using a continuous reading, fixed-slip technique.* West Conshohocken, PA: ASTM International.

ASTM E 2793-10e1. 2010. *Standard guide for the evaluation, calibration, and correlation of E274 friction measurement systems and equipment.* West Conshohocken, PA: ASTM International.

Austroads. 2007. *Austroads Test Method AG:AM/T008 – Validation and Repeatability Checks for a Deflectograph.* Sydney: Austroads Incorporated.

Austroads. 2011. *Austroads Test Method AG:AM/T007 – Pavement Deflection Measurement with a Deflectograph*. Sydney: Austroads Incorporated.

Baltzer S., D. Pratt, J. Weligamage, J. Adamsen, and G. Hildebrand. 2010. Continuous bearing capacity profile of 18,000 km Australian road network in 5 months. *Proceedings of the 24th ARRB Conference*. Melbourne, Australia.

Baus R. L. and W. Hong. 2004. Development of profiler-based rideability specifications for asphalt pavements and asphalt overlays, Report No. GT04-07. Department of Civil and Environmental Engineering. Columbia, SC: University of South Carolina.

Bay J.A. and K.H. Stokoe. 1998. *Development of a Rolling Dynamic Deflectometer for Continuous Deflection Testing of Pavements*. FHWA/TX-99/1422-3. Austin, TX: Center for Transportation Research, The University of Texas at Austin.

Bay J.A., K.H. Stokoe, B.F. McCullough, and D.R. Alexander. 1999. Profiling flexible highway pavement continuously with rolling dynamic deflectometer and at discrete points with falling weight deflectometer. *Transportation Research Record*, No. 1655, pp. 74–85. Washington, DC: TRB, National Research Council.

Bergström A., H. Åström, and R. Magnusson. 2003. Friction measurement on cycleways using a portable friction tester. *Journal of Cold Regions Engineering*, Vol. 17, No. 1, pp. 37–57.

Bowman B.L., B.P. Ellen, M. Stroup-Gardiner, and J. Fennell. 2002. *Evaluation of Pavement Smoothness Reduction and Pay Factor Determination for Alabama Department of Transportation*. Final Report 930-371. Auburn, Alabama: Highway Research Center and Department of Civil Engineering at Auburn University.

Bretonniere S. 1963. Etudes d'un deflectometer a boulet. *Bulletin de Liaison des Laboratoires Routiers*, Vol. 2, No. 2, pp. 43-5–43-16. Paris.

Brittain S. 2011. *Speed Correction for SCRIM Survey Machines*. Published Project Report PPR 587. Wokingham, UK: Transport Research Laboratory.

Brokaw M.P. 1973. *Development of the Road Meter*. Special Report 133. Washington, DC: Highway Research Board.

Caltrans. 2012. *Method of Test for Operation of California Profilograph and Evaluation of Profiles*. California Test 526. California: California Department of Transportation, Engineering Service Center.

Carey W.N., H.C. Huckins, and R.C. Leathers. 1962. *Slope Variance as a Measure of Roughness and the CHLOE Profilometer*. HRB Special Report 73. Washington, DC: Transportation Research Board.

Carey W.N. and P.E. Irick. 1960. The pavement serviceability-performance concept. *Highway Research Board Bulletin*, Vol. 250, p. 40.

CEDEX (Center for the Studies and Experimentation of Public Works). 2006. *Measuring of Pavement Deflections with a Curviameter*. NLT-333. Madrid, Spain: CEDEX.

CEN EN 13036-1. 2010. *Road and airfield surface characteristics – Test methods – Part 1: Measurement of pavement surface macro-texture depth using a volumetric patch technique*. Brussels: CEN.

CEN EN 13036-4. 2011. *Road and airfield surface characteristics – Test methods – Part 4: Method for measurement of slip/skid resistance of a surface – The pendulum test*. Brussels: CEN.

CEN EN 13036-6. 2008. *Road and airfield surface characteristics – Test methods – Part 6: Measurement of transverse and longitudinal profiles in the evenness and megatexture wavelength ranges*. Brussels: CEN.

CEN EN 13036-7. 2003. *Road and airfield surface characteristics – Test methods – Part 7: Irregularity measurement of pavement courses: The straightedge test*. Brussels: CEN.

CEN EN 13036-8. 2008. *Road and airfield surface characteristics – Test methods – Part 8: Determination of transverse unevenness indices*. Brussels: CEN.

CEN EN ISO 13471-1. 2004. *Characterization of pavement texture by use of surface profiles. Part 1: Determination of mean profile depth (ISO 13473-1: 1997)*. Brussels: CEN.

CEN/TS 13036-2. 2010. *Road and airfield surface characteristics – Test methods. Part 2: Assessment of the skid resistance of a road pavement surface by the use of dynamic measuring systems*. Brussels: CEN.

CEN/TS 15901-1. 2009. *Road and airfield surface characteristics – Part 1: Procedure for determining the skid resistance of a pavement surface using a device with longitudinal fixed slip ratio (LFCS): RoadSTAR*. Brussels: CEN.

CEN/TS 15901-2. 2009. *Road and airfield surface characteristics – Part 2: Procedure for determining the skid resistance of a pavement surface using a device with longitudinal controlled slip (LFCRNL): ROAR (road analyser and recorder of norsemeter)*. Brussels: CEN.

CEN/TS 15901-3. 2009. *Road and airfield surface characteristics – Part 3: Procedure for determining the skid resistance of a pavement surface using a device with longitudinal controlled slip (LFCA): The ADHERA*. Brussels: CEN.

CEN/TS 15901-4. 2009. *Road and airfield surface characteristics – Part 4: Procedure for determining the skid resistance of pavements using a device with longitudinal controlled slip (LFCT): Tatra runway tester (TRT)*. Brussels: CEN.

CEN/TS 15901-5. 2009. *Road and airfield surface characteristics – Part 5: Procedure for determining the skid resistance of a pavement surface using a device with longitudinal controlled slip (LFCRDK): ROAR (road analyser and recorder of norsemeter)*. Brussels: CEN.

CEN/TS 15901-6. 2009. *Road and airfield surface characteristics – Part 6: Procedure for determining the skid resistance of a pavement surface by measurement of the sideway force coefficient (SFCS): SCRIM®*. Brussels: CEN.

CEN/TS 15901-7. 2009. *Road and airfield surface characteristics – Part 7: Procedure for determining the skid resistance of a pavement surface using a device with longitudinal fixed slip ratio (LFCG): The GripTester®*. Brussels: CEN.

CEN/TS 15901-8. 2009. *Road and airfield surface characteristics – Part 8: Procedure for determining the skid resistance of a pavement surface by measurement of the sideway-force coefficient (SFCD): SKM*. Brussels: CEN.

CEN/TS 15901-9. 2009. *Road and airfield surface characteristics – Part 9: Procedure for determining the skid resistance of a pavement surface by measurement of the longitudinal friction coefficient (LFCD): DWWNL skid resistance trailer*. Brussels: CEN.

CEN/TS 15901-10. 2009. *Road and airfield surface characteristics – Part 10: Procedure for determining the skid resistance of a pavement surface using a device with longitudinal block measurement (LFCSK): The Skiddometer BV-8*. Brussels: CEN.

CEN/TS 15901-11. 2011. *Road and airfield surface characteristics – Part 11: Procedure for determining the skid resistance of a pavement surface using a device with longitudinal block measurement (LFCSR): The SRM*. Brussels: CEN.

CEN/TS 15901-12. 2011. *Road and airfield surface characteristics – Part 12: Procedure for determining the skid resistance of a pavement surface using a device with longitudinal controlled slip: The BV 11 and Saab friction tester (SFT)*. Brussels: CEN.

CEN/TS 15901-13. 2011. *Road and airfield surface characteristics – Part 13: Procedure for determining the skid resistance of a pavement surface by measurement of sideway force coefficient (SFCO): The Odoliograph*. Brussels: CEN.

Chong G.J., W.A. Phang, and G.A. Wrong. 1977. *Manual for Condition Rating of Rigid Pavements: Distress Manifestations*. Ontario Ministry of Transportation and Communications. Ontario: Downsview.

CNS Farnell Ltd (Hertfordshire, UK). 2014. Available at http://www.cnsfarnell.com.

Controls Srl controls@controls.it.

Cook A. 2011. *Performance Assessment of e-Spott Ground Penetrating Radar Device*. TRL PPR675. Crowthorne, UK: Transport Research Laboratory.

Cundill M.A. 1991. *The MERLIN Low-Cost Road Roughness Measuring Machine*. TRRL Report RR 301. Crowthorne, UK: Transport Research Laboratory.

Cundill M.A. 1996. *The MERLIN Road Roughness Machine: User Guide*. TRL Report 229. Crowthorne, UK: Transport Research Laboratory.

Descornet G. 1988. *Inventory of High-Speed Longitudinal and Transverse Road Evenness Measuring Equipment in Europe, FILTER Project*. FEHRL Technical Note 1999/01. Brussels: BRRC (Belgian Road Research Centre).

Douglas Equipment, Curtiss-Wright Flow Control Company. 2014. Available at http://www.douglas.cwfc.com.

Dunford A. 2010. *Grip Tester Trial – October 2009, Including SCRIM Comparison*. Published Project Report PPR 497. Wokingham, UK: Transport Research Laboratory.

Dynatest International A/S. 2014. Available at http://www.dynatest.com.

Eagleson B., S. Heisey, W.R. Hudson, A.H. Meyer, and K.H. Stokoe. 1982. *Comparison of the Falling Weight Deflectometer and the Dynaflect for Pavement Evaluation*. FHWA Research Report 256-1. Washington, DC: U.S. Department of Transportation, Federal Highway Administration.

Elseifi M., A.M. Abdel-Khalek, and K. Dasari. 2012. *Implementation of Rolling Wheel Deflectometer (RWD) in PMS and Pavement Preservation*. FHWA/11.492. Baton Rouge, LA: Department of Civil and Environmental Engineering, Louisiana State University.

ERI (Engineering and Research Int'l.) Inc., USA. 2014. Available at http://www.erikuab.com.

Euro Consulting Group. Available at http://www.euconsultinggroup.com.

Euroconsult. 2014. Available at http://www.roadsurveydevice.com.

FACE Construction Technologies Inc. 2014. Available at http://www.dipstick.com.

FHWA. 2011. *Rolling Wheel Deflectometer Network-Level Pavement Structural Evaluation, Measuring Deflection at Highway Speeds*. FHWA, U.S. Department of Transportation. Available at http://www.fhwa.dot.gov/pavement/management/rwd.

Findlay Irvine Ltd. Available at http://www.findlayirvine.com.

FKFS. Available at http://www.fkfs.de.

Flintsch G., S. Katicha, J. Bryce, B. Ferne, S. Nell, and B. Diefenderfer. 2013. *Assessment of Continuous Pavement Deflection Measuring Technologies*. SHRP 2 Report S2-R06F-RW-1. Washington, DC: TRB, National Academy of Sciences.

Furgo Roadware. Available at http://www.furgoroadware.com.

Gedafa D.S., M. Hossain, R. Miller, and D. Steele. 2008. Network level pavement structural evaluation using rolling wheel deflectometer. Paper No. 08-2648. *87th Transportation Research Board Annual Meeting*. Washington, DC.

Giles C.G., B. Sabey, and K. Cardew. 1964. *Development and Performance of the Portable Skid Resistance Tester*. Department of Industrial and Scientific Research. RRL Technical Paper No. 66. London: HMSO.

Gopaldas J., R.E. Lodge, and A. Wright. 2009. *Developing a New Method Measuring Pavement Layer Thickness*. TRL PPR390. Crowthorne, Wokingham, Berkshire, UK: Transport Research Laboratory.

Gorski M. 1999. Residual service life of flexible pavements and its impact on planning and selecting priorities for the structural strengthening of road networks. *XXI Congres Mondial de la Route (AIPCR)*. Kuala Lumpur, Malaisie.

Greenwood Engineering A/S. Available at http://www.greenwood.dk.

Grontmij A/S. Available at http://www.pavement-consultants.com.

GSSI Inc. Available at http://www.geophysical.com.

Hall J.W., K.L. Smith, L. Titus-Glover, J.C. Wambold, T.J. Yager, and Z. Rado. 2009. *Guide for Pavement Friction*. NCHRP Web only document 108. Transportation Research Board (TRB).

Hernán de Solminihac T., R. Salsilli, E. Köhler, and E. Bengoa. 2003. Analysis of pavement serviceability for the AASHTO design method: The Chilean case. *The Arabian Journal for Science and Engineering*, Vol. 28, No. 2B.

Highways Agency. Available at http://www.highways.gov.uk.

Highways Agency. 2004. *Design Manual for Roads and Bridges (DMRB), Volume 7: Pavement Design and Maintenance*. Section 3, Part 1, HD 28/04. London: Department for Transport, Highways Agency.

Highways Agency. 2008a. *The Manual of Contract Documents for Highway Works, Volume 1: Specification for Highway Works*. Series 900, Road Pavements-Bituminous bound materials. London: Department for Transport, Highways Agency.

Highways Agency. 2008b. *Design Manual for Roads and Bridges (DMRB), Volume 7: Pavement Design and Maintenance*. Section 3, Part 2, HD 29/08. London: Department for Transport, Highways Agency.

Highways Agency. 2008c. *Design Manual for Roads and Bridges (DMRB), Volume 7: Pavement Design and Maintenance.* Section 3, Part 3: HD 30/08. London: Department for Transport, Highways Agency.

Highways Agency. 2009. *The Manual of Contract Documents for Highway Works (MCDHW), Volume 1: Specification for Highway Works.* Series 700: Road pavements – General. London: Department for Transport, Highways Agency.

Hosking J.R. and G.C. Woodford. 1976a. *Measurement of Skidding Resistance, Part I, Guide to the Use of SCRIM.* Department of the Environment. TRRL Report, LR 737. Crowthorne, UK: Transport Research Laboratory.

Hosking J.R. and G.C. Woodford. 1976b. *Measurement of Skidding Resistance, Part III, Factors Affecting SCRIM Measurements.* Department of the Environment. TRRL Report, LR 739. Crowthorne, UK: Transport Research Laboratory.

Hosking R. 1992. *Road Aggregates and Skidding.* TRL, HMSO. London: Transport Research Laboratory.

Hveem F.N. 1955. Pavement deflections and fatigue failures. *Highway Research Board Bulletin*, Vol. 114. Highway Research Board.

Hveem F.N. 1960. Devices for recording and evaluating pavement roughness. *Highway Research Board Bulletin*, Vol. 264. Washington, DC: Highway Research Board.

Imine H., Y. Delanne, and N.K. M'Sirdi. 2006. Road profile input estimation in vehicle dynamics simulation. *Vehicle System Dynamics*, Vol. 44, No. 4. Taylor & Francis.

International Cybernetics Corp. Available at http://www.intlcybernetics.com.

Jacobs F.A. 1983. *M40 High Wycombe By-Pass: Results of a Bituminous Surface Texture Experiment.* Department of the Environment, Department of Transport. TRRL Report, LR 1065. Crowthorne, UK: Transport Research Laboratory.

Janoff M.S., J.B. Nick, P.S. Davit, and G.F. Hayhoe. 1985. *Pavement Roughness and Rideability.* NCHRP Report 275. National Research Council. Washington, DC: Transport Research Board.

JILS USA. Available at http://www.jilsfwd.com.

Jordan P.G. and J.C. Young. 1980. *Developments in the Calibration and Use of the Bump-Integrator for Ride Assessment.* TRRL Supplementary Report, SR 604. Crowthorne, Berkshire, UK: Transport Research Laboratory.

Keir WG. 1974. *Bump-Integrator Measurements in Routine Assessment of Highway Maintenance Needs.* TRRL SR26uc. Crowthorne, UK: Transport Research Laboratory.

Krarup J. 2012. *Traffic Speed Deflectometer (TSD).* Measuring systems for infrastructure – For roads and for railroads. Sacramento, CA: FWD User's Group.

LCPC, now IFSTTAR. Available at http://www.ifsttar.fr.

Less R.D. 1974. *Measurement of Surface Variations.* Final Report R-250. Iowa State Highway Commission, Materials Department.

Lucas J. and A. Viano. 1979. *Systematic Measurement of Evenness on the Road Network: High Output Longitudinal Profile Analyser.* LCPC Report No. 101. Paris: LCPC.

MALÅ. Available at http://www.malags.com.

Martin J.-M., F. Menant, and F. Lepert. 2012. Information relating to the measurement of the longitudinal evenness in France. *7th Symposium on Pavement Surface Characteristics: SURF 2012.* Virginia Tech Transportation Institute (VTTI).

Maurer P. 2004. The influence of different measuring tyre, measuring speed and macro-texture on skid resistance measurements. *Proceedings of SURF 2004.* Toronto, Canada.

McGhee K.H. 2004. *Automated Pavement Distress Collection Techniques.* NCHRP Synthesis 334. Washington, DC: Transportation Research Board.

MDRW (Ministère de la Région Wallonne). 2002. *54.26 Portance (déflexion du revêtement au passage d'un essieu), Chapitre 54 du Catalogue des Méthodes d'Essais, annexe au Cahier de Charges Type du Ministère de la Région wallonne.* Brussels, Belgium: Ministère de la Région Wallonne.

Mondal A., A.J. Hand, and D.R. Ward. 2000. *Evaluation of Lightweight Non-Contact Profilers.* Publication FHWA/IN/JTRP-2000/06. West Lafayette, Indiana: Joint Transportation Research Program, Indiana Department of Transportation and Purdue University.

Morosiuk G., M. Riley, and J.B. Odoki. 2004. *Modelling Road Deterioration and Works Effects.* Version 2, HDM-4. The Highway Development and Management Series, Vol. 6. Paris: World Road Association.

Moventor Oy Inc. Available at http://www.moventor.com.

Mrawira D. and R. Haas. 1996. Calibration of the TRRL's vehicle-mounted bump integrator. *Tanzania Engineer, Journal of the Institution of Engineers Tanzania*, Vol. 5, No. 5, pp. 13–24.

Muzet V., C. Heinkele, Y. Guillard, and J.M. Simonin. 2009. Surface deflection measurement using structured light. *7th International Conference on Non-destructive Testing in Civil Engineering (NDTCE'09)*. Nantes, France.

NCHRP. 2000. *Evaluation of Pavement Friction Characteristics.* NCHRP Synthesis Report 291. Transportation Research Board (TRB).

Nikolaides A., D. Evangelides, and G. Mintsis. 1992. Management of flexible pavement maintenance problem using PC software (PAVMAIN). *1st Panhellenic Congress of Bituminous Mixtures and Flexible Pavements*, p. 467. Thessaloniki, Greece.

Nikolaides A.F. 1978. Methods, factors, recommendations and materials in European skidding resistance. M.Sc. Thesis, University of Bradford, England.

Nippo Sangyo Co. Ltd. Available at http://www.nippou.com.

Norman P.J., R.A. Snowdon, and L.C. Jacobs. 1973. *Pavement Deflection Measurements and Their Application to Structural Maintenance and Overlay Design.* TRRL Report LR 571. Transport and Road Research Laboratory. Crowthorne, UK: Transport Research Laboratory.

NZTA. 2012. *Notes to Specification for State Highway Skid Resistance Management.* NZTA T10 Notes. New Zealand Transport Agency. Waka Kotahi.

OECD. 1984. *Road Transport Research, Road Surface Characteristics: The Interaction and Their Optimisation.* Road Transport and Intermodal Linkages Research Programme, Organisation for Economic Co-operation and Development. Paris: OECD.

Oliver J.W.H. 1980. *Temperature Correction of Skid Resistance Values Obtained with the British Portable Skid Resistance Tester.* AIR 314-2. Melbourne: ARRB.

Paquet J. 1978. *The CEBTP Curviameter – A New Instrument for Measuring Highway Pavement Deflection.* Paris, France: Centre Experimental de Recherches et d'Etudes du Batiment et des Travaux Publics.

Paterson W.D.O. 1987. *Road Deterioration and Maintenance Effects.* The John Hopkins University Press, Published for the World Bank.

Pavement Interactive. Available at http://www.pavementinteractive.org/article/deflection/.

PIARC. 1991. Surface characteristics. *Proceedings of the XIXth World Road Congress*, Technical Committee on the Surface Characteristics, Report 19.01B. Marrakech.

PipeHawk Plc. Available at http://www.pipehawk.com.

Priem H. 1989. *NAASRA Roughness Meter Calibration via the Road-Profile-Base International Roughness Index (IRI).* Research report ARR No. 164. Melbourne, Australia: Australian Road Research Board.

Ramos García J.A., Sánchez Domínguez, F. and Álvarez Loranca, R. 2013. Structural road surveys using a Curviameter device. Analysis of the temperature effect on asphalt mixes properties. *9th International Conference on the Bearing Capacity of Roads, Railways and Airfields (BCRRA)*. Trondheim, Norway, June.

Rainhart Co. 2014. Available at http://www.rainhart.com.

ROMDAS. 2014. Available at http://www.romdas.com.

RRL. 1969. *Instruction for Using the Portable Skid-Resistance Tester.* Road Note No. 27, Road Research Laboratory, Department of Science and Industrial Research. London: HMSO.

Rutherford M. 1993. *Road Rater Study. WA-RD 334.1 for WSDoT.* Springfield: National Technical Information Service.

Sabey B.E. 1966. *The Road Surface Texture and the Change in Skidding Resistance with Speed.* Ministry of Transport, Road Research Laboratory, Report LR 20. Harmondsworth (Crowthorne), UK: Transport Research Laboratory.

Sánchez D.F. 2008. New regulations for the measuring of deflections with a curviameter. *Symposium on the Measuring of Texture and Structural Condition of Pavements and Road Surface of Spain's Road Network, CEDEX.* Madrid, Spain, May.

Sarsys AB. Available at http://www.sarsys.se.

Sayers M.W., T.D. Gillespie, and W.D.O. Paterson. 1986b. *Guidelines for Conducting and Calibrating Road Roughness Measurements.* World Bank Technical Paper No. 46. Washington, DC: The World Bank.

Sayers M.W., T.D. Gillespie, and C.A.V. Queiroz. 1986a. *The International Road Roughness Experiment: Establishing Correlation and a Calibration Standard for Measurements.* World Bank Technical Paper No. 45. Washington, DC: The World Bank.

Sayers M.W. and S. Karamihas. 1996. *Interpretation of Road Roughness Profile Data.* UMTRI 96-19, Final Report prepared for Federal Highway Administration. FHWA Contract DTFH 61-92-C00143.

Sayers M.W. and S.M. Karamihas. 1998. *The Little Book of Profiling.* Michigan: The Regent of the University of Michigan.

Scala A.J. and Potter D.W. 1977. *Measurement of Road Roughness.* Technical Manual ATM 1. Australian Road Research Board.

Shahin M.Y. 2005. *Pavement Management for Airports, Roads and Parking Lots,* 2nd Edition. USA: Springer. Available at http://www.cecer.army.mil/paver.

Simonin J.-M., D. Lièvre, G. Hildebrand, and S. Rasmussen. 2005. Assessment of the Danish high speed deflectograph in France. *BCRA Conference.* Paris.

Steele D., J. Hall, R. Stubstad, A. Peekna, and R. Walker. 2009. Development of a high speed rolling wheel deflectometer. *88th Transportation Research Board Annual Meeting.* Washington, DC.

Still P.B., M.H. Burtwell, and J.C. Young. 1983. *Evaluation of a Microprocessor System for Bump-Integrators.* TRRL LR 1083. Crowthorne, Wokingham, Berkshire, UK: Transport Research Laboratory.

SSI (Surface Systems & Instruments) Inc. Available at http://www.smoothroad.com.

Swiss Road Standards. 1989–1992. *Swiss Road Standards for Road Maintenance Management System: No. 640 900: Basis, (1989), No. 640 901: System of objectives (1990), No. 640 902: Guide of implementation (1992), No. 640 925: Recording of state-data (1990).* Zurich: VSS (Union of Road Experts of Switzerland).

T&J Farnell Ltd. Available at http://www.tjfarnell.co.uk.

Texas A&M Transportation Institute (TTI). 2013. Using new technologies to solve old problems. *Texas Transportation Researcher,* Vol. 49, No. 4. Available at http://www.tti.tamu.edu.

TRL DFID. 1999. *A Guide to the Pavement Evaluation and Maintenance of Bitumen-Surfaced Roads in Tropical and Sub-Tropical Countries.* Overseas Road Note 18. Crowthorne, UK: Transport Research Laboratory.

TYROSAFE (Tyre and Road Surface Optimisation for Skid Resistance and Further Effects). 2008. *D04 Report on state-of-the-art of test methods.* FEHRL, 7th Framework Programme, December.

Ullidtz P. 1987. *Pavement Analysis.* Elsevier.

Van Geem V. 2010. Overview of interpretation techniques based on measurement of deflections and curvature radius obtained with the Curviameter. *6th European FWD User's Group Meeting.* Brussels.

Vectra France. Available at http://www.vectrafrance.com.

VTI, Sweden. Available at http://www.vti.se.

Walker R.S. and W.R. Hudson. 1973. *Method for Measuring Serviceability Index with Mays Road Meter.* Special Report 133. Washington, DC: Highway Research Board.

Wallman G. and H. Astrom. 2001. *Friction Measurement Methods and the Correlation between Road Friction and Traffic Safety-A Literature Review.* VTI Report 911A. Swedish National Road and Transport Research Institute.

Wang K. 2011. Prototyping automated distress survey based on 2D/3D laser images. *Proceedings of the 5th International Conference 'Bituminous Mixtures & Pavements'.* Thessaloniki, Greece.

Wang K., L. Li, and M. Moravec. 2011. Approaches to determining pavement texture with 1mm 3D laser images. *Proceedings of the 5th International Conference 'Bituminous Mixtures & Pavements'*. Thessaloniki, Greece.

Wang K.C.P. 2000. Design and implementation of automated systems for pavement surface distress Survey. *ASCE Journal of Infrastructure Systems*, Vol. 6, No. 1, pp. 24–32.

Wang K.C.P. and Gong W.G. 2005. The real-time automated system of cracking survey in a parallel environment. *ASCE Journal of Infrastructure System*, Vols. 11–13, pp. 154–164.

Wang M.C. 1985. Pavement response to Road Rater and axle loadings. *TRB Transportation Research Record*, No. 1043. Washington, DC: Transportation Research Board.

WayLink Systems Co. Available at http://www.waylink.com.

WDM Limited, UK. Available at http://www.wdm.co.uk.

Young J.C. 1977. *Calibration, Maintenance and Use of the Rolling Straight Edge*. TRL Supplementary Report, SR 290. Transport and Road Research Laboratory. Crowthorne, Wokingham, Berkshire, UK: Transport Research Laboratory.

Chapter 17

Pavement management

17.1 GENERAL

Ever since the construction of the first pavement, there was a need for periodic maintenance to keep it passable. This was the first form of pavement management, given that both works and finds required had to be estimated and programmed. At a later stage, the development of engine cars resulted in a dramatic increase in travelled speeds and in the necessity to construct safer and more resistant pavements.

The following additional factors have led to the necessity for more systematic approach of pavement maintenance works:

- The dramatic growth of the road network, its ageing and the need to keep it at a sufficient level of safety and functionality
- The increase of maintenance cost in conjunction with the difficulty to find the funds for maintenance when needed
- The realisation that social activities and environment are directly affected by the condition of the pavement
- The development of new methods and materials for pavement maintenance
- The imperative need for energy saving
- The necessity to re-use constructional materials to conserve natural resources
- The recognition that pavement condition affects user cost
- The development of devices for objective and fast evaluation of pavement condition
- The advanced development of computers and software
- The realisation of implementation of management tools and techniques in pavement engineering

The term that prevailed and began to be used in the late 1970s was *pavement management*.
Pavement management involves the implementation of the optimal solution at the most appropriate time, using the most appropriate materials and the most appropriate maintenance methods and techniques, aiming at the lowest possible maintenance cost and a longer service life for pavements. Pavement management is related to secondary capital needed for maintenance, after the main capital is spent for construction.

17.2 TERMS – DEFINITIONS

According to AASHTO (1985), *pavement management* is defined as 'The effective and efficient directing of the various activities involved in providing and sustaining pavements in a condition acceptable to the travelling public at the least life cycle cost'. Pavement management can be seen as the process of planning the maintenance and rehabilitation works to optimise pavement conditions.

The necessity for pavement management led to the development of the *pavement management system (PMS)*. PMS is a tool to aid pavement management. It may be seen as the process of managing the activities involved in pavement management in a systematic and coordinated way.

According to AASHTO (1993), PMS is defined as 'a set of tools or methods that assist decision makers in finding optimum strategies for providing, evaluating and maintaining pavements in a serviceable condition over a period of time'.

Apart from PMS, the term *pavement management program (PMP)* is also used by some agencies. PMP has the same structure and is similar to PMS but the financial aspect at the stage of selection and programming of works is perhaps considered to be of greater extent (budget predictions, source allocation, etc.).

It could be said that PMP provides a managerial tool of detailed network/project record, pavement condition record, maintenance history record, maintenance/rehabilitation strategies determination and long-term budget predictions and source allocations. PMP is absolutely essential for consortia undertaking highway projects by concession agreement, as they are obliged to manage the pavement in the most efficient way, providing an adequate and safe level of services (Nikolaides 2011).

In general, pavement management aims to manage pavement maintenance and rehabilitation in the most cost-effective way. This is achieved using a combination of low-cost treatments on pavements in good condition and more costly treatments on heavily distressed pavements.

Finally, a PMP may be manual or computerised. Computerised PMSs can manage and analyse more rapidly large databases derived from functional and structural pavement surveys, and most are able to forecast future pavement conditions.

PMSs may be developed for either a rural or an urban network; the principles for developing such systems or programs are the same. However, because of the differences in data reference system, functionality and priority policies, a PMS for a rural and an urban network should be developed separately. A differentiation may also exist among PMSs for motorway, national or provincial urban roads. In any case, a PMS is developed for the network level or project level.

Pavement management is a very wide subject and it is beyond the intention of this book to cover it in detail. The scope of this chapter is only to give the reader some useful general information.

17.3 PURPOSE OF PAVEMENT MANAGEMENT

The purpose of pavement management is to get financial, technical, organisational and administrative benefits (OECD 1987).

17.3.1 Financial benefits

Pavement management should aim to maximise net financial profits in relation to the financial restrictions imposed. This is achieved by the following:

a. Appropriate management of the available funds
b. Programming the maintenance/rehabilitation works in accordance with the available funds
c. Determination of the impact of various maintenance/rehabilitation alternative solutions to the cost of the proprietor and to the cost of the user
d. Determination of the impact of construction quality to the user cost
e. Objective evaluation/selection of the optimal choice, based solely on cost/benefit analysis

17.3.2 Technical benefits

Pavement management should offer technical benefits. To achieve technical benefits, a PMS should

a. Be composed of an extensive and integrated database (data bank) that should be constantly updated
b. Be reformed using the experience of the past and the technological advances of the present, by ameliorating the maintenance and construction techniques and avoiding the same mistakes
c. Choose the most appropriate maintenance/rehabilitation method
d. Use reliable forecast models of the pavement behaviour and reliable cost/benefit estimation models
e. Use criteria that aid decision making, such as desired level of pavement condition, warning level and intervention level

17.3.3 Organisational benefits

In pavement management, there should also be organisational benefits; to achieve this, a PMS should

a. Be able to reasonably determine the pavement condition at the network or project level
b. Plan and program both the present and future maintenance activities
c. Use the most effective and efficient methodology of systematic monitoring of the pavement condition
d. Predict the consequences that will result from different financing
e. Provide an objective basis for political decisions

17.4 LEVELS OF PAVEMENT MANAGEMENT ANALYSIS

PMSs used in decision making are distinguished into three reference levels: the *project* level, the *network* level and the *strategic* level.

The project-level PMS considers a specific section (or sections) of the road network. The network-level PMS considers the road network of a wide area such as a district or county. The strategic-level PMS considers the entire road network of a region or even of the whole country or state.

Some modern integrated PMSs are developed to operate at all levels depending on the requirements.

17.4.1 Pavement management at the project level

Pavement management at the project level involves decisions regarding the maintenance and rehabilitation of specific pavement sections, defining the 'project'. The decisions are made by operation engineers with good technical background and based on technical merits rather than on resource requirements and budget projections.

In a PMS at the project level, a detailed functional and structural evaluation of the pavement sections is conducted, the causes of deterioration are identified, followed by the selection of the appropriate intervention (routine maintenance, maintenance, rehabilitation or reconstruction). Figure 17.1 illustrates the elements contained in a PMS at the project level.

The functional evaluation of the pavement is usually carried out by a visual condition survey, coring and the use of necessary devices measuring surface characteristics such as

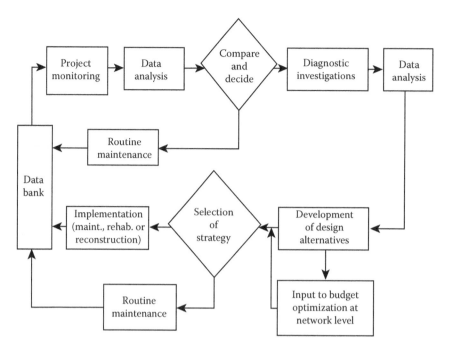

Figure 17.1 Elements of PMS at the project level. (From Peterson, D.E., *Pavement Management Practices, NCHRP Synthesis of Highway Practice 135*. National Research Council. Washington, DC: Transportation Research Board, 1987.)

skid resistance and transverse/longitudinal evenness. The decisions are usually based on the deterioration rate of the pavement section, assisted by historical maintenance and construction data. When alternative solutions are considered, life cycle cost analysis is applied. All collected data from the sections at the project level are stored in a main data bank in which data from other projects and from the road network are also stored.

In particular, a PMS at the project level

a. Considers all basic pavement design parameters such as subgrade strength, traffic volume, properties of materials, climatic conditions, cost of materials, age of pavement and remaining life
b. Deals with detailed pavement analysis, determines the cause for each individual distress and specifies corrective measures or suggests alternatives solutions
c. Applies life cycle cost analysis when alternatives are considered
d. Decides whether maintenance, rehabilitation or reconstruction will be implemented at each specific pavement section and establishes priorities
e. Selects the type of materials to be used and specifies rehabilitation thickness
f. Presents pavement condition, results and recommendations in tabular or graphical form
g. Feeds the main data bank with inventory data, pavement characteristics, pavement behaviour and corrective measures taken in the specific project sections

17.4.2 Pavement management at the network level

Pavement management at the network level deals with summary information related to the network concern, aiming at prioritisation of maintenance and rehabilitation works with respect to the amount of funding available.

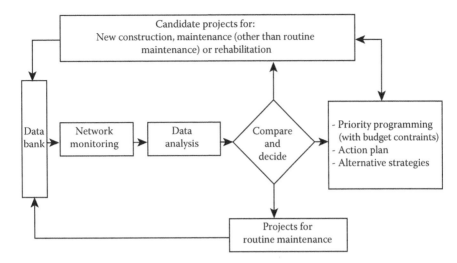

Figure 17.2 Elements of PMS at the network level. (From Peterson, D.E., *Pavement Management Practices, NCHRP Synthesis of Highway Practice 135.* National Research Council. Washington, DC: Transportation Research Board, 1987.)

The decisions are made by senior executives. As it is stated (Asphalt Institute 2009), they make decisions that play a part in determining pavement performance targets, distributing funds among regions or districts and establishing pavement preservation policies. Figure 17.2 illustrates the elements contained at the network level.

The evaluation of the functional and structural condition of the network pavements is carried out by traffic speed moving devices; limited coring and falling weight deflectometer (FWD) measurements are conducted only to confirm the findings from moving devices and obtain more structural details on pavement layers at selective locations.

All collected data, GIS referenced, are stored in the main data bank for future use.

A PMS at the network level

a. Depicts the current pavement condition of the network
b. Predicts and projects future needs
c. Identifies candidate projects for improvements
d. Prioritise the candidate projects
e. Determines budget requirements for short- and long-term needs
f. Estimates the consequences of the alternative fund investments on the future behaviour of pavement
g. Determines the final work plan, usually by an iterative process that involves moving neighbouring projects from one year to another, or combining similar actions, to gain economy of scale
h. Presents network pavement condition in map and tabular form
i. Feeds the main data bank with all data collected from the network and with the final decisions made regarding corrective measures

17.4.3 Pavement management at the strategic level

Pavement management is used nowadays on strategic decisions made by government officials, transportation boards, city councils or an agency's upper management. All are charged with long-term decision making based on pavement performance targets, fund requirements to achieve the performance targets, distribution of funds among regions or districts and pavement preservation policy.

Traditionally, strategic decisions have been less structured than decisions made at other levels and the information on which decisions are based is more speculative, requiring the ability to predict future conditions under a variety of scenarios. In the absence of reliable information to serve as the basis for sound business decisions, political priorities may prevail (AASHTO 2012).

17.5 PAVEMENT MANAGEMENT COMPONENTS

A PMS, irrespective of the analysis level, is composed of the following components:

a. Pavement inventory
b. Pavement condition information (survey)
c. Traffic data
d. History of post works
e. Database
f. Analysis module
g. Reporting module

The interrelation of the above components is explained in Figure 17.3.

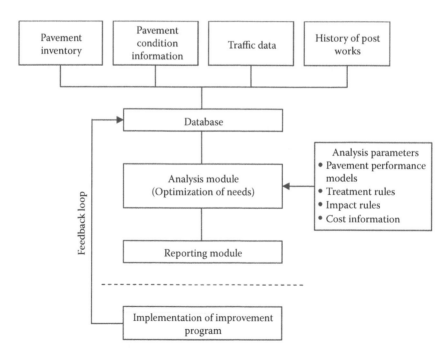

Figure 17.3 Basic pavement management components.

17.5.1 Pavement inventory

The pavement inventory has a supportive and informative character; however, it is absolutely necessary for pavement management. The minimum pavement characteristics to be recorded are as follows:

- Jurisdictional information, such as concessionaire, district, region or city.
- Location information, which includes the beginning and end point of each pavement segment.
- Road classification, for example, highway, principal, secondary or arterial road, cross-road, interchange and so on. A term often used is branch identification. A branch is an easily identifiable part of the pavement network with a distinct use.
- The type of pavement and type of shoulder (when it exists).
- The dimensions of the road (branch), for example, length, width and number of lanes, width of shoulders (when they exist) and so on.
- The historical construction data, for instance, year/month of construction, one stage or planned stage construction, maintenance/rehabilitation history, materials used, layer thickness, bearing capacity of formation layer material and anything else considered useful.
- Past traffic data.

Pavement inventory may also contain additional data such as bridge or underpass locations, road marking, streetlights, safety barriers, street furniture in general, accident records, construction, maintenance, rehabilitation cost, contracting companies' data, noise and air pollution measurements and anything else related to the road.

The abovementioned complementary data are not of course all necessary for a pavement management study. They are useful, however, as they provide an integrated picture of the road. The existence of all the abovementioned characteristics makes the data bank more complete and offers the opportunity to take other decisions being part of road management.

17.5.2 Pavement condition information

Pavement condition information is perhaps the most fundamental of all input data, as it will define the present and future needs for maintenance or rehabilitation works.

Pavement condition is assessed by pavement condition surveys carried out visually or by the use of appropriate static or mobile devices, measuring functional and structural pavement properties.

The results of pavement condition surveys are expressed in terms of condition indices. The techniques and means of measuring functional and structural condition of a pavement are discussed in Chapter 15.

The techniques range from visual-windscreen subjective surveys to mechanistic (use of automated devices operating at near-traffic speeds) objective surveys. It is up to the agency to decide which approach to use considering the length of pavement to be surveyed, availability of devices and running expenses. The agency should also consider the importance of updating the pavement condition frequently.

In all cases, survey procedures should be consistent from one survey to another in order for the information gathered to be comparable.

17.5.3 Traffic data

Historic traffic data and traffic counts are also very important because they are necessary to determine past traffic and predict future pavement condition and remaining pavement life.

Historic traffic data are relatively easy to be found in most countries. If not available, past traffic can be estimated from traffic counts executed during surveying period, the average annual increase of commercial vehicles from date of construction or last intervention and number of years elapsed.

17.5.4 History of post works

The history of post-construction works, if carried out, must be known in detail. This provides valuable information on estimating rates of deterioration for locally existent conditions. Initial construction details and history of post-construction works link the differences in pavement performance to pavement structure characteristics.

17.5.5 Database

The inventory, pavement condition, traffic and historic works data should be stored in a database. The data storage can range from simple spreadsheets to a rational computerised database. The latter is advantageous since retrieving, sorting and updating data become an easier task; it also provides access and the possibility of distant sharing by all agencies' offices.

A fundamental characteristic of a database is its location referencing system (LRS) and its segmentation.

Existing LRSs are almost exclusively linear and highway or street oriented. Figure 17.4 shows a simplified linear method based on the mile post system. The linear method based on interchange numbering has also been used (see Figure 17.5).

The emergence of global positioning systems and other spatial technologies drives the need for an LRS that can accommodate and integrate data expressed in multiple dimensions (AASHTO 2012).

More information regarding linear LRSs and multimodal LRSs used in the United States is provided by Vonderohe et al. (1997) and Adams et al. (2001).

The presentation of the content of a database may be conducted in various ways, such as

   a. Text and tabular format (the simplest format)
   b. Diagrammatic format (graphic representation of tables)

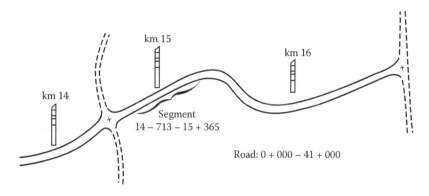

Figure 17.4 Linear LRS based on mile posts.

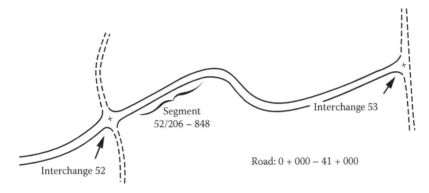

Figure 17.5 Linear LRS based on interchange numbering.

c. Road profile format (linear graphs showing information along the linear representation of the pavement)
d. Network maps format (specialised map drawings, may be GIS linked, showing pavement characteristic features)

The presentation of the content of the database should be simple and comprehensible by the user. For this reason, most of the PMSs use a combination of two or more data presentation formats.

17.5.6 Analysis module

The analysis module is the heart of a PMS since, after processing and analysing the data, the maintenance/rehabilitation program is optimised within given constrains, usually funding limitations. Project-level optimisation and network-level optimisation may share the same methods: ranking or benefit/cost analysis. However, the interface between project-level management and network-level management cannot be managed easily.

To support the analysis and prior optimisation, some parameters must be established. The most common ones are the pavement performance models, the treatment rules, the impact rules and cost information.

17.5.6.1 Pavement performance models

The *pavement performance* or *pavement deterioration models* predict future changes in the pavement condition. They are critical because they determine when maintenance/rehabilitation will be required, hence estimating needs for future funding. Thus, pavement performance prediction models should be as reliable as possible.

The above is not an easy task since the general term, *pavement performance*, is affected by many individual functional and structural pavement characteristics such as evenness, rutting, cracking, ravelling, potholing, skid resistance and so on. Each of these characteristics deteriorates differently depending on the existing conditions and type of pavement and hence requires a different way of treatment. To assist pavement management, for all pavement characteristic properties, there should be an individual unique deterioration model; an almost impossible task.

However, the World Bank initiated a study in 1969 that later became a large-scale program of collaborative research, known as Highway Design and Maintenance Standards

Study (HDM Study), from which prediction models were established for almost all pavement distresses. The prediction models developed were reported by Paterson (1987) and incorporated in the highway design and maintenance standards model, HDM-III model (Watanatada et al. 1987). These models improved upon the introduction of the HDM-4 models (Morosiuk et al. 2004). These distress models could assist pavement management.

Table 17.1 presents the prediction models used in HDW-4 for distresses considered in flexible and concrete pavements (excluding shoulders).

Most of the HDM-4 performance prediction models for flexible pavements have been calibrated to suit Australasia's conditions (Austroads 2008).

Distress prediction models have also been developed under the MEPDG (AASHTO 2008), which again after local calibration could assist pavement management. The distresses considered under MEPDG are shown in Table 17.2.

Although the national calibration–validation process was successfully completed for MEPDG in the United States (NCHRP 2004), further calibration and validation studies in accordance with local conditions were recommended by the MEPDG to improve the accuracy of MEPDG distress prediction models. Some of the evaluation studies have already been completed (Ceylan et al. 2013; FHWA 2010; Hall et al. 2011; Rahman et al. 2011).

A prediction model of the composite distress index PCI (pavement condition index), when properly developed, can also be used in PMSs. A successful example is the PCI prediction model developed and incorporated in the MicroPaver software (Shahin 2005).

Table 17.1 HDM-4 performance prediction models per distress type

Flexible pavements	Rigid pavements	
	JPCP and JRCP	CRCP
Distress type and unit of measurement		
Structural cracking (all and wide cracking)[a]	Transverse joint faulting (for slabs with or without dowels), mm	Roughness, PSR[h,i]
Reflection cracking[a]	Transverse joint spalling in JPCP, % of spalled joints	Failures on CRCP, total number of fails in the most trafficked lane
Transverse thermal cracking[b]	Transverse joint spalling in JRCP, % of spalled joints	Skid resistance (prediction only), SFC_{50}[g]
Ravelling[c]	Transverse cracking in PCP, % of cracked slabs	
Potholing[d]	Transverse cracks for JRCP, number/mile	
Rutting[e], in mm	Percentage of cracked slabs for JPCP, % of cracked slabs	
Roughness, in m/km IRI	Roughness for JPCP, m/km IRI	
Macrotexture, texture depth, in mm[f]	Roughness for JRCP, PSR[h,i]	
Microtexture, skid resistance, in SFC_{50}[g]		

[a] Initiation cracking, in years, and progression cracking, in % of total cwy area.
[b] Initiation cracking, in years, and progression cracking, in number/km.
[c] Initiation ravelling, in years, and progression ravelling, in % of total cwy area.
[d] Initiation ravelling, in years, and progression potholing, in total number of additional pothole units/km.
[e] Different model for rutting owing to structural deformation, plastic deformation or surface wear (studded tyres).
[f] Incremental change of sand patch texture depth during analysis year.
[g] SFC_{50}, sideway force coefficient at 50 km/h.
[h] PSR, present serviceability rating.
[i] To convert PSR to IRI use the following equation: $IRI = -3.67 \times \log_e(0.2 \times PSR)$.

Table 17.2 MEPDG performance prediction models per distress type

Flexible pavements and HMA overlays	Rigid pavements	
	JPCP	CRCP
Distress type and unit of measurement		
Rutting, in inches	Transverse cracking, % of cracked slab	Punchouts, number/mile
Load-related cracking: • Bottom/up cracking (alligator cracking), % of total lane area • Top/down cracking (longitudinal), ft/mile	Mean transverse joint faulting, inches	Smoothness (roughness), inches/mile IRI
Thermal cracking (transverse cracking), ft/mile	Smoothness (roughness), inches/mile IRI	
Reflection cracking in overlays, total reflected cracked area for month m[a]		
Smoothness (roughness), inches/mile IRI		

[a] Data from AASHTO, *Mechanistic-Empirical Pavement Design Guide (MEPDG)*. American Association of State Highway and Transportation Officials. Washington, DC: AASHTO, 2008.

Finally, when FWD measurements are available from which the developed strains at the critical interlayer can be determined, fatigue equations derived in relation to the tensile stain of the asphalt or the vertical strain of the subgrade criteria may also be used in PMS to predict the structural failure of the pavement, hence prolonging the pavement's life.

The above approach, assisted by pavement condition surveys, has been successfully used by the author for the development of PMSs on the project level; one project was 112 km long and another project was 230 km long (Nikolaides 2008, 2010).

In the same pavement management studies, it was found that the remaining life estimated from FWD measurements correlated well with the remaining life determined by back calculations using the design nomographs of AUTh pavement design methodology for flexible pavements (see Section 13.6). In this approach the provision of detailed past traffic data is essential.

17.5.6.2 Treatment rules

Treatment rules must be defined to describe the conditions under which intervention works (maintenance, rehabilitation or reconstruction) are considered to be feasible. The degree of sophistication for defining treatment rules depends on the agency.

The treatment rules are as follows:

a. For the restoration of skid resistance, apply micro-surfacing if the pavement is in good condition and daily traffic is less than a certain level, or alternatively, apply asphalt concrete for very thin layers (25 to 30 mm) (AC-VTL) if the pavement is in good condition irrespective of traffic volume.

b. Mill and overlay with 40 mm hot mix asphalt (HMA) (dense asphalt concrete) and an AC-VTL wearing course if the pavement is in fair condition.

c. Mill and overlay with appropriate thickness HMA and an AC-VTL wearing course if the pavement is in poor condition.

Similar rules may be defined for preventing maintenance (crack filling, ravelling, etc.), restoration of longitudinal evenness and transverse evenness (rutting) and so on.

17.5.6.3 Impact rules

A PMS should include impact rules for treatment selection.

An impact rule dictates how much an increase in pavement condition can be expected from the application of a certain treatment and how this treatment will perform with time. A treatment to be applied may restore pavement's condition to excellent, which can last for a number of years. Others may simply delay deterioration for a certain period. The duration or life expectancy for each treatment is critical and is best determined from past experience.

Consideration of the impact rules must be made in the cost analysis to assist in decision making.

17.5.6.4 Cost information

A PMS must provide cost information since, in most cases, it is the decisive parameter in selecting the treatment policy.

Cost information may be a simple budget requirement for each year in the analysis period, for the life cycle cost analysis for each alternative treatment solution or for each of the treatments or treatment categories.

The PMS may also provide quantities per material and works required (e.g. tons of hot or cold mixtures, volume or area of milling, area of micro-surfacing, length of crack filling, area of patching or full replacement, length or area of road marking, etc.). This provision is more commonly found in PMSs at the project level.

17.5.7 Reporting module

The results of a pavement management analysis and the information contained in the database can be reported in a number of different formats, including text, tables and graphics. Many agencies link their pavement management databases to geographic information systems (GISs) or other map packages to visually display the information. GIS displays of pavement management information are effective tools when reporting pavement conditions to decision makers (Asphalt Institute 2009).

A typical presentation of pavement management results can be found in all relevant bibliography; some examples are given in the AASHTO *Pavement Management Guide* (AASHTO 2012) and in the work of Shahin (2005).

17.5.8 Implementation of improvement program

After a final decision is made regarding the selection of the improvement program (preventing maintenance, maintenance, rehabilitation or reconstruction), its implementation follows. All information regarding completion dates, changes in pavement condition, materials used and any other construction details are fed back to the PMS database. This will be valuable information for future updates of PMS.

17.6 SOFTWARE FOR PAVEMENT MANAGEMENT

A great number of software programs for pavement management have been developed ever since the introduction of MicroPaver, perhaps the first integrated software developed in the early 1970s and is still in use (MicroPAVER 2013). Even a brief description of each PMS software program developed is an impossible task.

However, a catalogue of almost all PMS software programs developed in the United States from the public and private sector can be found in FHWA (2002); some of them are described and highlighted in Wolters et al. (2011). Similar PMS software programs have been developed in various other countries satisfying the requirements set by the agency.

In the United Kingdom, the Highways Agency's road network is managed using the specially developed PMS known as HAPMS (Highways Agency 2014).

In Greece, the first attempt to create a pavement maintenance software program was carried out in the early 1990s by the author and his research team. The program was known as PAVMAIN (Nikolaides et al. 1992).

The same research team developed a mainframe PMS software program known as ROADMAN (Nikolaides and Evangelides 1995).

REFERENCES

AASHTO. 1985. *Guidelines on pavement management*. AASHTO Joint Task Force on Pavements. Washington, DC: American Association of State Highway and Transportation Officials.

AASHTO. 1993. *AASHTO guide for design of pavement structures*, 4th Edition, with 1998 Supplement. Washington, DC: American Association of State Highway and Transportation Officials.

AASHTO. 2008. *Mechanistic-empirical pavement design guide (MEPDG)*. Washington, DC: American Association of State Highway and Transportation Officials.

AASHTO. 2012. *Pavement management guide*, 2nd Edition. Washington, DC: American Association of State Highway and Transportation Officials.

Adams T.M., N.A. Koncz, and A.P. Vonderohe. 2001. *Guidelines for the Implementation of Multimodal Transportation Location Referencing Systems*. NCHRP Report 460. National Research Council. Washington, DC: Transportation Research Board.

Asphalt Institute MS-16. 2009. *Asphalt in Pavement Preservation and Maintenance*, 4th Edition. Manual Series No. 16. Asphalt Institute.

Austroads. 2008. *Development of HDM-4 Road Deterioration (RD) Model Calibrations for Sealed Granular and Asphalt Roads*. Publication No. AP–T97/08. Sydney: Austroads Inc.

Ceylan H., S. Kim, K. Gopalakrishnan, and D. Ma. 2013. *Iowa Calibration of MEPDG Performance Prediction Models*. In Trans Project 11-401. Ames, IO: Institute for Transportation, Iowa State University.

FHWA. 2002. *Pavement Management Catalogue*, 2002 Edition. Washington, DC: Federal Highway Administration, Office of Asset Management.

FHWA. 2010. *Local Calibration of the MEPDG Using Pavement Management Systems*, Volume I, Final Report. Washington, DC: Federal Highway Administration.

Hall K.D., D.X. Xiao, and K.C.P. Wang. 2011. Calibration of the MEPDG for flexible pavement design in 1 Arkansas. Paper No. 13-2667-1. *TRB 90th Annual Meeting*. Washington, DC: Transportation Research Board.

Highways Agency. 2014. Available at http://www.highways.gov.uk.

MicroPAVER. 2013. Available at http://www.paverteam.com.

Morosiuk G., M.J. Riley, and J.B. Odoki. 2004. *HDM-4, Highway Development and Management, Volume Six, Modelling road deterioration and works effect, Version 2*. The Highway Development and Management Series. La Défence Cedex, France: The World Road Association (PIARC).

NCHRP. 2004. *Guide for Mechanistic-Empirical Design of New and Rehabilitated Pavement Structures*. National Cooperative Highway Research Program 1-37 A. Washington, DC: Transportation Research Board, National Research Council.

Nikolaides A. 2008. *Pavement Condition Evaluation and Pavement Management System of Eastern Sections of Egnatia Odos Motorway*. Technical Report. Thessaloniki, Greece: Egnatia Odos S.A., Maintenance Department.

Nikolaides A. 2010. *Pavement Management Program of Maliakos-Kleidi Motorway (All Sections except Interchanges)*. Technical Report. Goni, Greece: Agean Motorway S.A.

Nikolaides A. 2011. Pavement management program: A necessary tool to engineers, authorities and investors. *Proceedings of the 5th International Conference 'Bituminous Mixtures and Pavements'*. Thessaloniki, Greece.

Nikolaides A. and D. Evangelides. 1995. ROADMAIN – An integrated PMS using GPS. *1st Highway Engineering Conference, Technical Chamber of Greece*. Larisa, Greece.

Nikolaides A., D. Evangelidis, and G. Mintis. 1992. Pavement maintenance management of flexible pavements using computer software. *Proceedings 1st Panhellenic Conference 'Asphalt Concrete and Flexible Pavements'*. Thessaloniki, Greece: Aristotle University of Thessaloniki.

OECD. 1987. *Pavement Management Systems*. Road Transport Research. Paris: OECD.

Paterson W.D.O. 1987. *Road Deterioration and Maintenance Effects: Models for Planning and Management*. Washington, DC: World Bank.

Peterson D.E. 1987. *Pavement Management Practices, NCHRP Synthesis of Highway Practice 135*. National Research Council. Washington, DC: Transportation Research Board.

Rahman M.S., R.C. Williams, and T. Scholz. 2011. Mechanistic-empirical pavement design guide calibration for pavement rehabilitation in Oregon. Paper No. 13-4347-1. *TRB 90th Annual Meeting*. Washington, DC: Transportation Research Board.

Shahin M.Y. 2005. *Pavement Management for Airports, Roads and Parking Lots*, 2nd Edition. Springer, USA.

Vonderohe A.P., C.L. Chou, F. Sun, and T.M. Adams. 1997. *A Generic Data Model for Linear Referencing Systems*. NCHRP Research Results Digest 218. National Research Council, Washington, DC: Transportation Research Board.

Watanatada T., C.G. Harral, W.D.O. Paterson, A.M. Dhareshwar, A. Bhandari, and K. Tsunokawa. 1987. *The Highway Design and Maintenance Standards Model, Volume 1, Description of the HDM-III Model*. Washington, DC: The International Bank for Reconstruction and Development/ The World Bank.

Wolters A., K. Zimmerman, K. Schattler, and A. Rietgraf. 2011. *Implementing Pavement Management Systems for Local Agencies – State of the Art/State-of-Practice Synthesis*. FHWA-ICT-11-094. Urbana, IL: Illinois Center for Transportation.

Chapter 18

Pavement recycling

18.1 GENERAL

The construction industry, as well as other sectors, are always concerned about reduction of available funds, uninterrupted supply of raw materials, energy saving and environment protection.

To face the above concerns, there is a need to optimise the use of available funds, to optimise the properties of materials used, to minimise energy expenditure in all construction or maintenance/rehabilitation works and to find alternative solutions that will not damage, deteriorate or have a negative impact to the environment.

A major contribution to the above concerns is re-using or recycling the materials of old pavements, such as asphalt mixtures (asphalts), concrete or base/sub-base material.

The first form of pavement recycling appeared in the 1920s. However, the pavement manufacturing and maintenance industry started giving emphasis on recycling only after the energy crisis (1973), which caused a substantial cost increase in raw materials. Since 1973, and for a decade, the research and application of recycling have grown worldwide.

Despite the fall and normalisation of the cost of crude oil in the 1980s, recycling gained ground by the end of the 1980s. Nowadays, recycled materials are used in maintenance, rehabilitation, strengthening or even reconstruction works. Given the environmental and energy problem, which is constantly exacerbated, recycling will continue to gain more supporters.

Recycling may be applied to all layers of flexible and rigid pavements. This chapter emphasises on asphalt layer recycling and, to a lesser extent, on concrete slab recycling. Recycling of the unbound or hydraulically bound materials, although possible, is effectively restricted to pavement reconstruction works, which are not as common as maintenance and rehabilitation of asphalt layers. However, provided the unbound or cement-bound materials possess the required properties, the techniques and procedures followed are similar to those as in new pavement construction.

The material removed from the asphalt layers to be re-used, hence recycled, containing bitumen and aggregates, is called reclaimed asphalt (RA), under European Union terminology, or reclaimed asphalt pavement (RAP), under US terminology.

RA or RAP when properly milled or crushed and screened consists of high-quality well-graded aggregates coated by bitumen (asphalt cement). This material can be used in a number of highway applications such as aggregate substitute and bitumen supplement in hot or cold asphalts (bituminous mixtures), recycled asphalt mixtures, granular base or sub-base, stabilised base aggregate or as an embankment or fill material.

The term *RA* will be used from now in this chapter unless reference or description of a technique/procedure, used in the United States, is made where the term *RAP* is used.

As for the reclaimed concrete, after removing the steel, when properly crushed and screened, it can be used as aggregate replacement in hot mix production, as a granular base or sub-base or as an embankment or fill material.

18.2 WHAT PAVEMENT RECYCLING IS OFFERING

Pavement recycling offers conservation of natural resources and bitumen saving, energy saving, environmental protection and reduction of construction cost.

18.2.1 Conservation of natural resources and bitumen saving

The conservation of natural resources for road aggregate production, particularly of hard and durable aggregates for surfacing, is a serious matter particularly for countries that do not possess sufficient reserves of such materials. Even for countries that do possess sufficient reserves, the use of RA lengthens the exploitation period of the existing quarries and delays the opening of new ones; in some cases, it is a difficult task because of the strict environmental rules imposed.

Needless to say, for countries possessing no natural resources for road aggregate production, the use of RA is also beneficial since it reduces the cost of asphalts owing to the incorporation of less amount of expensive new aggregate in the mix.

The use of RA decreases bitumen consumption. An asphalt concrete usually requires 4.5% to 5.5% bitumen by weight of mix. A recycled asphalt concrete requires additional bitumen usually between 1% and 3%, by weight of mix. From the above, a certain amount of bitumen is saved, which is, on average, on the order of 3.0% by weight of mix, that is,

Table 18.1 Available quantities of RA in Europe and elsewhere in the year 2012

Country	Available RA ($\times 10^6$ tonnes)	% of available RA used in			% of the new hot and warm mix production that contains reclaimed material
		Hot/warm recycling	Cold recycling	Unbound layers	
Belgium	1.5	61	a	a	49
Czech Republic	1.4	22	30	15	10
Finland	1.0	a	a	a	65
France	6.5	62	a	a	>60
Germany	11.5	87	a	13	97
Great Britain	4.5	a	a	a	a
Italy	10.0	20	a	a	a
Netherlands	4.0	80	15	a	73
Sweden	1.0	75	5	10	70
Switzerland	1.6	52	17	9	24
Turkey	3.8	2	5	93	1
Total[b]	49.9	a	a	a	a
Australia	0.533	17	a	a	a
United States	64.0	95	–	5	a
Ontario, Canada	3.5	80	3	17	75

Source: EAPA, Asphalt in figures, http://www.eapa.org, 2012.

[a] No data available.
[b] Total of all European countries listed in EAPA (2012) figures including those with RA tonnage less than 1.0×10^6 (14 more countries).

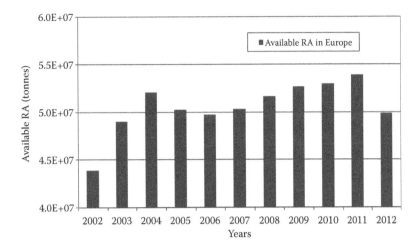

Figure 18.1 Annual tonnage of available RA in Europe from 2002 to 2012. (From EAPA, Asphalt in figures, http://www.eapa.org, 2012.)

30 L less bitumen per tonne of recycled asphalt concrete produced. Considering the annual production of asphalts, in each country and globally, the decrease in bitumen consumption is substantial.

Table 18.1 gives the annual quantities of RA resulting from milling or other means of removal of asphalt layers, called available RA quantities, in European countries in the year 2012. It is noted that only those with available RA tonnage of greater than 1 million tons are listed.

Table 18.1 also gives the percentage of available RA used in various highway applications (hot/warm recycling, cold recycling and unbound layers), as well as the percentage of new hot and warm mix production that contains RA. As it can be seen, the country with the most tonnage of RA available is Germany. It is also interesting to note that in the same country, almost all new hot/warm asphalt mixtures contain RA.

Figure 18.1 shows the annual available tonnage of RA in Europe from 2002 to 2012. As it can be seen, there was a distinct increase in available RA after the year 2003.

Similar statistical results concerning the annual tonnage of available RA in the United States can be found in a NAPA survey report (Hansen and Copeland 2013).

18.2.2 Energy saving

Energy saving is related primarily to the energy saved during the transportation of the constituent materials of asphalt (aggregates and bitumen). Additional energy saving may occur when cold recycling is used instead of hot recycling.

A recent study found that 33.273 MJ/km^2 of energy is required for the reconstruction of a 7 cm thick asphalt layer when cold in situ recycling is used, while 120.351 MJ/m^2 of energy is required when the conventional technique, that is, hot mix asphalt, is used (Mauduit et al. 2011).

Another study recommended that the energy factor should also be taken into consideration when life cycle cost analysis is employed (Uddin 2011).

18.2.3 Conservation of environment

Pavement recycling leads to conservation and protection of the environment, since the use of RA not only delays the opening of new quarries but also solves the environmental problem of deposition of reclaimed material.

The deposition of RA presupposes the existence of sufficient and appropriate deposited area, scarce to find in most cases, and if so, it causes changes to the landscape (landscape intrusion). In parallel, the concentration of hydrocarbons contained in the asphalt will pollute the soil and the water reservoir in the surrounding area.

Additionally, the effect of asphalt recycling is positive, particularly when cold recycling is used on carbon emissions. It has been found that carbon emissions were only 624 kg eq C/1000 m^2 during the reconstruction of a 7 cm thick asphalt layer using cold recycling, in contrast to 2381 kg eq C/1000 m^2 by way of the conventional technique with hot bituminous mixture (Mauduit et al. 2011).

18.2.4 Reduction of construction cost

Pavement recycling can also result in a noticeable reduction in maintenance, rehabilitation or reconstruction cost. This is derived from reduction in asphalt production cost and reduction in transport cost of the material, which is substantial when in situ recycling is selected.

The percentage of reduction in construction cost varies from project to project, since all factors contributing to the construction cost vary.

From the first recycling applications conducted in the United States, it was reported that the recycled asphalt production cost, when produced in a stationary plant, ranged between 10% and 39%, depending on the project (Finn 1980; TRB 1978). Similarly, when recycling was conducted in situ, the total reduction in construction cost was on the order of 33%.

Similar comparative studies in Europe gave lower reduction percentages, that is, lower than 30% (Chappat and Plaut 1982; Maraux et al. 1991). It was also found that, in some cases, there was no essential reduction in asphalt production cost when produced in a stationary plant.

Even though pavement recycling could result in the reduction of construction cost, this alone is not the decisive factor for selecting it. For most countries, conservation of energy and natural resources, environment protection and reduction of atmospheric pollution are the reasons why they embark on pavement recycling for all highway applications.

18.3 PAVEMENT RECYCLING METHODS

Pavement recycling methods differ from the flexible to the rigid type of pavement. In flexible pavements, two basic methods are distinguished: hot recycling (HR) and cold recycling (CR). Each one is further distinguished into in situ (or in-place) and in-plant (central plant) recycling.

As for the rigid pavement, there is only one method of recycling, the cold in-plant recycling.

In flexible and, to a certain extent, rigid pavements, cold planning (or milling) is considered as a recycling method.

Figure 18.2 shows the general pavement recycling tree.

Hot recycling in flexible pavements consists of removing the existing asphalt layer to a determined depth and replacing it with hot asphalt resulting from the same reclaimed material, with the possible addition of new material, that is, mixed in situ or new/recycled asphalt mixed in a central plant.

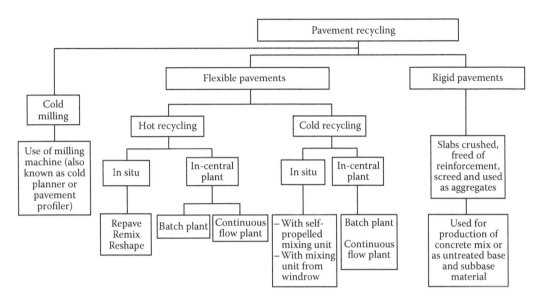

Figure 18.2 General categories of recycling.

Cold recycling in flexible pavements consists of removing the existing asphalt layer to a determined depth and replacing it with cold asphalt resulting from the same reclaimed material, cold mixed in situ or cold recycled asphalt mixed in a stationary (central) plant. Alternatively, all asphalt layers may be removed and the milled material is blended in situ to provide an upgraded homogeneous base material. This procedure is known as full-depth cold recycling or full-depth reclamation (FDR).

Rigid pavement recycling is the procedure of crushing and fragmentising the concrete slab and re-using the fragmented concrete as coarse aggregate to produce a new concrete mix or base/sub-base material (stabilised or non-stabilised material).

Cold milling (or planning) is the removal of the existing asphalt layer to a desired depth determined by the surface/pavement deficiency to be corrected, using special machinery called milling machines or planners. The generated reclaimed material is removed from the site and stockpiled for future use (production of recycled asphalt or other usage).

The advantages and disadvantages of each of the recycling methods and cold milling are summarised in Tables 18.2 through 18.5.

Table 18.2 Advantages and disadvantages of cold planning (milling)

Advantages	Disadvantages
• Correction of longitudinal evenness • Removal of rutting, surface bleeding, aged asphalt and other surface deficiencies • Improvement of surface skid resistance • Removal of asphalt for further recycling • Surface preparation prior to asphalt laying • Increased pavement efficiency • Less traffic and public disruption and higher productivity during asphalt works	• Some dust emission • Reduction of pavement structural capacity if milled area is not resurfaced • Milling machine is not so efficient with high-stiffness asphalt or asphalts containing large aggregate particles

Table 18.3 Advantages and disadvantages of hot recycling

	Advantages	Disadvantages
Hot in situ (in-place) recycling	• Conservation of materials and energy resources • Elimination of disposal problem of reclaimed asphalt • Minimisation of transport cost of material • Correction of all pavement distresses and cross slope • Correction of aggregate gradation may be possible • Pavement elevation is maintained • Economic savings	• Quality control not as effective as in in-plant hot recycling • High equipment cost • Equipment not suitable for urban projects with tight bends • Material variability almost impossible to be faced effectively
Hot in-central plant recycling	• As in hot in situ recycling • Deficiencies of reclaimed asphalt can be corrected • More effective correction of cross slope	• Increases transport cost of recycled material • Requirement to find stockpile area in the plant area

Table 18.4 Advantages and disadvantages of cold recycling

	Advantages	Disadvantages
Cold in situ recycling	• Conservation of materials and energy resources • Elimination of disposal problem of reclaimed asphalt • Minimisation of transport cost of materials • Correction of all pavement distresses and cross slope • Pavement elevation may be maintained • Ultimate energy savings • Minimisation of environmental pollution owing to reduction of smoke and gas emissions • Economic savings	• Quality control not as effective as in cold in-plant recycling • Delay in opening to traffic attributed to curing of cold asphalt • Requirement to place surface dressing or hot asphalt overlay (extra cost)
Cold in-plant recycling	• Conservation of materials and energy resources • Correction of all pavement distresses and cross slope • Correction of aggregate gradation and improvement of asphalt • Pavement elevation may be maintained • Energy savings • Minimisation of environmental pollution owing to less smoke and gas emissions • Better quality control of recycled cold asphalt • Economic savings	• Increases transport cost of recycled asphalt • Requirement to find stockpile area in the plant area • Requirement to place hot asphalt overlay or surface dressing

Table 18.5 Advantages and disadvantages of rigid pavement

Advantages	Disadvantages
• Minimises old slab disposal problem • Contributes to conservation of natural resources for aggregate production • Resistance to freeze and thaw is increased, in some cases	• Resulting recycled aggregates generally more porous • If used as sand, workability of the concrete mix is reduced • The existence of reinforcement increases slightly the recycled aggregate production cost

The selection of recycling method depends on various factors such as pavement condition, road category, traffic volume, age of pavement, past interventions, availability of equipment/ machinery, expected pavement performance after recycling, cost and energy saving achieved. Selecting the most appropriate recycling method is not an easy task, since many parameters, which vary from one project to another, are involved.

18.4 COLD MILLING (PLANNING)

Cold milling, or planning, consists of the removal of the existing pavement to a desired depth and depositing the RA for future use. Cold milling is used to correct corrugations, rutting, bitumen bleeding, shoving and cross slope or to restore surface skid resistance.

Cold milling of flexible pavements is carried out by milling machines also known as cold planners or pavement profilers. Cold milling of rigid pavements is carried out by similar machines usually called grinding machines.

The resulting textured surface can be used immediately or after treatment by an overlay consisting of recycled or new asphalt, provided the surface has been thoroughly cleaned and, in the case of overlay, tack coated.

18.4.1 Milling machines

The milling machine is used in all recycling methods of flexible pavements, while the grinding machine is used for surface skid resistance restoration of rigid pavements.

Milling machines are self-powered/self-propelled units moving on either wheels or crawler tracks; wheels provide better manoeuvrability, and usually small- to medium-size milling machines are equipped with wheels (milling width approximately up to 1.3 m and milling depth up to approximately 180 mm).

The milling machine has a rotary cutting drum equipped with replaceable cutting 'teeth' or 'tools' and a front-loading conveyor to load the milled material onto the truck.

The width of the cutting drum, hence milling width, usually varies from as narrow as 350 mm to as wide as 2500 mm, and the milling depth in a single pass may vary from 0 to 350 mm, depending on the model and the manufacturer. The travelling speed during milling is slow (a few meters per minute) and varies depending on the depth of milling the machine operates on. Modern milling machines are equipped with a high-precision automatic levelling system that optimises milling to specific elevations and grades. All milling machines are also equipped with a water spraying system to minimise or eliminate dust emission. Figure

Figure 18.3 Milling machine, W-200 model. (From Wirtgen GmbH, http://www.wirtgen.de, 2013.)

18.3 shows a typical modern milling machine. More information about cold milling can be found in Wirtgen (2013).

Milling usually takes place when the pavement surface is 'cold' (ambient temperatures). However, sometimes the surface is preheated by a heating unit travelling in front of the milling machine. Heating of the pavement surface prior to milling is required only in flexible pavement recycling.

18.5 HOT RECYCLING

Hot recycling is the most common method used worldwide for the rehabilitation of asphalt layers. As the name implies, heating is required during mixing and repaving of the reclaimed or the recycled asphalt.

Hot recycling may be carried out in situ or in a central mixing plant (in-plant). Hot in situ recycling is distinguished into *repave*, *remix* and *reshape* procedures. The basic steps followed in all versions of hot recycling are shown in Figure 18.4.

18.5.1 Hot in situ recycling

Hot in situ recycling is a quite common form of recycling in which the recycling processes of the RA are completed on site. The process consists of heating the pavement to a temperature such that the top asphalt layer can be scarified (or rotary milled) to the specific depth. The loosened asphalt is then mixed, placed and compacted. Depending on the procedure followed, new asphalt material may or may not be added prior to compaction.

The pavement is heated to temperatures above approximately 90°C using naked flame or infrared heating units (direct heating) or, for more effective in-depth heating, super-heated compressed air heaters (indirect heating). The typical treatment depth ranges from approximately 20 to 100 mm, depending on the procedure followed (repave, remix or reshape).

The hot in situ recycling process requires the use of a number of equipment including preheaters, heaters, scarifiers, mixers, pavers and compactors.

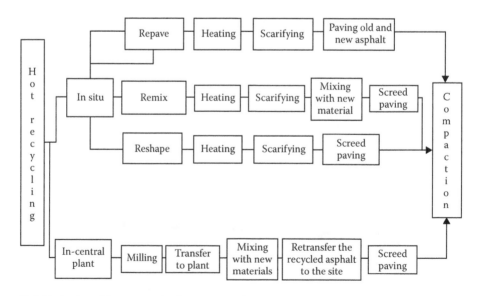

Figure 18.4 Methods and basic steps for hot in situ and in-central plant recycling.

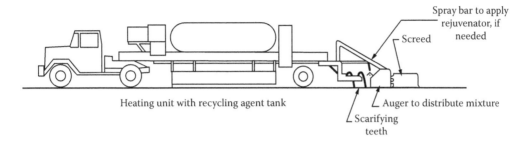

Figure 18.5 Reshape process. (From Kandhal, P.S. and Mallick, R.B., *Pavement Recycling Guidelines for State and Local Governments – Participant's Reference Book, FHWA-SA-97.* Washington, DC: Federal Highway Administration, 1997.)

In hot in situ recycling, it is vital to examine the old asphalt mix before commencing the work by taking representative samples to establish the areas where the composition of the existing material varies.

18.5.1.1 Reshape

Reshape, or surface recycling, is used to transform a partially failed surface layer to a 'good as new' layer without the addition of new hot mix asphalt or new aggregate material. Through this method, surface cracks are healed and surface profile is reshaped free of deformations.

The reshape process is conducted in one step consisting of heating the pavement surface, scarifying the top 25 to 50 mm of the wearing course, working and distributing the loose asphalt by augers and placing the worked mix by a screed. Recycling agent may be added, if needed, to the scarified mix before distribution. Compaction is carried out by pneumatic rollers or static/vibrating steel rollers. Figure 18.5 shows diagrammatically the hot in situ reshape process.

18.5.1.2 Remix

Remix is the hot in situ recycling process in which the pavement surface is heated, scarified and mixed; a rejuvenating agent is added if required; new hot asphalt material is added;

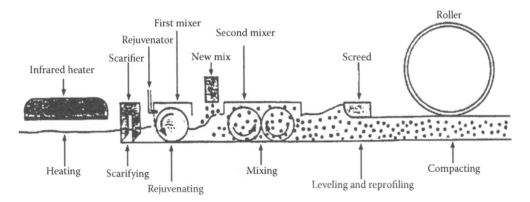

Figure 18.6 Schematic concept of remix process in a single pass. (From Kandhal, P.S. and Mallick, R.B., *Pavement Recycling Guidelines for State and Local Governments – Participant's Reference Book, FHWA-SA-97.* Washington, DC: Federal Highway Administration, 1997.)

and all materials are mixed again and screed. The homogeneous recycled mix is placed and compacted in one layer.

Remix is generally used when the properties of the existing top asphalt layer require some modification and ensure a fully integrated and blended layer. Remix is usually carried out in a single operation (stage).

Figure 18.6 shows the concept of a typical remix process in a single stage. The treatment depths in a single-stage procedure, or single-pass operation, usually range between 30 and 50 mm.

When remix is carried out in multiple passes of heating and scarifying, the scarified material is placed in a windrow. The material from the windrow is then carried to a pugmill mixer where it is mixed with new hot asphalt material; a rejuvenating agent may also be added to the mixer, if needed. Again, the recycled asphalt mix is paved and compacted in one homogeneous layer. The treatment depths in multiple passes increase and usually range between 40 and 80 mm.

The bitumen grade and the composition of the new asphalt material are selected to balance any defects in the composition of the existing material. When machinery permits, virgin aggregate material and new bitumen are added instead of new asphalt material.

18.5.1.3 Repave

The repave process consists of heating, scarifying and paving the scarified (reclaimed) asphalt and then paving the new asphalt mix. Both paved layers are compacted as one layer.

The repave process is carried out in a single pass using a purpose-built machine (see Figure 18.7). Alternatively, repaving may be carried out in multiple passes, where heating, scarifying and paving operations are carried out separately.

The scarified material functions as a levelling course and the new hot asphalt mix functions as a new wearing course (overlay). The addition of new hot mix asphalt inevitably increases the level of the pavement depending on the thickness of the new overlay.

Compaction of both scarified (reclaimed) and new asphalt material is carried out by a combination of double steel drum vibratory rollers and pneumatic-type rollers.

The repave process offers perhaps the ultimate rehabilitation/hot recycling results compared to the other two hot recycling in situ processes. In particular, its advantages over the other two in situ processes are as follows: (a) complete restoration of surface skid resistance since new asphalt material is used, (b) strengthening of the pavement to a certain extent and (c) minimisation of smoke emission.

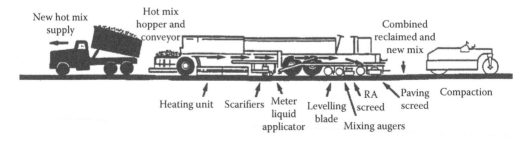

Figure 18.7 Single-pass repaving process. (From Kandhal, P.S. and Mallick, R.B., *Pavement Recycling Guidelines for State and Local Governments – Participant's Reference Book*, FHWA-SA-97. Washington, DC: Federal Highway Administration, 1997.)

Prerequisites for repave is that the pavement (a) should not show premature surface deformation (rutting, shoving and corrugations) caused by asphalt deficiency in deformability and (b) should not show extensive fatigue cracking.

Repave should be applied when the pavement is still in sound structural condition and should be seen as an economical process that restores the properties of the wearing course contributing to a certain extent to the structural capacity of the pavement.

18.5.2 Hot recycling in a central plant

The central plant hot recycling method, often called just hot recycling, is the oldest recycling method that emerged at the beginning of the 20th century. A lot of experience has since been gained, and today, it is one of the most widely used recycling methods.

This method gives the best possible quality mixture and ensures high construction quality. The advantages and disadvantages of this method are outlined in Table 18.3.

Hot recycling in a central plant consists of milling the old flexible pavement, transferring the reclaimed material to the plant, further crushing (if needed), sieving the RA and mixing the RA with new materials (bitumen and aggregate). The general steps of hot recycling in a central plant are outlined in Figure 18.4. The reclaimed material may be stored or recycled immediately.

Mixing is carried out in hot mixing batch plants or drum-mixing plants capable of incorporating RA. Details on hot mixing plants can be found in Section 8.2.

The percentage of RA or RAP used for the production of recycled asphalt is dictated by a number of factors such as (a) type of mixing plant used, (b) usage of recycled asphalt (i.e. surfacing, binder courses or asphalt base) and (c) suitability of constituent materials of the RA.

Given that the RA is suitable for recycling, higher percentages of RA are incorporated in drum-mixing plants than in conventional batch plants. Reports from various studies indicate that typically batch plants operate at RA ranging from 15% to 30% of the total recycled mix, while the respective percentage in drum-mixing plants may be much higher (more than 60% RA).

Generally, less RA is used for the production of recycled wearing course mixtures in comparison to RAs for binder or base course recycled mixtures.

In order to safeguard the performance of the recycled material and control emissions, some agencies may impose restriction on the percentage of RA to be used. It has been reported that the maximum amount of reclaimed material in central plant hot recycling was generally limited to approximately 60% (Highways Agency 1998), and the majority of US states imposed a 50% permissible limit regardless of hot mix recycling procedure (Eaton 1991). However, the improvement on drum-mixing technology, the use of rejuvenators and the requirement for more comprehensive testing on the constituents of the RA (particularly bitumen) and on the recycled asphalt make it possible today to incorporate even more than 70% reclaimed material in central plant hot recycling.

18.6 COLD RECYCLING

Cold recycling is a pavement recycling method in which no heating is required at any stage of work. The advantages and disadvantages of cold recycling are outlined in Table 18.4.

The binder used is exclusively bitumen emulsion and cold recycling may be carried out in situ or in a stationary plant (central plant). The in situ cold recycling is distinguished into full-depth recycling, also known as full-depth reclamation (FDR), and partial-depth cold recycling, also known as cold in situ recycling (CR).

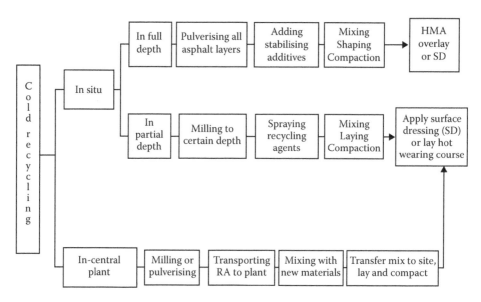

Figure 18.8 Methods and basic steps for cold in situ and stationary plant recycling.

The cold recycled mixture resulting from FDR is used as base and sub-base materials, while the cold recycled mixture resulting from in situ partial-depth recycling or cold mixing in a stationary plant is used as wearing course mixture. They may also be used as wearing course mixtures in sections with low traffic with the provision of surface dressing. In all cases, after compaction and curing, a hot asphalt overlay or surface dressing is applied, whichever is more appropriate considering the project conditions.

The basic steps followed for cold in situ and stationary plant recycling are shown in Figure 18.8.

18.6.1 In situ full-depth cold recycling

In situ full-depth cold recycling, also known as FDR or structural road recycling, is a cold recycling method where all asphalt layers and, in some cases, a predetermined portion of the underlying base material are uniformly pulverised and usually mixed with stabilising additives to produce a stabilised base course material. The pulverised reclaimed material may also be used as it is untreated for unbound base course material.

The stabilising additives, in dry or liquid form, can basically be ordinary Portland cement (OPC), lime, fly ash, cement kiln dust (dry of slurry), bitumen emulsion, foamed asphalt or combinations of two or more of these additives.

The above stabilising additives (binders) according to the RSTA Code of Practice for in situ structural recycling (RSTA 2012) are classified into five categories: (a) quick hydraulic binders, with OPC as the main hydraulic component; (b) slow hydraulic binders – binders such as pulverised fuel ash (PFA)/lime or granulated blast furnace slag/lime, excluding bituminous binders and OPC; (c) medium strength gain hydraulic binders containing both OPC and PFA; (d) quick viscoelastic bituminous binders but also including OPC; and (e) slow viscoelastic bituminous binder as the main component but excluding OPC.

The treatment depths vary depending on the thickness of the existing structure, which generally ranges between 100 and 300 mm (ARRA 2001; Kandhal and Mallick 1997).

FDR is an in-place rehabilitation process that can be used for reconstruction, lane widening, minor profile improvements and increased structural capacity by addressing the full range of pavement distresses (Stroup-Gardiner 2011).

The main steps of FDR are pulverisation, introduction of additive, addition of new aggregate (when necessary), shaping, compaction and application of a wearing course.

The basic equipment needed in FDR is a reclaimer, stabilising tankers, a motor grader and rollers.

The reclaimer is a high-horse-powered, self-propelled machine equipped with a pulverising/mixing drum and a stabilising additive delivery system. Depending on the type of reclaimer, the recycled material is placed by the paving screed integrated into the reclaimer or by motor graders.

The initial compaction is carried out by vibrating steel drum rollers or large-sized pneumatic rollers. Final shaping and compaction to the required longitudinal profile and cross slope are followed after the recycled material is cured. The curing period may range from 1 day to several days depending on the environmental conditions (temperature and wind) and the stabilising additives. During the curing period, construction traffic should be kept off the stabilised material.

After final compaction, the surface is covered by hot mix asphalt overlay or surface dressing (single or double), depending on the project conditions, constraints and requirements.

18.6.2 In situ partial-depth cold recycling

In situ partial-depth cold recycling, also known as in situ cold recycling, consists of milling the pavement surface to a certain depth (usually 50 to 100 mm), mixing the reclaimed material with a recycling agent and placing the recycled mix by screed and compaction. The reclaimed material may also be screened prior to mixing, if the equipment used allows it.

The recycling agent is bitumen emulsion or foamed bitumen, with the addition of very small amount of cement, lime or water–cement slurry, most probably a requirement for the acceleration of breaking and for strength development of the recycled cold mix.

Milling, mixing and laying may take place using a single-unit cold in situ recycling machine, also called cold recycler, or with a two-unit cold in situ recycling train, consisting of a milling machine and a mix paver.

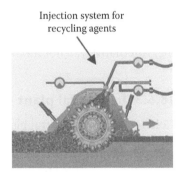

Milling drum   Augers and   Details of milling drum
paving screed

Injection system for recycling agents

Figure 18.9 Single unit cold in-place recycling machine, 2200 CR model. (From Wirtgen GmbH, http://www.wirtgen.de, 2013.)

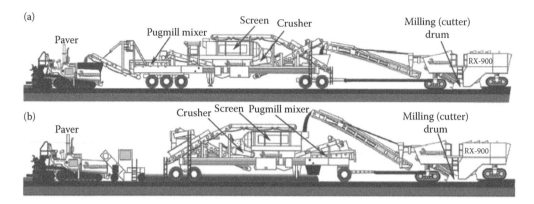

Figure 18.10 Modern two-plus-one cold in situ recycling equipment. (a) With paver-loading equipment. (b) With windrow equipment. (From ASTEC Industries Inc., http://www.astecinc.com [Roadtec Inc., http://www.roadtec.com].)

The cold recycler removes the old pavement surface by milling with a down cutting rotor. The recycling agents are injected into the milling drum to coat the reclaimed material. Figure 18.9 shows a typical single-unit cold recycler and a schematic illustration of the milling drum. This particular machine can also operate as a milling machine alone.

Apart from the two-unit cold in situ recycling train, a multi-unit train may also be used in cold in situ partial-depth recycling. The multi-unit train consists of a milling machine, a trailer-mounted crushing and screening unit and a trailer-mounted pugmill mixer and paver. Originally, all of the above machines were individual units. Today, the modern cold recycling equipment combines the crushing and screening unit and the pugmill mixer in one unit. These units discharge the cold recycled material in a windrow or directly to the paver. Figure 18.10 shows a three-unit cold in situ recycling train.

The two-unit, or multiple-unit, cold in situ recycling trains provide the ultimate process control and quality of recycled mix. This is because the milled material may be crushed further, achieving better particle distribution, the oversized materials can be removed from the reclaimed material and the amount of recycling agents is directly related to the volume of reclaimed material. The latter ensures application rates of recycling agents as much as required.

The initial compaction after rolling and the final compaction after rolling are carried out by vibrating steel drum rollers or pneumatic rollers.

After final compaction, the surface is usually treated with single or double surface dressing. Alternatively, a hot mix asphalt overlay may be selected to be applied.

18.6.3 In-central plant cold recycling

Cold recycling in a central plant is the process in which asphalt removed from the surface of an existing pavement is transported to a central cold mix plant where it may be stockpiled for future use or processed immediately.

The RA is obtained by cold milling or ripping and may need further crushing and screening. Mixing is carried out by stationary purposely built cold mixing plants, as described in Section 6.5.1. The cold recycling mix can be stockpiled for a short period or immediately transported to the site.

The recycling agent is bitumen emulsion with the addition of a very small amount of cement or lime. The type, grade and rate of bitumen emulsion and the amount of cement or lime are determined through evaluation of the RA and cold mix design process. The mix design process may dictate the addition and use of new aggregates.

The placement of the recycled cold asphalt is carried out with conventional hot asphalt mix pavers. However, recycled cold asphalt may be placed by motor graders.

The preliminary and final compaction of the recycled cold asphalt is carried out by pneumatic and vibrating steel drum rollers.

The compacted cold central plant recycled asphalt is generally overlaid with either surface dressing (single or double) or hot asphalt mix, depending on the traffic volume on the project site.

18.7 EVALUATION OF RA

Evaluation of RA is of vital importance since it enables the determination of its acceptability and the quantity and type of new material to be added (aggregate material and recycling agents) after performing the appropriate mix design.

The RA evaluation process consists of sampling and laboratory testing for the determination of

a. Binder properties
b. Aggregate grading
c. Binder content

Sampling of RA should be representative and may be carried out from existing pavement to be recycled, from a lorry load (hauling trucks) of RA or from stockpiles.

18.7.1 Sampling of RA

Sampling from existing pavement consists of core cutting (core diameter ≥150 mm) or sawing out slabs (slab size approximately 300 mm × 300 mm). Samples from each sampling location should be of sufficient mass for bitumen extraction, recovery and testing of bitumen (8 kg is usually sufficient). Sampling locations should be random and at an appropriate frequency; one every 500 m lane length is considered by many agencies to be sufficient. However, national relevant specifications should always be followed, such as CEN EN 12697-27 (2000), ASTM D 979 (2012) and ASTM D 5361 or AASHTO T 168 (2011), or other specifications.

Sampling of RA from a lorry load or stockpiles is carried out as in aggregate sampling and in accordance to relevant national specifications, such as CEN EN 932-1 (1996), ASTM D 75 (2009), AASHTO T 2 (2010) or other specifications. However, sampling from stockpiles (feedstock) requires that the number of samples should be sufficient; CEN EN 13108-8 (2005) recommends 1 for every 500 t of feedstock, rounded off upwards, with a minimum of 5.

18.7.2 Binder properties

Binder properties consist of the determination of penetration, the softening point or the viscosity of the binder extracted from the RA.

Binder is recovered by using the rotary evaporator method (CEN EN 12697-3 2013 or ASTM D 5404 2012), by fractionating column (CEN EN 12697-4 2005) or, similar to the fractionating column method, by the Abson method (AASHTO R 59 2011; ASTM D 1856 2009). Some details regarding the test methods are given in Section 9.6.2.

The determination of penetration and softening of the recovered binder is carried out as in virgin bitumen (see Sections 4.2 and 4.3).

The determination of viscosity is carried out by vacuum capillary viscometers at 60°C (see Section 4.8.2) or by a dynamic shear rheometer (DSR) (see Section 4.8.4.2).

The penetration (or softening point) of the recovered binder from the RA indirectly determines the degree of bitumen ageing and acts as a decisive parameter whether the RA can be recycled or should be used as a granular base/sub-base material.

The harder the RA binder, the less likely for the RA and the virgin binder to effectively blend. Although an extensive study found that blending of the RAP binder and virgin bitumen occurs to a significant extent (McDaniel et al. 2000) and the inclusion of the RA is not a 'black rock', another research indicated that mixing the virgin binder with the RAP results to some degree in a double coat of the old and the new binder (Mollenhauer and Gaspar 2012).

To ensure effective blending of RA with the virgin binder, some European countries have set limit values to the hardness of the RA binder in terms of penetration and softening point (see Table 18.6).

It is noted that CEN EN 13108-8 (2005) categorises the RA for recycling into penetration category P_{15}, or softening point category S_{70}, and into other RAs.

RA is categorised as P_{15} if the mean penetration value of at least five samples taken from feedstock is 15 dmm and none of the individual values obtained is less than 10 dmm.

Similarly, RA is categorised as S_{70} if the mean softening point value of at least five samples taken from feedstock is 70°C and none of the individual values obtained is less than 77°C.

For other RA categories, either the mean penetration values or the mean softening values is declared as category P_{dec} or S_{dec}.

It is noted that the European practice on using RA for the production of recycled asphalts requires the penetration or softening point of the binder in the resulting mixture, calculated from the penetrations or the softening points of the added binder and the recovered binder from the RA, to meet the penetration or softening point requirements of the selected grade, as if the works had to be conducted with virgin bitumen.

A prerequisite for the above is that the RA had been produced with paving grade bitumen and the binder to be added to the mixture is also paving grade bitumen.

The penetration or softening point in the resulting mixture can be calculated using the following equations, provided by the EN standards for hot asphalts:

Table 18.6 Limit values of recovered binder for RA to be appropriate for recycling

Country	Penetration (dmm)	Softening point (°C)
France	>5	<77
Belgium	>10	–
United Kingdom	>15	–
Germany, Ireland, Poland, Portugal	>15	<70
Slovakia	–	<70

Source: Ipavec, A. et al., *Synthesis of the European National Requirements and Practices for Recycling HMA and WMA (Direct_Mat Project)*. 5th Eurasphalt & Eurobitume Congress, Istanbul, Turkey, 2012.

For penetration:

$$a \times \lg \text{pen}_1 + b \times \lg \text{pen}_2 = (a + b) \times \lg \text{pen}_{\text{mix}},$$

where pen_{mix} is the calculated penetration of the binder in the mixture containing RA, pen_1 is the penetration of the binder recovered from the RA, pen_2 is the penetration of the added binder and a and b are portions by mass of the binder from the RA (a) and from the added binder (b) in the mixture; $a + b = 1$.

For softening point:

$$T_{\text{R\&Bmix}} = a \times T_{\text{R\&B1}} + b \times T_{\text{R\&B2}},$$

where $T_{\text{R\&Bmix}}$ is the calculated softening point of the binder in the mixture containing RA, $T_{\text{R\&B1}}$ is the softening point of the binder recovered from the RA, $T_{\text{R\&B2}}$ is the softening point of the added binder/softening point of the asphalt added and a and b are portions by mass of the binder from the RA (a) and from the added binder (b) in the mixture; $a + b = 1$.

The determination of the penetration or of the softening point of the binder of the resulting mix is necessary only when more than 10% by mass of the total mixture of RA in the recycled mixture for surface course is used or when more than 20% by mass of the total mixture of RA in the recycled mixture for regulating courses, binder courses and bases is used.

Hot recycling may be applied for the production of recycled asphalt concrete (CEN EN 13108-1 2008), asphalt concrete for very thin layers (CEN EN 13108-2 2008), hot rolled asphalt (CEN EN 13108-4 2008), SMA (CEN EN 13108-5 2008) and porous asphalt (CEN EN 13108-7 2008).

The viscosity of the extracted binder from the RA is used, by some mix design methodologies, for the determination of the percentage of the given virgin bitumen or recycling agent to be added to modify/rejuvenate the aged binder to the required level. This approach is described in ASTM D 4887 (2011).

Finally, the shear modulus, G^*, and the phase angle, s, measured by the DSR, in fact the ratio $G^*/\sin s$, should be used for blending new bitumen with recycled asphalt, when the Superpave PG bitumen classification system is used (ARRA 2001). Relevant information can be found in the same manual (ARRA 2001). A procedure to develop blending charts for RA and virgin binders using the Superpave PG classification is described by McDaniel et al. (2001).

With regard to Australian and New Zealand road agency specifications related to the use of RA, the criteria used by each agency are summarised in Austroads (2013). In the same reference, a proposed binder blend procedure is also described.

18.7.3 Aggregate grading and binder content

The aggregate grading determination of the RA and the binder content determination in the RA are conducted in the usual manner similar to hot or cold mix asphalt (see Sections 2.11.1 and 9.6, respectively).

This facilitates the determination of the mass of each portion of new aggregates and the mass of new bitumen to be added for the production of the recycled mix.

18.7.4 Other geometrical and physical tests on reclaimed aggregates

Other geometrical and physical tests on the aggregates of the RA (angularity, resistance to fragmentation, resistance to polishing polished stone value [PSV], resistance to surface abrasion aggregate abrasion value [AAV], particle density, water absorption, etc.) may also be necessary to be determined when comprehensive mix design is followed.

All tests are carried out in the usual manner as for virgin aggregates (see Section 2.10).

18.7.5 Homogeneity of RA

Homogeneity of the RA is also especially important. The homogeneity of the feedstock should be determined from the variability of the percentages of coarse and fine aggregates and of fines in the RA, the binder content of the RA and either the penetration, the softening point or the viscosity of the binder recovered from the RA (CEN EN 13108-8 2005). For ensuring homogeneity, a maximum range or standard deviation values should be set for each of the above properties of the RA.

In European countries with homogeneity requirements, the following RA properties were found to be taken into consideration (Ipavec et al. 2012):

- Percentage of aggregate <0.063 mm
- Percentage of aggregate between 0.063 and 2 mm
- Percentage of aggregate >2 mm
- Binder content
- Binder penetration

Some countries set tolerance limits for the above RA properties regardless of the RA content in the recycled mix. In some countries, like France and Germany, the maximum allowed RA content is set according to the homogeneity of the stockpile of RA (Ipavec et al. 2012).

18.8 SUITABILITY OF RA

Procedures for assessing the suitability of the RA to be recycled have also been developed by various agencies and vary from country to country.

Figure 18.11 shows an example of an assessment procedure used in the United Kingdom.

Another procedure for assessing the suitability of the RA for recycling into surface courses in particular can be found in Carswell et al. (2010).

18.9 HOT RECYCLING MIX DESIGN

Various mix design methods for hot recycled mix have been developed by various organisations worldwide. The main principles in all mix design procedure are the same, and the following are the general design steps followed:

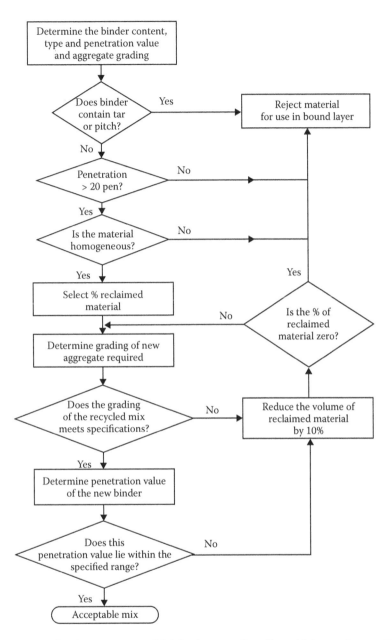

Figure 18.11 Flow chart for the assessment of RA for hot recycling. (From Highways Agency, *Design Manual for Roads and Bridges [DMRB], Volume 7: Pavement Design and Maintenance, Section 4, Part 1, HD31/94 [incorporating Amendment No. 2 February 1998]*. London: Highways Agency, 1998.)

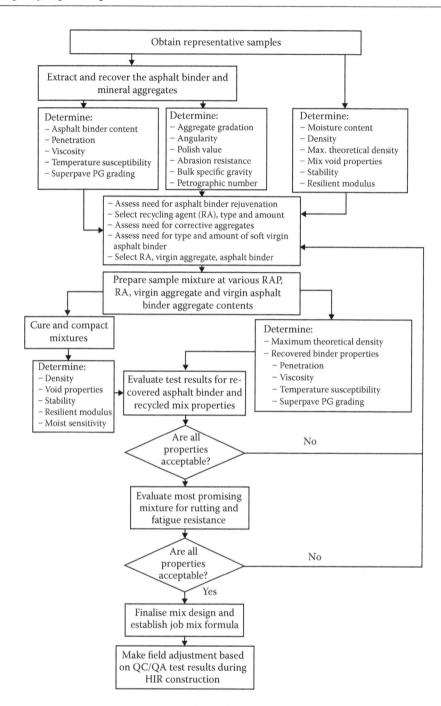

Figure 18.12 Comprehensive mix design flow chart of hot in situ recycled (HIR) asphalt mixture. (From ARRA, *Basic Asphalt Recycling Manual.* Asphalt Recycling & Reclaiming Association. Annapolis, MD: ARRA, 2001.)

a. Determination of properties of RA constituents (aggregate gradation, binder content, properties of aged binder, physical and geometrical properties of aggregates and chemical properties of aggregates, if required)
b. Determination of properties of virgin material (aggregate and bitumen)
c. Determination and selection of recycling agent (binder or rejuvenating agent)
d. Determination of aggregate gradation of recycled asphalt
e. Determination of properties of recycled asphalt base on Marshall, Superpave, Hveem or other mix design procedure
f. Determination of job mix formula
g. Or alternatively, determination of job mix formula on the basis of the performance-based characteristic properties of the recycled asphalt

A comprehensive/traditional hot in-place mix design method is described in the ARRA manual (ARRA 2001) and the mix design flow chart used is given in Figure 18.12. A similar mix design procedure is followed for hot in-plant mix design.

In the United Kingdom, a design method of RA for thin surfacing system has recently been developed. Details can be found in Carswell et al. (2010).

The current US practices on hot mix designs and other matters related to in-place cold recycling have been reported in a National Cooperative Highway Research Program (NCHRP) report (Stroup-Gardiner 2011).

18.10 COLD RECYCLING MIX DESIGN

The mix design procedure of cold recycled asphalts produced in situ (except full reclamation) or in a central plant is similar to and perhaps simpler than the one followed in hot mix recycling.

In cold recycling mix design, the type of recycling additive, usually bituminous emulsion (cationic or anionic, medium or slow setting) and sometimes foamed bitumen, does not depend on the penetration or the viscosity of the aged asphalt binder. Hence, the reclaimed material is seen as 'black aggregate' and the recycling additive simply coats the black aggregate. However, when a rejuvenator is incorporated, there is a certain softening of the aged binder. Cut-back bitumen used in the past is not favourable anymore for environmental and safety reasons.

The other difference with hot recycling mix design is the pre-mix moisture content and final moisture content of the recycled mixture. Moisture content determination is important in cold recycling mix design because it affects the workability of the material and can control the degree of compaction achieved.

In general, the steps of a cold recycling mix design consist of the following:

a. Determination of properties of RA constituents (aggregate gradation, binder content and, rarely, physical and geometrical properties of aggregates, as well as chemical properties of aggregates, if required)
b. Selection of type and grade of recycling additive
c. Determination of properties of virgin material (aggregate, bitumen emulsion or foamed bitumen, if used)
d. Selection of aggregate gradation of recycled cold asphalt (dense- or open-graded mixtures)
e. Determination of pre-mix moisture content and amount of recycling additive for adequate coating

f. Laboratory testing of the recycled mix for adequate mechanical and volumetric properties specified, using Marshall or gyratory (Superpave) compacted specimens, after determining the curing period

g. Determination of mix formula

h. In situ trial testing for minor adjustments

i. Establishing job mix formula, which may be re-adjusted if weather conditions change significantly from those during trial testing

There are various cold recycling mix design methods. Stroup-Gardiner (2011) provides statistical information and other details related to the cold recycling mix designs used in the United States.

One of the cold recycling mix design methods is described in detail in the ARRA manual (ARRA 2001).

Another mix design procedure of cold recycled material produced in situ including full-depth cold recycling (FDR) or in-plant (ex situ) using a wide range of binders and binder blends is also provided by Merrill et al. (2004).

The reader can also find very useful information regarding laboratory tests related to mix design of cold in situ recycling (FDR) and recommendations needed for test methods to make them applicable for the design of FDR materials in a Federal Highway Administration report (Bang et al. 2011).

18.11 PAVEMENT DESIGN USING RECYCLED ASPHALT

Pavement design using hot or cold recycled asphalts may be carried out by the same methodologies as the ones used for pavements with virgin hot or cold asphalt layers, provided the mechanical properties of the recycled materials are considered.

In addition to the pavement design methodologies covered in Chapter 12, the pavement design method developed by Merrill et al. (2004) specifically for pavements with cold recycled materials may also be used.

18.12 RECYCLING OF RIGID PAVEMENTS

The recycling of rigid pavements is relatively new compared to the recycling of flexible pavements. Recycling of rigid pavements started over the last 30–40 years mainly in countries that do not possess natural deposits for aggregate production.

Today, there are also other factors that affect decisions as regards recycling rigid pavements. The main factors are as follows: (a) preservation of natural deposits, (b) the lack of aggregate availability close to work site, (c) deposition of old concrete slabs, (d) need for cost reduction, (e) development of new methods and machinery for easy removal of reinforcement and (f) existence of great variety of additives for improving concrete strength and workability.

The reclaimed material from old concrete slabs (crushed and screened concrete particles) can be used as aggregate for (a) production of concrete, (b) stabilised or non-stabilised base and sub-base material, (c) construction of embankments or drainage filters, (d) construction of shoulders and (e) production of asphalts for base and binder courses.

During the recycling process of old rigid pavements (concrete slabs), the slabs are removed from site and transferred to a central disposal area where they are subjected to further crushing and screening after removal of reinforcement.

The major difficulty during recycling of rigid pavements is the initial crushing and bubbling carried out using specific hammers and the removal of reinforcement. Partial removal of the reinforcement may be conducted manually while the total removal of reinforcement is carried out by specific machinery using magnets. The crushers used for crushing further the concrete blocks are jaw- or hammer-type crushers. The crushed and sieved concrete material is stockpiled in the usual manner.

18.12.1 Properties of reclaimed aggregate from rigid pavements

With regard to the properties of the reclaimed aggregate from old slabs, the following can be summarised.

a. The reclaimed aggregates produced have good cubical shape, high porosity and lower bulk density compared to conventional crushed aggregates.
b. The incorporation of coarse aggregates for concrete production does not significantly affect the cement/water ratio or its workability.
c. When they are used as sand, the resulting concrete mixture is less workable and a higher water/cement ratio is required to achieve the same strength and workability. For this reason, many agencies do not use sand fraction for complete sand replacement. However, when used as sand replacement, the percentage usually incorporated is on the order of 30% maximum.
d. Provided that a small quantity of sand fraction is used, the concrete strength is similar to the strength of concretes with conventional aggregates regardless of the amount of coarse reclaimed material used.
e. In some cases, improvement on resistance to freeze/thaw can result.
f. Washing of the reclaimed aggregates from an old concrete slab does not seem to have any effect on concrete strength.

More information regarding reclaimed aggregate from old concrete slabs can be found in NCHRP Report 154 (Yrjanson 1989) and in NCHRP Report 435 (Stroup-Gardiner and Wattenberg-Komas 2013).

A good literature search on the use of recycled concrete aggregate on Portland cement concrete pavements can be found in a Washington State Department of Transport (WSDT) research report (Anderson et al. 2009).

To conclude, coarse aggregates, particularly those derived from recycled rigid pavements, are not inferior to conventional aggregates and they can be used as substitute to conventional aggregates. However, all necessary tests to ensure compliance with requirements should always be conducted prior to use.

REFERENCES

AASHTO R 59. 2011. *Recovery of asphalt binder from solution by Abson method*. Washington, DC: American Association of State Highway and Transportation Officials.
AASHTO T 2. 2010. *Sampling of aggregates*. Washington, DC: American Association of State Highway and Transportation Officials.

AASHTO T 168. 2011. *Sampling bituminous paving mixtures.* Washington, DC: American Association of State Highway and Transportation Officials.

Anderson K.W., J.S. Uhlmeyer, and M. Russell. 2009. *Use of Recycled Concrete Aggregate in PCCP: Literature Search.* Report No. WA-RD 726.1. Springfield, VA: National Technical Information Service.

ARRA. 2001. *Basic Asphalt Recycling Manual.* Asphalt Recycling & Reclaiming Association. Annapolis, MD: ARRA.

ASTEC Industries Inc. 2014. Available at http://www.astecinc.com (Roadtec Inc. Available at http://www.roadtec.com).

ASTM D 75/D 75M. 2009. *Standard practice for sampling aggregates.* West Conshohocken, PA: ASTM International.

ASTM D 979/D 979M. 2012. *Standard practice for sampling bituminous paving mixtures.* West Conshohocken, PA: ASTM International.

ASTM D 1856. 2009. *Standard test method for recovery of asphalt from solution by Abson method.* West Conshohocken, PA: ASTM International.

ASTM D 4887/D 4887M. 2011. *Standard practice for preparation of viscosity blends for hot recycled asphalt materials.* West Conshohocken, PA: ASTM International.

ASTM D 5404/D 5404M. 2012. *Standard practice for recovery of asphalt from solution using the rotary evaporator.* West Conshohocken, PA: ASTM International.

Austroads. 2013. *Maximising the Re-Use of Reclaimed Asphalt Pavement: Binder Blend Characterization.* AP-T245-13. Sydney, Australia: Austroads.

Bang S., W. Lein, B. Comes, L. Nehl, J. Anderson, P. Kraft, M. deStigter, C. Leibrock, L. Roberts, P. Sebaaly, D. Johnston, and D. Huft. 2011. *Quality Base Material Produced Using Full Depth Reclamation on Existing Asphalt Pavement Structure-Task 4: Development of FDR Mix Design Guide.* FHWA-HIF-12-015. Washington, DC: Federal Highway Administration.

Carswell I., J.C. Nicholls, I. Widyatmoko, J. Harris, and R. Taylor. 2010. *Best Practice Guide for Recycling into Surface Course.* TRL RN 43. Wokingham, Berkshire, UK: Transport Research Laboratory.

CEN EN 932-1. 1996. *Tests for general properties of aggregates – Part 1: Methods for sampling.* Brussels: CEN.

CEN EN 12697-3. 2013. *Bituminous mixtures – Test methods for hot mix asphalt – Part 3: Bitumen recovery: Rotary evaporator.* Brussels: CEN.

CEN EN 12697-4. 2005. *Bituminous mixtures – Test methods for hot mix asphalt – Part 4: Bitumen recovery: Fractionating column.* Brussels: CEN.

CEN EN 12697-27. 2000. *Bituminous mixtures – Test methods for hot mix asphalt – Part 27: Sampling.* Brussels: CEN.

CEN EN 13108-1:2006/AC. 2008. *Bituminous mixtures – Material specifications – Part 1: Asphalt Concrete.* Brussels: CEN.

CEN EN 13108-2:2006/AC. 2008. *Bituminous mixtures – Material specifications – Part 2: Asphalt Concrete for very thin layers.* Brussels: CEN.

CEN EN 13108-4:2006/AC. 2008. *Bituminous mixtures – Material specifications – Part 4: Hot Rolled Asphalt.* Brussels: CEN.

CEN EN 13108-5:2006/AC. 2008. *Bituminous mixtures – Material specifications – Part 5: Stone Mastic Asphalt.* Brussels: CEN.

CEN EN 13108-7:2008/AC. 2008. *Bituminous mixtures – Material specifications – Part 7: Porous asphalt.* Brussels: CEN.

CEN EN 13108-8. 2005. *Bituminous mixtures – Material specifications – Part 8: Reclaimed asphalt.* Brussels: CEN.

Chappat M. and J.F. Plaut. 1982. Recyclage des enrobés en centrale. *RGRA,* Vol. 584, No. 584, p. 11.

EAPA. 2012. Asphalt in figures. Available at http://www.eapa.org.

Eaton E. 1991. States fine tune asphalt recycling specs. Asphalt recycling and reclaiming '91. *Roads & Bridges,* Vol. 29, No. 10, pp. 26–29. Des Planes, IL: Scranton Gillette Communications.

Finn F.N. 1980. Seminar on asphalt pavement recycling: Overview of project selection. *Transportation Research Record,* Vol. 780, p. 8. Washington, DC: TRB.

Hansen K.R. and A. Copeland. 2013. *Annual Asphalt Pavement Industry Survey on Recycled Materials and Warm-Mix Asphalt Usage: 2009–2012.* Information Series 138. Lanham, MD: NAPA.

Highways Agency. 1998. *Design Manual for Roads and Bridges (DMRB), Volume 7: Pavement Design and Maintenance.* Section 4, Part 1, HD31/94 (incorporating Amendment No. 2 February 1998). London: Highways Agency.

Ipavec A., P. Marsac, and K. Mollenhauer. 2012. Synthesis of the European National requirements and practices for recycling HMA and WMA (Direct_Mat Project). *5th Eurasphalt & Eurobitume Congress.* Istanbul, Turkey.

Kandhal P.S. and R.B. Mallick. 1997. *Pavement Recycling Guidelines for State and Local Governments – Participant's Reference Book.* FHWA-SA-97. Washington, DC: Federal Highway Administration.

Maraux C., G. Glorie, and G. Van Heystraeten. 1991. Le recyclage d'enrobés asphaltique en central discontinue. Vol. CR32/91, p. 119. Brussels: Centre de Rechearches Routiéres.

Mauduit C., P.H. Carle, and M. Dauvergne. 2011. Comparison of carbon emission and energy consumption for two road maintenance solutions. *Proceedings of the 5th International Conference 'Bituminous Mixtures and Pavements'.* Thessaloniki, Greece.

McDaniel R., H. Soleymani, and R.M. Anderson. 2001. *Recommended Use of Reclaimed Asphalt Pavement in the Superpave Mix Design Method: Guidelines.* Research Results Digest 253. Washington, DC: Transportation Research Board.

McDaniel R., H. Soleymani, R.M. Anderson, P. Turner, and R. Peterson. 2000. *Recommended Use of Reclaimed Asphalt Pavement in the Superpave Mix Design Method.* NCHRP web document 30. Washington, DC: Transportation Research Board.

Merrill D., M. Nunn, and I. Carswell. 2004. *A Guide to the Use and Specifications of Cold Recycled Materials for the Maintenance of Road Pavements.* TRL Report TRL611. Wokingham, Berkshire, UK: TRL Ltd.

Mollenhauer K. and L. Gaspar. 2012. Synthesis of European knowledge on asphalt recycling: Options, best practices and research needs. *Proceedings of the 5th Eurasphalt & Eurobitume Congress.* Istanbul, Turkey.

RSTA. 2012. *Code of Practice for In-Situ Structural Road Recycling.* Wolverhampton, UK: Road Surface Treatments Association. Available at http://www.rsta-uk.org.

Stroup-Gardiner M. 2011. *Recycling and Reclamation of Asphalt Pavements Using In-Place Methods: A Synthesis of Highway Practice.* NCHRP Synthesis 421. Washington, DC: Transport Research Board.

Stroup-Gardiner M. and T. Wattenberg-Komas. 2013. *Recycled Materials and Byproducts in Highway Applications Volume 6: Reclaimed Asphalt Pavement, Recycled Concrete Aggregate, and Construction Demolition Waste.* NCHRP Synthesis 435. Washington, DC: TRB, National Academy of Science.

TRB. 1978. *Recycling Materials for Highways.* Transportation Research Board. National Research Council, NCHRP 54. Washington, DC: TRB.

Uddin W. 2011. Life cycle assessment of sustainable pavement systems. *Proceedings of the 5th International Conference 'Bituminous Mixtures and Pavements'.* Thessaloniki, Greece.

Wirtgen GmbH. 2014. Available at http://www.wirtgen.de.

Wirtgen. 2013. *Wirtgen Cold Milling Manual: Technology and Application.* Windhagen, Germany: Wirtgen GmbH.

Yrjanson W.A. 1989. *Recycling of Portland Cement Pavements.* National Research Council, NCHRP 154. Washington, DC: TRB.

Index

Page numbers followed by f and t indicate figures and tables, respectively.